DIGITAL DESIGN OF SIGNAL PROCESSING SYSTEMS

DIGITAL DESIGN OF SIGNAL PROCESSING SYSTEMS

A PRACTICAL APPROACH

Shoab Ahmed Khan

National University of Sciences and Technology (NUST), Pakistan

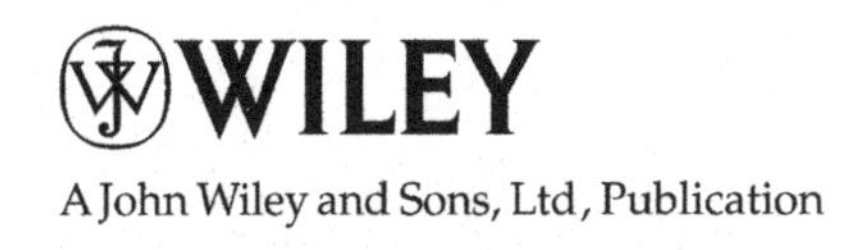

A John Wiley and Sons, Ltd, Publication

This edition first published 2011

Registered office
John Wiley & Sons Ltd, The Atrium, Southern Gate, Chichester, West Sussex, PO19 8SQ, United Kingdom

For details of our global editorial offices, for customer services and for information about how to apply for permission to reuse the copyright material in this book please see our website at www.wiley.com.

Library of Congress Cataloguing-in-Publication Data

Khan, Shoab Ahmed.
Digital design of signal processing systems : a practical approach / Shoab Ahmed Khan.
p. cm.
Includes bibliographical references and index.
ISBN 978-0-470-74183-2 (cloth)
1. Signal processing–Digital techniques. I. Title.
TK5102.9.K484 2010
621.382′2–dc22

2010026285

A catalogue record for this book is available from the British Library.

Print ISBN: 9780470741832 [HB]
ePDF ISBN: 9780470974698
oBook ISBN: 9780470974681
ePub ISBN: 9780470975251

Set in 9.5/11.5pt Times by Thomson Digital, Noida, India

Contents

Preface

Practising digital signal processing and digital system design for many years, and introducing and then developing the contents of courses at undergraduate and graduate levels, tempted me to write a book that would cover the entire spectrum of digital design from the signal processing perspective. The objective was to develop the contents such that a student, after taking the course, would be productive in an industrial setting in different roles. He or she could be a good algorithm developer, a digital designer and a verification engineer. An associated website (www.drshoabkhan.com) hosts RTL Verilog code of the examples in the book. Readers can also download PDF files of Microsoft PowerPoint presentations of lectures covering the material in the book. The lab exercises are provided for teachers' support.

The contents of this book show how to code algorithms in high-level languages in a way that facilitates their subsequent mapping on hardware-specific platforms. The book covers issues in implementing algorithms using fixed-point format. The ultimate conversion of algorithms developed in double-precision floating-point format to fixed-point is a critical design stage in system implementation. The conversion not only requires simple translation but in many cases also requires the designer to explore other structural options for mitigating quantization effects of fixed-point implementation. A number of commercially available system design and simulation tools provide support for fixed-point conversion and simulation. The MATLAB® fixed-point toolbox and utilities are important, and so is the support extended for fixed-point arithmetic in other high-level description languages such as SystemC. The issues of overflow, saturation, truncation and rounding are critical. The normalization and block floating-point option to optimize implementation should also be learnt. Chapter 3 covers all these issues and demonstrates the methodology with examples.

The next step in system design is to perform HW–SW partitioning. Usually this decision is made by an experienced designer. Chapters 1 and 3 give broad rules that help the designer to make this decision. The portion that is set aside for mapping in hardware is then explored for several architectural design options. Different ways of representing algorithms and their coding in MATLAB® are covered in Chapter 4. The chapter also covers mapping of the graphical representation of an algorithm on fully dedicated hardware.

Following the discussion on fully dedicated architectures, Chapter 5 lists designs of basic computational blocks. The chapter also highlights the architecture of embedded computational units in FPGAs and their effective use in the design. This discussion logically extends to algorithms that require multiplications with constants.

Chapter 6 gives an account of architectural optimization for designs that involve multiplications with constants. Depending on the throughput requirement and the target technology that constrains the clock rate, the architectural design decisions are made. Mapping an application written in a high-level language to hardware requires insight into the algorithm. Usually signal processing applications use nested loops. Unfolding and folding techniques are presented in Chapter 7. These techniques are discussed for code written in high-level languages and for algorithms that are described graphically as a dataflow graph (DFG). Chapter 4 covers the representation of algorithms as dataflow graphs. Different classes of DFGs are discussed. Many top-level design options are also discussed in Chapter 4 and Chapter 13. These options include a peer-to-peer KPN-connected network, shared bus-based design, and network-on-chip (NoC) based architectures. The top-level design is critical in overall performance, easy programmability and verification.

In Chapter 13 a complex application is considered as a network of connected processing elements (PEs). The PEs implement the functionality in an algorithm whereas the interconnection framework provides inter-PE communication. Issues of different scheduling techniques are discussed. These techniques affect the requirements of buffers between two connected nodes.

While discussing the hardware mapping of functionality in a PE, several design options are considered. These options include fully dedicated architecture (Chapter 4 and Chapter 6), parallel and unfolded architectures (Chapter 8), folded and time-shared architectures (Chapter 8 and Chapter 9) and programmable instruction set architectures (Chapter 10). Each architectural design option is discussed in detail with examples. Tradeoffs are also specified for the designer to gauge preferences of one over the other. Special consideration is given to the target platform. Examples of FGPAs with embedded blocks of multipliers with a fixed set of registers are discussed in Chapter 5.

Mapping of an algorithm in hardware must take into account the target technology. Novel methodologies for designing optimal architectures that meet stringent design constraints while keeping in perspective the target technology are elaborated. For a time-shared design, systolic and simple folded architectures are covered. Intricacies in folding a design usually require a dedicated controller to schedule operands on a shared HW resource. The controller is implemented as a finite state machine (FSM). FSM representations and designs are covered in Chapter 9. The chapter gives design examples. The testing of complex FSMs requires a lot of thought. Different coverage metrics are listed. Techniques are described that ensure maximum path coverage for effective testing of FSMs. For many complex applications, the designer has an option to define an instruction set that can effectively implement the application. A micro-programmed state machine design is covered in Chapter 10. Design examples are given to demonstrate the effectiveness of this design option. The designs are coded in RTL Verilog. The designer must know the coding rules and RTL guidelines from a synthesis perspective. Verilog HDL is covered, with mention of the guidelines for effective coding, in Chapter 2. This chapter also gives a brief description of SystemVerilog that primarily facilities testing and verification of the design. It also helps in modeling and simulating a system at higher levels of abstraction especially at transaction levels. Features of SystemVerilog that help in writing an effective stimulus are also given. For many examples, the RTL Verilog code is also listed with synthesis results. The book also provides an example of a communication receiver.

Two case studies of designs are discussed in detail. Chapter 11 presents an instruction set for implementing an adaptive algorithm for computationally intensive applications. Several architectural options are explored that trade off area with performance. Chapter 12 explores design options for a CORDIC-based DDFS algorithm. The chapter provides MATLAB® implementation of the basic CORDIC algorithm, and then explores fully parallel and folding architecture for implementing the CORDIC algorithm.

The book presents novel architectures for signal processing applications. Chapter 7 presents novel IIR filter-based decimation and interpolation designs. IIR filters are traditionally not used in these applications because for computing the current output sample they require previous output samples, so all samples need to be computed. This requires running the design at a faster clock for decimation and interpolation applications. The transformations are defined that only require an IIR filter to compute samples at a slower rate in decimation and interpolation applications.

In Chapter 10, the design of a DDFS based on the CORDIC algorithm is given. The chapter also presents a complete working of a novel design that requires only one stage of a CORDIC element and computes sine and cosine values. Then in Chapter 13 a novel design of time-shared and systolic AES architecture is presented. These architectures transform the AES algorithm to fit in an 8-bit datapath. Several innovative techniques are used to reduce the hardware complexity and memory requirements while enhancing the throughput performance of the design. Similarly novel architectures for massively parallel data compression applications are also covered.

The book can be adopted for a number of courses at senior undergraduate and graduate levels. It can be used for a senior undergraduate course on Advanced Digital Design and VLSI Signal Processing. Similarly the contents can be selected in a way to form a graduate level course in these two subjects.

Acknowledgments

I started my graduate studies at the Georgia Institute of Technology, Atlanta, USA, in January 1991. My area of specialization was digital signal processing. At the institute most of the core courses in signal processing were taught by teachers who had authored textbooks on the subjects. Dr Ronald W. Schafer taught "digital signal processing" using *Discrete Time Signal Processing* by Oppenheim, Schafer and Buck. Dr Monson H. Hayes taught "advanced signal processing" using his book, *Statistical Digital Signal Processing and Modeling*. Dr Vijay K. Madisetti taught from his book, *VLSI Digital Signal Processing*, and Dr Russell M. Mersereau taught from *Multi Dimensional Signal Processing* by Dudgeon and Mersereau. During the semester when I took a course with Dr Hayes, he was in the process of finalizing his book and would give different chapters of the draft to students as text material. I would always find my advisor, Dr Madisetti, burning the midnight oil while working on his new book.

The seed of desire to write a book in my area of interest was sowed in my heart in those days. After my graduation I had several opportunities to work on real projects in some of the finest engineering organizations in the USA. I worked for Ingersoll Rand, Scientific Atlanta, Picture Tel and Cisco Systems. I returned to Pakistan in January 1997 and started teaching in the Department of Computer Engineering at the College of Electrical and Mechanical Engineering (E&ME), National University of Sciences and Technology (NUST), while I was still working for Cisco Systems.

In September 1999, along with two friends in the USA, Raheel Ahmed Khan and Sherjil Ahmed, founded a startup company called Communications Enabling Technologies (or CET, later named Avaz Networks Inc. USA). Raheel Khan is a genuine designer of digital systems. Back in 1999 and 2000, we designed a few systems and technical discussions with him further increased my liking and affection, along with strengthening my comprehension of the subject. I was serving as CTO of the company and also heading the R&D team in Pakistan.

In 2001 we secured US$17 million in venture funding and embarked on developing what was at that time the highest density media processor system-on-chip (MPSoC) solution for VoIP carrier-class equipment. The single chip could process 2014 channels of VoIP, performing DTMF generation and detection, line echo cancellation (LEC), and voice compression and decompression on all these channels. I designed the top-level architecture of the chip and all instruction set processors, and headed a team of 160 engineers and scientists to implement the system. Designing, implementing and testing such a complex VLSI system helped me to understand the intricacies of the field of digital design. We were able to complete the design cycle in the short period of 10 months.

At the time we were busy in chip designing, I, with my friends at Avaz also founded the Center for Advanced Studies in Engineering (CASE), a postgraduate engineering program in computer engineering. I introduced the subject of "advanced digital design" at CASE and NUST. I would always teach this subject, and to compose the course contents I would collect material from research papers, reference books and the projects we were undertaking at Avaz. However, I could not find a book that did justice to this emerging field.

So it was in 2002 that I found myself compelled to write a textbook on this subject. In CASE all lectures are recorded on videos for its distance-learning program. I asked Sandeela Sameem, my student and an intern at Avaz, to make a Microsoft PowerPoint presentation from the design I drew and text I wrote on the whiteboard. These presentations served as the initial material for me to start writing the book. The course was offered once every year and in that semester I would write a little on a few topics, and would give the material to my students as reference.

In 2004, I with my core team of CET founded a research organization called the Center for Advanced Research in Engineering (CARE) and since its inception I have been managing the organization as CEO. At CARE, I have had several opportunities to participate in the design of machine vision, network analysis and digital communication systems. The techniques and examples discussed in this book are used in the award winning products from CARE. Software Defined Radio, 10 Gigabit VoIP monitoring system and Digital Surveillance equipment received APICTA (Asia Pacific Information and Communication Alliance) awards in September 2010 for their unique and effective designs.

My commitments as a professor at NUST and CASE, and as CEO of CARE did not allow me enough time to complete my dream of writing a book on the subject but my determination did not falter. Finally in March of 2008 I forwarded my proposal to John Wiley and Sons. In July 2008, I formally started work on the book.

I have been fortunate to have motivated students to help me in formatting the text and drawing the figures. Initially it was my PhD student, Fozia Noor Khan, who took pains to put the material in a good format and convert my hand-drawn figures into Visio images. Later this task was taken over by Hussnain Ali. He also read the text and highlighted areas that might need my attention. Finally it was my assistant, Shaista Zainab, who took over helping me in putting the manuscript in order. Also, many of my students helped in exploring areas. Among them were Zaheer Ahmed, Mohammad Mohsin Rahmatullah, Sheikh M. Farhan, Ummar Farooq and Rizwana Mehboob, who completed their PhDs under my co-supervision or supervision. There are almost 70 students who worked on their MS theses under my direct supervision working on areas related to digital system design. I am also deeply indebted to Dr Faisal Durbai for spending time reading a few chapters and suggesting improvements. My research associates, Usman Akram and Sajid, also extended their help in giving a careful reading of the text.

I should like to acknowledge a number of young and enthusiastic engineers who opted to work for us in Avaz and CARE, contributing to the development of several first-time-right ASICs and FPGA-based complex systems. Their hard work and dedication helped to enlighten my approach to the subject. A few names I should like to mention are Imran Qasir (IQ), Rahan Hameed, Nuaman, Mobeen, Hassan, Aeman Bukhari, Mahreen, Sadia, Hamza, Fahad Ali Mujahaid, Usman, Asim Munawar, Aysha Khalid, Alina Mufti, Hammood, Wajahat, Arsalan, Durdana, Shahbaz, Masood, Adnan, Khalid, Rabia Anwar and Javaria.

Thanks go to my collegues for keeping me motivated to complete the manuscript: Dr Habibullah Jamal, Dr Shamim Baig, Dr Abdul Khaliq, Hammad Khan, Dr Saad Rahman, Dr Muhammad Younis Javed, Asrar Ashraf, Dr Akhtar Nawaz Malik, Gen Muhammad Shahid, Dr Farrukh Kamran, Dr Saeed Ur Rahman, Dr Ismail Shah and Dr Sohail Naqvi.

Last but not least, my parents, brother, sisters and my wife Nuzhat who has given me support all through the years with love and compassion. My sons Hamza, Zaid and Talha grew from toddlers to teenagers seeing me taking on and then working on this book. My daughter Amina has consistently asked when I would finish the book!

1

Overview

No exponential is forever ... but we can delay "forever"

Gordon Moore

1.1 Introduction

This chapter begins from the assertion that the advent of VLSI (very large scale integration) has enabled solutions to intractable engineering problems. Gordon Moore predicted in 1965 the rate of development of VLSI technology, and the industry has indeed been developing newer technologies riding on his predicted curve. This rapid advancement has led to new dimensions in the core subject of VLSI. The capability to place billions of transistors in a small silicon area has tested the creativity of engineers and scientists around the world. The subject of digital design for signal processing systems embraces these new challenges. VLSI has revolutionized the commercial market, with products regularly appearing with increasing computational power, improved battery life and reduced physical size.

This chapter discusses several applications. The focus of the book is on applications primarily in areas of signal processing, multimedia, digital communication, computer networks and data security. Some of the applications are shown in Figure 1.1.

Multimedia applications have had a dramatic impact on our lives. Multimedia access on handheld devices such as mobile phones and digital cameras is a direct consequence of this technology.

Another area of application is high-data-rate communication systems. These systems have enormous real-time computational requirements. A modern mobile phone, for example, executes several complex algorithms, including speech compression and decompression, forward error-correction encoding and decoding, highly complex modulation and demodulation schemes, up-conversion and down-conversion of modulated and received signals, and so on. If these are implemented in software, the amount of real-time computation may require the power of a supercomputer. Advancement in VLSI technology has made it possible to conveniently accomplish the required computations in a hand-held device. We are also witnessing the dawn of new trends like wearable computing, owing much to this technology.

Digital Design of Signal Processing Systems: A Practical Approach, First Edition. Shoab Ahmed Khan.

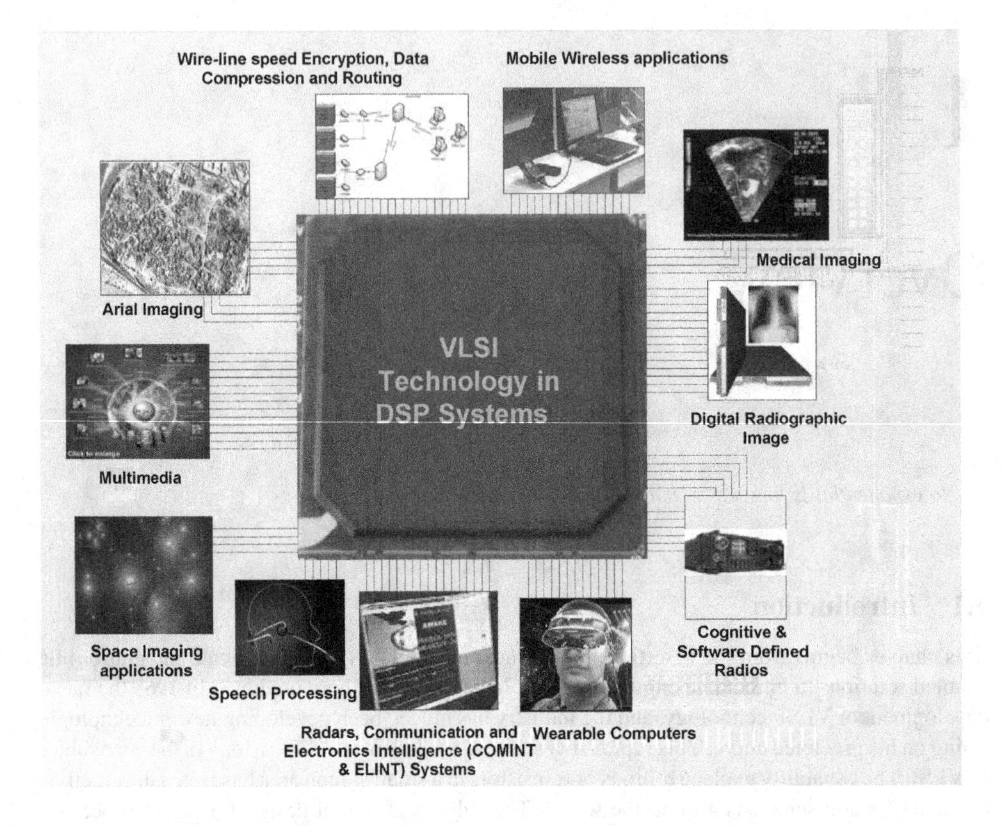

Figure 1.1 VLSI technology plays a critical role in realizing real-time signal processing systems

Broadband wireless access technology, processing many megabits of information per second, is another impressive display of the technology, enabling mobility to almost all the services currently running on desktop computers. The technology is also at work in spacecraft and satellites in space imaging applications.

The technology is finding uses in biomedical equipment, examples being digital production of radiographic and ultrasound images, and implantable devices such as the cardioverter defibrillator that acquires and digitizes heartbeats, detects any rhythmic abnormalities and symptoms of sudden cardiac arrest and applies an electric shock to help a failing heart.

This chapter selects a mobile communication system as an example to explain the design partitioning issues. It highlights that digital design is effective for mapping structured algorithms in silicon. The chapter also considers the design of a backplane of a high-end router to reveal the versatility of the techniques covered in this book to solve problems in related areas where performance is of prime importance.

The design process has to explore competing design objectives: speed, area, power, timing and so on. There are several mathematical transformations to help with this. Keeping in perspective the defined requirement specifications, transformations are applied that trade off less relevant design objectives against the other more important objectives. That said, for complex design problems these mathematical transformations are of less help, so an effective approach requires learning several 'tricks of the trade'. This book aims to introduce the transformations as well as giving tips for effective design.

The chapter highlights the impact of the initial ideas on the entire design process. It explains that the effect of design decisions diminishes as the design proceeds from concept to implementation. It establishes the rational for the system architect to positively impact the design process in the right direction by selecting the best option in the multidimensional design space. The chapter explores the spectrum of design options and technologies available to the designer. The design options range from the most flexible general-purpose computing machine like Pentium, to commercially available off-the-shelf digital signal processors (DSPs), to more application-specific instruction-set processors, to hard-wired application-specific designs yielding best performance without any consideration of flexibility in the solution. The chapter describes the target technologies on which the solution can be mapped, like general-purpose processors (GPPs), DSPs, application-specific integrated circuits (ASICs), and field-programmable gate arrays (FPGAs). It is established that, for complex applications, an optimal solution usually consists of a mix of these target technologies.

This chapter presents some design examples. The rationale for design decisions for a satellite burst modem receiver is described. There is a brief overview of the design of the backplane of a router. There is an explanation of the design of a network-on-chip (NoC) carrier-class VoIP media gateway. These examples follow a description of the trend from digital-only design to mixed-signal system-on-chips (SoCs). The chapter considers synchronous digital circuits where digital clocks are employed to make all components operate synchronously in implementing the design.

1.2 Fueling the Innovation: Moore's Law

Advancements in VLSI over a few decades have played a critical role in realizing the amazing electronic gadgets we live with today. Gordon Moore, founder of Intel, earlier predicted the rapid rate of these advancements. In 1965 he noted that the number of transistors on a chip was doubling every 18 to 24 months. Figure 1.2(a) shows the predicted curve known as Moore's Law from his original paper [1]. This 'law' has fueled innovation for five decades. Figure 1.2(b) shows Intel's response to his prediction.

Moore acknowledges that the trend cannot last forever, and he gave a presentation at an international conference, entitled "No exponential is forever, but we can delay 'forever'" [2]. Intel has plans to continue riding on the Moore's Law curve for another ten years and has announced a 2.9 billion-transistor chip for the second quarter of 2011. The chip will fit into an area the size of a fingernail and use 22-nanometer technology [3]. For comparison, the Intel 4004 microprocessor introduced in 1971 was based on a 10 000-nanometer process.

Integration at this scale promises enormous scope for designers and developers, and the development of design tools has matched the pace. These tools provide a level of abstraction so that the designer can focus more on higher level design concepts rather than low-level details.

1.3 Digital Systems

1.3.1 Principles

To examine the scope of the subject of digital design, let us consider an embedded signal processing system of medium complexity. Usually such a system consists of heterogeneous physical devices such as a GPP or micro-controller, multiple DSPs, a few ASICs, and FPGAs. An application implemented on such a system usually consists of numerous tasks of varying computational

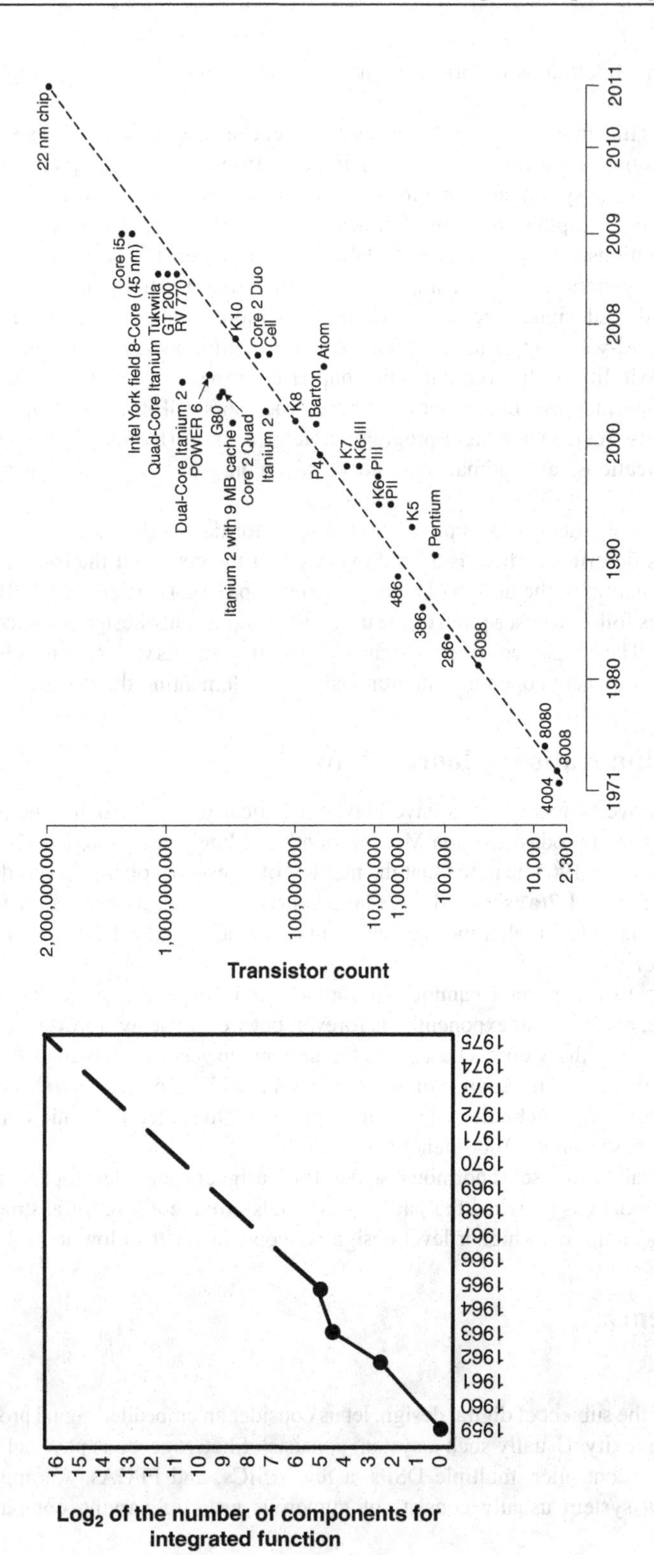

Figure 1.2 (a) The original prediction of Moore's Law. (b) Intel's response to Moore's prediction

complexity. These tasks are mapped on to the physical devices. The decision to implement a particular task on a particular device is based on the computational complexity of the task, its code density, and the communication it requires with other tasks.

The computationally intensive ('number-crunching') tasks of the application can be further divided into categories. The tasks for which commercial off-the-shelf ASICs are available are best mapped on these devices. ASICs are designed to perform specific functions of a particular application and are not programmable, as are GPPs. Based on the target technology, ASICs are of different types, examples of which are full-custom, standard-cell-based, gate-array-based, channeled gate array, channel-less gate array, and structured gate array. As these devices are application-specific they are optimized using integrated-circuit manufacturing process technology. These devices offer low cost and low power consumption. There are many benefits to using ASICs, but because of their fixed implementation a design cannot be made easily upgradable.

It is important to point out that several applications implement computationally intensive but non-standard algorithms. The designer, for these applications, may find that mapping the entire application on FPGAs is the only option for effective implementation. For applications that consist of standard as well as non-standard algorithms, the computationally intensive tasks are further divided into two groups: *structured* and *non-structured*. The tasks in the structured group usually consist of code that has loops or nested loops with a few instructions being repeated a number of times, whereas the tasks in the non-structured group implement more code-intensive components. The structured tasks are effectively mapped on FPGAs, while the non-structured parts of the algorithm are implemented on a DSP or multiple DSPs.

A field-programmable gate array comprises a matrix of configurable logic blocks (CLBs) embedded in an interconnected net. The FPGA synthesis tools provide a method of programming the configurable logic and the interconnects. The FPGAs are bought off the shelf: Xilinx [4], Altera [5], Atmel [6], Lattice Semiconductor [7], Actel [8] and QuickLogic [9] are some of the prominent vendors. Xilinx shares more than 50% of the programmable logic device (PLD) segment of the semiconductor industry.

FPGAs offer design reuse, and better performance than a software solution mapped on a DSP or GPP. They are, however, more expensive and give reduced performance and more power consumption compared with an equivalent ASIC solution if it exists. The DSP, on the other hand, is a microprocessor whose architecture is specially designed to support number-crunching signal processing applications. Usually a DSP can perform many multiplication and addition operations and supports special addressing modes that help in effective implementation of fast Fourier transform (FFT) and convolution algorithms.

The GPPs or microcontrollers are general-purpose computing machines. Types are 'complex instruction set computer' (CISC) and 'reduced instruction set computer' (RISC).

The tasks specific to user interfaces, control processes and other code-intensive protocols are usually mapped on GPPs or microcontrollers. For handling multiple concurrent tasks, events and interrupts, the microcontroller runs a real-time operating system. The GPP is also good at performing general tasks like configuring various devices in the system and interfacing with external devices. The microcontroller or GPP performs the job of a system controller. For systems of medium complexity, it is connected to a shared bus. The processor configures the ASICs and FGPAs, and also bootstraps the DSPs in the system. A high-speed bus like Amba High-speed Bus (AHB) is used in these systems [10]. The shared-bus protocol allows only one master to transfer the data. For designs that require parallel transfer of data, a multi-layer shared bus like Multi-Layer AHB (ML-AHB) is used [11]. The microcontroller also interfaces with the external displays and control panels.

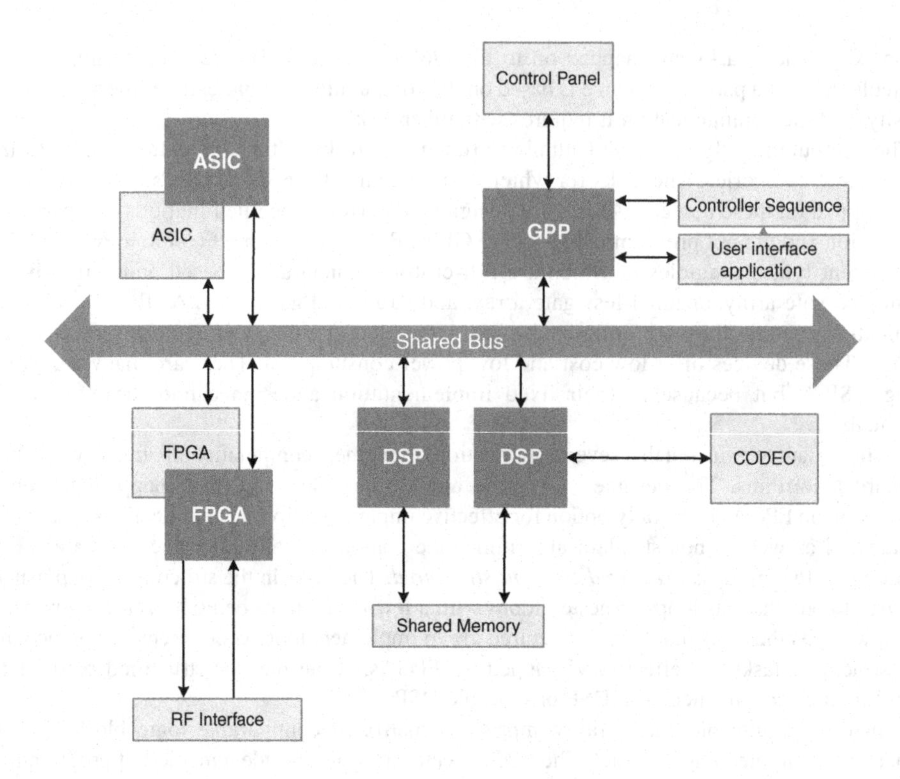

Figure 1.3 An embedded signal processing system with DSPs, FPGAs, ASICs and GPP

The digital design of a digital communication system interfaces with the RF front end. For voice-based applications the system also contains CODEC (more in Chapter 12) with associated analog interfaces. The FPAGs in the system also provide glue logic and interfaces with other devices in the system. There may also be dual-port RAM to provide shared memory to multiple DSPs in the system. A representative system is shown in Figure 1.3.

1.3.2 Multi-core Systems

Many applications are best mapped on general-purpose processors. As high-end computing applications demand more and more computational power in programmable devices, the vendors of GPPs are incorporating multiple cores of GPPs in a single SoC configuration. Almost all the vendors of GPPs, such as Intel, IBM, Sun Microsystems and AMD, are now placing multiple cores on a single chip for improved performance and high reliability. Examples are Intel's Yorkfield 8-core chip in 45-nm technology, Intel's 80-core teraflop processor, Sun's Rock 8-core CPU, Sun's UltraSPARC T1 8-core CPU, and IBM's 8-core POWER7. These multi-core solutions also offer the necessary abstraction, whereby the programmer need not be concerned with the underlying complex architecture, and software development tools have been produced that partition and map applications on these multiple cores. This trend is continuously adding complexity to digital design and software tool development. From the digital design perspective, multi processors based systems are required to communicate with each other, and inter-processors connections need to be scalable and expendable. The network-on-chip (NoC) design paradigm addresses issues of scalability of on-chip connectivity and inter-processor communication.

1.3.3 NoC-based MPSoC

Besides GPP-based multi-core SoCs for mapping general computing applications, there also exist other application-specific SoC solutions. An SoC integrates all components of a system in a single chipset. That includes microprocessor, application-specific accelerators, all interfaces to memory and peripheral devices, and so on.

Most high-end signal processing applications offer an inherent parallelism. To exploit this parallelism, these systems are mapped on multiple heterogeneous processors. Traditionally these processors are connected with shared memories on shared buses. As complex designs are integrating an increasing number of multi-processors on a single SoC (MPSoC) [12], designs based on a shared bus are not effective owing to complex arbitration, clock skews and latency issues. These designs require scalable and effective communication infrastructure. An NoC offers a good solution to these problems [13]. The NOC provides higher bandwidth, low latency, modularity, scalability, and a high level of abstraction to the system. The complex bus protocols route wires to connect various components, whereas an NOC uses packet-based protocols to provide connectivity among components. The NoC enables parallel transactions of data.

The basic architecture of an NOC is shown in Figure 1.4. Each processing element (PE) is connected to an on-chip router via a network interface (NI) controller.

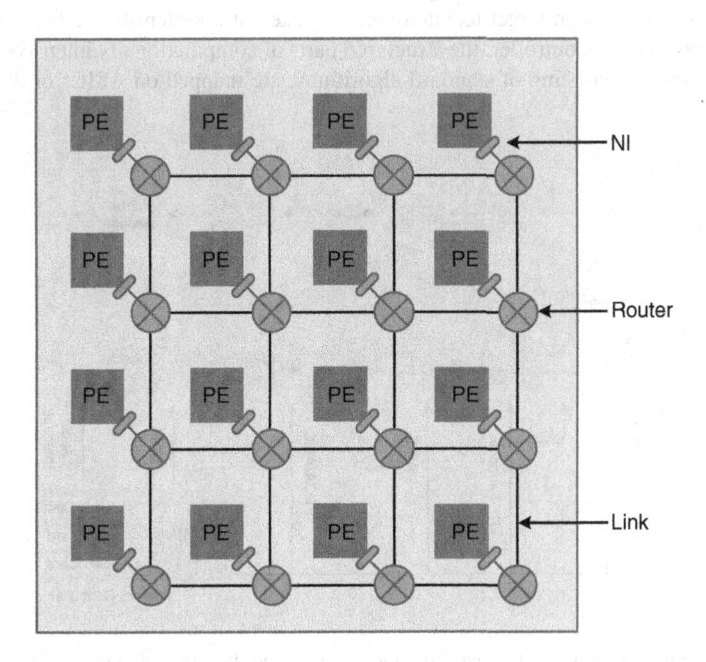

Figure 1.4 An NoC-based heterogeneous multi-core SoC design

Many vendors are now using NoC to integrate multiple PEs on a single chip. A good example is the use of NoC technology in the Play Station 3 (PS3) system by Sony Entertainment. A detailed design of an NoC-based system is given in Chapter 13.

1.4 Examples of Digital Systems

1.4.1 Digital Receiver for a Voice Communication System

A typical digital communication system for voice, such as a GSM mobile phone, executes a combination of algorithms of various types. A system-level block representation of these algorithms is shown in Figure 1.5. These algorithms fall into the following categories.

1. *Code-intensive algorithms.* These do not have repeated code. This category consists of code for phone book management, keyboard interface, GSM (Global System for Mobile) protocol stack, and the code for configuring different devices in the system.
2. *Structured and computationally intensive algorithms.* These mostly take loops in software and are excellent candidates for hardware mapping. These algorithms consist of digital up- and down-conversion, demodulation and synchronization loops, and forward error correction (FEC).
3. *Code-intensive and computationally intensive algorithms.* These lack any regular structure. Although their implementations do have loops, they are code-intensive. Speech compression is an example.

The GSM is an interesting example of a modern electronic device as most of these devices implement applications that comprise these three types of algorithm. Some examples are DVD players, digital cameras and medical diagnostic systems.

The mapping decisions on target technologies are taken at a system level. The code-intensive part is mapped on a microcontroller; the structured parts of computationally intensive components of the application, if consisting of standard algorithms, are mapped on ASICs or otherwise they

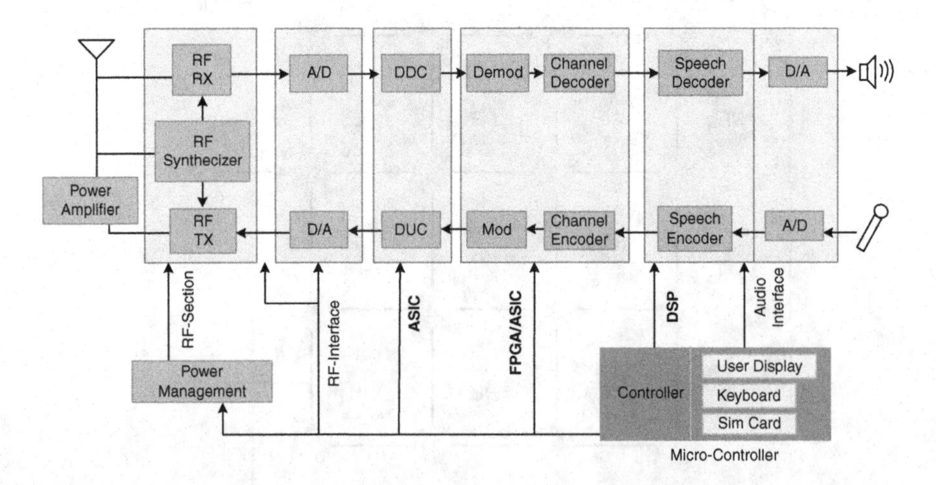

Figure 1.5 Algorithms in a GSM transmitter and receiver and their mapping on to conventional target technologies consisting of ASIC, FPGA and DSP

are implemented on FPGAs; and the computational and code-intensive parts are mapped on DSPs.

It is also important to note that only signals that can be acquired using an analog-to-digital (A/D) converter are implemented in digital hardware (HW) or software (SW), whereas the signal that does not meet the Nyquist sampling criterion can be processed only using analog circuitry. This sampling criterion requires the sampling rate of an A/D converter to be double the maximum frequency or bandwidth of the signal. A consumer electronic device like a mobile phone can only afford to have an A/D converter in the range 20 to 140 million samples per second (Msps). This constraint requires analog circuitry to process the RF signal at 900 MHz and bring it down to the 10–70 MHz range. After conversion of this to a digital signal by an A/D converter, it can be easily processed. A conventional mapping of different building blocks of a voice communication system is shown in Figure 1.5.

It is pertinent to mention that, if the volume production of the designed system are quite high, a mixed-signal SoC is the option of choice. In a mixed-signal SoC, the analog and digital components are all mapped on a single silicon device. Then, instead of placing physical components, the designer acquires soft cores or hard cores of the constituent components and integrates them on a single chip.

An SoC solution for the voice communication system is shown in Figure 1.6. The RF microcontroller, DSP and ASIC logic with on-chip RAM and requisite interfaces are all integrated on the same chip. A system controller controls all the interfaces and provides glue logic for all the components to communicate with each other on the device.

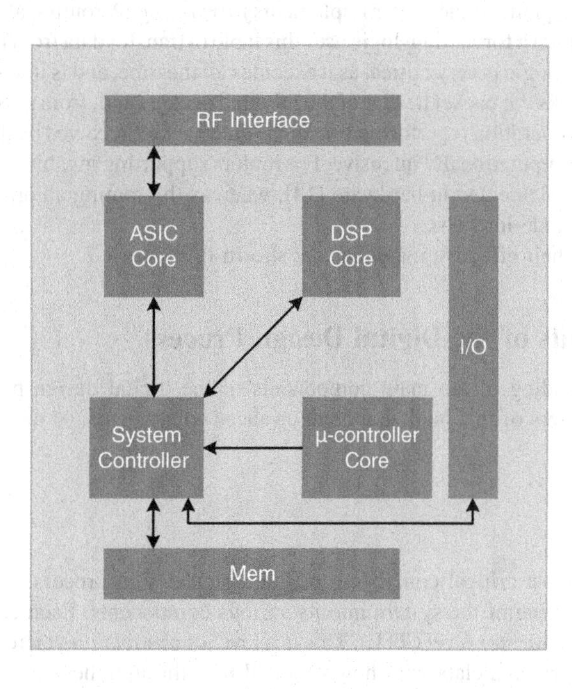

Figure 1.6 A system-on-chip solution

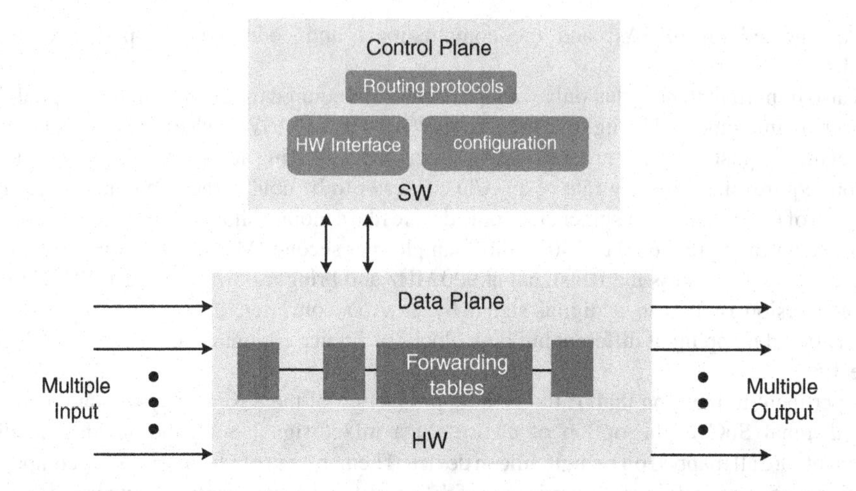

Figure 1.7 HW-SW partitioning of control and data plane in a router

1.4.2 The Backplane of a Router

A router consists mainly of two parts or planes, a control/management plane and a data plane. A code-intensive control or management plane implements the routing algorithms. These algorithms are executed only periodically, so they are not time-critical.

In contrast, the data plane of the router implements *forwarding*. A routing algorithm updates the routing table, after which a forwarding logic uses this table to transfer data from input ports to output ports. This forwarding logic is very critical as it executes all the time, and is implemented as the data plane. This plane checks the packet header of the inbound packets and, from a lookup table, finds its destination port. This operation is performed on all the data packets received by the router and is very well structured and computationally intensive. For routers supporting gigabit or multi-gigabit rates, this part is usually implemented in hardware [14], whereas the routing algorithms are mapped in software as they are code-intensive.

These planes and their effective mappings are shown in Figure 1.7.

1.5 Components of the Digital Design Process

A thorough understanding of the main components of the digital design process is important. The subsequent chapters of this book elaborate on these components, so they are discussed only briefly here.

1.5.1 Design

The 'design' is the most critical component of the digital design process. The *top-level design* highlights the partitioning of the system into its various components. Each component is further defined at the *register transfer level* (RTL). This is a level of abstraction where the digital designer specifies all the registers and elaborates how data will flow through these registers. The combinational logic between two sets of registers is usually described using high-level mathematical operations, and is drawn as a cloud.

1.5.2 Implementation

When the design has been described at RTL level, its implementation is usually a straightforward translation in a *hardware description language* (HDL) program. The program is then synthesized for mapping on an FPGA or ASIC implementation.

1.5.3 Verification

As the number of gates on a single silicon device increases, so do the challenges of verification. Verification is also critical in VLSI design as there is hardly any tolerance for bugs in the hardware. With application-specific integrated circuits, a bug may require a re-spin of fabrication, which is expensive, so it is important for an ASIC to be 'right first time'. Even bugs found in FPGA-based designs result in extended design cycles.

1.6 Competing Objectives in Digital Design

To achieve an effective design, a designer needs to explore the design space for tradeoffs of competing design objectives. The following are some of the most critical design objectives the designer needs to consider:

- area
- critical path delays
- testability
- power dissipation.

The *art* of digital design is to find the optimal tradeoff among these. These objectives are competing because, for example, if the designer tries to minimize area then the design may result in longer critical paths and may also affect the testability of the design. Similarly, if the design as synthesized for better timing means shorter critical paths, the design may result in a larger area. Better timing also means more power dissipation, which depends directly on the clock frequency. It is these competing objectives that make learning the techniques covered in this book very pertinent for designers.

1.7 Synchronous Digital Hardware Systems

The subject of digital design has many aspects. For example, the circuit may be *synchronous* or *asynchronous*, and it may be *analog* or *digital*. A digital synchronous circuit is always an option of choice for the designer. In synchronous digital hardware, all changes in the system are controlled by one or multiple clocks. In digital systems, all inputs/outputs and internal values can take only discrete values.

Figure 1.8 depicts an all-digital synchronous circuit in which all changes in the system are controlled by a global clock `clk`. A synchronous circuit has a number of registers, and values in these registers are updated at the occurrence of positive or negative edges of the clock signal. The figure shows positive-edge triggered registers. The output signal from the registers R_0 and R_1 are fed to the combinational logic. The signal goes through the combinational logic which consists of gates. Each gate causes some delay to the input signal. The accumulated delay on each path must be smaller than the time period of the clock, because the signal at the input of R_2 register must be stable before

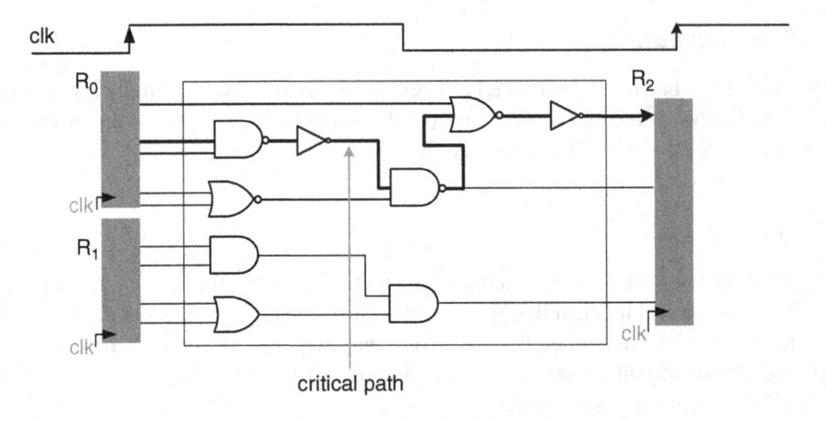

Figure 1.8 Example of a digital synchronous hardware system

the arrival of the next active edge of the clock. As there are a number of paths in any digital design, the longest path – the path that takes the maximum time for the signal to settle at the output – is called the *critical path*, as noted in Figure 1.8. The critical path of the design should be smaller than the permissible delay determined by the clock cycle.

1.8 Design Strategies

At the system level, the designer has a spectrum of design options as shown in Figure 1.9. It is very critical for the system designer to make good design choices at the conceptual level because they will have a deep impact on the rest of the design cycle.At the system design stage the designer needs only to draw a few boxes and take major design decisions like algorithm partitioning and target technology selection.

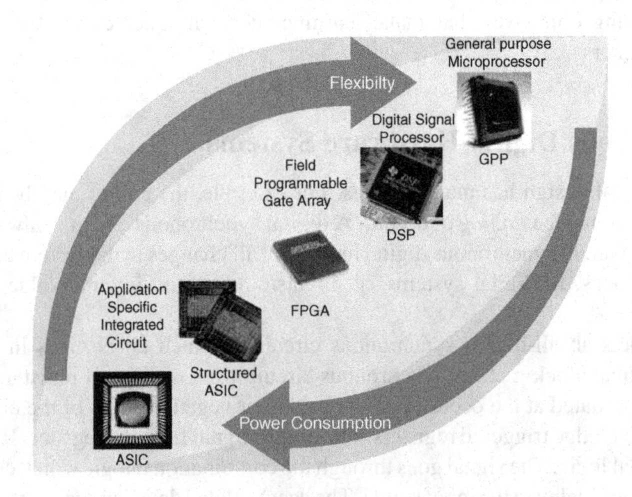

Figure 1.9 Target technologies plotted against flexibility and power consumption

If flexibility in programming is required, and the computational complexity of the application is low, and cost is not a serious consideration, then a general-purpose processor such as Intel's Pentium is a good option. In contrast, while implementing computationally intensive non-structured algorithms, flexibility in terms of programming is usually a serious consideration, and then a DSP should be the technology of choice.

In many applications the algorithms are computationally intensive but are also structured. This is usually the case in image and video processing applications, or a high-data-rate digital communication receiver. In these types of application the algorithms can be mapped on FPGAs or ASICs. While implementing algorithms on FPGAs there are usually two choices. One option is to design an *application-specific instruction-set processor* (ASIP). This type of processor is programmable but has little flexibility and can port only the class of applications using its application-specific instruction set. In the extreme case where performance is the only consideration and flexibility is not required, the designer should choose a second option, whereby the design is dedicated to that particular application and logic is hardwired without giving any consideration to flexibility. This book discusses these two design options in detail.

The performance versus flexibility tradeoff is shown in Figure 1.10. It is interesting to note that, in many high-end systems, usually all the design options are exercised. The code-intensive part of the application is mapped on GPPs, non-structured signal processing algorithms are mapped on DSPs, and structured algorithms are mapped on FPGAs, whereas for standard algorithms ASICs are used. This point is further elaborated in the design examples later.

These apparently simple decisions are very critical once the system proceeds along the design cycle. The decisions are especially significant for the blocks that are partitioned for hardware mapping. The algorithms are analyzed and architectures are designed. The designer either selects ASIP or dedicated hard-wired. The designer takes the high-level design and starts implementing the hardware. The digital design is implemented at RTL level and then it is synthesized and tools translate the code to gate level. The synthesized design is physically placed and routed. As the design

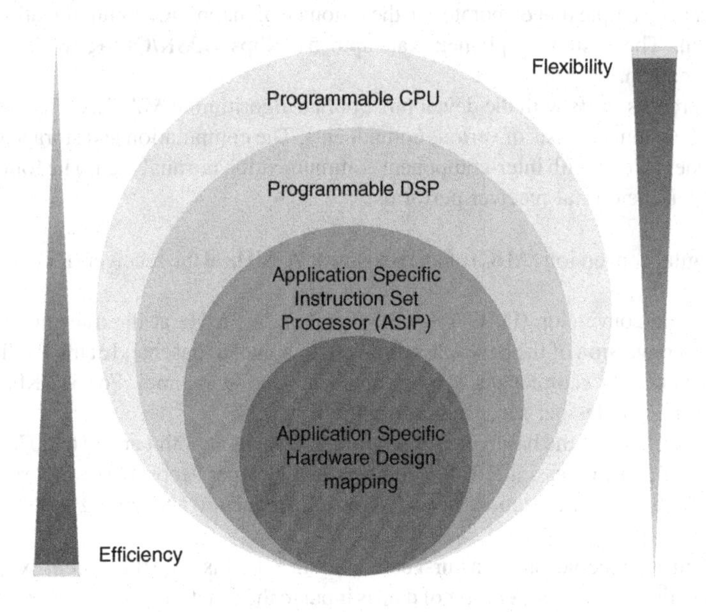

Figure 1.10 Efficiency verses flexibility tradeoff while selecting a design option

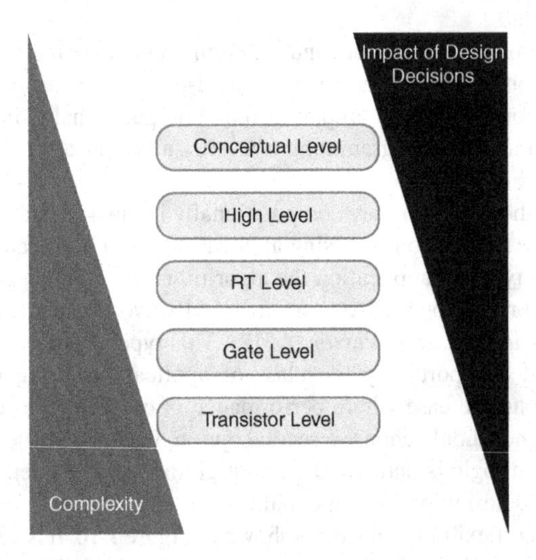

Figure 1.11 Design decision impact and complexity relationship diagram

goes along the design cycle, the details are very complex to comprehend and change. The designer at every stage has to make decisions, but as the design moves further along the cycle these decisions have less impact on the overall function and performance of the design. The relationship between the impact of the design decision and its associated complexity is shown in Figure 1.11.

1.8.1 Example of Design Partitioning

Let us consider an example that elaborates on the rationale of mapping a communication system on a hybrid platform. The system implements an upto 512Kbps BPSK/QPSK (phase-shift keying) satellite burst modem.

The design process starts with the development of an algorithm in MATLAB®. The code is then profiled. The algorithm consists of various components. The computation and storage requirements of each component along with inter-component communication are analyzed. The following is a list of operations that the digital receiver performs.

- Analog to digital conversion (ADC) of an IF signal at 70 MHz at the receiver (Rx) using band-pass sampling.
- Digital to analog conversion (DAC) of an IF signal at 24.5 MHz at the transmitter (Tx).
- Digital down-conversion of the band-pass digitized IF signal to baseband at the Rx. The baseband signal consists of four samples per symbol on each I and Q channel. For 512 Kbps this makes 2014 Ksps (kilo samples per second) on both the channels.
- Digital up-conversion of the baseband signal from 2014 ksps at both I and Q to 80 Msps at the Tx.
- Digital demodulator processing 1024 K complex samples per second. The algorithm at the Rx consists of: start of burst detection, header removal, frequency and timing loops and slicer.

In a burst modem, the receiver starts in burst detection state. In this state the system executes the start of the burst detection algorithm. A buffer of data is input to the function that computes some measure of presence of the burst. If the measure is greater than a threshold, 'start of burst' (SoB) is declared. In this state the system also detects the unique word (UW) in the transmitted burst and identifies the start of data. If the UW in the received burst is not detected, the algorithm transits back into the burst detection mode. When both the burst and the UW are detected, then the algorithm transits to the estimation state. In this state the algorithm estimates amplitude, timing, frequency and phase errors using the known header placed in the transmitted burst. The algorithm then transits to the demodulation state. In this state the system executes all the timing, phase and frequency error-correction loops. The output of the corrected signal is passed to the slicer. The slicer makes the soft and hard decisions. For forward error correction (FEC), the system implements a Viterbi algorithm to correct the bit errors in the slicer soft decision output, and generates the final bits [15]. The frame and end of frame are identified. In a burst, the transmitter can transmit several frames. To identify the end of the burst, the transmitter appends a particular sequence in the end of the last frame. If this sequence is detected, the receiver transits back to the SoB state. The state diagram of the sequence of operation in a satellite burst modem receiver is shown in Figure 1.12.

The algorithm is partitioned to be mapped on different components based on the nature of computations required in implementing the sub-components in the algorithm. The following mapping effectively implements the system.

- A DSP is used for mapping computationally intensive and non-regular algorithms in the receiver. These algorithms primarily consist of the demodulator and carrier and timing recovery loops.
- ASICs are used for ADC, DAC, digital down-conversion (DDC) and digital up-conversion (DUC). A direct digital frequency synthesis (DDFS) chip is used for cosine generation that mixes with the baseband signal.

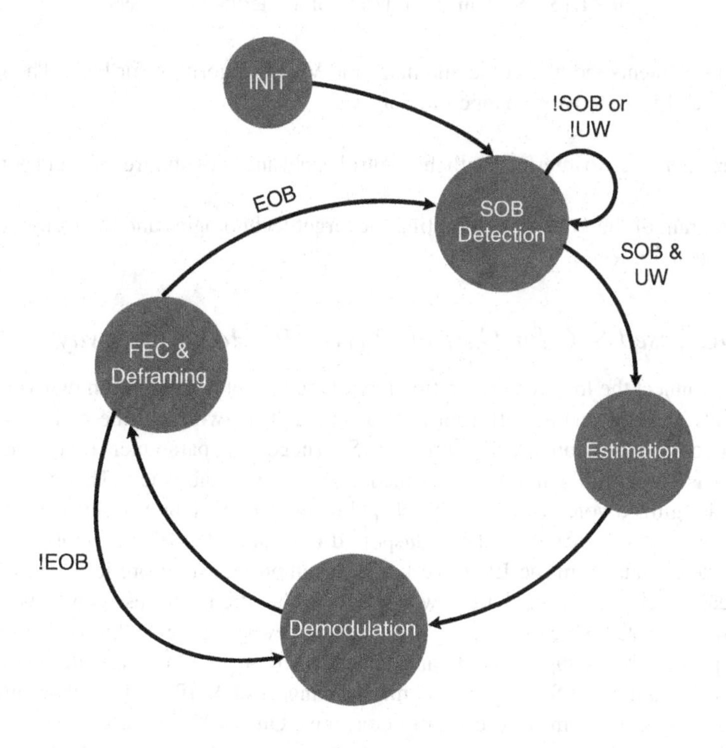

Figure 1.12 Sequence of operations in a satellite burst modem receiver

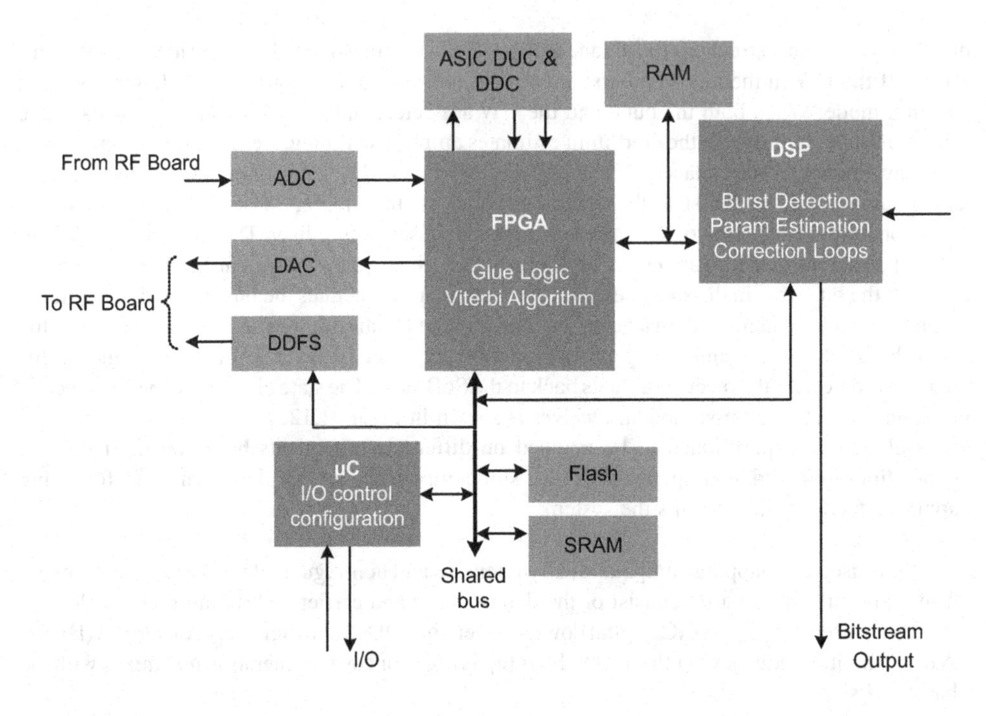

Figure 1.13 System-level design of a satellite burst receiver

- An FPGA implements the glue logic and maps the Viterbi algorithm for FEC. The algorithm is very regular and is effectively mapped in hardware.

A microcontroller is used to interface with the control panel and to configure different components in the system.

A block diagram of the system highlighting the target technologies and their interconnection is shown in Figure 1.13.

1.8.2 NoC-based SoC for Carrier-class VoIP Media Gateway

VoIP systems connect the legacy voice network with the packet network such that voice, data and associated signaling information are transported on the IP network. In the call setup stage, the signaling protocol (e.g. session initiation protocol, SIP) negotiates parameters for the media session. After the call is successfully initiated, the media session is established. This session takes the uncompressed digitized voice from the PSTN (public switched telephone network) interface and compresses and packages it before it is transported on a packet network. Similarly it takes the incoming packeted data from the IP network and decompresses it before it is sent on the PSTN network. A carrier-class VoIP media gateway processes hundreds of these channels.

The design of an SoC for a carrier-class VoIP media gateway is given in Figure 1.14. A matrix of application-specific processing elements are embedded in an NOC configuration on an SoC. In carrier-class application the SoC processes many channels of VoIP [16]. Each channel of VoIP requires the system to implement a series of algorithms. Once a VoIP call is in progress, the SoC needs to first process 'line echo cancellation' (LEC) and 'dual-tone multi-frequency' (DTMF)

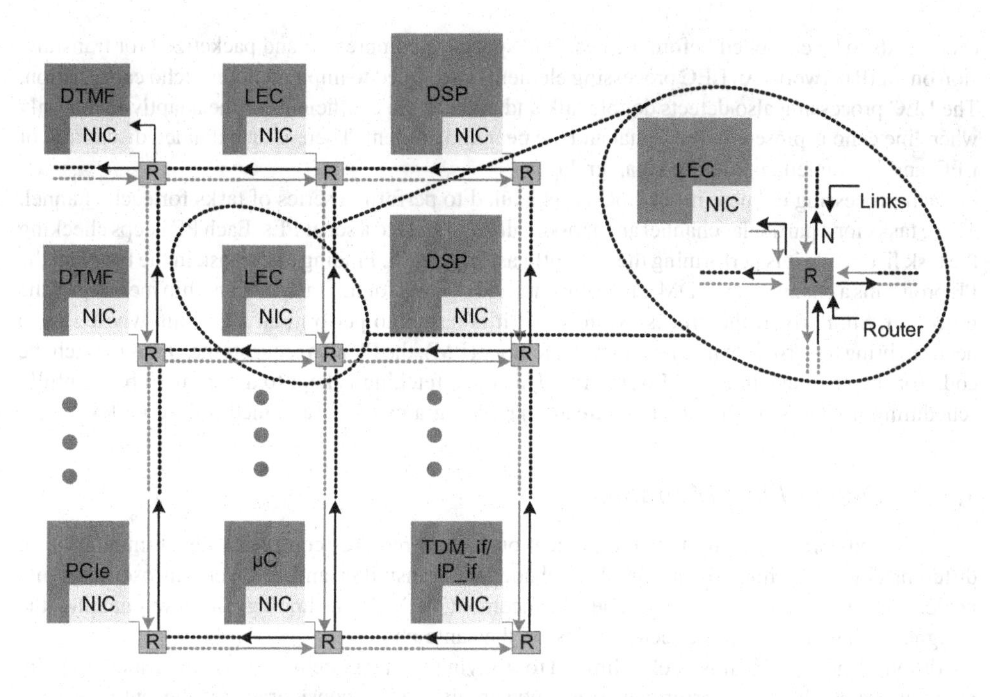

Figure 1.14 NoC-based SoC for carrier-class VoIP applications. Multiple layers of application-specific PEs are attached with an NoC for inter-processor communication

detection on each channel, and then it decompresses the packeted voice and compresses the time-division multiplex (TDM) voice. The SoC has two interfaces, one with the PSTN network and the other with the IP network. The interface with the PSTN may be an H.110 TDM interface. Similarly the interfaces on the IP side may be a combination of POS, UTOPIA or Ethernet. Besides these interfaces, the SoC may also have interfaces for external memory and PCI Express (PCIe). All these components on a chip are connected to a NOC for inter-component communication.

The design assumes that the media gateway controller and packet processor are attached with the media gateway SoC for complete functionality of a VoIP system. The packets received on the IP interface are saved in external memory. The data received on the H.110 interface is buffered in an on-chip memory before being transferred to the external memory. An on-chip RISC microcontroller is intimated to process an initiated call on a specified TDM slot by the host processor on a PCIe interface.

The microcontroller keeps a record of all the live calls, with associated information like the specification on agreed encoder and decoder between caller and callee. The microcontroller then schedules these calls on the array of multiprocessors by periodically assigning all the tasks associated with processing a channel that includes LEC, in-voice DTMF detection, encoding of TDM voice, and decoding of packeted voice. The PEs program external DMA for fetching TDM voice data for compression and packeted voice for decompression. The processor also needs to bring the context from external memory before it starts processing a particular channel. The context has the states of different variables and arrays saved while processing the last frame of data on a particular channel.

The echo is produced at the interface of 4-line to 2-line hybrid at the CO office. Owing to impedance mismatch in the hybrid, the echo of far-end speech is mixed in the near-end voice. This echo needs to be cancelled before the near-end speech is compressed and packetized for transmission on an IP network. An LEC processing element is designed to implement line echo cancellation. The LEC processing also detects double talk and updates the coefficients of the adaptive filter only when line echo is present in the signal and the near end is silent. There is an extended discussion of LEC and its implementation in Chapter 11.

Each processing element in the SoC is scheduled to perform a series of tasks for each channel. These tasks for a particular channel are periodically assigned to a set of PEs. Each PE keeps checking the task list, while it is performing the currently assigned task. Finding a new task in the task list, the PE programs a channel of the DMA to bring data and context for this task into on-chip memory of the processor. Similarly, if the processor finds that it is tasked to perform an algorithm where it also needs to bring the program into its program memory (PM), the PE also requests the DMA to fetch the code for the next task in the PM of the PE. This code fetching is kept to a minimum by carefully scheduling the tasks on the PEs that already have programs of the assigned task in its PM.

1.8.3 Design Flow Migration

As explained earlier, usually the communication system requires component-level integration of different devices to implement digital baseband, RF transmitter and receiver, RF oscillator and power management functionality. The advancement in VLSI technology is now enabling the designer to integrate all these technologies on the same chip.

Although the scope of this book is limited to studying digital systems, it is very pertinent to point out that, owing to cost, performance and power dissipation considerations, the entire system including the analog part is now being integrated on a single chip. This design flow migration is show in Figure 1.15. The ASICs and microcontroller are incorporated as intellectual property (IP) cores and reconfigurable logic (RL) of the FPGAs is also placed on the same chip. Along with digital components, RF and analog components are also integrated on the same chip. For example, a mixed-signal integrated circuit for a mobile communication system usually supports ADC and DAC for on-chip analog-to-digital and digital-to-analog conversion of baseband signals, phase-locked loops (PLLs) for generating clocks for various blocks, and codec components supporting PCM and other

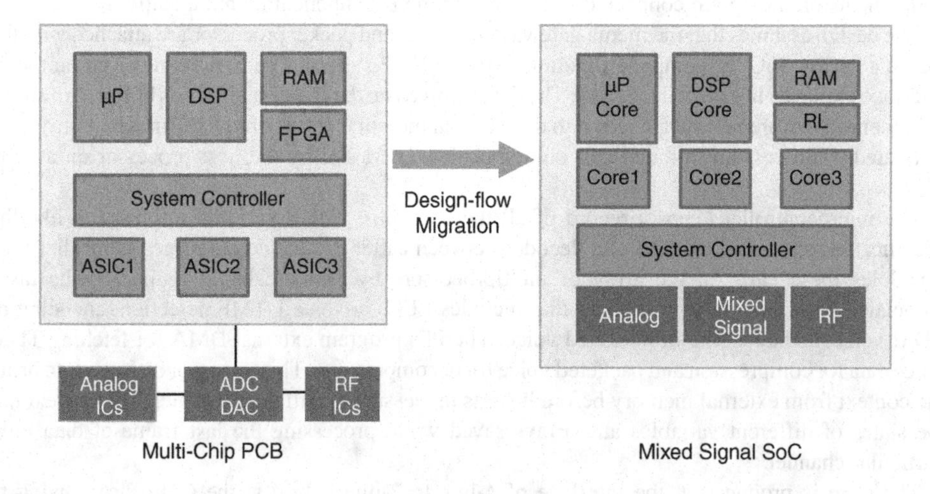

Figure 1.15 Mixed-signal SoC integrating all components on a multi-chip board on a single chip

standard formats [17]. There are even integrated circuits that incorporate RF and power management blocks on the same chip using deep sub-micron CMOS technology [18].

References

1. G. E. Moore, "Cramming more components onto integrated circuits," *Electronics*, vol. 38, no. 8, April 1965.
2. G. E. Moore, in Plenary Address at ISSCC, 2003.
3. S. Borkar, "Design perspectives on 22-nm CMOS and beyond," in *Proceedings of Design Automation Conference*, 2009, ACM/IEEE, pp. 93–94.
4. www.xilinx.com
5. www.altera.com
6. www.atmel.com
7. www.latticesemi.com
8. www.actel.com
9. www.quicklogic.com
10. www.arm.com/products/system-ip/amba/amba-open-specifications.php
11. A. Landry, M. Nekili and Y. Savaria, "A novel 2-GHz multi-layer AMBA high-speed bus interconnect matrix for SoC platforms," in *Proceedings of IEEE International Symposium on Circuits and Systems*, 2005, vol. 4, pp. 3343–3346.
12. W. Wolf, A. A. Jerraya and G. Martin, "Multiprocessor system-on-chip (MPSoC) technology," *IEEE Transactions on Computer-Aided Design of Integrated Circuits and Systems*, 2008, vol. 27, pp. 1701–1713.
13. S. V. Tota, M. R. Casu, M. R. Roch, L. Macchiarulo and M. Zamboni, "A case study for NoC-based homogeneous MPSoC architectures," *IEEE Transactions on Very Large Scale Integration Systems*, 2009, vol. 17, pp. 384–388.
14. R. C. Chang and B.-H. Lim, "Efficient IP routing Table VLSI design for multi-gigabit routers," *IEEE Transactions on Circuits and Systems I*, 2004, vol. 51, pp. 700–708.
15. S. A. Khan, M. M. Saqib and S. Ahmed, "Parallel Viterbi algorithm for a VLIW DSP," in *Proceedings of ICASSP*, 2000, vol. 6, pp. 3390–3393.
16. M. M. Rahmatullah, S. A. Khan and H. Jamal, "Carrier-class high-density VoIP media gateway using hardware/ software distributed architecture," *IEEE Transactions on Consumer Electronics*, 2007, vol. 53, pp. 1513–1520.
17. B. Baggini, "Baseband and audio mixed-signal front-end IC for GSM/EDGE applications," *IEEE Journal of Solid-State Circuits*, 2006, vol. 41, 1364–1379.
18. M. Hammes, C. Kranz and D. Seippel, "Deep submicron CMOS technology enables system-on-chip for wireless communications ICs," *IEEE Communications Magazine*, 2008, vol. 46, pp. 151–161.

2

Using a Hardware Description Language

2.1 Overview

This chapter gives a comprehensive coverage of Verilog and SystemVerilog. The focus is mostly on Verilog, which is a hardware description language (HDL).

The chapter starts with a discussion of a typical design cycle in implementing a signal processing application. The cycle starts with the requirements specification, followed by the design of an algorithm using tools like MATLAB®. To facilitate partitioning of the algorithm into hardware (HW) and software (SW), and its subsequent mapping on different platforms, algorithm design and coding techniques in MATLAB® are described. The MATLAB® code has to be structured so that the algorithm developers, SW designers and HW engineers can correlate various components and can seamlessly integrate, test and verify the design and can return to the original MATLAB® implementation if there are any discrepencies in the results.

The chapter then has a brief account of Verilog. As there are several textbooks available on Verilog [1–3], this chapter focuses primarily on design and coding guidelines and relevant rules. There is a particular emphasis on coding rules for keeping synthesis in perspective. A description of 'register transfer level' (RTL) Verilog is presented. RTL signifies the placement of registers in hardware while keeping an account of the movement of data among these registers.

SystemVerilog adds more features for modeling and verification. Although Verilog itself provides constructs to write test benches for verification, it lacks features that are required to verify a complex design. Traditionally verification engineers have resorted to other languages, such as Vera or e, or have used a 'program language interface' (PLI) to interface Verilog code with verification code written in C/C++. The use of PLI requires complex interface coding. SystemVerilog enhances some of the features of Verilog for hardware design, but more importantly adds powerful features that facilitate verification of more complex designs. Assertion, interface, package, coverage and randomization are examples of some of these features.

Digital Design of Signal Processing Systems: A Practical Approach, First Edition. Shoab Ahmed Khan.

2.2 About Verilog

2.2.1 *History*

Philip Moorby invented Verilog in1983/84. At that time he was with Gateway Design Automation. VHDL is another language used for designing hardware. It was with the advent of synthesis tools by Synopsys in1987 when Verilog and VHDL started to change the whole paradigm and spectrum of hardware design methodology. Within a few years, HDLs became the languages of choice for hardware design. In 1995, Open Verilog International (OVI) IEEE-1364 placed Verilog in the public domain to compete with VHDL [4].

It was critical for Verilog to keep pace with the high densities predicted by Moore's Law. Now the average process geometries are shrinking and billion-transistor chips are designed using 45-nm and smaller nanometer technologies.

The Verilog standard is still evolving. More and more features and syntax are being added that, on one hand, are providing higher level of abstraction, and on the other hand are helping the test designer to effectively verify an RTL design. Most of this advancement has been steered by the IEEE. Following the release of IEEE standard 1364-1995, in 1997 the IEEE formed another working group to add enhancements to the existing Verilog standard. The new standard was completed in 2001, and this variant of the language is called Verilog-2001 [5]. It provides additional support, flexibility and ease of programming to developers.

In 2001, a consortium (Accellara) of digital design companies and electronic design automation (EDA) tool vendors set up a committee to work on the next generation of extensions to Verilog. In 2003, the consortium released SystemVerilog 3.0, without ratification. In 2004, it released System-Verilog 3.1 [6] which augmented in Verilog-2001 many features that facilitated design and verification. In 2005, while still maintaining two sets of standards, the IEEE released Verilog-2005 [7] and SystemVerilog-2005 [8], the latter adding more features for modeling and verification.

2.2.2 *What is Verilog?*

Verilog is a hardware description language. Although it looks much like C, it is not a software programming language. It is very important for the Verilog programmer to understand hardware concepts. Each line of Verilog code in the design means one or more components in hardware.

Verilog is rich in constructs and functionality. Some of the constructs are specific to supporting verification and modeling and do not synthesize to infer hardware. The synthesis is performed using a *synthesis tool*, which is a compiler that translates Verilog into a gate-level design. The synthesis tool understands only a subset of Verilog, the part of Verilog called 'RTL Verilog'. All the other constructs are 'non-RTL Verilog'. These constructs are very helpful in testing, verification and simulation.

It is imperative for the designer to know at the register transfer level what is being coded in the design. The RTL signifies the placement of registers in the design and the flow of data among the registers. The complete Verilog is a combination of RTL and non-RTL constructs. A good hardware designer must have sound understanding of these differences and comprehensive command of RTL Verilog constructs. The programmer must also have a comprehension of the design to be coded in RTL Verilog.

Advancements in technology are allowing designers to realize ever more complex designs, posing real challenges for testing and verification engineers. The testing of a complex design requires creativity and ingenuity. Many features specific to verification are being added in SystemVerilog, which is a companion standard supported by most of the Verilog tool vendors. Verilog also provides a socket-level interface, known as 'programming language interface' (PLI), to be used with other programming environments such as C/C++, .NET, JavE and MATLAB®. This has extended the scope of hardware design verification from HW designers to SW engineers. Verification has become a challenging and exciting discipline. The author's personal experience in designing application-specific ICs with many million gates to complex systems on FPGAs has convinced him that the verification of many designs is even more challenging then designing the system itself. While designing, the designer needs to use genuine creativity and special abilities to make interesting explorations in the design space. The author believes that hardware design is an art, though techniques presented in this book provide excellent help; but coding a design in RTL is a well-defined and disciplined science. This chapter discusses Verilog coding with special focus on RTL and verification.

2.3 System Design Flow

Figure 2.1 shows a typical design flow of a design implementing a signal processing application. An explanation of this flow is given in Chapter 3. This section only highlights that a signal processing application is usually divided into software and hardware components. The hardware design is implemented in Verilog. The design is then mapped either on custom ASICs or FPGAs. This design needs to work with the rest of the software application. There are usually standard interfaces that enable the SW and HW components to transfer data and messages.

Architecture is designed to implement the hardware part of the application. The design contains all the requisite interfaces for communicating with the part implemented in software. The SW is mapped on a general-purpose processor (GPP) or digital signal processor (DSP). The HW design and the interfaces are coded in Verilog. This chapter focuses on RTL coding of the design and its verification for correct functionality. The verified design is synthesized on a target technology. The designer, while synthesizing the design, also constrains the synthesis tool either for timing or area. The tool generates a *gate-level netlist* of the design. The tool also reports if there are paths that are not meeting the timing constraints defined by the designer for running the HW at the desired clock speed. If that happens, the designer either makes the tool meet the timing by trying different synthesis options, or transforms the design by techniques described in this book. The modified design is re-coded in RTL and the process of synthesis is repeated until the design meets the defined timings. The gate-level netlist is then sent for a physical layout, and for custom ASICs the design is then 'taped-out' for fabrication. The field-programmable gate array tools provide an integrated environment for synthesis, layout and implementation of a bit stream to FPGA.

2.4 Logic Synthesis

The code written in RTL Verilog is synthesized for gate-level implementation. The synthesis process takes the RTL Verilog and translates it into an optimized gate-level netlist. For logic synthesis the user specifies design constraints and the target technology in the form of a standard *cell library*. The library has standard basic logic gates such as AND and OR, or macro cells like adders, multipliers, flip-flops, multiplexers and so on. The tool completely converts the design described in RTL hardware description language into a design that contains standard cells.

To optimally map the high-level description into real HW, the tool performs several steps. A typical flow of synthesis first converts the RTL description into non-optimized Boolean logic. It then performs several transformations to optimize the logic subject to user constraints. This optimization is independent of the target technology. Finally, the tool maps the optimized logic to technology-specific standard cells.

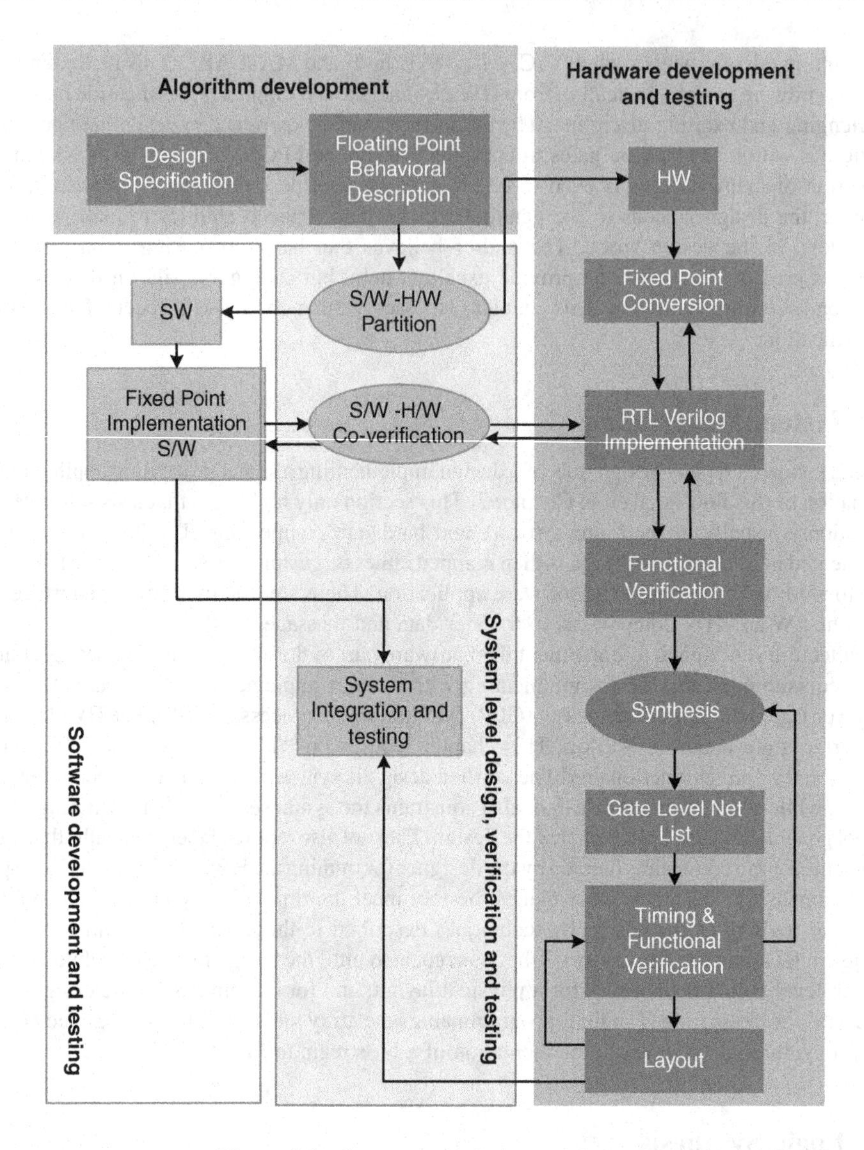

Figure 2.1 System-level design components

2.5 Using the Verilog HDL

2.5.1 Modules

A Verilog code has a top-level module, which may instantiate many other modules. The module is the basic building block in Verilog. Each module contains statements and instantiation of lower level modules. In RTL design this module, once synthesized, infers digital logic. The designer conceives a hardware design as hierarchically interconnecting lower level modules forming higher level modules. In the next level of hierarchy, the modules so constructed are further connected to design even high level modules. Thus the design has multiple layers of modules. At the top level the

```
module FA (<port declaration>);
.
.
.

.
endmodule
```

(a)

```
module FA(
    input a,
    input b,
    input c_in,
    output sum,
    output c_out);
    assign {c_out, sum}
    = a+b+c_in;
endmodule
```

(b)

Figure 2.2 Module definition (a) template (b) example

designer may also conceive the functionality of an application in terms of interconnected modules. Individual modules may also be incrementally synthesized to facilitate synthesis of large designs.

Modules are declared and instantiated like classes in C++, but module declarations cannot be nested. Instances of low-level modules are interconnected, and modules have *ports* for these interconnections.

Figure 2.2(a) shows a template of a module definition. A module starts with keyword `module` and ends with keyword `endmodule`. The ports of a module can be `input`, `output` or `in_out`. Figure 2.2(b) shows a simple example to illustrate the concept: the module `FA` has three input ports, `a`, `b` and `c_in`, and two output ports, `sum` and `c_out`.

2.5.2 Design Partitioning

2.5.2.1 Guidelines for RTL Design

A guide for effective RTL coding from the synthesis perspective is given in Figure 2.3 [9]. The partitioning of a digital design into a number of modules is important. A module should be neither too small nor too large. Where possible, the design should be partitioned in a way that module

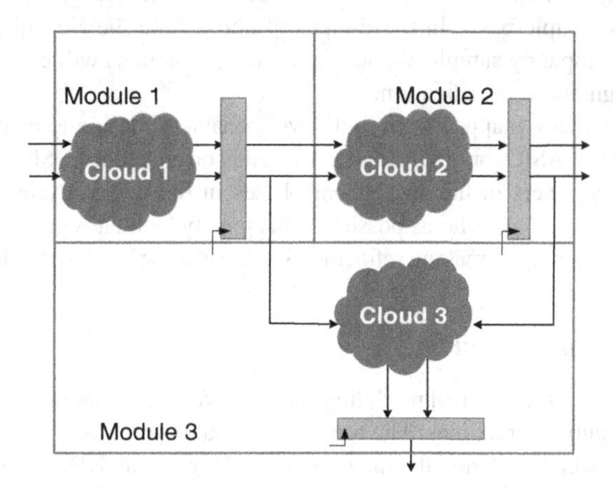

Figure 2.3 Design partitioning in number of modules with module boundaries on register outputs

boundaries reside at register outputs, as shown in the figure. This will make it easier to synthesize the top-level module or hierarchical synthesis at any level with timing constraints. The designer should also ensure that no combination cloud crosses module boundaries. This gives the synthesis tool more leverage to generate optimized logic.

2.5.2.2 Guidelines for System Level Design Flow

The design flow of a digital design process has been shown in Figure 2.1. A system designer first captures *requirements and specifications* (R&S) of the real-time system under design. Implementation of the algorithm in SW or HW needs to perform computations on the input data and produce output data at the specified rates. For example, for a multimedia processing system, the requirement can be in terms of processing P color or grayscale frames of $N \times M$ pixels per second. The processing may be compression, rendering, object recognition and so on. Similarly for digital communication applications, the requirement can be described in terms of data rates and the communication standard that modulates this data for transmission. An example is a design that supports up to a 54-Mbps OFDM-based communication system that uses a 64-QAM modulation scheme.

Algorithm development is one of the most critical steps in system design. Algorithms are developed using tools such as MATLAB®, Simulink or C/C++/C#, or in any high-level language. Functionally meeting R&S is a major consideration when the designer selects an algorithm out of several options. For example, in pattern matching the designer makes an intelligent choice out of many techniques including 'chamfer distance transform', 'artificial neural network' and 'correlation-based matching'.

Although meeting functional requirements is the major consideration, the developer must keep in mind the ultimate implementation of the algorithm on an embedded platform consisting of ASICs, FPGAs and DSPs. To ease design partitioning on a hybrid embedded platform, it is important for a system designer to define all the components of the design, clearly specifying the data flow among them. A component should implement a complete entity with defined functionality in the design. It is quite pertinent for the system designer to clearly define inputs and outputs and internal variables.

The program flow should be defined as it will happen in the actual system. For example, with hard real-time signal processing systems, the data is processed on a block by block basis. In this form, a buffer of input data is acquired and is passed to the first component in the system. The component processes this buffer of data and passes the output to the component next in execution order. Alternatively, in many applications, especially in communication receiver design, the processing is done on a sample by sample basis. In these types of application the algorithmic implementation should process data sample by sample. Adhering to these guidelines will ease the task of HW/SW partitioning, co-design and co-verification.

The design is sequentially mapped from high-level behavioral design to embedded system partitioning in HW mapped on ASICs or FPGAs and SW running on embedded DSPs or microcontrollers. It is important for the designers in the subsequent phases in the design cycle to stick to the same components and variable names as far as possible. This greatly facilitates going back and forth in the design cycle while the designer is making refinements and verifying its functionality and performance.

2.5.3 Hierarchical Design

Verilog works well with a hierarchical modeling concept. Verilog code contains a top-level module and zero or more instantiated modules. The top-level module is not instantiated anywhere. Several instantiations of a lower-level module may exist. Verilog is an HDL and, unlike with other programming languages, once synthesized each instantiation infers a physical copy of the HW with its own logic gates, registers and wires. Ports are used to interconnect instantiated modules.

```
module FA(a, b, c_in, sum,
c_out);
input a, b, c;
ouput sum, c_out;

assign {c_out, sum} = a+b+c_in;
endmodule
```

(a)

```
module FA(
input a, b, c_in,
output sum, c_out);

assign {c_out, sum} = a+b+c_in;
endmodule
```

(b)

Figure 2.4 Verilog FA module with input and output ports. (a) Port declaration in module definition and port listing follows the definition (b) Verilog-2001 support of ANSI style port listing in module definition

Figure 2.4 shows two ways of listing ports in a Verilog module. In Verilog-95, ports are defined in the module definition and then they are listed in any order. Verilog-2001 also supports ANSI-style port listing, whereby the listing is incorporated in the module definition.

Using the FA module of Figure 2.4(a), a 3-bit ripple carry adder (RCA) can be designed. Figure 2.5 shows the composition of the adder as three serially connected FAs. To realize this simple design in Verilog, the module RCA instantiates FA three times. The Verilog code of the design is given in Figure 2.6(a).

If ports are declared in Verilog-95 style, then the order of port declaration in the module definition is important but the order in which these ports are listed as `input`, `output`, `c_in` and `c_out` on the following lines has no significance. As Verilog-2001 lists the ports in the module boundary, their order should be maintained while instantiating this module in another module.

For modules having a large number of ports, this method of instantiation is error-prone and *should be avoided*. The ports of the instantiated module then should be connected by specifying names. In this style of Verilog, the ports can be connected in any order, as demonstrated in Figure 2.6(b).

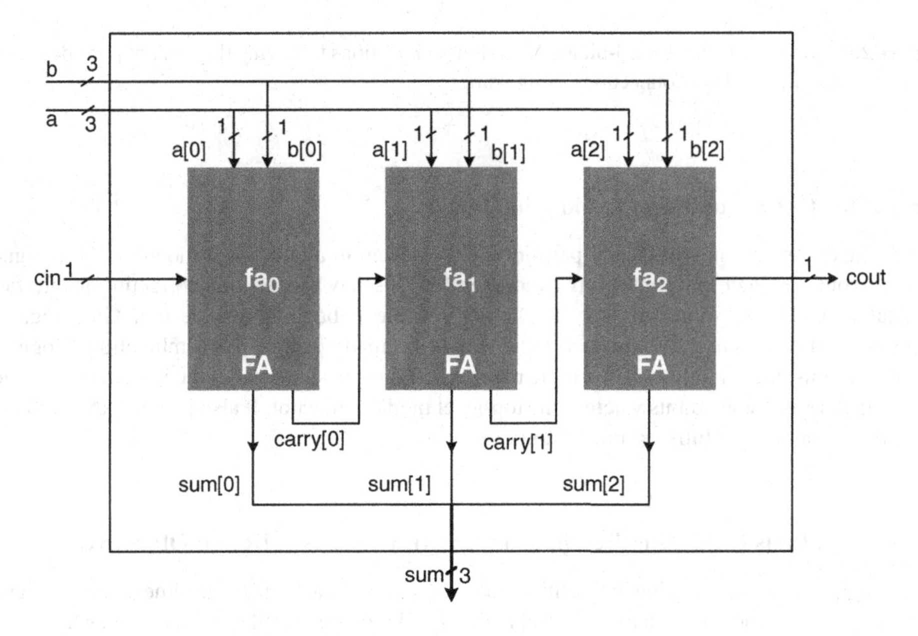

Figure 2.5 Design of a 3-bit RCA using instantiation of three FAs

```
module RCA(
    input [2:0] a, b,
    input c_in,
    output [2:0] sum,
    output c_out);

wire carry[1:0];

// module instantiation
FA fa0(a[0], b[0], c_in,
    sum[0], carry[0]);
FA fa1(a[1], b[1], carry[0],
    sum[1], carry[1]);
FA fa2(a[2], b[2], carry[1],
    sum[2], c_out);

endmodule
```

(a)

```
module RCA(
    input [2:0] a, b,
    input c_in,
    output [2:0] sum,
    output c_out);

wire carry[1:0];

// module instantiation
FA fa0(.a(a[0]),.b( b[0]),
.c_in(c_in),
.sum(sum[0]),
.c_out(carry[0]));
FA fa1(.a(a[1]), .b(b[1]),
.c_in(carry[0]),
.sum(sum[1]),
.c_out(carry[1]));
FA fa2(.a(a[2]), .b(b[2]),
.c_in(carry[1]),
.sum(sum[2]),
.c_out(c_out));

endmodule
```

(b)

Figure 2.6 Verilog module for a 3-bit RCA. (a) Port connections following the order of ports definition in the FA module. (b) Port connections using names

2.5.3.1 Synthesis Guideline: Avoid Glue Logic

While the designer is hierarchically partitioning the design in a number of modules, the designer should avoid *glue logic* that connects two modules [9]. This may happen after correcting an interface mismatch or adding some missing functionality while debugging the design. Glue logic is demonstrated in Figure 2.7. Any such logic should be made part of the combinational logic of one of the constituent modules. Glue logic may cause issues in synthesis as the individual modules may satisfy timing constraints whereas the top-level module may not. It also prevents the synthesis tool from generating a fully optimized logic.

2.5.3.2 Synthesis Guideline: Design Modules with Common Design Objectives

The designer must avoid placing time-critical and non-time-critical logic in the same module [9], as in Figure 2.8(a). The module with time-critical logic should be synthesized for best timing, whereas the module with non-time-critical logic is optimized for best area. Putting them in the same module will

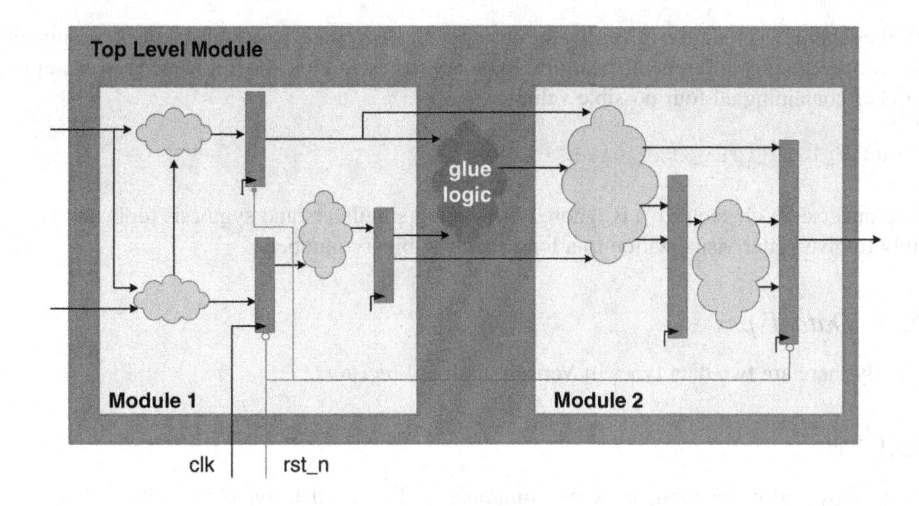

Figure 2.7 Glue logic at the top level should be *avoided*

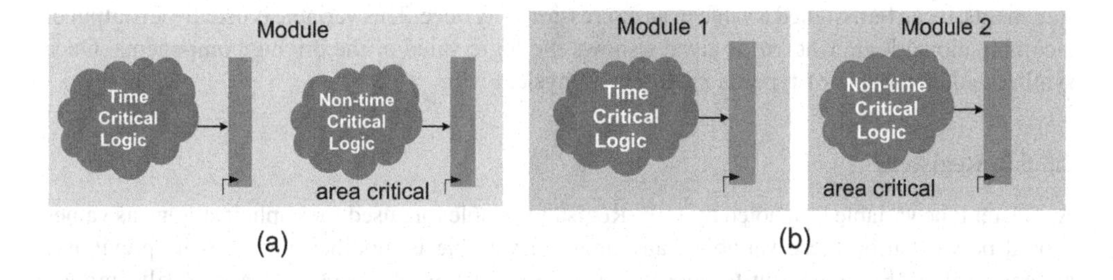

Figure 2.8 Synthesis guidelines. (a) A bad design in which time-critical and non-critical logics are placed in the same module. (b) Critical logic and non-critical logic placed in separate modules

produce a sub-optimal design. The logic should be divided and placed into two separate modules, as depicted in Figure 2.8(b).

2.5.4 Logic Values

Unlike with other programming languages, a *bit* in Verilog may contain one of four values, as given in Table 2.1. It is important to remember that there is no unknown value in a real circuit, and an 'x' in simulation signifies only that the Verilog simulator cannot determine a definite value of 0 or 1.

Table 2.1 Possible values a bit may take in Verilog

0	Zero, logic low, false, or ground
1	One, logic high, or power
x	Unknown
z	High impedance, unconnected, or tri-state port

While running a simulation in Verilog the designer may encounter a variable taking a combination of the above values at different bit locations. In binary representation, the following is an example of a number containing all four possible values:

```
20'b 0011_1010_101x_x0z0_011z
```

The underscore character (_) is ignored by Verilog simulators and synthesis tools and is used simply to give better visualization to a long string of binary numbers.

2.5.5 Data Types

Primarily there are two data types in Verilog, *nets* and *registers*.

2.5.5.1 Nets

Nets are physical connections between components. The net data types are `wire`, `tri`, `wor`, `trior`, `wand`, `triand`, `tri0`, `tri1`, `supply0`, `supply1` and `trireg`. An RTL Verilog code mostly uses the `wire` data type. A variable of type `wire` represents one or multiple bit values. Although this variable can be used multiple times on the right-hand side in different assignment statements, it can be assigned a value in an expression only once. This variable is usually an output of a combinational logic whereas it always shows the logic value of the driving components. Once synthesized, a variable of type `wire` infers a physical wire.

2.5.5.2 Registers

A register type variable is denoted by `reg`. Register variables are used for implicit storage as values should be written on these variables, and unless a variable is modified it retains its previously assigned value. It is important to note that a variable of type `reg` does not necessarily imply a hardware register; it may infer a physical wire once synthesized. Other register data types are `integer`, `time` and `real`.

A Verilog simulator assigns `'x'` as the default value to all uninitialized variables of type `reg`. If one observes a variable taking a value of `'x'` in simulation, it usually traces back to an uninitialized variable of type `reg`.

2.5.6 Variable Declaration

In almost all software programming languages, only variables with fixed sizes can be declared. For example, in C/C++ a variable can be of type `char`, `short` or `int`. Unlike these languages, a Verilog variable can take any width. The variable can be signed or unsigned. The following syntax is used for declaring a signed wire:

```
wire signed [<range>] <net_name> <net_name>*;
```

Here * implies optional and the range is specified as `[Most Significant bit (MSb) : Least Significant bit (LSb)]`. It is read as MSb down to LSb. If not specified, the default value of the range is taken as one bit width. A similar syntax is used for declaring a signed variable of type `reg`:

```
reg signed [<range>] <reg_name> <reg_name>*;
```

A memory is declared as a two-dimensional variable of type `reg`, the range specifies the width of the memory, and start and end addresses define its depth. The following is the syntax for memory declaration in Verilog:

```
reg [<range>] <memory_name> [<start_addr> : <end_addr>];
```

The Verilog code in the following example declares two 1-bit wide signed variables of type `reg` `(x1 and x2)`, two 1-bit unsigned variables of type wire (`y1` and `y2`), an 8-bit variable of type reg (`temp`), and an 8-bit wide and 1-Kbyte deep memory `ram-local`. Note that a double forward slanted bar is used in Verilog for comments:

```
reg signed x1, x2; // 1-bit signed variables of type reg x1 and x2
wire y1, y2; // 1-bit variables of type wire, y1 and y2
reg [7:0] temp; // 8-bit reg temp
reg [7:0] ram_local [0:1023]; //8-bit wide and 1-Kbyte deep memory
```

A variable of type `reg` can also be initialized at declaration as shown here:

```
reg x1 = 1'b0; // 1-bit reg variable x1 initialize to 0 at declaration
```

2.5.7 Constants

Like variables, a constant in Verilog can be of any size and it can be written in decimal, binary, octal or hexadecimal format. Decimal is the default format. As the constant can be of any size, its size is usually written with 'd', 'b', 'o' or 'h' to specify decimal, binary, octal or hexadecimal, respectively. For example, the number 13 can be written in different formats as shown in Table 2.2.

2.6 Four Levels of Abstraction

As noted earlier, Verilog is a hardware description language. The HW can be described at several levels of detail. To capture this detail, Verilog provides the designer with the following four levels of abstraction:

- switch level
- gate level
- dataflow level
- behavioral or algorithmic level.

A design in Verilog can be coded in a mix of levels, moving from the lowest abstraction of switch level to the highly abstract model of behavioral level. The practice is to use higher levels of

Table 2.2 Formats to represent constants

Decimal	13 or 4′d13
Binary	4′b1101
Octal	4′o15
Hexadecimal	4′hd

abstraction like dataflow and behavioral while designing logic in Verilog. A synthesis tool then translates the design coded using higher levels of abstraction to gate-level details.

2.6.1 Switch Level

The lowest level of abstraction is switch- or transistor-level modeling. This level is used to construct gates, though its use is becoming rare as CAD tools provide a better way of designing and modeling gates at the transistor level. A digital design in Verilog is coded at RTL and switch-level modeling is not used in RTL, so this level is not covered in this chapter. Interested readers can get relevant information on this topic from the IEEE standard document on Verilog [7].

2.6.2 Gate Level or Structural Modeling

Gate-level modeling is at a low level of abstraction and not used for coding design at RTL. Our interest in this level arises from the fact that the synthesis tools compile high-level code and generate code at gate level. This code can then be simulated using the stimulus earlier developed for the RTL-level code. The simulation at gate level is very slow compared with the original RTL-level code. A selective run of the code for a few test cases may be performed to derive confidence in the synthesized code. The synthesis tools have matured over the years and so are the coding guidelines. Gate-level simulation is also becoming rare.

Gate-level simulation can be performed with timing information in a *standard delay file* (SDF). The SDF is generated for pre-layout or post-layout simulation by, respectively, synthesis or place and route tools. The designer can run simulation using the gate-level netlist and the SDF. There is a separate timing calculator in all synthesis tools. The calculator provides timing violations if there are any. For synchronous designs the use of gate-level simulation for post-synthesis or layout timing verification is usually not required.

The code at gate level is built from *Verilog primitives*. These primitives are built-in gate-level models of basic functions, including `nand`, `nor`, `and`, `or`, `xor`, `buf` and `not`. Modeling at this level requires describing the circuit using logic gates. This description looks much like an implementation of a circuit in a basic logic design course. Delays can also be modeled at this level. A typical gate instantiation is

```
and #delay instance-name (out, in1, in2, in3)
```

The first port in the primitive, `out`, is always a 1-bit output followed by several 1-bit inputs (here `in1`, `in2` and `in3`); the `and` is a Verilog primitive that models functionality of an AND gate, while `#delay` specifies the delay from input to output of this gate.

Example 2.1

This example designs a 2:1 multiplexer at gate level using Verilog primitives. The design is given in Figure 2.9(a). The `sel` wire selects one of the two inputs `in1` and `in2`. If `sel` = 0, `in1` is selected, otherwise `in2` is selected. The implementation requires `and`, `not` and `or` gates, which are available as Verilog primitives. Figure 2.9(b) lists the Verilog code for the gate-level implementation of the design. Note `#5`, which models delay from input to output of the AND gate. This delay in Verilog is a unit-less constant. It gives good visualization once the

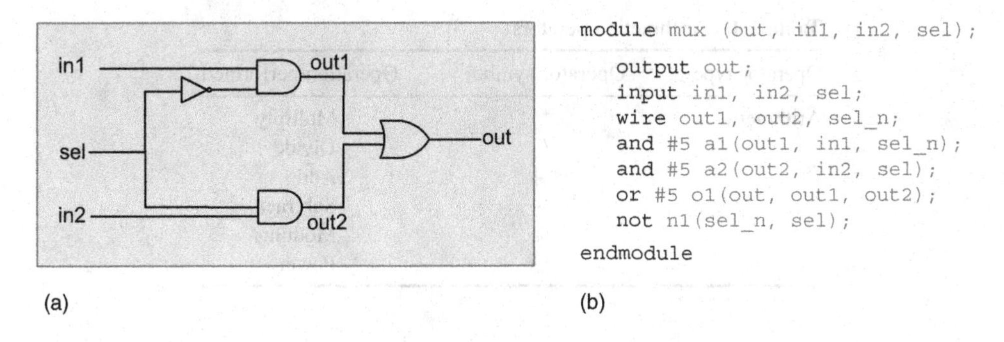

Figure 2.9 (a) A gate-level design for a 2 : 1 multiplexer. (b) Gate-level implementation of a 2 : 1 multiplexer using Verilog primitives

waveforms of input and output are plotted in a Verilog simulator. These delays are ignored by synthesis tools.

2.6.3 Dataflow Level

This level of abstraction is higher than the gate level. Expressions, operands and operators characterize this level. Most of the operators used in dataflow modeling are common to software programmers, but there are a few others that are specific to HW design. Operators that are used in expressions for dataflow modeling are given in Table 2.3. At this level every expression starts with the keyword `assign`. Here is a simple example where two variables `a` and `b` are added to produce `c`:

```
assign c = a + b;
```

The value on wire `c` is continuously driven by the result of the arithmetic operation. This assignment statement is also called 'continuous assignment'. In this statement the right-hand side must be a variable of type `wire`, whereas the operands on the left-hand side may be of type `wire` or `reg`.

Table 2.3 Operators for dataflow modeling

Type	Operators								
Arithmetic	+	−	=	*	/	%	**		
Binary bitwise	~	&	~&	|	~|	^	~^	^~	
Unary reduction	&	~&	|	~|	^	~^	+	-	
Logical	!	&&	||	==	===	!=	!==	==	===
Relational	<	>	<=	>=					
Logical shift	≫	≪							
Arithmetic shift	⋙	⋘							
Conditional	?:								
Concatenation	{ }								
Replication	{{ }}								

Table 2.4 Arithmetic operators

Operator type	Operator symbol	Operation performed
Arithmetic	*	Multiply
	/	Divide
	+	Add
	−	Subtract
	%	Modulus
	**	Power

2.6.3.1 Arithmetic Operators

The arithmetic operators are given in Table 2.4. It is important to understand the significance of using these operators in RTL Verilog code as each results in a hardware block that performs the operation specified by the operator. The designer should also understand the type of HW the synthesis tool generates once the code containing these operators is synthesized. In many circumstances, the programmer can specify the HW block from an already developed library to synthesis tools. Many FPGAs have build-in arithmetic units. For example, the Xilinx family of devices have embedded blocks for multipliers and adders. While writing RTL Verilog for targeting a particular device, these blocks can be instantiated in the design. The following code shows instantiation of two built-in 18×18 multipliers in the Virtex-II family of FPGAs:

```
// Xilinx 18x18 built-in multipliers are instantiated
MULT18X18 m1 ( out1, in1, in2);
MULT18X18 m2 ( out2, in3, in4);
```

The library from Xilinx also provides a model for `MULT18x18` for simulation. Adders and multipliers are extensively used in signal processing, and use of a divider is preferably avoided. Verilog supports both signed and unsigned operations. For signed operation the respective operands are declared as `signed wire` or `reg`.

The size of the output depends on the size of the input operands and the type of operation. The multiplication operator results in an output equal to the sum of sizes of both the operands. For addition and subtraction the size of the output is the size of the wider operand and a carry or borrow bit.

2.6.3.2 Conditional Operators

The conditional operator of Table 2.5 infers a multiplexer. A statement with the conditional operator is:

```
out = sel ? a : b;
```

Table 2.5 Conditional operator

Operator type	Operator symbol	Operation performed
Conditional	?:	Conditional

This statement is equivalent to the following decision logic:

```
if(sel)
    out = a;
else
    out = b;
```

The conditional operator can also be used to infer higher order multiplexers. The code here infers a 4:1 multiplexer:

```
out = sel[1] ? ( sel[0] ? in3 : in2 ) : ( sel[0] ? in1 : in0 );
```

2.6.3.3 Concatenation and Replication Operators

Most of the operators in Verilog are the same as in other programming languages, but Verilog provides a few that are specific to HW designs. Examples are concatenation and replication operators, which are shown in Table 2.6.

Example 2.2

Using a concatenation operator, signals or parts of signals can be concatenated to make a new signal. This is a very convenient and useful operator for the hardware designer, who can bring wires from different parts of the design and tag them with a more appropriate name.
In the example in Figure 2.10, signals `a[3:0]`, `b[2:0]`, `3'b111` and `c[2:0]` are concatenated together in the specified order to make a 13-bit signal, `p`.

Table 2.6 Concatenation and replication operators

Operator type	Operator symbol	Operation performed
Concatenation	{ }	Concatenation
Replication	{{ }}	Replication

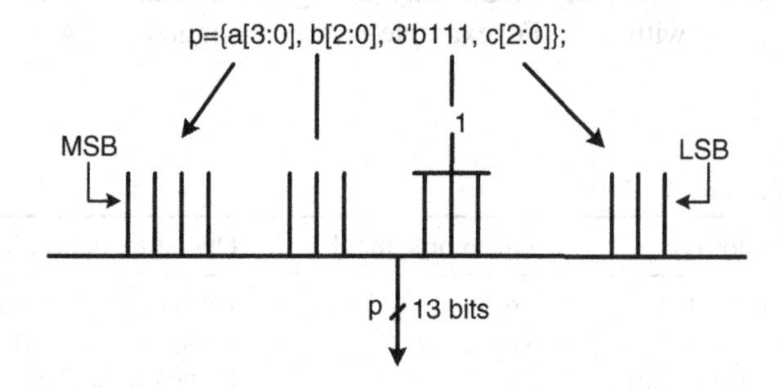

Figure 2.10 Example of a concatenation operator

Table 2.7 Logical operators

Operator type	Operator symbol	Operation performed
Logical	!	Logical negation
	||	Logical OR
	&&	Logical AND

Example 2.3

A replication operator simply replicates a signal multiple times. To illustrate the use of this, let

```
A = 2'b01;
B = {4{A}} // the replication operator
```

The operator replicates `A` four times and assigns the replicated value to `B`.
Thus `B` = `8' b 01010101`.

2.6.3.4 Logical Operators

These operators are common to all programming languages (Table 2.7). They operate on logical operands and result in a logical TRUE or FALSE. The logical negation operator (!) checks whether the operand is FALSE, then it results in logical TRUE; and vice versa. Similarly, if one or both of the operands is TRUE, the logical OR operator (||) results in TRUE; and FALSE otherwise. The logical AND operator is TRUE if both the logical operands are TRUE, and it is FALSE otherwise. When one of the operands is an `x`, then the result of the logical operator is also `x`.

The bitwise negation operator (~) is sometimes mistakenly used as a logical negation operator. In the case of a multi-bit operand, this may result in an incorrect answer.

2.6.3.5 Logic and Arithmetic Shift Operators

Shift operators are listed in Table 2.8. Verilog can perform logical and arithmetic shift operations. The logical shift is performed on `reg` and `wire`.

Example 2.4

Right shifting of a signal by a constant n drops n least significant bits of the number and appends the n most significant bits with zeros. For example, shift an unsigned `reg A = 6'b101111` by 2:

```
B = A >> 2;
```

Table 2.8 Shift operators

Operator type	Operator symbol	Operation performed
Logic shift	>>	Unsigned right shift
	<<	Unsigned left shift
Arithmetic shift	>>>	Signed right shift
	<<<	Signed left shift

This drops two LSBs and appends two zeros at the MSB position, thus:

```
B = 6'b001011
```

Example 2.5

Arithmetic shift right of an operand by *n* drops the *n* LSBs of the operand and fills the *n* MSBs with the sign bit of the operand. For example, shift right a wire A= 6'b101111 by 2:

```
B = A >>> 2;
```

This operation will drop two LSBs and appends the sign bit to two MSB locations. Thus B = 6'b111011.
Arithmetic and logic shift *left* by *n* performs the same operation, as both drop *n* MSBs of the operand without any consideration of the sign bit.

2.6.3.6 Relational Operators

Relational operators are also very common to software programmers and are used to compare two numbers (Table 2.9). These operators operate on two operands as shown below:

```
result = operand1 OP operand2;
```

This statement results in a logical value of TRUE or FALSE. If one of the bits of any of the operands is an x, then the operation results in x.

2.6.3.7 Reduction Operators

Reduction operators are also specific to HW design (Table 2.10). The operator performs the prescribed operation on all the bits of the operand and generates a 1-bit output.

Table 2.9 Relational operator

Relational operator	Operator symbol	Operation performed
	>	Greater than
	<	Less than
	>=	Greater than or equal to
	<=	Less than or equal to

Table 2.10 Reduction operators

Operator type	Operator symbol	Operation performed
Reduction	&	Reduction AND
	~&	Reduction NAND
	|	Reduction OR
	~|	Reduction NOR
	^	Reduction XOR
	^~ or ~^	Reduction XNOR

Example 2.6

Apply the & reduction operator to a 4-bit number A=4'b1011:

```
assign out = &A;
```

This operation is equivalent to performing a bitwise & operation on all the bits of A:

```
out = A[0] & A[1] & A[2] & A[3];
```

2.6.3.8 Bitwise Arithmetic Operators

Bitwise arithmetic operators are also common to software programmers. These operators perform bitwise operations on all the corresponding bits of the two operands. Table 2.11 gives all the bitwise operators in Verilog.

Example 2.7

This example performs bitwise | operation on two 4-bit numbers A = 4'b1011 and B=4'b0011. The Verilog expression computes a 4-bit C:

```
assign C = A | B;
```

performs the OR operation on corresponding bits of A and B and the operation is equivalent to:

```
C[0] = A[0] | B[0]
C[1] = A[1] | B[1]
C[2] = A[2] | B[2]
C[3] = A[3] | B[3]
```

2.6.3.9 Equality Operators

Equality operators are common to software programmers, but Verilog offers two flavors that are specific to HW design: case equality (===) and case inequality (!==). A simple equality operator (==) checks whether all the bits of the two operands are the same. If any operand has an x or z as one of its bits, the answer to the equality will be x. The === operator is different from == as it also matches x with x and z with z. The result of this operator is always a 0 or a 1. There is a similar difference between != and !==. The following example differentiates the two operators.

Table 2.11 Bitwise arithmetic operators

Operator type	Operator symbol	Operation performed
Bitwise	~	Bitwise negation
	&	Bitwise AND
	~&	Bitwise NAND
	|	Bitwise OR
	~|	Bitwise NOR
	^	Bitwise XOR
	^~ or ~^	Bitwise XNOR

Table 2.12 Equality operators

Operator type	Operator symbol	Operation performed
Equality	==	Equality
	!=	Inequality
	===	Case equality
	!==	Case Inequality

Example 2.8

While comparing `A=4'b101x` and `B=4'b101x` using == and ===, `out = (A==B)` will be `x` and `out = (A===B)` will be `1` (Table 2.12).

2.6.4 Behavioral Level

The behavioral level is the highest level of abstraction in Verilog. This level provides high-level language constructs like `for`, `while`, `repeat`, `if-else` and `case`. Designers with a software programming background already know these constructs.

Although the constructs are handy and very powerful, the programmer must know that each construct in RTL Verilog infers hardware. High-level constructs are very tempting to use, but the HW consequence of their inclusion must be well understood. For example, `for loop` to a software programmer suggests a construct that simply repeats a block of code a number of times, but if used in RTL Verilog the code infers multiple copies of the logic in the loop. There are behavioral-level synthesis tools that take a complete behavioral model and synthesize it, but the logic generated using these tools is usually not optimal. The tools are not used in those designs where area, power and speed are important considerations.

Verilog restricts all the behavioral statements to be enclosed in a *procedural block*. In a procedural block all variables on the left-hand side of the statements must be declared as of type `reg`, whereas operands on the right-hand side in expressions may be of type `reg` or `wire`.

There are two types of procedural block, `always` and `initial`.

2.6.4.1 `Always` and `Initial` Procedural Blocks

A procedural block contains one or multiple statements per block. An assignment statement used in a procedural block is called a procedural assignment. The `initial` block executes only once, starting at $t=0$ simulation time, whereas an `always` block executes continuously at $t=0$ and repeatedly thereafter.

The characteristics of an `initial` block are as follows.

- This block starts with the `initial` keyword. If multiple statements are used in the block, they are enclosed within `begin` and `end` constructs, as shown in Figure 2.11.
- This block is non-synthesizable and non-RTL. This block is used only in a stimulus.
- There are usually more than one `initial` blocks in the stimulus. All `initial` blocks execute concurrently in arbitrary order, starting at simulation time 0.
- The simulator kernel executes the `initial` block until the execution comes to a `#delay` operator. Then the execution is suspended and the simulator kernel places the execution of this block in the event list for `delay-time` units in the future.
- After completing `delay-time` units, the execution is resumed where it was left off.

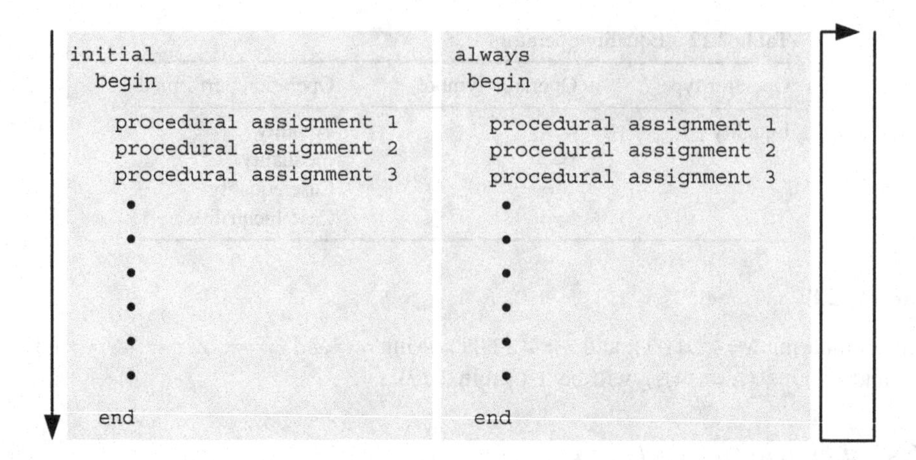

Figure 2.11 `Initial` and `always` blocks

An `always` block is synthesizable provided it adheres to coding guidelines for synthesis. From the perspective of its execution in a simulator, an `always` block behaves like an `initial` block except that, once it ends, it starts repeating itself.

2.6.4.2 Blocking and Non-blocking Procedural Assignments

All assignments in a procedural block are called procedural assignments. These assignments are of two types, blocking and non-blocking. A blocking assignment is a regular assignment inside a procedural block. These assignments are called blocking because each assignment blocks the execution of the subsequent assignments in the sequence. In RTL Verilog code, these assignments are used to model combinational logic. For the RTL code to infer combinational logic, the blocking procedural assignments are placed in an `always` procedural block.

There are several ways of writing what is called the *sensitivity list* in an `always` block. The sensitivity list helps the simulator in effective management of simulation. It executes an `always` block only if one of the variables in the sensitivity list changes. The classical method of sensitivity listing is to write all the inputs in the block in a bracket, where each input is separated by an 'or' tag. Verilog-2001 supports comma-separated sensitivity lists. It also supports just writing a '*' for the sensitivity list. The simulator computes the list by analyzing the block by itself.

The code in Figure 2.12 illustrates the use of a procedural block to infer combinational logic in RTL Verilog code. The `always` block contains two blocking procedural assignments. The sensitivity list includes the two inputs `x` and `y`, which are used in the procedural block. A list of inputs `x` and `y` to these assignments are placed with the `always` statement. This list is the sensitivity list. This procedural block once synthesized will infer combinational logic. The three methods of writing a sensitivity list are shown in Figure 2.12.

It should also be noted that, as the left-hand side of a procedural assignment must be of type `reg`, so `sum` and `carry` are defined as variables of type `reg`.

In contrast to blocking procedural assignments, non-blocking procedural assignments do not block other statements in the block and these statements execute in parallel. The simulator executes this functionality by assigning the output of these statements to temporary variables, and at the end of execution of the block these temporary variables are assigned to actual variables.

<pre>reg sum, carry;
 	
always @ (x or y)
 	
begin
 	
  sum = x^y;
 	
  carry = x&y;
 	
end</pre>
	<pre>reg sum, carry;
	
	always @ (x, y)
	
	begin
	
	  sum = x^y;
	
	  carry = x&y;
	
	end</pre>
		<pre>reg sum, carry;
		
		always @ (*)
		
		begin
		
		  sum = x^y;
		
		  carry = x&y;
		
		end</pre>

(a)	(b)	(c)

Figure 2.12 Blocking procedural assignment with three methods of writing the sensitivity list. (a) Verilog-95 style. (b) Verilog-2001 support of comma-separated sensitivity list. (c) Verilog-2001 style that only writes * in the list

The left-hand side of the non-blocking assignment must be of type `reg`. The non-blocking procedural assignments are primarily used to infer synchronous logic. Shown below is the use of a non-blocking procedural assignment that infers two registers, `sum_reg` and `carry_reg`:

```
reg sum_reg, carry_reg;
always @ (posedge clk)
begin
   sumreg <= x^y;
   carry_reg <= x&y;
end
```

Both of the non-blocking assignments are simultaneously executed by the simulator. The use of a blocking assignment in generating synchronous logic is further explained in the next section.

2.6.4.3 Multiple Procedural Assignments

From the simulation perspective, all procedural blocks simultaneously start execution at $t = 0$. The simulator, however, schedules their execution in an arbitrary order. Now a variable of data type `reg` can be assigned values at multiple locations in a module. Any such multiple assignments to a variable of type `reg` must always be placed in the same procedural block. If a variable is assigned values in different procedural blocks and the values are assigned on it at the same time, the value assigned to the variable depends on the order in which the `simulator` executes these blocks. This may cause errors in simulation and pre- and post-synthesis results may not match.

2.6.4.4 Time Control # and @

Verilog provides different timing constructs for modeling timing and delays in the design. The Verilog simulator works on unit-less timing for simulating logic. The simulated time at any instance in the design can be accessed using built-in variable `$time`. It is a unit-less integer. The timing at any instance in simulation can be displayed by using the `$display` as shown here:

```
$display ($time, "a=%d", a);
```

The programmer can insert delays in the code by placing #<number>. On encountering this statement, the simulation halts the execution of the statement until <number> of time units have passed. The control is released from that statement or block so that other processes can execute. Synthesis tools ignore this statement in RTL Verilog code, and the statements are mostly used in test code or to model propagation delay in combinational logic.

Another important timing control directive in Verilog is `@`. This directive models event-based control. It halts the execution of the statement until the event happens. This timing control construct is used to model combinational logic as shown here:

```
always @ (a or b)
   c = a^b;
```

Here the execution of the assignment statement `always @ (a or b)` will happen only if `a` or `b` changes value. These signals are listed in the sensitivity list of the block.

The time control `@` is used also to model synchronous logic. The code here models a positive-edge trigger flip-flop:

```
always @ (posedge clk)
    dout <= din;
```

It is important to note that, while coding at RTL, the non-blocking procedural assignment should be used only to model synchronous logic and the blocking procedural assignment to model combinational logic.

2.6.4.5 RTL Coding Guideline: Avoid Combinational Feedback

The designer must avoid any combinational feedback in RTL coding. Figure 2.13(a) demonstrates combinational feedback, as does the following code:

```
reg [15:0] acc;
always@(acc)
    acc = acc + 1;
```

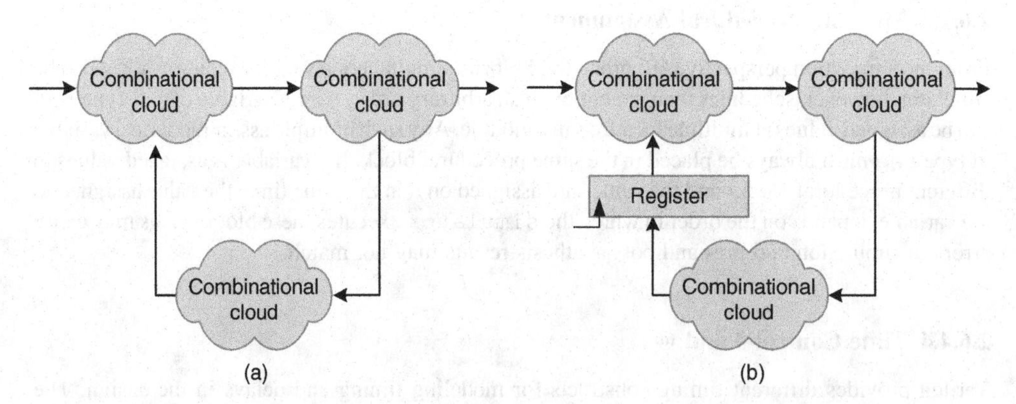

Figure 2.13 Combinational feedback must be voided in RTL Verilog code. (a) A logic with combinational feedback. (b) The register is placed in the feedback path of a combinational logic

Any such code does not make sense in design and simulation. The simulator will never come out of this block as the change in acc will bring it back into the procedural block. If logic demands any such functionality, a *register* should be used to break the combinational logic, as shown in Figure 2.13 (b) where a register is placed in the combinational feedback paths.

2.6.4.6 The Feedback Register

In many digital designs, a number of registers reside in feedback paths of combinational logic. Figure 2.14 shows digital logic of an accumulator with a feedback register. In designs with feedback registers there must be a reset, because a pre-stored value in a feedback register will corrupt all future computations. From the simulation perspective, Verilog assumes a logic value x in all register variables. If this value is not cleared, this x feeds back to a combinational cloud in the first cycle and may produce a logic value x at the output. Then, in all subsequent clock cycles the simulator – irrespective of the input data to the combinational cloud – may compute x and keep showing x at the output of the simulation.

A register can be initialized using a synchronous or an asynchronous reset. In both cases, an active-low or active-high reset signal can be used. An asynchronous active-low reset is usually used in designs because it is available in most technology libraries. Below are examples of Verilog code to infer registers with asynchronous active-low and active-high resets for the accumulator example;

```
// Register with asynchronous active-low reset
always @ (posedge clk or negedge rst_n)
begin
  if(!rst_n)
    acc_reg <= 16'b0;
  else
    acc_reg <= data+acc_reg;
end
// Register with asynchronous active-high reset
always @ (posedge clk or posedge rst)
begin
  if(rst)
    acc_reg <= 16'b0;
  else
    acc_reg <= data+acc_reg;
end
```

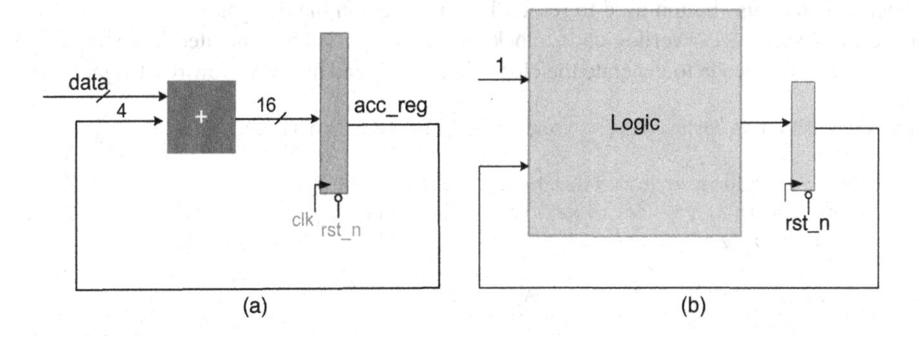

Figure 2.14 Accumulator logic with a feedback register

The negedge rst_n and posedge rst directives in the always statement and subsequently if(!rst_n) and if(rst) statements in each block, respectively, are used to infer these resets. To infer registers with synchronous reset, either active-low or active-high, the always statement in each block contains only the posedge clk directive. Given below are examples of Verilog code to infer registers with synchronous active-low and active-high resets:

```
// Register with asynchronous active-low reset
always @ (posedge clk)
begin
  if(!rst_n)
    acc_reg <= 16'b0;
  else
    acc_reg <= data+acc_reg;
end
// Register with asynchronous active-high reset
always @ (posedge clk)
begin
  if(rst)
    acc_reg <= 16'b0;
  else
    acc_reg <= data+acc_reg;
end
```

2.6.4.7 Generating Clock and Reset

The clock and reset that go to every flip-flop in the design are not generated inside the design. The clock usually comes from a crystal oscillator outside the chip or FPGA. In the place and route phase of the design cycle, clocks are specially treated and are routed using *clock trees*. These are specially designed routes that take the clocks to registers and flip-flops while causing minimum skews to these special signals. In FPGAs, the external clocks must be tied to one of the dedicated pins that can drive large nets. This is achieved by locking the clock signal with one of these pins. For Xilinx it is done in a 'user constraint file' (UCF). This file lists user constraints for placement, mapping, timing and bit generation [11].

Similarly, the reset usually comes from outside the design and is tied to a pin that is physically connected with a push button used to reset all the registers in the design.

To test and verify RTL Verilog code, clock and reset signals are generated in a stimulus. The following is Verilog code to generate the clock signal clk and an active-low reset signal rst_n:

```
initial // All the initializations should be in the initial block
begin
    clk = 0; // clock signal must be initialized to 0
    # 5 rst_n = 0; // pull active low reset signal to low
    # 2 rst_n=1; // pull the signal back to high
end
always // generate clock in an always block
     #10 clk=(~clk);
```

These blocks are incorporated in the stimulus module. From the stimulus these signals are input to the top-level module.

2.6.4.8 Case Statement

Like C and other high-level programming languages, Verilog supports switch and case statements for multi-way decision support. This statement compares a value with number of possible outcomes and then branches to its match.

The syntax in Verilog is different from the format used in C/C + +. The following code shows the use of the case statement to infer a 4:1 multiplexer:

```
module mux4_1(in1, in2, in3, in4, sel, out);
input [1:0] sel;
input [15:0] in1, in2, in3, in3;
output [15:0] out;
reg [15:0] out;
always @(*)
begin
    case (sel)
        2'b00: out = in1;
        2'b01: out = in2;
        2'b10: out = in3;
        2'b11: out = in4;
        default: out = 16'bx;
    endcase
end
endmodule
```

The select signal sel is evaluated, and the control branches to the statement that matches with this value. When the sel value does not match with any listed value, the default statement is executed. Two variants of case statements, casez and casex, are used to make comparison with the 'don't care' situation. The statement casez takes z as don't care, whereas casex takes z and x as don't care. These don't care bits can be used to match with any value. This provision is very handy while implementing logic where only a few of the bits are used to take a branch decision:

```
always @(op_code)
begin
    casez (op_code)
        4'b1???: alu_inst(op_code);
        4'b01??: mem_rd(op_code);
        4'b001?: mem_wr(op_code);
    endcase
end
```

This block compares only the bits that are specified and switches to one of the appropriate tasks. For example, if the MSB of the op_code is 1, the casez statement selects the first statement and the alu_inst task is called.

2.6.4.9 Conditional Statements

Verilog supports the use of conditional statements in behavioral modeling. The `if-else` statement evaluates the expression. If the expression is TRUE it branches to execute the statements in the `if` block, otherwise the expression may be FALSE, `0`, `x` or `z`, so the statements in `else` block are executed. The example below gives a simple use. If the `brach_flag` is non-zero, the PC is equated to `brach_addr`; otherwise if the `brach_flag` is `0`, `x` or `z`, the PC is assigned the value of `next_addr`.

```
if (brach_flag)
     PC = brach_addr
else
     PC = next_addr;
```

The `if-(else if)-else` conditional statement provides multi-way decision support. The expressions in `if-(else if)-else` statements are successively evaluated and, if any of the expressions is TRUE, the statements in that block are executed and the control exits from the conditional block. The code below demonstrates the working of multi-way branching using the `if-(else if)-else` statement:

```
always @(op_code)
begin
     if (op_code == 2'b00)
          cntr_sgn = 4'b1011;
     else if (op_code == 2'b01;
          cntr_sgn = 4'b1110;
else
          cntr_sgn = 4'b0000;
end
```

The code successively evaluates the `op_code` in the order specified in `if`, `else if` and `else` statements and, depending on the value of `op_code`, it appropriately assigns value to `cntr_sgn`.

2.6.4.10 RTL Coding Guideline: Avoid Latches in the Design

A designer must avoid any RTL syntax that infers latches in the synthesized netlist. A latch is a storage device that stores a value without the use of a clock. Latches are usually technology-specific and must be avoided in synchronous designs. To avoid latches the programmer must adhere to coding guidelines.

For decision statements, the programmer should either fully specify assignments or must use a default assignment. A variable in an `if` statement in a procedural block for combinational logic infers a latch if it is not assigned a value under all conditions. This is depicted in the following code:

```
input [1:0] sel;
reg [1:0] out_a, out_b;
always @ (*)
     begin
          if (sel == 2'b00)
          begin
               out_a = 2'b01;
```

```
                    out_b = 2'b10;
                end
                else
                    out_a = 2'b01;
end
```

As out_b is not assigned any value under else, the synthesis tool will infer a latch for storing the previous value of out_b in cases where an else condition is TRUE. To avoid this latch the programmer should either assign some value to out_b in the else block, or assign default values to *all* variables outside a conditional block. This is shown in the following code:

```
input [1:0] sel;
reg [1:0] out_a, out_b;

always @ (*)
     begin
          out_a = 2'b00;
          out_b = 2'b00;
          if (sel=2'b00)
          begin
               out_a = 2'b01;
               out_b = 2'b10;
          end
          else
              out_a = 2'b01;
end
```

The syntheses tool will also infer a latch when conditional code in the combinational block misses any one or more conditions. This scenario is depicted in the following code:.

```
input [1:0] sel;
reg [1:0] out_a, out_b;

always @*
     begin
          out_a = 2'b00;
          out_b = 2'b00;
          if (sel==2'b00)
          begin
              out_a = 2'b01;
              out_b = 2'b10;
end
else if (sel == 2'b01)
     out_a = 2'b01;
end
```

This code misses some possible values of sel and checks for only two listed values, 2'b01 and 2'b00. The synthesis tool will infer a latch for out_a and out_b to retain previous values in case any one of the conditions not covered occurs. This stype of coding must be avoided. In an if, else if, else block, the block must come with an else statement; and in scenarios where the case statement is used, either all conditions must be specified, and for each condition values should be assigned to all variables, or a `default` condition must be used and all variables must be assigned default values outside the conditional block. The correct way of coding is depicted here:

```
always @*
    begin
        out_a = 2'b00;
        out_b = 2'b00;
        if (sel==2'b00)
        begin
            out_a = 2'b01;
            out_b = 2'b10;
        end
        else if (sel == 2'b01)
             out_a = 2'b01;
        else
           out_a = 2'b00;
end
```

Here is the code showing the correct use of `case` statements:

```
always @*
     begin
         out_a = 2'b00;
         out_b = 2'b00;
         case (sel)
         2'b00:
         begin
             out_a = 2'b01;
             out_b = 2'b10;
         end
         2'b01:
            out_a = 2'b01;
         default:
            out_a = 2'b00;
     endcase
end
```

2.6.4.11 Loop Statements

Loop statements are used to execute a block of statements multiple times. Four types of loop statement are supported in Verilog: `forever`, `repeat`, `while` and `for`. The statement `forever` continuously executes its block of statements. The remaining three statements are commonly used to execute a block of statements *a fixed number of times*. Their equivalence is shown below. For `repeat`:

```
i=0;
repeat (5)
begin
    $display("i=%d\n", i);
    i=i+1;
end
```

For `while`:

```
i=0;
while (i<5)
begin
	$display("i=%d\n", i);
	i=i+1;
end
```

For `for`:

```
for (i=0; i<5; i=i+1)
begin
	$display("i=%d\n", i);
end
```

2.6.4.12 Ports and Data Types

In Verilog, input ports of a module are always of type `wire`. An output, if assigned in a procedural block, is declared as `reg`, and in cases where the assignment to the output is made using a continuous assignment statement, then the output is defined as a `wire`. The `inout` is always defined as a `wire`. The data types are shown in Figure 2.15.

The input to a module is usually the output of another module, so the figure shows that the output of $module_1$ is the input to $module_0$. The port declaration rules can be easily followed using the arrow analogy, whereby the head of the arrow drawn across the module must be defined as `wire` and the tail declared as `reg` or `wire` depending on whether the assignment is made inside a procedural block or in a continuous assignment.

2.6.4.13 Simulation Control

Verilog provides several system tasks that do not infer any hardware and are used for simulation control. All system tasks start with the sign $. Some of the most frequently used tasks and the actions they perform are listed here.

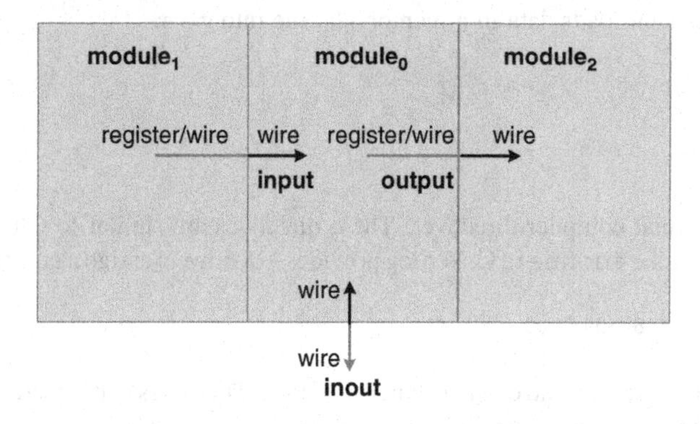

Figure 2.15 Port listing rules in Verilog. Head is always a wire. Tail may be a wire or reg based on whether it is, respectively, an assignment statement or a statement in a procedure block

- $finish makes the simulator exit simulation.
- $stop suspends the simulation and the simulator enters an interactive mode, but the simulation can be resume from the point of suspension.
- $display prints an output using a format similar to C and creates a new line for further printing.
- $monitor is similar to $display but it is active all the time. Only one monitor task can be active at any time in the entire simulation. This task prints at the end of the current simulation time the entire list when one of the listed values changes.

The following example gives the format of $monitor and $display which closely resemble the printf() function in C:

```
$monitor($time, "A=%d, B=%d, CIN=%b, SUM=%d, COUT=%d", A, B, CIN, SUM, COUT);
$display($time, "A=%d, B=%d, CIN=%b, SUM=%d, COUT=%d", A, B, CIN, SUM, COUT);
```

$time in these statements prints the simulation time at the time of execution of the statement. These statements display the values of A, B, CIN and COUT in decimal, binary, decimal and decimal number representations, respectively. The directives %d, %o, %h and %b are used to print values in decimal, octal, hexadecimal and binary formats, respectively.

$fmonitor and $fdisplay write values in a file. The file first needs to be open using $fopen. The code below shows the use of these tasks for printing values in a file:

```
modulator_vl = $fopen("modulator.dat");
if (modulator_vl == 0) $finish;
$fmonitor(modulator_vl,"data=%h bits=%h", data_values, decision_bits);
```

2.6.4.14 Loading Memory Data from a File

System tasks $readmemb and $readmemh are used to load data from a text file written in binary or hexadecimal, respectively, into specified memory. The example here illustrates the use of these tasks. First memory needs to be defined as:

```
reg [7:0] mem[0:63];
```

The following statement loads data in a memory.dat file into mem:

```
$readmemb ("memory.dat", mem);
```

2.6.4.15 Macros

Verilog supports several compiler directives. These directives are similar to C programming pre-compiler directives. Like #define in C, Verilog provides 'define to assign a constant value to a tag:

```
'define DIFFERENCE 6'b011001
```

The tag can then be used instead of a constant in the code. This gives better readability to the code. The use of the 'define tag is shown here:

```
if (ctrl == 'DIFFERENCE)
```

2.6.4.16 Preprocessing Commands

These are conditional pre-compiler directives used to selectively execute a piece of code:

```
'ifdef G723
$display ("G723 execution");
'else
$display ("other codec execution");
'endif
```

The `'include` directive works like `#include` in C and copies the contents in the file at the location of the statement. The statement

```
'include "filename.v"
```

copies the contents of filename.v at the location of the statement.

2.6.4.17 Comments

Verilog supports C-type comments. Their use is shown below:

```
reg a; // One-line comment
```

Verilog also supports block comments (as in C):

```
/* Multi-line comment that
reg acc;
results in the reg acc declaration being commented out */
```

Example 2.9

This example implements a simple single-tap infinite impulse response (IIR) filter in RTL Verilog and writes its stimulus to demonstrate coding of a design with feedback registers. The design implements the following equation:

$$y[n] = 0.5y[n-1] + x[n]$$

The multiplication by 0.5 is implemented by an arithmetic shift right by 1 operation. A register `y_reg` realizes $y[n-1]$ in the feedback path of the design, thus needing reset logic. The reset logic is implemented as an active-low asynchronous reset. The module has 16-bit data `x`, clock `clk`, reset `rst_n` as inputs and the value of `y` as output. The module `IIR` has two procedural blocks. One block models combinational logic and the other sequential. The block that models combinational logic consists of an adder and hard-wired shifter. The adder adds the input data `x` in shifted value of `y_reg`. The output of the combinational cloud is assigned to `y`. The sequential block latches the value of `y` in `y_reg`. The RTL Verilog code for the module `IIR` is given below:

```
module iir(
input signed [15:0] x,
input clk, rst_n,
output reg signed [31:0] y);
```

```
reg signed [31:0] y_reg;
  always @(*)     \\ combinational logic block
    y = (y_reg>>>1) + x;
    always @(posedge clk or negedge rst_n) \\ sequential logic block
    begin
      if (!rst_n)
        y_reg <= 0;
      else
        y_reg <= y;
    end
endmodule
```

The stimulus generates a clock and a reset signal. This reset is applied to the feedback register before the first positive edge of the clock. Initialization on clock and generation of reset is done in an `initial` block. Another `initial` block is used to give a set of input values to the module. These values are generated in a loop. The monitor statement prints the input and output with simulation time on the screen. The `$stop` halts the simulation after 60 time units and `$finish` ends the simulation. It is important to note that a simulation with clock input must be terminated using `$finish`, otherwise it never ends. The code for the stimulus is listed below:

```
module stimulus_irr;
reg [15:0] X;
reg CLK, RST_N;
wire [31:0] Y;
integer i;
iir IRR0(X, CLK, RST_N, Y); \\ instantiation of the module

initial
begin
  CLK = 0;
  #5 RST_N = 0; \\ resetting register before first posedge clk
  #2 RST_N = 1;
end
initial
begin
  X = 0;
    for(i=0; i<10; i=i+1) \\ generating input values every clk cycle
    #20 X = X + 1;
  $finish;
end
always \\ clk generation
  #10 CLK = CLK;
initial
$monitor($time, " X=%d, sum=%d, Y=%d", X, IRR0.y, Y);
initial
begin
  #60 $stop;
end
endmodule
```

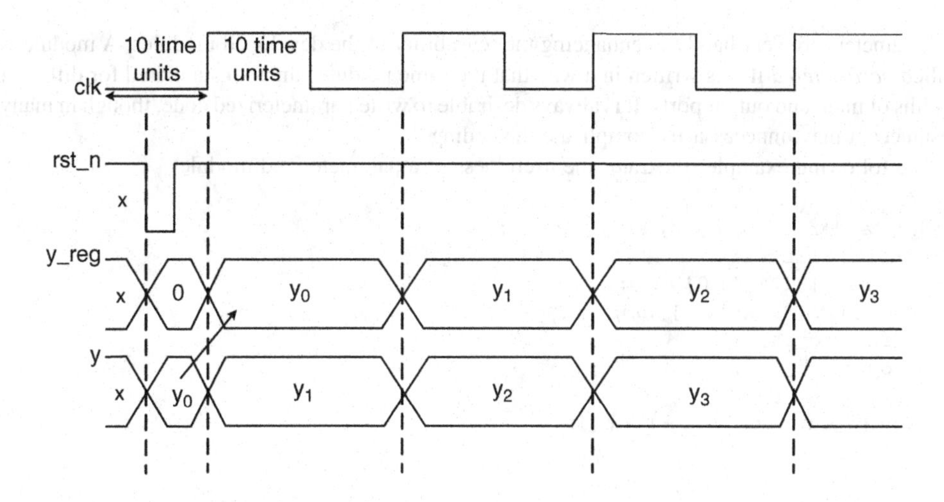

Figure 2.16 Timing diagram for the IIR filter design of example 2.9

2.6.4.18 Timing Diagram

In many instances before writing Verilog code and stimuli, it is quite useful to sketch a timing diagram. This is usually a great help in understanding the interrelationships of different logic blocks in the design. Figure 2.16 illustrates the timing diagram for the IIR filter design of the pervious subsection.

A clock is generated with time period of 20 units. The active-low reset is pulled low after 5 time units and then pulled high after 2 time units. As soon as the reset is pulled low, the `y_reg` is cleared and set to `0`. The first `posedge` of the clock after 10 time units latches the output of the combinational logic `y` into `y_reg`. The timing diagram should be drawn first and then accordingly coded in stimulus and checked in simulation for validity of results.

All Verilog simulators also provide waveform viewers that can show the timing diagram of selected variables in the simulation run. Figure 2.17 shows the screen output of the waveform viewer of ModelSim simulator for the IIR filter example above.

2.6.4.19 Parameters

Parameters are constants that are local to a module. A parameter is assigned a default value in the module and for every instance of this module it can be assigned a different value.

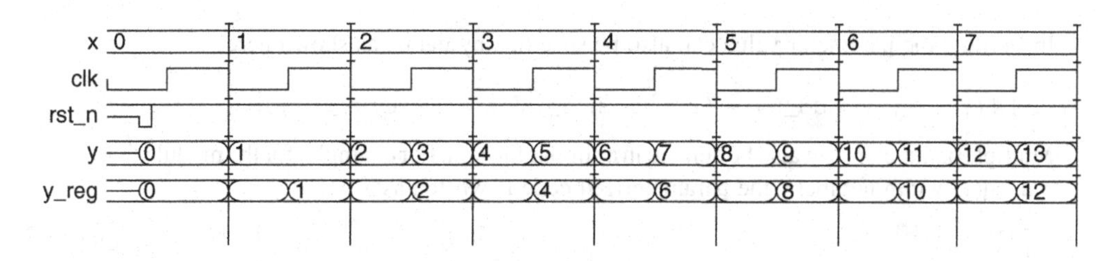

Figure 2.17 Timing diagram from the ModelSim simulator

Parameters are very handy in enhancing the reusability of the developed modules. A module is called *parametered* if it is written in a way that the same module can be instantiated for different widths of input and output ports. It is always desirable to write parameterized code, though in many instances it may unnecessarily complicate the coding.

The following example illustrates the usefulness of a parameterized module:

```
module adder (a, b, c_in, sum, c_out);
parameter SIZE = 4;
     input [SIZE-1: 0] a, b;
     output [SIZE-1: 0] sum;
     input c_in;
     output c_out;

assign {c_out, sum} = a + b + c_in;

endmodule
```

The same module declaration using ANSI-style port listing is given here:

```
module adder
#(parameter SIZE = 4)
     (input [SIZE-1: 0] a, b,
     output [SIZE-1: 0] sum,
     input c_in,
     output c_out);
```

This module now can be instantiated for different values of `SIZE` by merely specifying the value while instantiating the module. Shown below is a section of the code related to the instantiation of the module for adding 8-bit inputs, `in1` and `in2`:

```
module stimulus;
reg [7:0] in1, in2;
wire [7:0] sum_byte;
reg c_in;
wire c_out;

adder #8 add_byte (in1, in2, c_in, sum_byte, c_out);
.
.
endmodule
```

In Verilog, the parameter value can also be specified by name, as shown here:

```
adder #(.SIZE(8)) add_byte (in1, in2, c_in, sum_byte, c_out);
```

Multiple parameters can also be defined in a similar fashion. For example, for the module that adds two unequal width numbers, the parameterized code is written as:

```
module adder
#(parameter SIZE1 = 4, SIZE2=6)
     (input [SIZE1-1: 0] a,
     input [SIZE2-1: 0] b,
     output [SIZE2-1: 0] sum,
```

```
    input c_in,
    output c_out);
```

The parameter values can then be specified using one of the following two options:

```
adder #(.SIZE1(8), .SIZE2(10)) add_byte
        (in1, in2, c_in, sum_byte, c_out);
```

or, keeping the parameters in the same order as defined:

```
adder #(8,10) add_byte (in1, in2, c_in, sum_byte, c_out);
```

2.6.5 Verilog Tasks

Verilog `task` can be used to code functionality that is repeated multiple times in a module. A task has `input`, `output` and `inout` and can have its local variables. All the variables defined in the module are also accessible in the task. The task must be defined in the same module using `task` and `endtask` keywords.

To use a `task` in other modules, the task should be written in a separate file and the file then should be included using an `‘include` directive in these modules. The tasks are called from `initial` or `always` blocks or from other tasks in a module. The task can contain any behavioral statements including timing control statements. Like module instantiation, the order of `input`, `output` and `inout` declarations in a task determines the order in which they must be mentioned for calling. As tasks are called in a procedural block, the output must be of type `reg`, whereas the inputs may be of type `reg` or `wire`. Verilog-2001 adds a keyword `automatic` to the task to define a re-entrant task.

The following example designs a task `FA` and calls it in a loop four times to generate a 4-bit ripple carry adder:

```
module RCA(
    input [3:0] a, b,
    input c_in,
    output reg c_out,
    output reg [3:0] sum
);
reg carry[4:0];
integer i;

task FA(
    input in1, in2, carry_in,
    output reg out, carry_out);
        {carry_out, out} = in1 + in2 + carry_in;
endtask

always@*
begin
    carry[0]=c_in;
    for(i=0; i<4; i=i+1)
    begin
        FA(a[i], b[i], carry[i], sum[i], carry[i+1]);
    end
    c_out = carry[4];
end
endmodule
```

2.6.6 Verilog Functions

Verilog `function` is in many respects like `task` as it also implements code that can be called several times inside a module. A function is defined in the module using `function` and `endfunction` keywords. The function can compute only one output. To compute this output, the function must have at least one input. The output must be assigned to an implicit variable bearing the name and range of the function. The range of the output is also specified with the `function` declaration. A function in Verilog cannot use timing constructs like # or `@`. A function can be called from a procedural block or continuous assignment statement. It may also be called from other functions and tasks, whereas a function cannot call a task. A re-entrant function can be designed by adding the `automatic` keyword.

A simple example here writes a function to implement a `2:1` multiplexer and then uses it three times to design a `4:1` multiplexer:

```
module MUX4to1 (
  input [3:0] in,
  input [1:0] sel,
  output out);
  wire out1, out2;
  function MUX2to1;
    input in1, in2;
    input select;
    assign MUX2to1 = select ? in2:in1;
  endfunction
    assign out1 = MUX2to1(in[0], in[1], sel[0]);
    assign out2 = MUX2to1(in[2], in[3], sel[0]);
    assign out = MUX2to1(out1, out2, sel[1]);
endmodule
/* stimulus for testing the module MUX4to1 */
module testFunction;
  reg [3:0] IN;
  reg [1:0] SEL;
  wire OUT;
  MUX4to1 mux(IN, SEL, OUT);
  initial
begin
    IN = 1;
    SEL = 0;
    #5 IN = 7;
    SEL = 0;
    #5 IN = 2; SEL=1;
    #5 IN = 4; SEL = 2;
    #5 IN = 8; SEL = 3;
    end
  initial
    $monitor($time, " %b %b %b\n", IN, SEL, OUT);
endmodule
```

2.6.7 Signed Arithmetic

Verilog supports signed `reg` and `wire`, thus enabling the programmer to implement signed arithmetic using simple arithmetic operators. In addition to this, `function` can also return a signed value, and inputs and outputs can be defined as signed `reg` or `wire`. The following lines define signed `reg` and `wire` with keyword `signed`:

```
reg signed [63:0] data;
wire signed [7:0] vector;
input signed [31:0] a;
function signed [128:0] alu;
```

Verilog also supports type-casting using system functions `$signed` and `$unsigned` as shown here:

```
reg [63:0] data; // Unsigned data type
always @ (a)
begin
  out = ($signed(data) )>>> 2;// Type-cast to perform signed arithmetic
end
```

where >>> is used for the arithmetic shift right operation.

2.7 Verification in Hardware Design

2.7.1 Introduction to Verification

Verilog is especially designed for hardware modeling and lacks features that facilitate verification of complex digital designs. In these circumstances, designers resort to using other tools like `Vera` or `e` for verification [12]. To resolve this limited scope for verification in Verilog and to add more advanced features for HW design, the EDA vendors constituted a consortium. In 2005, the IEEE standardized Verilog and SystemVerilog languages [6, 8]. Many advanced features have been added in SystemVerilog. These relate to enhanced constructs for design and test-bench generation, assertion and direct programming interfaces (DPIs).

The EDA industry is trying to respond to increasing demands to elegantly handle chip design migration from the IC scale to the multi-core SoC scale. Verification is the greatest challenge, and for complex designs it is critical to plan for it right from the start. A *verification plan* (Vplan) should be developed by studying the function specification document.

As SoC involves several standard interfaces, it is possible that verification test-benches already exist for many components of the design in the form of *verification intellectual property* (VIP). Good examples are the test-benches developed for ARM, AMBA and PCI buses. Such VIPs usually consist of a test-bench, a set of assertions, and coverage matrices. An aggregated coverage matrix should always be computed to ensure maximum coverage. Guidelines have been published by Accellera that ensure interoperability and reuse of test-benches across design domains [13].

Simulators are very common in verifying an RTL design, but they are very slow in testing a design with many million gates. In many design instances, after the design is verified for a subset of test cases that includes the corner cases, more elaborate verification is performed using FPGA-based accelerators [14]. Finally, the verification engineers also plan verification of the first batches of ICs.

Many languages and tools have evolved for effective verification. Verilog, SystemVerilog, e and SystemC are some of the most used for test-bench implementation; usually a mix of these tools is used. *Open verification methodology* (OVM) enables these tools to coexist in an integrated verification environment [15]. The *OVM & Verification Methodology Manual* (VMM) has class libraries that verification engineers can use to enhance productivity [16].

Mixed-signal ICs add another level of complexity to verification. The design requires an integrated testing methodology to verify a mixed-signal design. Many vendors support mixed-signal verification in their offered solutions. The analog design is modeled in Verilog-AMS.

It is important to note that verification should be performed in a way that the code developed for verification is reusable and becomes a VIP. SystemVerilog is mostly the language of choice for developing VIPs, and vendors are adding complete functionality of the IEEE standard of System Verilog for verification in development tools. SystemVerilog supports constraint value generation that can be configured dynamically. It can generate constraint random stimulus sequences. It can also randomly select the control paths out of many possibilities. It also provides functional converge modeling: the model dynamically reactivates constrained random stimulus generation. SystemVerilog also supports coverage monitoring.

2.7.2 Approaches to Testing a Digital Design

2.7.2.1 Black-box Testing

This is testing against specifications when the internal structure of the system is unknown. A set of inputs is applied and the outputs are checked against the specification or expected output, without considering the inner details of the system (Figure 2.18). The design to be tested is usually called the 'device under test' (DUT).

2.7.2.2 White-box Testing

This tests against specifications while making use of the known internal structure of the system. It enables the developer to locate a bug for quick fixing. Usually this type of testing is done by the developer of the module.

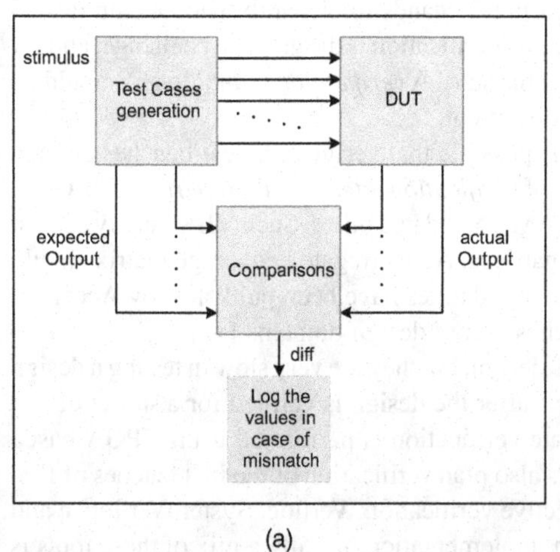

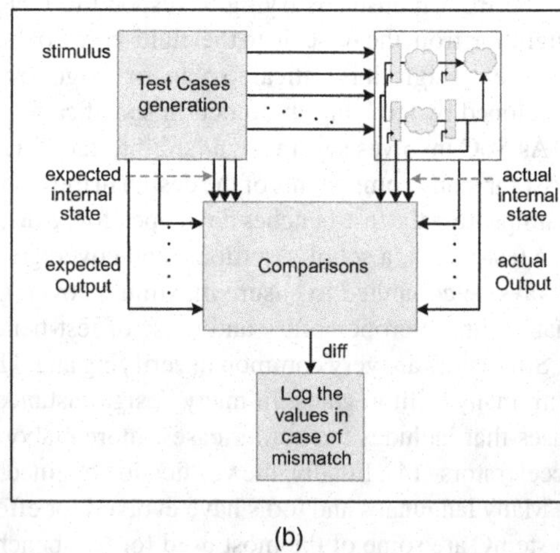

Figure 2.18 Digital system design testing using (a) the black-box technique and (b) the white-box technique

2.7.3 Levels of Testing in the Development Cycle

A digital design goes through several levels of testing during development. Each level is critical as an early bug going undetected is very costly and may lead to changes in other parts of the system. The testing phase can be broken down into four parts, described briefly below.

2.7.3.1 Module- and Component-level Testing

A component is a combination of modules. White-box testing techniques are employed. The testing is usually done by the developer of the modules, who has a clear understanding of the functionality and can use knowledge of the internal structure of the module to ease bug fixing.

2.7.3.2 Integration Testing

In integration testing, modules implemented as components are put together and their interaction is verified using test cases. Both black-box and white-box testing are used.

2.7.3.3 System-level Testing

This is conducted after integrating all the components of the system, to check whether the system conforms to specifications. Black-box testing is used and is performed by a test engineer. The testing must be done in an unbiased manner without any preconceptions or design bias. As the codings of different developers are usually integrated at the system level, an unbiased tester is important to identify faults and bugs and then assign responsibilities.

The first step is functional verification. When that is completed, the system should undergo performance testing in which the throughput of the system is evaluated. For example, an AES (advanced encryption standard) processor, after functional verification, should be tested to check whether it gives the required performance of encrypting data with a defined throughput.

After the system has been tested for specified functionality and performance, next comes stress testing. This stretches the system beyond the requirements imposed earlier on the design. For example, an AES processor designed to process a 2 Mbps link may be able to process 4 Mbps.

2.7.3.4 Regression Testing

Regression testing is performed after any change to the design is made as a consequence of bug fixing or any modification in the design. Regression tests are a sub-set of test vectors that the designer needs to run after any bug fixing or significant modification in an already tested design. Both black-box and white-box methodologies are used. Fixing a bug may resolve the problem under consideration but can disturb other parts of the system, so regression testing is important.

2.7.4 Methods for Generating Test Cases

There are several methods for generating test cases. The particular choice depends on the size of the design and the level at which the design is to be tested.

2.7.4.1 Exhaustive Test-vector Generation

For a small design or for module-level testing, the designer may want to exhaustively generate all possible scenarios. However, the time taken by testing increases exponentially with the size of the inputs. For example, testing a simple 16×16-bit multiplier requires $2^{16} \times 2^{16}$ test vectors. The simulators can spend hours or even days in exhaustive testing of designs of even moderate size. The designer therefore needs to test intelligently, choosing sample points and focusing especially on corner cases. For mathematical computations, the overflow and saturation logic cases are corner cases. Similarly for other designs, the inputs that test the maximum strength of the system should be applied.

2.7.4.2 Random Testing

For large designs, the designer may resort to random testing. The values of inputs are randomly generated out of a large pool of possible values. In many instances this random testing should be biased to cover stress points and corner cases, while avoiding redundancy and invalid inputs.

2.7.4.3 Constraint-based Testing

Constraint-based testing works with random testing, whereby the randomness is constrained to work in a defined range. In many instances, constraint testing makes use of symbolic execution of the model to generate an input sequence.

2.7.4.4 Tests to Locate a Fault

In many design instances, the first set of input sequences and test strategies are used only to identify faults. Based on the occurrence and type of faults, automatic test patterns are generated that localize the fault for easy debugging.

2.7.4.5 Model Checkers

The designer can make use of models for checking designs that implement standard protocols (e.g. interfaces). Appropriate checkers are placed in the design. The input is fed to the model as well as to the design. When there is non-conformity the checkers fire to identify the location of the bug.

2.7.5 Transaction-level Modeling

Many levels of modeling are used in hardware design. RTL and functional-level modeling have already been mentioned. For functional-level modeling, algorithms are implemented in tools like MATLAB®, and in many design instances a design that is functionally verified is directly converted into RTL. However, designs are becoming more and more complex. This is especially the case for SoC and MPSoC, where more and more components are being added on a single piece of silicon. The interworking of the processors or other components on the chip is also becoming ever more critical. This interworking at register transfer level is very complex as it deals with bus protocols or NoC protocols. While analyzing the interworking of these components, usually this level of detail is not required and interworking can only be studied by observing the physical links to make complex packet or data transactions.

Transaction-level modeling (TLM) deals with designs that have multiple components. These components communicate with each other on some medium. At TLM, the detailed RTL functionality of the components and RTL protocol are not important. TLM separately deals with communications as transactions and the behavior of each component at the functional level. Transaction-level modeling is easy to develop and fast to simulate, so enabling the designer to verify the functionality of the design at transaction level early in the development life cycle. RTL models, though, are developed in parallel but are very slow to simulate for verification and analysis of the design. For a complex SoC design the architects need to develop these three models: the functional model in the early stages, while the transaction-level and RTL are developed in parallel. Building three models of a system requires them to be consistent at different stages of the design cycle.

2.8 Example of a Verification Setup

A complete setup for testing a signal-processing based design in hardware is shown in Figure 2.19. A C++ environment generates constrained random test vectors to be input to the algorithm running in C++ and also to the translated design that is implemented in TLM. A *transactor block* converts the test vector into transactions, and these are input to the transaction-level model. The output of the model is also in terms of transactions. A transactor converts the transactions into results that can be compared with the output of the simulation result. A checker compares the two results to find functional equivalence. The input to the simulator is also fed to a *coverage block*. This block checks the level of coverage and can direct the dynamic constrained random generator to generate the input sample to maximize the coverage.

When the transaction model of Figure 2.19(a) is verified, the same setup can be used to test the RTL design as in Figure 2.19(b). A *driver block* is added that converts a transaction into an RTL detailed signal to be input to the device under test (DUT). The output of the RTL implementation of the DUT is also converted back to transactions. A *monitor* checks the cycle-by-cycle behavior of the DUT using assertions.

2.9 SystemVerilog

As designs become more complex in functionality, test-vector generation for appropriate coverage is also becoming critical. Verification engineers have been using tools specific to verification, such as `Vera` and `e`. Nevertheless there has been a need to have a unified language that supports both design and verification of complex designs. SystemVerilog (SV) is such an initiative that offers a unified language that is very powerful to model complex systems and provides advanced level constructs. These constructs facilitate concise writing of test-benches and the analysis of coverage. Most of the EDA tool vendors are continuously adding support for SV. The added features make SV a very powerful language for hardware modeling and verification.

2.9.1 Data Types

SystemVerilog supports additional data types `logic`, `int`, `bit`, `byte`, `longint` and `shortint`. The data type `reg` of Verilog is ambiguous because `reg` also means a physical register but once inferred may result in a physical wire or a register. A `logic` type is similar to a `reg` where all the bits in the variable of this type can take any one of four values: `0`, `1`, `x` and `z`. In the other data types each bit can be `0` or `1`. The variables of these types are automatically initialized to `0` at time zero. Table 2.13 shows all the additional data types in SystemVerilog and their related information.

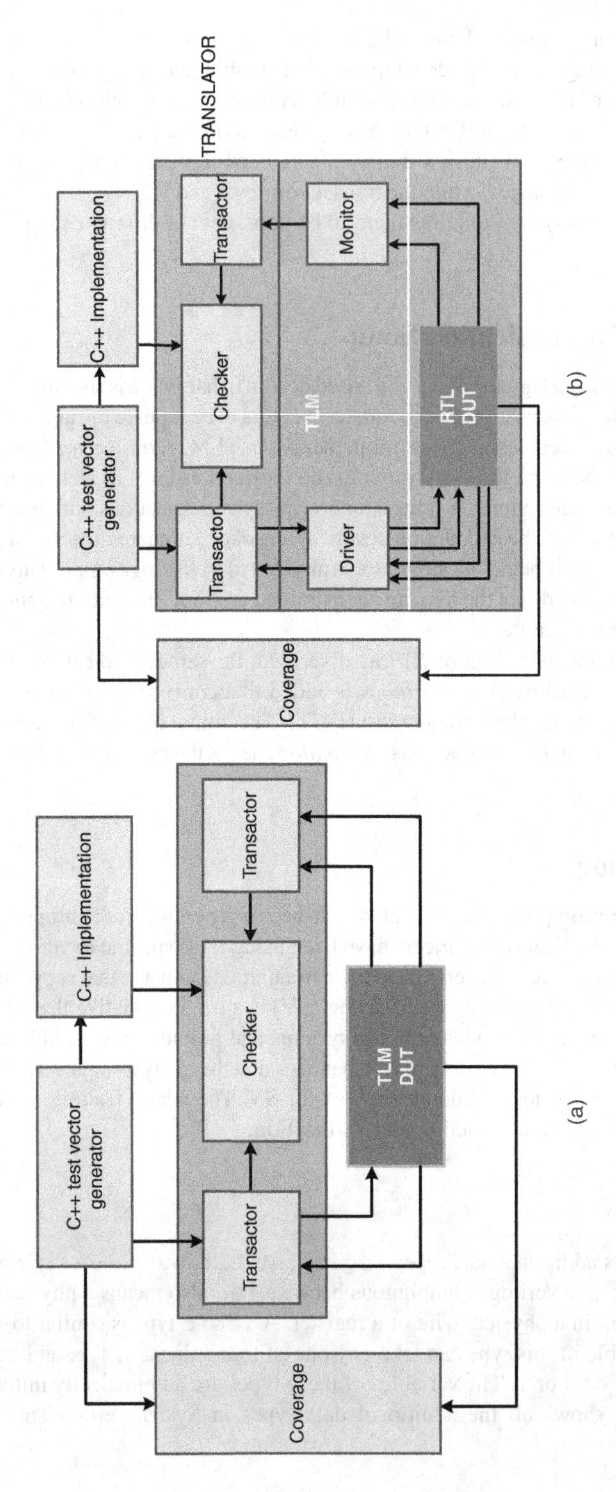

Figure 2.19 Test setups for verification of a complex design that is implemented in C++, TLM and RTL. (a) Setup with TLM. (b) Same verification setup with TLM replaced by RTL DUT, a driver and a monitor block

Table 2.13 Additional datatypes in SystemVerilog

Data type	Description	States	Example
logic	User-defined	Four states 0,1, x,z	logic [15:0] a,b;
int	32-bit signed	Two states 0,1	int num;
bit	User-defined	Two states 0,1	bit [5:0] in;
byte	8-bit signed	Two states 0,1	byte t;
longint	64-bit signed	Two states 0,1	longint p;
shortint	16-bit signed	Two states 0,1	shortint q;

There are two ways to define an array in SV: packed and unpacked. SystemVerilog can operate on an entire two-dimensional (2-D) array of packed data, whereas the unpacked arrays can be operated only on an indexed value. The unpacked 1-D and 2-D arrays are declared as:

```
bit up_data [15:0];
bit [31:0] up_mem [0:511];
```

For packed cases the same arrays are declared as:

```
bit [15:0] p_data;
bit [31:0] [0:511] p_mem1, p_mem2;
```

There are some constraints while operating on packed and unpacked arrays. The unpacked arrays can be sliced as:

```
slice_data = up_mem[2] [31:15];
                        // most significant byte at mem location 2
```

An operator can be applied on an entire packed array of data. An example is:

```
add_mem = p_mem1 + p_mem2;
```

Dynamic arrays can also be declared as:

```
bit [15:0] array[];
array = new[1023];
```

2.9.2 Module Instantiation and Port Listing

If the same names of ports are used in the instantiated module, the port names can be directly mentioned using .<name> or can be simply skipped while only ports having different names are mentioned. Consider a module defined as:

```
module FA(in1, in2, sum, clk, rest_n);
```

Assuming the instance has the first three ports with the same name, the instance can be written as:

```
FA ff (.in1, .sum, .in2, .clk(clk_global), .rest_n (rst_n));
```

or more concisely as:

```
FA ff (.*, .clk(clk_global), .rest_n (rst_n));
```

2.9.3 Constructs of the C/C++ Type

SV supports many C/C++ constructs for effective modeling.

2.9.3.1 `typedef, struct` and `enum`

The constructs `typedef`, `struct` and `enum` of C/C++ add descriptive power to SV. Their use is the same as in C. Examples of their use are:

```
typedef bit [15:0] addr;
typedef struct {
   addr src;
   addr dst;
   bit [31:0] data;
   }packet_tcp;
module packet (input packet_tcp packet_in,
   input clk,
   output packet_tcp packet_out);
always_ff @ (posedge clk)
begin
   packet_out.dst <= packet_in.src;
   packet_out.src~ packet_in.data;
end
endmodule
```

The `enum` construct can be used to define states of an FSM. It can be used in place of the Verilog `parameter` or `define`. The first constant gets a value of `0`. When a value is assigned to some constant, the following constants in the list are sequentially incremented. For example:

```
typedef enum logic [2:0]
{idle = 0,
read = 3,
dec, // = 4
exe // = 5} states;
states pipes;
```

The `enum` can also be directly defined as:

```
enum {idle, read=3, dec, exe} pipes;

case (pipes)
    idle: pc = pc;
    read: pc = pc+1;
.
.
endcase
```

2.9.3.2 Operators

The advanced features in SV enable it to model complex HW features in very few lines of code. For this, SV supports C-language like constructs such as:

$$\text{operand}_1\ OP = \text{operand}_2$$

where *OP* could be +, −, *,/, %, >>, <<, >>>, <<<, &, | or ^. For example, $x = x + 3$ can be written as:

```
x +=3;
```

SystemVerilog also supports post- and pre- increment and decrement operations ++x, −−x, x++ and x−−.

2.9.4 `for` and `do-while` Loops

SystemVerilog adds C/C++ type `for` and `do-while` loops. An example of the `for` loop is:

```
for(i=0, j=0, k=0; i+j+k<10; i++, j++, k++)
```

An example of the `do-while` loop is:

```
do
begin
if (sel_1 == 0)
    continue;
if (sel_2==3) break;
end
while (sel_2==0);
```

In this code, if `sel_1` is zero, `continue` makes the program jump to the start of the loop at `do`. When `sel_2` is 3, `break` makes the program exit the `do-while` loop, otherwise the loop is executed until the time `sel_2` is zero.

2.9.5 The `always` Procedural Block

SV helps in solving the issue of the sensitivity list. There are several variants of the `always` block that give distinct functionality for inferring combinational or sequential logic. For a combinational block, SV provides `always_comb`. Similarly `always_latch` infers a latch. and `always_ff` realizes synchronous logic:

```
module adder(input signed [3:0] in1, in2,
    input clk, rst_n,
    output logic signed [15:0] acc);
    logic signed [15:0] sum;

// Combinational block
always_comb
    begin: adder
    sum = in1 + in2 + acc;
    end: adder
```

```
// Sequential block
always_ff @(posedge clk or negedge rst_n)
    if (!rst_n)
        acc <= 0;
    else
        acc <= sum;
endmodule
```

2.9.6 *The* final *Procedural Block*

The final procedural block is like the initial block in that it too executes only once, but at the end of the simulation. It is good for displaying a summary of results:

```
final
begin
   $display($time, "simulation time, the simulation ends\n");
end
```

2.9.7 *The* unique *and* priority *Case Statements*

In Verilog, while synthesizing the code, the user may need to specify the type of logic intended to infer from a case statement. The synthesis directives full-case and full-case parallel-case are used to indicate, respectively, whether the user intends the logic to consider the first match it finds in a case statement if there is a possibility of finding more than one match, or that the user guarantees that all cases are handled in the coding and each case will only uniquely match with one of the selections. This behavior is very specific to synthesis and has no implication on simulation. SV provides equivalent directives, which are unique and priority, to guarantees the simulation behavior matches with the intended synthesis results. The examples below explain the two directives:

```
always @*
unique case (sel) //Equivalent to full-case parallel-case synthesis directive
  2'b00: out = in0;
  2'b01: out = in1;
  2'b10: out = in2;
  2'b11: out = in3;
  default: out = x;
endcase
```

The priority case is used in instances where the programmer intends to prioritize the selection and more than one possible match is possible:

```
always @*
priority case (1'b1) //equivalent to full-case synthesis directive

    irq1: out = in0;
    irq3: out = in1;
    irq2: out = in2;
```

```
    irq4: out = in3;
    default: out = 'x;
endcase
```

2.9.8 Nested Modules

SV supports nested modules, so that a module can be declared inside another module. For example:

```
module top_level;

module accumulator(input clk, rst_n, input [7:0] data, output bit [15:0] acc);

always_ff @ (posedge clk)
begin
    if (!rst_n)
        acc <= 0;
    else
        acc <= acc + data;
end
endmodule

logic clk=0;

always #1 clk = ~clk;
logic rst_n;
logic [7:0] data;
logic [15:0] acc_reg;

accumulator acc_inst(clk, rst_n, data, acc_reg);
initial
begin
    rst_n = 0;
    #10 rst_n = 1;
    data = 2;
    #200 $finish;
end

initial
$monitor($time, "%d, %d\n", data, acc_reg);
endmodule
```

2.9.9 Functions and Tasks

SV enhances Verilog functions and tasks with more features and flexibility. No `begin` and `end` is required to place multiple statements in functions and tasks. Unlike with a function in Verilog that always returns one value, SV functions can return a `void`. Use of the `return` statement is also added, whereby a function or a task returns a value before reaching the end. In SV, the input and output can also be passed by name; and, in a similar manner to module port listing, default arguments are also allowed.

The following example shows a function that returns a `void`:

```
function void expression (input integer a, b, c, output integer d);
    d = a+b-c;
endfunction: expression
```

Below is another example that illustrates a function returning before it ends:

```
function integer divide (input integer a, b);
    if (b)
        divide = a/b;
    else
    begin
        $display(''divide by 0\n'');
        return ('hx);
    end
// Rest of the function
.
.
endfunction: divide
```

2.9.10 The Interface

The interface is a major addition in SV. The interface encapsulates connectivity and replaces a group of ports and their interworking with a single identity that can be used in module definition. The interface can contain parameters, constants, variables, functions and tasks. The interface provides a higher level of abstraction to users for modeling and test-bench generation.

Consider two modules that are connected through an interface, as shown in Figure 2.20. The roles of the ports `input` and `output` change from one interconnection to the other. The `modport` configures the direction on ports in an interface to be an input or output.

```
interface local_bus (input logic clk);
bit rqst;
bit grant;
bit rw;
bit [4:0] addr;
wire [7:0] data;

modport tx      (input grant,
    output rqst, addr, rw,
    inout data,
    input clk);

modport rx (output grant,
    input rqst, addr, rw,
    inout data,
    input clk);

endinterface

module src (input bit clk,
     local_bus.tx busTx);
integer i;
```

```
logic [7:0] value = 0;
assign busTx.data = value;
initial
begin
    busTx.rw = 1;
    for (i=0; i<32; i++)
    begin
        #2 busTx.addr = i;
        value += 1;
end
busTx.rw = 0;
end
// Rest of the module details here

module dst ( input bit clk,
local_bus.rx busRx);
logic [7:0] local_mem [0:31];
always @ (posedge clk)

    if (busRx.rw)
    local_mem[busRx.addr] = busRx.data;
endmodule

// In the top-level module these modules are instantiated with interface
   declaration.
module local_bus_top;

logic clk = 0;
local_bus bus(clk); // the interface declaration

always #1 clk = ~clk;

    src SRC (clk, bus.tx);
    dst DST (clk, bus.rx);

initial
$monitor ($time, "\t%d %d %d %d\n", bus.rx.rw, bus.rx.addr,
                   bus.rx.data, DST.local_mem[bus.rx.addr]);

endmodule
```

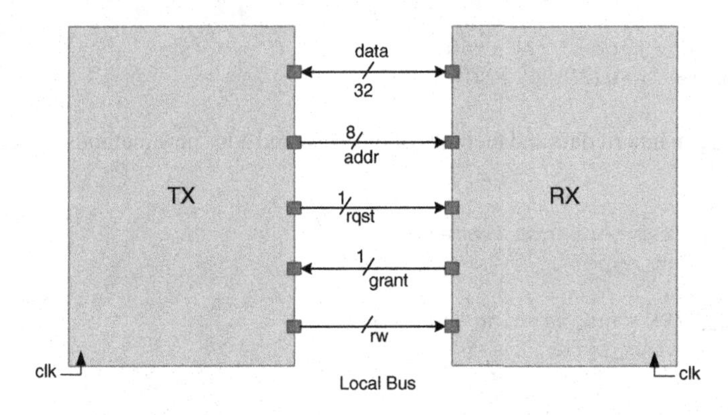

Figure 2.20 Local bus interface between two modules

2.9.11 Classes

In SV, as in C++, a class consists of data and methods. The methods are functions and tasks that operate on the data in the class. SV supports key aspects of object-oriented programming (OOP), including inheritance, encapsulation and polymorphism.

A class is declared with internal or external declared functions that operate on the data defined in the class. The example below defines a class with an internal and external declared method:

```
class frame{
byte dst_addr;
bit [3:0] set_frame_type;
data_struct payload;
function byte get_src_addr ()
    return src_addr;
endfunction
extern task assign_dst_addr_type (input byte addr, input bit[3:0] type);

endclass

task frame::assign_dst_addr(input byte addr, input bit [3:0] type);
    dst_addr = addr;
    frame_type = type;
endtask
```

The syntax only declares an object class of type `frame`. One or multiple instances of this class can be created as follows:

```
frame first_frame = new;
```

A class constructor can also be used to initialize data as:

```
class frame

function new (input byte addr, input [3:0] type)
     dst_addr = addr;
     frame_type = type;
endfunction
.
.
endclass
// Set the dst and type of the frame
frame msg_frame = new(8'h00, MSG);
```

Another class can inherit data and methods of this class and adds new methods and can change the existing methods.

```
class warning_frame extends frame;
bit [2:0] warning_type;

function MSG_TYPE send_warning ();
    return warning_type;
endfuction;
endclass
```

Using object-oriented programming (OOP), the classes defined in OVM and VMM can be extended for effective verification. For example, `vmm_data` class has many member functions, including `set_log ()`, `display()`, `copy()` and `compare()` [15]. These methods can be used and new methods can be added by extending this base class:

```
class burst_frame extends vmm_data;
```

SV restricts inheritance to single inheritance only. The derived class can override the methods of the parent class. The derived class can also use members of the parent class with a keyword `super`:

```
class abs_energy;
    integer amp;
function integer energy ();
    energy = amp*amp;
endfunction
endclass
```

A derived class can be declared that overrides the function `energy`:

```
class mod_energy extends abs_energy;
integer amp;
function integer energy();
    energy = 0.5*super.energy() + amp * super.amp;
endfunction
endclass
```

SV also supports data hiding and encapsulation. To restrict the scope of a member to the parent class only, the member is declared as local or protected. The local members are accessible only to methods defined inside the class, and these members cannot be inherited by derived classes as they are not visible to them. A protected method can be inherited by a derived class.

Virtual classes or methods in SV provide polymorphism. These classes are used to create a template. Using this template, real classes are derived. The virtual methods defined in a virtual class are overridden by the derived classes:

```
virtual class frame;
ip frame,
atm frame,
stm frame
virtual class frame;
virtual function integer send (bit [255:0] frame_data);
endfunction
endclass
```

The derived classes are:

```
class ethernet extends frame;
function integer send ( bit [255:0] frame_data);
// The contents of the function
.
.
endfuntion
endclass
```

```
class atm extends frame;

frame gateway_frame [10];
```

Frames of various types can be declared and assigned to the array:

```
ethernet frame_e = new;
atm frame_a = new;

gateway_frame [0] = frame_e;
gateway_frame [1] = frame_a;
```

The statement

```
gateway_frame [1].send();
```

makes the compiler finds out which frame will be sent.

2.9.12 Direct Programming Interface (DPI)

SV can directly access a function written in C using a DPI. Similarly a function or task written in SV can be exported to a C program. SV makes interworking of C and SV code very trivial and there is no need to use PLI. The C functions in SV are then called using import directive, while functions and tasks of SV to be used in a C function are accessible by using export DPI declaration. The illustration here shows DPI use:

```
// top-level module that instantiates a module that calls a C function
module top_level();

moduleCall_C Call_C (rst, clk, in1, in2, out1, ...;
.
.
.
endmodule
```

The instantiated module `Call_C` of type `moduleCall_C` uses an import directive for interfacing with a C program:

```
module moduleCall_C(rst, clk, in1, in2, out1,...);
.
.
import "DPI-C" context task fuctionC (....);

always@(posedge clk)
    functionC (rst,in1, in2, out1,....);

    export "DPI-C" task CallVeri1;
    export "DPI-C" task CallVeri2;


task CallVeri1 (input int addr1, output int data1);
.
.
    endtask
```

```
.
.
task CallVeri2 (input int addr2, output int data2);
.
.
    endtask
endmodule
```

The C function `functionC` is called from the SV module, and this function further calls `funct1 ()` and `funct2 ()`. These two functions use tasks `CallVeri1` and `CallVeri2` defined in SV:

```
// required header files

void fuctionC (int rst, ....)
{
.
.
    rest = rst;
.
    funct1(...);
    funct2(...);
.
.
}

void funct1 (void)
{
.
.
    CallVeri1(....);
.
}

void funct2 (void)
{
.
.
    CallVeri2(....);
.
}
```

2.9.13 *Assertion*

Assertion is used to validate the behavior of a design. It is used in the verification module or in an independent checker module. SV supports two types of assertion, immediate and concurrent. The immediate assertion is like an `if-else` statement. The expression in `assert` is checked for the desired behavior. If this expression fails, SV provides that one of the three severity system tasks can be called. These tasks are `$warning`, `$error` and `$fatal`. The user may also use `$info` where no severity on assertion is required. Below is an example:

```
assert(value>=5)
else $warning("Value above range");
```

Concurrent assertion checks the validity of a property. There are several ways to build properties; these may be compound expressions using logical operators or sequences:

```
assert property (request && !ready)
```

An example of a sequence is:

```
assert property (@posedge clk) req |-> ##[2:5] grant);
```

Here, the implication operator (|->) checks on every `posedge` of `clk` the assertion of `req`, and when it is asserted then the `grant` must be asserted in 2 to 5 following clock cycles.

2.9.14 Packages

SV has borrowed the concept of a package from VHDL. By using `package`, SV can share user-defined type definitions across multiple modules, interfaces, other programs and packages. The package can contain, for example, parameters, constants, type definitions, tasks, functions, import statements from other packages and global variables. Below is an example:

```
package FSM_types
// global typedef
typedef enum FSM{INVALID, READ, DECODE, EXECUTE, WRITE} pipelines;
        bit idle; // global variable initialize to 0
    task invalid_cycle (input [2:0] curret_state) //global task
        if (current_state == INVALID)
            $display("invalid state");
        $finish;
    endtask: invalid_cycle
endpackege
```

2.9.15 Randomization

SV supports unconstrained and constrained random value generation. The function `randomize` returns `1` if it successfully generates the constrained random value, otherwise it returns `0`.

```
bit [15:0] value1, value2;
bit valid;

initial
begin
    for(i=0; i<1024; i++)
        valid = randomize (value1, value2);
    end
end
```

The randomization can also be constrained by adding a `with` clause. The example given above can be constrained as:

```
    valid = randomize (value1, value2); with (value1>32; value1
```

2.9.16 *Coverage*

The coverage in SV gives a quantitative measure of the extent that the functioning of a DUT is verified is the simulation environment. The statistics are gathered using coverage groups. With a coverage group, the user lists variables as `converpoints`. The simulator collects statistics of the values these variables take in simulation. The simulator stores the values of these variables in a coverage database.

```
module stimulus;
logic [15:0] operand1, operand2;
.
.
covergroup cg_operands @ (posedge clk)
      o1: coverpoint = operand1;
      o2: coverpoint = operand2;
endgroup : cg_operands
.
.
.
cg_operands cover_ops = new( );
.
endmodule
```

Each coverage point contains a set of bins. These bins further refine the values the variable takes for each range.

```
covergroup cg_operands @ (posedge clk)
  o1: coverpoint = operand1 {
        bins low = {0,63};
        bins med = {64,127};
        bins high = {128,255};
        }
  o2: coverpoint = operand2 {
        bins low = {0,63};
        bins med = {64,127};
        bins high = {128,255};
        }
endgroup : cg_operands.
```

The coverage group can be used inside a module, class or interface.

Exercises

Exercise 2.1

Write RTL Verilog code to implement the design given in Figure 2.21. Generate the appropriate reset signal for the feedback register used in the design. Develop a test plan and write a stimulus to test the design for functional correctness. Also write a test case to count the number of cycles it takes for the register `out_reg` to overflow for `in1` and `in2` and `sel` set to `1`. Also, code the design and stimulus in SystemVerilog.

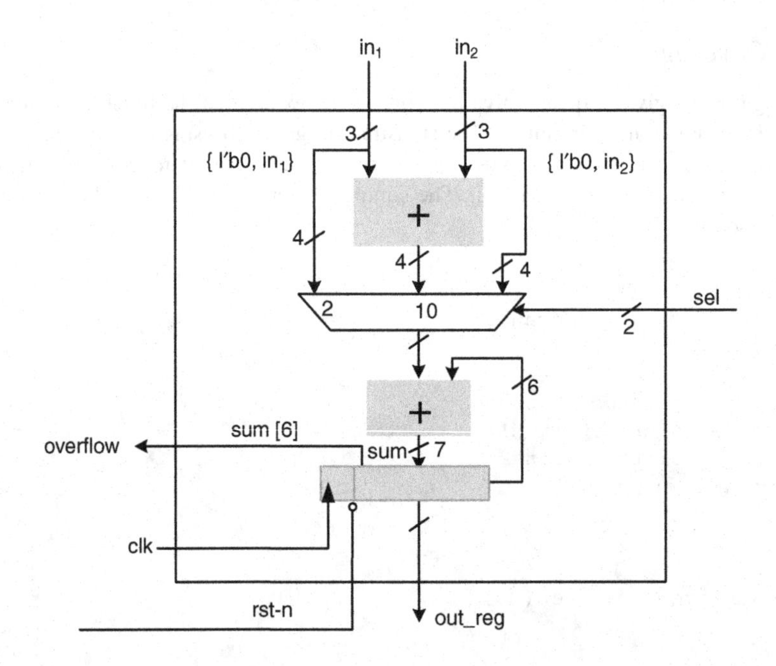

Figure 2.21 Digital design at register transfer level with a feedback register

Exercise 2.2

Design an ALU datapath that performs the following operations in parallel on two 16-bit signed inputs `A` and `B` and assigns the value of one of the outputs to 16-bit `C`. The selection of operation is based on a 2-bit selection line. Code the design in both Verilog and SystemVerilog. Write test vectors to verify the implementation.

```
C=A+B
C=A-B
C=A&B
C=A|B
```

Exercise 2.3

Code the logic of the following equations in RTL Verilog:

```
acc0 = acc1 + in1;
acc1= acc0 + in2;
out = acc0 + acc1;
```

where `in1` and `in2` are two 32-bit inputs, `out` is a 40-bit output, and `acco` and `acc1` are two 40-bit internal registers. At every positive edge of a clock the inputs are fed to the design and `out` is produced.

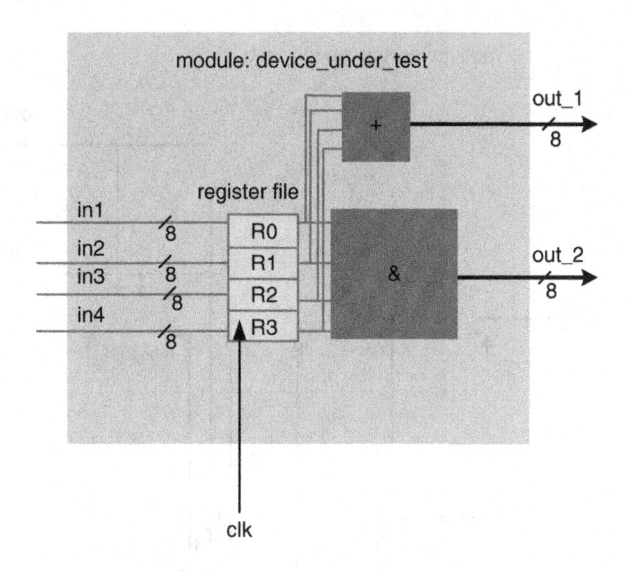

Figure 2.22 Digital design at register transfer level showing the combinational and synchronous components

Exercise 2.4

Write RTL Verilog code of the module `device_under_test` given in Figure 2.22. Four 8-bit inputs, `in1`, `in2`, `in3` and `in4`, are input to four 8-bit registers, `R0`, `R1`, `R2` and `R3`, respectively, at every positive edge of the clock. Four values in these registers are added, and bitwise AND operation is performed on the values stored in these registers to produce `out_1` and `out_2`.

Exercise 2.5

Write an RTL Verilog code and its stimulus to implement and test the module `device_under_test` given in Figure 2.23. The inputs and outputs of the module are shown. Generate a 20-time unit clock `clk` from the stimulus, as werll as a `rst_n` signal to reset the device before the first positive edge of the clock. The widths of each input and output signals are shown. Write test vectors by varying all the input signals in each test. Make sure each test vector is valid only for one clock period. Use the `monitor` statement to print the values of the inputs and outputs on the screen. Finally, rewrite the stimulus in SystemVerilog and use coverage to test the design for selective ranges of input values. Use datapaths of exercise 2.2 for the ALU.

Exercise 2.6

Design a datapath with three 8-bit accumulators. The first accumulator, `acc1`, adds a 4-bit input `data` in `acc1` in every clock cycle. The second accumulator, `acc2`, adds the first accumulator in itself, and the third accumulator, `acc3`, adds the first and second accumulators in itself in every clock cycle. Each accumulator has an asynchronous reset. Draw the RTL-level diagram and code the design in RTL Verilog.

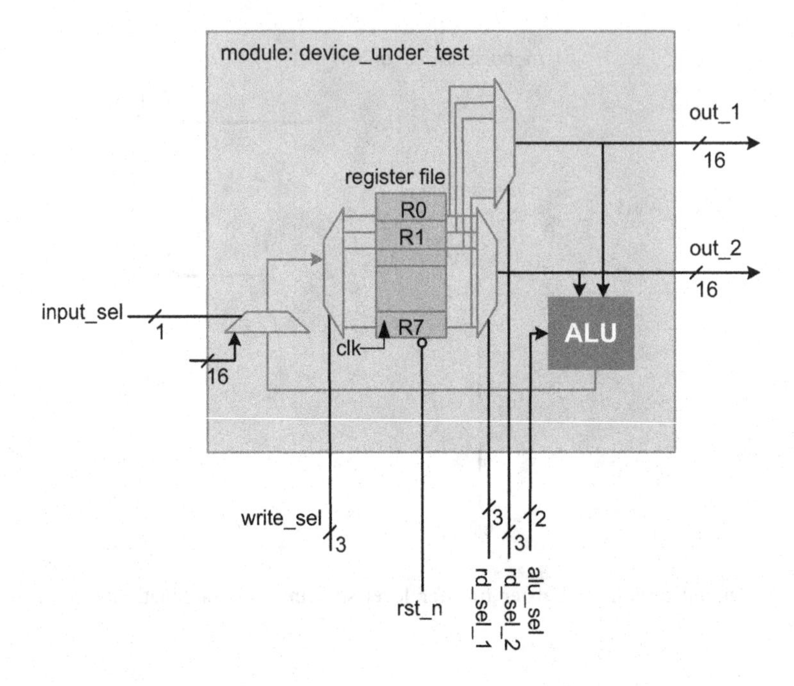

Figure 2.23 Digital design with multiple inputs and outputs

Exercise 2.7

Write RTL Verilog code to implement the design given in Figure 2.24. The feedback register needs to be reset using a negative-level asynchronous reset. Write a stimulus for the design. The `out` is the output of the module, and `in` is the input. Identify other signals that need to be defined as ports of the module.

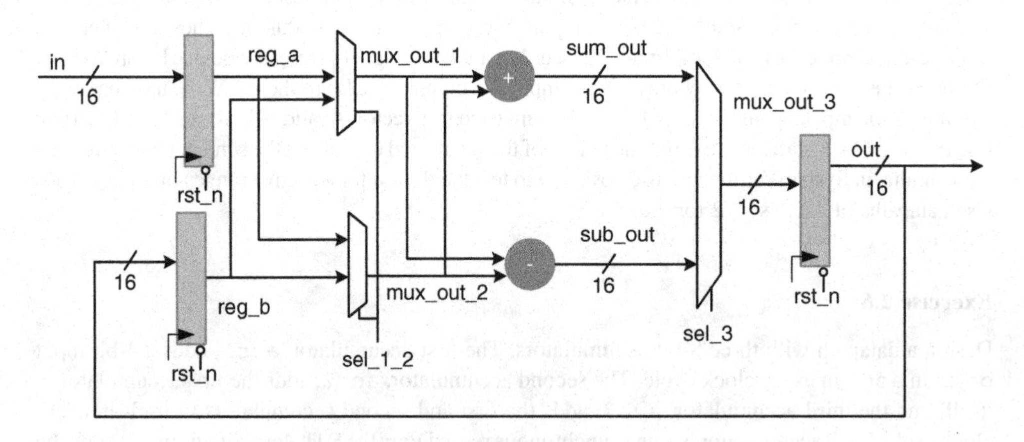

Figure 2.24 RTL design of a digital circuit

Exercise 2.8

Draw an RTL diagram for the following Verilog code. Clearly specify the data widths of all the wires, and show multiplexers, registers, reset and clock signals.

```
module test_module (input [31:0] x0,
input [1:0] sel, input clk,
          rst_n, output reg [31:0] y0);
reg [31:0] x1, x2, x3;
reg [31:0] y1;
wire [31:0] out;
assign out = (x0 + x1 + x2 + x3 + y1)>>>2;
always @(posedge clk or negedge rst_n)
begin
     if(!rst_n) begin
          x1 <= 0;
          x2 <= 0;
          x3 <= 0;
     end
     else if (sel==0) begin
          x3 <= x2;
          x2 <= x1;
          x1 <= x0;
     end
     else if (sel == 01) begin
          x3 <= x1;
          x2 <= x0;
          x1 <= x2;
     end
     else begin
          x3 <= x3;
          x2 <= x2;
          x1 <= x0;
     end
end
always @ (posedge clk or negedge
rst_n) begin
     if(!rst_n) begin
          y1 <= 0;
          y0 <= 0;
     end
     else begin
          y1 <= y0;
          y0 <= out;
end
     end
endmodule
```

Exercise 2.9

Partition the RTL-level design given in Figure 2.25 into two or three modules for better synthesis result. Write RTL Verilog code for the design. For the combinational cloud, write an empty function or a task to implement the interfaces.

Exercise 2.10

Design architecture, and implement it in RTL Verilog to realize the following difference equation:

$$y[n] = x[n] - x[n-1] + x[n-2] + x[n-3] + 0.5y[n-1] + 0.25y[n-2].$$

Implement multiplication with 0.5 and 0.25 by shift operations.

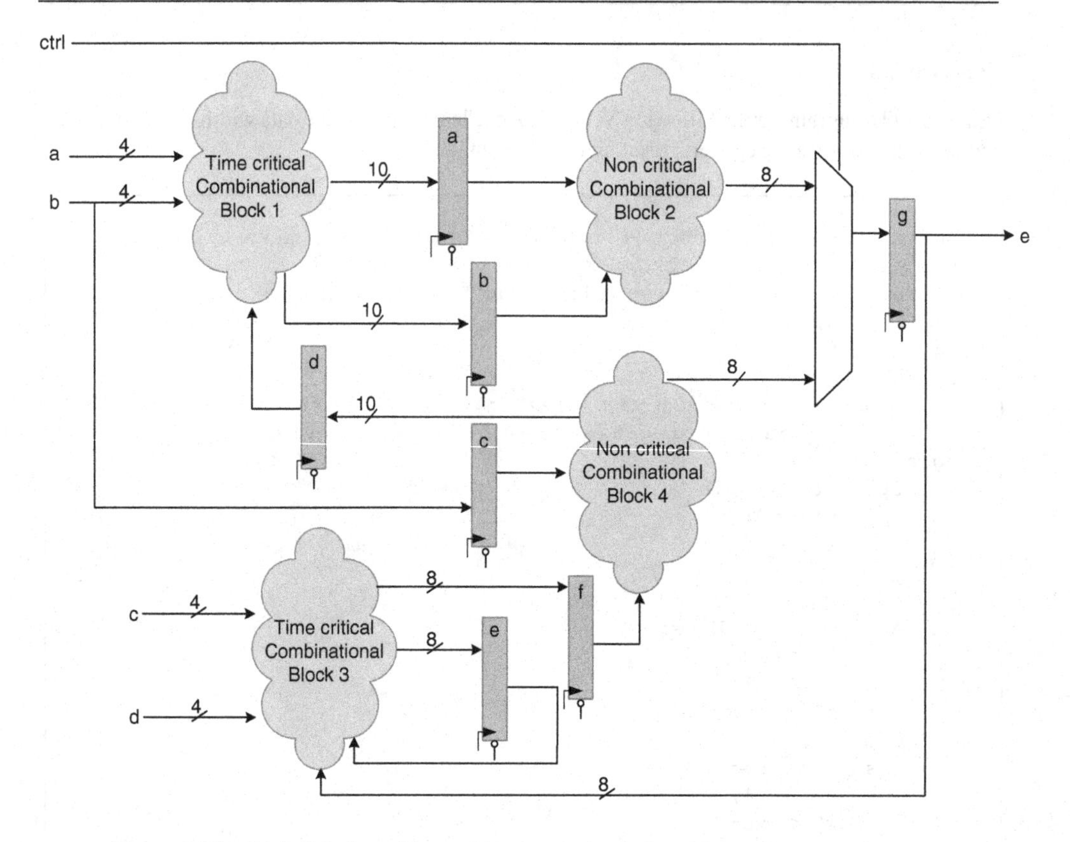

Figure 2.25 Digital design with combinational clouds for different design objectives

References

1. S. Palnitkar, *Verilog HDL*, 2nd edn, Prentice-Hall, 2003.
2. M. D. Ciletti, *Advanced Digital Design with the Verilog HD*, Prentice-Hall, 2003.
3. Z. Navabi, *Verilog Digital System Design: Register Transfer Level Synthesis, Testbench, and Verification*, McGraw-Hill, 2006.
4. IEEE, Standard 1364-1995: "Hardware description language based on Verilog," 1995.
5. IEEE, Standard 1364-2001: "Verilog hardware description language," 2001.
6. "SystemVerilog 3.1, ballot draft: Accellera's extensions to Verilog," April 2003.
7. IEEE, Standard 1364 -2005: "Verilog hardware description language," 2005.
8. www.systemverilog.org
9. M. Keating and P. Bricaud, *Reuse Methodology Manual for System-on-a-Chip Designs*, 3rd edn, Kluwer academic Publishers, 2002.
10. IEEE, Standard 1800-2005: "SystemVerilog: unified hardware design, specification and verification language," 2005.
11. www.xilinx.com
12. V. Berman, "IEEE P1647 and P1800: two approaches to standardization and language design," *IEEE Design & Testing of Computers*, 2005, vol. 22, pp. 283–285.
13. Accellera Standards, *Verification Intellectual Property (VIP) Recommended Practices*, vol. 1.1, August 2009.
14. A. Alimohamnnad, S. F. Fard and B. F. Cockburn, "FPGA-based accelerator for the verification of leading-edge wireless systems," in *Proceedings of Design Automation Conference*, 2009, ACM/IEEE, pp. 844–847.
15. www.ovmworld.org
16. www.vmmcentral.org

3

System Design Flow and Fixed-point Arithmetic

Pure mathematics is, in its way, the poetry of logical ideas.

Albert Einstein

3.1 Overview

This chapter describes a typical design cycle in the implementation of a signal processing application. The first step in the cycle is to capture the requirements and specifications (R&S) of the system. The R&S usually specify the sampling rate, a quantitative measure of the system's performance, and other application-specific parameters. The R&S constrain the designer to explore different design options and algorithms to meet them in the most economical manner. The algorithm exploration is usually facilitated by MATLAB®, which is rich in toolsets, libraries and functions. After implementation and analysis of the algorithm in MATLAB®, usually the code is translated into higher level programming languages, for example, C/C++ or C#.

This requires the chapter to focus on *numbering systems*. Representing signed numbers in two's complement format is explained. In this representation, the most significant bit (MSB) has negative weight and the remainder of the bits carry positive weights. Although the two's complement arithmetic greatly helps addition and subtraction, as subtraction can be achieved by addition, the negative weight of the sign bit influences multiplication and shifting operations. As the MSB of a signed number carries negative weight, the multiplication operation requires different handling for different types of operand. There are four possible combinations for multiplication of two numbers: unsigned–unsigned, signed–unsigned, unsigned–signed and signed–signed. The chapter describes two's complement multiplication to highlight these differences. Scaling of signed number is described as it is often needed while implementing algorithms in fixed-point format.

Characteristics of two's complement arithmetic from the hardware (HW) perspective are listed with special emphasis on corner cases. The designer needs to add additional logic to check the corner cases and deal with them if they occur. The chapter then explains floating-point format and builds the rationale of using an alternate fixed-point format for DSP system implementation. The chapter also

Digital Design of Signal Processing Systems: A Practical Approach, First Edition. Shoab Ahmed Khan.

justifies this preference in spite of apparent complexities and precision penalties. Cost, performance and power dissipation are the main reasons for preferring fixed-point processors and HW for signal processing systems. The fixed-point implementations are widely used for signal processing systems whereas floating-point processors are mainly used in feedback control systems where precision is of paramount importance. The chapter also highlights that currently integrated high gate counts on FPGAs encourages designers to use floating-point blocks as well for complex DSP designs in hardware.

The chapter then describes the equivalent format used in implementing floating-point algorithms in fixed-point form. This equivalent format is termed the `Qn.m` format. All the floating-point variables and constants in the algorithm are converted to Q*n*.*m* format. In this format, the designer fixes the place of an implied decimal point in an N-bit number such that there are n bits to the left and m bits to the right. All the computations in the algorithm are then performed on fixed-point numbers.

The chapter gives a systematic approach to converting a floating-point algorithm in MATLAB® to its equivalent fixed-point format. The approach involves the steps of levelization, scalarization, and then computation of ranges for specifying the Q*n*.*m* format for different variables in the algorithm. The fixed-point MATLAB® code then can easily be implemented. The chapter gives all the rules to be followed while performing Q*n*.*m* format arithmetic. It is emphasized that it is the developer's responsibility to track and manage the decimal point while performing different arithmetic operations. For example, while adding two different Q-format numbers, the alignment of the decimal point is the responsibility of the developer. Similarly, while multiplying two Q-format signed numbers the developer needs to throw away the redundant sign bit. This discussion leads to some critical issues such as overflow, rounding and scaling in implementing fixed-point arithmetic. Bit growth is another consequence of fixed-point arithmetic, which occurs if two different Q-format numbers are added or two Q-format numbers are multiplied. To bring the result back to pre-defined precision, it is rounded and truncated or simply truncated. Scaling or normalization before truncation helps to reduce the quantization noise. The chapter presents a comprehensive account of all these issues with examples.

For communication systems, the noise performance of an algorithm is critical, and the finite precision of numbers also contributes to noise. The performance of fixed-point implementation needs to be tested for different ratios of signal to quantization noise (SQNR). To facilitate partitioning of the algorithm in HW and SW and its subsequent mapping on different components, the chapter describes algorithm design and coding guidelines for behavioral implementation of the system. The code should be structured such that the algorithm developers, SW designers and HW engineers can correlate different implementations and can seamlessly integrate, test and verify the design and can also fall back to the original algorithm implementation if there is any disagreement in results and performance. FPGA vendors like Xilinx and Altera in collaboration with Mathworks are also providing Verilog code generation support in several Simulink blocks. DSPbuilder from Altera [1] and System Generator from Xilinx [2] are excellent utilities. In a Simulink environment these blocksets are used for model building and simulation. These blocks can then be translated for HW synthesis. A model incorporating these blocks also enables SW/HW co-simulation. This co-simulation environment guarantees bit and cycle exactness between simulation and HW implementation. MATLAB® also provides connectivity with ModelSim though a Link to ModelSim utility.

Logic and arithmetic shifts of numbers are discussed. It is explained that a full analysis should be performed while converting a recursive digital system to fixed-point format. Different computational structures exist for implementing a recursive system. The designer should select a structure that is least susceptible to quantization noise. The MATLAB® filter design toolbox is handy to do the requisite analysis and design space exploration. The chapter explains this issue with the help of an example. The chapter ends by giving examples of floating-point C code and its equivalent fixed-point C conversion to clarify the differences between the two implementations.

To explain different topics in the chapter several numerical examples are given. These examples use numbers in binary, decimal and hexadecimal formats. To remove any confusion among these representations, we use different ways of distinguishing the formats. A binary number is written with a base 2 or b, such as 1001_b and 100_2. The binary numbers are also represented in Verilog format, for example $3'b1001$. Decimal numbers are written without any base or with a base 10, like −0.678 and -0.678_{10}. The hexadecimal representation is 0×23df.

3.2 System Design Flow

3.2.1 Principles

This section looks at the design flow for an embedded system implementing a digital signal processing application. Such a system in general consists of hybrid target technologies. In Chapter 1 the composition of a representative system is discussed in detail. While designing such an embedded system, the application is partitioned into hardware and software components.

Figure 3.1 depicts the design flow for an embedded signal processing application. Capturing R&S is the first step in the design cycle. The specifications are usually in terms of requirements on the sampling rate, a quantitative measure of performance in the presence of noise, and other application-specific parameters. For example, for a communication system a few critical requirements are:

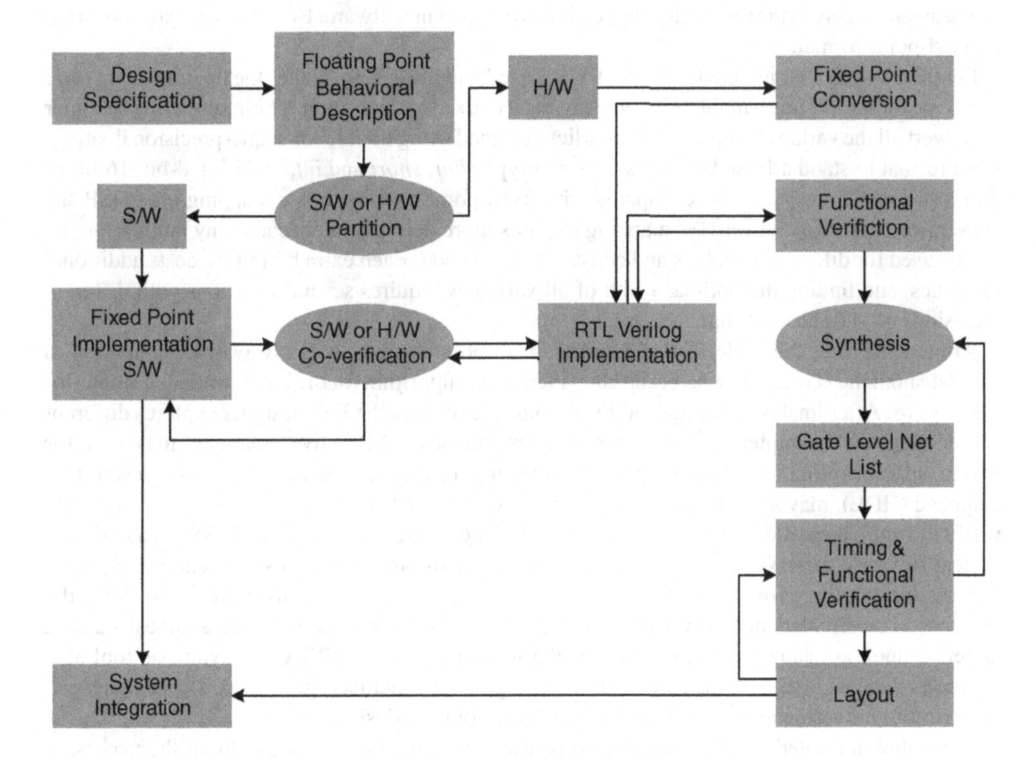

Figure 3.1 System-level design components

maximum supported data rate in bits per second (bps) at the transmitter, the permissible bit error rate (BER), the channel bandwidth, the carrier and intermediate frequencies, the physical form factor, power rating, user interfaces, and so on. The R&S summary helps the designer to explore different design options and flavors of algorithms.

The requirements related to digital design are forwarded to algorithm developers. The algorithm developers take these requirements and explore digital communication techniques that can meet the listed specifications. These algorithms are usually coded in behavioral modeling tools with rich pre-designed libraries of functions and graphical aids. MATLAB® has become the tool of choice for algorithm development and exploration. It is up to an experienced designer to finalize an algorithm based on its ease of implementation and its compliance to the R&S of the design. After the implementation of the algorithm in MATLAB®, usually the code is translated into high level language, for example, C/C++ or C#. Although MATLAB® has been augmented with fixed-point tools and a compiler that can directly convert MATLAB® code to C or C++, but in many instances a hand-coded implementation in C/C++ is preferred for better performance. In many circumstances it is also important for the designer to comprehend the hidden complexities of MATLAB® functions: first to make wise HW/SW partitioning decisions, and then to implement the design on an embedded platform consisting of GPP, DSPs, ASICs and FPGSs.

Partitioning of the application into HW/SW is a critical step. The partitioning is driven by factors like performance, cost, form factor, power dissipation, time to market, and so on. Although an all-software implementation is easiest to code and implement, in many design instances it may not be feasible. Then the designer needs to partition the algorithm written in C/C++ into HW and SW parts. A well-structured, computationally intensive and easily separable code is preferably mapped on hardware, whereas the rest of the application is mapped in software. Both the parts are converted to fixed-point format.

For the SW part a target fixed-point DSP is selected and the designer takes the floating-point code that is meant for SW implementation and converts it to fixed-point format. This requires the designer to convert all the variables and constants earlier designed using double- or single-precision floating-point format to standard fixed-point variables of type *char, short* and *int,* requiring 8-bit, 16-bit or 32-bit precisions, respectively. Compared with fixed-point conversion for mapping on a DSP, the conversion to fixed-point for HW mapping requires more deliberation because any number of bits can be used for different variables and constants in the code. Each extra bit in HW costs additional resources, and finding the optimal width of all variables requires several iterations with different precisions of variables settings.

After converting the code to the requisite fixed-point format, it is always recommended to simulate both the HW and SW parts and characterize the algorithm for different simulated signal-to-noise ratios. After final partitioning and fixed-point conversion, the HW designer explores different architectures for the implementation. This book presents several effective techniques to be used for design space exploration. The designed architecture is then coded in a hardware description language (HDL), maybe Verilog or VHDL. The designer needs to write *test vectors* to thoroughly verify the implementation. In instances where part of the application is running in SW on a DSP, it is critical to verify the HW design in a co-verification environment. The designer can use System-Verilog for this purpose. Several commercially available tools like MATLAB® also help the designer to co-simulate and co-verify the design. The verified RTL code is then synthesized on a target technology using a synthesis tool. If the technology is an FPGA, the synthesis tool also generates the final layout. In the case of an ASIC, a gate-level netlist is generated. The designer can perform timing verification and then use tools to lay out the design.

In parallel, a printed circuit board (PCB) is designed. The board, along with all the necessary interfaces like PCI, USB and Ethernet, houses the hybrid technologies such as DSP, GPPs, ASICs

Table 3.1 General requirement specification of a UHF radio

Characteristics	Specification
Output power	2W
Spurious emission	< 60 dB
Harmonic suppression	> 55 dB
Frequency stability	2 ppm or better
Reliability	> 10,000 hours MTBF minimum < 30 minutes MTTR
Handheld	12V DC nickel metal hydride, nickel cadmium or lithium-ion (rechargeable) battery pack

and FPGAs. Once the board is ready the first task is to check all the devices on the board and their interconnections. The application is then mapped where the HW part of it runs on ASICs or FPGAs and the SW executes on DSPs and GPPs.

3.2.2 *Example: Requirements and Specifications of a UHF Software-defined Radio*

Gathering R&S is the first part of the system design. This requires a comprehensive understanding of different stakeholders. The design cycle starts with visualizing different components of the system and then breaking up the system into its main components. In the case of a UHF radio, this amounts to breaking the system down into an analog front end (AFE) and the digital software-defined radio (SDR). The AFE consists of an antenna unit, a power amplifier and one or multiple stages of RF mixer, whereas the digital part of the receiver takes the received signal at intermediate frequency (IF) and its corresponding transmitter part up-converts the baseband signal to IF and passes it to AFE. Table 3.1 lists general requirements on the design. These requirements primarily list the power output, spurious emissions and harmonic suppression outside the band of operation, and the frequency stability of the crystal oscillator in parts per million (ppm). A frequency deviation of 2 ppm amounts to 0.0002% over the specified temperature range with respect to the frequency at ambient temperature 25 °C. This part also lists mean time between failure (MTBF) and mean time to recover (MTTR). Table 3.2, lists transmitter and receiver specifications. These are important

Table 3.2 Transmitter and receiver characteristics

Characteristic	Specification
Frequency range BW	420 MHz to 512 MHz
Data rate	Up to 512 kbps multi-channel non-line of sight
Channel	Multi-path with 15 μs delay spread and 220 km/h relative speed between transmitter and receiver
Modulation	OFDM supporting BPSK, QPSK and QAM
FEC	Turbo codes, convolution, Reed–Solomon
Frequency hopping	> 600 hops/s, frequency hopping on full hopping band
Waveforms	Radio works as SDR and should be capable of accepting additional waveforms

because the algorithm designer has to explore digital communication techniques to meet them. The specification lists OFDM to be used supporting BPSK, QPSK and QAM modulations. The technique caters for multipath effects and fast fading due to Doppler shifts as the transmitter and receiver are expected to move very fast relative to each other. The communication system also supports different FEC techniques at 512 kbps. Further to this the system needs to fulfill some special requirements. Primarily these are to support frequency hopping and provide flexibility and programmability in the computing platform for mapping additional waveforms. All these specifications pose stringent computational requirements on the digital computing platform and at the same time require them to be programmable; that is, to consist of GPP, DSP and FPGA technologies.

The purpose of including this example is not to entangle the reader's mind in the complexities of a digital communication system. Rather, it highlights that a complete list of specifications is critical upfront before initiating the design cycle of developing a signal processing system.

3.2.3 Coding Guidelines for High-level Behavioral Description

The algorithm designer should always write MATLAB® code for easy translation into C/C ++ and its subsequent HW/SW partitioning, where HW is implementation is in RTL Verilog and SW components are mapped on DSP. Behavioral tools like MATLAB® provide a good environment for the algorithm designer to experiment with different algorithms without paying much attention to implementation details. Sometimes this flexibility leads to a design that does not easily go down in the design cycle. It is therefore critical to adhere to the guidelines while developing algorithms in MATLAB®. The code should be written in such a way that eases HW/SW partitioning and fixed-point translation. The HW designer, SW developer and verification engineer should be able to conveniently fall back to the original implementation in case of any disagreement. The following are handy guidelines.

1. The code should be structured in distinct components defined as MATLAB® functions with defined interfaces in terms of input and output arguments and internal data storages. This helps in making meaningful HW/SW partitioning decisions.
2. All variables and constants should be defined and packaged in data structures. All user-defined configurations should preferably be packaged in one structure while system design constants are placed in another structure. The internal states must be clearly defined as structure elements for each block and initialized at the start of simulation.
3. The code must be designed for the processing of data in chunks. The code takes a predefined block of data from an input FIFO (first-in/first-out) and produces a block of data at the output, storing it in an output FIFO, mimicking the actual system where usually data from an analog-to-digital converter is acquired in a FIFO or in a ping-pong buffer and then forwarded to HW units for processing. This is contrary to the normal tendency followed while coding applications in MATLAB®, where the entire data is read from a file for processing in one go. Each MATLAB® function in the application should process data in blocks. This way of processing data requires the storing of intermediate states of the system, and these are used by the function to process the next block of data.

An example MATLAB® code that works on chunks of data for simple baseband modulation is shown below. The top-level module sets the user parameters and calls the initialization functions and main modulation function. The processing is done on a chunk-by-chunk basis.

```
% BPSK = 1, QPSK = 2, 8PSK = 3, 16QAM = 4
% All-user defined parameters are set in structure USER_PARAMS
USER_PARAMS.MOD_SCH = 2; %select QPSK for current simulation
USER_PARAMS.CHUNK_SZ = 256; %set buffer size for chunk by chunk processing
USER_PARAMS.NO_CHUNKS = 100;% set no of chunks for simulation
% generate raw data for simulation
raw_data = randint(1, USER_PARAMS.NO_CHUNKS*USER_PARAMS.CHUNK_SZ)
% Initialize user defined, system defined parameters and states in respective
% Structures
PARAMS = MOD_Params_Init(USER_PARAMS);
STATES = MOD_States_Init(PARAMS);
mod_out = [];
% Code should be structured to process data on chunk-by-chunk basis
for iter = 0:USER_PARAMS.NO_CHUNKS-1
  in_data = raw_data
  (iter*USER_PARAMS.CHUNK_SZ+1:USER_PARAMS.CHUNK_SZ*(iter+1));
  [out_sig,STATES]= Modulator(in_data,PARAMS,STATES);
  mod_out = [mod_out out_sig];
end
```

The parameter initialization function sets all the parameters for the modulator.

```
% Initializing the user defined parameters and system design parameters
% In PARAMS
function PARAMS = MOD_Params_Init(USER_PARAMS)
% Structure for transmitter parameters
PARAMS.MOD_SCH = USER_PARAMS.MOD_SCH;
PARAMS.SPS = 4; % Sample per symbol
% Create a root raised cosine pulse-shaping filter
PARAMS.Nyquist_filter = rcosfir(.5 , 5, PARAMS.SPS, 1);
% Bits per symbol, in this case bits per symbols is same as mod scheme
PARAMS.BPS = USER_PARAMS.MOD_SCH;
% Lookup tables for BPSK, QPSK, 8-PSK and 16-QAM using gray coding
BPSK_Table = [(-1 + 0*j) (1 + 0*j)];
QPSK_Table = [(-.707 - .707*j) (-.707 + .707*j)...
             (.707 - .707*j) (.707 + .707*j)];
PSK8_Table = [(1 + 0j) (.7071 + .7071i) (-.7071 + .7071i) (0 + i)...
             (-1 + 0i) (-.7071 - .7071i) (.7071 - .7071i) (0 - 1i)];
QAM_Table = [(-3 + -3*j) (-3 + -1*j) (-3 + 3*j) (-3 + 1*j) (-1 + -3*j)...
             (-1 + -1*j) (-1 + 3*j) (-1 + 1*j) (3 + -3*j) (3 + -1*j)...
             (3 + 3*j) (3 + 1*j) (1 + -3*j) (1 + -1*j) (1 + 3*j) (1 + 1*j)];
% Constellation selection according to bits per symbol
if(PARAMS.BPS == 1)
  PARAMS.const_Table = BPSK_Table;
elseif(PARAMS.BPS == 2)
  PARAMS.const_Table = QPSK_Table;
elseif(PARAMS.BPS == 3)
  PARAMS.const_Table = PSK8_Table;
elseif(PARAMS.BPS == 4)
  PARAMS.const_Table = QAM_Table;
```

```
else
  error(ERROR!!! This constellation size not supported)
end
```

Similarly the state initialization function sets all the states for the modulator for chunk-by-chunk processing of data. For this simple example, the symbols are first zero-padded and then they are passed through a Nyquist filter. The delay line for the filter is initialized in this function. For more complex applications this function may have many more arrays.

```
function STATES = MOD_States_Init(PARAMS)
% Pulse shaping filter delayline
STATES.filter_delayline = zeros(1,length(PARAMS.Nyquist_filter)-1);
```

And finally the actual modulation function performs the modulation on a block of data.

```
function [out_data, STATES] = Modulator(in_data, PARAMS, STATES);
% Bits to symbols conversion

sym = reshape(in_data,PARAMS.BPS,length(in_data)/PARAMS.BPS)';
% Binary to decimal conversion
sym_decimal = bi2de(sym);
% Bit to symbol mapping
const_sym = PARAMS.const_Table(sym_decimal+1);
% Zero padding for up-sampling
up_sym = upsample(const_sym,PARAMS.SPS);
% Zero padded signal passed through Nyquist filter
[out_data, STATES.filter_delayline] =
            filter(PARAMS.Nyquist_filter,1,up_sym,
            STATES.filter_delayline);
```

This MATLAB® example defines a good layout for implementing a real-time signal processing application for subsequent mapping in software and hardware. In complex applications some of the MATLAB® functions may be mapped in SW while others are mapped in HW. It is important to keep this aspect of system design from inception and divide the implementation into several components, and for each component group its data of parameters and states in different structures.

3.2.4 *Fixed-point versus Floating-point Hardware*

From a signal processing perspective, an algorithm can be implemented using fixed- or floating-point format. The floating-point format stores a number in terms of mantissa and exponent. Hardware that supports the floating point-format, after executing each computation, automatically scales the mantissa and updates the exponent to make the result fit in the required number of bits in a defined way. All these operations make floating-point HW more expensive in terms of area and power than fixed-point HW.

A fixed-point HW, after executing a computation, does not track the position of the decimal point and leaves this responsibility to the developer. The decimal point is fixed for each variable and is predefined. By fixing the point a variable can take only a fixed range of values. As the variable is bounded, if the result of a calculation falls outside of this range the data is lost or corrupted. This is known as *overflow*.

There are various solutions for handling overflows in fixed-point arithmetic. Handling overflows requires saturating the result to its maximum positive or minimum negative value that can be assigned to a variable defined in fixed-point format. This results in reduced performance or accuracy. The programmer can fix the places of decimal points for all variables such that the arrangement prevents any overflows. This requires the designer to perform testing with all the possible data and observe the ranges of values all variables take in the simulation. Knowing the ranges of all the variables in the algorithm makes determination of the decimal point that avoids overflow very trivial.

The implementation of a signal processing and communication algorithm on a fixed-point processor is a straigtforward task. Owing to their low power consumption and relative cheapness, fixed-point DSPs are very common in many embedded applications. Whereas floating-point processors normally use 32-bit floating-point format, 16-bit format is normally used for fixed-point implementation. This results in fixed-point designs using less memory. On-chip memory tends to occupy the most silicon area, so this directly results in reduction in the cost of the system. Fixed-point designs are widely used in multimedia and telecommunication solutions.

Although the floating-point option is more expensive, it gives more accuracy. The radix point floats around and is recalculated with each computation, so the HW automatically detects whether overflow has occurred and adjusts the decimal place by changing the exponent. This eliminates overflow errors and reduces inaccuracies caused by unnecessary rounding.

3.3 Representation of Numbers

3.3.1 *Types of Representation*

All signal processing applications deal with numbers. Usually in these applications an analog signal is first digitized and then processed. The discrete number is represented and stored in N bits. Let this N-bit number be $\boldsymbol{a} = a_{N-1} \ldots a_2 a_1 a_0$. This may be treated as *signed* or *unsigned.*

There are several ways of representing a value in these N bits. When the number is unsigned then all the N bits are used to express its magnitude. For signed numbers, the representation must have a way to store the sign along with the magnitude.

There are several representations for signed numbers [3–7]. Some of them are one's complement, sign magnitude [8], canonic sign digit (CSD), and two's complement. In digital system design the last two representations are normally used. This section gives an account of two's complement representation, while the CSD representation is discussed in Chapter 6.

3.3.2 *Two's Complement Representation*

In two's complement representation of a signed number $\boldsymbol{a} = a_{N-1} \ldots a_2 a_1 a_0$, the most significant bit (MSB) a_{N-1} represents the sign of the number. If $\boldsymbol{a}$ is positive, the sign bit a_{N-1} is zero, and the remaining bits represent the magnitude of the number:

$$\boldsymbol{a} = \sum_{i=0}^{N-2} a_i 2^i \quad \text{for } a \geq 0 \tag{3.1}$$

Therefore the two's complement implementation of an N-bit positive number is equivalent to the $(N-1)$-bit unsigned value of the number. In this representation, 0 is considered as a positive number. The range of positive numbers that can be represented in N bits is from 0 to $(2_{N-1}-1)$.

For negative numbers, the MSB $a_{N-1}=1$ has a negative weight and all the other bits have positive weight. A closed-form expression of a two's complement negative number is:

$$\boldsymbol{a} = -2^{N-1} + \sum_{i=0}^{N-2} a_i 2^i \quad \text{for } a < 0 \tag{3.2}$$

Combining (3.1) and (3.2) into (3.3) gives a unified representation to two's complement numbers:

$$\boldsymbol{a} = -2^{N-1} a_{N-1} + \sum_{i=0}^{N-2} a_i 2^i \tag{3.3}$$

It is also interesting to observe an unsigned equivalent of negative two's complement numbers. Many SW and HW simulation tools display all numbers as unsigned numbers. While displaying numbers these tools assign positive weights to all the bits of the number. The equivalent unsigned representation of negative numbers is given by:

$$2^N - |a| \tag{3.4}$$

where $|\boldsymbol{a}|$ is equal to the absolute value of the negative number $\boldsymbol{a}$.

Example: −9 in a 5-bit two's complement representation is 10111. The equivalent unsigned representation of −9 is $2^5 - 9 = 23$.

3.3.3 Computing Two's Complement of a Signed Number

The two's complement of a signed number refers to the negative of the number. This can be computed by inverting all the bits and adding 1 to the least significant bit (LSB) position of the number. This is equivalent to inverting all the bits while moving from LSB to MSB and leaving the least significant 1 as it is. From the hardware perspective, adding 1 requires an adder in the HW, which is an expensive preposition. Chapter 5 covers interesting ways of avoiding the adder while computing the two's complement of a number in HW designs.

Example: The 4-bit two's complement representation of −2 is $4'b1110$, its $2'$s complement is obtained by inverting all the bits, into $4'b0001$, and adding 1 to that inverted number giving 4′b0010.

Table 3.3 lists the two's complement representations of 4 bit numbers and their unsigned equivalent numbers. In two's complement, the representation of a negative number does look odd because after 7 comes −8, but it facilitates the hardware implementation of many basic arithmetic operations. For this reason, this representation is used widely for executing arithmetic operations in special algorithms and general-purpose architectures.

It is apparent from (3.3) above that the MSB position plays an important role in two's complement representation of a negative number and directly affects many arithmetic operations. While implementing algorithms using two's complement arithmetic, a designer needs to deal with the issue of overflow. An overflow occurs if a calculation generates a number that cannot be represented using the assigned number of bits in the number. For example, 7 + 1 = 8, and 8 cannot be represented as a 4-bit signed number. The same is the case with −9, which may be produced as (−8) − 1.

Table 3.3 Four-bit representation of two's complement number and its equivalent unsigned number

Decimal number	Two's complement representation				Equivalent unsigned number
	-2^3	2^2	2^1	2^0	
0	0	0	0	0	0
+1	0	0	0	1	1
+2	0	0	1	0	2
+3	0	0	1	1	3
+4	0	1	0	0	4
+5	0	1	0	1	5
+6	0	1	1	0	6
+7	0	1	1	1	7
−8	1	0	0	0	8
−7	1	0	0	1	9
−6	1	0	1	0	10
−5	1	0	1	1	11
−4	1	1	0	0	12
−3	1	1	0	1	13
−2	1	1	1	0	14
−1	1	1	1	1	15

3.3.4 Scaling

While implementing algorithms using finite precision arithmetic it is sometimes important to *avoid* overflow as it adds an error that is equal to the complete dynamic range of the number. For example, the case $7 + 1 = 8 = 4'b1000$ as a 4-bit signed number is −8. To avoid overflow, numbers are *scaled down.* In digital designs it is sometimes also required to sign extend an N-bit number to an M-bit number for $M > N$.

3.3.4.1 Sign Extension

In the case of a signed number, without affecting the value of the number, $M - N$ bits are sign extended. Thus a number can be extended by any number of bits by copying the sign bit to extended bit locations. Although this extension does not change the value of the number, its unsigned equivalent representation will change. The new equivalent number will be:

$$2^M - |a| \tag{3.5}$$

Example: The number −2 as a 4-bit binary number is $4'b1110$. As an 8-bit number it is $8'b11111110$.

3.3.4.2 Scaling-down

When a number has redundant sign bits, it can be scaled down by dropping the redundant bits. This dropping of bits will not affect the value of the number.

Example: The number – 2 as an 8-bit two's complement signed number is $8'b11111110$. There are six redundant sign bits. The number can be easily represented as a 2-bit signed number after throwing away six significant bits. The truncated number in binary is $2'b10$.

3.4 Floating-point Format

Floating-point representation works well for numbers with large dynamic range. Based on the number of bits, there are two representations in IEEE 754 standard [9]: 32-bit single-precision and 64-bit double-precision. This standard is almost exclusively used across computing platforms and hardware designs that support floating-point arithmetic. In this standard a normalized floating-point number x is stored in three parts: the sign s, the excess exponent e, and the significand or mantissa m, and the value of the number in terms of these parts is:

$$x = (-1)^s \times 1 \times m \times 2^{e-b} \tag{3.6}$$

This indeed is a sign magnitude representation, s represents the sign of the number and m gives the normalized magnitude with a 1 at the MSB position, and this implied 1 is not stored with the number. For normalized values, m represents a fraction value greater than 1.0 and less than 2.0. This IEEE format stores the exponent e as a biased number that is a positive number from which a constant bias b is subtracted to get the actual positive or negative exponent.

Figure 3.2 shows this representation for a single-precision floating point number. Such a number is represented in 32 bits, where 1 bit is kept for the sign s, 8 bits for the exponent e and 23 bits for the mantissa m. For a 64-bit double-precision floating-point number, 1 bit is kept for the sign, 11 bits for the exponent and 52 bits for the mantissa. The values of bias b are 127 and 1023, respectively, for single-and double-precision floating-point formats.

Example: Find the value of the following 32-bit binary string representing a single-precision IEEE floating-point format: 0_10000010_11010000_00000000_0000000. The value is calculated by parsing the number into different fields, namely sign bit, exponent and mantissa, and then computing the value of each field to get the final value, as shown in Table 3.4.

Example: This example represents -12.25 in single-precision IEEE floating-point format. The number -12.25 in sign magnitude binary is -00001100.01. Now moving the decimal point to bring it into the right format: $-1100.01 \times 2^0 = -1.10001 \times 2^3$. Thus the normalized number is -1.10001×2^3.

$$\begin{aligned}
&\text{Sign bit (s)} = 1\\
&\text{Mantissa field(m)} = 10001000_00000000_0000000\\
&\text{Exponent field(e)} = 3 + 127 = 130 = 82\,\text{h} = 1000\,0010
\end{aligned}$$

So the complete 32-bit floating-point number in binary representation is:

1_10000010_10001000_00000000_0000000

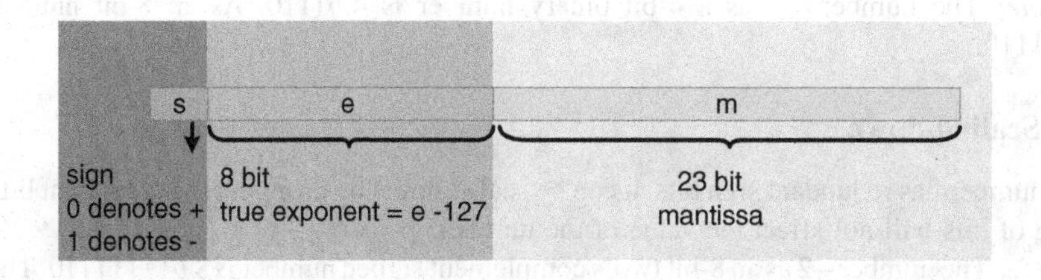

Figure 3.2 IEEE format for single-precision 32-bit floating point number

Table 3.4 Value calculated by parsing and then computing the value of each field to get the final value, $+14.5_{10}$ (see text)

Sign bit	Exponent		Mantissa
0	10000010		11010000_00000000_0000000
$(-1)^0$	$\times 2^{(130-127)}$	$\times$	$(1\,.\,1\,1\,0\,1)_2$
1	$\times 2^{(3)}$	$\times$	$(1 + 0.5 + .25 + 0 + 0.0625)_{10}$
$(+1)$	$\times 2^{(3)}$	$\times$	$(1.8125)_{10}$

A floating-point number can also overflow if its exponent is greater or smaller than the maximum or minimum value the e-field can represent. For IEEE single-precision floating point format, the minimum to maximum exponent ranges from 2^{-126} to 2^{127}. On the same account, allocating 11 bits in the e-field as opposed to 8 bits for single-precision numbers, minimum and maximum values of a double-precision floating-point number ranges from 2^{-1022} to 2^{1023}.

Although the fields in the format can represent almost all numbers, there are still some special numbers that require unique handling. These are:

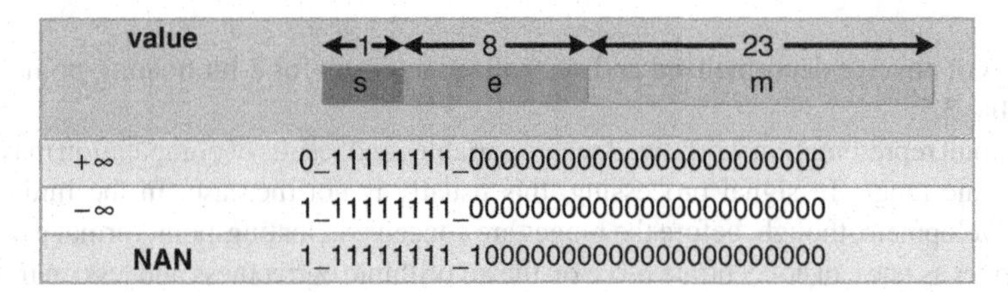

A $\pm\infty$ may be produced if any floating-point number is divided by zero, so $1.0/0.0 = +\infty$. Similarly, $-1.0/0.0 = -\infty$. Not A Number (NAN) is produced for an invalid operations like 0/0 or $\infty - \infty$.

3.4.1 Normalized and Denormalized Values

The floating-point representation covers a wide dynamic range of numbers. There is an extent where the number of bits in the mantissa is enough to represent the exact value of the floating-point number and zero placed in the exponent. This range of values is termed *denormalized.* Beyond this range, the representation keeps normalizing the number by assigning an appropriate value in the exponent field. A stage will be reached where the number of bits in the exponent is not enough to represent the normalized number and results in $+\infty$ or $-\infty$. In instances where a wrong calculation is conducted, the number is represented as NAN. This dynamic range of floating-point representation is shown in Figure 3.3.

In the IEEE format of floating-point representation, with an implied 1 in the mantissa and a nonzero value in the exponent, where all bits are neither 0 nor 1, are called *normalized* values. *Denormalized* values are values where all exponent bits are zeros and the mantissa is non-zero.

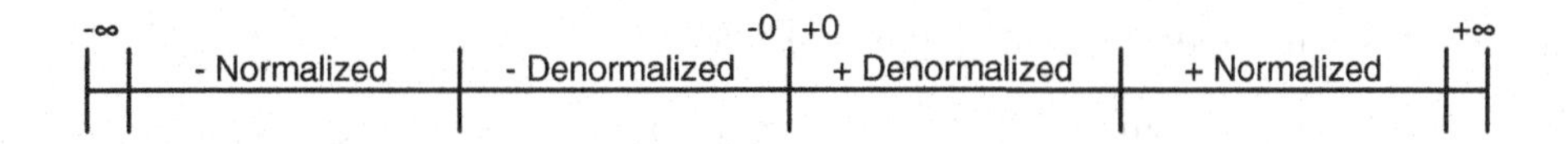

Figure 3.3 Dynamic range of a floating-point number

These values represent numbers in the range of zero and smallest normalized number on the number line:

$$\text{value} = \begin{cases} (-1)^s \cdot (2^{e-127}) \cdot (1.m) & \text{normalized}, 0 < e < 255 \\ (-1)^s \cdot (2^{e-126}) \cdot (0.m) & \text{denormalized}, \ e = 0, m > 0. \end{cases}$$

The example below illustrates normalized and denormalized numbers.

Example: Assume a floating-point number is represented as an 8-bit number. There is one sign bit and 4 and 3 bits, respectively, are allocated to store exponent and mantissa. By traversing through the entire range of the number, limits for denormalized and normalized values can be easily marked. The value where the e-field is all zeros and the m-field is non-zero represents denormalized values. This value does not assume an implied 1 in the mantissa. For a normalized value an implied 1 is assumed and a bias of $+7$ is added in the true exponent to get the e-field. Therefore for a normalized value this bias is subtracted from the e-field to get the actual exponent. These numbers can be represented as:

$$\text{value} = \begin{cases} (-1)^s \cdot (2^{e-7}) \cdot (1.m) & \text{normalized}, 0 < e < 7 \\ (-1)^s \cdot (2^{-6}) \cdot (0.m) & \text{denormalized}, e = 0, m > 0. \end{cases}$$

The ranges of positive denormalized and normalized numbers for 8-bit floating-point format are given in Table 3.5.

Floating-point representation works well where variables and results of computation may vary over a large dynamic range. In signal processing, this usually is not the case. In the initial phase of algorithm development, though, before the ranges are conceived, floating-point format gives comfort to the developer as one can concentrate more on the algorithmic correctness and less on implementation details. If there are no strict requirements on numerical accuracy of the computation, performing

Table 3.5 Various 8-bit floating-point numbers

	s	E	m	Exp	Value		
Denormalized numbers	0	0000	000	-	0		
	0	0000	001	−6	0.125×2^{-6}	←	Closest to zero
	0	0000	010	−6	0.5×2^{-6}		
	...						
	0	0000	110	−6	0.75×2^{-6}		
	0	0000	111	−6	0.875×2^{-6}	←	Largest denormalized value
	0	0001	000	−6	1.0×2^{-6}	←	Smallest normalized value
Normalized numbers	0	0001	001	−6	1.125×2^{-6}		
	...						
	0	0110	110	−1	1.75×2^{-1}		
	0	0110	111	−1	1.875×2^{-1}		
	0	0111	000	0	1		
	0	0111	001	0	1.125×2^{0}		
	...						
	0	1110	111	7	1.875×2^{7}	←	Largest normalized value
	0	1111	000		∞		

floating-point arithmetic in HW is an expensive preposition in terms of power, area and performance, so is normally avoided. The large gate densities of current generation FPGAs and ASICs allow designers to map floating-point algorithms in HW if required. This is especially true for more complex signal processing applications where keeping the numerical accuracy intact is considered critical [10].

Floating-point arithmetic requires three operations: exponent adjustment, mathematical operation, and normalization. Although no dedicated floating-point units are provided on FPGAs, embedded fast fixed-point adder/subtractors, multipliers and multiplexers are used in effective floating-point arithmetic units [11].

3.4.2 Floating-point Arithmetic Addition

The following steps are used to perform floating-point addition of two numbers:

S0 Append the implied bit of the mantissa. This bit is 1 or 0 for normalized and denormalized numbers, respectively.
S1 Shift the mantissas from S0 with smaller exponent e_s to the right by $e_l - e_s$, where e_l is the larger of the two exponents. This shift may be accomplished by providing a few guard bits to the right for better precision.
S2 If any of the operands is negative, take two's complement of the mantissa from S1 and then add the two mantissas. If the result is negative, again takes two's complement of the result.
S3 Normalize the sum back to IEEE format by adjusting the mantissa and appropriately changing the value of the exponent e_l. Also check for underflow and overflow of the result.
S4 Round or truncate the resultant mantissa to the number of bits prescribed in the standard.

Example: Add the two floating-point numbers below in 10-bit precision, where 4 bits are kept for the exponent, 5 bits for the mantissa and 1 bit for the sig. Assume the bias value is 7:

0_1010_00101
0_1001_00101

Taking the bias +7 off from the exponents and appending the implied 1, the numbers are:

$$1.00101_b \times 2^3 \text{ and } 1.00101_b \times 2^2$$

S0 As both the numbers are positive, there is no need to take two's complements. Add one extra bit as guard bit to the right, and align the two exponents by shifting the mantissa of the number with the smaller exponent accordingly:

$$1.00101_b \times 2^3 \rightarrow 1.001010_b \times 2^3$$
$$1.00101_b \times 2^2 \rightarrow 0.100101_b \times 2^3$$

S1 Add the mantissas:

$$1.001010_b + 0.100101_b = 1.101111_b$$

S2 Drop the guard bit:

$$1.10111_b \times 2^3$$

S3 The final answer is $1.10111_b \times 2^3$, which in 10-bit format of the operands is 0_1010_10111.

3.4.3 Floating-point Multiplication

The following steps are used to perform floating-point multiplication of two numbers.

S0 Add the two exponents e_1 and e_2. As inherent bias in the two exponents is added twice, subtract the bias once from the sum to get the resultant exponent e in correct format.
S1 Place the implied 1 if the operands are saved as normalized numbers. Multiply the mantissas as unsigned numbers to get the product, and XOR the two sign bits to get the sign of the product.
S2 Normalize the product if required. This puts the result back to IEEE format. Check whether the result underflows or overflows.
S3 Round or truncate the mantissa to the width prescribed by the standard.

Example: Multiply the two numbers below stored as 10-bit floating-point numbers, where 4 bits and 5 bits, respectively, are allocated for the exponent and mantissa, 1 bit is kept for the sign and 7 is used as bias:

$$3.5 \rightarrow 0\_1000\_11000$$
$$5.0 \rightarrow 0\_1001\_01000$$

S0 Add the two exponents and subtract the bias 7 (=0111) from the sum:

$$1000 + 1001 - 0111 = 1010$$

S1 Append the implied 1 and multiply the mantissa using unsigned × unsigned multiplication:

$$\begin{array}{l} 1.11 \\ \underline{1.01} \\ 111 \\ \\ 00\mathrm{x} \\ \underline{111\mathrm{xx}} \\ 10.0011 \end{array}$$

S2 Normalize the result:

$$(10.0011)_b \times 2^3 \rightarrow (1.00011)_b \times 2^4$$
$$(1.00011)_b \times 2^4 \rightarrow (17.5)_{10}$$

S3 Put the result back into the format of the operands. This may require dropping of bits but in this case no truncation is required and the result is $(1.00011)_b \times 2^4$ stored as 0_1011_00011.

3.5 Qn.m Format for Fixed-point Arithmetic

3.5.1 Introducing Qn.m

Most signal processing and communication systems are first implemented in double-precision floating-point arithmetic using tools like MATLAB®. While implementing these algorithms the main focus of the developer is to correctly assimilate the functionality of the algorithm. This MATLAB® code is then converted into floating-point C/C++ code. C++ code usually runs much faster than MATLAB® routines. This code conversion also gives more understanding of the algorithm as the designer might have used several functions from MATLAB® toolboxes. Their

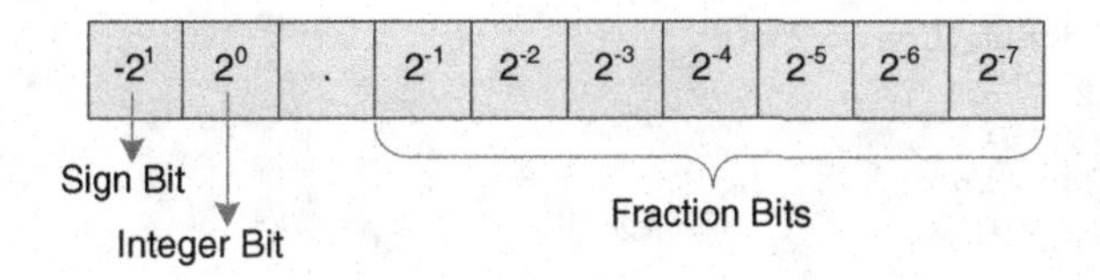

Figure 3.4 Fields of the bits and their equivalent weights for the text example

understanding is critical for porting these algorithms in SW or HW for embedded devices. After getting the desired performance from the floating-point algorithm, this implementation is converted to fixed-point format. For this the floating-point variables and constants in the simulation are converted to Q*n.m* fixed-point format. This is a fixed positional number system for representing floating-point numbers.

The Q*n.m* format of an *N*-bit number sets *n* bits to the left and *m* bits to the right of the binary point. In cases of signed numbers, the MSB is used for the sign and has negative weight. A two's complement fixed-point number in Q*n.m* format is equivalent to $b = b_{n-1}b_{n-2}...b_1b_0b_{-1}b_{-2}...b_{-m}$, with equivalent floating point value:

$$-b_{n-1}2^{n-1} + b_{n-2}2^{n-2} + \ldots + b_1 2^1 + b_0 + b_{-1}2^{-1} + b_{-2}2^{-2} + \ldots b_{-m}2^{-m} \tag{3.7}$$

Example: Compute the floating-point equivalent of 01_1101_0000 in signed Q2.8 format. The fields of the bits and their equivalent weights are shown in Figure 3.4.

Assigning values to bit locations gives the equivalent floating-point value of the Q-format fixed-point number:

$$01\_1101\_0000 = 0 + 1 + \frac{1}{2} + \frac{1}{4} + \frac{1}{16} = 01.8125$$

This example keeps 2 bits to store the integer part and the remaining 8 bits are for the decimal part. In these 10 bits the number covers −2 to +1.9961.

3.5.2 *Floating-point to Fixed-point Conversion of Numbers*

Conversion requires serious deliberation as it results in a tradeoff between performance and cost. To reduce cost, shorter word lengths are used for all variables in the code, but this adversely affects the numerical performance of the algorithm and adds quantization noise.

A floating-point number is simply converted to Q*n.m* fixed-point format by brining *m* fractional bits of the number to the integer part and then dropping the rest of the bits with or without rounding. This conversion translates a floating-point number to an integer number with an implied decimal. This implied decimal needs to be remembered by the designer for referral in further processing of the number in different calculations:

$$\text{num\_fixed} = \text{round}(\text{num\_float} \times 2^{\text{m}})$$
$$\text{or num\_fixed} = \text{fix}\,(\text{num\_float} \times 2^{\text{m}})$$

The result is *saturated* if the integer part of the floating-point number is greater than *n*. This is simply checked if the converted fixed-point number is greater than the maximum positive or less than the minimum negative *N*-bit two's complement number. The maximum and minimum values of an *N*-bit two's complement number are $2^{N-1} - 1$ and -2^{N-1}, respectively.

```
num_fixed = round(num_float ×2^m)
if (num_fixed > 2^(N-1) -1)
    num_fixed = 2^(N-1) -1
elseif (num_fixed < -2^(N-1))
    num_fixed = -2^(N-1)
```

Example: Most commercially available off-the-shelf DSPs have 16-bit processors. Two examples of these processors are the Texas Instruments TMS320C5514 and the Analog Devices ADSP-2192. To implement signal processing and communication algorithms on these DSPs, Q1.15 format (commonly known as Q15 format) is used. The following pseudo-code describes the conversion logic to translate a floating-point number `num_float` to fixed-point number `num_fixed_Q15` in Q1.15 format:

```
num_fixed_long = (long)(num_float × 2^15)
if (num_fixed_long > 0x7fff)
   num_fixed_long = 0x7fff
elseif (num_fixed_long < 0xffff8000)
   num_fixed_long =0xffff8000
num_fixed_Q15 = (short)(num_fixed_long & 0xffff))
```

Using this logic, the following lists a few floating-point numbers (`num_float`) and their representation in Q1.15 fixed-point format (`num_fixed_Q15`):

```
0.5 → 0x4000
-0.5 → 0xC000
0.9997 → 0x7FF6
0.213 → 0x1B44
-1.0 → 0x8000
```

3.5.3 Addition in Q Format

Addition of two fixed-point numbers a and b of $Qn_1.m_1$ and $Qn_2.m_2$ formats, respectively, results in a $Qn.m$ format number, where n is the larger of n_1 and n_2 and m is the larger of m_1 and m_2. Although the decimal is implied and does not exist in HW, the designer needs to align the location of the implied decimal of both numbers and then appropriately sign extend the number that has the least number of integer bits. As the fractional bits are stored in least significant bits, no extension of the fractional part is required. The example below illustrates Q-format addition.

Example: Add two signed numbers a and b in Q2.2 and Q4.4 formats. In Q2.2 format a is 1110, and in Q4.4 format b is 0111_0110. As n_1 is less than n_2, the sign bit of a is extended to change its format from Q2.2 to Q4.2 (Figure 3.5). The extended sign bits are shown with bold letters. The numbers are also placed in such a way that aligns their implied decimal point. The addition results in a Q4.4 format number.

3.5.4 Multiplication in Q Format

If two numbers a and b in, respectively, $Qn_1.m_1$ and $Qn_2.m_2$ formats are multiplied, the multiplication results in a product in $Q(n_1+n_2)\cdot(m_1+m_2)$ format. If both numbers are signed two's complement numbers, we get a redundant sign bit at the MSB position. Left shifting the product

implied decimal

Qn1.m1	1	1	1	1 •	1	0			= Q4.2 = -2+1+0.5 = -0.5
Qn2.m2	0	1	1	1	0	1	1	0	= Q4.4 = 1+2+4+025+0.125 = 7.375
Qn.m	0	1	1	0	1	1	1	0	= Q4.4 = 2+4+0.5+0.25+0.125 = 6.875

Figure 3.5 Example of addition in Q format

by 1 removes the redundant sign bit, and the format of the product is changed to $Q(n_1 + n_2 - 1)\cdot(m_1 + m_2 + 1)$. Below are listed multiplications for all possible operand types.

3.5.4.1 Unsigned by Unsigned

Multiplication of two unsigned numbers in Q2.2 and Q2.2 formats results in a Q4.4 format number, as shown in Figure 3.6. As both numbers are unsigned, no sign extension of partial products is required. The partial products are simply added. Each partial product is sequentially generated and successively shifted by one place to the left.

3.5.4.2 Signed by Unsigned

Multiplication of a signed multiplicand in Q2.2 with an unsigned multiplier in Q2.2 format results in a Q4.4 format signed number, as shown in Figure 3.7. As the multiplicand is a signed number, this

```
        1 1 0 1 = 11.01 in Q2.2 = 3.25
        1 0 1 1 = 10.11 in Q2.2 = 2.75
        -------
        1 1 0 1
      1 1 0 1 X
    0 0 0 0 X X
  1 1 0 1 X X X
---------------
1 0 0 0 1 1 1 1= 1000.1111 in Q4.4 i.e.8.9375
```

Figure 3.6 Multiplication, unsigned by unsigned

```
        1 1 0 1 = 11.01 in Q2.2 = -0.75
        0 1 0 1 = 01.01 in Q2.2 = 1.25
1 1 1 1 1 1 0 1 extended sign bits shown in bold
0 0 0 0 0 0 0 X
1 1 1 1 0 1 X X
0 0 0 0 0 X X X
---------------
1 1 1 1 0 0 0 1 = 1111.0001 in Q4.4 i.e.-0.9375
```

Figure 3.7 Multiplication, signed by unsigned

```
             1 0 0 1 = 10.01 in Q2.2 = 2.25 (unsigned)
             1 1 0 1 = 11.01 in Q2.2 = -0.75 (signed)
             -------
             1 0 0 1
           0 0 0 0 X
         1 0 0 1 X X
     1 0 1 1 1 X X X 2's compliment of the positive multiplicand 01001
     ---------------
     1 1 1 0 0 1 0 1  = 1110.0101 in Q4.4 i.e.-1.6875
```

Figure 3.8 Multiplication, unsigned by signed

requires sign extension of each partial product before addition. In this example the partial products are first sign-extended and then added. The sign extension bits are shown in bold.

3.5.4.3 Unsigned by Signed

Multiplication of an unsigned multiplicand in Q2.2 with a signed multiplier in Q2.2 format results in a Q4.4 format signed number, as shown in Figure 3.8. In this case the multiplier is signed and the multiplicand is unsigned. As all the partial products except the final one are unsigned, no sign extension is required except for the last one. For the last partial product the two's complement of the multiplicand is computed as it is produced by multiplication with the MSB of the multiplier, which has negative weight. The multiplicand is taken as a positive 5-bit number, 01001, and its two's complement is 10111.

3.5.4.4 Signed by Signed

Multiplication of a signed number in Q1.2 with a signed number in Q1.2 format results in a Q1.5 format signed number, as shown in Figure 3.9. Sign extension of partial products is again necessary. This multiplication also produces a redundant sign bit, which can be removed by left shifting the result. Multiplying a Qn_1.m_1 format number with a Qn_2.m_2 number results in a Q$(n_1 + n_2)\cdot(m_1 + m_2)$ number, which after dropping the redundant sign bit becomes a Q$(n_1 + n_2 - 1)\cdot(m_1 + m_2 + 1)$ number.

The multiplication in the example results in a redundant sign bit shown in bold. The bit is dropped by a left shift and the final product is 111000 in Q1.5 format and is equivalent to −0.25.

When the multiplier is a negative number, while calculating the last partial product (i.e. multiplying the multiplicand with the sign bit with negative weight), the two's complement of the multiplicand is taken for the last partial product, as shown in Figure 3.10.

```
          1 1 0 = Q1.2 = -0.5 (signed)
          0 1 0 = Q1.2 = 0.5 (signed)
          -----
    0 0 0 0 0 0
    1 1 1 1 0 X
    0 0 0 0 X X
    -----------
    1 1 1 1 0 0 = Q1.5 format 1_11000=-0.25
```

Figure 3.9 Multiplication, signed by signed

```
        1 1. 0 1 = -0.75 in Q2.2 format
          1. 1 0 1 = -0.375 in Q1.3 format
    ---------------
    1 1 1 1 1 1 0 1
    0 0 0 0 0 0 0 X
    1 1 1 1 0 1 X X
    0 0 0 1 1 X X X
    ---------------
    0 0 0 0 1 0 0 1 = shifting left by one 00.010010 in Q2.6 format is 0.28125
```

Figure 3.10 Multiplication, signed by signed

3.5.5 *Bit Growth in Fixed-point Arithmetic*

Bit growth is one of the most critical issues in implementing algorithms using fixed-point arithmetic. Multiplication of an N-bit number with an M-bit number results in an $(N + M)$-bit number, a problem that grows in recursive computations. For example, implementing the recursive equation:

$$y[n] = ay[n-1] + x[n] \tag{3.8}$$

requires multiplication of a constant a with a previous value of the output $y[n-1]$. If the first output is an N-bit number in $Qn_1.m_1$ format and the constant a is an M-bit number in $Qn_2 \cdot m_2$ format, in the first iteration the product $ay[n-1]$ is an $(N + M)$-bit number in $Q(n_1 + n_2 - 1)\cdot(m_1 + m_2 + 1)$ format. In the second iteration the previous value of the output is now an $N + M$ bits, and once multiplied by a it becomes $N + 2M$ bits. The size will keep growing in each iteration. It is therefore important to curtail the size of the output. This requires *truncating* the output after every mathematical operation. Truncation is achieved by first rounding and then dropping a defined number of least significant bits or without rounding dropping these bits.

The bit growth problem is also observed while adding two different Q-format numbers. In this case the output format of the number will be:

$$Qn.m = Q\max(n_1, n_2).\max(m_1, m_2).$$

Assigning an appropriate format to the output is very critical and one can use Schwarz's inequality of (3.9) to find an upper bound on the output values of a linear time-invariant (LTI) system:

$$|y[n]| \leq \sqrt{\sum_{n=-\infty}^{\infty} h^2[n] \sum_{n=-\infty}^{\infty} x^2[n]} \tag{3.9}$$

Here $h[n]$ and $x[n]$ are impulse response and input to an LTI system, respectively. This helps the designer to appropriately assign bits to the integer part of the format defined to store output values. Assigning bits to the fraction part requires deliberation as it depends on the tolerance of the system to quantization noise.

Example: Adding an 8-bit Q7.1 format number in an 8-bit Q1.7 format number will yield a 14-bit Q7.7 format number.

```
            ↙truncation
0 1 1 1_0 1 1 1 in Q4.4 is 7.4375
              1  ←— rounding
_______________
0 1 1 1_1 0 0 1
0 1 1 1_1 0 0 = 7.5
```

Figure 3.11 Rounding followed by truncation

3.5.5.1 Simple Truncation

In multiplication of two Q-format numbers, the number of bits in the product increases. The precision is sacrificed by dropping some low-precision bits of the product: $Qn_1.m_1$ is truncated to $Qn_1.m_2$, where $m_2 < m_1$.

Example:

0111_0111 in Q4.4 is 7.4375
Truncated to Q4.2 gives 0111_01 = 7.25

3.5.5.2 Rounding Followed by Truncation

Direct truncation of numbers biases the results, so in many applications it is preferred to round before trimming the number to the desired size. For this, 1 is added to the bit that is at the right of the position of the point of truncation. The resultant number is then truncated to the desired number of bits. This is shown in the example in Figure 3.11. First rounding and then truncation gives a better approximation; in the example, simple truncation to Q4.2 results in a number with value 7.25, whereas rounding before truncation gives 7.5 – which is closer to the actual value 7.4375.

3.5.6 *Overflow and Saturation*

Overflow is a serious consequence of fixed-point arithmetic. Overflow occurs if two positive or negative numbers are added and the sum requires more than the available number of bits. For example, in a 3-bit two's complement representation, if 1 is added to 3 (= $3'b011$), the sum is 4 (= $4'b0100$). The number 4 thus requires four bits and cannot be represented as a 3-bit two's complement signed number as $3'b100$ (= −4). This causes an error equal to the full dynamic range of the number and so adversely affects subsequent computation that uses this number.

Figure 3.12 shows the case of an overflow for a 3-bit number, adding an error equal to the dynamic range of the number. It is therefore imperative to check the overflow condition after performing arithmetic operations that can cause a possible overflow. If an overflow results, the designer should set an *overflow flag*. In many circumstances, it is better to curtail the result to the maximum positive or minimum negative value that the defined word length can represent. In the above example the value should be limited to $3'b011$.

Thus, the following computation is in 3-bit precision with an overflow flag set to indicate this abnormal result:

$$3+1=3 \text{ and overflow flag} = 1$$

Similarly, performing subtraction with an overflow flag set to ! is:

$$-4-1=-4 \text{ overflow flag} = 1$$

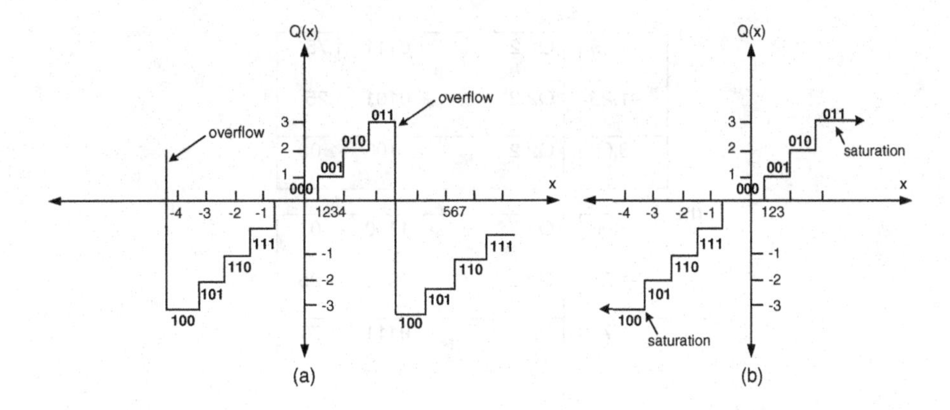

Figure 3.12 Overflow and saturation. (a) Overflow introduces an error equal to the dynamic range of the number. (b) Saturation clamps the value to a maximum positive or minimum negative level

Figure 3.12 shows the overflow and saturation mode for 3-bit two's complement numbers, respectively. The saturation mode clamps the value of the number to the maximum positive or minimum negative.

The designer may do nothing and let the overflow happen, as shown in Figure 3.13(a). For the same calculation, the designer can clamp the answer to a maximum positive value in case of an overflow, as shown in Figure 3.13(b).

3.5.7 Two's Complement Intermediate Overflow Property

In an iterative calculation using two's complement arithmetic, if it is guaranteed that the final result will be within the precision bound of the assigned fixed-point format, then any amount of intermediate overflows will not affect the final answer.

This property is illustrated in Figure 3.14(a) with operands representing 4-bit precision. The first column shows the correct calculation but, owing to 4-bit precision, the result overflows and the

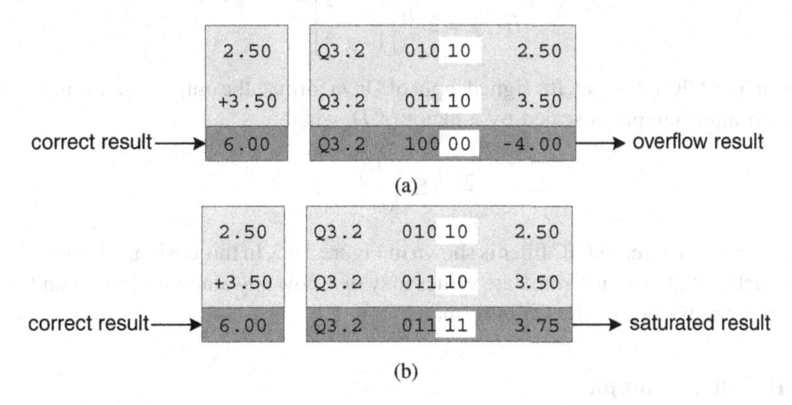

Figure 3.13 Handling overflow: (a) doing nothing; (b) clamping

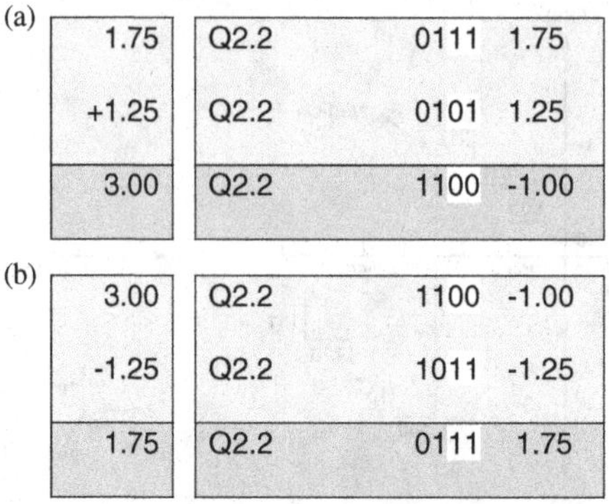

(a)

1.75	Q2.2	0111	1.75
+1.25	Q2.2	0101	1.25
3.00	Q2.2	1100	-1.00

(b)

3.00	Q2.2	1100	-1.00
-1.25	Q2.2	1011	-1.25
1.75	Q2.2	0111	1.75

Figure 3.14 Two's complement intermediate overflow property: (a) incorrect intermediate result; (b) correct final result

calculation yields an incorrect answer as shown in the third column. If the further calculation uses this incorrect answer and adds a value that guarantees to bring the final answer within the legal limits of the format, then this calculation will yield the correct answer even though one of the operands in the calculation is incorrect. The calculation in Figure 3.14(b) shows that even using the incorrect value from the earlier calculation the further addition will yield the right value.

This property of two's complement arithmetic is very useful in designing architectures. The algorithms in many design instances can guarantee the final output to fit in a defined format with possible intermediate overflows.

3.5.7.1 CIC Filter Example

The above property can be used effectively if the bounds on output can be determined. Such is the case with a CIC (cascaded integrator-comb) filter which is an integral part of all digital down-converters. A CIC filter has M stages of integrator executing at sampling rate f_s, followed by a decimator by a factor of L, followed by a cascade of M stages of Comb filter running at decimated sampling rate f_s/L. The transfer function of such a CIC filter is:

$$H(z) = 2^{-D}\left[\frac{1-z^{-L}}{1-z^{-1}}\right]^{M} \tag{3.10}$$

It is proven for a CIC filter that, for signed input of Q$n.m$ format, the output also fits in Q$n.m$ format provided each input sample is scaled by a factor of D, where:

$$2^{-D} \leq \left(\frac{1}{L}\right)^{M}$$

M is the order of the filter. A CIC filter is shown in Figure 3.15. In this design, provided the input is scaled by a factor of D, then intermediate values may overflow any amount of times and the output remains within the bounds of the Q$n.m$ format.

3.5.7.2 FIR Filter Example

If all the coefficients of an FIR (finite impulse response) filter are scaled to fit in Q1.15 format, and the filter design package ensures that the sum of the coefficients is 1, then fixed-point implementation of

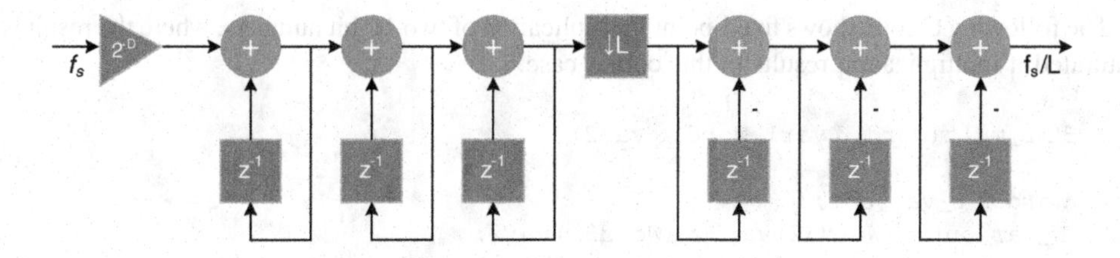

Figure 3.15 An *M*th-order CIC filter for decimating a signal by a factor of *L* (here, $M=3$). The input is scaled by a factor of *D* to avoid overflow of the output. Intermediate values may still overflow a number of times but the output is always correct

the filter does not overflow and uses the complete dynamic range of the defined format. The output can be represented in Q1.15 format. MATLAB® filter design utilities guarantee that the sum of all the coefficients of an FIR filter is 1; that is:

$$\sum_{n=0}^{L-1} h[n] = 1$$

where L is the length of the filter. This can be readily checked in MATLAB® for any amount of coefficients and cutoff frequency, as is done below for a 21-coefficient FIR filter with cutoff frequency π/L:

```
>> sum(fir1(20,.1))
ans =
1.0000
```

3.5.8 Corner Cases

Two's complement arithmetic has one serious shortcoming, in that in *N*-bit representation of numbers it does not have an equivalent opposite of -2^{N-1}. For example, for a 4-bit signed number there is no equivalent opposite of -16, the maximum 4-bit positive number being $+15$. This is normally referred to as a corner case, and the digital designer needs to be concerned with the occurrence of this in computations. Multiplying two signed numbers and dropping the redundant sign bit may also result in this corner case.

For example, consider multiplying -1 by -1 in Q1.2 format, as shown in Figure 3.16. After throwing away the redundant sign bit the answer is 100000 ($=-1$) in Q1.5 format, which is incorrect. The designer needs to add an exclusive logic to check this corner case, and if it happens *saturate* the result with a maximum positive number in the resultant format.

```
          1 0 0 = Q1.2 = -1
          1 0 0 = Q1.2 = -1
    ------------
    0 0 0 0 0 0
    0 0 0 0 0 X
    0 1 0 0 X X    2's complement
    ------------
    0 1 0 0 0 0
    ------------
    1 0 0 0 0 0 = After dropping redundant sign bit
```

Figure 3.16 Multiplying -1 by -1 in Q1.2 format

The following C code shows fixed-point multiplication of two 16-bit numbers, where the result is saturated if multiplication results in this corner case.

```
Word32 L_mult(Word16 var1,Word16 var2)
{
    Word32 L_var_out;
    L_var_out = (Word32)var1 * (Word32)var2;
                                    // Sign*sing multiplication
/*
Check corner case. If before throwing away redundant sign bit the answer
is 0x40000000, it is the corner case, so set the overflow flag and clamp the
value to maximum 32-bit positive number in fixed-point format.
*/
    if (L_var_out != (Word32)0x40000000L) // 8000H*8000H=40000000Hex
    {
    L_var_out *= 2; //Shift the redundant sign bit out
    }
    else
    {
    Overflow = 1;
    L_var_out = 0x7fffffff;
                              //If overflow then clamp the value to MAX_32
    }
    return(L_var_out);
}
```

3.5.9 *Code Conversion and Checking the Corner Case*

The C code below implements the dot-product of two arrays h_float and x_float:

```
float dp_float (float x[], float h[], int size)
{
    float acc;
    int i;
    acc = 0;
    for( i = 0; i < size; i++)
    {
        acc += x[i] * h[i];
    }
    return acc;
}
```

Let the constants in the two arrays be:

```
float h_float[SIZE] = {0.258, -0.309, -0.704, 0.12};
float x_float[SIZE]= {-0.19813, -0.76291,0.57407,0.131769);
```

As all the values in the two arrays h_float and x_float are bounded within $+1$ and -1, an appropriate format for both the arrays is Q1.15. The arrays are converted into this format by multiplying by 2^{15} and rounding the result to change these numbers to integers:

```
short h16[SIZE] = {8454, -10125, -23069, 3932};
short x16[SIZE] = {-6492, -24999, 18811, 4318};
```

The following C code implements the same functionality in fixed-point format assuming `short` and `int` are 16-bit and 32-bit wide, respectively:

```
int dp_32 (short x16[], short h16[], int size)
{
  int prod32;
  int acc32;
  int tmp32;
  int i;
  acc32 = 0;
  for( i = 0; i < size; i++ )
  {
    prod32 = x16[i] * h16[i]; // Q2.30 = Q1.15 * Q1.15
    if (prod32 == 0x40000000) // saturation check
      prod32 = MAX_32;
    else
      prod32 ≪= 1; // Q1.31 = Q2.31 ≪ 1;
    tmp32 = acc32 + prod32; // accumulator
  // +ve saturation logic check
    if( ( (acc32>0) 0 ) )
      acc32 = MAX_32;
  // Negative saturation logic check
    else if( ( (acc32<0) && (prod32<0) ) && ( tmp32>0 ) )
      acc32 = MIN_32;
  // The case of no overflow
    else
      acc32 = tmp32;
  }
}
```

As is apparent from the C code, fixed-point implementation results in several checks and format changes and thus is very slow. In almost all fixed-point DSPs the provision of shifting out redundant sign bits in cases of sign multiplication and checking of overflow and saturating the results are provided in hardware.

3.5.10 Rounding the Product in Fixed-point Multiplication

The following C code demonstrates complex fixed-point multiplication for Q1.15-format numbers. The result is rounded and truncated to Q1.15 format:

```
typedef struct COMPLEX_F
{
  short real;
  short imag;
}
COMPLEX_F;
COMPLEX_F ComplexMultFixed (COMPLEX_F a, COMPLEX_F b)
```

```
{
  COMPLEX_F out;
  int L1, L2, tmp1, tmp2;
  L1 = a.real * b.real;
  L2 = a.imag * b.imag;
  // Rounding and truncation
  out.real=(short)(((L1 - L2)+0x00004000)>>15);
  L1 = a.real * b.imag;
  L2 = a.imag * b.real;
  // Rounding and truncation
  out.imag=(short)(((L1 + L2)+0x00004000)>>15
  return (out);
}
```

3.5.10.1 Unified Multiplier

From the hardware design perspective, placing four separate multipliers to cover general multiplication is an expensive proposition. A unified multiplier can handle all the four types of multiplication. For $N_1 \times N_2$ multiplication, this is achieved by appending one extra bit to the left of the MSB of both of the operands. When an operand is unsigned, the extended bit is zero, otherwise the extended bit is the sign bit of the operand. This selection is made using a multiplexer (MUX). The case for a 16×16 unified multiplier is shown in Figure 3.17.

After extending the bits, $(N_1 + 1) \times (N_2 + 1)$ signed by signed multiplication is performed and only $N_1 \times N_2$ bits of the product are kept for the result. Below are shown four options and their respective modified operands for multiplying two 16-bit numbers, a and b:

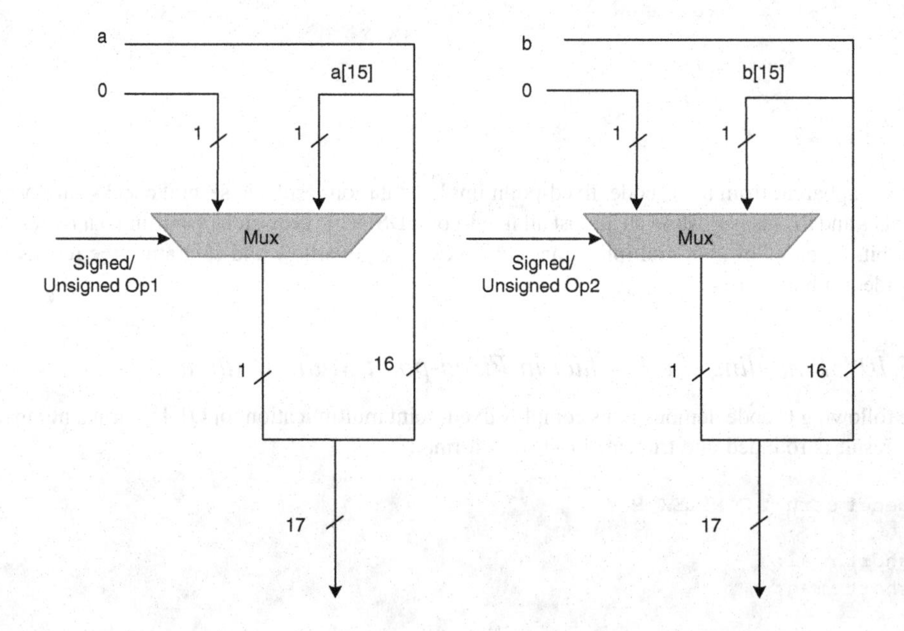

Figure 3.17 One-bit extension to cater for all four types of multiplications using a single signed by signed multiplier

```
unsigned by unsigned: {1'b0, a}, {1'b0, b}
unsigned by signed: {1'b0, a}, {b[15], b}
signed by unsigned: {a[15], a}, {1'b0, b}
signed by signed: {a[15], a}, {b[15], b}.
```

Further to this, there are two main options for multiplications. One option multiplies integer numbers and the second multiplies Q-format numbers. The Q-format multiplication is also called *fractional multiplication*. In fractional signed by signed multiplication, the redundant sign bit is dropped, whereas for integer sign by sign multiplication this bit is kept in the product.

A unified multiplier that takes care of all types of signed and unsigned operands and both integer and fractional multiplication is given in Figure 3.18. The multiplier appropriately extends the bits of

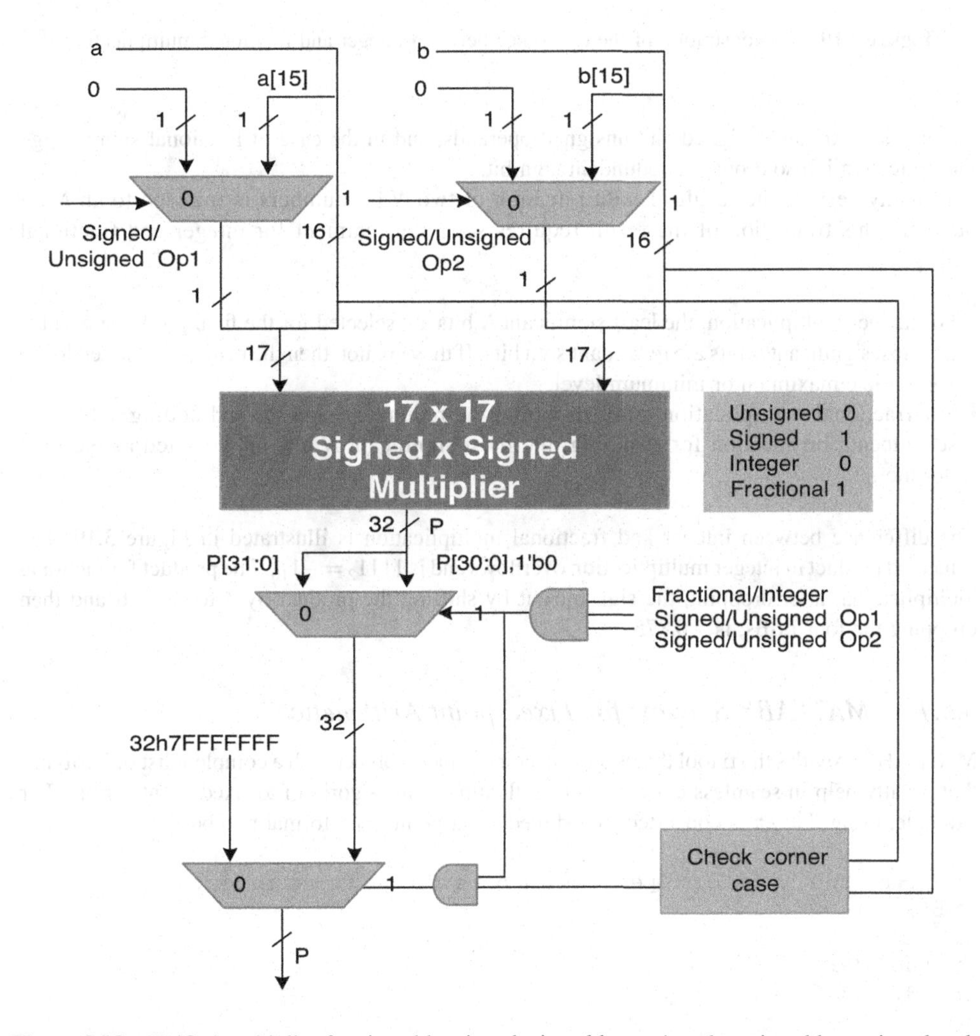

Figure 3.18 Unified multiplier for signed by signed, signed by unsigned, unsigned by unsigned and unsigned by signed multiplication for both fraction and integer modes, it also cater for the corner case

4-bit integer arithmetic

```
        1 0 0 1 = -7₁₀
        0 1 1 1 = +7₁₀
1 1 1 1 1 0 0 1
1 1 1 1 0 0 1 X
1 1 1 0 0 1 X X
1 1 0 0 1 1 1 1 = -49₁₀

1 1 0 0 1 1 1 1 → 1111
      overflow
```

Q1.3 fractional arithmetic

```
      1. 0 0 1 = -0.875₁₀
      0. 1 1 1 = 0.875₁₀
1 1. 1 1 1 0 0 1
1 1. 1 1 0 0 1 X
1 1. 1 0 0 1 X X
1 1. 0 0 1 1 1 1 = 1.0011110
                 = - 0.7656₁₀

1. 0 0 1 1 1 1 0 → 1. 0 0 1 = -0.875₁₀
```

Figure 3.19 Demonstration of the difference between integer and fractional multiplication

the operands to handle signed and unsigned operands, and in the case of fractional sign by sign multiplication it also drops the redundant sign bit.

In many designs the result of multiplication of two N-bit numbers is rounded to an N-bit number. This truncation of the result requires separate treatment for integer and fractional multipliers.

- For integer multiplication, the least significant N bits are selected for the final product provided the most significant N bits are redundant sign bits. If they are not, then the result is saturated to the appropriate maximum or minimum level.
- For fractional multiplication, after dropping the redundant sign bit and adding a 1 to the seventeenth bit location for rounding, the most significant N bits are selected as the final product.

The difference between integer and fractional multiplication is illustrated in Figure 3.19. The truncated product in integer multiplication overflows and is $1111_2 = -1_{10}$. The product for factional multiplication, after dropping the sign bit out by shifting the product by 1 to the left and then dropping the four LSBs, is -0.875.

3.5.11 MATLAB® Support for Fixed-point Arithmetic

MATLAB® provides the fi tool that is a fixed-point numeric object with a complete list of attributes that greatly help in seamless conversion of a floating-point algorithm to fixed-point format. For example, using fi(), π is converted to a signed fixed-point Q3.5 format number:

```
>> PI = fi(pi, 1, 3+5, 5);% Specifying N bits and the fraction part
>> bin(PI)
01100101
>> double(PI)
   3.1563
```

All the available attributes that a programmer can set for each fixed-point object are given in Figure 3.20.

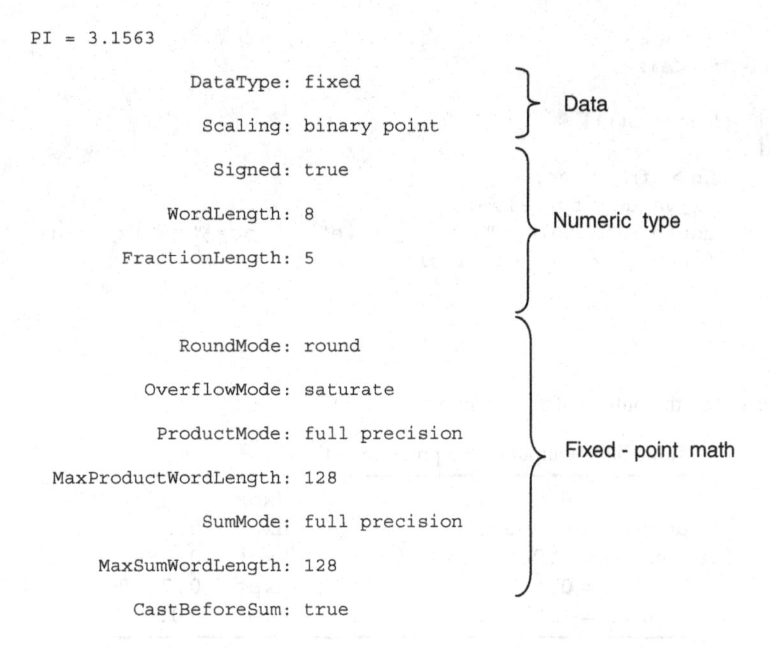

Figure 3.20 MATLAB® support for fixed-point arithmetic

3.5.12 SystemC Support for Fixed-point Arithmetic

As with MATLAB®, once a simulation is working in SystemC using double-precision floating-point numbers, by merely substituting or redefining the same variable with fixed-point attributes the same code can be used for fixed-point implementation. This simple conversion greatly helps the developer to explore a set of word lengths and rounding or truncation options for different variables by changing the attributes.

Here we look at an example showing the power of SystemC for floating-point and fixed-point simulation. This program reads a file with floating-point numbers. It defines a fixed-point variable `fx_value` in Q6.10 with simple truncation and saturation attributes. The program reads a floating-point value in a `fp_value` variable originally defined as of type double, and a simple assignment of this variable to `fx_value` changes the floating-point value to Q6.10 format by rounding the number to 16 bits; and when the integer part is greater than what the format can support the value is saturated. SystemC supports several options for rounding and saturation.

```
int sc_main (int argc , char *argv[])
{
sc_fixed <16, 6, SC_TRN, SC_SAT> fx_value;
double fp_value;
int i, size;
    ofstream fout("fx_file.txt");
    ifstream fin("fp_file.txt");
    if (fin.is_open())
        fin >> size;
    else
        cout << "Error opening input file!\n";
        cout << "size = " << size << endl;
```

```
    for (i=0; i<size; i++)
    {
        if (!fin.eof())
        {
            fin >> fp value;
            fx_value = fp_value;
            cout << "double = " << fp_value"\t fixpt = " << fx_value<< endl;
            fout << fx_value<< endl;
        }
    }
}
```

Table 3.6 shows the output of the program.

Table 3.6 Output of the program in the text

double = 0.324612	fixpt = 0.3242
double = 0.243331	fixpt = 0.2432
double = 0.0892605	fixpt = 0.0889
double = 0.8	fixpt = 0.7998
double =−0.9	fixpt = −0.9004

3.6 Floating-point to Fixed-point Conversion

Declaring all variables as single- or double-precision floating-point numbers is the most convenient way of implementing any DSP algorithm, but from a digital design perspective the implementation takes much more area and dissipates a lot more power than its equivalent fixed-point implementation. Although floating-point to fixed-point conversion is a difficult aspect of algorithm mapping on architecture, to preserve area and power this option is widely chosen by designers.

For fixed-point implementation on a programmable digital signal processor, the code is converted to a mix of standard data types consisting of 8-bit char, 16-bit short and 32-bit long. Hence, defining the word lengths of different variables and intermediate values of all computations in the results, which are not assigned to defined variables, is very simple. This is because in most cases the option is only to define them as char, short or long. In contrast, if designer's intention is to map the algorithm on an application-specific processor then a run of optimization on these data types can further improve the implementation. This optimization step tailors all variables to any arbitrary length variables that just meet the specification and yields a design that is best from area, power and speed.

It is important to understand that a search for an optimal world length for all variables is an NP-hard problem. An exhaustive exploration of world lengths for all variables may take hours for a moderately complex design. Here, user experience with an insight into the design along with industrial practices can be very handy. The system-level design tools are slowly incorporating optimization to seamlessly evaluate a near-optimal solution for word lengths [12].

Conversion of a floating-point algorithm to fixed-point format requires the following steps [6].

S0 Serialize the floating-point code to separate each floating-point computation into an independent statement assigned to a distinct variable var_i.

S1 Insert range directives after each serial floating-point computation in the serialized code of S0 to measure the range of values each distinct variable takes in a set of simulations.

S2 Design a top layer that in a loop runs the design for all possible sets of inputs. Each iteration executes the serialized code with range directives for one possible set of inputs. Make these directives keep a track of maximum `max_var_i` and minimum `min_var_i` values of each variable `var_i`. After running the code for all iterations, the range directives return the range that each variable `var_i` takes in the implementation.

S3 To convert each floating-point variable `var_i` to fixed-point variable `fx_var_i` in its equivalent $Qn_i.m_i$ fixed-point format, extract the integer length n_i using the following mathematical relation:

$$\mathrm{n_i} = \log_2(\max|(\mathrm{max\_val_i},\ \mathrm{min\_val_i})|) + 1 \tag{3.11}$$

S4 Setting the fractional part m_i of each fixed-point variable `fx_var_i` requires detailed technical deliberation and optimization. The integer part is critical as it must be correctly set to avoid any overflow in fixed-point computation. The fractional part, on the other hand, determines the computational accuracy of the algorithm as any truncation and rounding of the input data and intermediate results of computation appear as quantization noise. This noise, generated as a consequence of throwing away of valid part of the result after each computation, propagates in subsequent computations in the implementation. Although a smaller fractional part results in smaller area, less power dissipation and improved clock speeds, it adds quantization noise in the results.

Finding an optimal word length for each variable is a complex problem. An analytical study of the algorithm can help in determining an optimal fractional part for each variable to give acceptable ratio of signal to quantization noise (SQNR) [7]. In many design instances an analytical study of the algorithm may be too complex and involved.

An alternative to analytical study is an exhaustive search of the design space considering a pre-specified minimum to maximum fraction length of each variable. This search, even in simple design problems, may require millions of iterations of the algorithm and thus is not feasible or desirable. An intelligent trial and error method can easily replace analytical evaluation or an infeasible exhaustive search of optimal word length. Exploiting the designer's experience, known design practices and published literature can help in coming up with just the right tailoring of the fractional part. Several researchers have also used optimization techniques like 'mixed integer programming' [8], 'genetic algorithms' [9], and 'simulating annealing' [10]. These techniques are slowly finding their way into automatic floating-point to fixed-point conversion utilities. The techniques are involved and require intensive study, so interested readers are advised to read relevant publications [7–12]. Here we suggest only that designer should intelligently try different word lengths and observe the trend of the SQNR, and then settle for word lengths that just give the desirable performance.

Figure 3.21 lists a sequence of steps while designing a system starting from R&S, development of an algorithm in floating point, determination of the integer part by range estimation, and then iteratively analyzing SQNR on fixed-point code to find the fractional part of $Qn.m$ format for all numbers, and finally mapping the fixed-point algorithm on HW and SW components.

3.7 Block Floating-point Format

Block floating-point format improves the dynamic range of numbers for cases where a block of data goes though different stages of computation. In this format, a block of data is represented with a common exponent. All the values in the block are in fixed-point format. This format, before truncating the intermediate results, intelligently observes the number of redundant sign bits of all the values in the block, to track the value with the minimum number of redundant sign bits in the

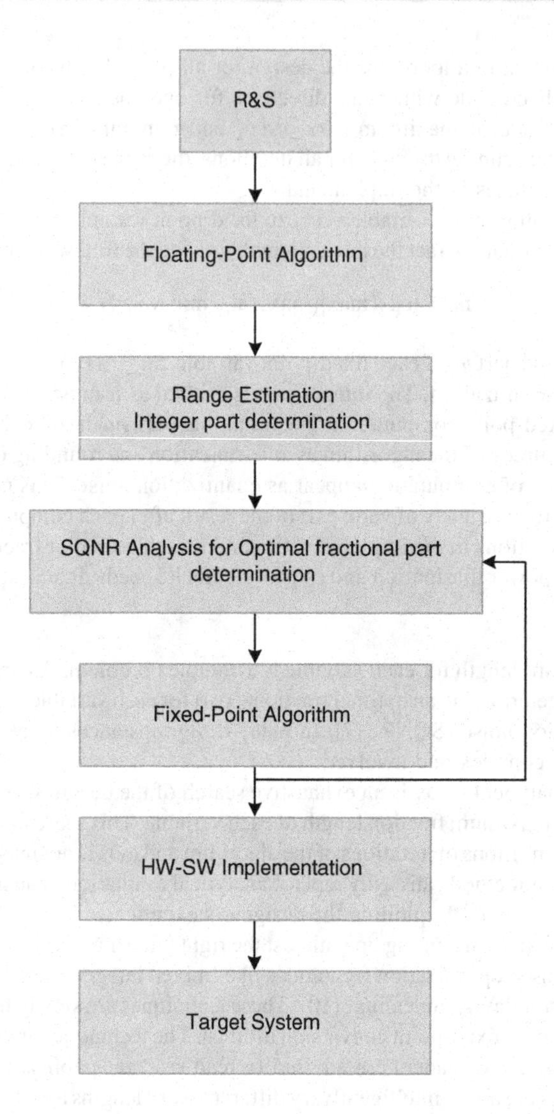

Figure 3.21 Steps in system design starting from gathering R&S, implementing algorithms in floating-point format and the algorithm conversion in fixed-point while observing SQNR and final implementation in HW and SW on target platforms

block. This can be easily extracted by computing the value with maximum magnitude in the block. The number of redundant sign bits in the value is the block exponent for the current block under consideration. All the values in the block are shifted left by the block exponent. This shifting replaces redundant sign bits with meaningful bits, so improving effective utilization of more bits of the representation. Figure 3.22 shows a stream of input divided into blockswith block exponent value.

Block floating-point format is effectively used in computing the fast Fourier transform (FFT), where each butterfly stage processes a block of fixed-point numbers. Before these numbers are fed to the next stage, a block exponent is computed and then the intermediate result is truncated to be fed to

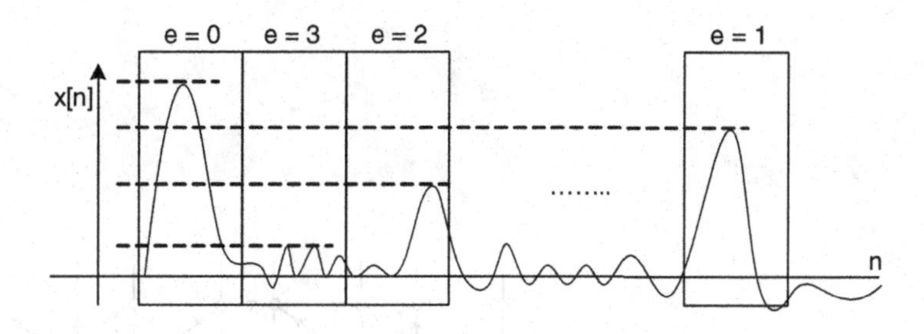

Figure 3.22 Applying block floating-point format on data input in blocks to an algorithm

the next block. Every stage of the FFT causes bit growth, but the block floating-point implementation can also cater for this growth.

The first two stages of the Radix-2 FFT algorithm incur a growth by a factor of two. Before feeding the data to these stages, two redundant sign bits are left in the representation to accommodate this growth. The block floating-point computation, while scaling the data, keeps this factor in mind. For the remainder of the stages in the FFT algorithm, a bit growth by a factor of four is expected. For this, three redundant sign bits are left in the block floating-point implementation. The number of shifts these blocks of data go through from the first stage to the last is accumulated by adding the block exponent of each stage. The output values are then readjusted by a right shift by this amount to get the correct values, if required. The block floating point format improves precision as more valid bits take part in computation.

Figures 3.23(a) and (b) illustrate the first stage of a block floating-point implementation of an 8-point radix-2 FFT algorithm that caters for the potential of bit growth across the stages. While observing the minimum redundant sign bits (four in the figure) and keeping in consideration the bit growth in the first stage, the block of data is moved to the left by a factor of two and then converted to the requisite 8-bit precision. This makes the block exponent of the input stage to be 2. The output of the first stage in 8-bit precision is shown. Keeping in consideration the potential bit growth in the second stage, the block is shifted to the right by a factor of two. This shift makes the block exponent equal to zero. Not shown in the figure are the rest of the stages where the algorithm caters for bit growth and also observes the redundant sign bits to keep adjusting the block floating-point exponent.

3.8 Forms of Digital Filter

This section briefly describes different forms of FIR and IIR digital filters.

3.8.1 Infinite Impulse Response Filter

The transfer function and its corresponding difference equation for an IIR filter are given in (3.12) and (3.13), respectively:

$$H(z) = \frac{Y(z)}{X(z)} = \frac{\sum_{k=0}^{N} b_k z^{-k}}{1 + \sum_{k=1}^{M} a_k z^{-k}} \tag{3.12}$$

$$y[n] = \sum_{k=0}^{N} b_k \, x[n-k] - \sum_{k=1}^{M} a_k \, y[n-k] \tag{3.13}$$

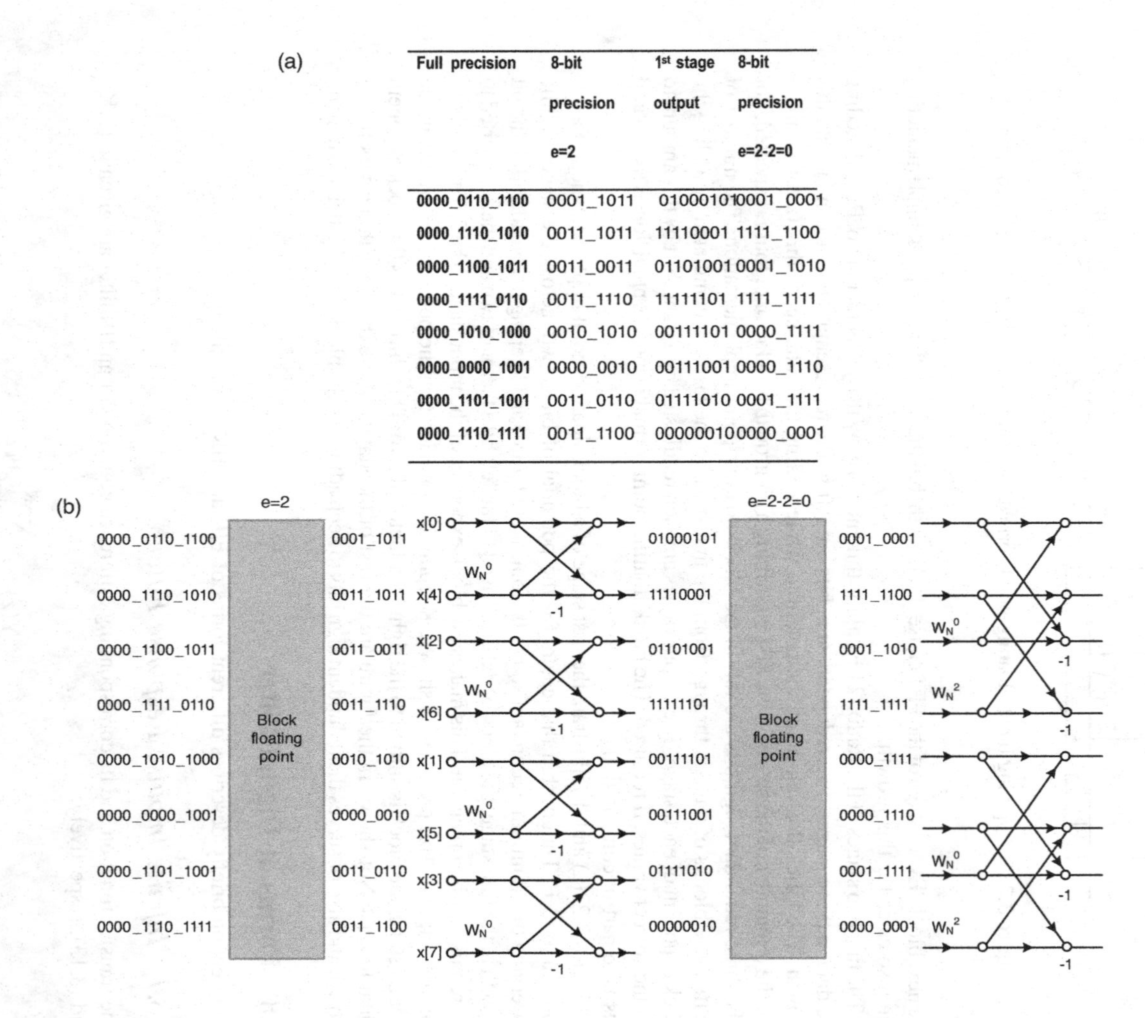

(a)

Full precision	8-bit precision e=2	1st stage output	8-bit precision e=2-2=0
0000_0110_1100	0001_1011	01000101	0001_0001
0000_1110_1010	0011_1011	11110001	1111_1100
0000_1100_1011	0011_0011	01101001	0001_1010
0000_1111_0110	0011_1110	11111101	1111_1111
0000_1010_1000	0010_1010	00111101	0000_1111
0000_0000_1001	0000_0010	00111001	0000_1110
0000_1101_1001	0011_0110	01111010	0001_1111
0000_1110_1111	0011_1100	00000010	0000_0001

Figure 3.23 FFT butterfly structure employing block floating-point computation while keeping the bit growth issue in perspective

There are several mathematical ways of implementing the difference equation. The differences between these implementations can be observed if they are drawn as dataflow graphs (DFGs). Figure 3.24 draws these analytically equivalent DFGs for implementing Nth-order IIR filter.

Although all these implementations are analytically equivalent, they do vary in terms of hardware resource requirements and their sensitivity to coefficient quantization. While implementing a signal processing algorithm in double- precision floating-point format, these varying forms have little practical significance; but for fixed-point implementation, it is important to understand their relative susceptibilities to quantization noise. It is important for a digital designer to understand that just converting a floating-point implementation to fixed-point using any optimization technique of choice may not yield desirable results and the system may even become unstable. It is therefore essential always to select a form that has minimum sensitivity to coefficient quantization before the format conversion is performed. Thus the first design objective for a fixed-point implementation of an algorithm is to choose a form that is least sensitive to quantization noise; and then, subsequent to this, the designer should try to minimize HW resources for the selected implementation.

3.8.2 *Quantization of IIR Filter Coefficients*

Fixed-point conversion of double-precision floating-point coefficients of an IIR filter moves the pole–zero location and thus affects the stability and frequency response of the system. To illustrate this effect, consider the design of an eighth-order system in MATLAB®, with pass-band ripple of 0.5 dB, stop-band attenuation of 50 dB and normalized cutoff frequency $\omega_c = 0.15$. The coefficients of the numerator **b** and denominator **a** obtained using MATLAB® function `ellip` with these specifications are given below:

$$\mathbf{b} = 0.0046{-}0.0249 \quad 0.0655{-}0.1096 \quad 0.1289{-}0.1096 \quad 0.0655{-}0.0249 \quad 0.0046$$
$$\mathbf{a} = 1.0000{-}6.9350 \quad 21.5565{-}39.1515 \quad 45.3884{-}34.3665 \quad 16.5896{-}4.6673 \quad 0.5860$$

The maximum values of the magnitude of the coefficients in **b** and **a** are 0.1289 and 45.3884, respectively. When converting the coefficients to fixed-point format, these values require an integer part of 1 and 7 bits, respectively. The word length of the fractional part is application-dependent.

Before any attempt is made to choose appropriate Q-formats for filter implementation, let us observe the effect of quantization on stability and frequency response of the system. The pole–zero plot of the filter transfer function with coefficients in double-precision floating-point numbers is shown in the top left part of Figure 3.25. Now the filter coefficients in arrays **b** and **a** are quantized to 24-bit, 16-bit and 12-bit precisions, and the Q-formats of these precisions are Q1.23 and Q7.17, Q1.15 and Q7.9, and Q1.11 and Q7.6, respectively. The pole–zero and the frequency response of the system for all these four cases are plotted in Figures 3.26 and 3.27, respectively.

It is evident from the pole–zero plots that the filter is unstable for 16-bit and 12-bit quantization cases as some of its poles move outside the unit circle. If an IIR filter is meant for fixed-point implementation, it is important to first select an appropriate form for the implementation before coding the design in high-level languages for its subsequent conversion to fixed-point format. The same filter has been implemented using second-order-sections (SoS), and pole–zero plots show that

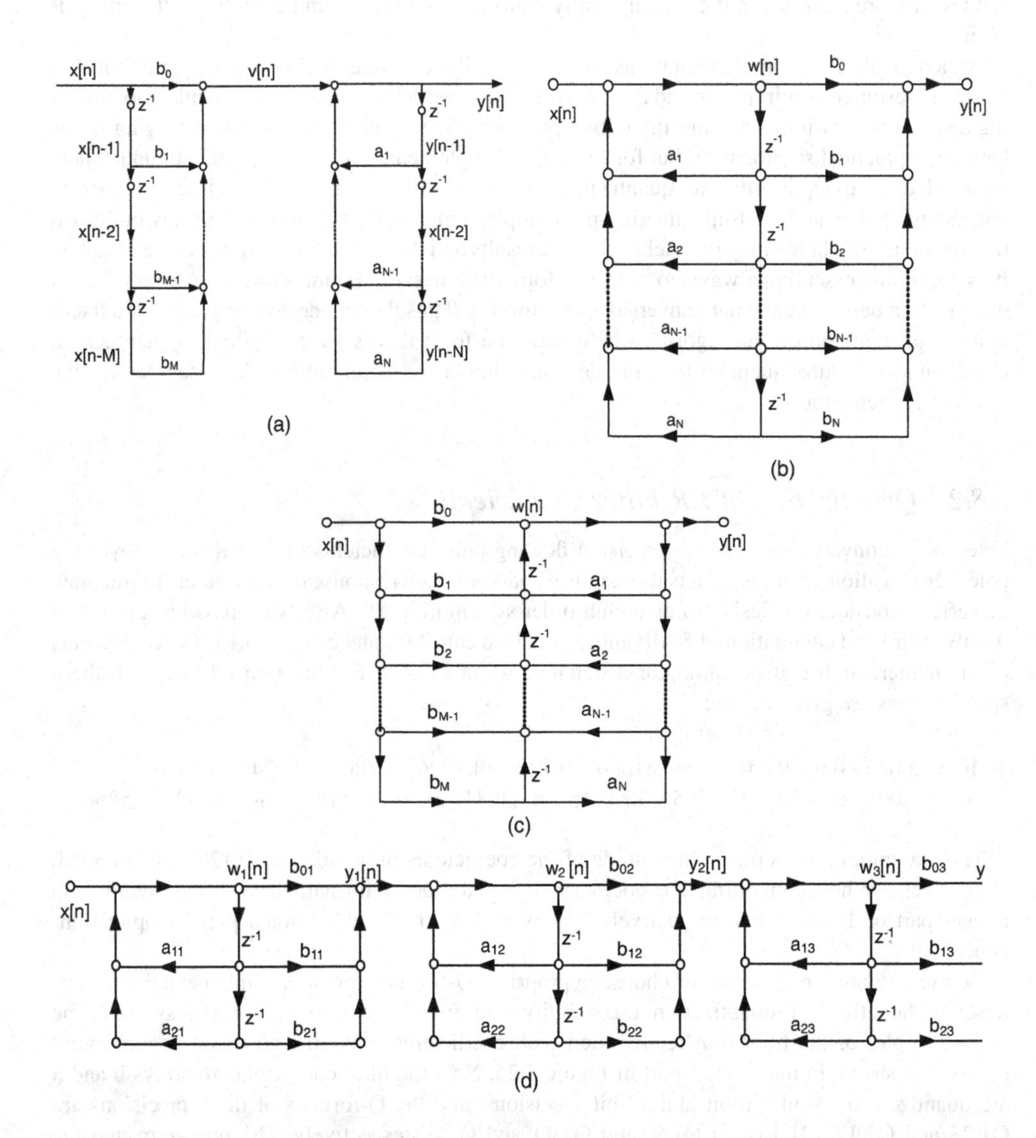

Figure 3.24 Equivalent implementation for realizing an Nth-order IIR system. (a) Direct Form-I. (b) Direct Form-II. (c) Transposed Direct Form II. (d) Cascade form using DF-II second-order sections. (e) Parallel form using DF-II second-order sections

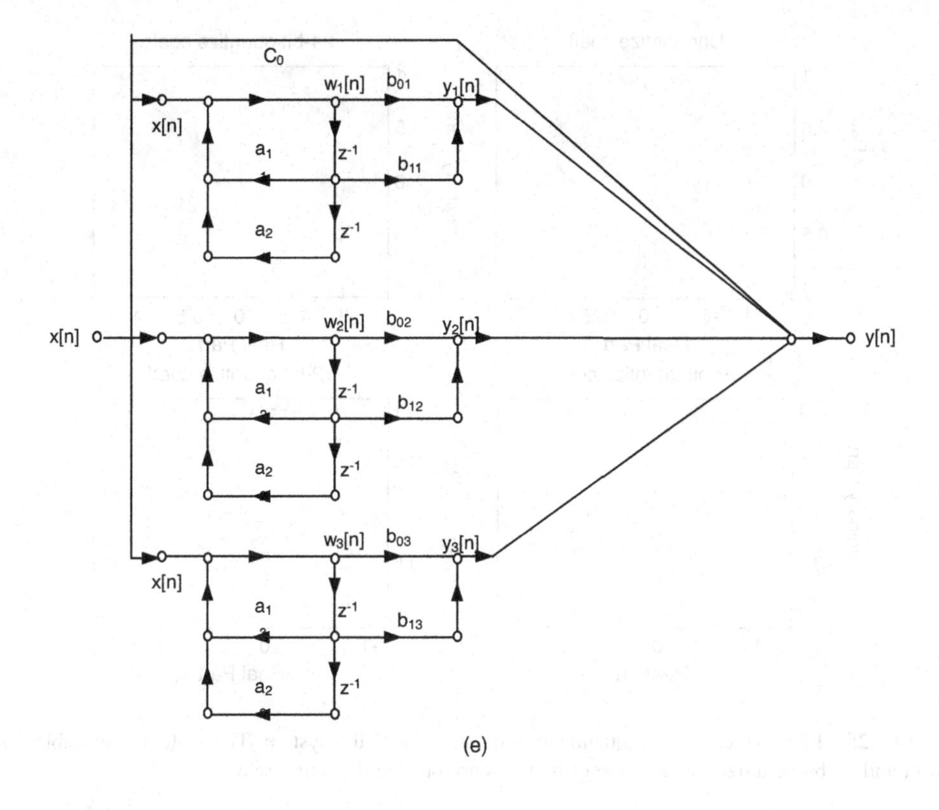

Figure 3.24 *Continued*

the filter is stable for all the above cases as all its poles remain inside the unit circle. The matrix of coefficients for all the four sections from MATLAB® is:

0.0046	−0.0015	0.0046	1	−1.6853	0.7290
1	−1.5730	1	1	−1.7250	0.8573
1	−1.7199	1	1	−1.7542	0.9488
1	−1.7509	1	1	−1.7705	0.9882

Each row lists three coefficients of **b** and three coefficients of **a** for its respective section. Based on the maximum of absolute values of the coefficients for each section, 2 bits are required for the integer part of respective Q-format. The filter is analyzed for 12-bit and 8-bit precision, and all the coefficients are converted into Q2.10 and Q2.6 format for the two formats. The pole–zero plots for these four sections for both the cases are shown in Figure 3.27. In both cases the designs using cascaded SoS are stable, so the overall system remains stable even for 8-bit quantization. It is therefore important to first analyze the stability of the system while selecting the word lengths for fixed-point implementation of IIR filters.

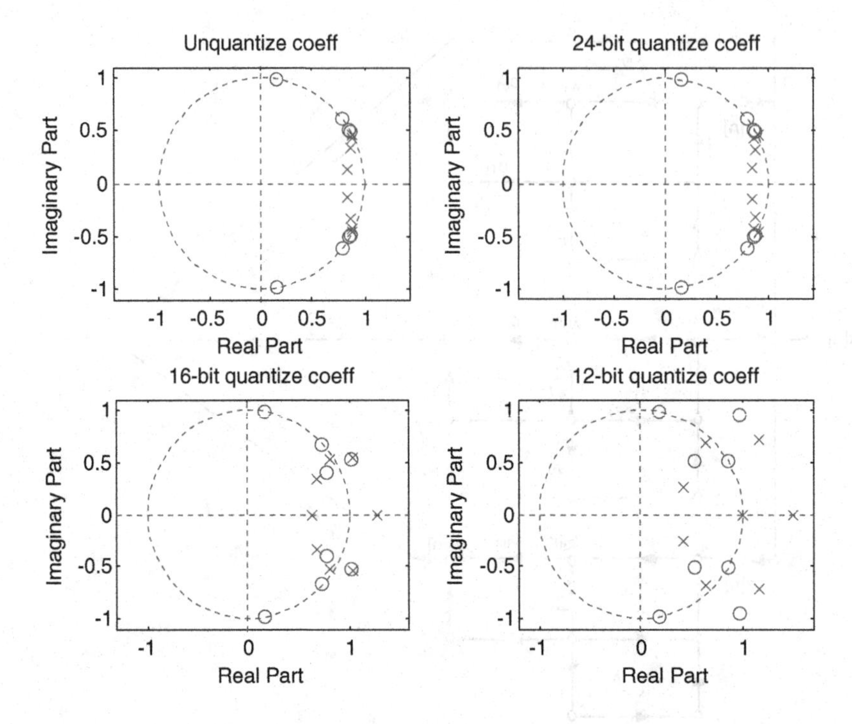

Figure 3.25 Effect of coefficient quantization on stability of the system. The system is unstable for 16-bit and 12-bit quantization as some of its poles are outside the unit circle

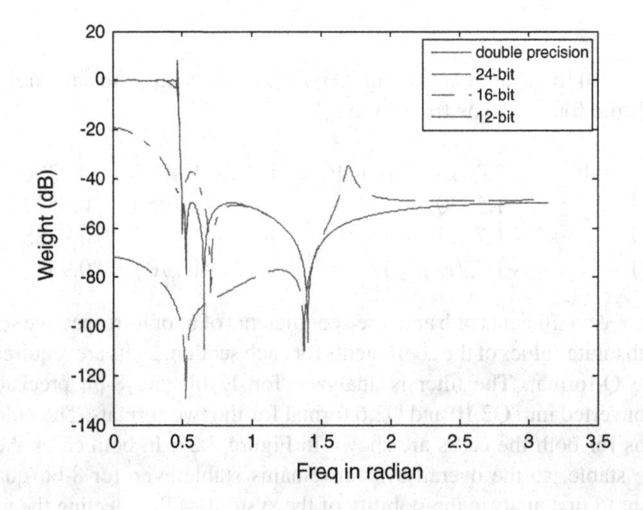

Figure 3.26 Frequency response of the eighth-order system for different quantization levels

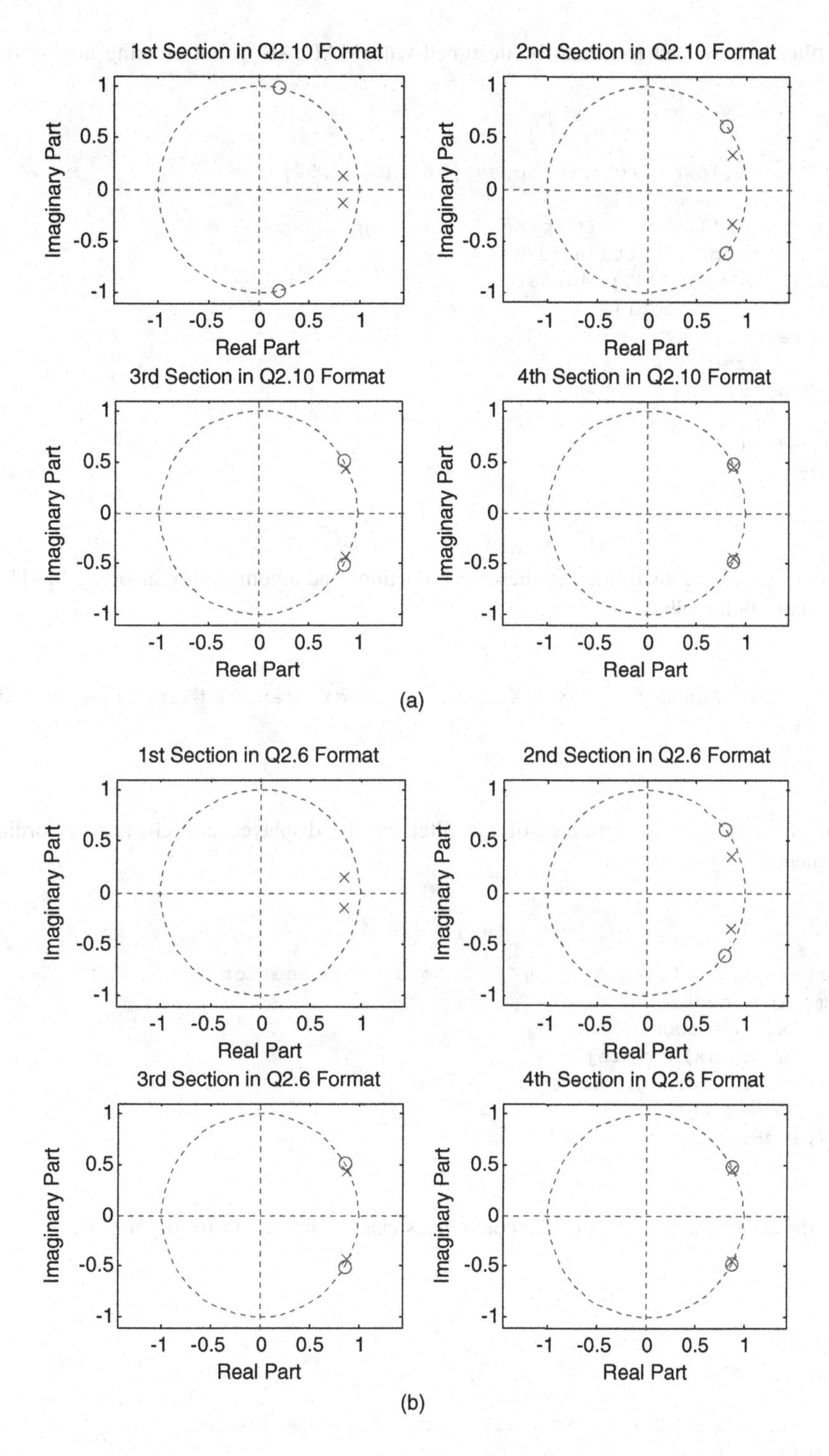

Figure 3.27 (a) Filter using cascaded SoSs with coefficients quantized in Q2.10 format. (b) Filter using cascaded SoSs with coefficients quantized in Q2.6 format

The filter in cascaded form can be designed with MATLAB®, first creating a filter design object:

```
>> d = fdesign.lowpass('n,fp,ap,ast',8,.15,.5,50)
d =
ResponseType: 'Lowpass with passband-edge specifications and
          stopband attenuation'
SpecificationType: 'N,Fp,Ap,Ast'
Description: {4x1 cell}
NormalizedFrequency: true
Fs: 'Normalized'
FilterOrder: 8
Fpass: 0.1500
Apass: 0.5000
Astop: 50
```

Then design an elliptical filter on these specifications and automatically invoking MATLAB® filter visualization toolbox `fvtool`:

```
>> ellip(d); % Automatically starts fvtool to display the filter in second order
sections.
```

By using a *handle*, all attributes of the filter can be displayed and changed according to requirements:

```
>> hd=ellip(d)
hd =
FilterStructure: 'Direct-Form II, Second-Order Sections'
Arithmetic: 'double'
sosMatrix: [4x6 double]
ScaleValues: [5x1 double]
ResetBeforeFiltering: 'on'
States: [2x4 double]
NumSamplesProcessed: 0
```

Now the arithmetic precision of the coefficients can be changed to fixed-point by:

```
>> set(hd,'arithmetic','fixed')
>> hd
hd =
FilterStructure: 'Direct-Form II, Second-Order Sections'
Arithmetic: 'fixed'
```

```
sosMatrix: [4x6 double]
ScaleValues: [5x1 double]
ResetBeforeFiltering: 'on'
States: [2x4 embedded.fi]
NumSamplesProcessed: 0
CoeffWordLength: 16
CoeffAutoScale: true
Signed: true
InputWordLength: 16
InputFracLength: 15
StageInputWordLength: 16
StageInputAutoScale: true
StageOutputWordLength: 16
StageOutputAutoScale: true
OutputWordLength: 16
OutputMode: 'AvoidOverflow'
StateWordLength: 16
StateFracLength: 15
ProductMode: 'FullPrecision'
AccumMode: 'KeepMSB'
AccumWordLength: 40
CastBeforeSum: true
RoundMode: 'convergent'
OverflowMode: 'wrap'
```

While exploring different design options for optimal word length, any of the attributes of `hd` can be changed using set directive.

Filter Design and Analysis Tool (FDATOOL) is a collection of tools for design, analysis, conversion and code generation. The tool designs FIR and IIR digital filters in both floating-point and fixed-point formats. The tool provides a great degree of control with a user-friendly interface. Figure 3.28 shows two snapshots of the GUI depicting all the design options available to the designer for designing, analyzing, fixed-point conversion and code generation. The elliptical filter can also be conveniently designed by setting different options in FDATOOL.

3.8.3 Coefficient Quantization Analysis of a Second-order Section

Even in second-order section the quantization effects can be further mitigated. The conversion in fixed-point format of the denominator coefficient as given in (3.14) creates a non-uniform grid:

$$H(z) = \frac{1}{1 + a_1 z^{-1} + a_2 z^{-2}} \tag{3.14}$$

The grid is shown in Figure 3.29 for a_1 and a_2 converted to Q3.3 and Q2.4 formats, respectively. A filter design package, depending on the specifications, may place a double-precision pole at any point inside the unit circle. The conversion of the coefficients to fixed-point

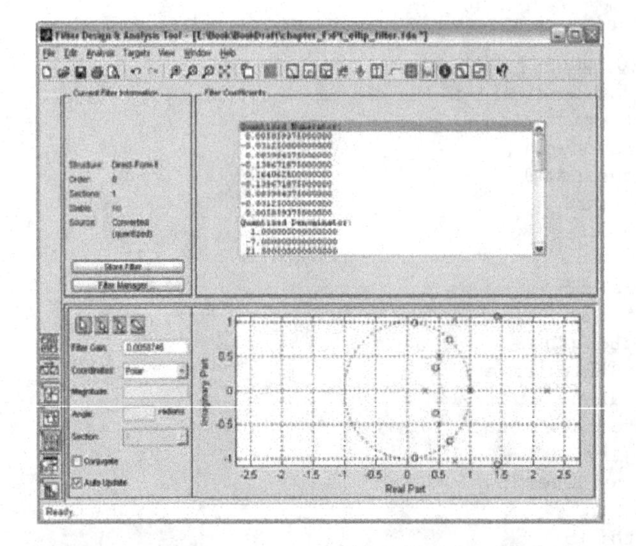

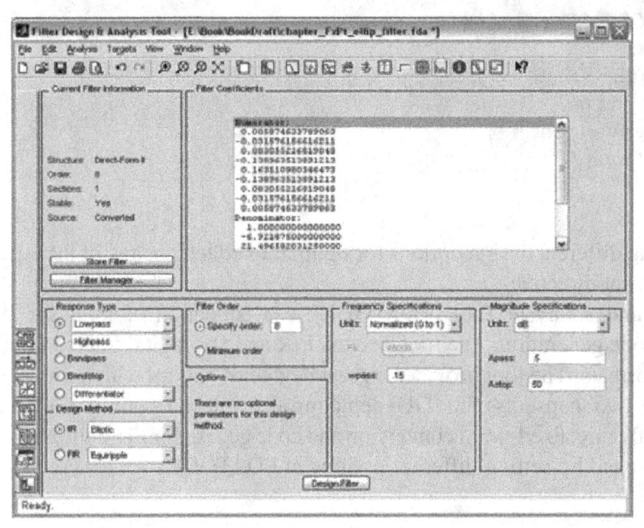

Figure 3.28 The MATLAB® filter design and analysis toolbox offers varying options to design, quantize, analyze and generate Verilog or C code of a designed filter according to the given specifications

format moves the respective poles to quantized locations and makes the system experience more quantization noise at places with wider gaps. The structure given in Figure 3.30 creates a uniform rectangular grid and helps the designer to exactly model the quantization noise independent of location of its double-precision poles and zeros (r and θ are the radius and angle of the complex conjugate poles).

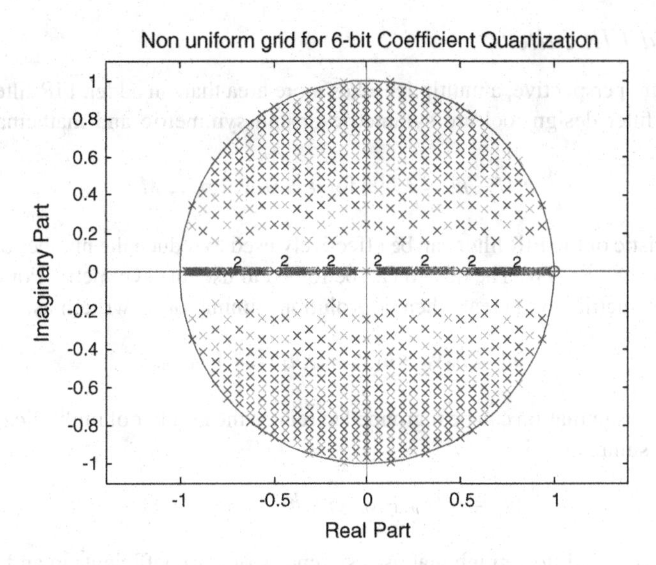

Figure 3.29 Possible pole locations when the coefficients of the denominator of an SoS are quantized to 6-bit fixed-point numbers

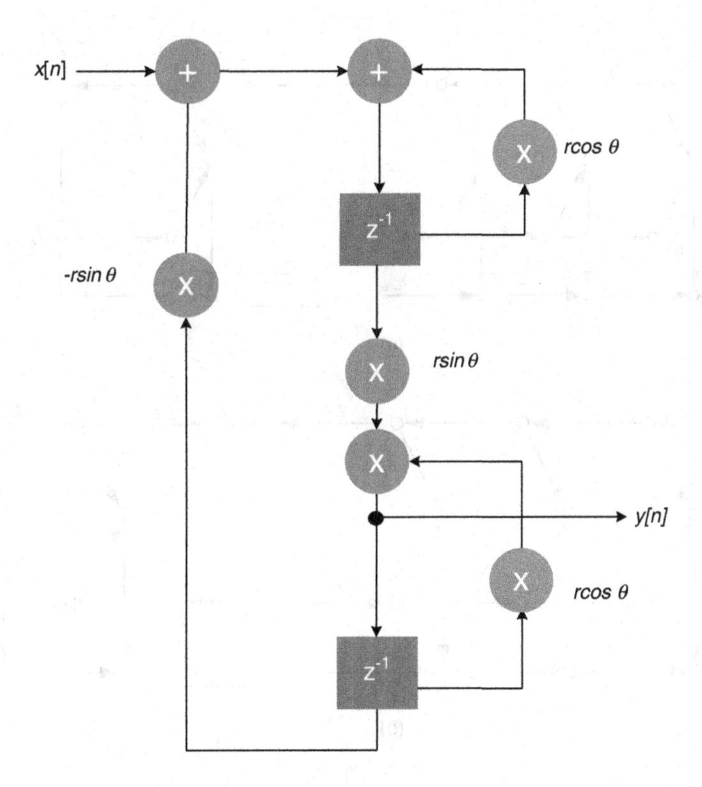

Figure 3.30 Filter structure to place quantized pole-pair $re^{\pm j\theta}$ at intersections of evenly placed horizontal and vertical lines inside the unit circle

3.8.4 Folded FIR Filters

From the hardware perspective, a multiplier takes more area than an adder. FIR filter design using the MATLAB® filter design tool are symmetric or anti-symmetric and mathematically can be represented as:

$$h[M-n] = \pm h[n] \text{ for } n = 0, 1, 2, \ldots, M$$

This characteristic of the FIR filter can be effectively used to reduce the number of multipliers in mapping these designs in HW. The design can be folded to use this symmetry. For example, if the filter has four symmetric coefficients then convolution summation is written as:

$$y[n] = h_0 x_n + h_1 x_{n-1} + h_1 x_{n-2} + h_0 x_{n-3}$$

The terms in this summation can be regrouped to reduce the number of multiplications from four to two. The new summation is:

$$y[n] = h_0(x_n + x_{n-3}) + h_1(x_{n-1} + x_{n-2})$$

The generalized folded flow graph that uses symmetry of the coefficients in an FIR is shown in Figure 3.31 for even and odd number of coefficients.

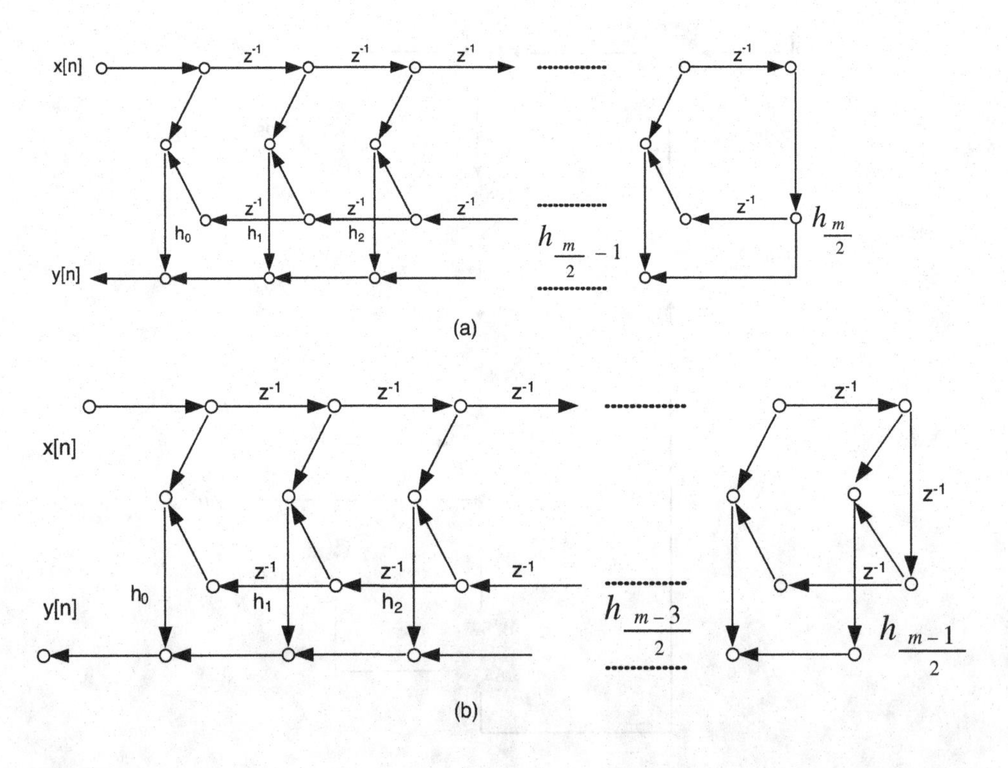

Figure 3.31 Symmetry of coefficients reduces the number of multipliers to almost half. (a) Folded design with odd number of coefficients. (b) Folded design with even number of coefficients

3.8.5 Coefficient Quantization of an FIR Filter

As FIR systems are always stable, stability is not a concern when quantizing double-precision floating-point coefficients of an FIR system to any fixed-point format. The coefficient quantization primarily adds an undesirable additional frequency response into the frequency response of the original system with double-precision floating-point coefficients. The mathematical reasoning of this addition is given in these expressions:

$$h_Q[n] = h[n] + \Delta h[n] \tag{3.15}$$

$$H_Q(e^{j\omega}) = \sum_{n=0}^{M}(h[n] + \Delta h[n])e^{j\omega n} \tag{3.16}$$

$$H_Q(e^{j\omega}) = H(e^{j\omega}) + \sum_{n=0}^{M}\Delta h[n]e^{j\omega n} \tag{3.17}$$

Thus quantization of coefficients of an FIR filter adds a frequency response equal to the frequency response of $\Delta h[n]$ caused due to dropping of bits of $h[n]$. Consider a filter designed in MATLAB® using:

```
>> h=fir1(10,.1);
```

The coefficients are converted to Q1.7 format using the fi utility as:

```
>> hQ = fi(h, 1, 8,7);
```

$\Delta h[n]$ is computed as:

```
>> hQ-h
```

and the respective values of these variables are:

h = 0.0100 0.0249 0.0668 0.1249 0.1756 0.1957 0.1756 0.1249 0.0668 0.0249 0.0100
h_Q = 0.0078 0.0234 0.0703 0.1250 0.1719 0.1953 0.1719 0.1250 0.0703 0.0234 0.0078
$\Delta h[n]$ = −0.0022 −0.0014
0.0035 0.0001 −0.0037 −0.0004 −0.0037 0.0001 0.0035 −0.0014 −0.0022

The quantization of coefficients obviously changes the frequency response of the filter. There is a possibility that the quantized filter may no longer satisfy the original design specifications. Figure 3.32 shows the original filter and modified filter, clearly demonstrating degradation in the pass-band ripples. The designer may need to over-design a filter to let the quantization effects use this leverage for meeting the specifications.

A more optimal solution to this problem is to design a filter in the fixed-point domain. Researchers have tended to focus more on conversion of signal processing applications originally designed using floating-point format to fixed-point implementation, but very few have investigated designing in the fixed-point format. A direct design in fixed-point gives more control to the designer and takes his or her worry away from meeting a specification while finding an optimal word length for the coefficients. These techniques take the system

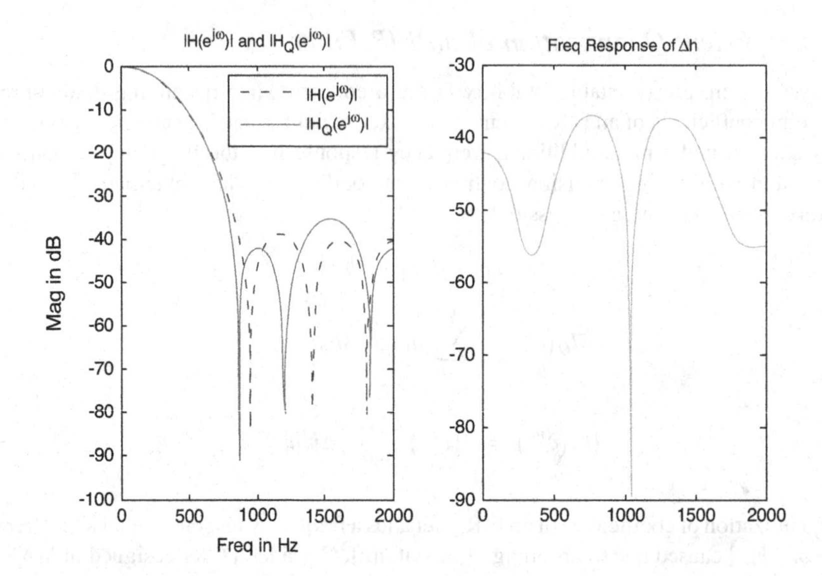

Figure 3.32 Change in magnitude of frequency response due to floating-point to fixed-point conversion

specification and then directly design the system in fixed-point format. Optimization techniques are exploited, but these techniques have yet not found a place in commercial toolboxes primarily because of their computational complexity. Finding computationally feasible techniques requires more deliberation and effort from researchers working in areas of signal processing and digital design.

Exercises

Exercise 3.1

Design a 13-bit floating-point multiplier. The floating-point number has 1, 4 and 8 bits for, respectively, its sign s, exponential e and mantissa m. Assume a bias value of 7 for the representation. Use an 8×8-bit unsigned multiplier to multiply two mantissas, and a 4-bit adder and a subtractor to add the two exponents and then subtract the bias to take the effect of twice added bias from the addition. Normalize the multiplication and add its effect in the computed exponential. Draw an RTL diagram and code the design in Verilog. Write a stimulus to check your design for normalized and denormalized values. Finally, check the design for multiplication by $\pm\infty$.

Exercise 3.2

Add the following two floating-point numbers using 32-bit IEEE floating-point format. Show all the steps in the addition that should help in designing a floating-point adder. The numbers are:

$$x = -23.175$$
$$y = 109.5661$$

Exercise 3.3

Design a floating-point multiply accumulator (MAC) that first multiplies two 32-bit numbers and then adds the accumulator into the product. The MAC unit should only normalize the result once after multiplication and addition. Code the design in RTL and write a stimulus to test the design.

Exercise 3.4

Convert the floating-point implementation of Section 3.2.3 to 16-bit fixed-point format. Simulate the floating point and fixed-point implementation. Design a receiver that demodulates the signal. Count bit errors. Add AWGN (additive white Gaussian noise) and count bit errors for different signal-to-noise ratios.

Exercise 3.5

Multiply the following 8-bit numbers A and B by considering them as U × U, U × S, S × U and S × S, where U and S stands for unsigned and signed fraction numbers:

$$A = 8'b1011\_0011$$
$$B = 8'b1100\_0101$$

Also multiply the numbers considering both of these numbers as signed integers and then signed fractional numbers in Q1.7 format. Also multiply them by considering A as a signed integer and B as a signed Q1.7 format number. Design the HW and implement it in RTL Verilog.

Exercise 3.6

Add rounding and truncation logic in the 17 × 17-bit unified multiplier of Figure 3.18. The design should generate a 16-bit product and also generate a flag if truncation of the integer multiplication results in overflow.

Exercise 3.7

Design HW logic to implement a block floating-point module that computes a block exponential of an 8-element 16-bit array of numbers before truncating them to 8-bit numbers. Test your design for multiple blocks of data with a block exponent varying from 0 to 7. Realize the design in RTL Verilog.

Exercise 3.8

Floating-point implementation of a component of a communication algorithm is given below:

```
double Mixer_Out[3];
double Buff_Delayline[3];
// Filter coefficients in double precision
double HF_DEN [] = {1.0000, -1.0275, 0.7265};
double HF_NUM [] = {0.1367, 0.5, -0.1367};
// Input values for testing the code
#define LEN_INPUT 10
double input[LEN_INPUT] = {
    0.2251, 0.4273, 0.6430, 0.8349, 1.0089, 0.6788, 0.3610,
                                    -0.3400, -0.6385, -0.8867};
```

```
// Section of the code for floating to fixed-point conversion, the code to be
written in a function for simulation
for(int i=0; i<LEN_INPUT; i++)
{
    Mixer_Out[2] = Mixer_Out[1];
    Mixer_Out[1] = Mixer_Out[0];
    Mixer_Out[0] = input[i];
    Buff_Delayline[2] = Buff_Delayline[1];
    Buff_Delayline[1] = Buff_Delayline[0];
    Buff_Delayline[0]=-Buff_Delayline[1]*HF_DEN[1]
    -Buff_Delayline[2]*HF_DEN[2]
    +Mixer_Out[0]*HF_NUM[0]
    +Mixer_Out[1]*HF_NUM[1]
    +Mixer_Out[2] * HF_NUM[2];
}
```

1. Simulate the communication algorithm using input samples from the floating-point array `input[]`.
2. Convert the algorithm into fixed-point format using 16-bit precision by appropriately converting all the variables into Q$n.m$ format, where $n+m=16$.
3. Compare your fixed-point result with the floating-point implementation for the same set of inputs in fixed-point format.
4. Draw RTL design of computations that are performed in the loop.

Exercise 3.9

To implement the difference equation below in fixed-point hardware, assume the input and output are in Q1.7 format:

$$y[n] = -0.9821y[n-1] + x[n]$$

First convert the coefficients to appropriate 8-bit fixed-point format, use rounding followed by truncation and saturation logic. Simulate the floating-point and fixed-point implementation in MATLAB®. Code the design in RTL Verilog. Write stimulus and compare the results with the MATLAB® simulation result.

Exercise 3.10

Implement the following equation, where $x[n]$ and $y[n]$ are in Q1.7 and Q2.6 formats, respectively, and $z[n]$ is an unsigned number in Q0.8 format. Design architecture for computing:

$$w[n] = x[n]y[n] + y[n]z[n] + x[n] + y[n] + z[n]$$

Give the Q format of $w[n]$.

Exercise 3.11

The C code below implements this difference equation for 12 input values $y[n] = -2.375x[n] + 1.24y[n-1]$:

```
float y[12], x[12];
y[0] = 0.0
for (n =1; n< 12; n++ )
{
y[n] = -2.375 x[n] + 1.24 y[n-1];
}
```

The code defines floating-point arrays $x[.]$ and $y[.]$. Convert the code to appropriate fixed-point format. Specify the format and simulate the implementation for given values of input. Assume $|x[n]| < 1$. Compare your result with the floating-point implementation result. Calculate the mean squared error by computing the mean of squares of differences of the two output values for the test vector given below:

$$x[n] = \{0.5, 0.5, -0.23, 0.34, 0.89, 0.11, -0.22, 0.13, 0.15, 0.67, -0.15, -0.99\}$$

Exercise 3.12

Write RTL Verilog code to implement a four-coefficient FIR filter given in Figure 3.33. Assume all the coefficients $h_0 \ldots h_3$ and input data $x[n]$ are in Q1.15 format. Appropriately truncate the result from multiplication units to use 18-bit adders, and define the appropriate format for $y[n]$. Optimize the design assuming the coefficients of the FIR filter are symmetric.

Exercise 3.13

Using the MATLAB® filter toolbox, design an IIR filter with the following specifications: $\delta_1 = 0.001$; $\delta_2 = 0.0001$; $w_p = 0.2\pi$; $w_s = 0.25\pi$. Convert the filter into multiple cascaded second-order sections, and then quantize the coefficients as 16-bit signed numbers using an appropriate Q$n.m$ format. Draw the dataflow of the design and write RTL Verilog code to implement the design using signed arithmetic.

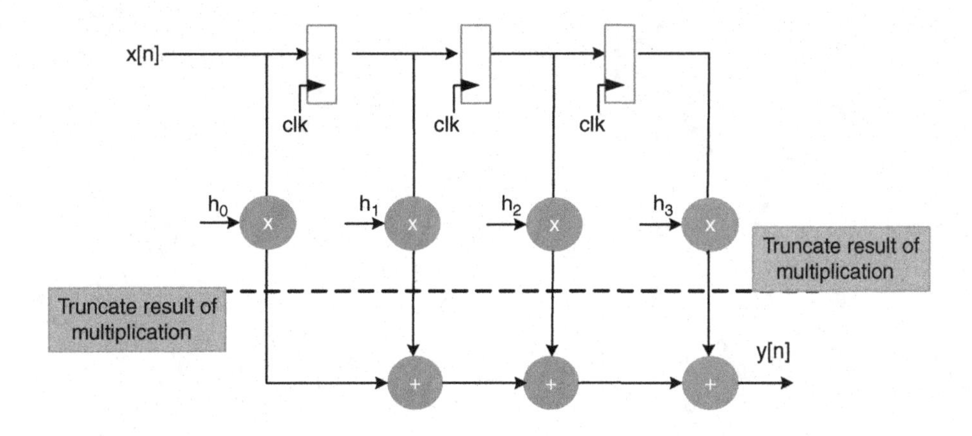

Figure 3.33 Four-coefficient FIR filter

References

1. www.altera.com/products/software/products/dsp/dsp-builder.html
2. www.xilinx.com/tools/sysgen.htm
3. J. J. F. Cavanagh, *Digital Computer Arithmetic*, 1984, McGraw-Hill.
4. V. C. Hamachar, Z. G. Vranesic and S. G. Zaky, *Computer Organization*, 1990, McGraw-Hill.
5. M. M. Mano, *Computer Engineering*, 1988, Prentice-Hall.
6. J. F. Wakerly, *Digital Design Principles und Practices*, 1990, Prentice-Hall.
7. S. Waserand M. J. Flynn, *Introduction to Arithmetic for Digital Systems Designers*, 1982, Holt, Rinehart & Winston.
8. R. C. Dorf, *Computers, Software Engineering, and Digital Devices*, 2005, CRC Press.
9. ANSI/IEEE, Standard 754-1985: "Binary floating-point arithmetic," 1985.
10. P. Karlström, A. Ehliarand D. Liu, "High-performance low-latency FPGA-based floating-point adder and multiplier units in a Virtex 4," in *Proceedings of the 24th Norchip Conference*, 2006, pp. 31–34, Linkoping, Sweden.
11. G. Govindu, L. Zhuo, S. Choi, P. Gundalaand V. K. Prasanna,"Area and power performance analysis of a floating-point based application on FPGAs," in *Proceedings of the 7th Annual Workshop on High-performance Embedded Computing*, 2003, Lincoln Laboratory, MIT, USA.
12. K. Han and B. L. Evans, "Optimum word-length search using a complexity-and-distortion measure," *EURASIP Journal of Applied Signal Processes*, 2006, no. 5, pp. 103–116.

4

Mapping on Fully Dedicated Architecture

4.1 Introduction

Although there are many applications that are commonly mapped on field-programmable gate arrays (FPGAs) and application-specific integrated circuits (ASICs), the primary focus of this book is to explore architectures and design techniques that implement real-time signal processing systems in hardware. A high data rate, secure wireless communication gateway is a good example of a signal processing system that is mapped in hardware. Such a device multiplexes several individual voice, data and video channels, performs encryption and digital modulation on the discrete signal, up-converts the signal, and digitally mixes it with an intermediate frequency (IF) carrier.

These types of digital design implement complex algorithms that operate on large amounts of data in real time. For example, a digital receiver for a G703 compliant E3 microwave link needs to demodulate 34.368 Mbps of information that were modulated using binary phase-shift keying (BPSK) or quadrature phase shift keying (QPSK). The system further requires forward error-correction (FEC) decoding and decompression along with several other auxiliary operations in real time.

These applications can be conveniently conceived as a network of interconnected components. For effective HW mapping, these components should be placed to work independently of other components in the system. The Kahn Process Network (KPN) is the simplest way of synchronizing their interworking. Although the implementation of a system as a KPN theoretically uses an unbounded FIFO (first-in/first-out) between two components, several techniques exist to optimally size these FIFOs. The execution of each component is blocked on a read operation on FIFOs on incoming links. A KPN also works well to implement event-driven and multi-rate systems. For signal processing applications, each component in the KPN implements one distinct block of the system. For example, the FEC in a communication receiver is perceived as a distinct block. This chapter gives a detailed account of the KPN and demonstrates its effectiveness and viability for digital system design with a case study.

In addition to this, for implementing logic in each block, the algorithm for the block can first be described graphically. There are different graphical ways of representing signal processing algorithms. The dataflow graph (DFG), signal flow graph, block diagram and data dependency graph (DDG) are a few choices for the designer. These representations are quite similar but have subtle differences. This chapter gives an extensive account of DFG and its variants. Graphical

Digital Design of Signal Processing Systems: A Practical Approach, First Edition. Shoab Ahmed Khan.

representation of the algorithm in any of these formats gives convenient visualization of algorithms for architecture mapping.

For blocks that need to process data at a very high data rate, the fastest achievable clock frequency is almost the same as the number of samples the block needs to process every second. For example, a typical BPSK receiver needs four samples per symbol to decode bits; for processing 36.5 Mbps of data, the receiver must process 146 million samples every second. The block implementing the front end of such a digital receiver is tasked to process even higher numbers of samples per symbol. In these applications the designer aims to achieve a clock frequency that matches the sampling frequency. This matching eases the digital design of the system, because the entire algorithm then requires one-to-one mapping of algorithmic operations to HW operators. This class of architecture is called 'fully dedicated architecture' (FDA). The designer, after mapping each operation to a HW operator, may need to appropriately place pipeline registers to bring the clock frequency equal to the sampling frequency.

The chapter describes this one-to-one mapping and techniques and transformations for adding one or multiple stages of pipelining for better timing. The scope of the design is limited to synchronous systems where all changes to the design are mastered by either a global clock or multiple clocks. The focus is to design digital logic at register transfer level (RTL). The chapter highlights that the design at RTL should be visualized as a mix of combinational clouds and registers, and the developer then optimizes the combinational clouds by using faster computational units to make the HW run at the desired clock while constraining it to fit within a budgeted silicon area. With feedforward algorithms the designer has the option to add pipeline registers in slower paths, whereas in feedback designs the registers can be added only after applying certain mathematical transformations. Although this chapter mentions pipelining, a detail treatment of pipelining and retiming are given exclusive coverage in Chapter 7.

4.2 Discrete Real-time Systems

A discrete real-time system is constrained by the sampling rate of the input signal acquired from the real world and the amount of processing the system needs to perform on the acquired data in real time to produce output samples at a specified rate. In a digital communication receiver, the real-time input signal may be modulated voice, data or video and the output is the respective demodulated signal. The analog signal is converted to a discrete time signal using an analog-to-digital (A/D) converter.

In many designs this real-time discrete signal is processed in fixed-size chunks. The time it takes to acquire a chunk of data and the time required to process this chunk pose a hard constraint on the design. The design must be fast enough to complete its processing before the next block of data is ready for its turn for processing. The size of the block is also important in many applications as it causes an inherent delay. A large block size increases the delay and memory requirements, whereas a smaller block increases the block-related processing overhead. In many applications the minimum block size is constrained by the selected algorithm.

In a communication transmitter, a real-time signal – usually voice or video – is digitized and then processed by general-purpose processors (GPPs), ASICs or FPGAs, or any combination of these. This processed discrete signal is converted back to an analog signal and transmitted on a wired or wireless medium.

A signal processing system may be designed as single-rate or multiple-rate. In a single-rate system, the numbers of samples per second at the input and output of the system are the same, and the number of samples per second does not change when the samples move from one block to another for processing. Communication systems are multi-rate systems: data is processed at different rates in different blocks. For each block the number of samples per second is specified. Depending on whether the system is a transmitter or a receiver, the number of samples per second may respectively increase or decrease for subsequent processing. For example, in a digital front end of a receiver the samples go through multiple stages of decimation, and in each stage the number of samples per second decreases. On the transmitter side, the number of samples per second *increases* once the data moves from one block to another. In a digital up-converter, this increase in number of samples is achieved by passing the signal through a series of interpolation stages. The number of samples each block needs to process every second imposes a throughput constraint on the block. For speech the data acquisition rate varies from 8 kHz to 16 kHz, and samples are quantized from 12-bit to 16-bit precision. For voice signals, like CD-quality music, the sampling rate goes to 44.1 kHz and each sample is quantized using 16-bit to 24-bit precision. For video signals the sampling rate further increases to 13.4 MHz with 8-bit to 12-bit of precision. Figure 4.1 illustrates the periodic sampling of

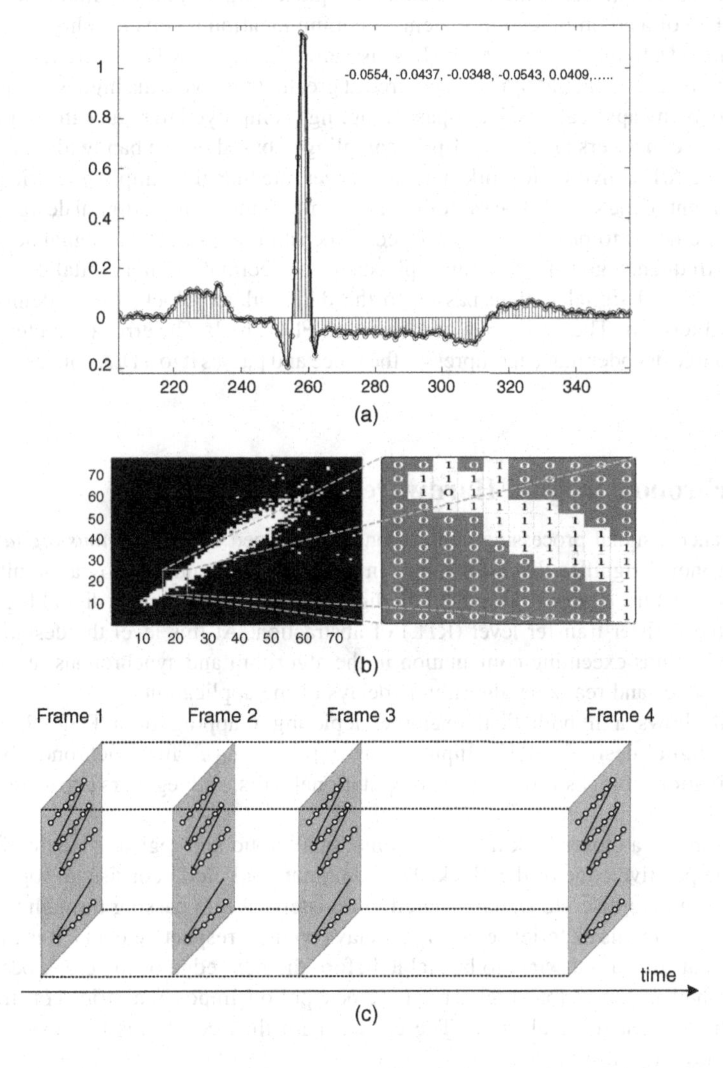

Figure 4.1 Periodic sampling of (a) 1-D signal; (b) 2-D signal; (c) 3-D signal

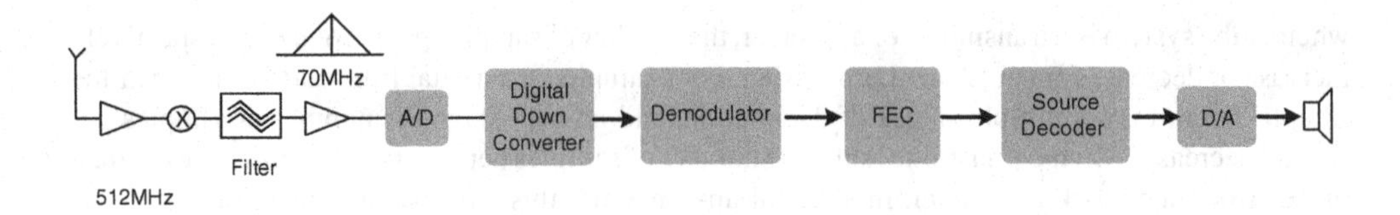

Figure 4.2 Real-time communication system

a one-dimensional electrocardiogram signal (ECG), a 2-D signal of a binary image, and 3-D signal of a video.

Another signal of interest is the intermediate frequency signal that is output from the analog front end (AFE) of a communication system. A communication receiver is shown in Figure 4.2. This IF signal is at a frequency of 70 MHz. It is discretized using an A/D converter. Adhering to the Nyquist sampling criterion, the signal is acquired at greater than twice the highest frequency of the signal at IF. In many applications, band-pass sampling is employed to reduce the requirement for high-speed A/D converters [1, 2]. Band-pass sampling is based on the bandwidth of an IF signal and requires the A/D converter to work at a much lower rate than the sampling at twice the highest frequency content of the signal. The sampling rate is important from the digital design perspective as the hardware needs to process data acquired at the sampling rate. This signal acquired at the intermediate frequency goes through multiple stages of decimation in a digital down-converter. The down-converted signal is then passed to the demodulator block, which demodulates the signal and extracts bits. These bits then go through an FEC block. The error-corrected data is then passed to a source decoder that decompresses the voice and passes it to a D/A converter for playing on a speaker.

4.3 Synchronous Digital Hardware Systems

In many instances, signal processing applications are mapped on *synchronous digital logic*. The word 'synchronous' signifies that all changes in the logic are controlled by a circuit clock, and 'digital' means that the design deals only with digital signals. Synchronous digital logic is usually designed at the register transfer level (RTL) of abstraction. At this level the design consists of combinational clouds executing computation in the algorithm and synchronous registers storing intermediate values and realizing algorithmic delays of the application.

Figure 4.3 shows a hypothetical example depicting mapping of a DSP algorithm as a synchronous digital design at RTL. Input values $x_1[n] \ldots x_4[n]$ are synchronously fed to the logic. Combinational blocks 1 to 5 depict computational units, and registers a to g are algorithmic delays.

Figure 4.4 shows a closer look into a combinational cloud and registers where all inputs are updated at the positive edge of the clock. The combinational cloud consists of logic gates. The input to the cloud are discrete signals, stored in registers. While passing through the combinational logic these signals experience different delays on their respective paths. It is important for all the signals at the output wires to be stable before the next edge of the clock occurs and the output is latched in the output register. The clock period imposes a strict constraint on the longest or the slowest (critical) path. The critical path thus constrains the fastest achievable clock frequency of a design.

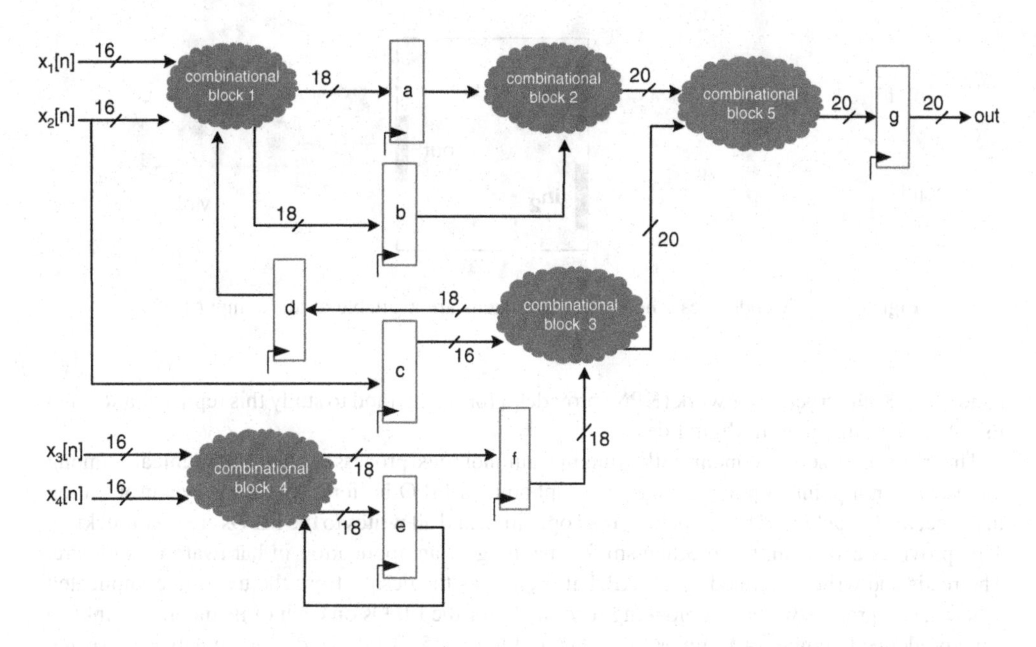

Figure 4.3 Register transfer level design consisting of combinational blocks and registers

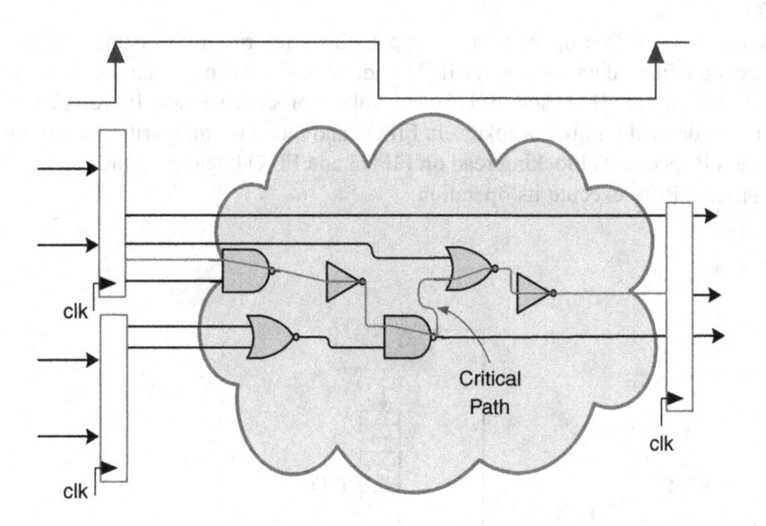

Figure 4.4 Closer look at a combinational cloud and registers depicting the critical path

4.4 Kahn Process Networks

4.4.1 Introduction to KPN

A system implementing a streaming application is best represented by autonomously running components taking inputs from FIFOs on input edges and generating output in FIFOs on the output

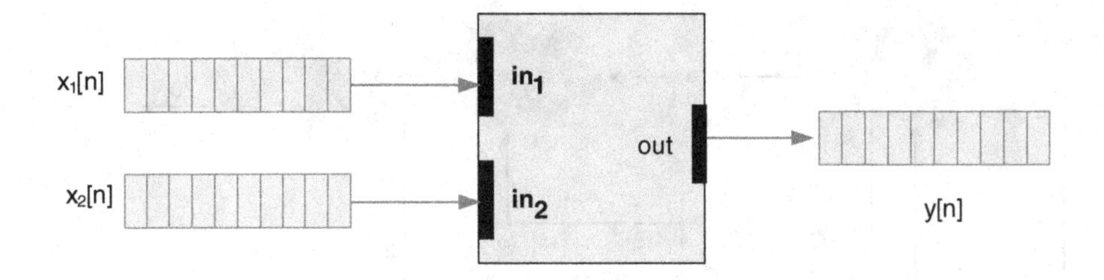

Figure 4.5 A node fires when sufficient tokens are available at all its input FIFOs

edges. The Kahn Process Network (KPN) provides a formal method to study this representation and its subsequent mapping in digital design.

The KPN is a set of concurrently running autonomous processes that communicate among themselves in a point-to-point manner over unbounded FIFO buffers, where the synchronization in the network is achieved by a blocking read operation and all writes to the FIFOs are non-blocking. This provides a very simple mechanism for mapping of an application in hardware or software. The reads and writes confined to the KPN also elevates the design from the use of a complicated scheduler. A process waits in a blocking read mode for the FIFOs on each of its incoming links to get a predefined number of samples, as shown in Figure 4.5. All the nodes in the network execute after their associated input FIFOs have acquired enough data. This execution of a node is called *firing*, and samples are called *tokens*. Thus, firing produces tokens and they are placed in respective output FIFOs.

Figure 4.6 shows a KPN implementing a hypothetical algorithm consisting of four processes. Process P_1 gets the input data stream in FIFO1 and, after processing a defined number of tokens, writes the result in output FIFO2 and FIFO3 on its links to processes P_2 and P_3, respectively. Process P_2 waits for a predefined number of tokens in FIFO2 and then fires and writes the output tokens in FIFO4. Process P_3 performs blocking read on FIFO3 and FIFO4, and then fires and writes data in FIFO5 for process P_4 to execute its operation.

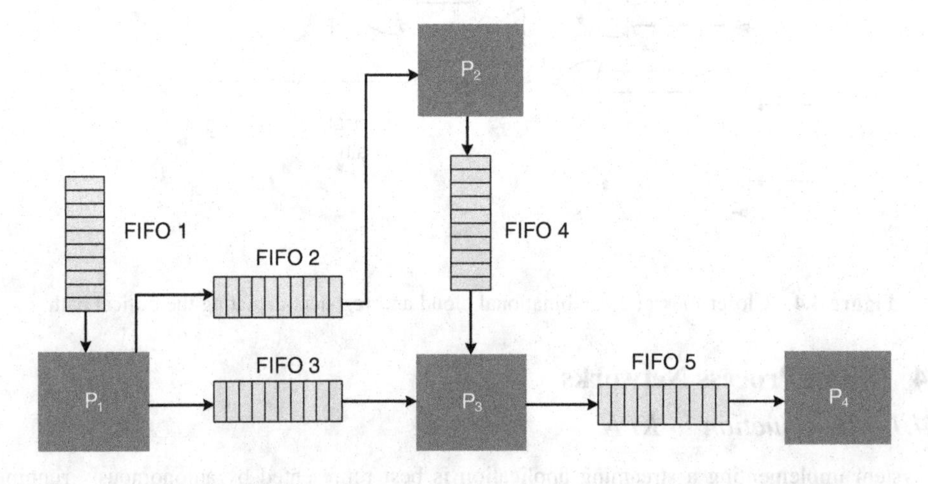

Figure 4.6 Example of a KPN with four processes and five connected FIFOs

Confining the buffers of the FIFOs to minimum size without affecting the performance of the network is a critical problem attracting the interest of researchers. A few solutions have been proposed [4, 5].

The KPN can also be implemented in software, where each process executes in a separate thread. The process, in a sequence of operations, waits on a read from a FIFO and, when the FIFO has enough samples, the thread performs a read operation and executes. The results are written into output FIFO. The KPN also works well in the context of mapping of a signal processing algorithm on reconfigurable platforms.

4.4.2 KPN for Modeling Streaming Applications

To map a streaming application as a KPN, it is preferable to first implement the application in a high-level language. The code should be broken down into distinguishable blocks with clearly identified streams of input and output data. This formatting of the code helps a designer to map the design as a KPN in hardware. The MATLAB® code below implements a hypothetical application:

```
N = 4*16;
K = 4*32;

% source node
for i=1:N
    for j=1:K
        x(i,j)= src_x ();
    end
end

% processing block 1
for i=i:N
    for j=1:K
        y1(i,j)=func1(x(i,j));
    end
end

% processing block 2
for i=1:N
    for j=1:K
        y2(i,j)=func2(y1(i,j));
    end
end

% sink block
m=1;
for i=1:N
    for j=1:K
        y(m)=sink (x(i,j), y1(i,j), y2(i,j));
        m=m+1;
    end
end
```

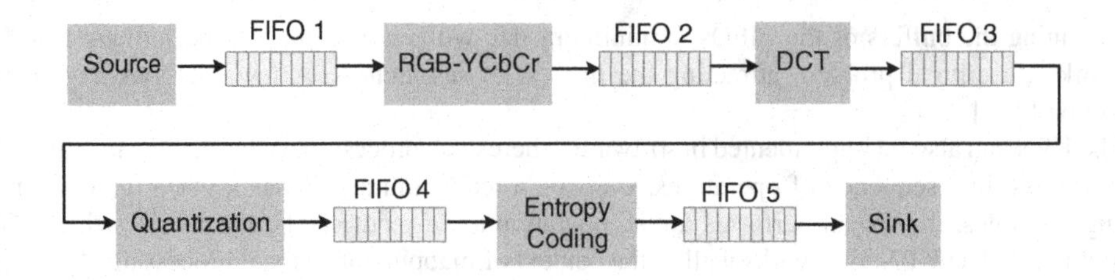

Figure 4.7 KPN implementing JPEG compression

4.4.2.1 Example: JPEG Compression Using KPN

Although KPN best describes distributed systems, it is also suited well to model streaming applications. In these, different processes in the system, in parallel or in a sequence, incrementally transform a stream of input data. Here we look at an implementation of the Joint Photographic Experts Group (JPEG) algorithm to demonstrate the effectiveness of a KPN for modeling streaming applications.

A raw image acquired from a source is saved in FIFO1, as shown in Figure 4.7. The algorithm is implemented on a block-by-block basis. This block is read by node 1 that transforms each pixel of the image from RGB to YCbCr and stores the transformed block in FIFO2. Now node 2 fires and computes the discrete cosine transform (DCT) of the block and writes the result in FIFO3. Subsequently node 3 and then node 4 fire and compute quantization and entropy coding and fill FIFO4 and FIFO5, respectively.

To implement JPEG, MATLAB® code is first written as distinguishable processes. Each process is coded as a function with clearly defined data input and output. The designer can then easily implement FIFOs for concurrent operation of different processes. This coding style helps the user to see the application as distinguishable and independently running processes for KPN implementation. Tools such as Compaan can exploit the code written in this subset of MATLAB® to perform profiling, HW/SW partitioning and the SW code generation. Tools such as Laura can be used subsequently for HW code generation for FPGAs [6, 7].

A representative top-level code in MATLAB® is given below (the code listing for functions `zigzag_runlength` and `VLC_huffman` are not given and can be easily coded in MATLAB®:

```
Q=[8  36 36 36 39 45 52 65;
36 36 36 37 41 47 56 68;
36 36 38 42 47 54 64 78;
36 37 42 50 59 69 81 98;
39 41 47 54 73 89 108 130;
45 47 54 69 89 115 144 178;
53 56 64 81 108 144 190 243;
65 68 78 98 130 178 243 255];
BLK_SZ = 8; % block size

imageRGB = imread('peppers.png'); % default image
figure, imshow(imageRGB);
title('Original Color Image');
ImageFIFO = rgb2gray(RGB);
figure, imshow(ImageFIFO);
title('Original Gray Scale Image');
```

```
fun = @dct2;
DCT_ImageFIFO = blkproc(ImageFIFO, [BLK_SZ BLK_SZ], fun);
figure, imshow(DCT_ImageFIFO);
title('DCT Image');
QuanFIFO = blkproc(DCT_FIFO, [BLK_SZ BLK_SZ], 'round(x./P1)', Q);
figure, imshow(QuanFIFO);

fun = @zigzag_runlength;
zigzagRLFIFO = blkproc(QuanFIFO, [BLK_SZ BLK_SZ], fun);
% Variable length coding using Huffman tables
JPEGoutFIFO = VLC_huffman(zigzagRLFIFO, Huffman_dict);
```

This MATLAB® code can be easily mapped as a KPN. A graphical illustration of this KPN mapping is shown in Figure 4.8. To conserve FIFO sizes for hardware implementation, the processing is performed on a block-by-block basis.

4.4.2.2 Example: MPEG Encoding

This section gives a top-level design of a hardware implementation of the MPEG video compression algorithm. Steps in the algorithm are given in Figure 4.9.

Successive video frames are coded as I and P frames. The coding of an I frame closely follows JPEG implementation, but instead of using a quantization table the DCT coefficients are quantized by a single value. This coded I frame is also decoded following the inverse processes of quantization and DCT. The decoded frame is stored in motion video (MV) memory.

For coding a P frame, first the frame is divided into macro blocks. Each block (called a target block) is searched in the previously decoded and saved frame in MV memory. The block that matches best with the target block is taken as a reference block and a difference of target and

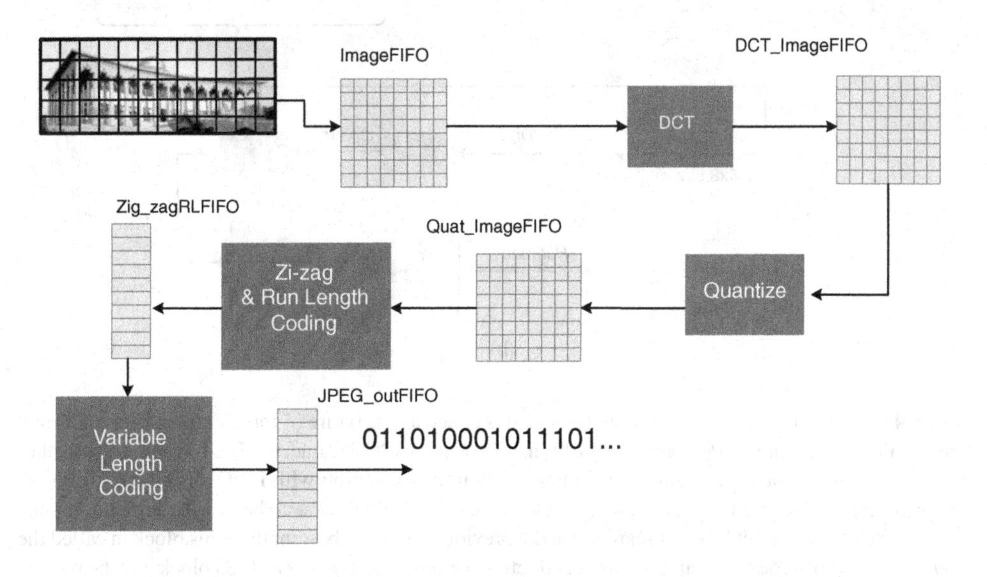

Figure 4.8 Graphical illustration of a KPN implementing the JPEG algorithm on block by block basis

reference block is computed. This difference block is then quantized and Huffman-coded for transmission. The motion vector refereeing the block in the reference frame is also transmitted. At the receiver the difference macro-block is decoded and then added in the reference block for recreation of the target block.

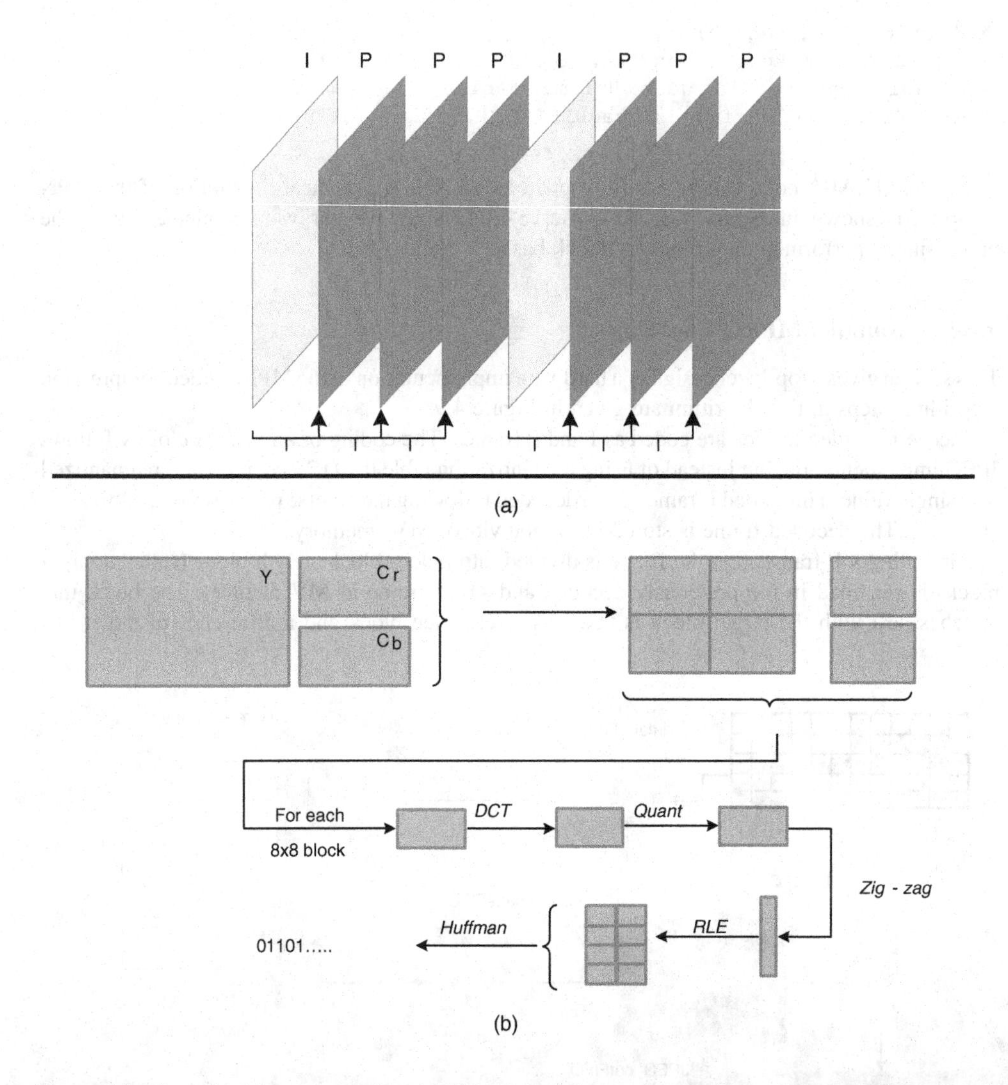

Figure 4.9 Steps in video compression. (a) Video is coded as a mixture of Intra-Frames (I-Frames) and Inter- or Predicted Frames (P-Frames). (b) I-Frame is coded as JPEG frame. (c) The I-Frame and all other frames are also decoded for coding of P-Frame, KPN implementation with FIFO's is given. (d) For P-Frame coding, the frame is divided into sub-blocks and each block is searched in the previously self-decoded frame. The target block is searched in the previous block for best match. This block in called the *reference block.* (e) The difference of target and reference block is coded as a JPEG block and the motion vector is also coded and transmitted with the coded block

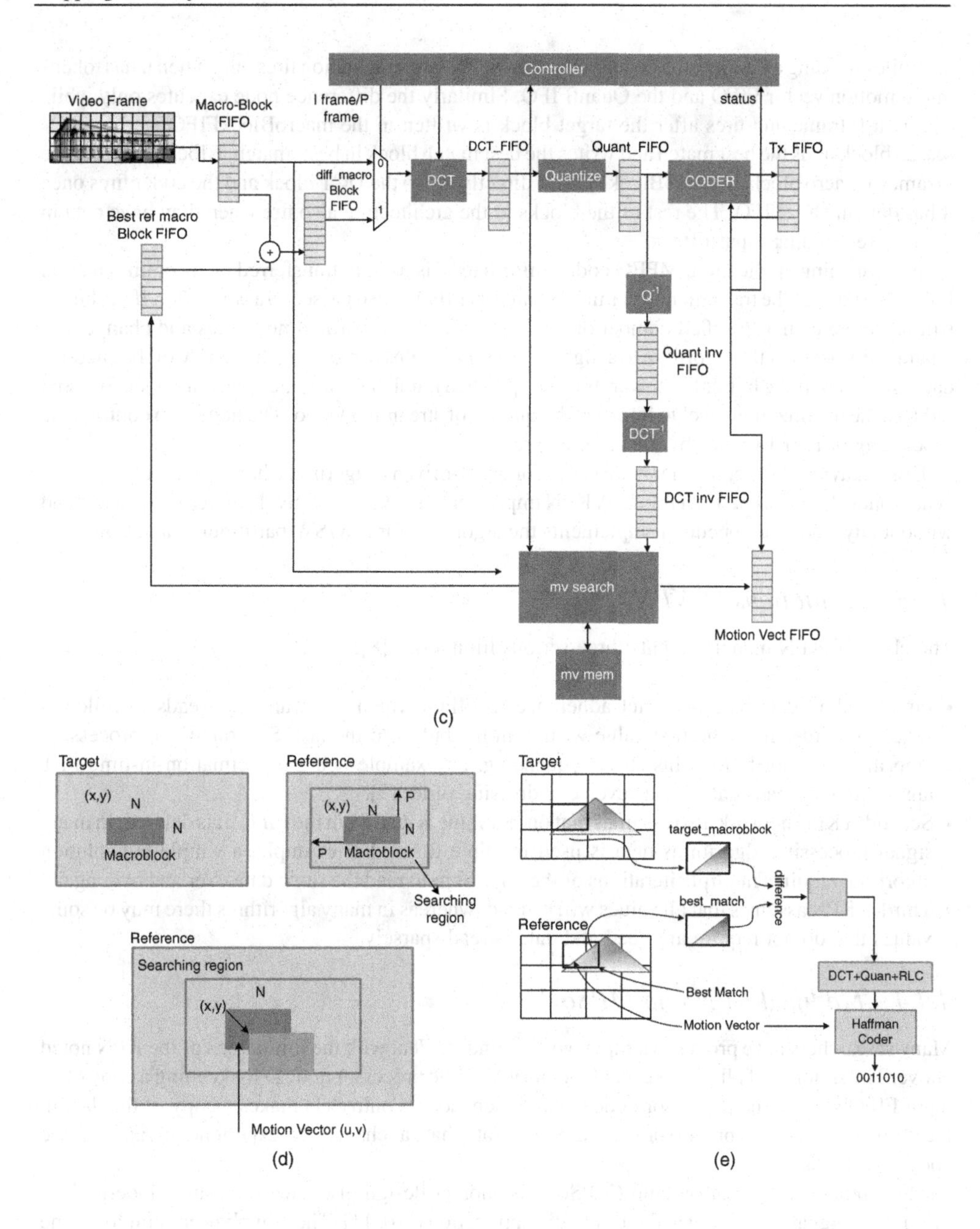

Figure 4.9 (*Continued*)

While encoding a P frame, the coder node in the KPN implementation fires only when it has tokens in the motion vector FIFO and the QuantFIFO. Similarly the difference node executes only while coding a P frame and fires after the target block is written in the macroBlockFIFO, and the MV search block finds the best match and writes the best match block in best_match_BlockFIFO. For the I frame, a macro block in marcoBlockFIFO is directly fed to the DCT block and the coder fires once it has data in DCTFIFO. The rest of the blocks in the architecture also fire when they have data in their corresponding input FIFOs.

In a streaming application, MPEG coded information is usually transferred on a communication link. The status of the transmit buffer implemented as FIFO is also passed to a *controller*. If the buffer still has more than a specified number of bits to transfer, the controller node fires and changes the quantization level of the algorithm to a higher value. This helps in reducing the bit rate of the encoded data. Similarly, if the buffer has fewer than the specified number of bits, the controller node fires and reduces the quantization level to improve the quality of streaming video. The node also controls the processing of P or I frames by the architecture.

This behavior of the controller node sets it for event-driven triggering, where the rest of the nodes synchronously fire to encode a frame. A KPN implementation very effectively mixes these nodes and without any elaborate scheduler implements the algorithm for HW/SW partitioning and synthesis.

4.4.3 Limitations of KPN

The classical KPN model presents three serious limitations [8]:

- First, reading data requires strict adherence to FIFO, which constrains the reads to follow a sequential order from the first value written in the buffer to the last. Several signal processing algorithms do not follow this strict sequencing, an example being a decimation-in-time FFT algorithm that reads data in bit-reverse addressing order [9].
- Second, a KPN network also assumes that once a value is read from the FIFO, it is deleted. In many signal processing algorithms data is used multiple times. For example, a simple convolution algorithm requires multiple iterations of the algorithm to read the same data over and over again.
- Third, a KPN assumes that all values will be read, whereas in many algorithms there may be some values that do not require any read and data is read sparsely.

4.4.4 Modified KPN and MPSoC

Many researchers have proposed simple workarounds to deal with the limitations of the KPN noted above. The simplest of all is to use local memory M in the processor node D for keeping a copy of its input FIFO's data. The processor node (or an independent controller) makes a copy of the data in local memory of the processor, in case the data has a chance of experiencing any of the above limitations.

Multi-processor system-on-chip (MPSoC) is another design of choice for many modern high-throughput signal processing and multimedia applications [10, 11]. These appliations employ some form of KPN to model the problem, and then the KPN, in many design methodologies, is automatically translated into MPSoC. A representative design is shown in Figure 4.10. Each processor has its local memory M, a memory controller (MC) and requisite FIFOs. It is connected to the FIFOs of other processors through a cross bar switch, a P2P network, a shared bus or a more elaborate network-on-chip (NOC) fabric. Each processor has a communciation controller (CC) that implements arbitration logic for sending data to FIFOs of other processors. The implementation is structured as tiles of these basic units. A detailed discussion on NoC design is given in Chapter 13.

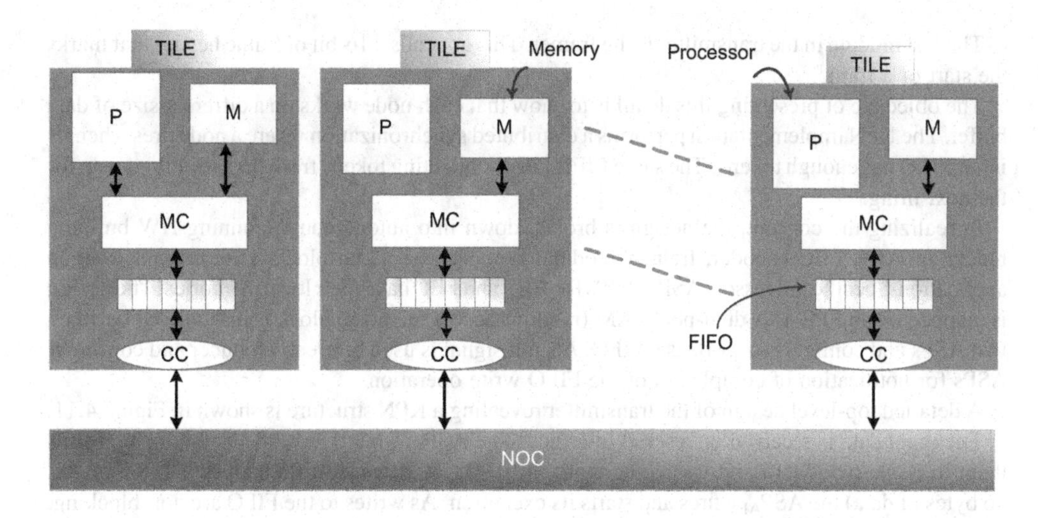

Figure 4.10 Modified KPN eliminating key limitations of the classical KPN configuration

4.4.5 Case Study: GMSK Communication Transmitter

The KPN configuration is used here in the design of a communication system comprising a transmitter and a receiver. As there is similarity in the top-level design of transmitter and receiver is the same, this section describes only the top-level design of the transmitter.

At the top level the KPN very conveniently models the design by mapping different blocks in the transmitter as autonomously executing hardware building blocks. FIFOs are used between two adjacent blocks and local memory in each block is used to store intermediate results. A producer block, after completing its execution, writes output data in its respective FIFO and notifies its respective consumer block to fire.

The transmitter system implements baseband digital signal processing of a Gaussian minimum shift-keying (GMSK) transmitter. The system takes the input data and modulates it on to an intermediate frequency of 21.4 MHz. The input to the system is an uninterrupted stream of digitized data. The input data rate is variable from 16 kbps up to 8 Mbps. In the system the data incrementally undergoes a series of transformations, such as encryption, forward error correction, framing, modulation and quadrature mixing.

The encryption module performs 256-bit encryption on a 128-bit input data block. A 256-bit key is input to the block. The encryption module expands the key for multiple rounds of encryption operation. The FEC performs 'block turbo code' (BTC) for a block size of $m \times n$ with values of m and n to be 11, 26 and 57. For different options of m and n the block size ranges from 121 to 3249 bits. The encoder adds redundant bits for error correction and changes the size of the data to $k \times l$, where k and l can take values 16, 32 or 64. For example, the user may configure the block encoder to input 11×57 data and encode it to 16×64. The user can also enable the interleaver that takes the data arranged row-wise and then reads it column-wise.

As AES (advanced encryption standard) block does not fall in the integer multiple of FEC boundaries, so two levels of marker are inserted. One header marks the start of the AES block and the other indicates the start of the FEC block. For example, each 128-bit AES block is appended with an 8-bit marker. Similarly, each FEC block is independently appended with a 7-bit marker.

The last module in the transmitter is the framer. This appends a 16-bit of frame header that marks the start of a frame.

The objective of presenting this detail is to show that each node works on a different size of data buffer. The KPN implementation performs a distributed synchronization where a node fires when its input FIFO has enough tokens. The same FIFO keeps collecting tokens from the preceding node for the next firing.

In realizing the concept, the design is broken down into autonomously running HW building blocks of AES, FEC encoder, framer, modulator and mixer. Each block is implemented as an application-Specific processor (ASP). ASPs for AES and FEC have their local memories. The design is mapped on an FPGA. A dual-port RAM (random-access memory) block is instantiated between two ASPs and configure to work as a FIFO. A 1-bit signal is used between producer and consumer ASPs for notification of completion of the FIFO write operation.

A detailed top-level design of the transmitter revealing a KPN structure is shown in Figure 4.11. The input data is received on a serial interface. The ASP_{AES} interface collects these bits, forms them in a one-byte word and writes the word in $FIFO_{AES}$. After the FIFO collects 128 bits (i.e. 16 bytes of data) the ASP_{AES} fires and starts its execution. As writes to the FIFO are non-blocking, thus the interface keeps collecting bits, forms them in one-byte words and writes the words in the FIFO. The encryption key is expended at the initialization and is stored in a local memory of ASP_{AES} labeled as MEM_{AES}. The internal working of the ASP_{AES} is not highlighted in this chapter as the basic aim of the section is to demonstrate the top-level modeling of the design as KPN. A representative design of ASP_{AES} is given in Chapter 13.

The next processor in the sequence is ASP_{FEC}. The processor fires after its FIFO stores sufficient number of samples based on the mode selected for BTE. For example, the 11×11 mode requires 121 bits. Each firing of ASP_{AES} produces 128 bits and an 8-bit marker is attached to indicate the start of the AES block. ASP_{FEC} fires once and uses 121 bits. The remaining 15 bits are left in the FIFO. In the next firing these 15 bits along with 106 bits from current AES frame are used and 30 bits are left in the FIFO. The number of remaining bits keeps increasing, and after a number of iterations they make a complete FEC block that requires an additional firing of the ASP_{FEC}. The FEC block is a multi-rate block as it adds additional bits for error correction at the receiver. For the case in consideration, 121 bits generate 256 bits at the output.

At the completion of encoding, the ASP_{FEC} notifies the ASP_{Framer}. The ASP_{Framer} also waits to collect the defined number of bits before it fires. It also appends a 16-bit header to the frame of data from ASP_{FEC} for synchronization at the receiver. The framer writes a complete frame of data in a buffer. From here onward the processing is done on a bit-by-bit basis. The GMSK modulator first filters the non-return-to-zero (NRZ) data using a Gaussian low-pass filter and the resultant signal is then frequency modulated (FM). The bandwidth of the filter is chosen in such a way that it yields a narrowband GMSK signal given by:

$$x(t) = \sqrt{2P_c} \cos\left(2\pi f_c t + \varphi_s(t)\right)$$

where f_c is the carrier frequency, P_c is the carrier power and $\varphi_s(t)$ is the phase modulation given by:

$$\varphi_s(t) = 2\pi f_d \int_{-\infty}^{t} \sum_{k=-\infty}^{\infty} a_k g(v - kT) dv$$

where f_d is the modulation index and is equal to 0.5 for GMSK, a_k is the binary data symbol and is equal to ± 1, and $g(.)$ is the Gaussian pulse. A block diagram of the GMSK baseband modulator is shown in Figure 4.12.

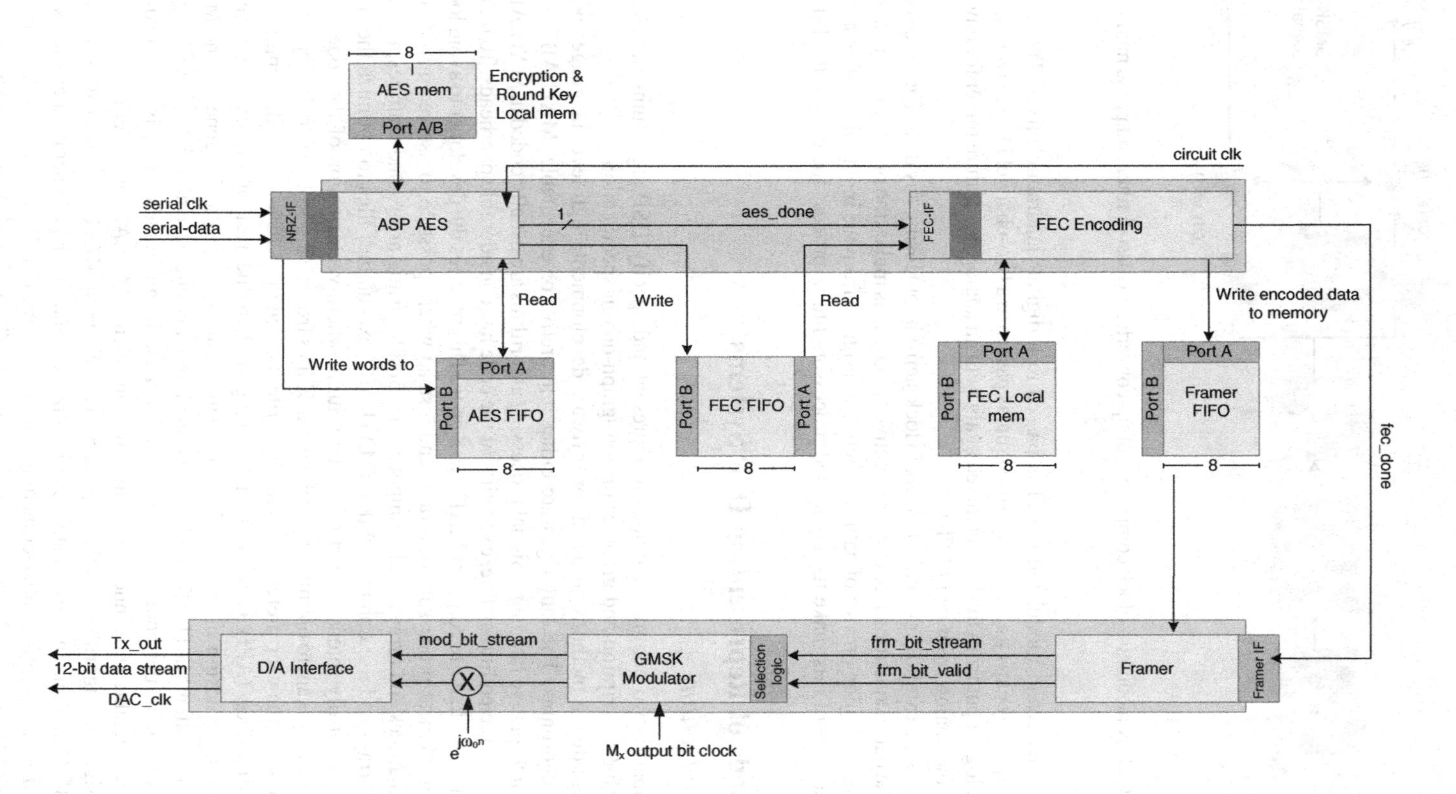

Figure 4.11 Detailed top-level design of the transmitter modeled as a KPN

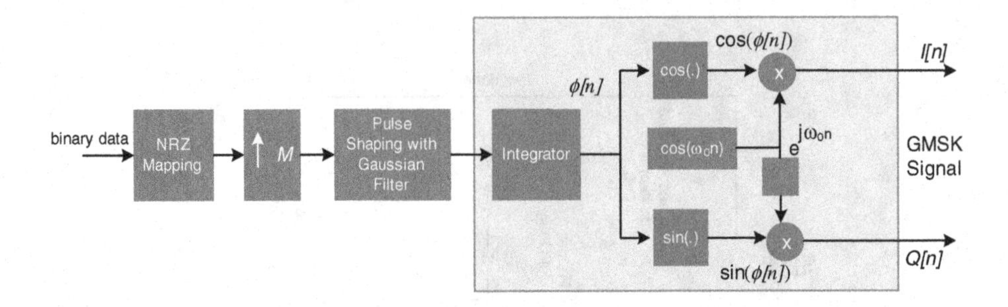

Figure 4.12 GMSK modulator block comprising of up-converter, Gaussian filter and phase modulator

The complex baseband modulated signal is passed to a digital quadrature mixer. The mixer multiplies the signal by $\exp(j\omega_o n)$. The mixed signal is passed to a two-channel D/A converter. The quadrature analog output from the D/A is mixed again with an analog quadrature mixer for onward processing by the analog front end (AFE).

The processors are carefully designed to work in lock step without losing any data. The processing is faster than the data acquisition rate at the respective FIFOs. This makes the processors wait a little after completing execution on a set of data. The activation signals from the input FIFOs after storing the required number of bits make the connected processors fire to process the data in the FIFO.

4.5 Methods of Representing DSP Systems

4.5.1 Introduction

There are primarily two common methods to represent and specify DSP algorithms: language-driven executable description and graphics or flow-graph-driven specification.

The language-driven methods are used for software development, high-level languages being used to code algorithms. The languages are either interpretive or executable. MATLAB® is an example of an interpretive language. Signal processing algorithms are usually coded in MATLAB®. As the code is interpreted line by line, execution of the code has considerable overheads. To reduce these it is always best to write compact MATLAB® code. This requires the developer to avoid loops and element-by-element processing of arrays, and instead where possible to use vectorized code that employs matrices and arrays in all computations. Besides using arrays and matrices, the user can predefine arrays and use compiled MEX (MATLAB® executable) files to optimize the code.

For a computationally intensive part of the algorithm, where vectorization of the code is not possible, the user may want to optimize by creating MEX files from C/C++ code or already written MATLAB® code. If the user prefers to manually write the code in C/C++, a MEX application programming interface (API) is used for interfacing it with the rest of the code written in MATLAB®. In many design instances it is more convenient to automatically generate the MEX file from already written MATLAB® code. For computationally intensive algorithms, the designer usually prefers to write the code in C/C++. As in these languages the code is compiled for execution, the executable runs much faster than its equivalent MATLAB® simulation.

In many designs the algorithm developer prefers to use a mix of MATLAB® and C/C++ using the MATLAB® Complier. In these instances visualization and other peripheral calculations are performed in MATLAB® and number-crunching routines are coded in C/C++, compiled and then called from the MATLAB® code. The exact semantics of MATLAB® and C/C++ are not of much interest in hardware mapping. The main focus of this chapter is graphical representation of DSP algorithms, so the description of exact semantics of procedural languages is left for the readers to learn from other resources. It is important to point out that there are tools such as Compaan that convert a code written in a subset of MATLAB® or C constructs to HW description [12].

Although language-driven executable specifications have wider acceptability, graphical methods are gaining ground. The graphical methods are especially convenient for HW mapping and understanding the working and flow of the algorithm. A graphical representation is the method of choice for developing optimized hardware, code generation and synthesis.

Another motivation to learn about the graphical representation comes from its use in many commercially available signal processing tools, including Simulink from Mathworks [13], Advanced Design Systems (ADS) from Agilent [14], Signal Processing Worksystem (SPW) from Cadence [15], Cocentric System Studio from Synopsys [16], Compaan from Leiden University [12], LabVIEW from National Instruments [17], Grape from K. U. Leuven, PeaCE from Seoul National University [18], SteamIT from MIT [19], DSP Station from Mentor Graphics [20], Hypersignal from Hyperception [21], and open-source software like Ptolemy-II [22] from the University of California at Berkeley and Khoros [23] from the University of New Mexico. All these tools at a coarser level use some form of KPN and each process in the representation uses an executable flow graph. It is important to point out that alternatively in many of these tools, the nodes of a graphical representation may also run a program written in one of more conventional programming language. Although the node behavior is still captured in procedural programming languages, overall system interpretation as a graph allows parallel mapping of the nodes on multiple processors or HW.

The graphical methods also support structural and hierarchical design flow. Each node in the graph is hierarchically built and may encompass a graph in itself. The specifications are simulatable and can also be synthesized. These methods also emphasize component-based architecture design. The components may be parameterized to be reused in a number of design instances. Each component can further be described at different levels of abstraction. This helps in HW/SW co-simulation and exploration of the design space for better HW/SW partitioning.

The concept is illustrated in Figure 4.13. Process P3 is implemented in RTL Verilog and optimized assembly language targeting a particular DSP processor. The idea is to explore both options of HW and SW implementation and then, based on design objectives, seamlessly map P3 either on the DSP processor or on the FPGA.

4.5.2 *Block Diagram*

A block diagram is a very simple graphical method that consists of functional blocks connected with directed edges. A connected edge represents flow of data from source block to destination block. Figure 4.14 shows a block diagram representation of a 3-coefficients FIR filter implementing this equation:

$$y[n] = h_0x[n] + h_1x[n-1] + h_2x[n-2] \tag{4.1}$$

In the figure, the functional blocks are drawn to represent multiplication, addition and delay operations. These blocks are connected by directed edges showing the precedence of operations and data flow in the architecture. It is important to note that the DSP algorithms are iterative. Each iteration of the algorithm implements a series of tasks in a specific order. A delay element stores the intermediate result produced by its source node. This result is used by the destination node on the same edge in the next iteration of the algorithm. Figure 4.15 shows source node U performing

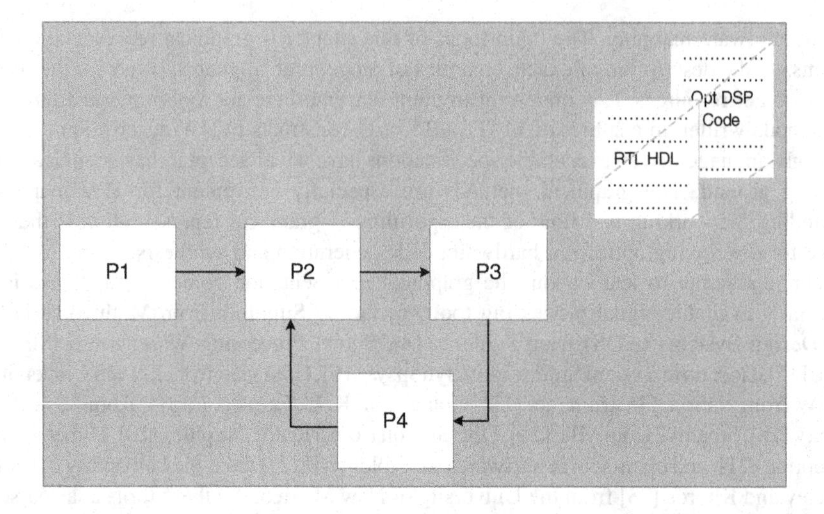

Figure 4.13 Graphical representation of a signal processing algorithm where node P3 is described in RTL and optimized assembly targeting an FPGA or a particular DSP respectively

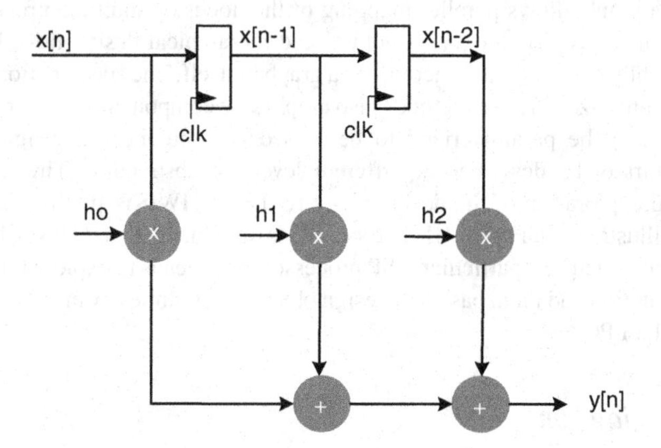

Figure 4.14 Dataflow graph representation of a 3-coefficient FIR filter

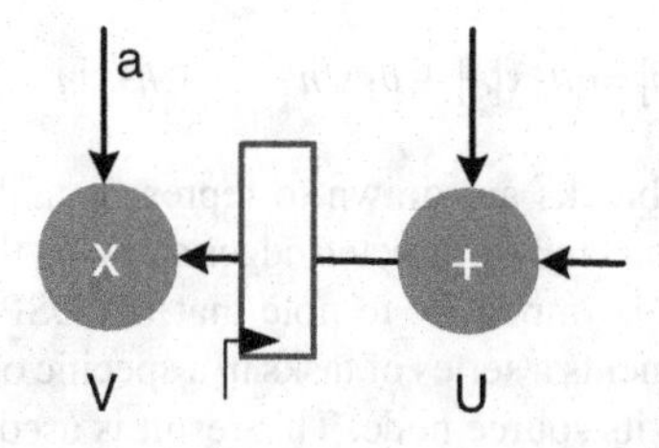

Figure 4.15 Source and destination nodes with a delay element on the joining edge

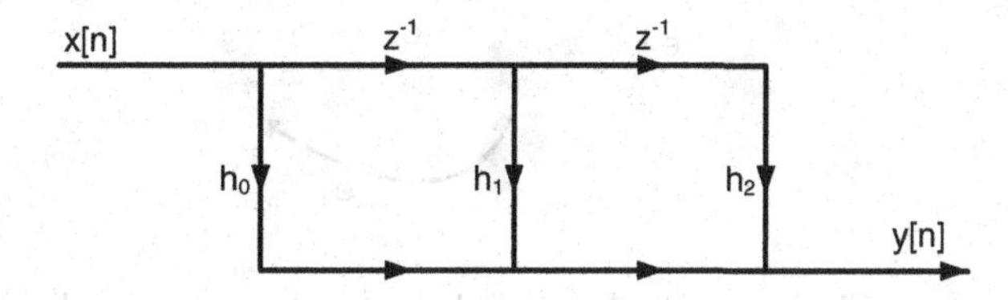

Figure 4.16 Signal flow graph representation of a 3-coefficient FIR filter

the addition operation and storing the value in the register on the edge, whereas the node V uses the stored value of the last iteration in the register for its multiplication by a constant *a*.

4.5.3 Signal Flow Graph

A signal flow graph (SFG) is a simplified version of a block diagram. The operation of multiplication with a constant and delays are represented by edges, whereas nodes represent addition, subtraction and input and output (I/O) operations.

Figure 4.16 shows an SFG implementation of equation (4.1) above. The edges with h_0, h_1 and h_2 represent multiplication of the signal at the input of the edges by respective constants, z^{-1} on the arrow represents a delay element, the node where two arrows meet shows the addition operation, and a node with one incoming and two outgoing edges means that the data on the input edge is broadcast to output edges.

SFGs are primarily used in describing DSP algorithms. Their use in representing architectures is not very attractive.

4.5.4 Dataflow Graph or Data Dependency Graph

In a DFG representation, a signal processing algorithm is described by a directed graph $G = \langle V, E \rangle$, where a node $v \in V$ represents a computational unit or, in a hierarchical design, a sub-graph already described subscribing to the rules of DFG. A directed edge $e \in E$ from a source node to a destination node represents either a FIFO buffer or just precedence of execution. It is also used to represent algorithmic delays introduced to data while it moves from source node to destination node. A destination node can fire only if it has on all its input edges the predefined number of tokens. Once a node executes, it consumes the defined number of tokens from its input edges and generates a defined number of tokens on all its output edges.

A DFG representation is of special interest to hardware designers as it captures the data-driven property of a DSP algorithm. A DFG representation of a DSP algorithm exposes the hidden concurrency among different parts of the algorithm. This concurrency then can be exploited for parallel mapping of the algorithm in HW. A DFG can be used to represent synchronous, asynchronous and multi-rate DSP algorithms. For an asynchronous algorithm, an edge 'e' represents a FIFO buffer.

DFG representation is very effective in designing and synthesizing DSP systems. It motivates the designer to think in terms of components. A component-based design, if defined carefully, helps in reuse of the components in other design instances. It also helps in module-level optimization, testing and verification. The system can be easily profiled and the profiling results along with the structure of the DFG facilitate HW/SW partitioning and subsequent co-design, co-simulation and co-verification.

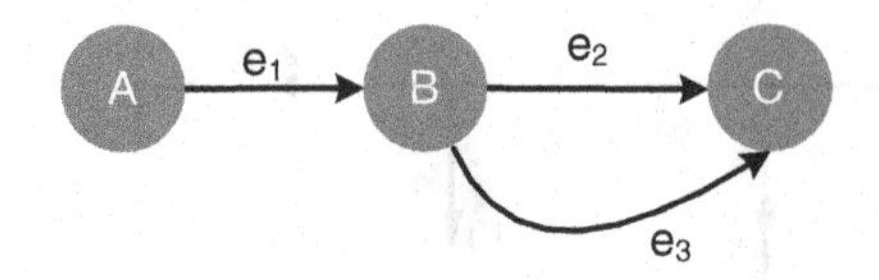

Figure 4.17 Hypothetical DFG with three nodes and three edges

Each node in the graph may represent an atomic operation like multiplication, addition or subtraction. Alternatively, each node may coarsely define a computational block such as FIR filter, IIR filter or FFT. In these design instances, a predesigned library for each node may already exist. The designer needs to exactly compute the throughput for each node and the storage requirement on each edge. While mapping a cascade of these nodes to HW, a controller can be easily designed or automatically generated that synchronizes the operation of parallel implementation of these nodes.

A hypothetical DFG is shown in Figure 4.17, where each node represents a computational block defined at coarse granularity and edges represent connectivity and precedence of operation. For HW mapping, appropriate HW blocks from a predesigned library are selected or specifically designed with a controller synchronizing the nodes to work in lock step for parallel or sequential implementation.

Figure 4.18 shows the scope of different sub-classes of graphical representations. These representations are described in below sections. The designer needs to select, out of all these representations, an appropriate representation to describe the signal processing algorithm under consideration. The KPN is the most generalized representation and can be used for implementing a wide range of signal processing systems. In many design instances, the algorithm can be defined more precisely at a finer level with details of number of cycles each node takes to execute its computation and the number of tokens it consumes at firing, and as a result the number of tokens it produces at its output edges. These designs can be represented using cyclo-static DFG (CSDFG), synchronous DFG (SDFG) and homogenous SDFG (HSDFG) – in reducing order of generality.

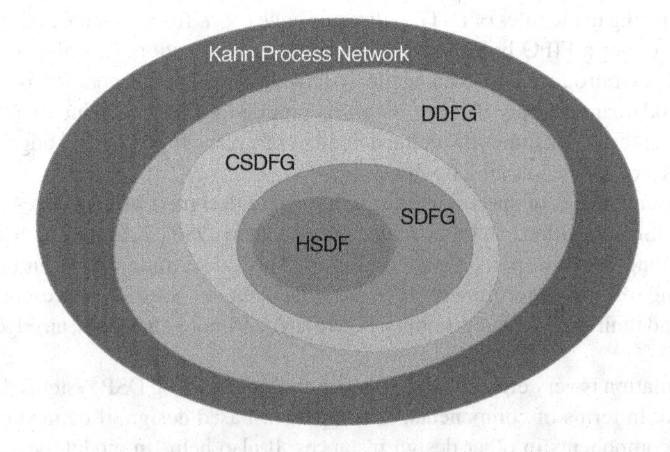

Figure 4.18 Scope of different subclasses of graphical representation, moving inward from the most generalized KPN to dynamic DFG (DDFG), to CSDFG, to SDFG, and ending at the most constrained HSDFG

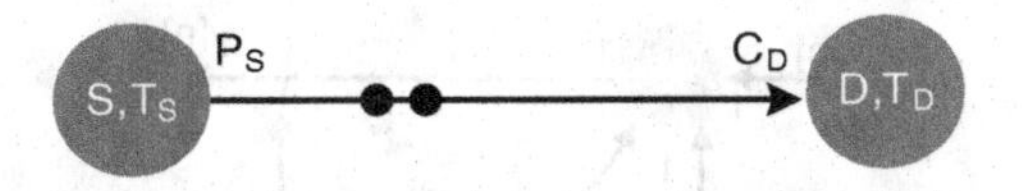

Figure 4.19 SDFG with nodes S and D

When the number of cycles each node takes to execute and the number of tokens produced and consumed by the node are not known *a priori*, these design cases can be modeled as dynamic DFG (DDFG).

4.5.4.1 Synchronous Dataflow Graph

A typical streaming application continuously processes sampled data from an A/D converter and the processed data is sent to a D/A converter. An example of a streaming application is a multimedia processing system consisting of processes or tasks where each task operates on a predefined number of samples and then produces a fixed number of output values. These tasks are periodically executed in a defined order.

A *synchronous* dataflow graph (SDFG) best describes these applications. Here the number of tokens consumed by a node on each of its edges, and as a result of its firing the number of tokens it produces on its output edges, are known *a priori*. This leads to efficient HW mapping and implementation.

Nodes are labeled with their names and number of cycles or time units they take in execution, as demonstrated in Figure 4.19. An edge is labeled at tail and head with production and consumption rates, respectively. The black dots on the edges are the algorithm delays between two nodes. Figure 4.19 shows source and destination nodes S and D taking execution time T_S and T_D units, respectively. The data or token production rate of S is P_S and the data consumption rate of node D is C_D. The edge shows two algorithmic delays between the nodes specifying that node D uses two iterations-old data for its firing.

An example of an SDFG is shown in Figure 4.20(a). The graph consists of three nodes A, B and C with execution times of 2, 1 and 1 time units, respectively. The production and consumption rates of each node are also shown, with the number of required initial delays represented by filled circles on the edges.

An SDFG can also be represented by a *topology matrix*, as shown in Figure 4.20(b). Each column of the matrix represents a node and each row corresponds to an edge. The three columns represent nodes A, B and C in the same order. The first row depicts edge e_1 connecting nodes A and B. Similarly the second and third rows show the connections on edges e_2 and e_3, respectively.

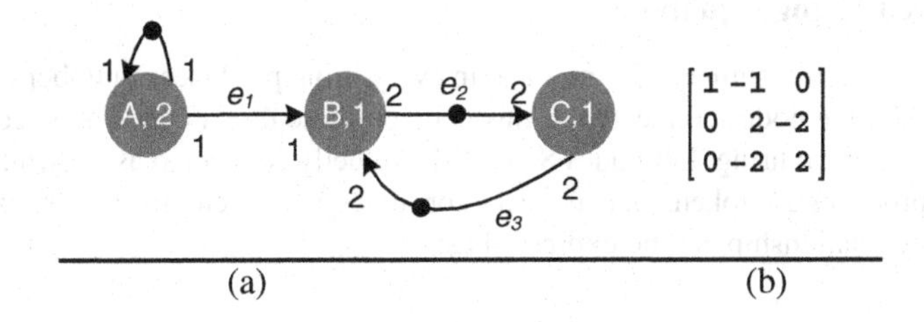

Figure 4.20 (a) Example of an SDFG with three nodes A, B and C. (b) Topology matrix for the SDFG

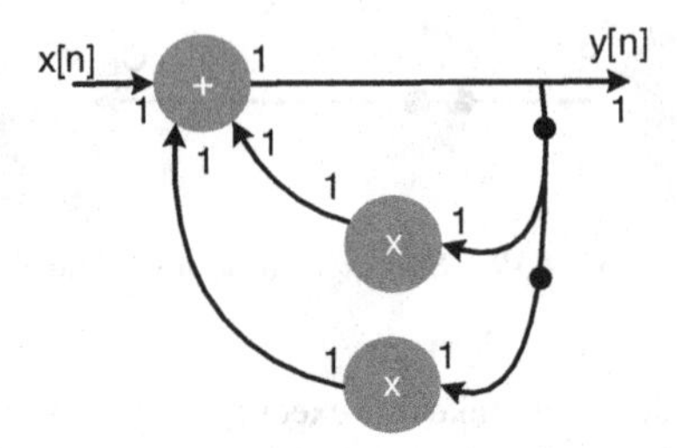

Figure 4.21 SDFG implementing the difference equation (4.2)

The production and consumption rates are represented by positive and negative numbers, respectively.

4.5.4.2 Example: IIR Filter as an SDFG

Draw an SDFG to implement the difference equation below:

$$y[n] = x[n] + a_1 y[n-1] + a_2 y[n-2] \tag{4.2}$$

The SDFG modeling the equation is given in Figure 4.21. The graph consists of two nodes for multiplications and one node for addition, each taking one token as input and producing one token at the output. The feedback paths from the output to the two multipliers each requires a delay. These delays are shown with black dots on the respective edges.

4.5.4.3 Consistent and Inconsistent SDFG

An SDFG is described as *consistent* if it is correctly constructed. A consistent SDFG, once it executes, neither starves for data nor requires any unbounded FIFOs on its edges. These graphs represent both single- and multiple-rate systems. The consistency of an SDFG can be easily confirmed by computing the rank of its topology matrix and checking whether that is one less than the number of nodes in the graph [24].

An SDFG is described as *inconsistent* and experiences a deadlock in a streaming application if some of its nodes starve for data at its input to fire, or on some of its edges it may need unbounded FIFOs to store an unbounded production of tokens.

4.5.4.4 Balanced Firing Equations

A node in an SDFG fires a number of times, and in every firing produces a number of tokens on its outgoing edges. All the nodes connected to this node on these direct edges must consume all the tokens in their respective firing. Let nodes S and D be directly connected as shown in Figure 4.19, where node S produces P_S tokens and node D consumes C_D tokens in their respective firings. A balanced firing relationship can be expressed as:

$$f_s P_S = f_D C_D \tag{4.3}$$

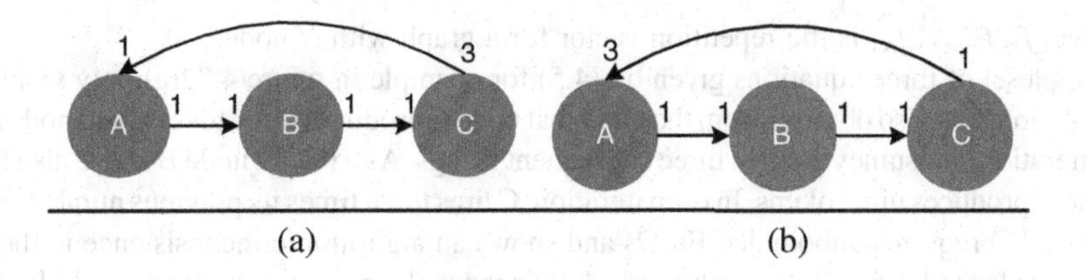

Figure 4.22 Examples of inconsistent DFGs resulting in only the trivial solution of a set of balanced-firing equations where no node fires. (a) DFG requires unbounded FIFOs. (b) Node A will starve for data as it requires three tokens to fire and only gets one from node B

where f_S and f_D are non-zero finite minimum positive values. This definition can be hierarchically extended for an SDFG with N nodes, where node 1 is the input and node N is the output:

$$\begin{aligned} f_1 P_1 &= f_2 C_2 \\ f_2 P_2 &= f_3 C_3 \\ &\vdots \\ f_{N-1} P_{N-1} &= f_N C_N \end{aligned} \tag{4.4}$$

The values $f_1 \ldots f_N$ form a vector called the *repetition vector*. It can be observed that this vector is obtained by computing a non-trivial solution to the set of equations given in (4.4). The repetition vector requires each node to fire f_i times. This makes all the nodes in an SDFG fire synchronously, producing enough tokens for the destination nodes and consuming all the data values being produced by its source nodes.

A set of balanced equations for an SDFG greatly helps a digital designer to design logic in each node such that each node takes an equal number of cycles for its respective firing. The concept of balanced equations and their effective utilization in digital design is illustrated later in this chapter. It is pertinent to note that there are design instances where a cyclo-static schedule is required where a node may take different time units and generate a different number of tokens in each of its firings and then periodically repeat the pattern of firing.

Example: Find a solution of balanced equations for the SDFG given in Figure 4.22(a). The respective balanced equations for the links A → B, B → C and C → A are:

$$\begin{aligned} f_A 1 &= f_B 1 \\ f_B 1 &= f_C 1 \\ f_C 3 &= f_A 1 \end{aligned}$$

No non-trivial solution exists for these set of equations as the DFG is inconsistent.

4.5.4.5 Consistent and Inconsistent SDFGs

If no non-trivial solution with non-zero positive values exists for the set of balanced firing equations, then it is inferred that the DFG is an inconsistent SDFG. This is equivalent to the check on the rank of the topology matrix **T**, as any set of balanced equations for an SDFG can be written as:

$$\mathbf{Tf} = 0 \tag{4.5}$$

where $\mathbf{f} = [f_1, f_2, \ldots f_N]$ is the repetition vector for a graph with N nodes.

Solving a set of three equations given by (4.5) for example in figure 4.22(a) only yields a non-trivial solution. It is also obvious from the DFG that node C produces three tokens, and node A, in the second iteration, consumes them in three subsequent firings. As a result, node B and C also fire three times and C produces nine tokens. In each iteration, C fires three times its previous number of firings. Thus this DFG requires unbounded FIFOs and shows an algorithmic inconsistence in the design. Using the balanced firing relationship of (4.3), produce/consume equations for links A $\rightarrow$ B, B $\rightarrow$ C and C $\rightarrow$ A are written as:

$$f_A 1 = f_B 1$$

$$f_B 1 = f_C 1 \tag{4.6}$$

$$f_C 3 = f_A 1$$

The only solution for the set of equations is the trivial $f_A = f_B = f_C = 0$. Another equally relevant case is shown in Figure 4.22(b) where, following the sequential firing of different nodes, node A will starve for data in the second iteration as it requires three tokens for its firing and will only get one from node B.

4.5.5 Self-timed Firing

In a self-timed firing, a node fires as soon as it gets the requisite number of tokens on its incoming edges. For implementing SDFG on dedicated hardware, either self-timed execution of nodes can be used or a repetition vector can be first calculated using a balanced set of equations. This vector calculates multiple firings of each node to make the SDFG consistent. This obviously requires a larger buffer size at the output of each edge to accumulate tokens resulting from multiple firing of a node. A self-timed SDFG implementation usually results in a transient phase, and after that the firing sequence repeats itself in a periodic phase. Self-timed SDFG implementation requires minimum buffer sizes as a node fires as soon as the buffers on its inputs have the requisite number of samples.

Example: Figure 4.23(a) shows an SDFG with three nodes, A, B and C, having execution time of 1, 2 and 2 time units, respectively. The consumption and production rates for each link are also shown. In a self-timed configuration, an implementation goes through a transient phase where nodes A and B fire three times to accumulate three token at the input of C. Node B takes two time units to complete its execution. In Figure 4.23(b) the first time unit is shown with capital letter 'B' and the second time unit is shown with small letter 'b'. This initial firing brings the self-timed implementation in periodic phase where C fires and generates three tokens at its output. This makes A fire three times, and so does B, generating three tokens at input of C – and the process is repeated in a periodic fashion.

4.5.6 Single-rate and Multi-rate SDFGs

In a Single-Rate SDFG, the consumption rate r_c (the number of tokens consumed on each incoming edge) is same as the production rate r_p (the number of tokens produced at the outgoing edges). In a Multi-Rate SDFG, these rates are not equal and one is a rational multiple of the other. For a decimation multi-rate system, $r_c > r_p$. An interpolation system observes a reverse relationship, $r_p < r_c$.

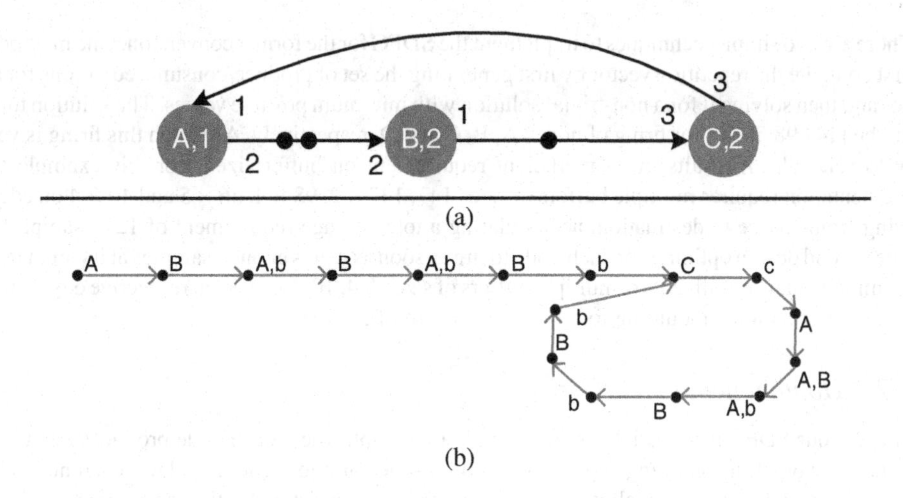

Figure 4.23 Example of self-timed firing where a node fires as soon as it has enough tokens on its input. (a) An SDFG. (b) Firing pattern using self-timed firing

An algorithm described as a synchronous DFG makes it very convenient for a designer or an automation tool to synthesize the graph and generate SW or HW code for the design. This representation also helps the designer or the tools to apply different transformations on the graph to make it optimized for a set of design constraints and objective functions. The representation can also be directly mapped on dedicated HW, or time-multiplexed HW. For time-multiplexed HW, several techniques have been evolved to generate an optimal schedule and HW architecture for the design. Time-multiplexed architectures are covered in Chapters 8 and 9.

Example: A good example of a multi-rate DFG is a format converter to store a CD-quality audio sampled at 44.1 kHz to a digital audio tape (DAT) that records a signal sampled at 48 kHz. The following is the detailed workout of the format conversion stages [25]:

$$\frac{f_{DAT}}{f_{CD}} = \frac{480}{441} = \frac{160}{147} = \frac{2}{1} \times \frac{4}{3} \times \frac{5}{7} \times \frac{4}{7}$$

The first stage of rate change is implemented as an interpolator by 2, and each subsequent stage is implemented as a sampling rate change node by a rational factor P/Q, where P and Q are interpolation and decimation factors. This sampling rate change is implemented by an interpolation by a factor of P, and then passing the interpolated signal through a low-pass FIR filter with cutoff frequency π/max (P,Q). The signal from the filter is decimated by a factor of Q [9].

Figure 4.24 shows the SDFG that implements the format converter. The SDFG consists of processing nodes A, B, C and D implementing sampling rate conversion by factors of 2/1, 4/3, 5/7 and 4/7, respectively. In the figure CD and DAT represent source and destination nodes.

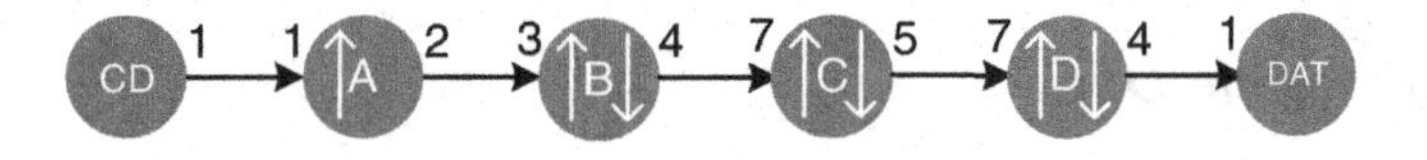

Figure 4.24 SDFG implementing CD to DAT format converter

There are two simple techniques to implement the SDFG for the format conversion. One method is to first compute the repetition vector by first generating the set of produce/consume equations for all edges and then solving it for a non-trivial solution with minimum positive values. The solution turns out to be [147 98 56 40] for firing of nodes A, B, C and D, respectively. Although this firing is very easy to schedule, it results in an inefficient requirement on buffer sizes. For this example the implementation requires multiple buffers of sizes 147, 147×2, 98×4, 56×5 and 40×4 on edges moving from source to destination, accumulating a total storage requirement of 1273 samples.

The second design option is for each node to fire as soon as it gets enough samples at its input port. This implementation will require multiple buffers of sizes 1, 4, 10, 11 and 4 on respective edges from source to destination, amounting to a total buffer size of 30 words.

4.5.7 *Homogeneous SDFG*

Homogeneous SDFG is a special case of a single-rate graph where each node produces one token or data value on all its outgoing edges. Extending this account to incoming edges, each node also consumes one data value from all its incoming edges. Any consistent SDFG can be converted into HSDFG. The conversion gives an exact measure of throughput, although the conversion may result in an exponential increase in the number of nodes and thus may be very complex for interpretation and implementation.

The simplest way to convert a consistent SDFG to HSDFG is to first find the repetition vector and then to make copies of each node as given in the repetition vector and appropriately draw edges from source nodes to the destination nodes according to the original SDFG.

Example: Convert the SDFG given in Figure 4.25(a) to its equivalent HSDFG. This requires first finding the repetition vector from a set of produce/consume equations for the DFG. The repetition vector turns out to be [2 4 3] for set [A B C]. The equivalent HSDFG is drawn by following the repetition vector where nodes A, B and C are drawn 2, 4 and 3 times, respectively. The parallel nodes are then connected according to the original DFG. The equivalent HSDFG is given in Figure 4.25(b).

Example: Section 4.4.2.1 considered a JPEG realization as a KPN network. It is interesting to observe that each process in the KPN can be described as a dataflow graph. Implementation of a two-dimensional DCT block is given here. There are several area-efficient algorithms for realizing a 2-D DCT in hardware [26–29], but let us consider the simplest algorithm that computes 2-D DCT by successively computing two 1-D DCTs while performing transpose before the first DCT computation. This obviously is a multi-rate DFG, as shown in Figure 4.26.

The first node in the DFG takes 64 samples as an 8×8 block B, and after performing transpose of the input block it produces 64 samples at the output. The next node computes a 1-D DCT C_i of each of the rows of the block, MATLAB® notation `B^T(i,:)` is used to represent each row $i\,(=0,\ldots,7)$ of block B^T. This node, in each of its firings, takes a row comprising 8 samples and computes 8 samples at the output. The next FIFO at the output keeps collecting these rows, and the next node fires only when all 8 rows consisting of 64 elements of C are computed and saved in the FIFO. This node computes the transpose of C. In the same manner the next node computes 8 samples of the one-dimensional DCT F_j of each column j of C^T.

This DFG can be easily converted into HSDFG by replicating the 1-D DCT computation 8 times in both the DCT computation nodes.

4.5.8 *Cyclo-static DFG*

In a cyclo-static DFG, the number of tokens consumed by each node, although varying from iteration to iteration, exhibits periodicity and repeats the pattern after a fixed number of iterations. A CS-DFG

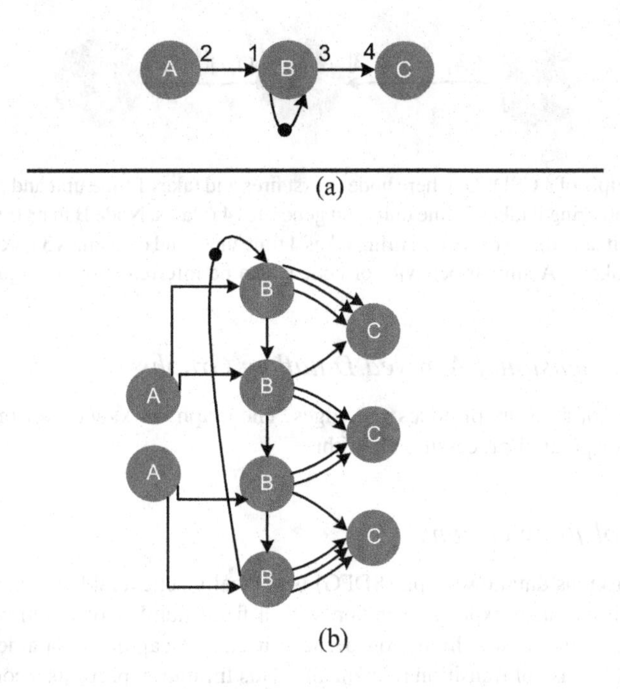

Figure 4.25 SDFG to HSDF conversion. (a) SDFG consisting of three nodes A, B and C. (b) An equivalent HSDG

is more generic and is suited to modeling several signal processing applications because it provides flexibility of varying production and consumption rates of each node provided the pattern is repeated after some finite number of iterations. This representation also works well for designs where a periodic sequence of functions is mapped on same HW block. Modeling it as CS-DFG, a node, in a periodic pattern, fires and executes a different function in a sequence and then repeats the sequence of calls after a fixed number of function calls. Each function observes different production and consumption rates, so the number of tokens in each firing of a node is different.

Figure 4.27 shows a CS-DFG where nodes A, B and C each executes two functions with execution times of [1, 3], [2, 4] and [3, 7] time units, respectively. In each of their firings the nodes produce and consume different numbers of tokens. This is shown in the figure on respective edges.

To generalize the execution in a CS-DFG node, in a periodic sequence of calls with period N, an iteration executes a function out of a sequence of functions $f_0(.), \ldots f_{(N-1)}$. Other nodes also call functions in a sequence. Each call consumes and produces a different number of tokens.

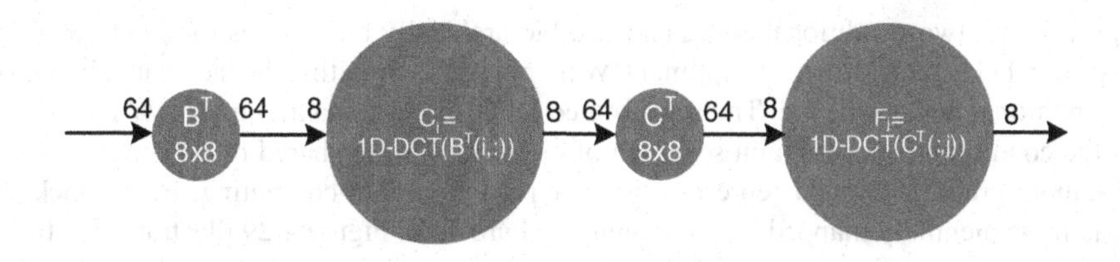

Figure 4.26 Computing a two-dimensional DCT in a JPEG algorithm presents a good example to demonstrate a multi-rate DFG

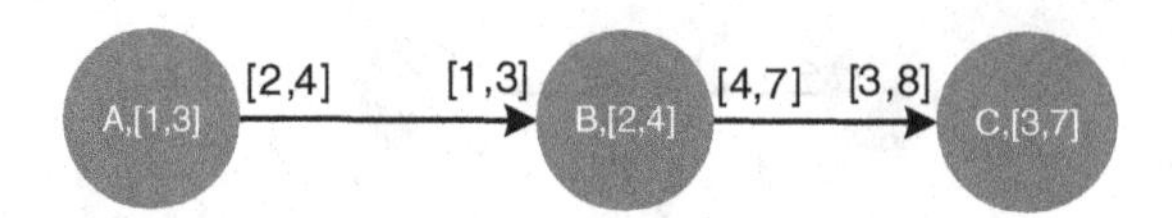

Figure 4.27 Example of a CSDFG, where node A first fires and takes 1 time unit and produces 2 tokens, and then in its second firing it takes 3 time units and generates 4 tokens. Node B in its first firing consumes 1 token in 2 time unit, and the in its second firing takes 4 time units and consumes 3 tokens; it respectively produces 4 and 7 tokens. A similar behavior of node C can be inferred from the figure

4.5.9 Multi-dimensional Arrayed Dataflow Graphs

This graph consists of an array of nodes and edges. The graph works well for multi-dimensional signal processing or parallel processing algorithms.

4.5.10 Control Flow Graphs

Although a synchronous dataflow graph (SDFG) is statically schedulable owing to it predictable communications, it can also express recursion with a fixed number of iterations. However, the representation has an expressive limitation because it cannot capture a data-dependent flow of tokens unless it makes use of transition refinements. This limitation prevents a concise representation of conditional flows incorporating `if-else-then` and `do-while` loops and data-dependent recursions.

As opposed to a DFG which is specific to a dataflow algorithm, a CFG is suitable to process a control algorithm. These algorithms are usually encountered in implementing communication protocols or controller designs for datapaths.

A control dataflow graph (CDFG) combines data-driven and control-specific functionality of an algorithm. Each node of the DFG represents a mathematical operation, and each edge represents either precedence or a data dependency between the operations. A CDFG may change the number of tokens produced and consumed by the nodes in different input settings. A DFG with varying rates of production and consumption is called a *dynamic* dataflow graph (DDFG).

Example: The example here converts a code consisting of decision statements to CDFG. Figure 4.28 shows a CDFG that implements the following code:

```
if (a==0)
    c=a+b×e;
else
    c=a-b;
```

The figure shows two conditional edges and two hierarchically built nodes for conditional firing. This type of DFG is transformed for optimal HW mapping by exploiting the fact that only one of the many conditional nodes will fire. The transformed DFG shares a maximum of HW resources and moves the conditional execution on selection of operands for the shared resource.

A dataflow graph is a good representation for reconfigurable computing. Each block in the DFG can be sequentially mapped on reconfigurable hardware. Figure 4.29 illustrates the fact that, in a JPEG implementation, sequential firing of nodes can be mapped on the same HW. This sharing of HW resources at run time reduces the area required, at the cost of longer execution times.

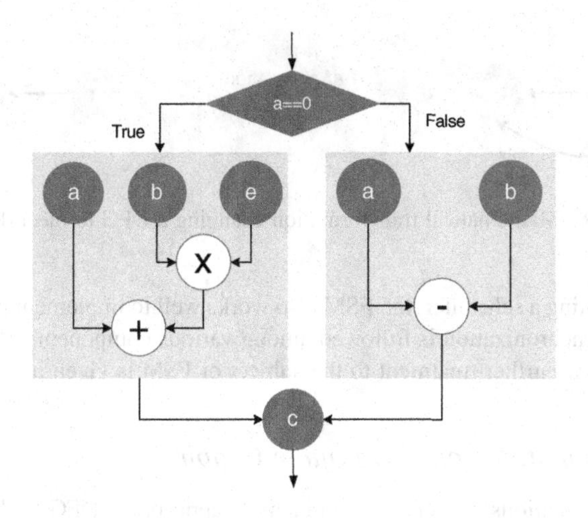

Figure 4.28 CDFG implementing conditional firing

4.5.11 Finite State Machine

A finite state machine (FSM) is used to provide control signals to a signal processing datapath for executing a computation with a selection of operands from a set of operands. An FSM is useful once a DFG is folded and mapped on a reduced HW, thus requiring a scheduler to schedule multiple nodes of the graph on a HW unit in time-multiplexed fashion. An FSM implements a scheduler in these designs.

A generic FSM assumes a system to be in one of a finite number of states. An FSM has a current state and, based on an internal or external event, it then computes the next state. The FSM then transitions into the next state. In each state, usually the datapath of the system executes a part of the algorithm. In this way, once the FSM transits from one state to another, the datapath keeps implementing different portions of the algorithm. After finishing the current iteration of the algorithm, the FSM sets its state to an initial value, and the datapath starts executing the algorithm again on a new set of data.

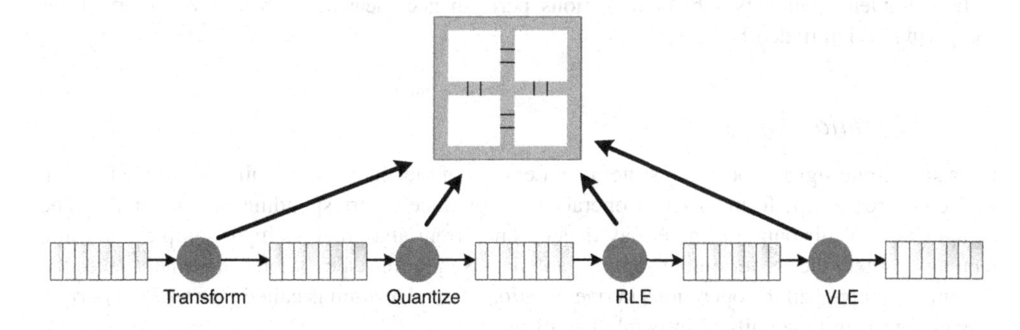

Figure 4.29 A dataflow graph is a good representation for mapping on a run-time reconfigurable platform

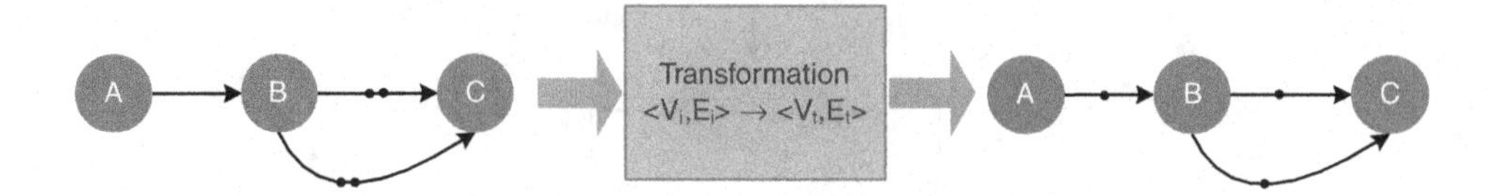

Figure 4.30 Mathematical transformation changing a DFG to meet design goals

Besides implementing a scheduler, the FSM also works well to implement protocols where a set of procedures and synchronization is followed among various components of the system, as with shared-bus arbitration. Further treatment to the subject of FSM is given in Chapter 9.

4.5.12 Transformations on a Dataflow Graph

Mathematical transformations convert a DFG to a more appropriate DFG for hardware implementation. These transformations change the implementation of the algorithm such that the transformed algorithm better meets specific goals of the design. Out of a set of design goals the designer may want to minimize critical path delay or the number of registers. To achieve this, several transformations can be applied to a DFG. Retiming, folding, unfolding and look-ahead are some of the transformations commonly used. These are covered in Chapter 7.

A transformation as shown in Figure 4.30 takes the current representation of $DFG_i = \langle V_i, E_i \rangle$ and translates it to another $DFG_t = \langle V_t, E_t \rangle$ with the same functionality and analytical behavior but different implementation.

4.5.13 Dataflow Interchange Format (DIF) Language

Dataflow interchange format is a standard language for specifying DSP systems in terms of executable graphs. Representation of an algorithm in DIF textually captures the execution model. An algorithm defined in DIF format is extensible and can be ported across platforms for simulation and for high-level synthesis and code generation. More information on DIF is given in [25].

4.6 Performance Measures

A DSP implementation is subject to various performance measures. These are important for comparing design tradeoffs.

4.6.1 Iteration Period

For a single-rate signal processing system, an iteration of the algorithm acquires a sample from an A/D converter and performs a set of operations to produce a corresponding output sample. The computation of this output sample may depend on current and previous input samples, and in a recursive system the earlier output samples are also used in this calculation. The time it takes the system to compute all the operations in one iteration of an algorithm is called the *iteration period.* It is measured in time units or in number of cycles.

For a generic digital system, the relationship between the sampling frequency f_s and the circuit clock frequency f_c is important. When these are equal, the iteration period is determined by the *critical path*. In designs where $f_c > f_s$, the iteration period is measured in terms of the number of clock cycles required to compute one output sample. This definition can be trivially extended for multi-rate systems.

4.6.2 Sampling Period and Throughput

The *sampling period* T_s is defined as the average time between two successive data samples. The period specifies the number of samples per second of any signal. The sampling rate or frequency ($f_s = 1/T_s$) requirement is specific to an application and subsequently constrains the designer to produce hardware that can process the data that is input to the system at this rate. Often this constraint requires the designer to minimize critical path delays. They can be reduced by using more optimized computational units or by adding *pipelining delays* in the logic (see later). The pipelining delays add latency in the design. In designs where $f_s < f_c$, the digital designer explores avenues of resource sharing for optimal reuse of computational blocks.

4.6.3 Latency

Latency is defined as the time delay for the algorithm to produce an output $y[n]$ in response to an input $x[n]$. In many applications the data is processed in batches. First the data is acquired in a buffer and then it is input for processing. This acquisition of data adds further latency in producing corresponding outputs for a given set of inputs.

Beside algorithmic delays, pipelining registers (see later) are the main source of latency in an FDA. In DSP applications, minimization of the critical path is usually considered to be more important than reducing latency. There is usually an inverse relationship between critical path and latency. In order to reduce the critical path, pipelining registers are added that result in an increase in latency of the design. Reducing a critical path helps in meeting the sampling requirement of a design.

Figure 4.31 shows the sampling period as the time difference between the arrivals of two successive samples, and latency as the time for the HW to produce a result $y[n]$ in response to input $x[n]$.

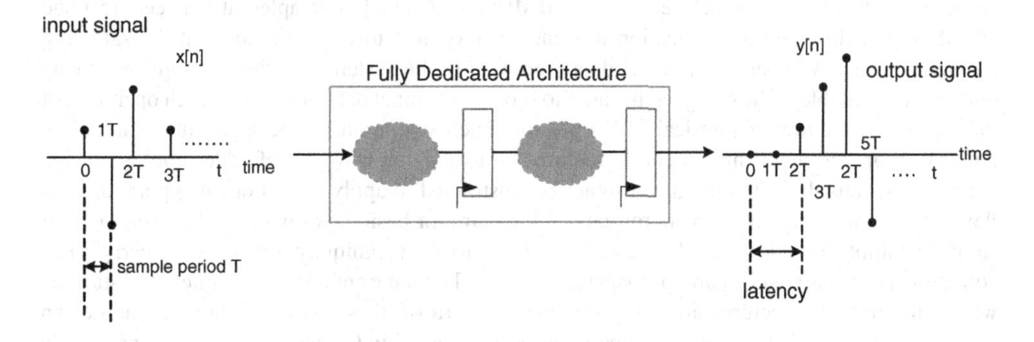

Figure 4.31 Latency and sampling period

4.6.4 Power Dissipation

4.6.4.1 Power

Power is another critical performance parameter in digital design. The subject of designing for low power use is gaining more importance with the trend towards more handheld computing platforms as consumer devices.

There are two classes of power dissipation in a digital circuit, static and dynamic. Static power dissipation is due to the leakage current in digital logic, while dynamic power dissipation is due to all the switching activity. It is the dynamic power that constitutes the major portion of power dissipation in a design. Dynamic power dissipation is design-specific whereas static power dissipation depends on technology.

In an FPGA, the static power dissipation is due to the leakage current through reversed-biased diodes. On the same FPGA, the use of dynamic power depends on the clock frequency, the supply voltage, switching activity, and resource utilization. For example, the dynamic power consumption P_d in a CMOS circuit is:

$$P_d = \alpha C V_{DD}^2 f$$

where α and C are, respectively, switching activity and physical capacitance of the design, V_{DD} is the supply voltage and f is the clock frequency. Power is a major design consideration especially for battery-operated devices.

At register transfer level (RTL), the designer can determine the portions of the design that are not performing useful computations and can be shut down for power saving. A technique called 'gated clock' [30–32] is used to selectively stop the clock in areas that are not performing computations in the current cycle. A detailed description is given in Chapter 9.

4.7 Fully Dedicated Architecture

4.7.1 The Design Space

To design an optimal logic for a given problem, the designer explores the design space where several options are available for mapping real-time algorithms in hardware. The maximum achievable clock frequency of the circuit, f_c, and the required sampling rate of data input to the system, f_s, play crucial roles in determining and selecting the best option.

Digital devices like FPGAs and custom ASICs can execute the logic at clock rates in the range of 30 MHz to at least 600 MHz. Digital front end of a software defined radio (SDR) requires sampling and processing of an IF signal centered at 70 MHz. The signal is sampled at between 150 and 200 MHz. For this type of application it is more convenient to clock the logic at the sampling frequency of the A/D converter or the data input rate to the system. For these designs not many options are available to the designer, because to process the input data in real time each operation of the algorithm requires a physical HW operator. Each operation in the algorithm – addition, multiplication and algorithmic delay – requires an associated HW unit of adder, multiplier and register, respectively. Although the designer is constrained to apply one-to-one mapping, there is flexibility in the selection of an appropriate architecture for basic HW operators. For example, for simple addition, an adder can be designed using various techniques: ripple carry, carry save, conditional sum, carry skip, carry propagate, and so on. The same applies to multipliers and shifters, where different architectures are being designed to perform these basic operations. The design options for these basic mathematical operations are discussed in Chapter 5. These options usually trade off area against timing.

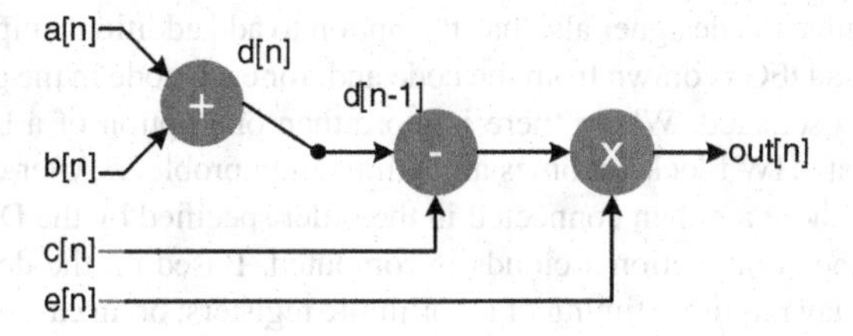

Figure 4.32 Mapping to a dataflow graph of the set of equations given in the text example

After one-to-one mapping of the DFG to architecture, the design is evaluated to meet the input data rate of the system. There will be cases where, even after selecting the optimal architectures for basic operations, the synthesized design does not meet timing requirements. The designer then needs to employ appropriate mathematical transformations or may add pipeline registers to get better timing (see later).

Example: Convert the following statements to DFG and then map it to fully dedicated architecture.

$$d[n] = a[n] + b[n] \tag{4.7}$$

$$out[n] = (d[n-1]-c[n])e[n] \tag{4.8}$$

This is a very simple set of equations requiring addition, a delay element, a subtraction and a multiplication. The index n represents the current iteration, and $n-1$ represents the result from the previous iteration. The conversion of these equations to a synchronous DFG is shown in Figure 4.32.

After conversion of an algorithm to DFG, the designer may apply mathematical transformations like retiming and unfolding to get a DFG that better suits HW mapping. Mapping of the transformed DFG to architecture is the next task. For an FDA the conversion to an initial architecture is very trivial, involving only replacement of nodes with appropriate basic HW building blocks.

Figure 4.33 shows the initial architecture mapping of the DFG of (4.7) and (4.8). The operations of the nodes are mapped to HW building blocks of an adder, subtractor, multiplier and a register.

4.7.2 Pipelining

Mapping of a transformed dataflow graph to fully dedicated architecture is trivial, because each operation is mapped to its matching hardware operator. In many cases this mapping may consist of paths with combinational logic that violates timing constraints. It is, therefore, imperative to break these combinational clouds with registers. Retiming transformation, which is discussed in Chapter 7, can also be used for effectively moving algorithmic registers in the combinational logic that violates timing constraints.

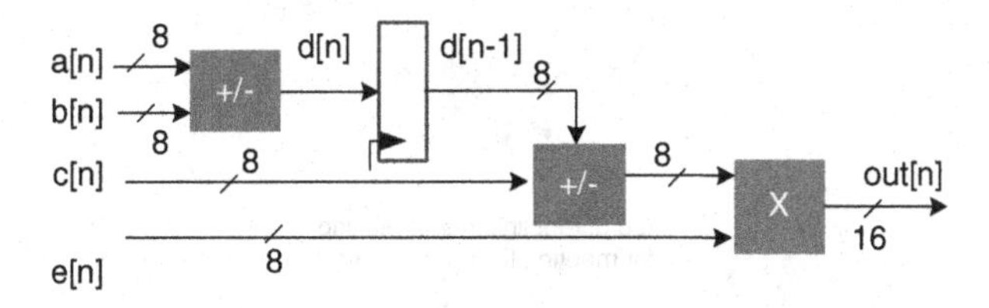

Figure 4.33 DFG to FDA mapping

For feedforward paths, the designer also has the option to add additional pipeline registers in the path. As the first step, a DFG is drawn from the code and, for each node in the graph, an appropriate HW building block is selected. Where there is more than one option of a basic building block, selecting an appropriate HW block becomes an optimization problem. After choosing appropriate basic building blocks,these are then connected in the order specified by the DFG. After this initial mapping, timing of the combinational clouds is computed. Based on the desired clock rate, the combinational logic may require retiming of algorithmic registers; or, in case of feedforward paths, the designer may need to insert additional pipeline registers.

Figure 4.34 shows the steps in mapping a DFG to fully dedicated architecture while exercising the pipelining option for better timing.

Maintaining coherency of the data in the graph is a critical issue in pipelining. The designer needs to make sure that the datapath traced from any primary input to any primary output passes through the same number of pipeline registers. Figure 4.35 shows that to partition the combinational cloud of

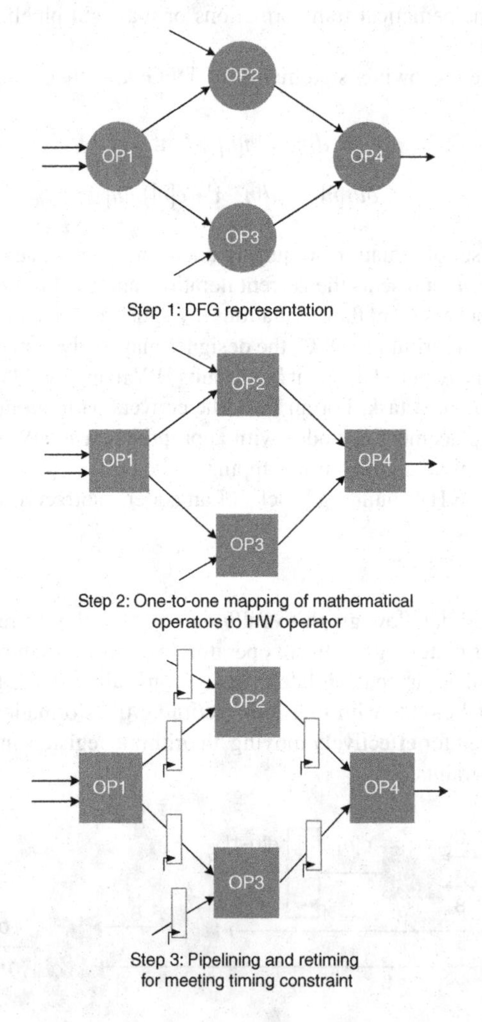

Figure 4.34 Steps in mapping DFG on to FDA with pipeline registers

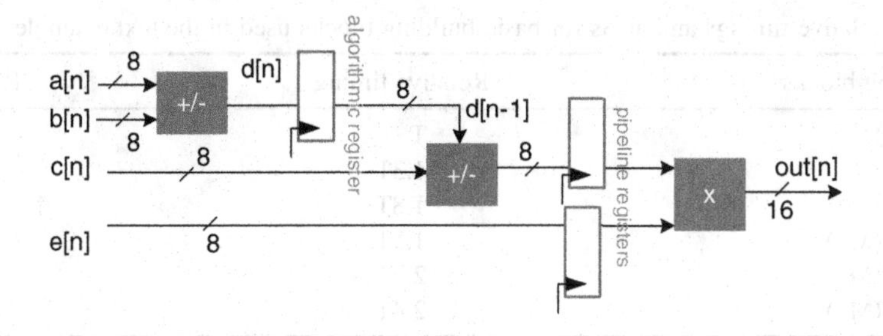

Figure 4.35 Pipelining while maintaining coherency in the datapath

Figure 4.33, a pipeline register needs to be inserted between the multiplier and the subtractor; but, to maintain coherency of input $e[n]$ to the multiplier, one register is also added in the path of the input $e[n]$ to the multiplier.

When some of the feedforward paths of the mapped HW do not meet timings, the designer can add pipeline registers to these paths. In contrast, there is no simple way of adding pipeline registers in *feedback* paths. The designer should mark any feedback paths in the design because these require special treatment for addition of any pipeline registers. These special transformations are discussed in Chapter 7.

4.7.3 Selecting Basic Building Blocks

As an example, a dataflow graph representing a signal processing algorithm is given in Figure 4.36(a). Let there be three different types of multiplier and adder in the library of predesigned basic building blocks. The relative timing and area of each building block is given in Table 4.1. This example designs an optimal architecture with minimum area and best timing while mapping the DFG to fully dedicated architecture.

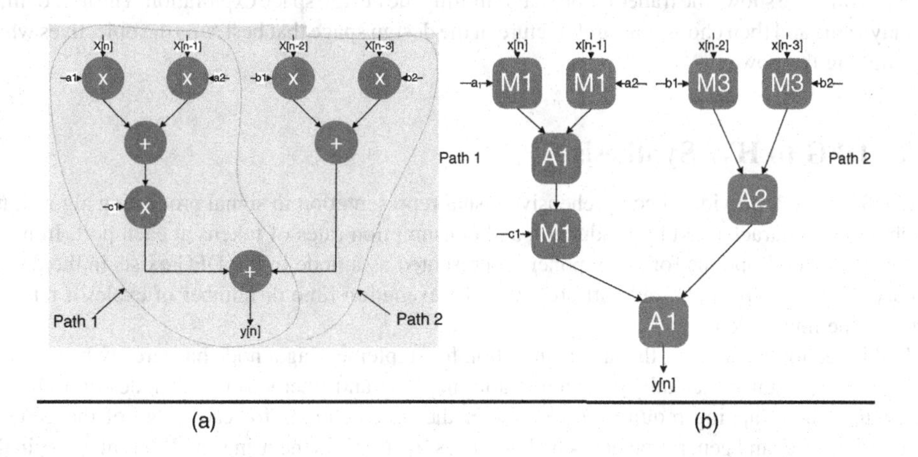

Figure 4.36 One-to-one mathematical operations. (a) Dataflow graph. (b) Fully dedicated architecture with optimal selection of HW blocks

Table 4.1 Relative timings and areas for basic building blocks used in the text example

Basic building blocks	Relative timing	Relative area
Adder 1 (A1)	T	2.0A
Adder 2 (A2)	1.3T	1.7A
Adder 3 (A3)	1.8T	1.3A
Multiplier 1 (M1)	1.5T	2.5A
Multiplier 2 (M2)	2.0T	2.0A
Multiplier 3 (M3)	2.4T	1.7A

The DFG of Figure 4.36(a) is optimally mapped to architecture in (b). The objective is to clock the design with the best possible sampling rate without addition of any pipeline register. The optimal design selects appropriate basic building blocks from the library to minimize timing issues and area.

There are two distinct sections in the DFG. The critical section requires the fastest building blocks for multiplications and additions. For the second sections, first the blocks with minimum area should be allocated. If any path results in a timing violation, an appropriate building block with better timing while still smaller area should be used to improve the timing. For complex designs the mapping problem can be modeled as an optimization problem and can be solved for an optimal solution.

4.7.4 Extending the Concept of One-to-one Mapping

The concept can be extended to graphs that are hierarchically built where each node implements an algorithm such as FIR or FFT. Then there are also several design options to realize these algorithms. These options are characterized with parameters of power, area, delay, throughput and so on. Now what is needed is a design tool that can iteratively evaluates the design by mapping different design options and then settling for the one that best suits the design objectives. A system design tool to work in this paradigm needs to develop different design options for each component offline and to place them in the design library.

Figure 4.37 demonstrates what is involved. The dots in the figure show different design instances and the solid line shows the tradeoff boundary limiting the design space exploration. The user defines a delay value and then chooses an architecture in the design space that best suits the objectives while minimizing the power.

4.8 DFG to HW Synthesis

A dataflow graph provides a comprehensive visual representation to signal processing algorithms. Each node is characterized by production and consumption rates of tokens at each port. In many instances, several options for a component represented as a node in the DFG exists in the design library. These components trade off area with the execution time or number of cycles it takes to process the input tokens.

In this section we assume that a design option for implementing a node has already been made. The section presents a technique for automatic mapping and interworking of nodes of a DFG in hardware. This mapping requires convertion of the specifications for each edge of the DFG to appropriate HW, and generation of a scheduler that synchronizes the firings of different nodes in the DFG [33].

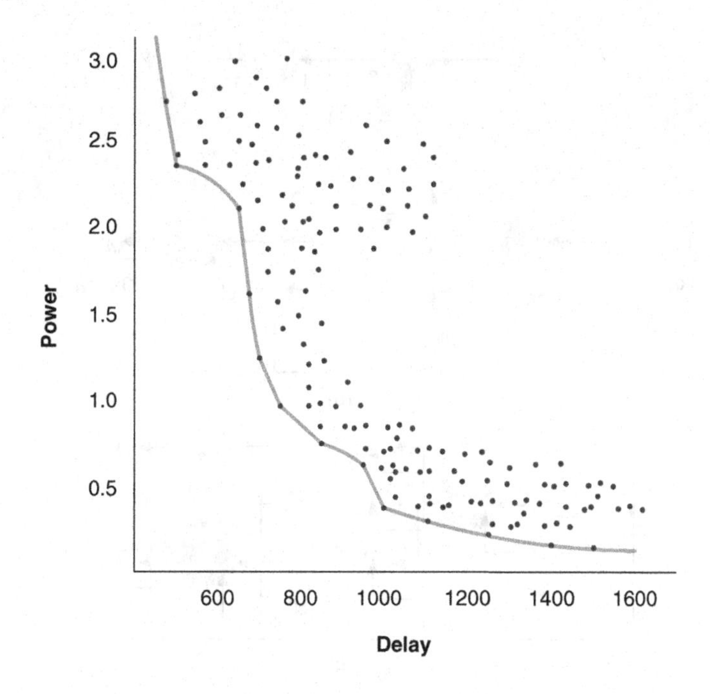

Figure 4.37 Power–delay tradeoff for a typical digital design

4.8.1 Mapping a Multi-rate DFG in Hardware

Each edge is considered with its rates of production and consumption of tokens. For an edge requiring multi-rate production to single-rate consumption, either a parallel or a sequential implementation is possible. A parallel implementation invokes each destination node multiple times, whereas a sequential setting stores the data in registers and sequentially inputs it to a single destination node.

For example, Figure 4.38(a) shows an edge connecting nodes A and B with production and consumption rates of 3 and 1, respectively. The edge is mapped to fully parallel and sequential designs where nodes are realized as HW components. The designs are shown in Figures 4.38(b) and (c).

4.8.1.1 Sequential Mapping

The input to component A is fed at the positive edge of clk_G, whereas component B is fed with the data at three times faster clock speed. Based on the HW implementation, A and B can execute at independently running clocks clk_A and clk_B. In design instances where a node implements a complete functionality of an algorithm, these clocks are much faster than clk_G. There may be design instances where, in a hierarchical fashion, each node also represents a DFG and the graph is mapped on fully dedicated architecture. Then clk_A and clk_B are set to be clk_G and $3 \times clk_G$, respectively.

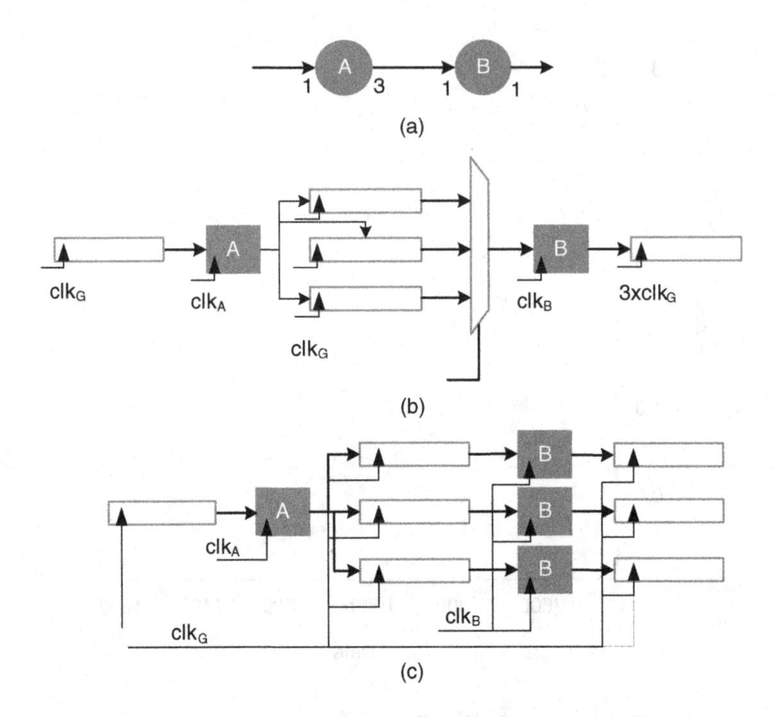

Figure 4.38 (a) Multi-rate to single-rate edge. (b) Sequential synthesis. (c) Parallel synthesis

4.8.1.2 Parallel Mapping

A parallel realization uses the same clock for both the inputs to components A and B as shown in Figure 4.38(c). Further to this, if the design is also mapped as FDA then the entire design would run on global clk_G.

In a multi-rate system where the consumption and production rates are different, the buffering design becomes more complex. This requires either a self-timed implementation or balanced-equation based multiple firing realizations. In a self-timed implementation, the firing of the destination node may happen a variable number of times in a period.

Example: Figure 4.39(a) shows a multi-rate edge with rate of production and consumption 3 and 4, respectively. Every time a source node fires it produces three tokens of N bits each, and writes them in a ping-pong pattern in two sets of three registers, {R1, R2, R3} and {R4, R5, R6}. The destination node B consumes the produced tokens in a firing sequence in relation to the firing of node A as [0 1 1 1]. Node A fires in the first cycle and stores the output values in registers {R1, R2, R3}. In the second cycle node B does not fire as it requires four tokens, but node A fires again and three new tokens are produced in registers {R4, R5, R6}. As the total number of tokens produced becomes six, so B fires and consumes four tokens in {R1, R2, R3, R4}. Node A also fires in this cycle and again saves the three tokens in registers {R1, R2, R3}, and total number of unconsumed tokens becomes five. In sequential firing, the node B fires again and consumes tokens in registers {R5, R6, R1, R2}, and A also fires and stores tokens in {R4, R5, R6}. The unconsumed tokens are four now, so B fires again and consumes four tokens in registers {R3, R4, R5, R6}. This pattern is repeated thereafter. Figure 4.39(b) shows the HW design that supports this production and consumption of tokens.

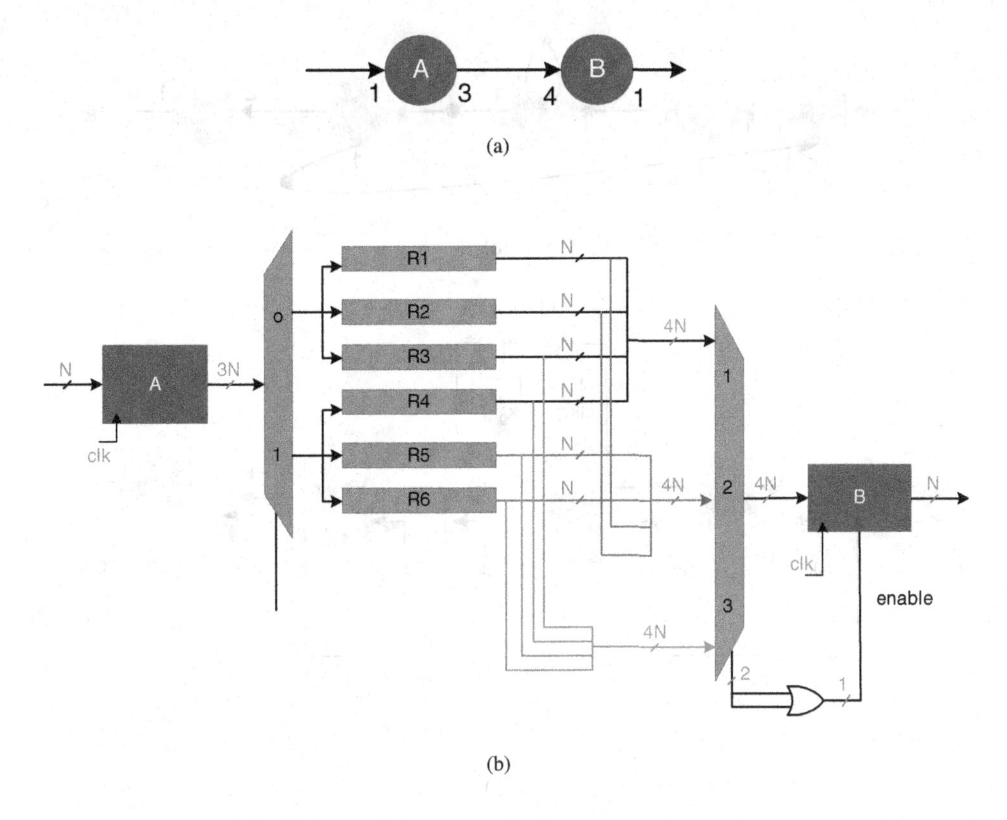

Figure 4.39 (a) A multi-rate DFG. (b) Hardware realization

4.8.2 Centralized Controller for DFG Realization

Chapter 9 is dedicated to state-machine based controller design, but in this section a brief introduction to a centralized controller for a DFG realization in HW is given. Several automation tools use a similar methodology to add a controller while generating a HW design for a DFG [33, 34].

The first step in implementing a central controller for a DFG is to classify all the nodes in the DFG as one of the following: a combination node, or a sequential node that executes and generates an output in every sample clock cycle, or a node that executes and generates output tokens in predefined number of circuit clock cycles, or a node that takes a variable number of circuit clock cycles to complete its execution. A sample clock cycle is the clock that latches a new sample into the logic, whereas a circuit clock is usually much faster than a sample clock and executes sequential logic in each node.

An automation tool either picks an appropriate implementation of a node from a design library or produces HDL code using a code-generation utility. A node containing combinational logic or sequential logic executing on the sample clock does not require any control signal except a reset signal that must be used to reset any feedback register in the design. A node with predefined number of circuit clock cycles requires a start signal from a controller, whereas a dynamic node gets a start signal from a controller and then, after completing its execution, it generates a *done* signal to notify the controller for further asserting control signals to subsequent nodes in the DFG.

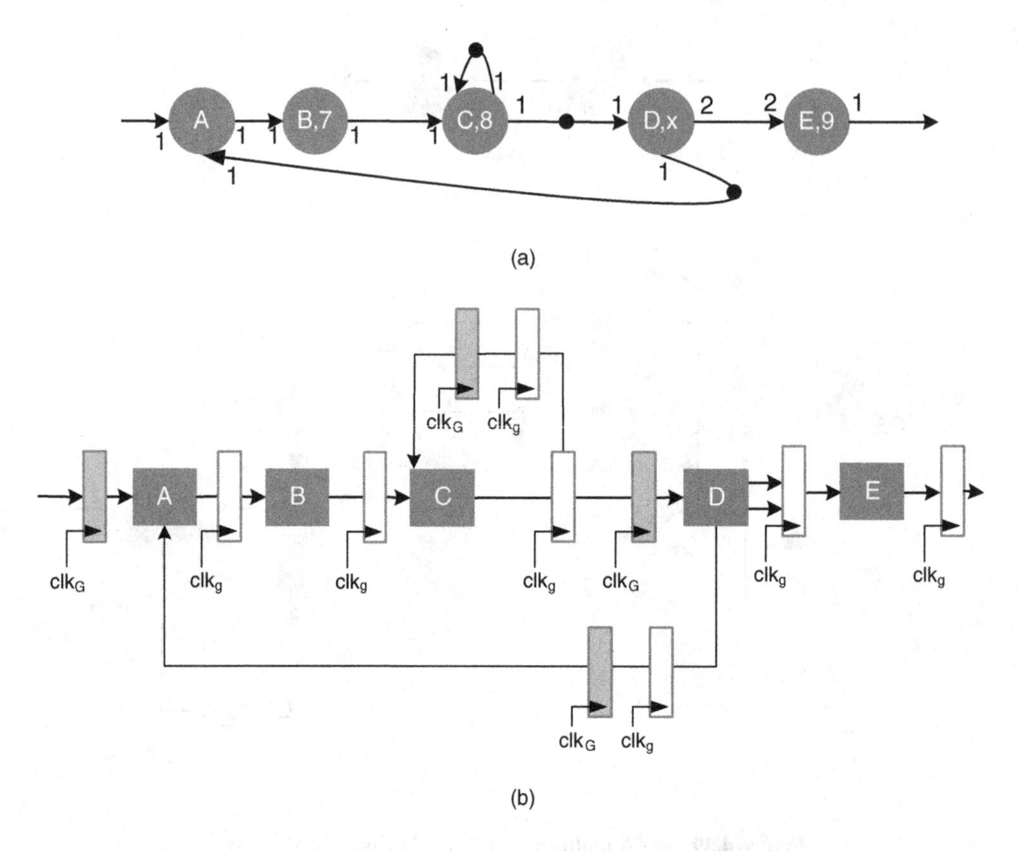

Figure 4.40 (a) Hypothetical DFG with different types of node. (b) Hardware realization

Figure 4.40(a) shows a DFG with nodes A, B, C, D and E. Node A is a combinational logic and nodes B, C and E take 7, 8 and 9 predefined number of circuit clock cycles, respectively. Node D dynamically executes and takes a variable number of cycles. Each black dot shows an algorithmic delay where data from previous iteration is used in subsequent nodes. This dot in actual HW is realized as a register that is clocked by the sampling clock. Now, having the graph and the information about each node, a centralized controller can be easily designed or automatically generated. Such a controller is shown in Figure 4.40(b). Each node with predefined number of cycles needs a start signal from the controller, and then the controller counts the number of cycles for the node to complete its execution. The controller then asserts an output *enable* signal to the register at the output of each such node. In the figure, the output from each node is latched in a register after assertion of an *enable* signal, whereas the register is clocked by the circuit clock. In the case of the dynamic node D, the controller not only notifies the node to start its execution; the node also after completing its execution asserts a *done* signal.

For the design in question, a `done_D` is asserted and the controller then asserts `en_out_D` to latch the output from node D in a register. The input and output to the DFG and the dots on the edges are replaced by registers clocked by sample clock clk_G, whereas the rest of the logic in the nodes and the registers at the output of each node are clocked by a circuit clock clk_g. All the feedback register are resetable using a global `rst_n` signal. It is important to ensure that for a synchronous DFG that implements a streaming application, a new sample to the DFG is fed in every sample clock cycle. The design must be constructed in a manner that all the nodes in the DFG complete their respective executions before the arrival of the next sample clock. Even for the dynamic node, the worst-case condition must be considered and the design must guarantee completion before the arrival of the next sample clock. All the nodes that can do so start and execute in parallel. Nodes A, B and C execute in a sequence, and D and E work in parallel and use the output of C from the previous iteration. If x, the execution of time of dynamic node D, is bounded by 10, then the execution of D and E takes a maximum of 19 circuit clocks, whereas nodes A, B and C take 16 clocks. Thus the circuit clock should be at least 19 times faster than the sampling clock.

Exercises

Exercise 4.1

The code below lists one iteration of an algorithm implementing a block in a digital communications receiver. The algorithm processes a complex input `InputSample` and gives it a phase correction. The phase correction is computed in this code. Draw DFG, and map it on to an FDA. Write RTL Verilog code of the design using 16-bit fixed point arithmetic. Write a stimulus to verify the RTL code.

```
Mixer_out = InputSample*(-j*Phase);
// a number of algorithmic registers
Mixer_Out[2] = Mixer_Out[1];
Mixer_Out[1] = Mixer_Out[0];
Mixer_Out[0] = Mixer_out;
Delayline[2] = Delayline[1];
Delayline[1] = Delayline[0];
// feedback loop, where a and b are arrays of 16-bit constants
// Kp and Ki are also 16-bit constant numbers
Delayline[0]= - Delayline[1]*a[1]
              - Delayline[2]*a[2]
              + Mixer_Out[0]*b[0]
              + Mixer_Out[1]*b[1]
              + Mixer_Out[2]*b[2];
offset = real(Delayline[2]*Delayline[0]);
offset_Delayline[0] = offset_Delayline[1];
offset_Delayline[1] = offset;
 phase_corr =  Kp*offset_Delayline[0]
             + Ki*offset_Delayline[1]
             + phase_corr
             - Kp*offset_Delayline[1];
Phase = Phase + phase_corr;
```

Exercise 4.2

Convert the C code shown below to its equivalent fixed-point listing in C. Consider `acc` to be in Q8.32 format and all the other variables to be Q1.15. Check corner cases and saturate the result if overflow occurs. Draw a DFG of the design and map the DFG as FDA. Write RTL Verilog code of the design. Simulate the design for a stream of input values in Q1.15 format.

```
float recursion (float input, float x[], float y[])
{
    float acc, out;
    int i;
    acc = 0;
    x[2]=x[1];
    x[1]=x[0];
    x[0]=input;
    for( i = 0; i < 2; i++)
    {
    acc += b[i] * x[i] + a[i] * y[i];
    }
      y[2]=y[1];
      y[1]=y[0];
      y[0]=acc;
}
```

The constants are:

```
float b[] = {0.258, -0.309, -0.704};
float a[] = {-0.123, -0.51, 0.223};
```

Exercise 4.3

Draw an HSDFG to implement a 4-tap FIR filter in transposed direct-form configuration.

Exercise 4.4

Design a sampling-rate converter to translate a signal sampled at 66 MHz to 245 MHz. Compute the buffer requirement if the system is designed to use fixed pattern computation using a repetition vector. Also compute the minimum buffer requirement if the system is implemented to execute on self-timed firing.

Exercise 4.5

Determine balanced equations for the graph shown in Figure 4.41. If the graph is inconsistent, mention the edge and the type of inconsistency and convert the graph to a consistent graph by suggesting modifications of consumption and production rates of any one of the nodes?

Exercise 4.6

Determine balanced equations and compute a non-trivial solution with minimum positive non-zero values for each firing for the SDFG shown in Figure 4.42. Design hardware consisting of registers and multiplexers to implement the graph.

Exercise 4.7

Production and consumption rates in the form of a 2-D array of data are given in Figure 4.43.

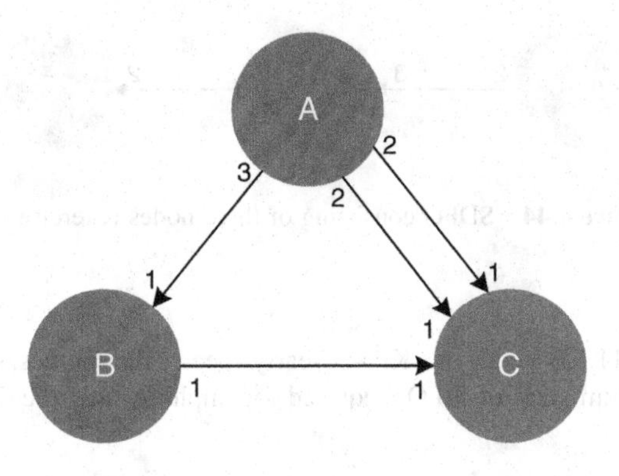

Figure 4.41 SDFG for computing balanced equations (exercise 4.5)

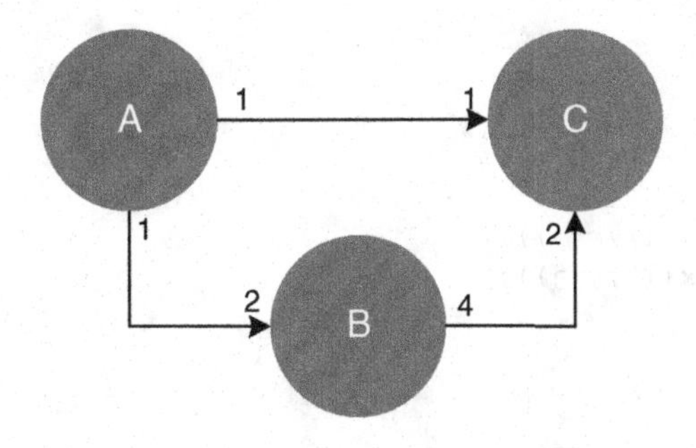

Figure 4.42 SDFG for computing a non-trivial solution (exercise 4.6)

1. Write a balanced equation for both the dimensions. Solve the equation to find a solution to establish the number of firings of B that is required to consume all the tokens produced by A.
2. Draw a 2-D array of data produced by A, and suggest a buffering arrangement with multiplexers to feed the required data to B for its respective firings.

Exercise 4.8

Compute the repetition vector for the SDFG shown in Figure 4.44. Translate the SDFG to HSDFG.

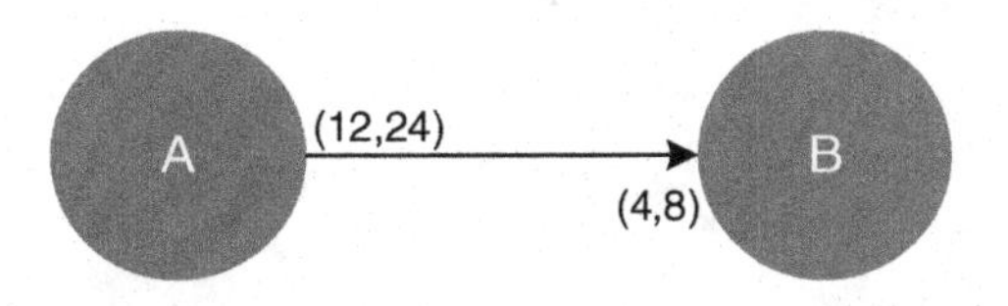

Figure 4.43 Nodes processing a 2-D array of data (exercise 4.7)

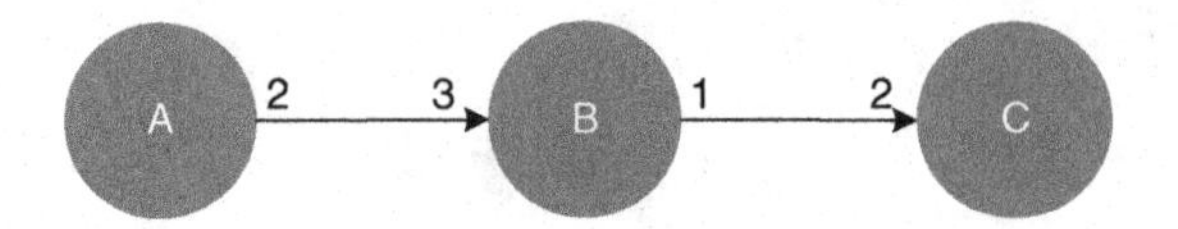

Figure 4.44 SDFG consisting of three nodes (exercise 4.8)

Exercise 4.9

Map the following MATLAB® code to a KPN. Clearly specify the sources, sink and the processing nodes. State the minimum sizes of FIFOs required for implementing the design in hardware.

```
N = 16;
K = 32;
for i=1:N
  x(i) = src_x ();
end
for j=1:K
  y(j) = src_y ();
end
for i=1:N
  for j=1:K
    x1(i,j)=func1(x(i),y(j));
    y1(i,j)=func2(x(i),y(j));
  end
end
for i=1:N
  for j=1:K
    z(i,j)=func3(x1(i,j),y1(i,j));
  end
end
m=1;
  for i=1:N
  for j=1:K
    y(m)=sink(x1(i,j), y1(i,j), z(i,j));
    m=m+1;
  end
end
```

Exercise 4.10

Map the following MATLAB® code to a KPN. The functions F and G should be mapped to different processing nodes. Clearly identify the data dependencies.

```
N = 8*16;
K = 8*32;
for i=1:N
  for j=1:K
      x(i,j)= src_x ();
  end
```

```
end
for i=1:8:N
  for j=1:8:N;
    for k=i:(i+8-1)
      for l=j:(j+8-1)
        y1(k,l)=func1(x(k,l));
      end
    end
    for k=i:(i+8-1)
      for l=j:(j+8-1)
        y2(k,l)=func2(y1(k,l));
      end
    end
    for k=i:(i+8-1)
     for l=j:(j+8-1)
       y3(k,l)=func3(y2(k,l), y1(k,l), x(k,l));
     end
    end
      for k=i:(i+8-1)
      y4(k,j)=0;       for l=j+1:(j+8-1)
        y4(k,l)=func4(y4(k,l-1), y1(k,l), x(k,l));
      end
     end
   end
end
for i=1:N
  for j=1:K
    out(i,j)=sink (y4(i,j), x(i,j));
  end
end
```

Exercise 4.11

Map the KPN shown in Figure 4.45 to a hardware architecture showing each processor tile with the processor, local memory, memory controller, communication controller, and network on chip block along with associated FIFOs.

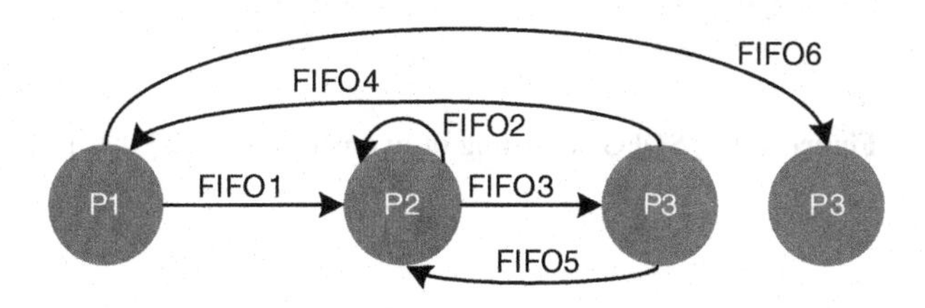

Figure 4.45 KPN with multiple nodes and channels (exercise 4.11)

Exercise 4.12

Check whether the DFG given in Figure 4.46 is consistent. If it is not, what minimum changes in the production or consumption rates can convert it into a consistent DFG. Give a self-timed firing sequence of a consistent DFG.

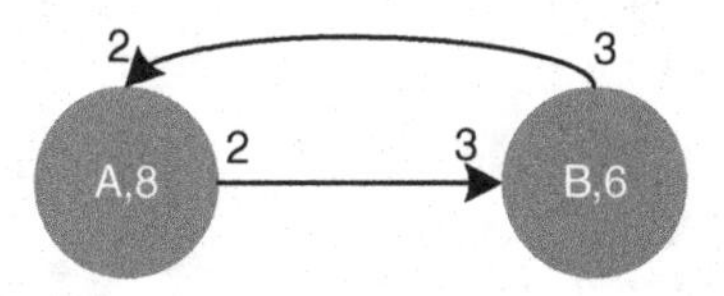

Figure 4.46 DFG consisting of two nodes (exercise 4.12)

Exercise 4.13

Convert the SDFG shown in Figure 4.47 into HSDFG. Compute the throughput of the graph and also give a hardware realization of the DFG.

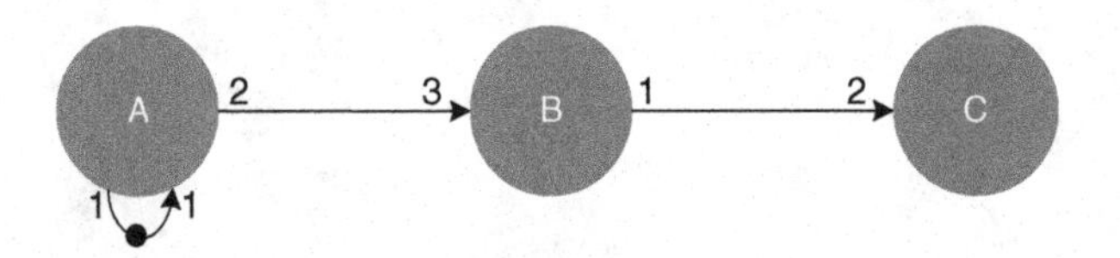

Figure 4.47 SDFG consisting of three nodes (exercise 4.13)

Exercise 4.14

Convert the SDFG given in Figure 4.48 into HSDFG. Give a synthesis realization of the DFG.

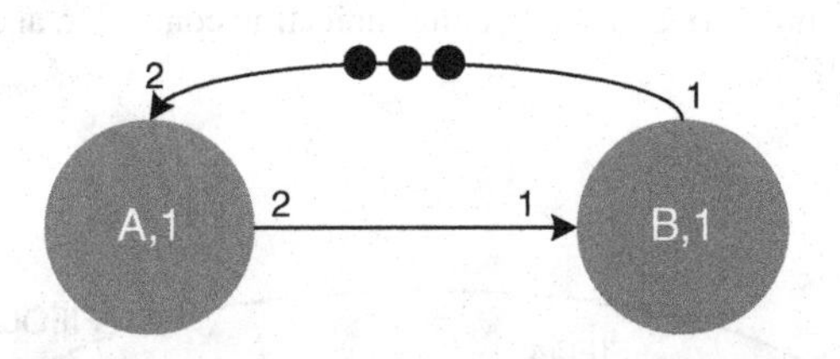

Figure 4.48 SDFG consisting of two nodes (exercise 4.14)

Exercise 4.15

The DFG given in Figure 4.49 implements an M-JPEG algorithm. Draw a KPN showing FIFOs.

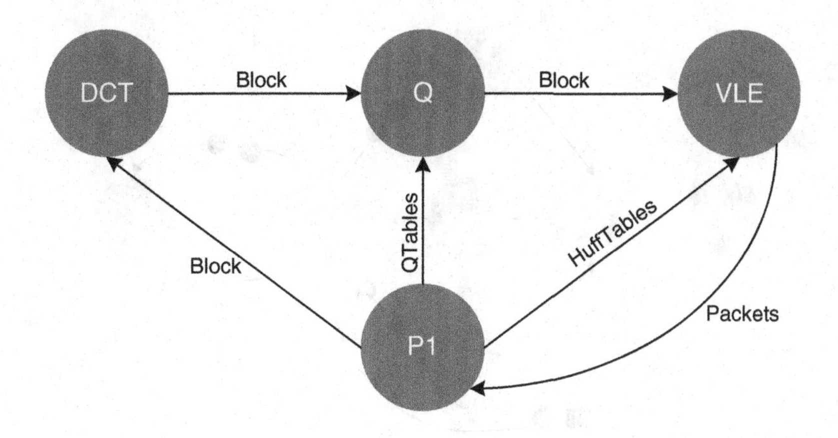

Figure 4.49 DFG implementing M-JPEG compression algorithm (exercise 4.15)

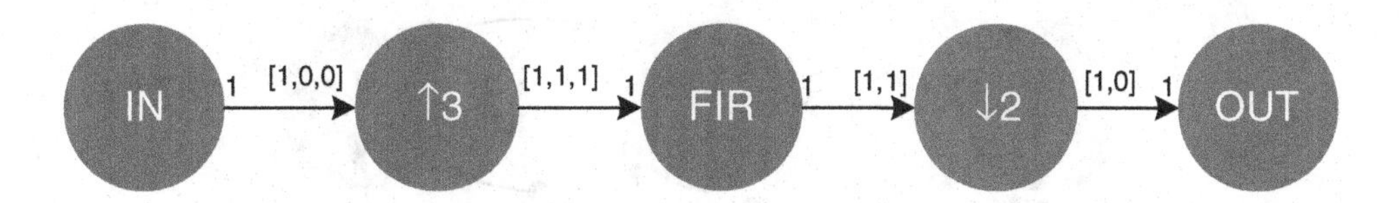

Figure 4.50 Cyclo-static DFG for 3/2 sampling rate change (exercise 4.16)

Exercise 4.16

A digital up/down converter can be realized as a cyclo-static DFG. The production and consumption rate as vectors in the DFG given in Figure 4.50 implement a 3/2 digital up/down converter. Design the registers and multiplexer to realize the design in HW.

Exercise 4.17

Draw a multi-dimensional DFG implementing a level-3 discrete wavelet transform based on sub-band decomposition of a 256×256 gray image. State the consumption and production rates on each edge. Also specify the buffer size requirements on the edges.

Exercise 4.18

For the HSDF given in Figure 4.51, compute the following:

1. a self-timed schedule for the graph;
2. a repetition vector based on the solution of balanced equations;
3. a hardware synthesis of the graph.

Exercise 4.19

For the SDFG given in Figure 4.52:

1. Write its topology matrix.
2. By computing the rank of the matrix computed in (1), determine whether the graph represents a consistent or an inconsistent SDFG.

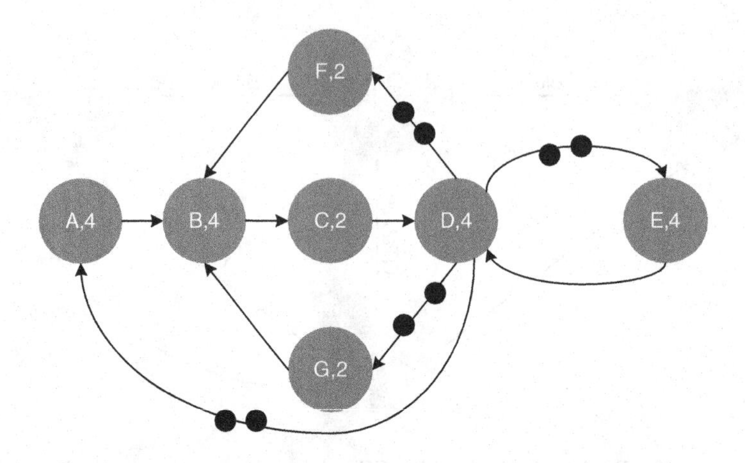

Figure 4.51 Graph implementing an HSDFG (exercise 4.18)

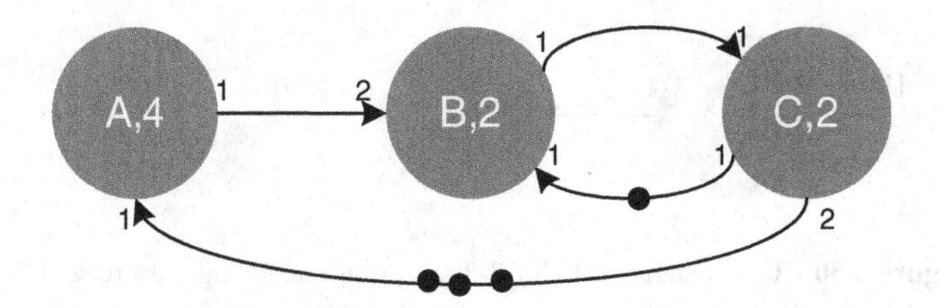

Figure 4.52 SDFG with three nodes (exercise 4.19)

3. Compute the repetition vector of the graph.
4. Map the DFG to hardware, assuming node A gets one sample of the input data at every sample clock. Clearly draw registers realizing delays and registers at cross-node boundaries clocked with the circuit clock.
5. Convert the graph to an HSDFG.

Exercise 4.20

The flow graph of Figure 4.53 shows a multi-rate DFG.

1. Design a sequential HW realization of the graph showing all the multiplexers, demultiplexers, registers and clocks in the design.
2. Write RTL Verilog code to implement the design in hardware.
3. Write balanced equations for the graph and solve the equations to find a parallel HW realization of the design. Draw the realization.

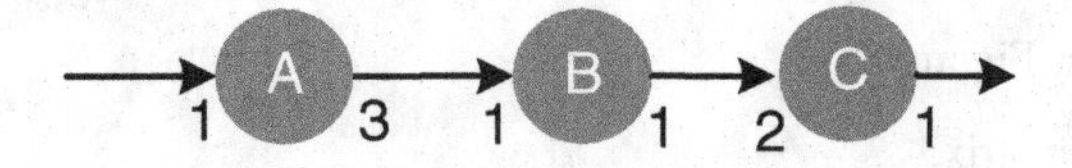

Figure 4.53 A multi-rate DFG (exercise 4.20)

References

1. R. G. Vaughan, N. L. Scott and D. R. White, "The theory of bandpass sampling," *IEEE Transactions on Signal Processing*, 1991, vol. 39, pp. 1973–1984.
2. C.-H. Tseng and S.-C. Chou, "Direct down-conversion of multiband RF signals using bandpass sampling," *IEEE Transactions on Wireless Communications*, 2006, vol. 5, pp. 72–76.
3. G. Kahn, "The semantics of a simple language for parallel programming," *Information Processing*, 1974, pp. 471–475.
4. E. A. Lee and T. M. Parks, "Dataflow process networks," *in Proceeding of the IEEE*, 1995, vol. 83, pp. 773–799.
5. E. Cheung, H. Hsieh and F. Balarin, "Automatic buffer sizing for rate-constrained KPN applications on multiprocessor system-on-chip," in *Proceedings of IEEE International High-level Design Validation and Test Workshop*, 2007, pp. 37–44.
6. T. Stefanov, C. Zissulescu, A. Turjan, B. Kienhuis and E. Deprettere, "System design using Kahn process networks: the Compaan/Laura approach," in *Proceedings of Design, Automation and Testing in Europe Conference*, Paris, 2004, pp. 340–345, IEEE Computer Society, Washington, USA.
7. S. Derrien, A. Turjan, C. Zissulescu, B. Kienhuis and E. Deprettere, "Deriving efficient control in process networks with Compaan/Laura," *International Journal of Embedded Systems*, 2008, vol. 3,no. 3, pp. 170–180.
8. K. Huang, D. Grunert and L. Thiele, "Windowed FIFOs for FPGA-based multiprocessor systems," in *Proceedings of 18th IEEE International Conference on Application-specific Systems*, pp. 36–41, Zurich, 2007.
9. A. V. Oppenheim, R. W. Schafer and J. R. Buck, *Discrete-time Signal Processing*, 2nd edn, 1999, Prentice-Hall.
10. H. Nikolov, T. Stefanov and E. Deprettere, "Systematic and automated multiprocessor system design, programming and implementation," *IEEE Transactions on Computer-Aided Design of Integrated Circuits and Systems*, 2008, vol. 27, pp. 542–555.
11. R. Ohlendorf, M. Meitinger, T. Wild and A. Herkersdorf, "A processing path dispatcher in network processor MPSoCs," *IEEE Transactions on Very Large Scale Integration Systems*, 2008, vol. 16, pp. 1335–1345.
12. Leiden University. Compaan: www.liacs.nl/∼cserc/compaan
13. www.mathworks.com/products/simulink
14. http://eesof.tm.agilent.com/products/ads_main.html
15. www.cadence.com
16. www.synopsys.com
17. National Instruments. NI LabVIEW: www.ni.com/labview/
18. Seoul National University. PeaCE (codesign environment): http://peace.snu.ac.kr/research/peace/
19. StreamIt: groups.csail.mit.edu/cag/streamit/
20. M. G. Corporation. Mentor Graphics: www.mentor.com
21. SIGNALogic. Hypersignal: www.signalogic.com
22. University of Berkeley. Ptolemy: http://ptolemy.berkeley.edu
23. University of New Mexico. Khoros: www.unm.edu
24. E. A. Lee and D. G. Messerschmitt, "Static scheduling of synchronous data-flow programs for digital signal processing," *IEEE Transactions on Computing*, 1987, vol. 36, pp. 24–35.
25. C.-J. Hsu,"Dataflow integration and simulation techniques for DSP system design tools," PhD thesis, University of Maryland, College Park, USA, 2007.
26. W. Khan and V. Dimitrov, "Multiplierless DCT algorithm for image compression applications," *International Journal of Information Theories and Applications*, 2004, vol. 11, no. 2, pp. 162–169.
27. M. T. Sun, T. C. Chen and A. M. Gottlieb, "VLSI implementation of a 16 × 16 discrete cosine transform," *IEEE Transactions on Circuits and Systems*, 1989, vol. 36, pp. 610–617.
28. L. V. Agostini, I. S. Silva and S. Bampi, "Pipelined fast 2-D DCT architecture for JPEG image compression," in *Proceedings of 14th Symposium on Integrated Circuits and Systems Design*, 2001, pp. 226–231, IEEE Computer Society, USA.
29. I. K. Kim, H. J. Cho and J. J. Cha, "A design of 2-D DCT/IDCT for real-time video applications," in *Proceedings of 6th International Conference on VLSI and CAD*, 1999, pp. 557–559, Seoul, South Korea.
30. L. Benini, P. Siegel and G. De Micheli, "Automatic synthesis of gated clocks for power reduction in sequential circuits," *IEEE Design and Test of Computers*, 1994, vol. 11, pp. 32–40.

31. L. Benini and G. De Micheli, "Transformation and synthesis of FSMs for low-power gated clock implementation," *IEEE Transactions on Computer-Aided Design of Integrated Circuits and Systems*, 1996, vol. 15, pp. 630–643.
32. L. Benini, G. De Micheli, E. Macii, M. Poncino and R. Scarsi, "Symbolic synthesis of clock-gating logic for power optimization of control-oriented synchronous networks," in *Proceedings of European Design and Test Conference*, Paris, 1997, pp. 514–520, IEEE Computer Society, Washington, USA.
33. H. Jung, H. Yang and S. Ha,"Optimized RTL code generation from coarse-grain dataflow specification for fast HW/SW cosynthesis," *Journal of VLSI Signal Processing,* June 2007 [online].
34. H. Jung, K. Lee and S. Ha, "Efficient hardware controller synthesis for synchronous data flow in system level design," *IEEE Transactions on Very Large Scale Integration Systems*, 2002, vol. 10, pp. 423–428.

5

Design Options for Basic Building Blocks

5.1 Introduction

A detailed description of system-level design of signal processing algorithms and their representation as dataflow graphs (DFGs) is given in Chapter 4. A natural sequel is to discuss design options for the fundamental operations that are the building blocks for hardware mapping of an algorithm. These blocks constitute the datapath of the design that primarily implements the number-crunching. They perform addition, subtraction, multiplication and arithmetic and logic shifts.

This chapter covers design options for parallel adders, multipliers and barrel shifters. Almost all vendors of field-programmable gate arrays (FPGAs) are now also embedding hundreds of basic building blocks. The incorporation of fast adders and multipliers along with hard and soft micros of microcontrollers has given a new dimension to the subject of digital design. The hardware can run in close proximity to the software running on embedded processors. The hardware can be designed either as an extension to the instruction set or as an accelerator to execute computationally intensive parts of the application while the code-intensive part executes on the embedded processor.

This trend of embedding basic computational units on an FPGA has also encouraged designers to think at a higher level of abstraction while mapping signal processing algorithms in HW. There are still instances where a designer may find it more optimal to explore all the design options to further optimize the design by exercising the architecture alternatives of the basic building blocks. This chapter gives a detailed account of using these already embedded building blocks in the design. The chapter then describes architectural design options for basic computational blocks.

5.2 Embedded Processors and Arithmetic Units in FPGAs

FPGAs have emerged as exciting devices for mapping high-throughput signal processing applications. For these applications the FPGAs outperform their traditional competing technology of digital signal processors (DSPs). No matter how many MACs the DSP vendor can place on a chip, still it cannot compete with the availability of hundreds of these units on a high-end FPGA device. The modern day FPGAs come with embedded processors, standard interfaces and signal processing building blocks consisting of multipliers, adders, registers and multiplexers. Different devices in a

Digital Design of Signal Processing Systems: A Practical Approach, First Edition. Shoab Ahmed Khan.

family come with varying numbers of these embedded units, and it is expected that the number and type of these building blocks on FPGAs will see an upward trend.

For example, the Xilinx Virtex™-4 and Virtex™-5 devices come with DSP48 and DSP48e blocks, respectively [1, 2]. The DSP48 block has one 18 × 18-bit two's complement multiplier followed by three 48-bit multiplexers, one 3-input 48-bit adder/subtractor and a number of registers. This block is shown in Figure 5.1(a). The registers in the block can be effectively used to add pipeline stages to the multiplier and adder. Each DSP48 block is independently configurable and has 40 different modes of operation. These modes are dynamically reconfigurable and can be changed in every clock cycle.

Similarly the Virtex™-II, Virtex™-II Pro, Spartan™ 3 and Spartan™ 3E devices are embedded with a number of two's complement 18 × 18-bit multipliers as shown in Figure 5.1(d). Figures 5.1(b) and (c) show arithmetic blocks in Altera and QuickLogic FPGAs.

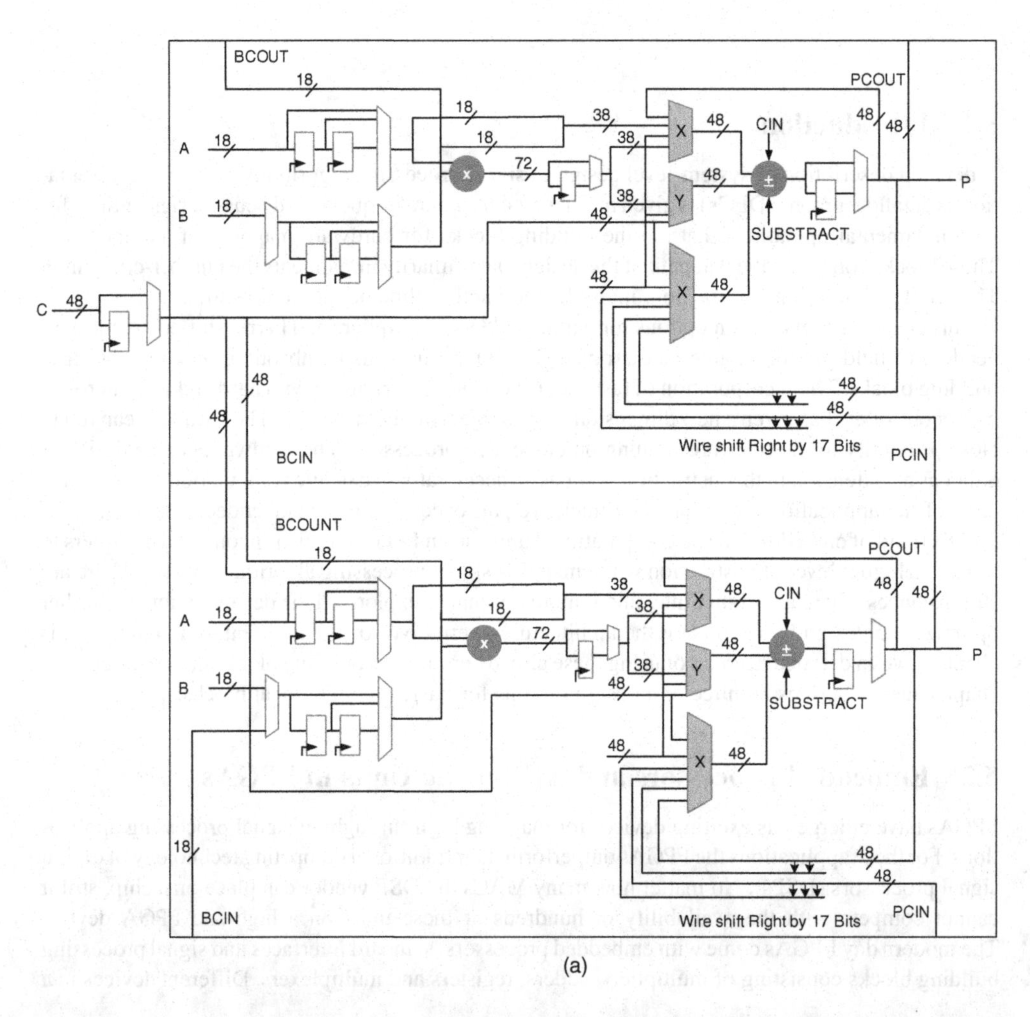

Figure 5.1 Dedicated computational blocks. (a) DSP48 in Virtex™-4 FPGA (derived from Xilinx documentation). (b) 18 × 18 multiplier and adder in Altera FPGA. (c) 8 × 8 multiplier and 16-bit adder in Quick Logic FPGA. (d) 18 × 18 multiplier in Virtex-II, Virtex-II pro and Spartan™-3 FPGA

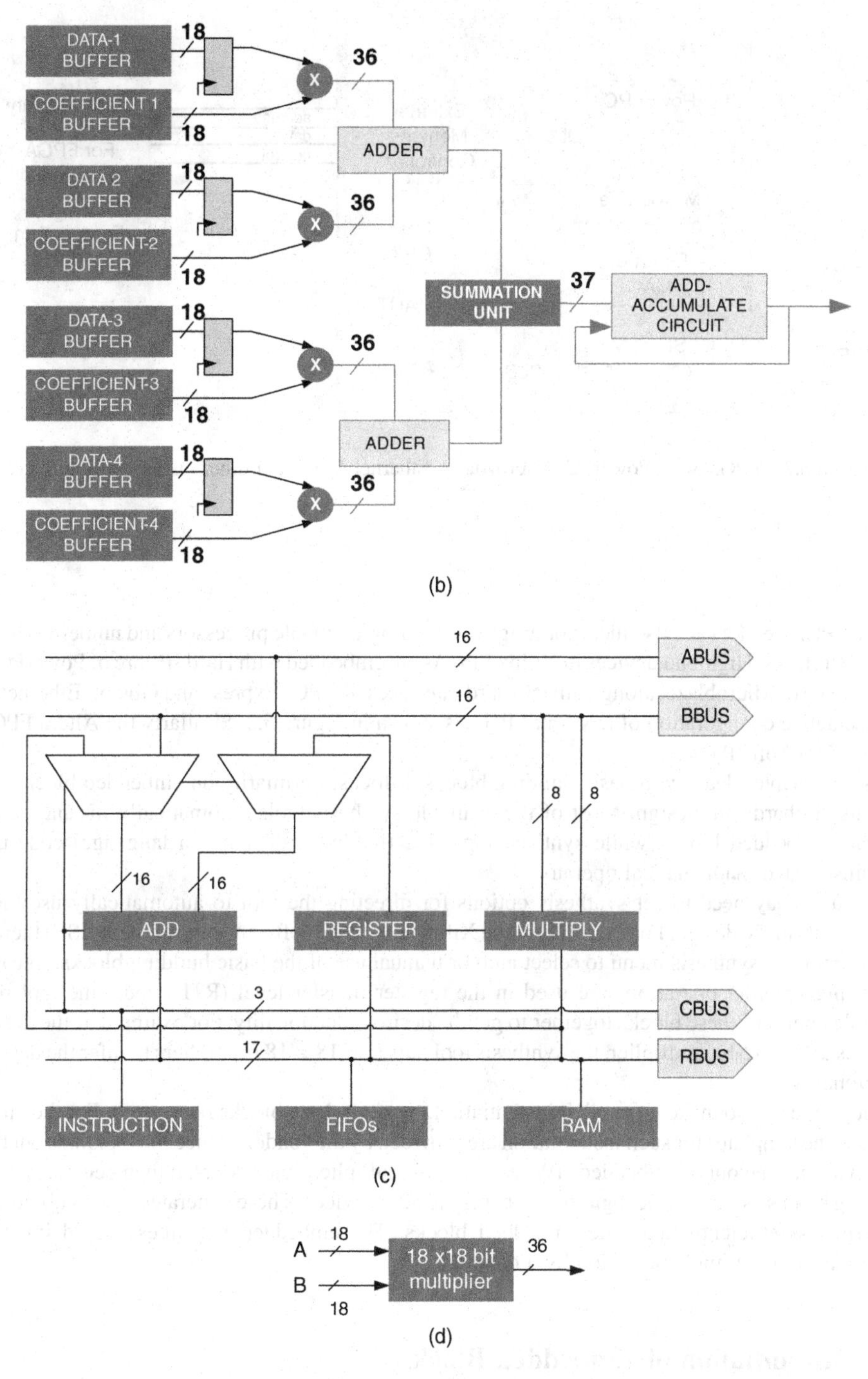

Figure 5.1 (*Continued*)

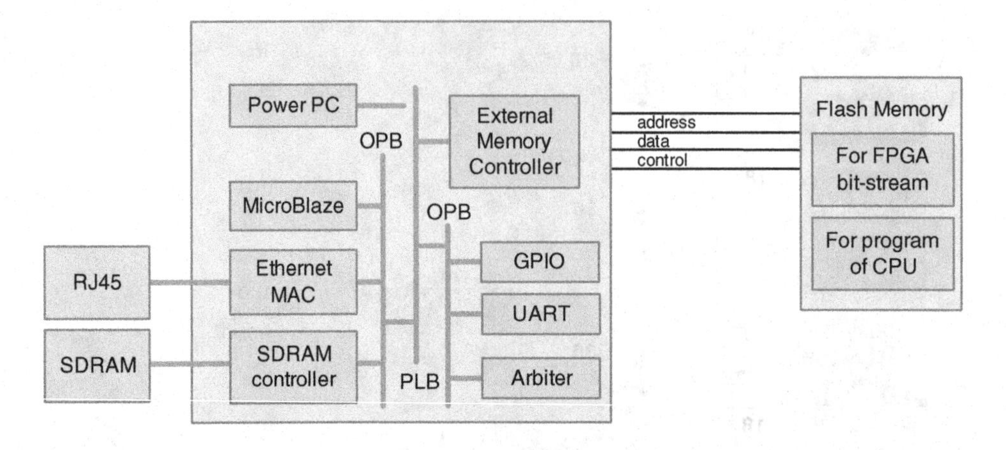

Figure 5.2 FPGA with PowerPC, MicroBlaze, Ethernet MAC and other embedded interfaces

The FPGA vendors are also incorporating cores of programmable processors and numerous high-speed interfaces. High-end devices in Xilinx FPGAs are embedded with Hard IP core of PowerPC or Soft IP core of Microblaze, along with standard interfaces like PCI Express and Gigabit Ethernet. A representative configuration of a Xilinx FPGA is shown in Figure 5.2. Similarly the Altera FPGA offers ARM Soft IP cores.

As this chapter deals with basic building blocks, it focuses primarily on embedded blocks and their use in hardware design. Most of the available synthesis tools automatically instantiate the relevant embedded blocks while synthesizing HDL (hardware description language) code that contains related mathematical operators.

The user may need to set synthesis options for directing the tool to automatically use these components in the design. For example, in the Xilinx integrated software environment (ISE) there is an option in the synthesis menu to select auto or manual use of the basic building blocks. In cases where higher order operations are used in the register transfer level (RTL) code, the tool puts multiple copies of these blocks together to get the desired functionality. For example, if the design requires a 32×32-bit multiplier, the synthesis tool puts four 18×18 multipliers to infer the desired functionality.

The user can also make an explicit instantiation to one of these blocks if required. For the target devices, the templates for such instantiation are provided by the vendors. Once the tool finds out that the device has run out of embedded HW resources of multipliers and adders, it then generates these building blocks using generic logic components on these devices. These generated blocks obviously perform less efficiently than the embedded blocks. The embedded resources indeed improve performance of an implementation by a quantum.

5.3 Instantiation of Embedded Blocks

To demonstrate the effectiveness of embedded blocks in HW design, an example codes a second-order infinite impulse response (IIR) filter in Direct Form (DF)-II realization. The block diagram of

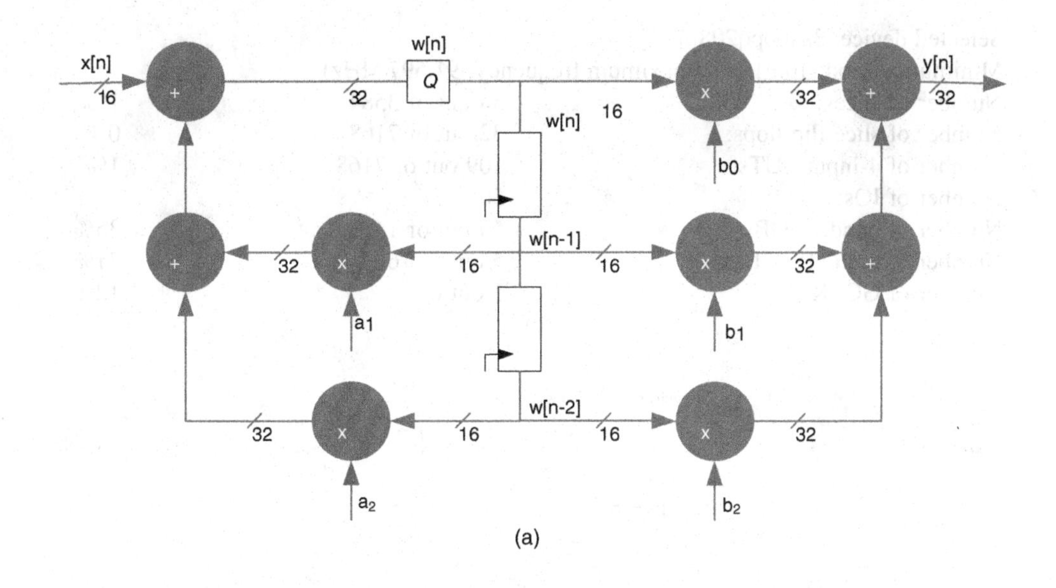

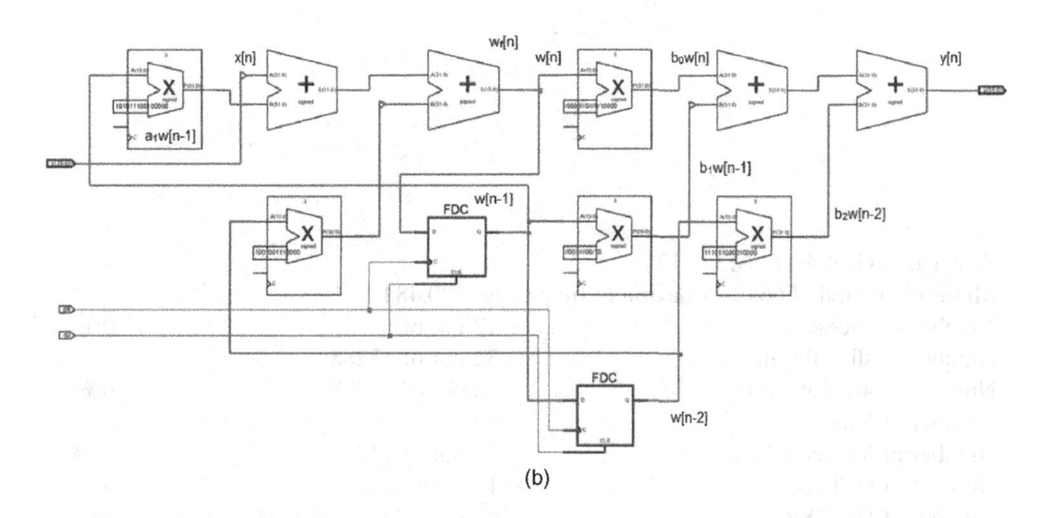

Figure 5.3 Hardware implementation of a second-order IIR filter on FPGA. (a) Block diagram of a second-order IIR filter in Direct Form-II realization. (b) RTL schematic generated by Xilinx's Integrated Software Environment (ISE). Each multiplication operation is mapped on an embedded 18×18 multiplier block of Xiliinx Spartan™ 3 FPGA. (c) Synthesis summary report of the design on Spartan™-3 FPGA. (d) RTL schematic generated by Xilinx ISE for Virtex™ 4 target device. The multiplication and addition operations are mapped on DSP48 multiply accumulate (MAC) embedded blocks. (e) Synthesis summary report for the RTL schematic of (d)

Selected device: 3s400pq208-5		
Minimum period: 10.917 ns (maximum frequency: 91.597 MHz)		
Number of slices:	58 out of 3584	1%
Number of slice flip-flops:	32 out of 7168	0%
Number of 4-input LUTs:	109 out of 7168	1%
Number of IOs:	50	
Number of bonded IOBs:	50 out of 141	35%
Number of multi 18 × 18s:	5 out of 16	31%
Number of GCLKs:	1 out of 8	12%

(c)

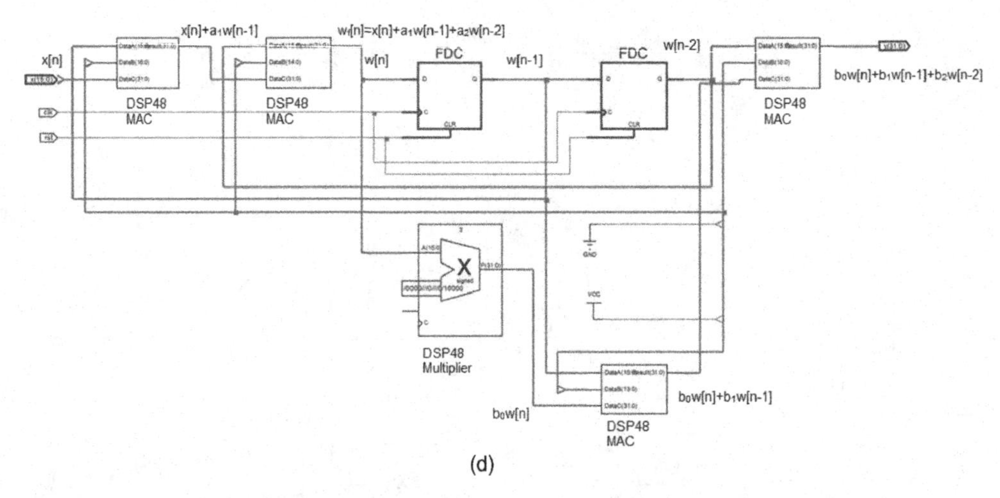

(d)

Selected device: 4vlx15sf363-12		
Minimum period: 7.664 ns (maximum frequency: 130.484 MHz)		
Number of slices:	17 out of 6144	0%
Number of slice flip-flops:	32 out of 12288	0%
Number of 4-input LUTs:	16 out of 12288	0%
Number of IOs:	50	
Number of bonded IOBs:	50 out of 240	20%
Number of GCLKs:	1 out of 32	3%
Number of DSP48s:	5 out of 32	15%

(e)

Figure 5.3 (*Continued*)

the filter is shown in Figure 5.3, to implement the difference equations:

$$w[n] = a_1 w[n-1] + a_2 w[n-2] + x[n] \tag{5.1a}$$

$$y[n] = b_0 w[n] + b_1 w[n-1] + b_2 w[n-2]. \tag{5.1b}$$

RTL Verilog code of a fixed-point implementation of the design is listed here:

```
module iir(xn, clk, rst, yn);

// x[n] is in Q1.15 format
input signed [15:0] xn;
input clk, rst;
output reg signed [31:0] yn;     // y[n] is in Q2.30 format

wire signed [31:0] wfn;          // Full precision w[n] in Q2.30 format
wire signed [15:0] wn;           // Quantized w[n] in Q1.15 format
reg signed [15:0] wn_1, wn_2;  // w[n-1]and w[n-2] in Q1.15 format

// all the coefficients are in Q1.15 format
wire signed [15:0] b0 = 16'h0008;
wire signed [15:0] b1 = 16'h0010;
wire signed [15:0] b2 = 16'h0008;
wire signed [15:0] a1 = 16'h8000;
wire signed [15:0] a2 = 16'h7a70;

assign wfn = wn_1*a1+wn_2*a2;    // w[n] in Q2.30 format with one redundant
                                    sign bit
/* through away redundant sign bit and keeping
   16 MSB and adding x[n] to get w[n] in Q1.15 format */

assign wn = wfn[30:16]+xn;
//assign yn = b0*wn + b1*wn_1 + b2*wn_2;  // computing y[n] in Q2.30 format
                                             with one redundant sign bit
always @(posedge clk or posedge rst)
begin
   if(rst)
   begin
      wn_1 <= #1 0;
      wn_2 <= #1 0;
      yn  <= #1 0;
   end
   else
   begin
      wn_1 <= #1 wn;
      wn_2 <= #1 wn_1;
      yn  <= #1 b0*wn + b1*wn_1 + b2*wn_2;     // computing y[n] in Q2.30
                                                 format with one redundant
                                                 sign bit

   end
end
endmodule
```

The design assumes that $x[n]$ and all the coefficients of the filter are in Q1.15 format. To demonstrate the usability of embedded blocks on FPGA, the code is synthesized for one of the Spartan™-3 family of devices. These devices come with embedded 18×18 multiplier blocks. In the synthesis option, the Xilinx ISE tool is directed to use the embedded multiplier blocks. The post-synthesis report shows the design is realized using five MULT 18×18 blocks. As there is no embedded adder in the Spartan™-3 family of devices, the tool creates a 32-bit adder using the fast carry chain logic

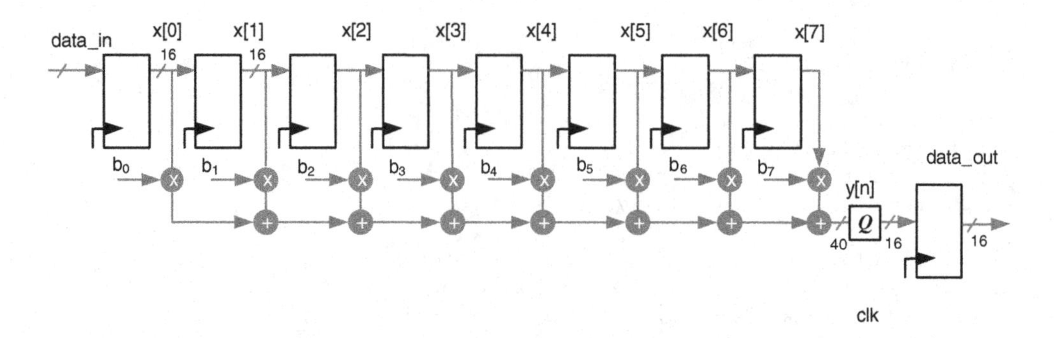

Figure 5.4 An 8-tap direct form (DF-I) FIR filter

provided in the LUT. A summary of the synthesis report listed in Figure 5.3(c) and schematic diagram given in (b) provide details of the Spartan™-3 resources being used by the design.

To observe the effectiveness of DSP48 blocks in Virtex™-4 FPGAs, the design is synthesized again after changing the device selection. As the tool is still directed to use DSP48 blocks, the synthesis result shows the use of five DSP48 blocks in the design, as is obvious from Figures 5.3(d) and (e) showing the schematic and a summary of the synthesis report of the design. Both the schematic diagrams are labeled with the actual variables and operators to demonstrate their relevance with the original block diagram. Virtex™-4 is clearly a superior technology and gives better timing for the same design. It is pertinent to point out that the synthesis timings given in this chapter are post-synthesis; although they give a good estimate, the true timings come after placement and routing of the design.

5.3.1 Example of Optimized Mapping

The Spartan™-3 family of FPGAs embeds dedicated 18 × 18-bit two's complement multipliers to speed up computation. The family offers different numbers of multipliers ranging from 4 to 104 in a single device. These dedicated multipliers help in fast and power-efficient implementation of DSP algorithms. Here a simple example illustrates the simplicity of embedding dedicated blocks in a design.

The 8-tap FIR filter of Figure 5.4 is implemented in Verilog. The RTL code is listed below:

```
// Module: fir_filter
// Discrete-Time FIR Filter
// Filter Structure : Direct-Form FIR
// Filter Order : 7
// Input Format: Q1.15
// Output Format: Q1.15
module fir_filter (
  input clk,
  input signed [15:0] data_in, //Q1.15
  output reg signed [15:0] data_out //Q1.15
  );
// Constants, filter is designed using Matlab FDATool, all coefficients are in
// Q1.15 format
parameter signed [15:0] b0 = 16'b1101110110111011;
parameter signed [15:0] b1 = 16'b1110101010001110;
parameter signed [15:0] b2 = 16'b0011001111011011;
```

```
parameter signed [15:0] b3 = 16'b0110100000001000;
parameter signed [15:0] b4 = 16'b0110100000001000;
parameter signed [15:0] b5 = 16'b0011001111011011;
parameter signed [15:0] b6 = 16'b1110101010001110;
parameter signed [15:0] b7 = 16'b1101110110111011;
reg signed [15:0] xn [0:7]; // input sample delay line
wire signed [39:0] yn; // Q8.32
// Block Statements
  always @ (posedge clk)
  begin
      xn[0] <= data_in;
      xn[1] <= xn[0];
      xn[2] <= xn[1];
      xn[3] <= xn[2];
      xn[4] <= xn[3];
      xn[5] <= xn[4];
      xn[6] <= xn[5];
      xn[7] <= xn[6];
      data_out <= yn[30:15];        // bring the output back in Q1.15
                                    //format
  end
  assign yn = xn[0] * b0 + xn[1] * b1 + xn[2] * b2 + xn[3] * b3 + xn[4] * b4 +  xn[5]
    * b5 +xn[6] * b6 + xn[7] * b7;
endmodule
```

The tool automatically infers eight of the embedded 18×18-bit multipliers. The synthesis report of Table 5.1(a) and schematic of Figure 5.5 clearly show the use of embedded multipliers in the design. The developer can also use an explicit instantiation of the multiplier by appropriately selecting the correct instance in the Device Primitive Instantiation option of the Language Template icon

Table 5.1 Synthesis reports: (a) Eight 18×18-bit embedded multipliers and seven adders from generic logic blocks are used on a Spartan™-3 family of FPGA. (b) Eight DSP48 embedded blocks are used once mapped on a Vertix™-4 family of FPGA

(a)			(b)		
Selected device: 3s200pq208-5			Selected device: 4vlx15sf363-12		
Minimum period: 23.290 ns			Minimum period: 16.958 ns		
(Maximum frequency:42.936 MHz)			(Maximum frequency: 58.969 MHz)		
Number of slices:	185 out of 1920	9%	Number of Slices:	9 out of 6144	0%
Number of Slice Flip Flops:	144 out of 3840	3%	Number of Slice Flip Flops:	16 out of 12288	0%
Number of 4 input LUTs:	217 out of 3840	5%	Number of IOs:	33	
Number of IOs:	33		Number of bonded IOBs:	33 out of 240	13%
Number of bonded IOBs:	33 out of 141	23%	Number of GCLKs:	1 out of 32	3%
Number of MULT18×18s:	8 out of 12	66%	Number of DSP48s:	8 out of 32	25%
Number of GCLKs:	1 out of 8	12%			

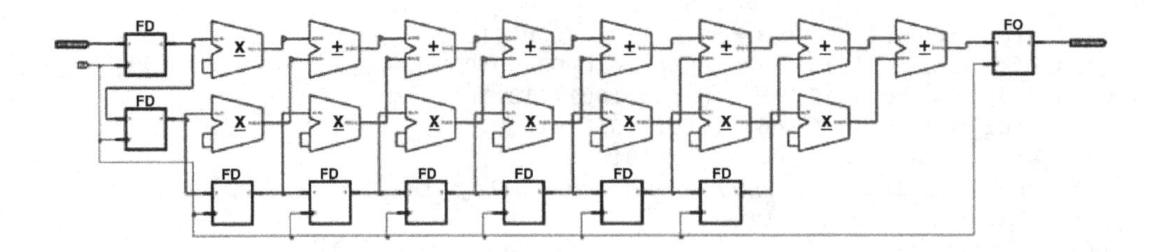

Figure 5.5 Schematic displaying the use of eight 18 × 18 embedded multiplier of Spartan™-3 FPGA

provided in the ISE toolbar. The code for the instance from the ISE-provided template is given here:

```
// MULT18X18: 18 x 18 signed asynchronous multiplier
// Virtex™-II/II-Pro, Spartan™-3
// Xilinx HDL Language Template, version 9.1i
MULT18X18 MULT18X18_inst (
     .P(P), // 36-bit multiplier output
     .A(A), // 18-bit multiplier input
     .B(B) // 18-bit multiplier input
     );
// End of MULT18X18_inst instantiation
```

As there is no embedded adder in the Spartan™-3 family of FPGAs, the tool crafts an adder using general-purpose logic. If in the device options we select a device in the Vierter™-4 family of FPGAs, which has several DSP48 blocks, then the tool uses MACs in the design and that obviously outperforms the Spartan™ design. The synthesis report of this design option is given in Table 5.1(b).

5.3.2 Design Optimization for the Target Technology

Although a digital design and its subsequent RTL Verilog implementation should be independent of technology, in many cases it is imperative to understand the target technology – especially when the device embeds dedicated arithmetic blocks. A design can effectively use the dedicated resources to their full potential and can improve performance many-fold without any additional HW cost.

To substantiate this statement, the FIR filter implementation of Figure 5.4 is redesigned for optimized mapping on devices with embedded DSP48 blocks. The objective is to use the potential of the DSP48 blocks. Each can add two stages of pipelining in the MAC operation. If the code does not require any pipelining in the MAC operation, these registers are bypassed. The design in Figure 5.4 works without pipelining the MAC and registers in the DSP48 blocks are not used. A pipeline implementation for FIR filter is shown in Figure 5.6. This effectively uses the pipeline registers in the embedded blocks. The RTL Verilog code is listed below:

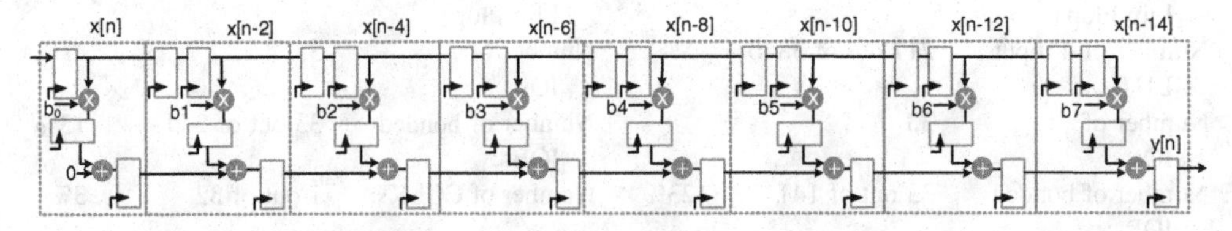

Figure 5.6 Pipeline implementation of 8-tap FIR filter optimized for mapping on FPGAs with DSP48 embedded blocks

```
module fir_filter_pipeline (
   input clk;
   input signed [15:0] data_in; //Q1.15
   output signed [15:0] data_out; //Q1.15
// Constants, filter is designed using Matlab FDATool, all coefficients are in
//Q1.15 format
parameter signed [15:0] b0 = 16'b1101110110111011;
parameter signed [15:0] b1 = 16'b1110101010001110;
parameter signed [15:0] b2 = 16b0011001111011011;
parameter signed [15:0] b3 = 16b0110100000001000;
parameter signed [15:0] b4 = 16b0110100000001000;
parameter signed [15:0] b5 = 16b0011001111011011;
parameter signed [15:0] b6 = 16b1110101010001110;
parameter signed [15:0] b7 = 16b1101110110111011;
reg signed [15:0] xn [0:14] ; // one stage pipelined input sample delay line
reg signed [32:0] prod [0:7]; // pipeline product registers in Q2.30 format
wire signed [39:0] yn; // Q10.30
reg signed [39:0] mac [0:7]; // pipelined MAC registers in Q10.30 format
integer i;
always @ ( posedge clk)
   begin
  xn[0] <= data_in;
  for (i=0; i<14; i=i+1)
 xn[i+1] = xn[i];
   data_out <= yn[30:15]; // bring the output back in Q1.15 format
end
always @ ( posedge clk)
begin
   prod[0] <= xn[0] * b0;
   prod[1] <= xn[2] * b1;
   prod[2] <= xn[4] * b2;
   prod[3] <= xn[6] * b3;
   prod[4] <= xn[8] * b4;
   prod[5] <= xn[10] * b5;
   prod[6] <= xn[12] * b6;
   prod[7] <= xn[14] * b7;
end
always @ (posedge clk)
begin
   mac[0] <= prod[0];
   for (i=0; i<7; i=i+1)
  mac[i+1] <= mac[i]+prod[i+1];
end
assign yn = mac[7];
endmodule
```

The design is synthesized and the synthesis report in Table 5.2 reveals that, athough the design uses exactly the same amount of resources, it results in a ~9-fold improvement in timing. This design can run at 528.82 MHz, compared to its DF-I counterpart that compiles for 58.96 MHz of best timing.

Table 5.2 Synthesis report of mapping pipelined 8-tap FIR filter on Virtex™-4 family of FPGA

Selected device: 4vlx15sf363-12		
Minimum period: 1.891 ns		
(Maximum frequency: 528.821 MHz)		
Number of slices:	9 out of 6144	0%
Number of slice flip-flops:	16 out of 12288	0%
Number of I/Os:	33	
Number of bonded IOBs:	33 out of 240	13%
Number of GCLKs:	1 out of 32	3%
Number of DSP48s:	8 out of 32	25%

It is important to point out that the timing for the DF-I realization increases linearly with the length of the filter, whereas the timing for the pipeline implementation is independent of the length of the filter.

Several other effective techniques and structures for implementing FIR filters are covered in Chapter 6.

5.4 Basic Building Blocks: Introduction

After the foregoing discussion of the use of dedicated multipliers and MAC blocks, it is pertinent to look at the architectures for the basic building blocks. This should help the designer to appreciate the different options for some of the very basic mathematical operations, and elucidate the tradeoffs in the design space exploration. This should encourage the reader to always explore design alternatives, no matter how simple the design. It is also important to understand that a designer should always prefer to use the dedicated blocks.

Several architectural options are available for selecting an appropriate HW block for operations like addition, multiplication and shifting. The following sections discuss some of these design options.

5.5 Adders

5.5.1 Overview

Adders are used in addition, subtraction, multiplication and division. The speed of any digital design of a signal processing or communication system depends heavily on these functional units. The *ripple carry adder* (RCA) is the slowest in adder family. It implements the traditional way of adding numbers, where two bits and a carry of addition from the previous bit position are added and a sum and a carry-out is computed. This carry is propagated to the next bit position for sequential addition of the rest of the significant bits in the numbers. Although the carry propagation makes it the slowest adder, its simplicity gives it the minimum gate count.

To cater for the slow carry propagation, fast adders are designed. These make the process of carry generation and its propagation faster. For example, in a *carry look-ahead adder* the carry-in for all the bit positions are generated simultaneously by a carry look-ahead generator logic [3, 4].

This helps in the computation of a sum in parallel with the generation of carries without waiting for their propagation. This results in a constant addition time independent of the length of the adder. As the word length increases, the hardware organization of the carry generation logic gets complicated. Hence adders with a large number of elements are implemented in two or three levels of carry look-ahead stages.

Another fast devive is the *carry select adder* (CSA) [5]. This partitions the addition in K groups; to gain speed the logic is replicated and for each group addition is performed assuming carry-in 0 and 1. The correct sum and carry-out from each group is selected by carry-out from the previous group. The selected carry-out is used to select sum and carry-out of the next adjacent group. For adding large numbers a hierarchical CSA is used; this divides the addition into multiple levels.

A *conditional sum adder* can be regarded as a CSA with maximum possible levels. The carry select operation in the first level is performed on 1-bit groups. In the next level two adjacent groups are merged to give the result of a 2-bit carry select operation. Merging of two adjacent groups is repeated until the last two groups are merged to generate the final sum and carry-out. The conditional sum adder is the fastest in adder family [6]. It is a log-time adder and the size of the adder can be doubled by the addition of one 2:1 multiplexer delay.

This chapter presents architectural details and lists RTL Verilog code for a few designs. It is important to highlight that mapping an adder on an FPGA may not yield expected performance results. This is due primarily to the fact that FPGAs are embedded with blocks that favor certain designs over others. For example, a fast carry chain in many FPGA families of devices greatly helps an RCA; so, on these FPGAs, an RCA to a certain bit width is the best design option for area and time.

5.5.2 Half Adders and Full Adders

A *half adder* (HA) is a combinational circuit used for adding two bits, a_i and b_i, without a carry-in. The sum s_i and carry output c_i are given by:

$$s_i = a_i \oplus b_i \tag{5.2a}$$

$$c_i = a_i b_i \tag{5.2b}$$

The critical path delay is *one* gate delay, and it corresponds to the length of any one of the two paths.

A *full adder* (FA) adds three bits. A 3-bit adder is also called a 3:2 compressor. One way to implement a full adder is to use the following equations:

$$s_i = a_i \oplus b_i \oplus c_i \tag{5.3a}$$

$$c_{i+1} = (a_i \oplus b_i)c_i + a_i b_i \tag{5.3b}$$

There are several gate-level designs to implement a full adder, some of which are shown in Figure 5.7. These options to implement a simple mathematical operation of adding three bits also emphasize our initial assertion that the designer must avoid gate-level modeling and dataflow modeling using bitwise operators. It is not possible for an RTL designer to know the optimal design without knowing the target libraries. In many instances it is preferred that the design should be at RTL and is technology-independent. In design instances where the target technology is known *a priori*, then the designer can craft an optimal design by effectively using the optimized components from the technology library. The earlier example of Section 5.3.2 provides a better comprehension of

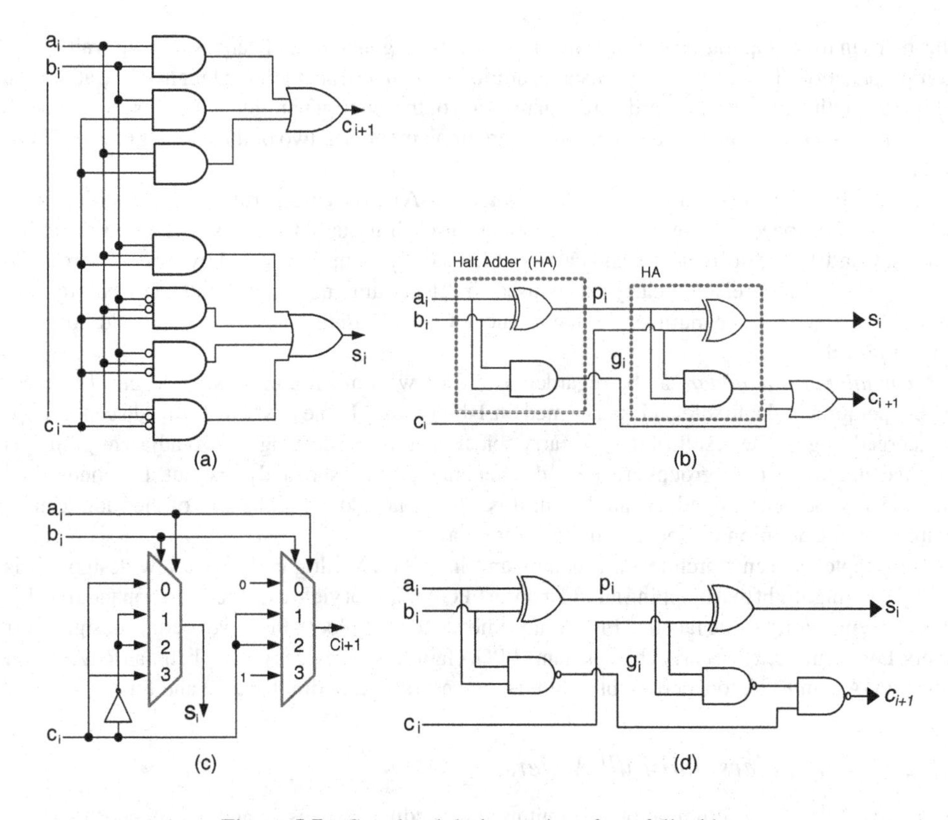

Figure 5.7 Gate-level design options for a full adder

this design rule where DSP48 blocks in the Virtex™-4 family of devices are effectively used to gain a nine times better performance without adding any additional resources compared to a technology-independent RTL implementation.

5.5.3 Ripple Carry Adder

The emphasis of this text is on high-speed architecture, and an RCA is perceived to be the slowest adder. This perception can be proven wrong if the design is mapped on an FPGA with embedded carry chain logic. An RCA takes minimum area and exhibits a regular structure. The structure is very desirable especially in the case of an FPGA mapping as the adders fit easily into a 2-dimensional layout. An RCA can also be pipelined for improved speed.

A ripple adder that adds two N-bit operands requires N full adders. The speed varies linearly with the word length. The RCA implements the conventional way of adding two numbers. In this architecture the operands are added bitwise from the least significant bits (LSBs) to the most significant (MSBs), adding at each stage the carry from the previous stage. Thus the carry-out from the FA at stage i goes into the FA at stage $(i + 1)$, and in this manner carry ripples from LSB to MSB (hence the name of ripple carry adder):

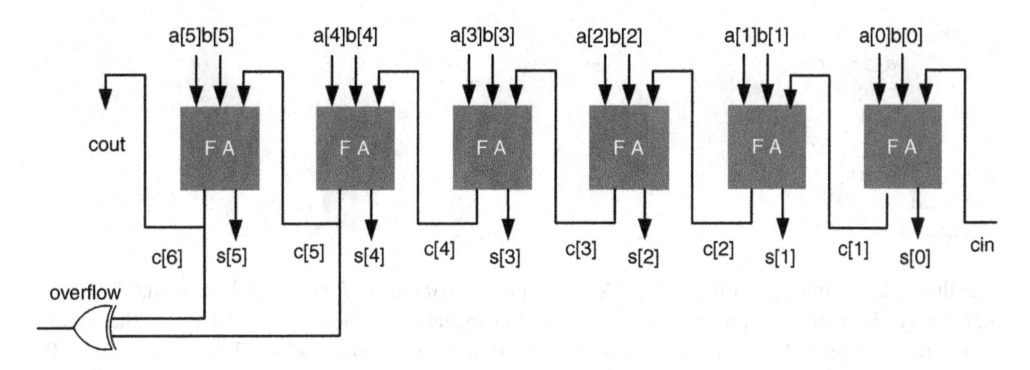

Figure 5.8 A 6-bit ripple carry adder

$$c_{i+1}, s_i \leftarrow a_i + b_i + c_i \tag{5.4}$$

A 6-bit RCA is shown in Figure 5.8. The overflow condition is easily computed by simply performing an XOR operation on the carry-out from the last two full adders. This overflow computation is also depicted in the figure.

The critical path delay of a ripple carry adder is:

$$T_{\mathrm{RCA}} = (N-1)T_{\mathrm{FA}} + T_m \tag{5.5}$$

where T_{RCA} is the delay of an N-bit ripple carry adder, while T_{FA} and T_m are delays of an FA and the carry generation logic of the FA, respectively.

It is evident from (5.5) that the performance of an RCA is constrained by the rippling of carries from the first FA to the last. To speed the propagation of the carry, most FPGAs are supported with fast carry propagation and sum generation logic. The carry propagation logic provides fast paths for the carry to go from one block to another. This helps the designer to implement fast RCA on the FPGA. The user can still implement parallel adders discussed in this chapter, but their mapping on FGPA logic may not generate adequate speed-ups as expected by the user when compared with the simplest RCA performance. In many design instances the designer may find an RCA as the fastest adder compared to other parallel adders discussed here.

Verilog implementation of a 16-bit RCA through dataflow modeling is given below. This implementation, besides using registers for input and output signals, simply uses addition operators, and most synthesis tools will (if not directed otherwise) infer an RCA from this statement:

```
module ripple_carry_adder #(parameter W=16)
(input clk,
    input [W-1:0] a, b,
    input cin,
    output reg [W-1:0] s_r,
    output reg cout_r);
wire [W-1:0] s;
wire cout;
reg [W-1:0] a_r, b_r;
reg cin_r;
assign {cout,s} = a_r + b_r + cin_r;
always@(posedge clk)
```

```
    begin
        a_r<=a;
        b_r<=b;
        cin_r<=cin;
        s_r<=s;
        cout_r<= cout;
    end
endmodule
```

If the code is synthesized for an FPGA that supports *fast carry chain logic*, like Vertix™-II pro, then the synthesis tool will infer this logic for fast propagation of the carry signal across the adder.

A Vertix™-II pro FPGA consists of a number of configurable logic blocks (CLBs), and each CLB has four 'slices'. Each slice further has two look-up tables (LUTs) and dedicated logic to compute generate (g) and propagate (p) functions. These functions are used by the synthesis tool to infer carry logic for implementing a fast RCA. The equations of g_i and p_i and logic it uses for generating fast carry-out c_{i+1} in a chain across an RCA are:

$$c_{i+1} = g_i + p_i c_i$$

$$p_i = a_i \oplus b_i$$

$$g_i = a_i b_i$$

This is shown in Figures 5.9(a) and (b). When the Verilog code of this section is synthesized on the device, the synthesized design uses fast carry logic in cascade to tie four CLBs in a column to implement a 16-bit RCA. The logic of 16-bit and 64-bit adders is shown in Figures 5.10(c) and (d), respectively.

5.5.4 Fast Adders

If not mapped on an FPGA with fast carry chain logic, an RCA usually is the slowest adder as each full adder requires carry-out from the previous onefor its sum and carry-out computation. Several alternative architectures have been proposed in the literature. All these architectures somehow accelerate the generation of carries for each stage. This acceleration results in additional logic. For FGPA implementation the designer needs to carefully select a fast adder because some have carry acceleration techniques well suited for FPGA architectures while others do not. As already discussed, an RCA makes the most optimal use of carry chains, although all the full adders need to fit in a column for effective use of this logic. This usually is easily achieved. A CSA also replicates RCA blocks, so each block still makes an effective use of fast carry chain logic with some additional logic for the selection of one sum out of two sums computed in parallel.

5.5.5 Carry Look-ahead Adder

A closer inspection of the carry generation process reveals that a carry does not have to depend explicitly on the preceding carries. In a carry look-ahead adder the carries entering all the bit positions of the adder are generated simultaneously by a carry look-ahead (CLA) generator; that is, computation of carries takes place in parallel with sum calculation. This results in a constant addition time independent of the length of the adder. As the word length increases, the hardware organization

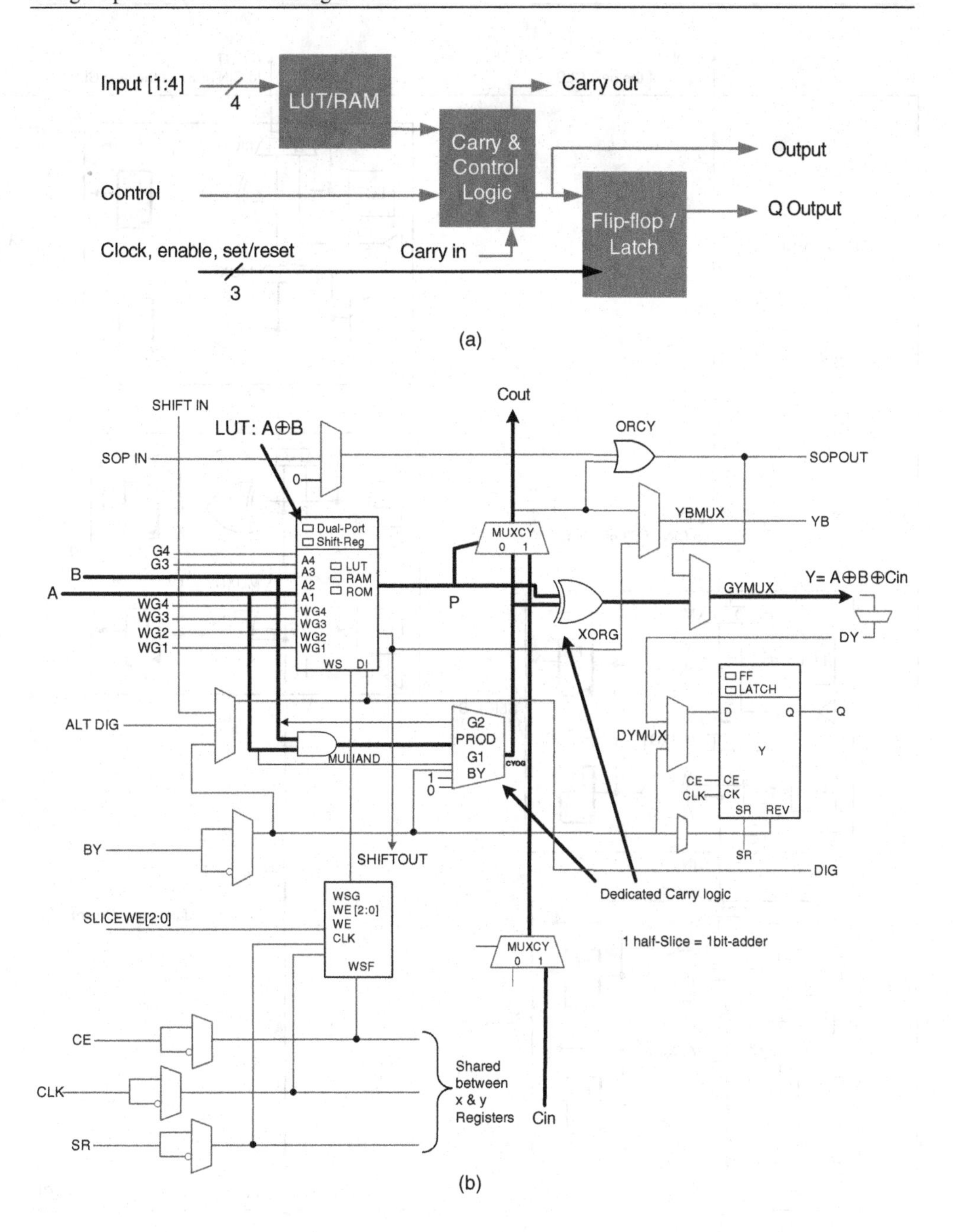

Figure 5.9 (a) Fast-carry logic blocks. (b) Fast-carry logic in Vertix™-II pro FPGA slice. (c) One CLB of Vertix™-II pro inferring two 4-bit adders, thus requiring four CLBs to implement a 16-bit adder. (d) A 64-bit RCA using fast-carry chain logic (derived from Xilinx documentation)

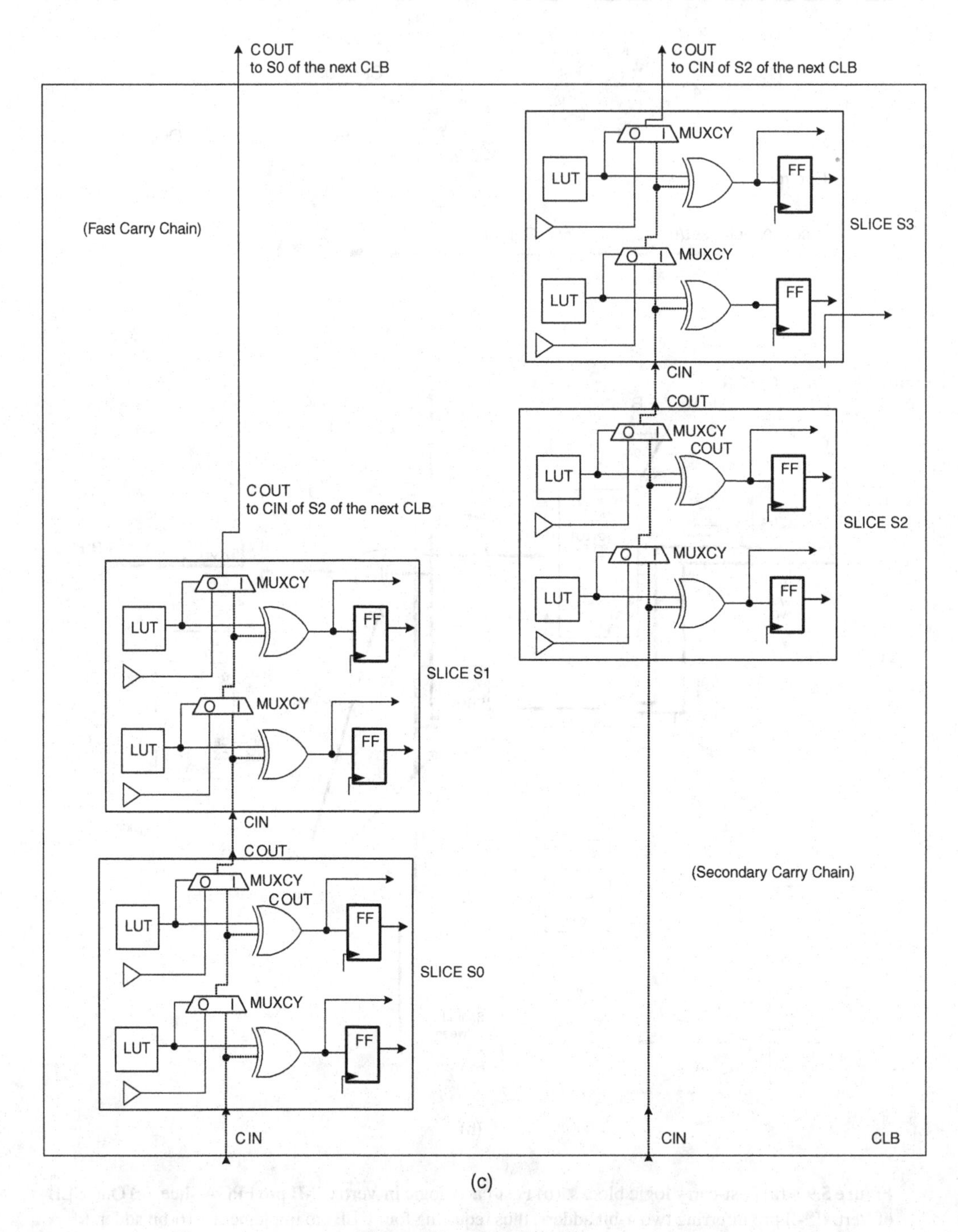

(c)

Figure 5.9 *(Continued)*

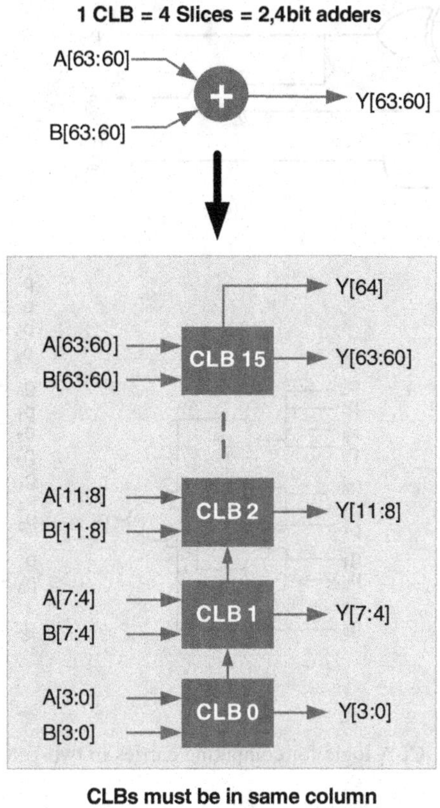

(d)

Figure 5.9 *(Continued)*

of the addition technique gets complicated. Hence adders with a large number of elements may require two or three levels of carry look-ahead stages.

A simple consideration of full adder logic identifies that a carry c_{i+1} is generated if $a_i = b_i = 1$, and a carry is propagated if either a_i or b_i is 1. This can be written as:

$$g_i = a_i b_i \tag{5.6a}$$

$$p_i = a_i \oplus b_i \tag{5.6b}$$

$$c_{i+1} = g_i + p_i c_i \tag{5.6c}$$

$$s_i = c_i \oplus p_i \tag{5.6d}$$

Thus a given stage generates a carry if g_i is TRUE and propagates a carry-in to the next stage if p_i is TRUE. Using these relationships, the carries can be generated in parallel as:

$$c_1 = g_0 + p_0 c_0 \tag{5.7a}$$

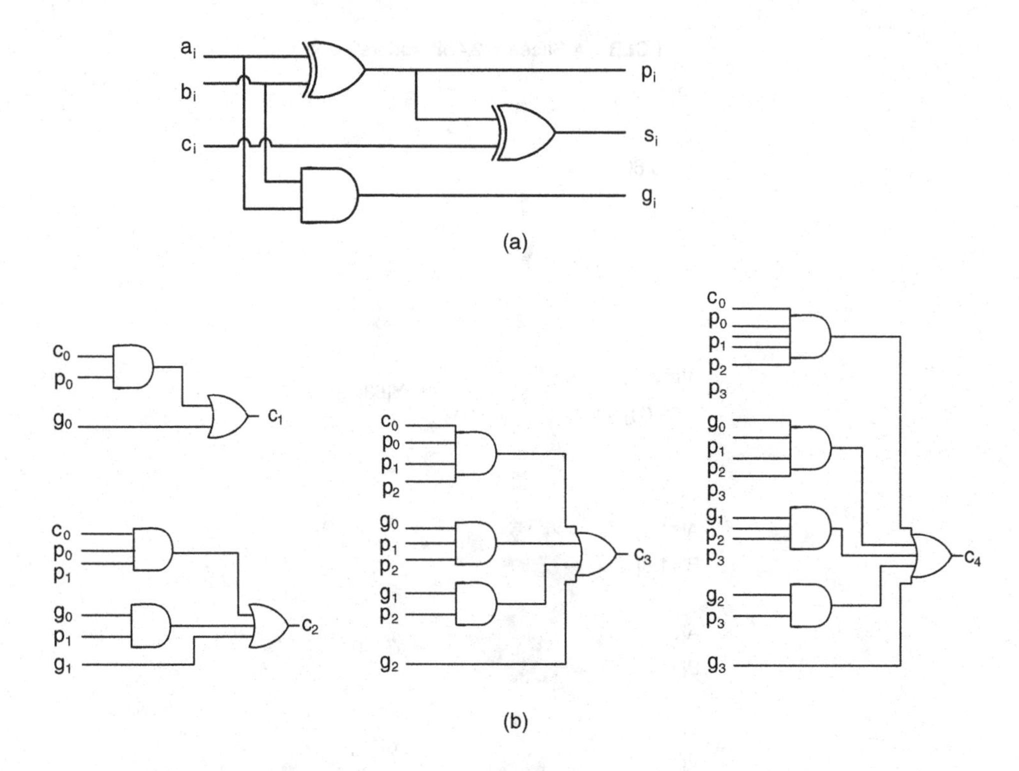

Figure 5.10 CLA logic for computing carries in two-gate delay time

$$c_2 = g_1 + p_1 c_1 \tag{5.7b}$$

$$c_2 = g_1 + p_1(g_0 + p_0 c_0) \tag{5.7c}$$

$$c_2 = g_1 + p_1 g_0 + p_0 p_1 c_0 \tag{5.7d}$$

$$c_3 = g_2 + p_2 g_1 + p_2 p_1 g_0 + p_2 p_1 p_0 c_0 \tag{5.7e}$$

Figure 5.10 shows that all carries can be computed in a time of two gate delays. The block in carry look-ahead architecture can be of any size, but the number of inputs to the logic gates increases with the size. In closed form, we can write the carry at position i as:

$$c_i = g_{i-1} + \sum_{j=0}^{i-2}\left(\prod_{k=j+1}^{i-1} p_j\right) g_j + \prod_{k=0}^{i-1} p_j c_0 \tag{5.8}$$

This requires $i + 1$ gates with a fan-in of $i + 1$. Thus each additional position will increase the fan-in of the logic gates.

Industrial practice is to use 4-bit wide blocks. This limits the computation of carries until c_3, and c_4 is not computed. The first four terms in c_4 are grouped as G_0 and the product $p_3p_2p_1p_0$ in the last term is tagged as P_0 as given here:

$$c_4 = g_3 + p_3g_2 + p_3p_2g_1 + p_3p_2p_1g_0 + p_3p_2p_1p_0c_0$$

Let:

$$G_0 = g_3 + p_3g_2 + p_3p_2g_1 + p_3p_2p_1g_0$$

$$P_0 = p_3p_2p_1p_0$$

Now the group G_0 and P_0 are used in the second level of the CLA to produce c_4 as:

$$c_4 = G_0 + P_0c_0$$

Similarly, bits 4 to 7 are also grouped together and c_5, c_6 and c_7 are computed in the first level of the CLA block using c_4 from the second level of CLA logic. The first-level CLA block for these bits also generates G_1 and P_1.

Figure 5.11 shows a 16-bit adder using four carry look-ahead 4-bit wide blocks in the first level. Each block also computes its G and P using the same CLA as used in the first level. Thus the second level generates all the carries c_4, c_8 and c_{12} required by the first-level CLAs. In this way the design is hierarchically broken down for efficient implementation. The same strategy is further extended to build higher order adders. Figure 5.12 shows a 64-bit carry look-ahead adder using three levels of CLA logic hierarchy.

5.5.6 Hybrid Ripple Carry and Carry Look-ahead Adder

Instead of hierarchically building a larger carry look-ahead adder using multiple levels of CLA logic, the carry can simply be rippled between blocks. This hybrid adder is a good compromise as it yields an adder that is faster than RCA and takes less area than a hierarchical carry look-ahead adder. A 12-bit hybrid ripple carry and carry look-ahead adder is shown in Figure 5.13.

5.5.7 Binary Carry Look-ahead Adder

The BCLA works in a group of two adjacent bits, and then from LSB to MSB successively combines two groups to formulate a new group and its corresponding carry. The logic for the carry generation of an N-bit adder is:

$$g_i = a_ib_i \tag{5.9a}$$

$$p_i = a_i \oplus b_i \tag{5.9b}$$

$$(G_i, P_i) = (g_i, p_i) \cdot (g_{i-1}, p_{i-1}) \cdot \ldots (g_1, p_1) \cdot (g_0, p_0) \tag{5.9c}$$

The problem can be solved recursively as:

$$(G_0, P_0) = (g_0, p_0)$$

for $i = 1$ to $N-1$

$$(G_i, P_i) = (g_i, p_i) \cdot (G_{i-1}, P_{i-1})$$

$$c_i = G_i + P_ic_0$$

end

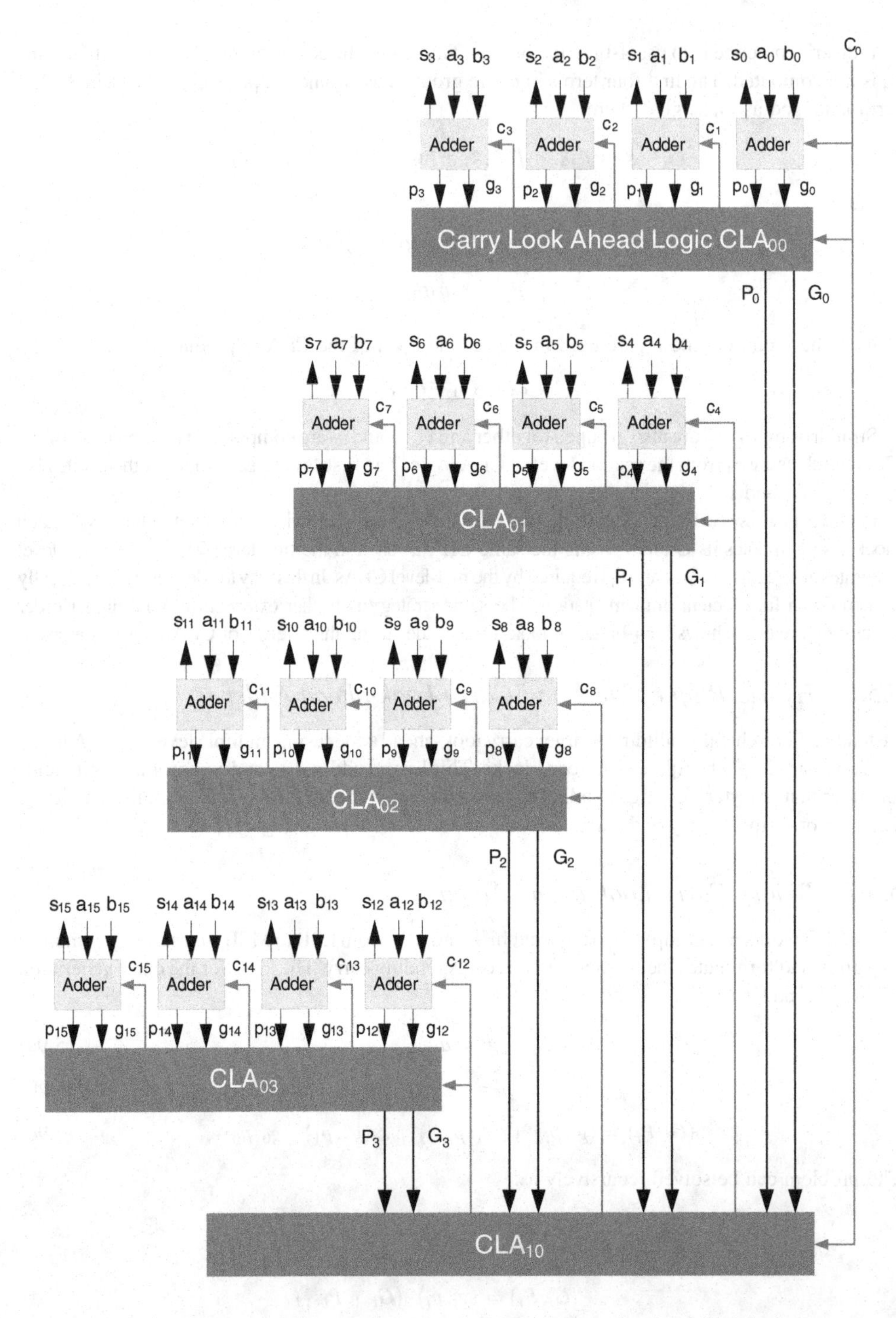

Figure 5.11 A 16-bit carry look-ahead adder using two levels of CLA logic

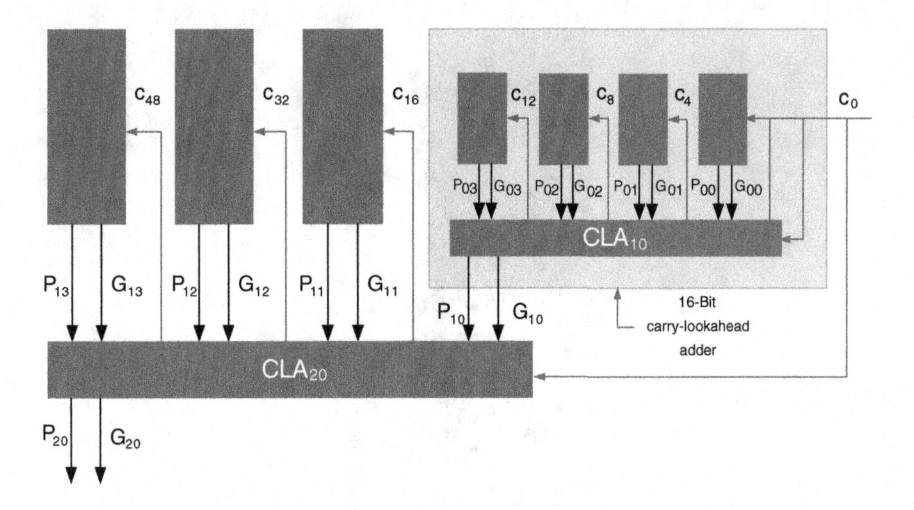

Figure 5.12 A 64-bit carry look-ahead adder using three levels of CLA logic

where the dot operator $\cdot$ is given as:

$$(G_i, P_i) = (g_i, p_i) \cdot (G_{i-1}, P_{i-1}) = (g_i + p_i G_{i-1}, p_i P_{i-1}) \tag{5.10}$$

There are several ways to logarithmically break the recursion into stages. The objective is to implement (5.9) for each bit position. Moving from LSB to MSB, this requires successive application of the $\cdot$ operator on two adjacent bit positions. For an N-bit adder, a serial implementation will require $N-1$ stages of operator implementation, as shown in Figure 5.14(a). Several optimized implementations are reported in the literature that improve on the serial realization [7, 8]. These are shown in Figures 5.14(b)-(e). That in (b) is referred to as a Brent–Kung Adder and results in a regular layout, shown in (f) where the tree depicted in black color is forward tree and calculates carry-out to the adder in minimum time. The RTL Verilog code and stimulus of a 16-bit linear BCLA is given here:

```
module BinaryCarryLookaheadAdder
# (parameter N = 16)
(input [N-1:0] a,b,
input c_in,
output reg [N-1:0] sum,
output reg c_out);
reg [N-1:0] p, g, P, G;
reg [N:0] c;
integer i;

always@(*)
begin
    for (i=0;i<N;i=i+1)
    begin
        // Generate all ps and gs
        p[i]= a[i] ^ b[i];
        g[i]= a[i] & b[i];
    end
end
```

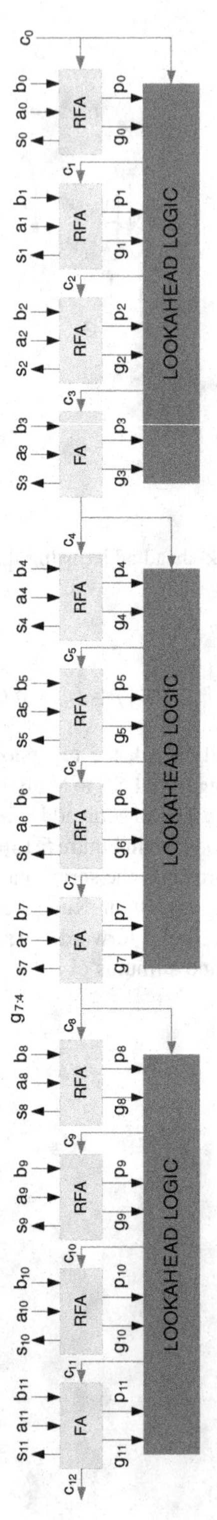

Figure 5.13 A 12-bit hybrid ripple carry and carry look-ahead adder

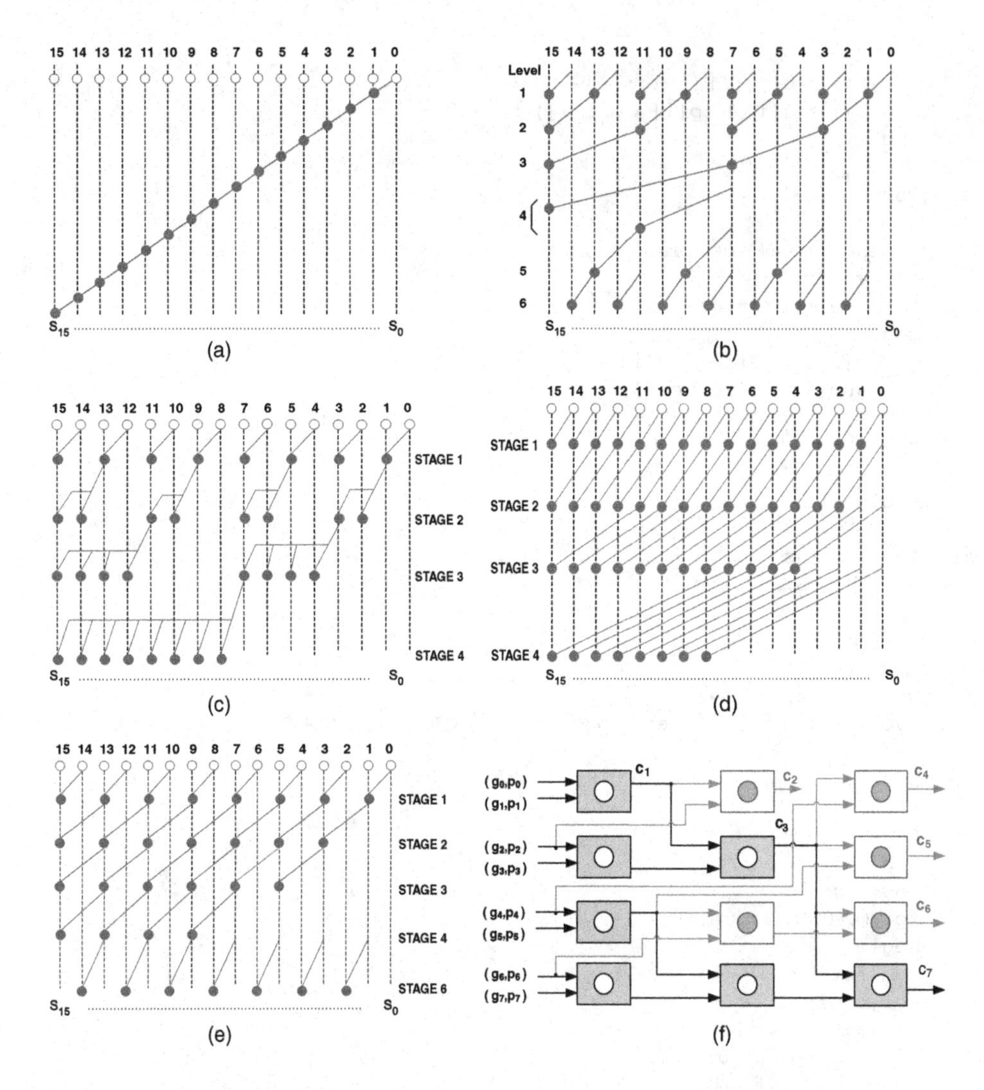

Figure 5.14 Binary carry look-ahead adder logic for generating all the carries in a tree of logic moving from LSB to MSB. (a) Serial implementation. (b) Brent–Kung adder. (c) Ladner–Fischer parallel prefix adder. (d) Kogge–Stone parallel prefix adder. (e) Han–Carlson parallel prefix adder. (f) Regular layout of an 8-bit Brent–Kung adder

```
always@(*)
begin
    // Linearly apply dot operators
    P[0] = p[0];
    G[0] = g[0];
    for (i=1; i<N; i=i+1)
    begin
```

```
        P[i]= p[i] & P[i-1];
        G[i]= g[i] | (p[i] & G[i-1]) ;
    end
end
always@ (*)
begin
    //Generate all carries and sum
    c[0]=c_in;
    for(i=0;i<N;i=i+1)
    begin
        c[i+1] = G[i] | (P[i] & c[0]);
        sum[i] = p[i] ^ c[i];
    end
    c_out = c[N];
end
endmodule

module stimulus;
reg [15:0] A, B;
reg CIN;
wire COUT;
wire [15:0] SUM;
integer i, j, k;
BinaryCarryLookaheadAdder #16 BCLA(A, B, CIN, SUM, COUT);
initial
begin
    A = 0;
    B = 0;
    CIN = 0;
    #5 A= 1;
    for (k=0; k<2; k=k+1)
    begin
        for (i=0; i<5; i=i+1)
        begin
            #5 A = A + 1;
            for (j=0; j<6; j=j+1)
                begin
                    #5 B = B+1;
                end
        end
        CIN=CIN+1;
    end
    #10 A = 16'hFFFF; B=0; CIN = 1;
end
initial
    $monitor ($time, " A=%d, B=%d, CIN=%b, SUM=%h, Error=%h,
    COUT=%b\n", A, B, CIN, SUM, SUM-(A+B+CIN), COUT);
endmodule
```

5.5.8 Carry Skip Adder

In an N-bit carry skip adder, N bits are divided into groups of k bits. The adder propagates all the carries simultaneously through the groups [9]. Each group i computes group P_i using the following relationship:

$$P_i = p_i p_{i+1} p_{i+2} \cdots p_{i+k-1}$$

where p_i is computed for each bit location i as:

$$p_i = a_i \oplus b_i$$

The strategy is that, if any group generates a carry, it passes it to the next group; but if the group does not generate its own carry owing to the arrangements of individual bits in the block, then it simply bypasses the carry from the previous block to its next block. This bypassing of a carry is handled by P_i. The carry skip adder can also be designed to work on unequal groups. A carry skip adder is shown in Figure 5.15.

For a 16-bit adder divided into groups of 4-bits each, the worst case is when the first group generates its carry-out and the next two subsequent groups due to the arrangements of bits do not generate their own carries but simply skip the carry from the first group to the last group. The last group makes use of this carry and then generates its own carry. The worst-case carry delay of a carry skip adder is less than the corresponding carry delay of an equal-width RCA.

5.5.9 Conditional Sum Adder

A conditional sum adder is implemented in multiple levels. At level 1, sum and carry bits for the input carry bits 0 and 1 are computed for each bit position:

$$s0_i = a_i \oplus b_i \tag{5.11a}$$

$$s1_i = a_i \sim \oplus b_i \tag{5.11b}$$

$$c0_i = a_i b_i \tag{5.11c}$$

$$c1_i = a_i + b_i \tag{5.11d}$$

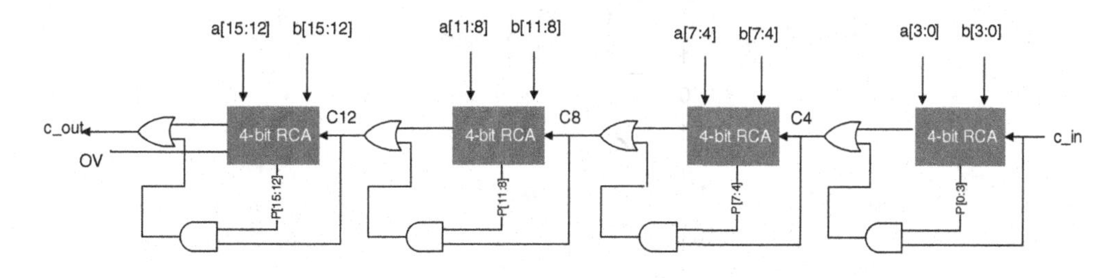

Figure 5.15 A 16-bit equal-group carry skip adder

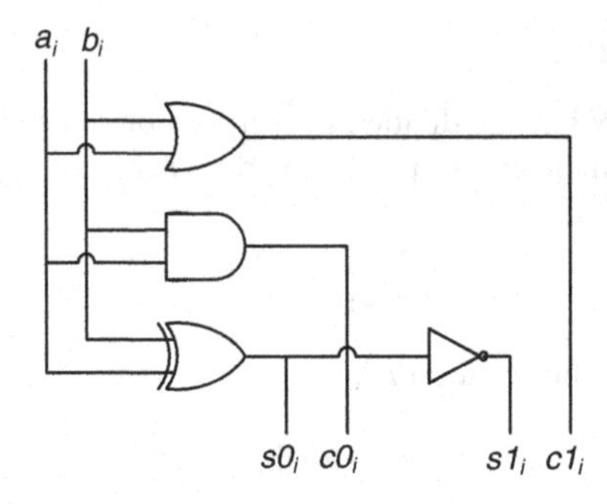

Figure 5.16 Conditional cell (CC)

Here, a_i and b_i are the ith bits of operands a and b, $s0_i$ and $c0_i$ are the sum and carry-out bits at location i calculated by assuming 0 as carry-in, and similarly $s1_i$ and $c1_i$ are the sum and carry-out bits at location i that are computed by taking carry-in as 1. These equations are implemented by a logic cell called a *condition cell* (CC). A representative implementation of a CC is shown in Figure 5.16.

At level 2, the results from level 1 are merged together. This merging is done by pairing up consecutive columns at level 1. For an N-bit adder, the columns are paired as $(i, i+1)$ for $i = 0, 1, \ldots, N-2$. The least significant carry (LSC) bits at bit location i (i.e $c0_i$ and $c1_i$) in each pair selects the sum and carry bits at bit location $i+1$ for the next level of processing. If the LSC bit is 0, then the bits computed for carry 0 are selected, otherwise the bits computed for carry 1 are selected for the next level. This level generates results as if a group of two bits are added assuming carry-in 0 and 1, as done in a carry select adder.

The example of Figure 5.17 illustrates the CSA by adding 3-bit numbers. In the first level each group consists of one bit each. The level adds bits at location i assuming carry-in of 0 and 1. In the next level, the three columns are divided into two groups. The column at bit location 0 forms the first group with the carry-in to the adder, and the rest of the two columns at bit location 1 and 2 form the second group. These two groups are separated with a dotted line in the figure. The selection of the

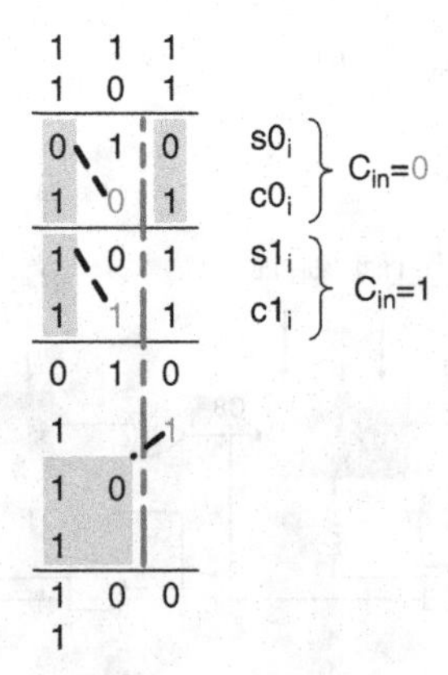

Figure 5.17 Addition of three bit numbers using a conditional sum adder

appropriate bits in each group by the LSC is shown with a diagonal line. The LSC from each group determines which one of the two bits in the next column will be brought down to the next level of processing. If the LSC is 0, the upper two bits of the next column are selected, otherwise the lower two bits are selected for the next level. For the second group, the sum bit of the first column is also dropped down to the next level. The LSCs in the first level are shown with bold fonts. Finally in the next level the two groups formed in the previous level are merged and the LSC selects one of the two 3-bit group to the next level. In this example, as the LSC is one, it selects the lower group.

Example: The conditional sum adder technique will be further illustrated using a 16-bit addition. Figure 5.18 shows all the steps in computing the sum of two 16-bit numbers. Figure 5.19 lays out the architecture of the adder. At level 0, 16 CCs compute sum and carry bits assuming carry-in 0 and 1.

		a_i	1	0	0	1	1	0	0	1	1	1	0	0	1	1	0	1	a_i
		b_i	0	0	1	1	0	1	1	0	1	1	0	1	0	1	1	0	b_i
Group width	Group carry-in	Group sum and block carry out																
			15	14	13	12	11	10	9	8	7	6	5	4	3	2	1	0	i
1	0		1	0	1	0	1	1	1	1	0	0	0	1	1	0	1	1	$s0_i$
			0	0	0	1	0	0	0	0	1	1	0	0	0	1	0	0	$c0_i$
	1		0	1	0	1	0	0	0	0	1	1	1	0	0	1	0		$s1_i$
			1	0	1	1	1	1	1	1	1	1	0	1	1	1	1		$c1_i$
2	0		1	0	0	0	1	1	1	1	1	0	0	1	0	0	1	1	
			0		1		0		0		1		0		1		0		
	1		1	1	0	1	0	0	0	0	1	1	1	0	0	1			
			0		1		1		1		1		0		1				
4	0		1	1	0	0	1	1	1	1	1	0	0	1	0	0	1	1	
			0				0				1				1				
	1		1	1	0	1	0	0	0	0	1	0	1	0					
			0				1				1								
8	0		1	1	0	0	1	1	1	1	1	0	1	0	0	0	1	1	
			0								1								
	1		1	1	0	1	0	0	0	0									
			0																
16	0		1	1	0	1	0	0	0	0	1	0	1	0	0	0	1	1	
			0																
	1																		

Figure 5.18 Example of a 16-bit conditional sum adder

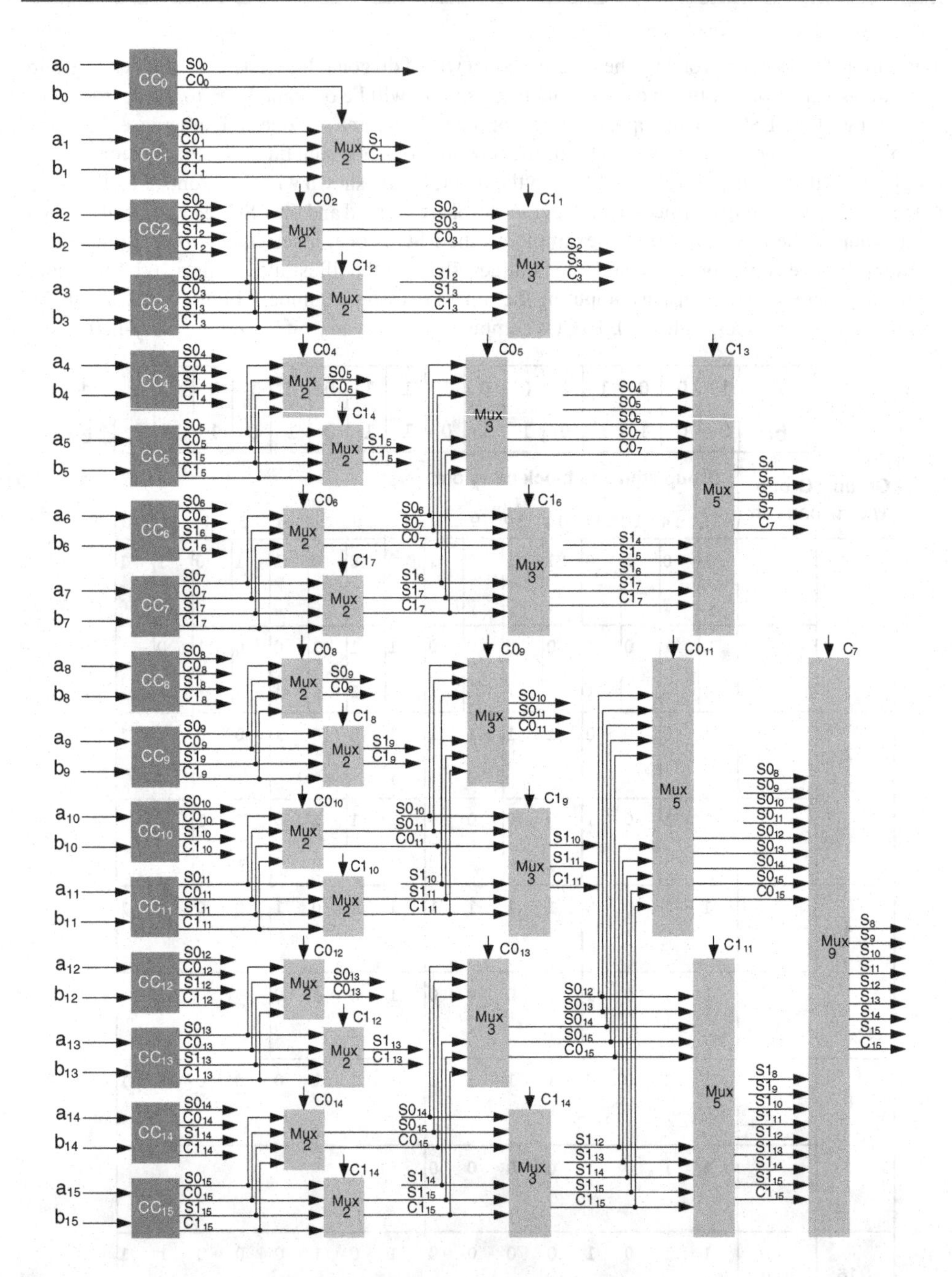

Figure 5.19 A 16-bit conditional sum adder

For this example the design assumes carry-in to the adder is 0, so CC_0 is just a half adder that takes LSBs of the inputs, a_0 and b_0 and only computes sum and carry bits $s0_0$ and $c0_0$ for carry-in 0. The rest of the bits require addition for both carry-in 0 and 1. CC_1 takes a_1 and b_1 as inputs and computes sum and carry bits for both carry-in 0 ($s0_1, c0_1$) and 1 ($s1_1, c1_1$). Similarly, for each bit location i (= 1, ..., 15), CC_i computes sum and carry for both carry-in 0 ($s0_i$, $c0_i$) and 1 ($s1_i$, $c1_i$). For the next level, two adjacent groups of bit locations i and $i + 1$ are merged to form a new group. The LSC in each group is used to select one of the pairs of its significant bits for the next level. The process of merging the groups and selecting bits in each group based on LSC is repeated $\log_2(16) = 4$ times to get the final answer.

Verilog code for an 8-bit conditional sum adder is given here:

```
// Module for 8-bit conditional sum adder
module conditional_sum_adder
#(parameter W = 8)
// Inputs declarations
(input [W-1:0] a, b,   // Two inputs a and b with a carry in cin
    input cin,
// Outputs declarations
output reg [W-1:0] sum,   // Sum and carry cout
output reg cout);
// Intermediate wires
wire s1_0, c2_0, s2_0, c3_0, s3_0, c4_0, s4_0, c5_0, s5_0, c6_0, s6_0,
         c7_0, s7_0, c8_0;
wire s1_1, c2_1, s2_1, c3_1, s3_1, c4_1, s4_1, c5_1, s5_1, c6_1, s6_1, c7_1,
         s7_1, c8_1;
// Intermediate registers
reg fcout;
reg s3_level_1_0, s3_level_1_1, s5_level_1_0, s5_level_1_1, s7_level_1_0,
               s7_level_1_1;
reg c4_level_1_0, c4_level_1_1, c6_level_1_0, c6_level_1_1, c8_level_1_0,
               c8_level_1_1;
reg c2_level_1;
reg c4_level_2;
reg s6_level_2_0, s6_level_2_1, s7_level_2_0, s7_level_2_1, c8_level_2_0,
               c8_level_2_1;

// Level 0
always @*
{fcout,sum[0]} = a[0] + b[0] + cin;
// Conditional cells instantiation
conditional_cell  c1( a[1], b[1], s1_0, s1_1, c2_0, c2_1);
conditional_cell  c2( a[2], b[2], s2_0, s2_1, c3_0, c3_1);
conditional_cell  c3( a[3], b[3], s3_0, s3_1, c4_0, c4_1);
conditional_cell  c4( a[4], b[4], s4_0, s4_1, c5_0, c5_1);
conditional_cell  c5( a[5], b[5], s5_0, s5_1, c6_0, c6_1);
conditional_cell  c6( a[6], b[6], s6_0, s6_1, c7_0, c7_1);
conditional_cell  c7( a[7], b[7], s7_0, s7_1, c8_0, c8_1);

// Level 1 muxes
always @*
    case(fcout) // For first mux
        1'b0: {c2_level_1, sum[1]} = {c2_0, s1_0};
```

```
   1'b1: {c2_level_1, sum[1]} = {c2_1, s1_1};
  endcase
always @* // For 2nd mux
    case(c3_0)
        1'b0: {c4_level_1_0, s3_level_1_0} = {c4_0, s3_0};
        1'b1: {c4_level_1_0, s3_level_1_0} = {c4_1, s3_1};
    endcase
always @* // For 3rd mux
    case(c3_1)
        1'b0: {c4_level_1_1, s3_level_1_1} = {c4_0, s3_0};
        1'b1: {c4_level_1_1, s3_level_1_1} = {c4_1, s3_1};
    endcase
always @* // For 4th mux
    case(c5_0)
        1'b0: {c6_level_1_0, s5_level_1_0} = {c6_0, s5_0};
        1'b1: {c6_level_1_0, s5_level_1_0} = {c6_1, s5_1};
    endcase
always @* // For 5th mux
    case(c5_1)
        1'b0: {c6_level_1_1, s5_level_1_1} = {c6_0, s5_0};
        1'b1: {c6_level_1_1, s5_level_1_1} = {c6_1, s5_1};
    endcase
always @* // For 6th mux
    case(c7_0)
        1'b0: {c8_level_1_0, s7_level_1_0} = {c8_0, s7_0};
        1'b1: {c8_level_1_0, s7_level_1_0} = {c8_1, s7_1};
    endcase
always @* // For 7th mux
    case(c7_1)
        1'b0: {c8_level_1_1, s7_level_1_1} = {c8_0, s7_0};
        1'b1: {c8_level_1_1, s7_level_1_1} = {c8_1, s7_1};
    endcase

// Level 2 muxes
always @* // First mux of level2
    case(c2_level_1)
        1'b0: {c4_level_2, sum[3], sum[2]} = {c4_level_1_0,
                              s3_level_1_0, s2_0};
        1'b1: {c4_level_2, sum[3], sum[2]} = {c4_level_1_1,
                              s3_level_1_1, s2_1};
    endcase
always @* // 2nd mux of level2
    case(c6_level_1_0)
        1'b0: {c8_level_2_0, s7_level_2_0, s6_level_2_0}
                             = {c8_level_1_0, s7_level_1_0, s6_0};
        1'b1: {c8_level_2_0, s7_level_2_0, s6_level_2_0}
                              ={c8_level_1_1, s7_level_1_1, s6_1};
    endcase
always @* // 3rd mux of level2
    case(c6_level_1_1)
        1'b0: {c8_level_2_1, s7_level_2_1, s6_level_2_1}
                              ={c8_level_1_0, s7_level_1_0, s6_0};
        1'b1: {c8_level_2_1, s7_level_2_1, s6_level_2_1}
                              ={c8_level_1_1, s7_level_1_1, s6_1};
```

```
    endcase

// Level 3 mux
always @*
    case(c4_level_2)
        1'b0: {cout,sum[7:4]} = {c8_level_2_0, s7_level_2_0,
                             s6_level_2_0, s5_level_1_0, s4_0};
        1'b1: {cout,sum[7:4]} = {c8_level_2_1, s7_level_2_1,
                             s6_level_2_1, s5_level_1_1, s4_1};
    endcase

endmodule

// Module for conditional cell
module conditional_cell(a, b, s_0, s_1, c_0, c_1);
input a,b;
output s_0, c_0, s_1, c_1;
assign s_0 = a^b;                  // sum with carry in 0
assign c_0 = a&b;                  // carry with carry in 0
assign s_1 = ~s_0;                  // sum with carry in 1
assign c_1 = a | b;                // carry with carry in 1
endmodule
```

5.5.10 Carry Select Adder

The carry select adder (CSA) is not as fast as the carry look-ahead adder and requires considerably more hardware if mapped on custom ASICs, but it has a favorable design for mapping on an FPGA with fast carry chain logic. Te CSA partitions an N-bit adder into K groups, where:

$$k = 0, 1, 2, \ldots, K-1 \tag{5.12a}$$

$$n_0 + n_1 + \ldots + n_{k-1} = N \tag{5.12b}$$

$$n_0 \leq n_1 \leq \ldots \leq n_{K-1} \tag{5.12c}$$

where n_k represents the number of bits in group k. The basic idea is to place two n_k-bit adders at each stage k. One set of adders computes the sum by assuming a carry-in 1, and the other a carry-in 0. The actual sum and carry are selected using a 2-to-1 MUX (multiplexer) based on the carry from the previous group.

Figure 5.20 shows a 16-bit carry select adder. The adder is divided into four groups of 4-bits each. As each block is of equal width, their outputs will be ready simultaneously. In an unequal-width CSA the block size at any stage in the adder is set larger than the block size at its less significant stage. This helps in reducing delay further as the carry-outs in less significant stages are ready to select the sum and carry of their respective next stages.

The CSA can be broken down into more than one stage. Figure 5.21 shows a two-stage CSA. The N-bit adder is divided into two groups of size $N/2$ bits. Each stage is further divided into two subgroups of $N/4$ bits. In the first stage each subgroup computes sums and carries for carry-in 0 and 1 and two subgroups are merged inside a group. Then, in the second stage, two groups are merged and final sums and carry-out are generated. Interestingly if we keep breaking the groups into more levels until each group contains one bit each, the adder architecture will be the same as for a conditional sum adder. This reveals that a conditional sum adder is a special case of a hierarchical CSA. The RTL Verilog code for a 16-bit hierarchical CSA is given here:

```
module HierarchicalCSA(a, b, cin, sum, c_out);
input [15:0] a,b;
```

```
input cin;
output c_out;
output [15:0] sum;
wire c4, c8, c8_0, c8_1, c12_0, c12_1, c16_0, c16_1, c16L2_0, c16L2_1;
wire [15:4] sumL1_0, sumL1_1;
wire [15:12] sumL2_0, sumL2_1;

// Level one of hierarchical CSA
assign {c4,sum[3:0]} = a[3:0] + b[3:0] + cin;
assign {c8_0, sumL1_0[7:4]}= a[7:4] + b[7:4] + 1'b0;
assign {c8_1, sumL1_1[7:4]}= a[7:4] + b[7:4] + 1'b1;
assign {c12_0,sumL1_0[11:8]}= a[11:8] + b[11:8] + 1'b0;
assign {c12_1,sumL1_1[11:8]}= a[11:8] + b[11:8] + 1'b1;
assign {c16_0, sumL1_0[15:12]}= a[15:12] + b[15:12] + 1'b0;
assign {c16_1, sumL1_1[15:12]}= a[15:12] + b[15:12] + 1'b1;

// Level two of hierarchical CSA
assign c8 = c4 ? c8_1 : c8_0;
assign sum[7:4] = c4 ? sumL1_1[7:4]: sumL1_0[7:4];

// Selecting sum and carry within a group
assign c16L2_0 = c12_0 ? c16_1 : c16_0;
assign sumL2_0 [15:12] = c12_0? sumL1_1[15:12] : sumL1_0[15:12];
assign c16L2_1 = c12_1 ? c16_1 : c16_0;
assign sumL2_1 [15:12] = c12_1? sumL1_1[15:12]: sumL1_0[15:12];

// Level three selecting the final outputs
assign c_out = c8 ? c16L2_1 : c16L2_0;
assign sum[15:8] = c8 ? {sumL2_1[15:12], sumL1_1[11:8]} :
                        {sumL2_0[15:12], sumL1_0[11:8]};
endmodule
```

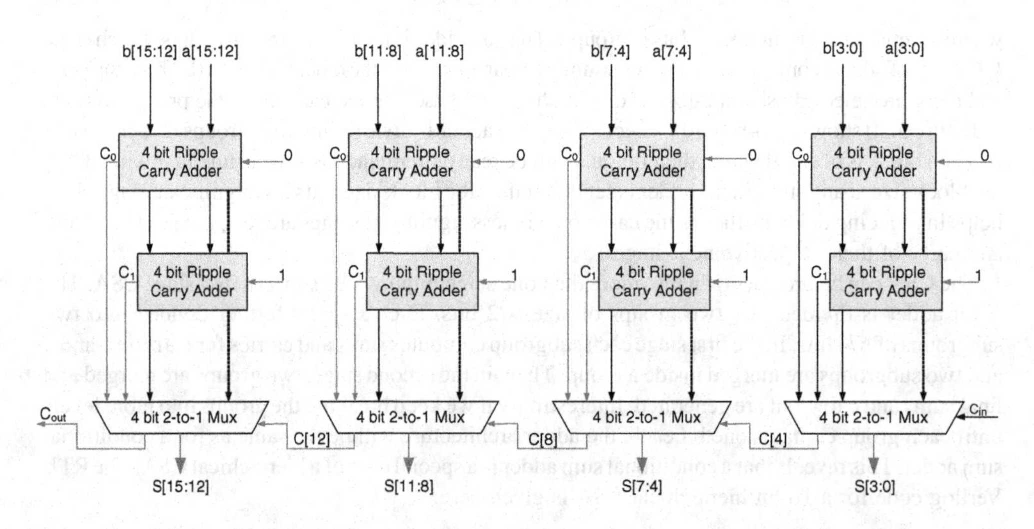

Figure 5.20 A 16-bit uniform-groups carry select adder

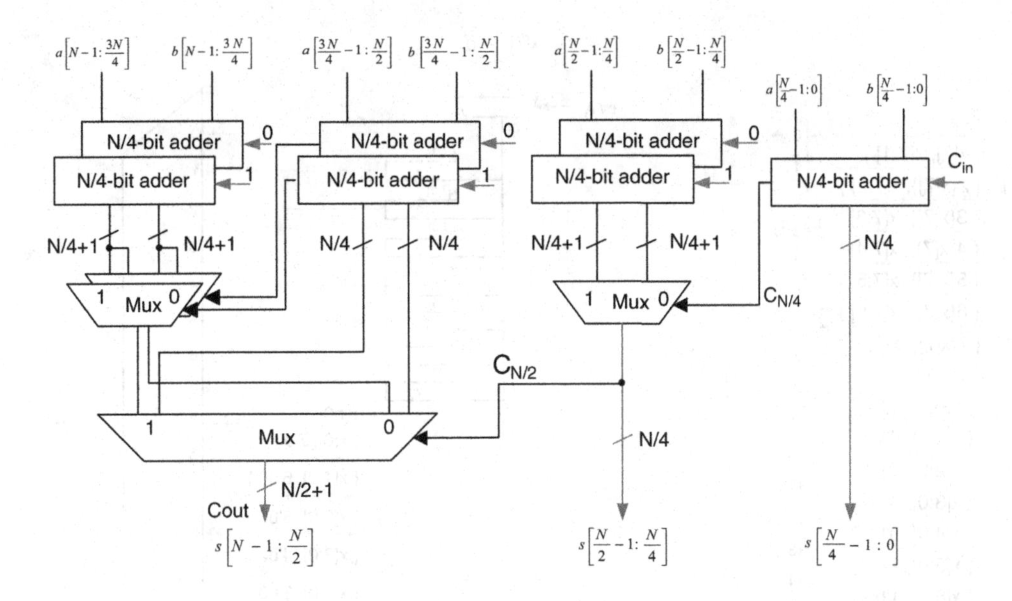

Figure 5.21 Hierarchical CSA

5.5.11 Using Hybrid Adders

A digital designer is always confronted with finding the best design option in area–power–time tradeoffs. A good account of these tradeoffs is given in [10]. The designer may find it appropriate to divide the adder into multiple groups and use different adder architecture for each group. This may help the designer to devise an optimal adder architecture for the design. This design methodology is discussed in [10] and [11].

5.6 Barrel Shifter

A single-cycle N-bit logic shifter implementing $x \gg s$, where s is a signed integer number, can be implemented by hardwiring all the possible shift results as input to a multiplexer and then using s to select the appropriate option at the output. The shifter performs a shift left operation for negative values of s. For example, $x \gg -2$ implies a shift left by 2.

The design of the shifter is shown in Figure 5.22(a), where x is the input operand and all possible shifts are pre-performed as input to the MUX and s is used as a select line to the MUX. The figure clearly shows that, for negative values of s, its equivalent positive number will be used by the MUX for selecting appropriate output to perform a shift left operation.

The design can be trivially extended to take care of arithmetic along with the logic shifts. This requires first selecting either the sign bit or 0 for appropriately appending to the left of the operand for shift right operation. For shift left operation, the design for both arithmetic and logic shift is same. When there are not enough redundant sign bits, the shift left operation results in overflow. The design of an arithmetic and logic shifter is given in Figure 5.22(b).

Instead of using one MUX with multiple inputs, a barrel shifter can also be hierarchically constructed. This design can also be easily pipelined. The technique can work for right as well as for left shift operations. For $x \gg s$, the technique works by considering s as a two's complement signed number where the sign bit has negative weight and the rest of the bits carry positive weights. Moving

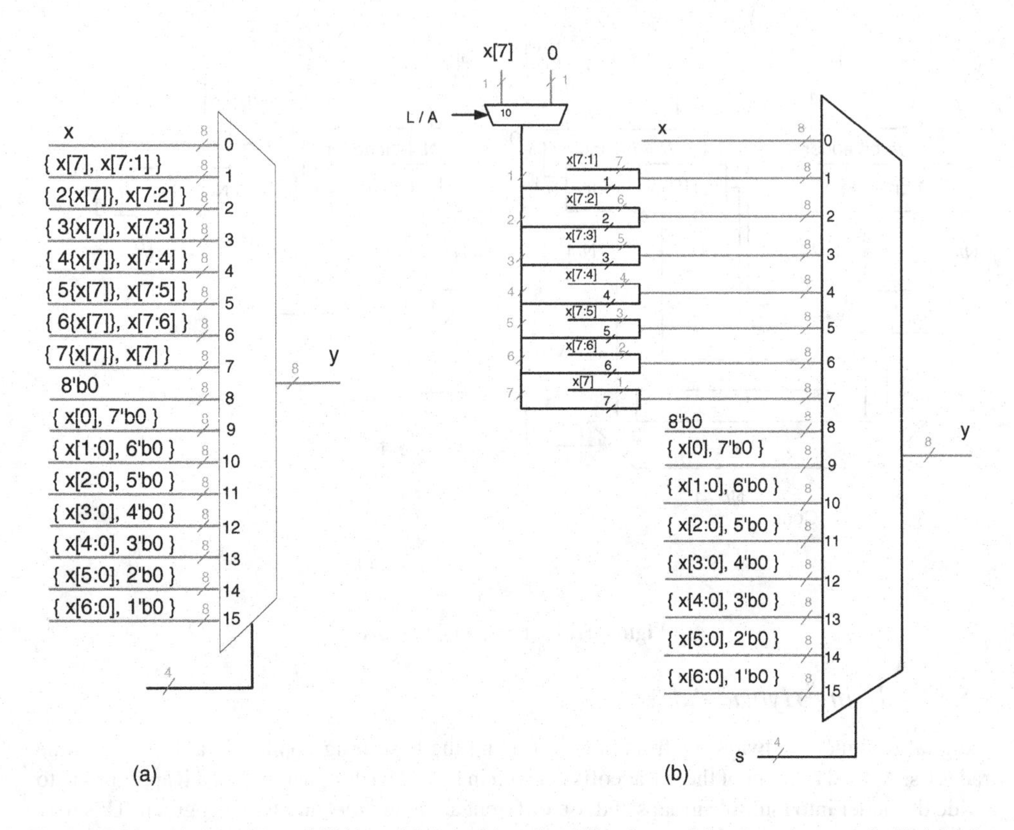

Figure 5.22 Design of Barrel Shifters (a) An arithmetic shifter for an 8-bit signed operand. (b) A logic and arithmetic shifter for an 8-bit operand

from MSB to LSB, each stage of the barrel shifter only caters for one bit and performs the requisite shift equal to the weight of the bit.

Figure 5.23 shows a barrel shifter for shifting a 16-bit number x by a 5-bit signed number s. Thus the barrel shifter can perform shifts from 0 to 15 to the right and 1 to 16 to the left in five stages or levels. First the shifter checks whether the logic or arithmetic shift is required and appropriately selects 0 or the sign bit of the operand for further appending for shift right operation. The barrel shifter then checks the MSB of s, as this bit has negative weight. Therefore, if $s[4]=1$, the shifter performs a shift left by 16 and keeps the result as a 31-bit number; otherwise, for $s[4]=0$, the number is appropriately extended to a 31-bit number for the next levels to perform appropriate shifts. For the rest of the bits the logic performs a shift right operation equal to the weight of the bit under consideration, and the design keeps reducing the number of bits to the width required at the output of each stage. The design of this hierarchical barrel shifter is shown in Figure 5.23(a) and its RTL Verilog code is given here:

```
module BarrelShifter(
input [15:0] x,
input signed [4:0] s,
input A_L,
output reg [15:0]y);
reg [30:0] y0;
```

```
reg [22:0] y1;
reg [18:0] y2;
reg [16:0] y3;
reg [14:0] sgn;

always @(*)
begin
// A_L =1 for Arithmetic and 0 for Logical shift
     sgn = (A_L) ? {15{x[15]}} : 15'b0;
     y0 = (s[4]) ? {x[14:0],16'b0} : {sgn[14:0], x[15:0]};
     y1 = (s[3]) ? y0[30:8] : y0[22:0];
     y2 = (s[2]) ? y1[22:4] : y1[18:0];
     y3 = (s[1]) ? y2[18:2] : y2[16:0];
     y = (s[0]) ? y3[16:1] : y3[15:0];
end
endmodule
```

This design can be easily pipelined by placing registers after any number of multiplexers and appropriately delaying the respective selection bit for coherent operation. A fully pipelined design of

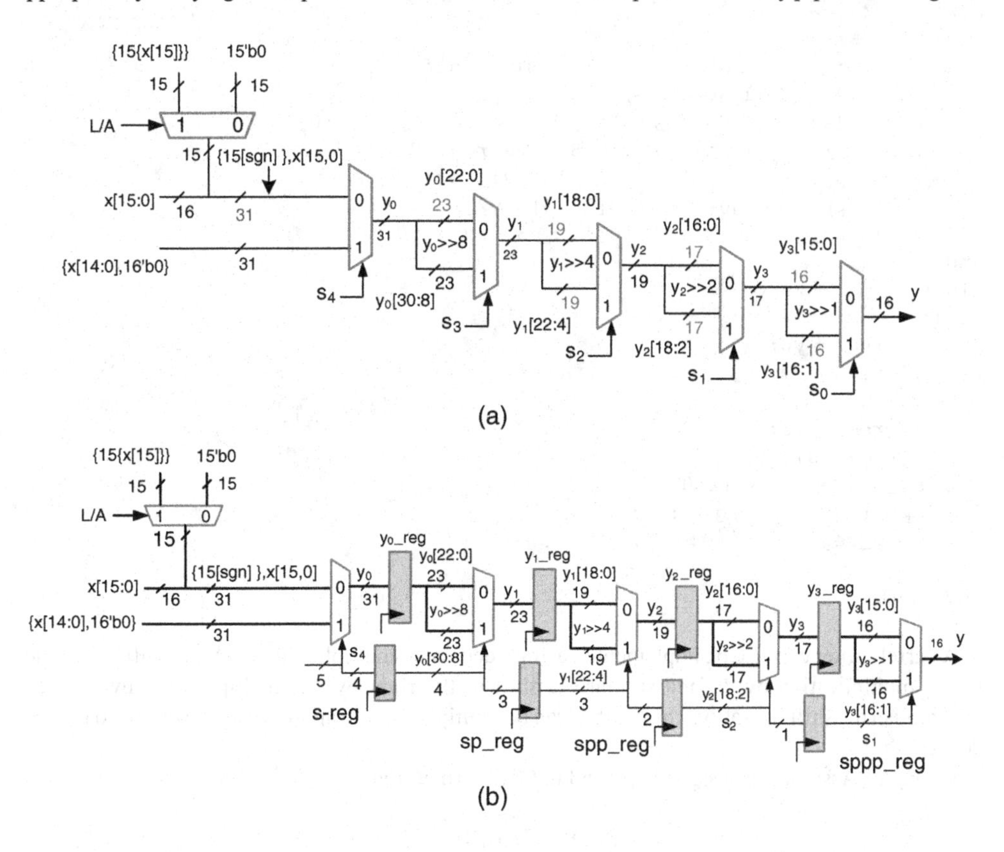

Figure 5.23 Design of a barrel shifter performing shifts in multiple stages. (a) Single-cycle design. (b) Pipelined design

a barrel shifter is given in Figure 5.23(b), and the code for RTL Verilog implementation of the design is listed here:

```
module BarrelShifterPipelined(
input clk,
input [15:0] x,
input signed [4:0] s,
input A_L,
output reg [15:0]y);
reg [30:0] y0, y0_reg;
reg [22:0] y1, y1_reg;
reg [18:0] y2, y2_reg;
reg [16:0] y3, y3_reg;
reg [14:0] sgn;
reg [3:0] s_reg;
reg [2:0] sp_reg;
reg [1:0] spp_reg;
reg sppp_reg;

always @(*)
begin
// A_L =1 for arithmetic and 0 for logical shift
    sgn = (A_L) ? {15{x[15]}} : 15'b0;
    y0 = (s[4]) ? {x[14:0],16'b0} : {sgn[14:0], x[15:0]};
    y1 = (s_reg[3]) ? y0_reg[30:8] : y0_reg[22:0];
    y2 = (sp_reg[2]) ? y1_reg[22:4] : y1_reg[18:0];
    y3 = (spp_reg[1]) ? y2_reg[18:2] : y2_reg[16:0];
    y = (sppp_reg) ? y3_reg[16:1] : y3_reg[15:0];
end
always @ (posedge clk)
begin
    y0_reg <= y0;
    y1_reg <= y1;
    y2_reg <= y2;
    y3_reg <= y3;
    s_reg <= s[3:0];
    sp_reg <= s_reg [2:0];
    spp_reg <= sp_reg [1:0];
    sppp_reg <= spp_reg [0];
end
endmodule
```

A barrel shifter can also be implemented using a dedicated multiplier in FPGAs. A shift by s to the left is multiplication by 2^s, and similarly a shift to the right by s is multiplication by 2^{-s}. To accomplish the shift operation, the number can be multiplied by an appropriate power of 2 to get the desired shift.

Example: Assume x is a signed operand in Q1.7 format. Let:

$$x = 8'1001\_0101$$

$x \gg 3$ can be performed by multiplying x by $y = 2^{-3}$, which in Q1.7 format is equivalent to $y = 8'b0001\_0000$. The fractional multiplication of x by y results in a Q2.14 format number:

14′b1111_1001_0101_0000. On dropping the redundant sign bit and 7 LSBs, as is done in fractional signed multiplication, the shift results in an 8-bit number:

$$z = 8'b1111\_0010$$

Following the same rationale, both arithmetic and logic shifts can be performed by appropriately assigning the operands to a dedicated multiplier embedded in FPGAs.

5.7 Carry Save Adders and Compressors

5.7.1 Carry Save Adders

This chapter has discussed various architectures for adding numbers. These adders, while adding two operands, propagate carries from one bit position to the next in computing the final sum and are collectively known as *carry propagate adders* (CPAs). Although three numbers can be added in a single cycle by using two CPAs, a better option is to use a carry save adder (CSA) that first reduces the three numbers to two and then any CPA adds the two numbers to compute the final sum. From the timing and area perspective, the CSA is one of the most efficiently and widely used techniques for speeding up digital designs of signal processing systems dealing with multiple operands for addition and multiplication. Several dataflow transformations are reported that extract and transform algorithms to use CSAs in their architectures for optimal performance [13].

A CSA, while reducing three operands to two, does not propagate carries; rather, a carry is saved to the next significant bit position. Thus this addition reduces three operands to two without carry propagation delay. Figure 5.24 illustrates addition of three numbers. As this addition reduces three numbers to two numbers, the CSA is also called a 3:2 *compressor.*

5.7.2 Compression Trees

There are a host of techniques that effectively use CSAs to add more than three operands. These techniques are especially attractive for reducing the partial products in multiplier design. Section 5.8 covers the use of these techniques in multiplier architectures, and then mentions their use in optimizing many signal processing architectures and applications.

5.7.3 Dot Notation

Dot notation is used to explain different reduction techniques. In this notation each bit in a multi-bit operands addition is represented by a dot. Figure 5.25 shows four dots to represent four bits of PP[0] in a 4 × 4-bit multiplier. The dots appearing in a column are to be added for computing the final

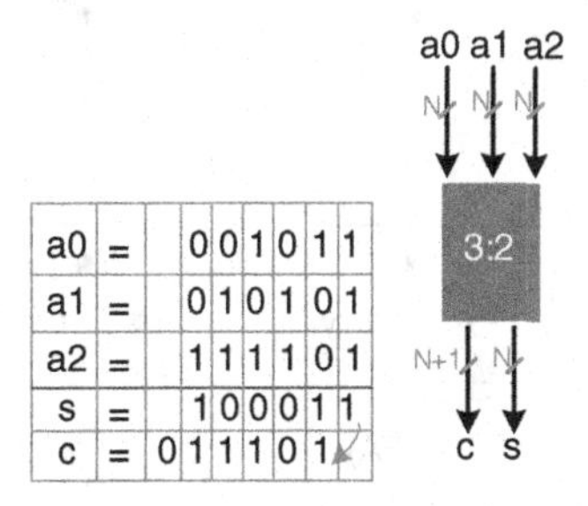

Figure 5.24 Carry save addition saves the carry at the next bit location

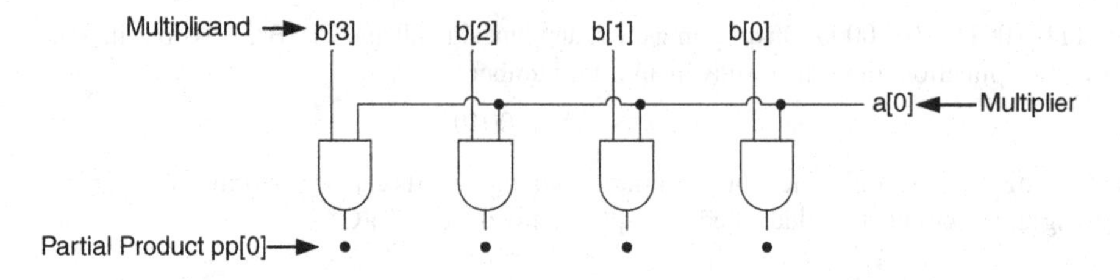

Figure 5.25 Dots are used to represent each bit of the partial product

product. A reduction technique based on a 3:2 compressor reduces three layers of PPs to two. The technique considers the number of dots in each column. If an isolated dot appears in a column, it is simply dropped down to the next level of logic. When there are two dots in a column, then they are added using a half adder, where the dot for the sum is simply dropped down in the same column and the dot for the carry is placed in the next significant column. Use of a half adder in this reduction is also known as a 2:2 compressor. The three dots in a column are reduced to two using a full adder. The dot for the sum is placed at the same bit location and the dot for the carry is moved to the next significant bit location. Figure 5.26 shows these three cases of dot processing.

5.8 Parallel Multipliers

5.8.1 Introduction

Most of the fundamental signal processing algorithms use multipliers extensively. Keeping in perspective their importance, the FPGA vendors are embedding many dedicated multipliers. It is still very important to understand the techniques that are practised for optimizing the implementation of these multipliers. Although a sequential multiplier can be designed that takes multiple cycles to compute its product, multiplier architectures that compute the product in one clock cycle are of interest to designers for high-throughput systems. This section discusses parallel multipliers.

A CSA is one of the fundamental building blocks of most parallel multiplier architectures. The partial products are first reduced to two numbers using a CSA tree. These two numbers are then added to get the final product. Many FPGAs have many dedicated summers. The summer can add three operands. The reduction trees can reduce the number of partial products to three instead of two to make full use of this block.

Any parallel multiplier architecture consists of three basic operations: partial product generation, partial product reduction, and computation of the final sum using a CPA, as depicted in Figure 5.27. For each of these operations, several techniques are used to optimize the HW of a multiplier.

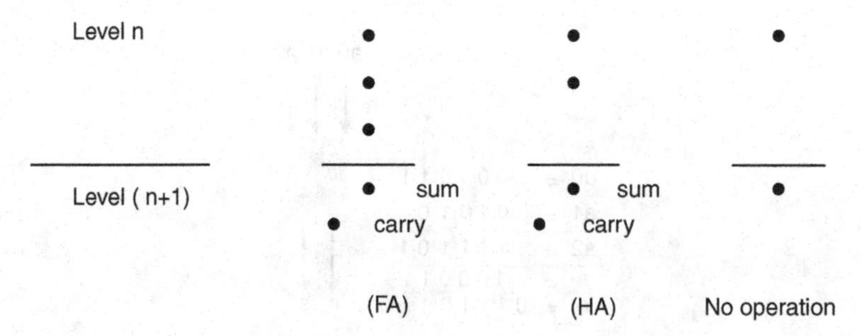

Figure 5.26 Reducing the number of dots in a column

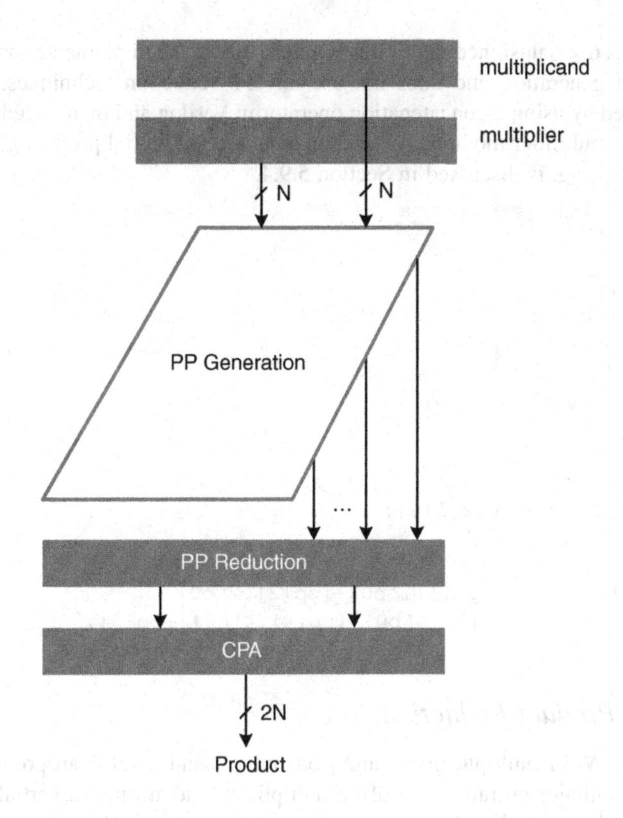

Figure 5.27 Three components of a multiplier

5.8.2 Partial Product Generation

While multiplying two N-bit unsigned numbers a and b, the partial products (PPs) are generated either by using an ANDing method or by implementing a modified Booth recoding algorithm. The first method generates partial product $PP[i]$ by ANDing each bit a_i of the multiplier with all the bits of the multiplicand b. Figure 5.28 shows the generation of partial products for a 6×6 multiplier. Each $PP[i]$ is shifted to the left by i bit positions before the partial products are added column-wise to produce the final product.

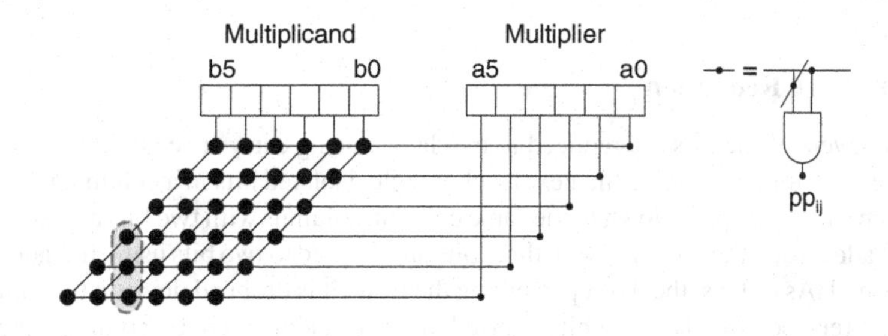

Figure 5.28 Partial-product generation for a 6×6 multiplier

Verilog code for a 6 × 6 unsigned multiplier is given below. The implementation only highlights the partial product generation and does not use any PP reduction techniques. These PPs are appropriately shifted by using a concatenation operator in Verilog and then added to complete the functionality of the multiplier module. The second technique of partial product generation, called modified Booth recoding, is discussed in Section 5.9.4.

```
module multiplier
(
input [5:0] a,b,
output [11:0] prod);
integer i;
reg [5:0] pp [0:5];//6 partial products
always@*
begin
    for(i=0; i<6; i=i+1)
        begin
            pp[i] = b & {6{a[i]}};
        end
end
    assign prod = pp[0]+{pp[1],1'b0}+{pp[2],2'b0} +
                        {pp[3],3'b0}+{pp[4],4'b0}+{pp[5],5'b0};
endmodule
```

5.8.3 Partial Product Reduction

While multiplying an N_1-bit multiplier a with an N_2-bit multiplicand b, N_1 PPs are produced by ANDing each bit $a[i]$ of the multiplier with all the bits of the multiplicand and shifting the partial product $PP[i]$ to the left by i bit positions. Using dot notations to represent bits, all the partial products form a parallelogram array of dots. These dots in each column are to be added to compute the final product.

For a general $N_1 \times N_2$ multiplier, the following four techniques are generally used to reduce N_1 layers of the partial products to two layers for their final addition using any CPA:

- carry save reduction
- dual carry save reduction
- Wallace tree reduction
- Dadda tree reduction.

Although the techniques are described here for 3:2 compressors, the same can be easily extended for other compressors and counters.

5.8.3.1 Carry Save Reduction

The first three layers of the PPs are reduced to two layers using carry save addition (CSA). In this reduction process, while generating the next level of logic, isolated bits in a column, in the selected three layers, are simply dropped down to the same column, columns with two bits are reduced to two bits using half adders and the columns with three bits are reduced to two bits using full adders. While adding bits using HAs or FAs, the dot representing the sum bit is dropped down in the same column whereas the dot representing the carry bit is placed in the next significant bit column. Once the first three partial products are reduced to two layers, the fourth partial product in the original composition is grouped with them to make a new group of three layers. These three layers are again reduced to two layers using the CSA technique. The process is repeated until the entire array is reduced to two layers of numbers. This scheme results in a few least significant product bits (called free product bits), and the rest of the bits appear in two layers, which are then added using any CPA to get the rest of the product bits.

Figure 5.29 shows carry save reduction scheme for reducing twelve PPs resulting from multiplication of two 12-bit numbers to two layers. At level 0, the first three PPs are selected. The first and the last columns in this group contain only 1 bit each. These bits are simply dropped down to level 1 without performing any operation. The two bits of the second and the second last columns are reduced using HAs. As the rest of the columns contain three bits each, FAs are used to reduce these bits to 2 bits each, where sum bits are placed in the same columns and carry bits are placed in the next columns of level 1. This placement of carries in the next columns is shown by diagonal lines. The use of an HA is shown by a cross line. In level 1, two least significant final product bits are produced (free product bits). In level 1 the two layers produced by reduction of level 0 are further grouped with the fourth PP and the process of reduction is repeated. For 12 layers of PPs, it requires 10 levels to reduce the PPs to two layers.

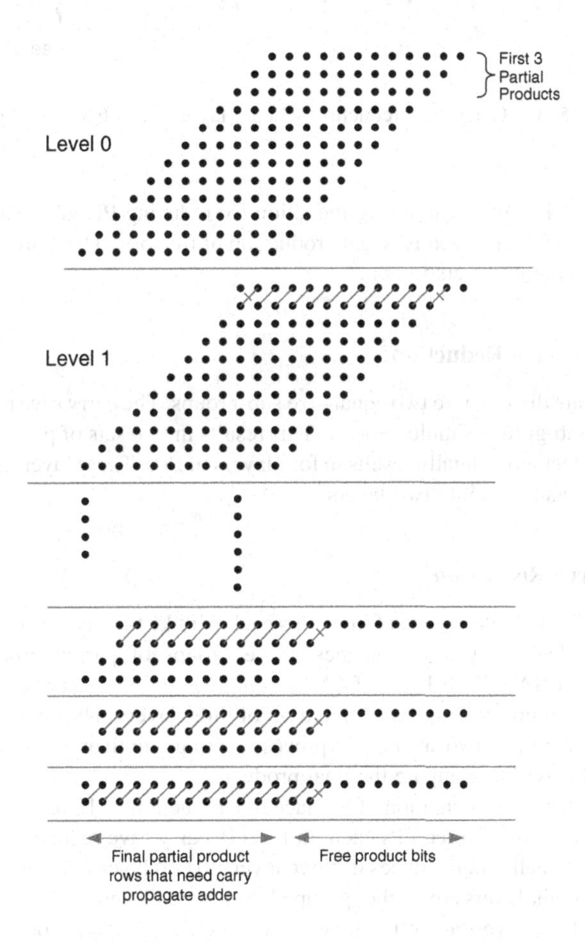

Figure 5.29 PP reduction for a 12×12 multiplier using a carry save reduction scheme

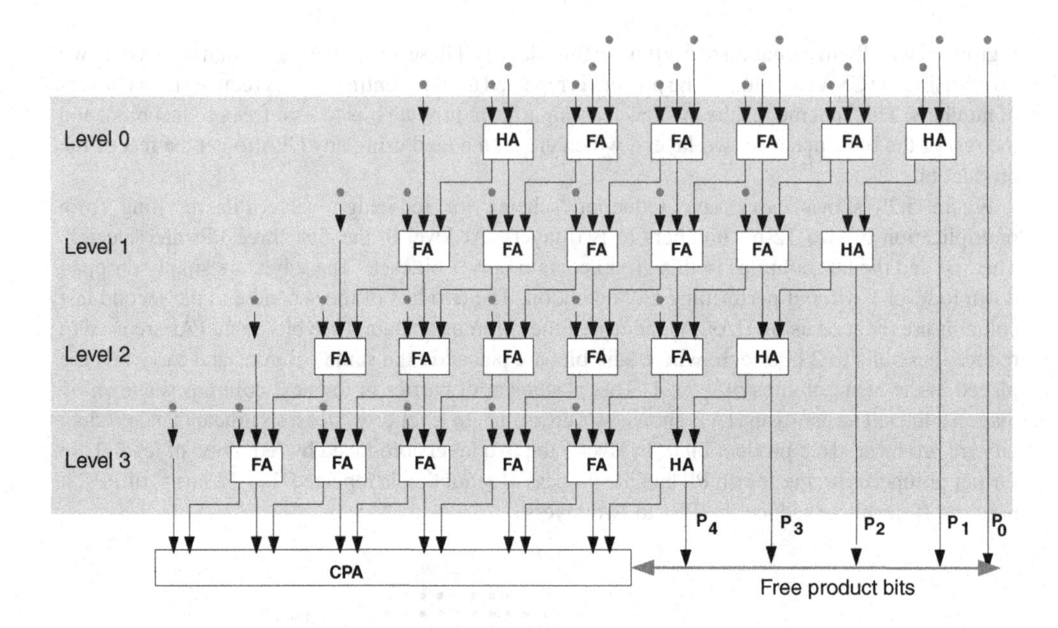

Figure 5.30 Carry save reduction scheme layout for a 6 × 6 multiplier

Figure 5.30 shows a layout of carry save reduction for reducing PPs of a 6 × 6 multiplier. The layout clearly shows use of HAs and FAs and production of free bits. There are four levels of logic and each level reduces three layers to two.

5.8.3.2 Dual Carry Save Reduction

The partial products are divided into two equal-size sub-groups. The carry save reduction scheme is applied on both the sub-groups simultaneously. This results in two sets of partial product layers in each sub-group. The technique finally results in four layers of PPs. These layers are then reduced as one group into three, and then into two layers.

5.8.3.3 Wallace Tree Reduction

Partial products are divided into groups of three PPs each. Unlike the linear time array reduction of the carry save and dual carry save schemes, these groups of partial products are reduced simultaneously using CSAs. Each layer of CSAs compresses three layers to two layers. These two layers from each group are re-grouped into a set of three layers. The next level of logic again reduces three-layer groups into two layers. This process continues until only two rows are left. At this stage any CPA can be used to compute the final product.

Figure 5.31 shows the implementation of Wallace tree reduction of 12 partial products. The PPs are divided into four groups of three PPs each. In level 0, carry save reduction is applied on each group simultaneously. Each group reduces its three layers to two layers, and as a result eight layers are produced. These eight layers are further grouped into three PPs each; this forms two groups of three layers each, and two layers are left as they are. In level 1, these two groups of three layers are

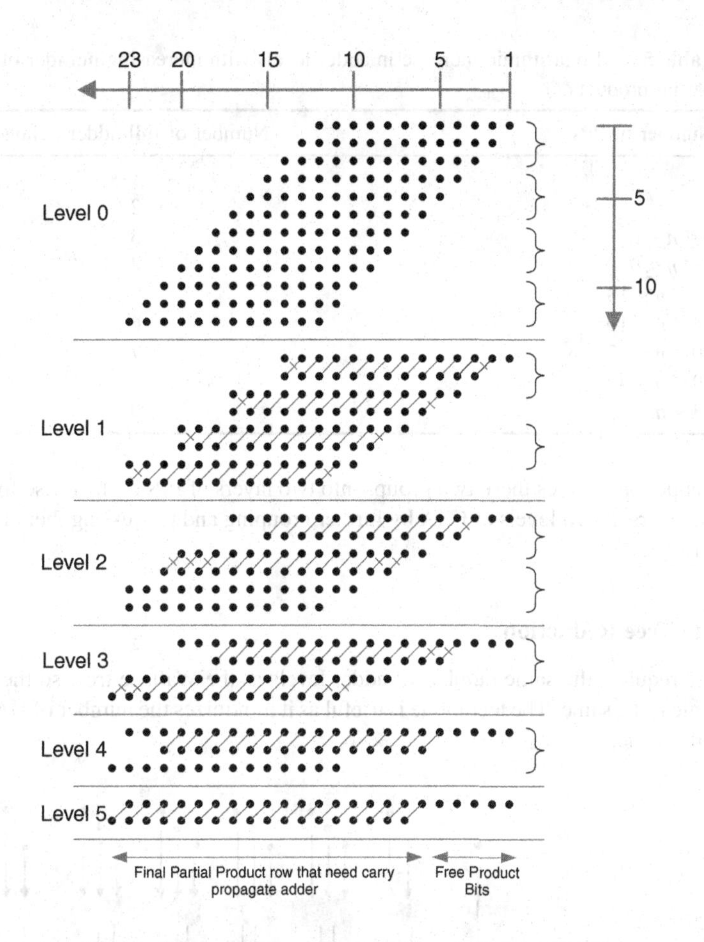

Figure 5.31 Wallace reduction tree applied on 12 PPs

further reduced using carry save addition, and this produces four layers. These four layers are combined with the two layers that are not processed in the previous level. Now in level 2 there are six layers; they form two groups of three layers and are reduced to four layers. Only one group is formed at this stage. Repeating the reduction process twice will reduce the number of layers to three and then two. The final two layers can then be added using any CPA to produce the final product.

Wallace reduction is one of the most commonly used schemes in multiplier architecture. It falls into the category of *log time array multiplier* as the reduction is performed in parallel in groups of threes and this results only in a logarithmic increase in the number of adder levels as the number of PPs increases (i.e. the size of the multiplier increases). The number of adder levels accounts for the critical path delay of the combinational cloud. Each adder level incurs one FA delay in the path. Table 5.3 shows the logarithmic increase in the adder levels as the number of partial products increases. As the Wallace reduction works on groups of three PPs, the adder levels are the same for a range of number of PPs. For example, if the number of PPs is five or six, it will require three adder levels to reduce the PPs to two layers for final addition.

Example: Figure 5.32 shows the layout of the Wallace reduction scheme on PP layers for a 6×6 multiplier of the reduction logic. In level 0, the six PPs are divided into two groups of three PPs each.

Table 5.3 Logarithmic increase in adder levels with increasing number of partial products

Number of PPs	Number of full-adder delays
3	1
4	2
$5 \leq n \leq 6$	3
$7 \leq n \leq 9$	4
$10 \leq n \leq 13$	5
$14 \leq n \leq 19$	6
$20 \leq n \leq 28$	7
$29 \leq n \leq 42$	8
$43 \leq n \leq 63$	9

The Wallace reduction reduces these two groups into two layers of PPs each. These four layers are reduced to two layers in two levels of CSA by further grouping and processing them through carry save addition twice.

5.8.3.4 Dadda Tree Reduction

The Dadda tree requires the same number of adder levels as the Wallace tree, so the critical path delay of the logic is the same. The technique is useful as it minimizes the number of HAs and FAs at each level of the logic.

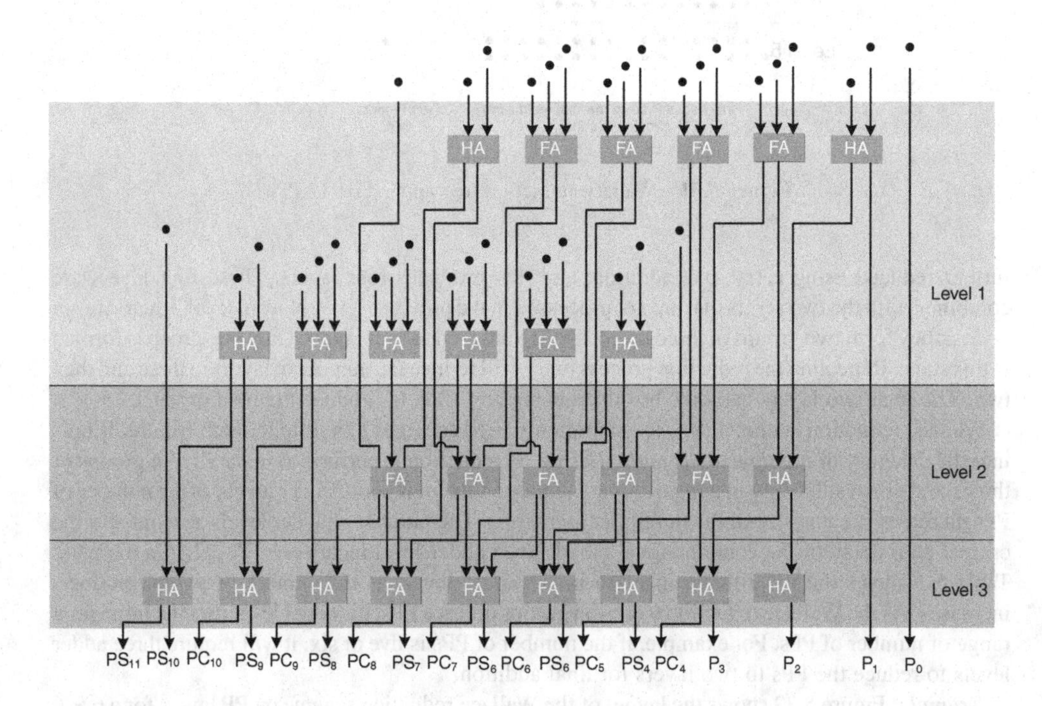

Figure 5.32 Wallace reduction tree layout for a 6 × 6 array of PPs

Consider the following sequence from a Wallace tree reduction scheme given as the upper limit of column 1 in Table 5.3:

$$2, 3, 4, 6, 9, 13, 19, 28, \ldots.$$

Each number represents the maximum number of partial products at each level that requires a fixed number of adder levels. The sequence also depicts that two partial products can be obtained from at most three partial products, three can be obtained from four, four from six, and so on.

The Dadda tree reduction considers each column separately and reduces the number of logic levels in a column to the maximum number of layers in the next level. For example, reducing PPs in a 12×12-bit multiplier, Wallace reduction reduces 12 partial products to eight whereas the Dadda scheme first reduces them to the maximum range in next the group, and this is nine as reducing twelve layers to eight will require the same number of logic levels as eight but results in less hardware.

In the Dadda tree each column is observed for the number of dots. If the number of dots in a column is less than the maximum number of PPs required to be reduced in the current level, they are simply dropped down to the next level of logic without any processing. Those columns that have more dots than the required dots for the next level are reduced to take the maximum layers in the next level.

Example: Figure 5.33 shows Dadda reduction on an 8×8 partial-product array. From Table 5.3 it is evident that eight PPs should be reduced to six PPs. In level 0, this reduction scheme leaves the columns that have less than or equal to six dots and only apply reduction on columns with more than six

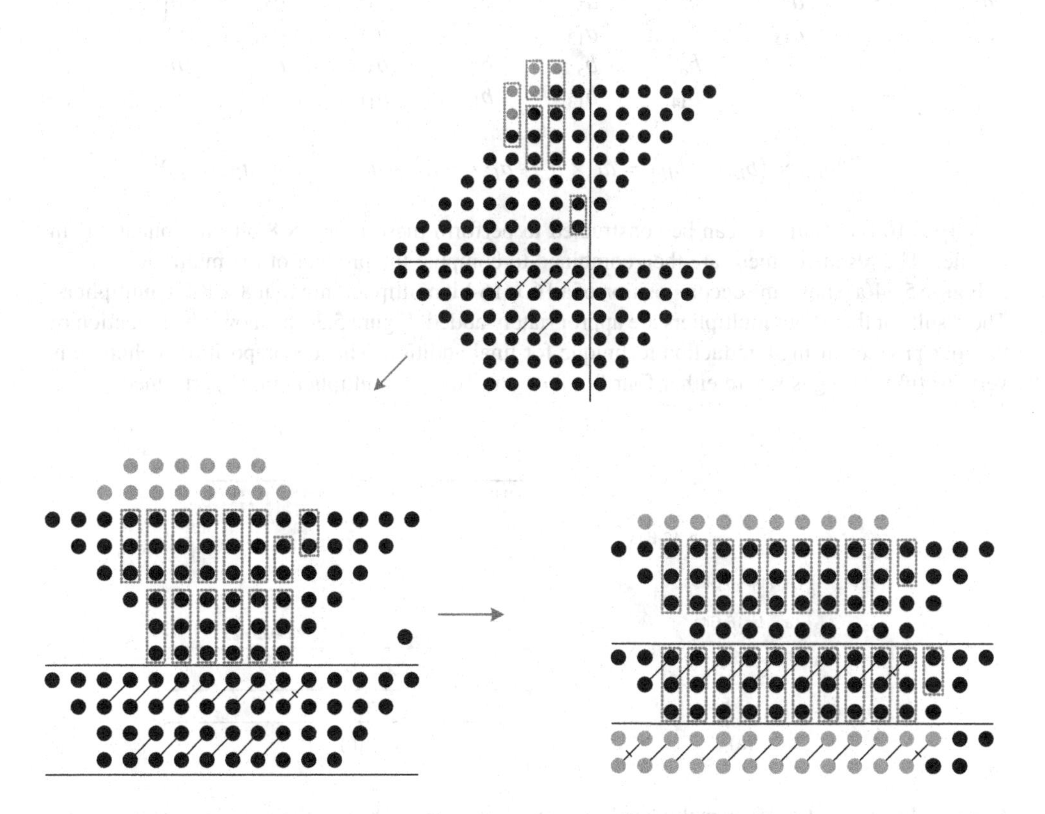

Figure 5.33 Dadda reduction levels for reducing eight PPs to two

dots. The columns having more than six dots are reduced to six. Thus the first six columns are dropped down to the next level without any reduction. Column 7 has seven dots; they are reduced to six by placing a HA. This operation generates a carry in column 8, which is shown in gray in the figure. Column 8 will then have nine dots. An HA and an FA reduce these nine dots to six and generate two carries in column 9. Column 9 will have nine dots, which are again reduced to six using an HA and an FA. (HAs and FAs are shown in the figure with diagonal crossed and uncrossed lines, respectively.) Finally, column 10 will reduce eight dots to six using an FA. The maximum number of dots in any column in the next level is six. For the next level, Table 5.3 shows that six PPs should be reduced to four. Each column is again observed for potential reduction. The reduction is only applied if the total number of dots, including dots coming from the previous column, is four or more, and these dots are reduced to three. The process is repeated to reduce three dots to two. This process is shown in Figure 5.33.

5.8.4 A Decomposed Multiplier

A multiplication can be decomposed into a number of smaller multiplication operations. For example, 16 × 16-bit multiplication can be performed by considering the two 16-bit operands a and b as four 8-bit operands, where a_H and a_L are eight MSBs and LSBs of a, and b_H and b_L are eight MSBs and LSBs of b. Mathematical decomposition of the operation is given here:

$$
\begin{array}{llllllllll}
a_L & = & a_7 & a_6 & a_5 & a_4 & a_3 & a_2 & a_1 & a_0 \\
a_H & = & a_{15} & a_{14} & a_{13} & a_{12} & a_{11} & a_{10} & a_9 & a_8 \\
b_L & = & b_7 & b_6 & b_5 & b_4 & b_3 & b_2 & b_1 & b_0 \\
b_H & = & b_{15} & b_{14} & b_{13} & b_{12} & b_{11} & b_{10} & b_9 & b_8
\end{array}
$$

$$(a_L + 2^8 a_H) \times (b_L + 2^8 b_H) = a_L \times b_L + a_L \times b_H 2^8 + a_H \times b_L 2^8 + a_H \times b_H 2^{16}$$

A 16 × 16-bit multiplier can be constructed to perform these four 8 × 8-bit multiplications in parallel. The results of these are then combined to compute the product of the multiplier.

Figure 5.34(a) shows the decomposition of a 16 × 16-bit multiplier into four 8 × 8-bit multipliers. The results of these four multipliers are appropriately added. Figure 5.34(b) shows the reduction of the four products using a reduction technique for final addition. This decomposition technique is very useful in designs where either four 8 × 8 or one 16 × 16 multiplication is performed.

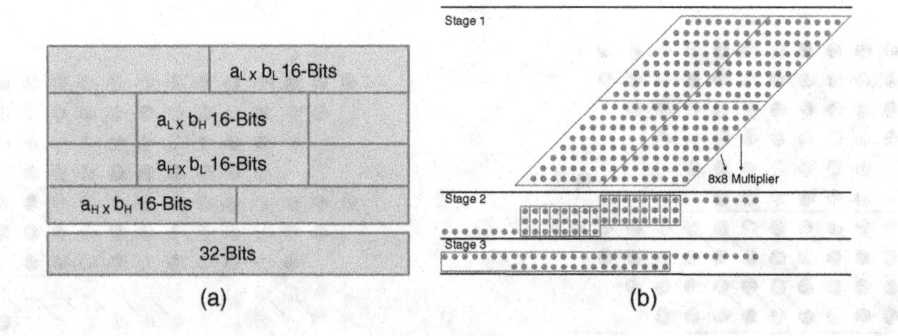

Figure 5.34 (a) A 16 × 16-bit multiplier decomposed into four 8 × 8 multipliers. (b) The results of these multipliers are appropriately added to get the final product

5.8.5 *Optimized Compressors*

Based on the concept of CSA, several other compressor blocks can be developed. For example a 4:2 compressor takes four operands and 1 bit for the carry-in and reduces them to 2 bits in addition to a carry-out bit. A candidate implementation of the compressor is shown in Figure 5.35(a). While compressing multiple operands the compressor works in cascade and creates an extended tile, as shown in Figure 5.35(b).

The use of this compressor in Wallace and Dadda reduction trees for an 8×8-bit multiplier is shown in Figures 5.35(c) and (d). This compressor, by using carry-chain logic, is reported to exhibit better timing and area performance compared with a CSA-based compression on a Virtex FGPA [14]. Similarly, a 5:3 bit counter reduces 5 bits to 3 bits with a carry-in and carry-out bit and is used in designing multiplier architectures in [15].

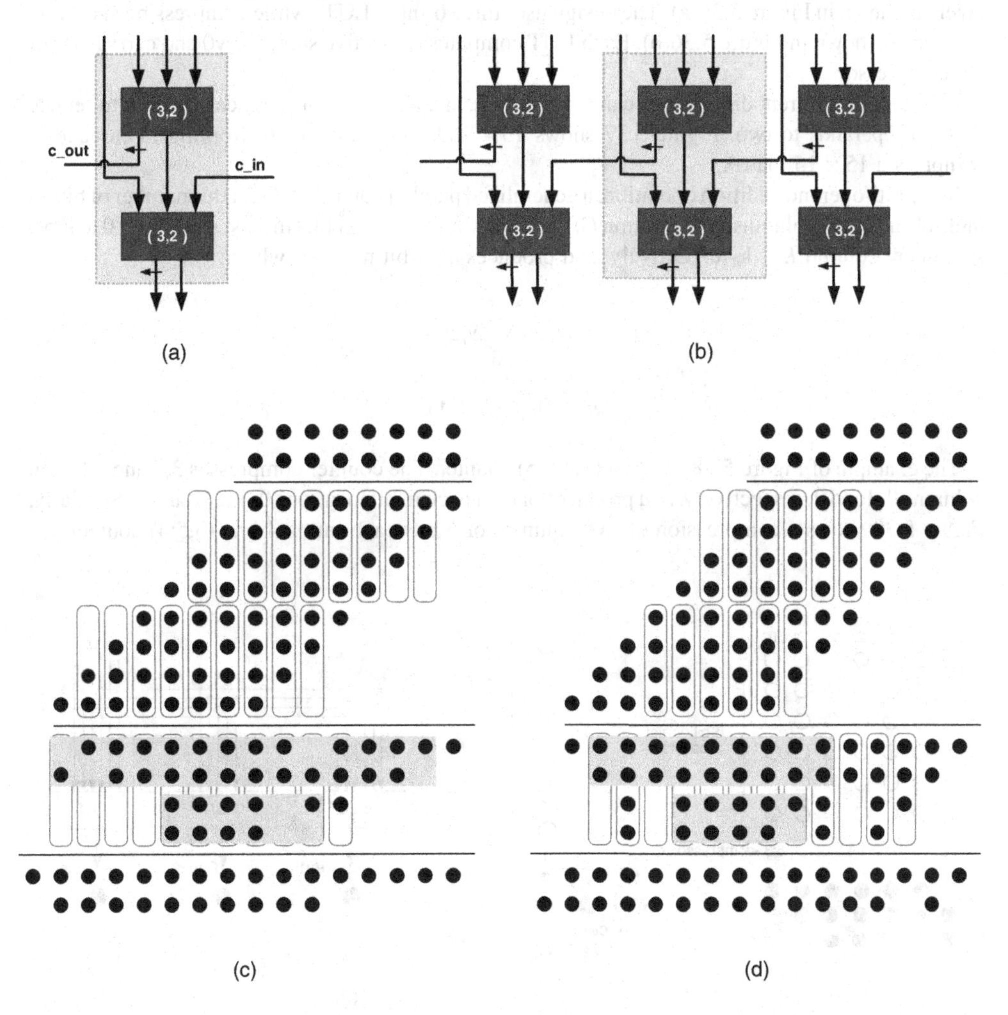

Figure 5.35 (a) Candidate implementation of a 4:2 compressor. (b) Concatenation of 4:2 compression to create wider tiles. (c) Use of a 4:2 compressor in Wallace tree reduction of an 8×8 multiplier. (d) Use of a 4:2 compressor in an 8×8 multiplier in Dadda reduction

5.8.6 Single- and Multiple-column Counters

Multi-operand additions on ASIC are traditionally implemented using CSA-based Wallace and Dadda reduction trees. With LUTs and carry chain-based FPGA architecture in mind, the counter-based implementation offers a better alternative to CSA-based multi-operand addition on FPGAs [16]. These counters add all bits in single or multiple columns to best utilize FGPA resources.

A single-column N:n counter adds all N bits in a column and returns an n-bit number, where $n = \log_2(N+1)$. Early FPGA architectures were 4-input LUT-based designs. Recently, 6-input LUTs with fast carry-chain logic have appeared in the Vertix™-4 and Virtex™-5 families. A design that effectively uses these features is more efficient than the others. A 6:3 counter fully utilizing three 6-input LUTs and carry-chain logic has been shown to outperform other types of compressors for the Virtex™-5 family of FPGAs [17]. A 6:3 counter compressing six layers of multi operands to three layers is shown in Figure 5.36(a). The design uses three 6-input LUTs while compressing six layers to three, as shown in Figure 5.36(b). Each LUT computes respective sum, carry0 and carry1 bits of the compressor.

Counters of different dimensions can also be built, and a mixture of these can be used to reduce multiple operands to two. Figure 5.37 shows 15:4, 4:3 and 3:2 counters working in cascade to compress a 15 × 15 matrix.

In a multi-operand addition operation, a generalized parallel counter (GPC) adds number of bits in multiple adjacent columns. A K-column GPC adds $N_0, N_1, \ldots, N_{K-1}$ bits in least significant 0 to most significant column $K-1$, respectively, and produces an n-bit number, where:

$$N = \sum_{i=0}^{K-1} N_i 2^i$$

$$n = \lceil \log_2(N+1) \rceil$$

The example of Figure 5.38 shows a (3,4,5:5) counter. The counter compresses 3,4 and 5 bits in columns 0, 1and 2, respectively, and produces one bit each in bit locations 0, 1, 2, 3 and 4. Similarly, Figure 5.39 shows a compression of two columns of 5 bits each into a 4-bits (5,5:4) counter.

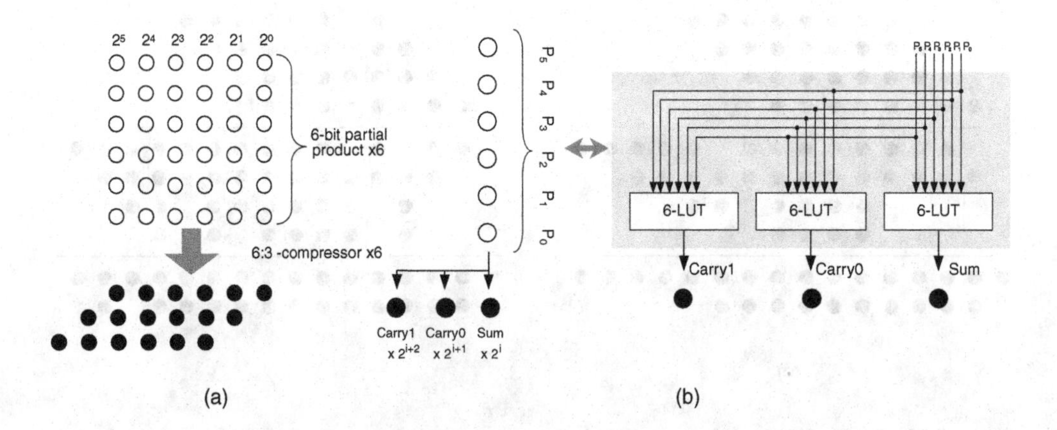

Figure 5.36 Single column counter (a) A 6:3 counter reducing six layers of multiple operands to three. (b) A 6:3 counter is mapped on three 6-input LUTs for generating sum, carry 0 and carry 1 output

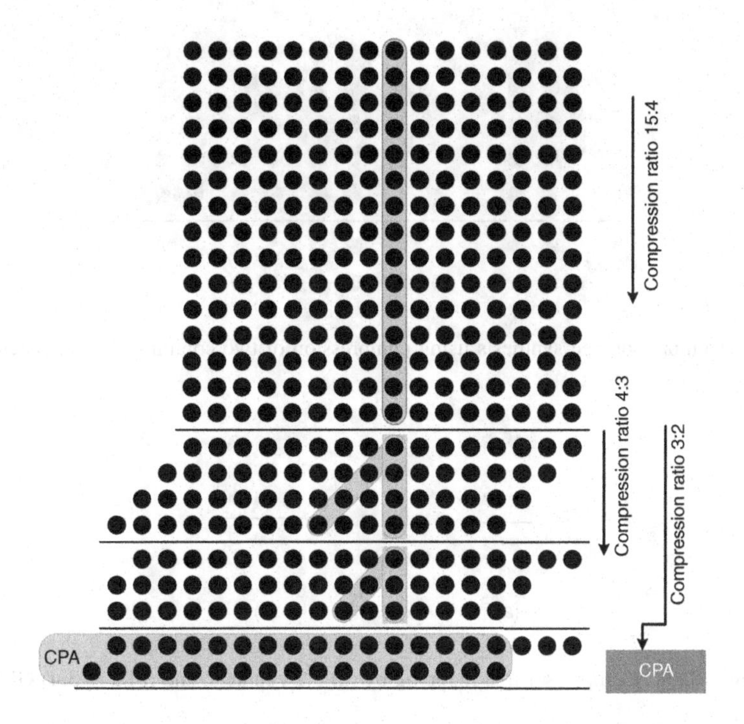

Figure 5.37 Counters compressing a 15 × 15 matrix

GPC offers flexibility once mapped on to a FPGA. The problem of configuring dimensions of GPC for mapping on FPGA for multi-operand addition is an NP-complete problem. The problem is solved using 'integer programming', and the method is reported to outperform adder tree-based implementation from the area and timing perspectives [18].

As stated earlier, FGPAs are best suited for counters and GPC-based compression trees. To fully utilize 6-LUT-based FPGAs, it is better that each counter or GPC should have six input bits and three (or preferably four) output bits, as shown in Figures 5.40 and 5.41. The four output bits are favored as

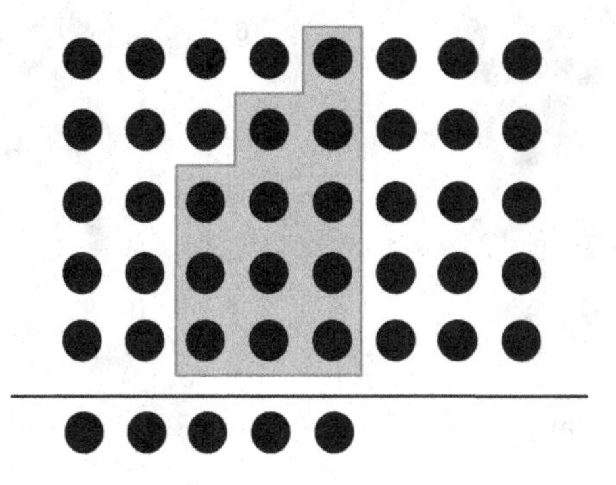

Figure 5.38 A (3,4,5:5) GPC compressing three columns with 3, 4 and 5 bits to 5 bits in different columns

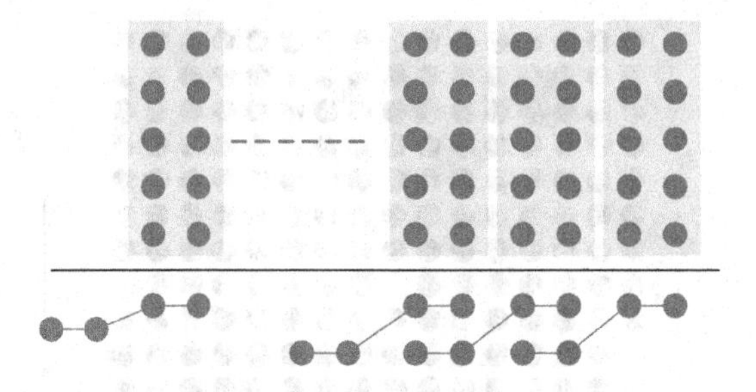

Figure 5.39 Compressor tree synthesis using compression of two columns of 5 bits each into 4-bit (5,5;4) GPCs

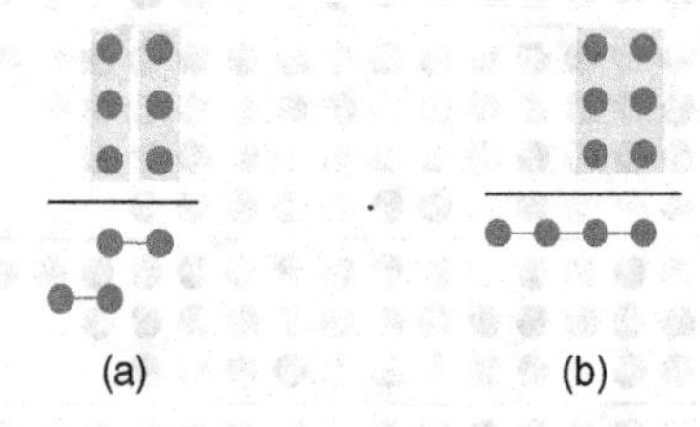

Figure 5.40 Compressor tree mapping by (a) 3:2 counters, and (b) a (3,3;4) GPC

the LUTs in many FPGAs come in groups of two with shared 6-bit input, and a 6:3 GPC would waste one LUT in every compressor, as shown in Figure 5.41(a).

5.9 Two's Complement Signed Multiplier

5.9.1 Basics

This section covers architectures that implement signed number multiplications. Recollecting the discussion in Chapter 3 on two's complement numbers, an N-bit signed number x is expressed as:

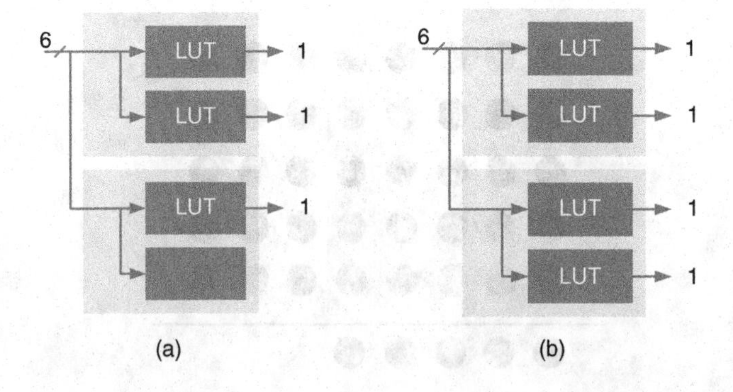

Figure 5.41 (a) The Altera FPGA adaptive logic module (ALM) contains two 6-LUTs with shared inputs; 6-input 3-output GPC has 3/4 logic utilization (b) A 6-input 4-output GPC has full logic utilization

$$x = -x_{n-1}2^{N-1} + \sum_{i=0}^{N-2} x_i 2^i$$

where x_{n-1} is the MSB that carries negative weight. When we multiply an N_1-bit signed multiplier a with an N_2-bit signed multiplicand b, we get N_1 partial products. For the first N_1-1 partial products, the $PP[i]$ is obtained by ANDing bit a_i of a with b and shifting the result by i positions to the left, This implements multiplication of b with $a_i 2^i$:

$$PP[i] = (a_i 2^i)\left(-b_{n-1}2^{N_2-1} + \sum_{i=0}^{N_2-2} b_i 2^i\right) \text{ for } i = 0, 1, \ldots, N_1-2$$

The $PP[i]$ in the above expression is just the signed multiplicand that is shifted by i to the left, so MSBs of all the partial products have negative weights. Furthermore, owing to the shift by i, all these PPs are unequal-width signed numbers. All these numbers are needed to be left-aligned by sign extension logic before they are added to compute the final product.

Now we need to deal with the last PP. As the MSB of a has negative weight, the multiplication of this bit results in a PP that is two's complement of the multiplicand. $PP[N_1 - i]$ is computed as

$$PP[N_1-1] = (-a_{N_1-1}2^{N_1-1})\left(-b_{n-1}2^{N_2-1} + \sum_{i=0}^{N_2-2} b_i 2^i\right)$$

All these N_1 partial products are appropriately sign extended and added to get the final product.

Example: Figure 5.42 shows 4 × 4-bit signed by signed multiplication. The sign bits of the first three PPs are extended and shown in bold. Also note that the two's complement of the last PP is taken to cater for the negative weight of the MSB of the multiplier. As can be seen, if the signed by signed multiplier is implemented as in this example, the sign extension logic will take significant area of the compression tree. It is desired to somehow remove this logic from the multiplier.

5.9.2 Sign Extension Elimination

A simple observation in a sign extended number leads us to an effective technique for elimination of sign extension logic. An equivalent of the sign extended number is computed by flipping the sign bit and adding a 1 at the location of the sign bit and extending the number with all 1s. Figure 5.43(a) explains the computation on a positive number. The sign-bit 0 is flipped to 1 and a 1 is added at the

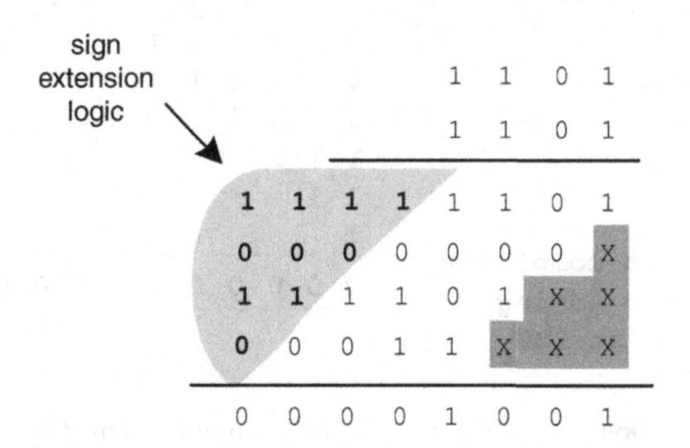

Figure 5.42 Showing 4 × 4-bit signed by signed multiplication

(a)

```
        extend all 1s      flip the sign bit
B = 0 0 0 0 0 0.1 1 0 1 0 1 1
B = 1 1 1 1 1 0̄.1 1 0 1 0 1 1
 +            1 → add 1 at the location of sign bit
    0 0 0 0 0 0.1 1 0 1 0 1 1
```

(b)

```
     extend all 1s      flip the sign bit
B = 1 1 1 1 1 1.1 1 0 1 0 1 1
B = 1 1 1 1 1 1̄.1 1 0 1 0 1 1
 +            1 → add 1 at the location of sign bit
    1 1 1 1 1 1.1 1 0 1 0 1 1
```

Figure 5.43 Sign-extension elimination

sign-bit location and extended bits are replaced by all 1s. This technique equivalently working on negative numbers is shown in Figure 5.43(b). The sign-bit 1 is flipped to 0 and a 1 is added to the sign-bit location and the extended bits are all 1s. Thus, irrespective of the sign of the number, the technique makes all the extended bits into 1s. Now to eliminate the sign extension logic, all these 1s are added off-line to form a correction vector.

Figure 5.44 illustrates the steps involved in sign-extension elimination logic on a 11×6-bit signed by signed multiplier. First the MSB of all the PPs except the last one are flipped and a 1 is added at the sign-bit location, and the number is extended by all 1s. For the last PP, the two's complement is computed by flipping all the bits and adding 1 to the LSB position. The MSB of the last PP is flipped again and 1 is added to this bit location for sign extension. All these 1s are added to find a *correction vector* (CV). Now all the 1s are removed and the CV is simply added and it takes care of the sign extension logic.

Example: Find the correction vector for a 4×4-bit signed multiplier and use the CV to multiply two numbers 0011 and 1101. In Figure 5.45(a), all the 1s for sign extension and two's complement are added and $CV = 0001\_0000$. Applying sign-extension elimination logic and adding CV to the PPs, the multiplication is performed again and it gives the same result, as shown in Figure 5.45(b). As the correction vector has just one non-zero bit, the bit is appended with the first PP (shown in gray).

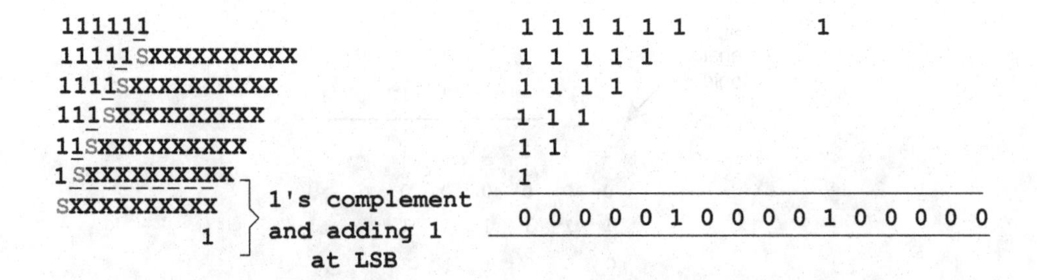

Figure 5.44 Sign-extension elimination and CV formulation for signed by signed multiplication

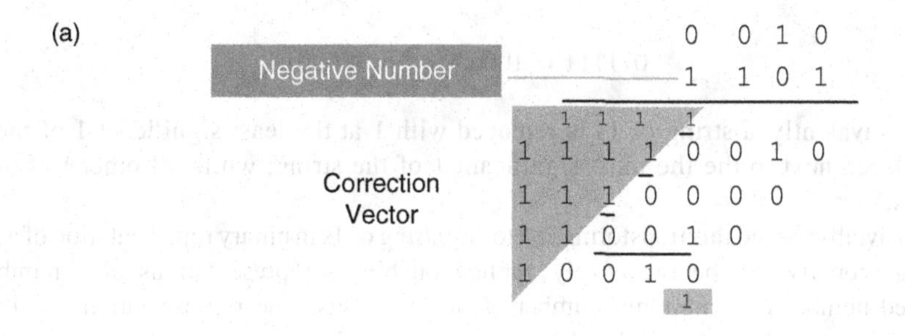

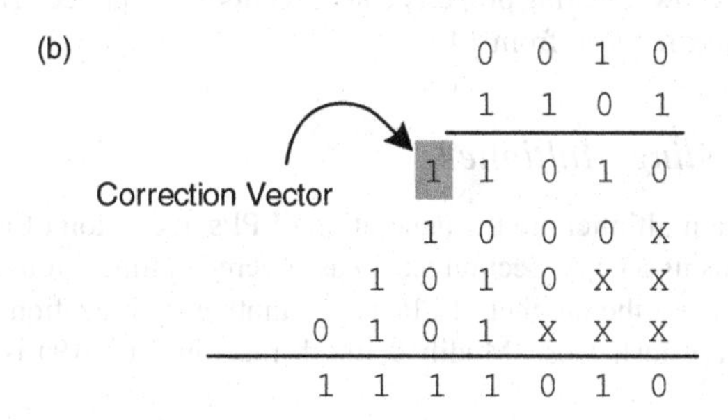

Figure 5.45 Multiplying two numbers, 0011 and 1101

Hardware implementation of a signed by signed multiplication using the sign-extension elimination technique saves area and thus is very effective.

5.9.3 String Property

So far we have represented numbers in two's complement form, where each bit is either a 0 or 1. There are, of course, other ways to represent numbers. A few of these are effective for hardware design for signal processing systems. Canonic signed digit (CSD) is one such format [19–21]. In CSD, a digit can be a 1, 0 or −1. The representation restricts the occurrence of two consecutive non-zeros in the number, so it results in a unique representation with minimum number of non-zero digits.

The CSD of a number can be computed using the string property of numbers. This property, while moving from LSB to MSB, successively observes strings of 1s and replaces each string with an equivalent value, using 1, 0 or −1. Consider the number 7. This can be written as 8 − 1, or in CSD representation:

$$0111 = 1000{-}1 = 100\bar{1}$$

The bit with a bar over has negative weight and the others have positive weights. Similarly, 31 can be written as 32 − 1, or in CSD representation:

$$011111 = 100000 - 1 = 10000\bar{1}$$

Thus, equivalently, a string of 1s is replaced with $\bar{1}$ at the least significant 1 of the string, and a 1 placed next to the the most significant 1 of the string, while all other bits are filled with zeros.

We can trivially extend this transformation to any string of 1s in binary representation of a number. The string property can be recursively applied on binary representations of a number. The transformed number has minimum number of non-zero bits. The representation and its use in digital design are further discussed in Chapter 6.

The example in Figure 5.46 shows how the string property can be recursively applied. The number of non-zero digits in the example has reduced from 14 to 6.

5.9.4 *Modified Booth Recoding Multiplier*

The three basic building blocks of a multiplier are the generation of PPs, reduction of PPs to two layers, and addition of these layers using a CSA. Section 5.8.3 has covered optimization techniques for partial products reduction. Reducing the number of PPs is yet another optimization technique that can be exploited in many design instances. 'Modified Booth recoding' (MBR) is one such technique.

While multiplying two N-bit signed numbers a and b, the technique generates one PP each by pairing all bits of b in 2-bit groups. The technique, while moving from LSB to MSB of b, pairs two bits together to make a group for recoding using the MBR algorithm. The two bits in a group can possibly be 00_2, 01_2, 10_2 and 11_2. Multiplication by 00_2, 01_2 and 10_2 simply results in 0, a and $2a = a \ll 1$, respectively where each PP is appropriately computed as one number. The fourth possibility of $11_2 = 3$ is $2 + 1$, and a simple shift would not generate the requisite PP; rather this multiplication will results in two PPs and they are a and $2a$. This means in the worst case the multiplier ends up having N partial products.

This problem of 11_2 generating two PPs is resolved by using the Booth recoding algorithm. This algorithm still works on groups of 2 bits each but recodes each group to use one of the five equivalent values: 0, 1, 2, −1, −2. Multiplication by all these digits results in one PP each. These equivalent values are coded by indexing into a look-up table. The look-up table is computed by exploiting the string property of numbers (see Section 5.9.3). The string property is observed on each pair of two bits while moving from LSB to MSB. To check the string property, the MSB of the previous pair is also required along with the two bits of the pair under consideration. For the first pair a zero is appended to the right. Table 5.4 shows the string property working on all possible 3-bit numbers for generating a table that is then used for recoding.

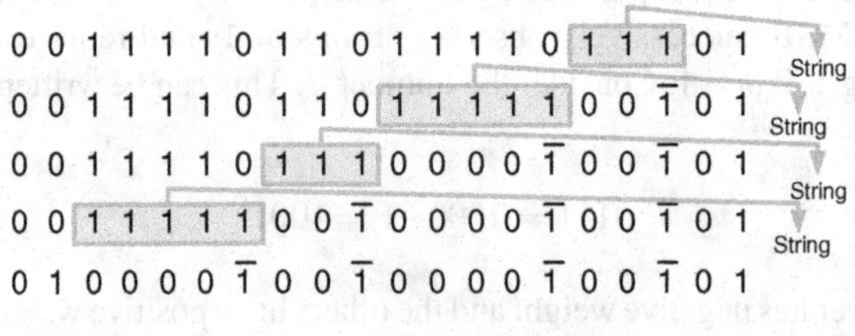

$$0\;0\;1\;1\;1\;1\;0\;1\;1\;0\;1\;1\;1\;1\;0\;1\;1\;1\;0\;1$$
$$0\;0\;1\;1\;1\;1\;0\;1\;1\;0\;1\;1\;1\;1\;1\;0\;0\;\bar{1}\;0\;1$$
$$0\;0\;1\;1\;1\;1\;0\;1\;1\;1\;0\;0\;0\;0\;\bar{1}\;0\;0\;\bar{1}\;0\;1$$
$$0\;0\;1\;1\;1\;1\;1\;0\;0\;\bar{1}\;0\;0\;0\;0\;\bar{1}\;0\;0\;\bar{1}\;0\;1$$
$$0\;1\;0\;0\;0\;0\;\bar{1}\;0\;0\;\bar{1}\;0\;0\;0\;0\;\bar{1}\;0\;0\;\bar{1}\;0\;1$$

String
String
String
String

Figure 5.46 Application of the string property

Table 5.1 Modified Booth recoding using the string property of numbers

2^1 2^0	String property implemented	Numeric computations	Recoded value
000	No string	0	0
001	End of string at bit location 0	2^0	1
010	Isolated 1	1	1
011	End of string at bit location 0	2^1	2
100	Start of string at bit location 1	-2^1	−2
101	End and start of string at bit locations 0 and 1, respectively	-2^0-2^1	−1
110	Start of string at bit location 0	-2^0	−1
111	Middle of string	0	0

Example: This example multiplies two 8-bit signed numbers 10101101 and 10001101 using the Booth recoding algorithm. The technique first pairs the bits in the multiplier in 2-bit groups. Then, on checking the MSB of the previous group, it recodes each group using Table 5.4. A zero is assumed as MSB of the previous group for the least significant group as there is no previous group. The eight groups with the MSB of the previous group are:

100 001 110 010.

From the table the groups are recoded as −2, 1, −1 and 1, and the corresponding four PPs are generated as shown in Figure 5.47. For a 1 recoded digit in the multiplier, the multiplicand is simply copied. As each PP is generated for a pair of 2 bits of the multiplier, the ith PP is shifted by $2i$ places to the left. For the second recoded digit of the multiplier that is −1, the two's complement of the multiplicand is copied. The PPs are generated for all the recoded digits; for the last digit of −2 the two's complement of the PP is further shifted left by one bit position to cater for multiplication by 2. All the four PPs are sign extended and then added to get the final product. Sign-extension elimination logic can also be used to reduce the logic for HW implementation.

A representative HW design for implementing an 8×8-bit multiplier is given in Figure 5.48. The BR0, BR1, BR2 and BR3 recode by considering groups of 2 bits and the MSB of the previous group into one of the five recoded options. Based on the recoded value, one of the five PPs $2a$, a, 0, $-a$ and $-2a$ are generated. All the four PPs are then input to a compressor. The compression generates two layers of PPs that are then added using a CPA and a 16-bit product is computed.

```
                10    10    11    01
                -2    +1    -1     1
-------------------------------------
11111111   |  10 |  10 |  11 |  01
00000001   |  01 |  00 |  11 |
11111010   |  11 |  01 |     |
00101001   |  1  |     |     |
-------------------------------------
00100101   |  01 |  00 |  10 |  01
```

Figure 5.47 Generation of four PPs

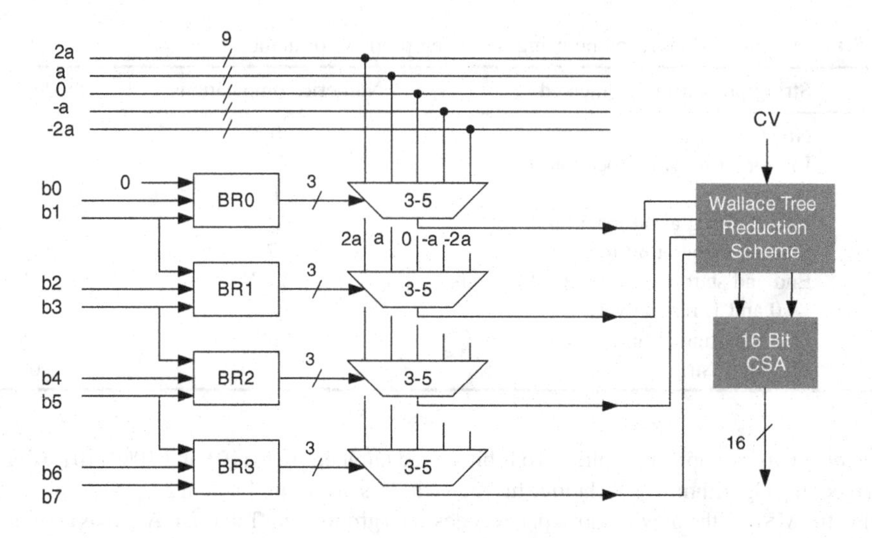

Figure 5.48 An 8 × 8-bit modified Booth recoder multiplier

5.9.5 Modified Booth Recoded Multiplier in RTL Verilog

The example here implements a 6 × 6-bit MBR multiplier. The three components in the design are Booth recoder, CV generation and PPs accumulation. The design is implemented in Verilog and code is listed at the end of this chapter. A Verilog function `RECODERfn` implements the recoding part. It divides the 6 bits of the multiplier into three groups of 2 bits each. `RECODERfn` takes three bits as input, consisting of two bits of each group and the MSB of the previous group, and generates one of the five values 0, 1, −1, 2 or −2 as output. A Verilog task, `GENERATE_PPtk`, generates three PPs with sign-extension elimination and a CV for the design. In cases where the recoded value is 2 or −2, the PP is generated by shifting the multiplicand or its compliment by 1 to the left, and this requires the PP to be 7 bits wide. For all other cases the seventh bit is the sign bit of the multiplicand. Sign-extension elimination logic is implemented. This requires flipping the MSB of the PP. For the cases of multiplication by −1 or −2, two's complement of the PP is computed by flipping all the bits and adding 1 to the LSB location. The addition of 1 is included in the CV, and the vector is appended by $2'b01$ for multiplication by −1 and $2'b10$ for multiplication by −2, as in this case the LSB is shifted by one bit position to the left. The output of the task is three PPs and six LSBs of the CV. The sign-extension elimination logic for the three PP is precalculated. By adding all the sign-extension elimination bits we get the five MSBs of the CV as $5'b01011$ as shown in Figure 5.49.

The code below illustrates the implementation of Booth recoding and sign-extension elimination logic in RTL Verilog. As the focus of the example is to illustrate the MBR technique, it simply adds all the PPs and the CV to compute the product. As the PPs and CV form four layers, for optimized

$$
\begin{array}{ccccc}
1 & 0 & 1 & 0 & 1 \\
1 & 1 & 1 & 1 & \bar{s} \\
1 & 1 & \bar{s} & & \\
\bar{s} & & & & \\
\hline
0 & 1 & 0 & 1 & 1
\end{array}
$$

Figure 5.49 Pre-calculated part of the CV

implementation a carry save reduction tree should be used to reduce these four layers to two, and then two layers should be added using a fast CPA.

```
module BOOTH_MULTIPLIER(
  input [5:0] multiplier,
  input [5:0] multiplicand,
  output [10:0] product);
  parameter WIDTH = 6;

  reg [6:0] pps [0:2];
  reg [10:0] correctionVector;
  reg [2:0] recoderOut[2:0];
  wire [6:0] a, a_n;
  wire [6:0] _2a, _2a_n;
  //integer i;

  /*
  1 x multiplicand, sign extend the multiplicand, and flip the sign bit
  2 x multiplicand, shift a by 1 and flip the sign bit
  1 x multiplicand, sign extend multiplicand and then flip all bits except the
sign bit
  2 x multiplicand, shift a by 1, flip all the bits except the sign bit
  */

  assign a_n  = { multiplicand[WIDTH-1], ~multiplicand};
  assign a    = { ~multiplicand[WIDTH-1], multiplicand};
  assign _2a_n = {multiplicand[WIDTH-1], ~multiplicand[WIDTH-2:0], 1'b0};
  assign _2a  = { ~multiplicand[WIDTH-1], multiplicand[WIDTH-2:0], 1'b0};

  // simply add all the PPs and CV, to complete the functionality
of the multiplier,
  // for optimized implementation, a reduction tree should be used
for compression

  assign product = pps[0] + {pps[1],2'b00} + {pps[2],4'b0000} +
correctionVector;

 always@*
 begin
   // compute booth recoded bits
   recoderOut[0] = RECODERfn ({multiplier[1:0],1'b0});
   recoderOut[1] = RECODERfn (multiplier[3:1]);
   recoderOut[2] = RECODERfn (multiplier[5:3]);

   // generate pps and correction vector
   GENERATE_PPtk (recoderOut[0], a, _2a, a_n, _2a_n, pps[0],
correctionVector[1:0]);
   GENERATE_PPtk (recoderOut[1], a, _2a, a_n, _2a_n, pps[1],
 correctionVector[3:2]);
   GENERATE_PPtk (recoderOut[2], a, _2a, a_n, _2a_n, pps[2],
 correctionVector[5:4]);

   // pre-computed CV for sign extension elimination
 correctionVector[10:6] = 5'b01011;
```

```
 //correctionVector[10:6] = 5'b00000;
end

/*
*****************************************************
* task:                 GENERATE_PPtk
* input:                multiplicand, multiplicand one's complement
*                       recoderOut: output from bit-pair recoder
* output:               correctionVector: addbits for 2's complement correction
* output:               ppi: ith partial product
*****************************************************
*/
task GENERATE_PPtk;
input [2:0] recoderOut;
input [WIDTH:0] a;
input [WIDTH:0] _2a;
input [WIDTH:0] a_n;
input [WIDTH:0] _2a_n;
output [WIDTH:0] ppi;
output [1:0] correctionVector;
reg  [WIDTH-1:0] zeros;
begin
  zeros = 0;
  case(recoderOut)
    3'b000:
    begin
      ppi              = {1'b1,zeros};
      //ppi              = {1'b0,zeros};
      correctionVector = 2'b00;
    end
    3'b001:
    begin
      ppi = a;
      correctionVector = 2'b00;
    end
    3'b010:
    begin
      ppi = _2a;
      correctionVector = 2'b00;
    end
    3'b110:
     begin
      ppi = _2a_n;
      correctionVector = 2'b10;
      //correctionVector = 2'b00;
     end
     3'b111:
     begin
     ppi = a_n;
     correctionVector = 2'b01;
     //correctionVector = 2'b00;
   end
   default:
   begin
```

```
      ppi = 'bx;
      correctionVector = 2'bx;
     end
   endcase
  end
 endtask
 /*
*****************************************************
 * Function: RECODERfn
 * input:one bit pair of multiplier with high order bit of previous pair, for
 *first pair a zero is appended as previous bit
 * output:Booth recoded output in radix-4 format, according to the following
table
 * ***************************************************** */
 function [2:0] RECODERfn;
  input [2:0] recoderIn;
  begin
   case(recoderIn)
   3'b000: RECODERfn = 3'b000;
   3'b001: RECODERfn = 3'b001;
   3'b010: RECODERfn = 3'b001;
   3'b011: RECODERfn = 3'b010;
   3'b100: RECODERfn = 3'b110;
   3'b101: RECODERfn = 3'b111;
   3'b110: RECODERfn = 3'b111;
   3'b111: RECODERfn = 3'b000;
   default: RECODERfn = 3'bx;
  endcase
  end
 endfunction
endmodule
```

5.10 Compression Trees for Multi-operand Addition

Although several devices in the FPGA families offer embedded multipliers, compression trees are still critical in many applications. The compression tree is the first building block in reducing the requirement on the number of CPAs for multi-operand addition. This is explained here with the help of an example.

The example adds five signed operands in Q1.5, Q5.3, Q4.7 and Q6.6 formats. Different bits in each operand can be represented with dots and be aligned with respect to the place of the decimal in their respective Q-format. The sign-extension logic is first constructed which is then eliminated by computing a correction vector and adding it as the sixth layer in dot representation. The placement of dots on a grid as shown in Figure 5.50 requires a compression tree to reduce the number of dots in each column to two. Any reduction technique can be used to demonstrate the effectiveness of the methodology. Here the Dadda reduction is employed. The two operands are then input to a CPA for final addition.

5.11 Algorithm Transformations for CSA

The CSA plays a key role in implementing high-throughput DSP applications in hardware. As a first step, while mapping a dataflow graph to architecture, the graph is observed to exhibit any potential

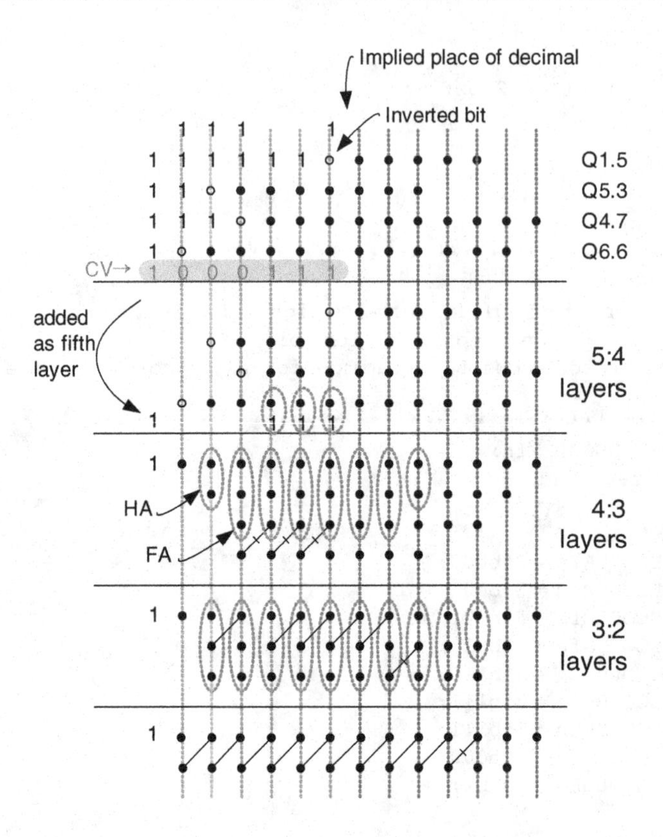

Figure 5.50 Example illustrating use of a compression tree in multi-operand addition

use of CSAs and the graph is modified accordingly. For example, consider implementing the following equations:

$$d[n] = a[n] + b[n] + c[n] \tag{5.13}$$

$$y[n] = d[n-1]e[n] \tag{5.14}$$

The equations are converted to the DFG of Figure 5.51(a). This is modified to use a CSA for compressing $a[n] + b[n] + c[n]$ into two numbers, which then are added using a CPA. The transformed DFG is shown in Figure 5.51(b).

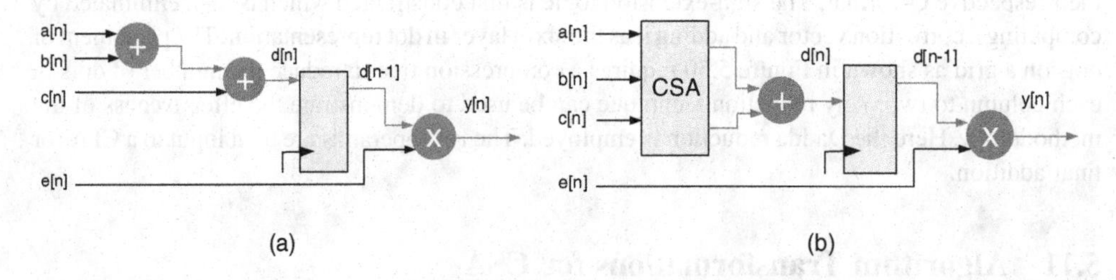

Figure 5.51 (a) FSFG with multi-operand addition. (b) Modified FSFG reducing three operands to two

This technique of extracting multi-operand addition can be extended to dataflow graphs where the graphs are observed to exhibit any potential use of CSA and compression trees. The graphs are then first transformed to optimally use the compression trees and then are mapped in HW. Such transformations are proven to significantly improve HW design of signal processing applications [13].

Multiple addition operations are the easiest of all the transformations. The compression tree can also be placed in the following add–compare–select operation:

$$
\begin{aligned}
&sum1 = op1 + op2;\\
&sum2 = op3 + op4;\\
&if(sim1 > sum2)\\
&\qquad sel = 0;\\
&else\\
&\qquad sel = 1;
\end{aligned}
$$

To transform the logic for optimal use of a compression tree, the algorithm is modified as:

$$sign(op1 + op2 - (op3 + op4)) = sign(op1 + op2 - op3 - op4)$$

$$sign(op1 + op2 + op3' + 1 + op4' + 1) = sign(op1 + op2 + op3' + op4' + 2)$$

This compression tree transformation on the equivalent DFG is shown in Figure 5.52. Similarly the following add and multiply operation is represented with the equivalent DFG:

$$op1 \times (op2 + op3)$$

The DFG can be transformed to effectively use a compression tree. A direct implementation requires one CPA to perform $op2 + op3$, and the result of this operation is then multiplied by $op1$. A multiplier architecture comprises a compression tree and a CPA. Thus to implement the computation two CPAs are required. A simple transformation uses the distributive property of the multiplication operator:

$$op1 \times op2 + op1 \times op3 \tag{5.15}$$

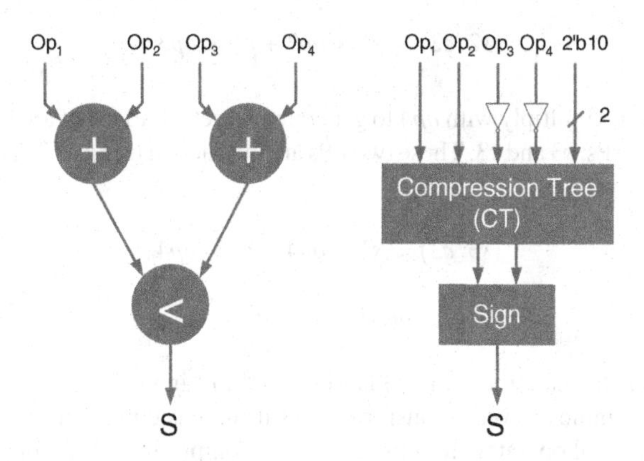

Figure 5.52 Compression tree replacement for an add compare and select operation

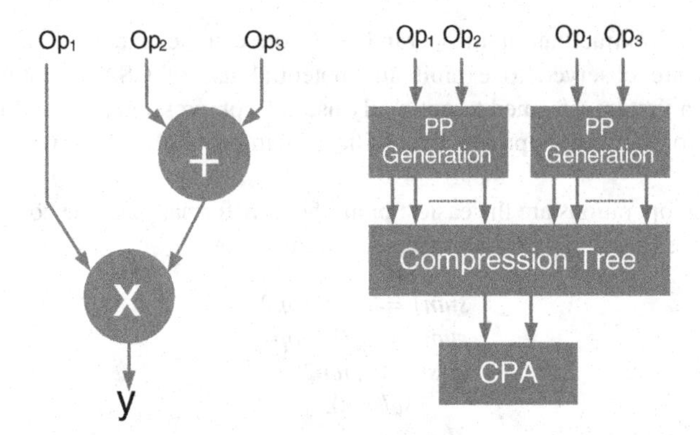

Figure 5.53 Transforming the add and multiply operation to use one CPA and a compression tree

This representation of the expression now requires one compression tree that is then followed by one CPA to compute the final value. The associated DFG and the transformation are shown in Figure 5.53.

Extending the technique of generating partial sums and carries can optimize hardware implementation of a cascade of multiplications as well:

$$prod = op1 \times op2 \times op3 \times op4$$

The transformation first generates PPs for $op1 \times op2$ and reduces them to two PPs, $s1$ and $c1$:

$$(s1, c1) = op1 \times op2$$

These two PPs independently multiply with $op3$, which generates two sets of PPs that are again reduced to two PPs, $s2$ and $c2$, using a compression tree:

$$(s2, c2) = s1 \times op3 + c1 \times op3$$

These two PPs further multiply with $op4$ to generate two sets of PPs that are again compressed to compute the final two PPs, $s3$ and $c3$. These two PPs are then added using a CPA to compute the final product:

$$(s3, c3) = s2 \times op4 + c2 \times op4$$

$$prod = s3 + c3$$

The equivalent transformation on a DFG is illustrated in Figure 5.54.

Following these examples, several transformations using basic mathematical properties can be used to accumulate several operators for effective use of compression trees for optimized hardware mapping.

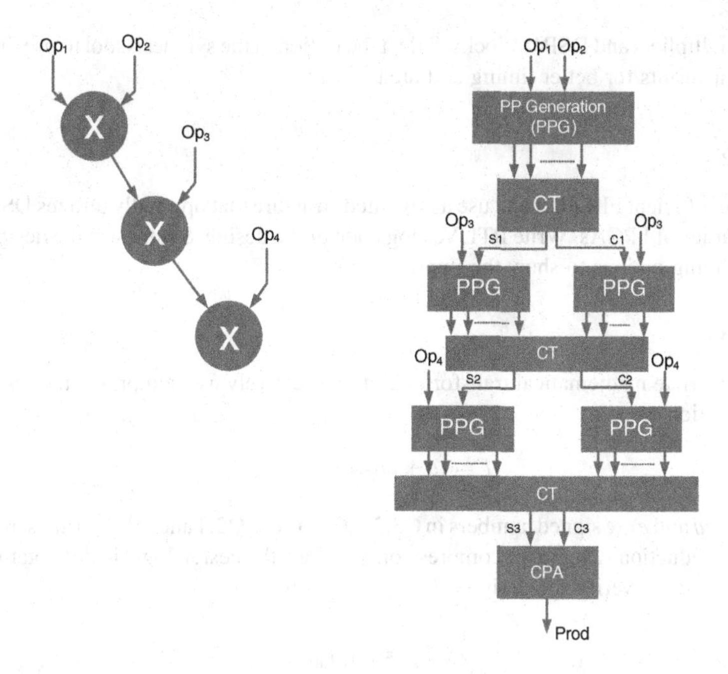

Figure 5.54 Transformation to use compression trees and a single CPA to implement a cascade of multiplication operations

Exercises

Exercise 5.1

An eighth-order IIR filter is designed using the `fdatool` of MATLAB®. The four second-order parallel sections of the filter are given below. The coefficients of the first-order numerators are:

1	0.6902
1	−0.1325
1	1.7434
1	0.0852

The coefficients of the corresponding second-order denominator of the four second-order filters are:

1	−1.0375	0.7062
1	−0.6083	0.9605
1	−1.3740	0.5431
1	−0.7400	0.8610

Convert the coefficients in appropriate 16-bit fixed-point format. Code the design using FDA in RTL Verilog. Use mathematical operators for multiplication and addition in the code. Test your design for functional correctness. Synthesize your design of two families of FPGAs that support

18×18-bit multipliers and DSP48 blocks. Select the option in the synthesis tool to use these blocks. Compare your results for better timing and area.

Exercise 5.2

Design an 8-coefficient FIR filter and use a pipelined structure that optimally utilizes DSP48 blocks of Xilinx families of FPGAs. Write RTL Verilog code of the design. Synthesize the design with and without pipelining options to show the improvements.

Exercise 5.3

Perform appropriate mathematical transformation to effectively use compression trees to map the following equation:

$$f = (a+2b)(c+d)-e.$$

Assume a, b, c, d and e are signed numbers in Q3.2, Q3.2, Q1.3, Q2.4 and Q1.3 formats, respectively. Use a Wallace reduction scheme for compression and draw the design logic in dot notation. Use the following numbers to verify the design:

$$a = 5'b011_{10}$$

$$b = 5'b111_{01}$$

$$c = 4'b0_{101}$$

$$d = 6'b01_{1100}$$

$$e = 4'b0_{011}$$

Exercise 5.4

Design an optimal hardware to implement the following equation:

$$d = a \times b - c$$

where a, b and c are in Q4.4, Q3.5 and Q2.6 format unsigned complex numbers, respectively. Verify your design for $a = 11.5 + j4.23$, $b = 2.876 + j1.23$ and $c = 1.22 + j3.32$. First convert these numbers in specified Q format and then design the logic, showing compression trees as blocks in the design.

Exercise 5.5

Design a 32-bit hybrid adder that combines different adder blocks. Use 4-bit RCA, 8-bit CSA, 8-bit conditional sum adder, 8-bit Brent–Kung and 4-bit of carry look-ahead adders in cascade. Write RTL Verilog code of the design and synthesize your design on any Spartan™-3 family of FPGAs. Now design different 32-bit adders exclusively using the above techniques. Synthesize all the adders on the same FPGA and compare your results for area and timing.

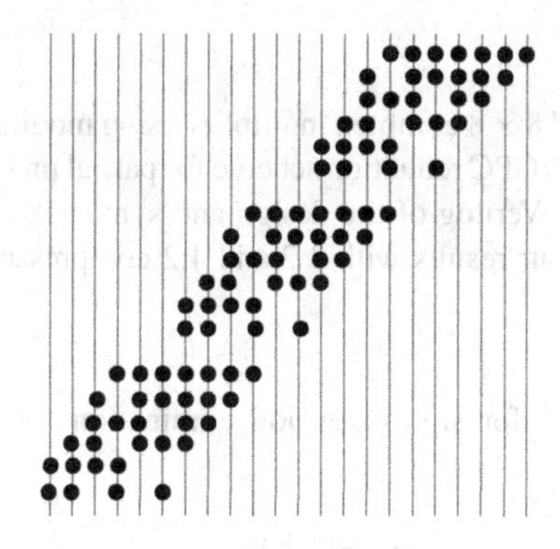

Figure 5.55 Bit array for Dadda reduction

Exercise 5.6

Add the following two numbers using conditional sum addition techniques:

1001_1101

1101_1011

Exercise 5.7

Design and code in RTL Verilog a 16-bit conditional sum adder with two stages of pipeline (add one register to break the combinational cloud). Assume `C_in` is not known prior to addition.

Exercise 5.8

Multiply a 5-bit signed multiplicand a with a 5-bit unsigned multiplier b by applying a sign-extension elimination technique, compute the correction vector for multiplication. Use the computed correction vector to multiply the following:

$$a = 11011$$
$$b = 10111$$

Exercise 5.9

Reduce the bit array shown in Figure 5.55 using a Dadda reduction scheme. Specify the number of FAs and HAs required to implement the scheme.

Exercise 5.10

Design and draw (using dot notation) an optimal logic to add twenty 1-bit numbers. Use a Dadda reduction tree for the compression.

Exercise 5.11

Design a single-stage pipelined 8 × 8 complex multiplier using modified Booth recoding for partial product generation and a (3,3:4) GPC reduction scheme for partial product reduction. Use minimum CPA in the design. Write RTL Verilog of the design and synthesize the design for a Spartan™-3 family of FPGAs. Compare your results with 3:2 and 4:2 compressor-based designs.

Exercise 5.12

Compute the correction vector for sign extension elimination for the following mathematical expression:

$$w = x + yz + y$$

where x, y and z are Q1.3, Q2.3 and Q3.3 signed numbers, respectively. Use the correction vector to evaluate the expression for $x = 1011$, $y = 11011$ and $z = 101111$. Give equivalent floating-point values for x, y, z and w.

Exercise 5.13

Compute the correction vector for sign extension elimination for computing the following equation:

$$\text{out} = \text{in1} + \text{in2} + \text{in3} - \text{in4} + \text{in5} - \text{in6}$$

where in1, in2, in3, in4, in5 and in6 are in Q3.5, Q7.8, Q1.6, Q8.2, Q2.4 and Q3.9 formats, respectively.

Exercise 5.14

Design and draw (using dot notation) an optimal logic to implement the following code:

```
if (a+b+c+d+e+f > 4)
   x=4;
elseif (2a+2b+c+d+e+2f > 0)
   x=6;
else
   x=0;
```

a, b, c, d, e and f are 6-bit signed numbers, x is a 4-bit register. Use a Wallace compression tree and minimum number of carry propagate adders (CPAs) to implement the design.

Exercise 5.15

Reduce the following equation using Wallace reduction tree, and count how many full adders and half adders are required to compute the expression in two layers:

$$z = x_0 + x_1 + x_2 + x_3 + x_4 + x_5$$

with

x_0	Q3.5
x_1	Q1.7
x_2	Q4.8
x_3	Q2.7
x_4	Q3.8
x_5	Q2.7

Use dot notation to represent the bits of each number, and assume all numbers are signed. Also compute the correction vector to eliminate sign-extension logic.

Exercise 5.16

Multiply the following two signed numbers using a modified Booth recoding technique:

$$a = 10011101$$
$$b = 10110011$$

References

1. Xilinx.Virtex-4 multi-platform FPGAs: www.xilinx.com/products/virtex4/index.htm
2. Xilinx.User guide: Virtex-5 FPGA XtremeDSP Design Considerations.
3. H. T. Kung and R. P. Brent, "A regular layout for parallel adders," *IEEE Transactions on Computers*, 1982, vol. 31, pp. 260–264.
4. T. Han and D. A. Carlson, "Fast area-efficient VLSI adders," in *Proceedings of 8th IEEE Symposium on Computer Arithmetic*, 1987, pp. 49–56.
5. O. J. Bedriji, "Carry select adder," *IRE Transactions on Electronic Computers*, 1962, vol. 11, pp. 340–346.
6. J. Sklansky, "Conditional-sum addition logic," *IRE Transactions on Electronic Computers*, 1960, vol. 9, pp. 226–231.
7. W. Schardein, B. Weghaus, O. Maas, B. J. Hosticka and G. TriSster, "A technology-independent module generator for CLA adders," in *Proceedings of 18th European Solid State Circuits Conference*, 1992, pp. 275–278, Copenhagen, Denmark.
8. V. Kantabutra, "A recursive carry look-ahead carry-select hybrid adder," *IEEE Transactions on Computers*, 1993, vol. 42, pp. 1495–1499.
9. M. Lehman and N. Burla, "Skip techniques for high-speed carry propagation in binary arithmetic circuits," *IRE Transactions on Electronic Computers*, 1961, vol. 10, pp. 691–698.
10. C. Nagendra, M. J. Irwin and R. M. Owens, "Area–time tradeoffs in parallel adders," *IEEE Transactions on Circuits and Systems-11*, 1996, vol. 43, pp. 689–702.
11. J.-G. Lee, J.-A. Lee and D.-Y. Lee, "Better area–time tradeoffs in an expanded design space of adder architecture by parameterizing bit-width of various carry propagated sub-adder-blocks," *Proceedings of 6th IEEE International Workshop on System-on-Chip for Real-Time Applications*, 2006, pp. 10–14.
12. Xilinx. DSP: Designing for Optimal Results – High-performance DSP using Virtex-4 FPGAs, 10th edn, 2005.
13. A. K. Verma, P. Brisk and P. Ienne, "Data-flow transformations to maximize the use of carry-save representation in arithmetic circuits," *IEEE Transactions on Computer-Aided Design of Integrated Circuits and Systems*, 2008, vol. 27, pp. 1761–1774.
14. J. Poldre and K. Tammemae, "Reconfigurable multiplier for Virtex FPGA family," in *Proceedings of International Workshop on Field-programmable Logic and Applications*, Glasgow, 1999, pp. 359–364, Springer-Verlag, London, UK.

15. O. Kwon, K. Nowka and E. E. Swartzlander, "A 16-bit by 16-bit MAC design using fast 5:3 compressor cells," *Journal of VLSI Signal Procsesing*, 2002, vol. 31, no. 2, pp. 77–89.
16. H. Parandeh-Afshar, P. Brisk and P. Ienne, "Efficient synthesis of compressor trees on FPGAs," in *proceedings of Asia and South Pacific Design Automation Conference*, 2008, pp. 138–143, IEEE Computer Society Press, USA.
17. K. Satoh, J. Tada, K. Yamaguchi and Y. Tamura, "Complex multiplier suited for FPGA structure," in *Proceedings of 23rd International Conference on Circuits/Systems, Computers and Communications*, 2008, pp. 341–344, IEICE Press, Japan.
18. H. Parandeh-Afshar, P. Brisk and P. Ienne, "Improving synthesis of compressor trees on FPGAs via integer linear programming," *Design, Automation and Test in Europe*, 2008, pp. 1256–1261.
19. A. Avizienis, "Signed-digit number representation for fast parallel arithmetic," *IRE Transactions on Electronic Computers*, 1961, vol. 10, pp. 389–400.
20. F. Xu, C. H. Chang and C. C. Jong, "Design of low-complexity FIR filters based on signed-powers-of-two coefficients with reusable common sub expressions," *IEEE Transactions on Computer Aided Design of Integrated Circuits and Systems*, 2007, vol. 26, pp. 1898–1907.
21. S.-M. Kim, J.-G. Chung and K. K. Parhi, "Low-error fixed-width CSD multiplier with efficient sign extension," *IEEE Transactions on Circuits and Systems II*, 2003, vol. 50, pp. 984–993.

6

Multiplier-less Multiplication by Constants

6.1 Introduction

In many digital system processing (DSP) and communication algorithms a large proportion of multiplications are by constant numbers. For example, the finite impulse response (FIR) and infinite impulse response (IIR) filters are realized by difference equations with constant coefficients. In image compression, the discrete cosine transform (DCT) and inverse discrete cosine transform (IDCT) are computed using data that is multiplied by cosine values that have been pre-computed and implemented as multiplication by constants. The same is the case for fast Fourier transform (FFT) and inverse fast Fourier transform (IFFT) computation. For fully dedicated architecture (FDA), where multiplication by a constant is mapped on a dedicated multiplier, the complexity of a general-purpose multiplier is not required.

The binary representation of a constant clearly shows the non-zero bits that require the generation of respective *partial products* (PPs) whereas the bits that are zero in the representation can be ignored for the PP generation operation. Representing the constant in canonic sign digit (CSD) form can further reduce the number of partial products as the CSD representation of a number has minimum number of non-zero bits. All the constant multipliers in an algorithm are in double-precision floating-point format. These numbers are first converted to appropriate fixed-point format. In the case of hardware mapping of the algorithm as FDA, these numbers in fixed-point format are then converted into CSD representation.

The chapter gives the example of an FIR filter. This filter is one of the most commonly used algorithmic building blocks in DSP and digital communication applications. An FIR filter is implemented by a convolution equation. To compute an output sample, the equation takes the dot product of a tap delay line of the inputs with the array of filter coefficients. The coefficients are predesigned and are double-precision floating-point numbers. These numbers are first converted to fixed-point format and then to CSD representation by applying the string property on their binary representation. A simple realization generates the PPs for all the multiplication operations in the dot product and reduces them using any reduction tree discussed in Chapter 5. The reduction reduces the PPs to two layers of sum and carry, which are then added using any carry propagate adder (CPA). The combinational cloud of the reduction logic can be pipelined to reduce the critical path delay of

Digital Design of Signal Processing Systems: A Practical Approach, First Edition. Shoab Ahmed Khan.

the design. Retiming is applied on an FIR filter and the transformed filter becomes a *transposed direct form* (TDF) FIR filter.

The chapter then describes techniques for complexity reduction. This further reduces the complexity of design that involves multiplication by constants. These techniques exploit the multiple appearances of common sub-expressions in the CSD representation of constants. The techniques are also applicable for designs where a variable is multiplied by an array of constants, as in a TDF implementation of an FIR filter.

The chapter discusses mapping a signal processing algorithm represented as a dataflow graph (DFG) on optimal hardware. The optimization techniques extensively use compression trees and avoid the use of CPAs, because from the area and timing perspectives a fast CPA is one of the most expensive building blocks in FDA implementation. The DFG can be transformed to avoid or reduce the use of CPAs. The technique is applied on IIR systems as well. These systems are recursive in nature. All the multipliers are implemented as compression trees that reduce all PPs to two layers of carry and sum. These two layers are not added inside the feedback loop, rather they are fed back as a partial solution to the next block. This helps in improving the timing of the implementation.

6.2 Canonic Signed Digit Representation

CSD is a radix-2 signed-digit coding. It codes a constant using signed digits 1, 0 and −1 [1, 2]. An N-bit constant C is represented as:

$$C = \sum_{i=0}^{N-1} s_i 2^i \text{ for } s_i \in \{-1, 0, 1\} \tag{6.1}$$

The expression implies that the constant is coded using signed digits 1, 0 or −1, where each digit si contributes a weight of 2^i to the constant value. The CSD representation has the following properties:

- No two consecutive bits in CSD representation of a number are non-zero.
- The CSD representation of a number uses a minimum number of non-zero digits.
- The CSD representation of a number is unique.

CSD representation of a number can be recursively computed using the string property. The number is observed to contain any string of 1s while moving from the least significant bit (LSB) to the most significant (MSB). The LSB in a string of 1s is changed to $\bar{1}$ that represents −1, and all the other 1s in the string are replaced with zeros, and the 0 that marks the end of the string is changed to 1. After replacing a string by its equivalent CSD digits, the number is observed again moving from the coded digit to the MSB to contain any further string of 1s. The newly found string is again replaced by its equivalent CSD representation. The process is repeated until no string of 1s is found in the number.

Example: Converting 16′*b*0011_1110_1111_0111 to CSD representation involves the following recursion. Find a string while moving from LSB to MSB and replace it with its equivalent CSD representation:

$$\begin{array}{c} 0011111011110111 \\ 001111101111100\bar{1} \end{array}$$

The newly formed number is observed again for any more string of 1s to be replaced by its equivalent CSD representation:

$$0011111011111 00\bar{1}$$
$$0011111100 00\bar{1}00\bar{1}$$

The process is repeated until all strings of 1s are replaced by their equivalent CSD representations:

$$0011111100 00\bar{1}00\bar{1}$$
$$0100000\bar{1}0000\bar{1}00\bar{1}$$

All these steps can be done simultaneously by observing isolated strings or a set of connected strings with one 0 in between. All the isolated strings with more than one 0 in between are replaced by their equivalent CSD representations, and for each connected string all the 0s connecting individual strings are changed to $\bar{1}$, and all the 1s in the strings are all changed to 0. The equivalent CSD representation computed in one step is:

$$0011111011110111$$
$$0100000\bar{1}0000\bar{1}00\bar{1}$$

6.3 Minimum Signed Digit Representation

MSD drops the condition of the CSD that does not permit two consecutive non-zero digits in the representation. MSD adds flexibility in representing numbers and is very useful in HW implementation of signal processing algorithms dealing with multiplication with constants. In CSD representation a number is unique, but a number can have more than one MSD representation with minimum number of non-zero digits. This representation is used in later in the chapter for optimizing the HW design of algorithms that require multiplication with constant numbers.

Example: The number 51 (=7′*b*0110011) has four non-zero digits, and the number is CSD representation, $10\bar{1}010\bar{1}$, also has four non-zero digits. The number can be further represented in the following MSD format with four non-zero digits:

$$011010\bar{1}$$
$$10\bar{1}0011$$

6.4 Multiplication by a Constant in a Signal Processing Algorithm

As stated earlier, in many DSP and digital communication algorithms a large proportion of multiplications are by constants. For FDA the complexity of a general-purpose multiplier is not required as PPs for only non-zero bits of the multiplier are generated. For example, when implementing the difference equation given by:

$$y[n] = 0.916y[n-1] + x[n] \tag{6.2}$$

$y[n-1]$ is multiplied by 0.916. In Q1.15 format the number is:

16’b0111_0101_0011_1111

As 11 bits in the constant are non-zero, PPs for these bits should be generated, whereas a general-purpose multiplier requires the generation of 16 PPs. Representing the constant in CSD form further reduces the number of PPs.

Here the constant in the equation is transformed to CSD representation by recursively applying the string property on the binary representation of the number:

$$16\text{'b}\,0111\_0101\_0011\_1111$$
$$=$$
$$0111\_0101\_0100\_000\bar{1}$$
$$=$$
$$100\bar{1}\_0101\_0100\_000\bar{1}$$
$$=$$
$$2^0 - 2^{-3} + 2^{-5} + 2^{-7} + 2^{-9} - 2^{-15}$$

The CSD representation of the constant thus reduces the number of non-zero digits from 11 to 6. For multiplication of 0.916 with $y[n-1]$ in FDA implementation thus requires generating only 6 PPs. These PPs are generated by appropriately shifting $y[n-1]$ by weight of the bits in Q1.15 format representation of the constant as CSD digits. The PPs are:

$$y[n-1] - y[n-1]2^{-3} + y[n-1]2^{-5} + y[n-1]2^{-7} + y[n-1]2^{-9} - y[n-1]2^{-15}$$

The PPs are generated by hardwired right shifting of $y[n-1]$ by 0,3,5,7,9 and 15 and adding or subtracting these PPs according to the sign of CSD digits at these locations. The architecture can be further optimized by incorporating $x[n]$ as the seventh PP and adding CV for sign extension elimination logic as the eighth PP in the compression tree. All these inputs to the compression tree are mathematically shown here:

$$y[n] = y[n-1] - y[n-1]2^{-3} + y[n-1]2^{-5} + y[n-1]2^{-7} + y[n-1]2^{-9} - y[n-1]2^{-15} + x[n] + CV$$

The CV is calculated by adding the correction for multiplication by 1 and $\bar{1}$ in the CSD representation of the number. The CV calculation follows the method given in Chapter 5. For multiplication by 1 the CV is computed by flipping the sign bit of the PP and adding 1 at the location of the sign bit and then extending the number by all 1s. Whereas for multiplication by $\bar{1}$ the PP is generated by taking the one's complement of the number. This sign extension and two's complement in this case require flipping of all the bits of the PP except the sign bit, adding 1 at the locations of the sign bit and the LSB, and then extending the number by all 1s. Adding contribution of all the 1s from all PPs gives us the CV:

$$1110\_1010\_1011\_1111\_0010\_0000\_0000\_001$$

Figure 6.1 shows the CV calculation. The filled dots show the original bits and empty dots show flipped bits of the PPs. Now these eight PPs can be reduced to two layers of partial sums and carries using any compression method described in Chapter 5.

Figure 6.2 shows the use of Wallace tree in reducing the PPs to two. Finally these two layers are added using any fast CPA. The compression is shown by all bits as filled dots.

6.5 Optimized DFG Transformation

Several architectural optimization techniques like carry save adder (CSA), compression trees and CSD multipliers can be used to optimize DFG for FDA mapping. From the area and timing

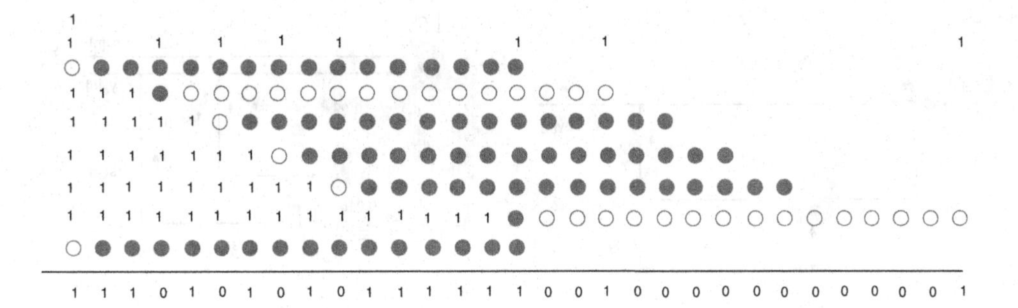

Figure 6.1 CV calculation

perspectives, a CPA is one of the most expensive building blocks in HW implementation. In FDA implementation, simple transformations are applied to minimize the use of CPAs.

Figure 6.3(a) shows the DFG mapping of (6.1). If implemented as it is, the critical path of the design consists of one multiplier and one adder. Section 6.4 showed the design for FDA by using a compression tree that reduces all the PPs consisting of terms for CSD multiplication, $x[n]$ and CV for sign extension elimination. The design then uses a CPA and reduces the two layers from the

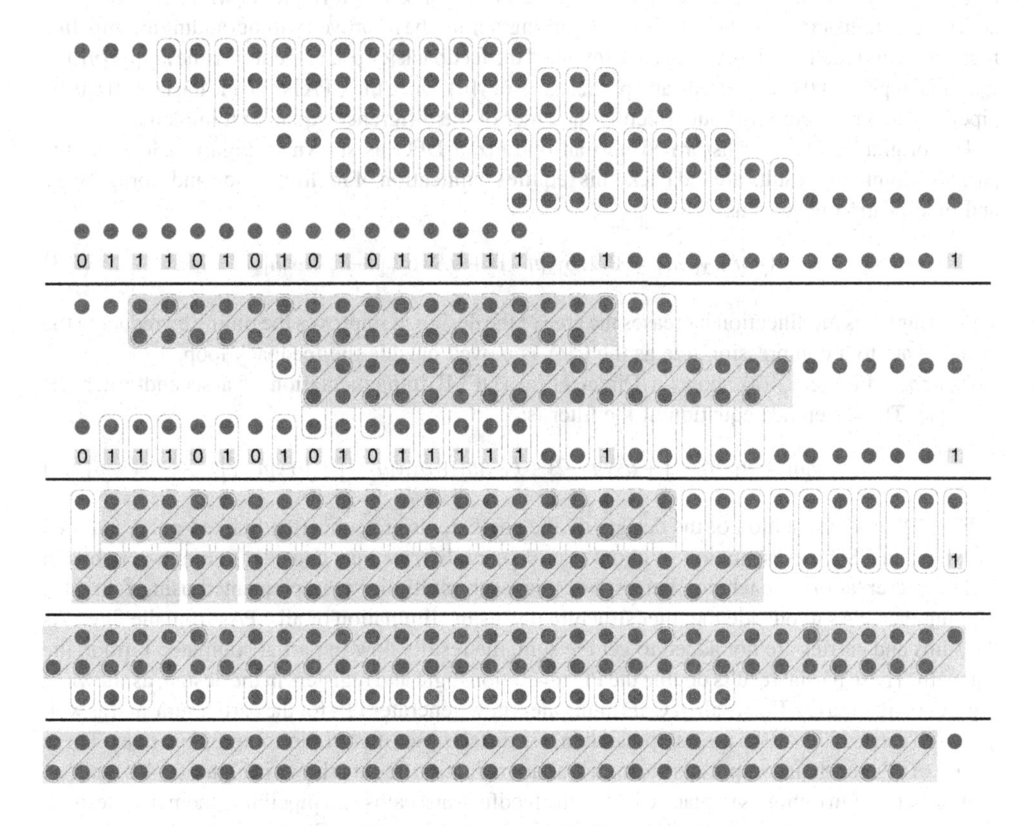

Figure 6.2 Use of the Wallace tree for PP reduction to two layers

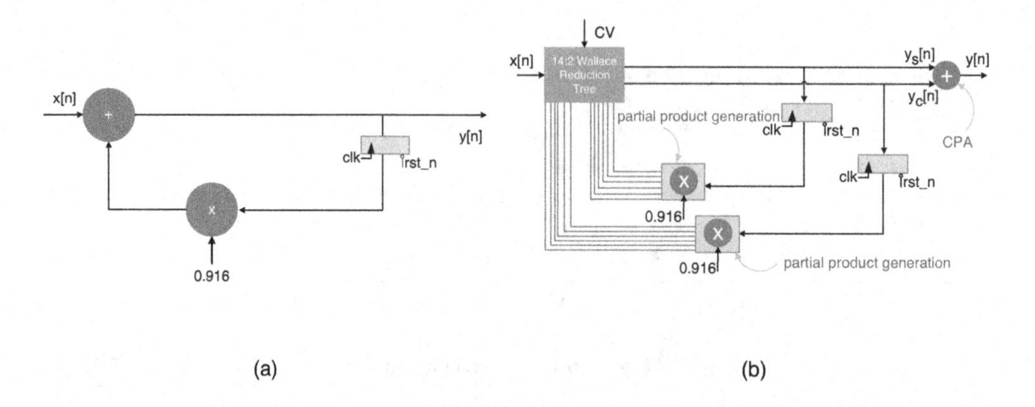

Figure 6.3 First-order IIR filter. (a) DFG with one adder and one multiplier in the critical path. (b) Transformed DFG with Wallace compression tree and CPA outside the feedback loop

reduction tree to compute $y[n]$. In instances where further improvement in the timing is required, this design needs modification.

It is important to note that there is no trivial way of adding pipeline registers in a feedback loop as it changes the order of the difference equation. There are complex transformations, which can be used to add pipeline registers in the design [3–5]. The timing can also be improved without adding any pipeline registers. This is achieved by taking the CPA out of the feedback loop, and feeding in the loop $y_s[n]$ and $y_c[n]$ that represent the partial sum and partial carry of $y[n]$. Once the CPA is outside the loop, it can be pipelined to any degree without affecting the order of the difference equation it implements.

The original DFG is transformed into an optimized DFG as shown in Figure 6.3(b). As the partially computed results are fed back, this requires duplication of multiplication and storage logic and modification in (6.1) as:

$$\{y_s[n], y_c[n]\} = 0.916y_s[n-1] + 0.916y_c[n-1] + x[n] \tag{6.3}$$

Although this modification increases the area of the design, it improves the timing by reducing the critical path to a compression tree as the CPA is moved outside the feedback loop.

Example: Figure 6.4(a) shows a Direct Form (DF)-II implementation of a second-order IIR filter [6]. The difference equation of the filter is:

$$y[n] = a_1y[n-1] + a_2y[n-2] + b_0x[n] + b_1x[n-1] + b_2x[n-2] \tag{6.4}$$

For HW implementation of the design, all the coefficients are converted to fixed-point and then translated to CSD representation. Without loss of generality, we may consider four non-zero bits in CSD representation of each coefficient. For this, each multiplier is implemented using four PPs. Moving the CPAs to outside the filter structure results in elimination of all CPAs from the filter. As the sums and carries are not added to get the sum, the results flow in partial computed form in the datapath. These partial results double the PP generation logic and registers in the design, as shown in Figures 6.4(b) and (c). Each pair of CSD multipliers now generates PPs for the partial sum, and as well as partial carries These PPs are compressed in two different settings of the compression trees. These two set of PPs and their compression using a Wallace reduction tree are shown in Figures 6.4(b) and (c).

Another design option is to place CPAs in the feedforward paths and pipelining them if so desired. This option is shown in Figure 6.4(d). For each pair of multipliers a CV is calculated, which takes

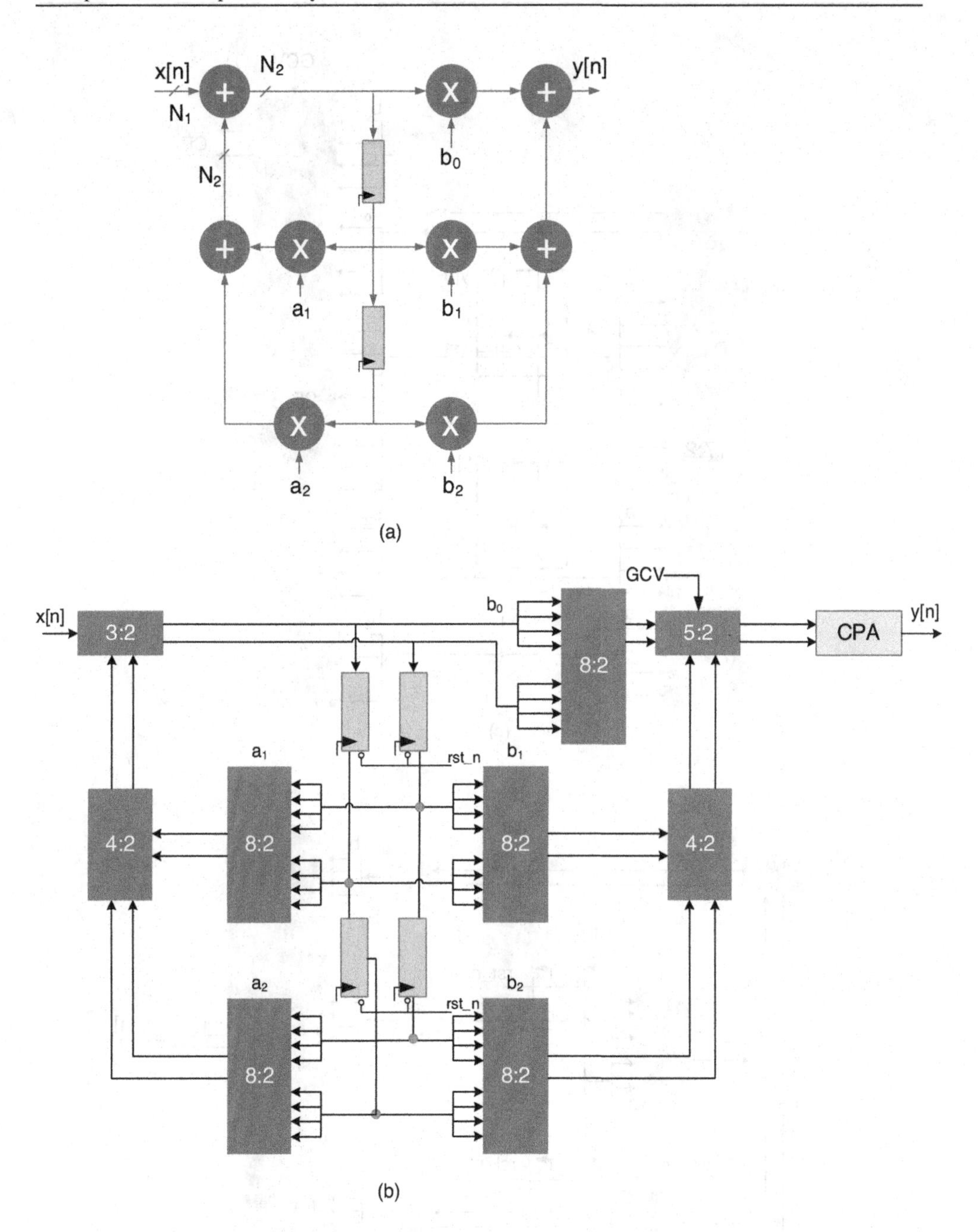

Figure 6.4 Second-order IIR filter. (a) DF-II dataflow graph. (b) Optimized implementation with CSD multipliers, compression trees and CPA outside the IIR filter. (c) Using unified reduction trees for the feedforward and feedback computations and CPA outside the filter. (d) CPA outside the feedback loop only

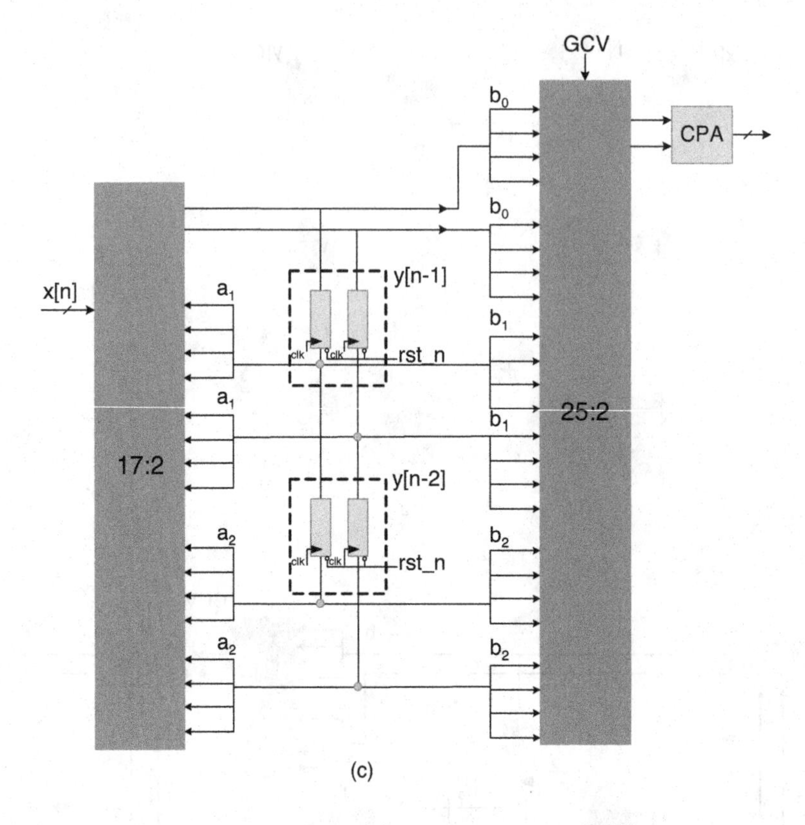

(c)

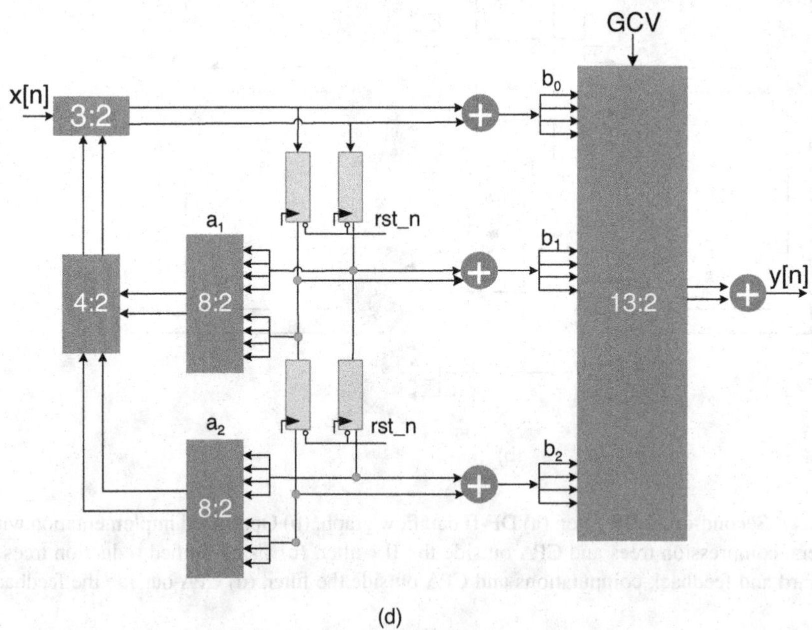

(d)

Figure 6.4 (*Continued*)

care of sign-extension elimination and two's complement logic. A global CV (GCV) is computed by adding all the CVs. The GCV can be added in any one of the compression trees. The figure shows the GCV added in the compression tree that computes final sum and carry, $y_s[n]$ and $y_c[n]$. A CPA is placed outside the loop. The adder adds the partial results to generate the final output. Once the CPA is outside the loops, it can be pipelined to any degree without affecting the order of the difference equation it implements.

If there are more stages of algorithm in the application, to get better timing performance the designer may chose to pass the partial results without adding them to the next stages of the design.

6.6 Fully Dedicated Architecture for Direct-form FIR Filter

6.6.1 *Introduction*

The FIR filter is very common in signal processing applications. For example it is used in the digital front end (DFE) of a communication receiver. It is also used in noise cancellation and unwanted frequency removal from received signals in a noisy environment. In many applications, the signal may be sampled at a very high rate and the samples are processed by an FIR filter. The application requires an FDA implementation of a filter executing at the sampling clock.

The FIR filter is implemented using the convolution summation given by:

$$y[n] = \sum_{k=0}^{L-1} h[k]x[n-k] \tag{6.4}$$

where the $h[k]$ represent coefficients of the FIR filter, and $x[n]$ represents the current input sample and $x[n-k]$ is the kth previous sample. For simplicity this chapter sometimes writes indices as subscripts:

$$y_n = \sum_{k=0}^{L-1} h_k x_{n-k} \tag{6.5}$$

A block diagram of an FIR filter for $L = 5$ is shown in Figure 6.5(a). This structure of the FIR filter is known as *direct-form* (DF) [6]. There are several other ways of implementing FIR filters, as discussed later.

In a DF implementation, to compute output sample y_n, the current input sample x_n is multiplied with h_0, and for each k each previous x_{n-k} sample is multiplied by its corresponding coefficient h_k and finally all the products are added to get the final output. An FDA implementation requires all these multiplications and additions to execute simultaneously, requiring L multipliers and $L-1$ adders. The multiplication with constant coefficients can exploit the simplicity of the CSD multiplier [7].

Each of these multipliers, in many design instances, is further simplified by restricting the number of non-zero CSD digits in each coefficient to four. One non-zero CSD digit in a coefficient approximately contributes 20 dB of stop-band attenuation [8], so the four most significant non-zero CSD digits in each coefficient attains around 80 dB of stop-band attenuation. The stop band attenuation is a measure of effectiveness of a filter and it represents how successfully the filter can stop unwanted signals.

Figure 6.5(b) shows each CSD multiplier with four non-zero digits generating four PPs. One approach is to compute the products by accumulating four PPs for each multiplication and then sum these products to get the final answer. A better approach is to reduce all the PPs generated for all multiplications and the GCV to two layers using a Wallace or Dadda reduction scheme. The two layers then can be summed using any fast CPA. The GCV is computed by adding CVs for each multiplication. Figure 6.5(b) shows the optimized architecture for a 5-coefficient FIR filter. More optimized solutions for FIR architectures are proposed in the literature. These solutions work on complexity reduction and are discussed in a later section.

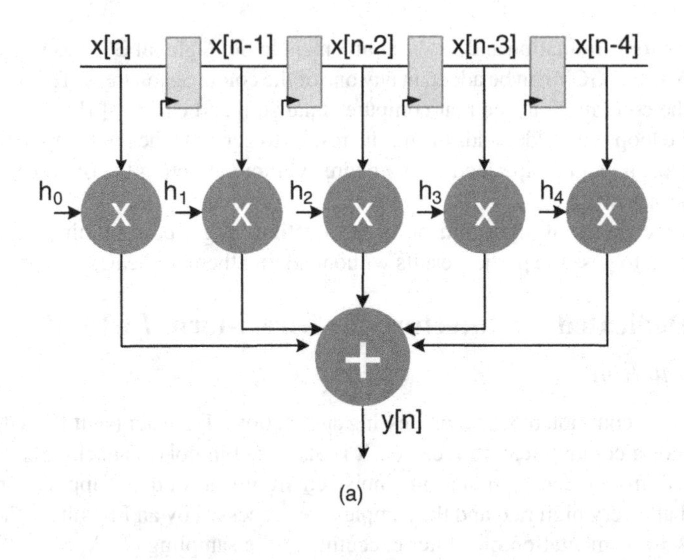

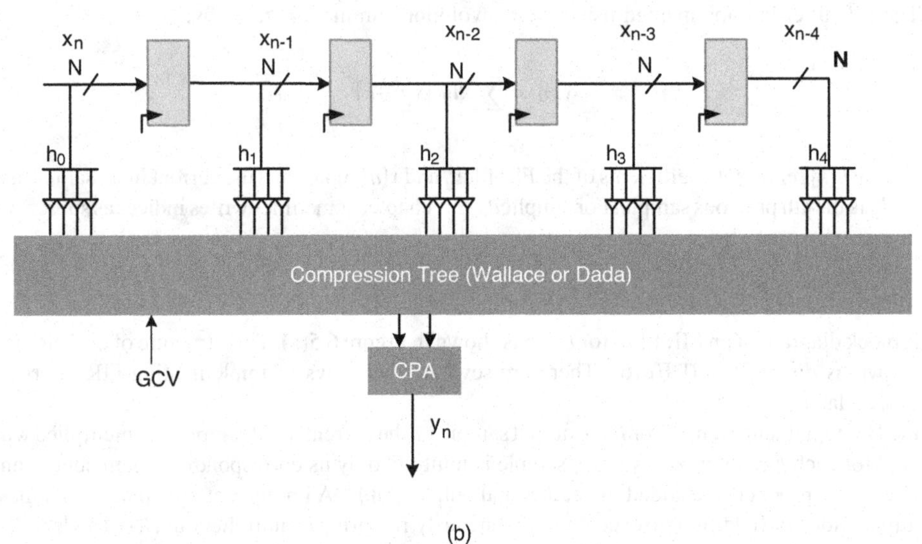

Figure 6.5 Five-coefficient FIR filter. (a) DF structure. (b) All multiplications are implemented as one compression tree and a single CPA

6.6.2 Example: Five-coefficient Filter

Consider the design of a 5-coefficient FIR filter with cutoff frequency $\pi/4$. The filter is designed using the `fir1()` function in MATLAB®. Convert $h[n]$ to Q1.15 format. Compute the CSD representation of each fixed-point coefficient. Design an optimal DF FDA architecture by keeping a maximum of four most significant non-zero digits in the CSD representation of the filter coefficients. Generate all PPs and also incorporate CV in the reduction tree. Use the Wallace reduction technique to reduce the PPs to two layers for addition using any fast CPA.

Solution: The five coefficients of the filter in double-precision floating-point using the MATLAB® `fir1` function are:

$$h[n] = [0.024553834015017 \\ 0.234389464237986 \\ 0.482113403493995 \\ 0.234389464237986 \\ 0.024553834015017]$$

Converting $h[n]$ into Q1.15 format gives:

$$h[n] = \text{round}(h[n]{*}2^{15}) = [805\ \ 7680\ \ 15798\ \ 7680\ \ 805]$$

Binary representation of the coefficients is:

$$16'b0000\_0011\_0010\_0101 \\ 16'b0001\_1110\_0000\_0000 \\ 16'b001\_1110\_1101\_10110 \\ 16'b0001\_1110\_0000\_0000 \\ 16'b0000\_0011\_0010\_0101$$

Converting the coefficients into CSD representation gives:

$$0000010\bar{1}0\_0100\_101 \\ 001000\bar{1}00\_0000\_000 \\ 010000\bar{1}0\_0\bar{1}00\_\bar{1}0\bar{1} \\ 001000\bar{1}00\_0000\_000 \\ 0000010\bar{1}0\_0100\_101$$

Keeping a maximum of four non-zero CSDs in each coefficient results in:

$$0000010\bar{1}001001 \\ 001000\bar{1}0000000000 \\ 010000\bar{1}00\bar{1}00\bar{1} \\ 001000\bar{1}000000000 \\ 0000010\bar{1}001001$$

The sixteen PPs are as follows:

$$\begin{aligned} y[n] &= \\ &(x[n]2^{-5} - x[n]2^{-7} + x[n]2^{-10} + x[n]2^{-13}) \\ &+ (x[n-1]2^{-2} - x[n]2^{-6}) \\ &+ (x[n-2]2^{-1} - x[n-2]2^{-6} - x[n-2]2^{-9} - x[n-2]2^{-12}) \\ &+ (x[n-3]2^{-2} - x[n-3]2^{-6}) \\ &+ (x[n-4]2^{-5} - x[n-4]2^{-7} + x[n-4]2^{-10} + x[n-4]2^{-13}) \end{aligned}$$

For sing-extension elimination, five CVs for each multiplier are computed and added to form the GCV. The computed CV_0 for the first coefficient is shown in Figure 6.6, and so:

$$CV_0 = 32'b1111\_1010\_1101\_1100\_0000\_0010\_0000\_0000$$

					1		1			1			1								
1	1	1	1	1	$\bar{x}_{15}$	x_{14}	x_{13}	x_{12}	x_{11}	x_{10}	x_9	x_8	x_7	x_6	x_5	x_4	x_3	x_2	x_1	x_0	
1	1	1	1	1	1	1	x_{15}	$\bar{x}_{14}$	$\bar{x}_{13}$	$\bar{x}_{12}$	$\bar{x}_{11}$	$\bar{x}_{10}$	$\bar{x}_9$	$\bar{x}_8$	$\bar{x}_7$	$\bar{x}_6$	$\bar{x}_5$	$\bar{x}_4$	$\bar{x}_3$	$\bar{x}_2$	$\bar{x}_1$
1	1	1	1	1	1	1	1	1	1	$\bar{x}_{15}$	x_{14}	x_{13}	x_{12}	x_{11}	x_{10}	x_9	x_8	x_7	x_6	x_5	x_4
1	1	1	1	1	1	1	1	1	1	1	1	1	$\bar{x}_{15}$	x_{14}	x_{13}	x_{12}	x_{11}	x_{10}	x_9	x_8	x_7

Figure 6.6 Computation for CV_0 (see text)

The computed CV_1 for the second coefficient is shown in Figure 6.7, and so:

$$CV_1 = 32'b1101\_1110\_0000\_0000\_0000\_0100\_0000\_0000$$

Following the same procedure, the CV_2 computes to:

$$CV_2 = 32'b1011\_1101\_1011\_1000\_0000\_0100\_1001\_0000$$

Because of symmetry of the coefficients, CV_3 and CV_4 are the same as CV_1 and CV_0, respectively. All these correction vectors are added to get the GCV:

$$GCV = 32'b01101111011100000001000010010000$$

This correction vector is added for sign-extension elimination.

The implementation of three designs in RTL Verilog is given in this section. The module `FIRfilter` uses multipliers and adders to implement the FIR filter, whereas module `FIRfilterCSD` converts the filter coefficients in CSD format while considering a maximum of four non-zero CSD digits for each coefficient. There are a total of 16 PPs that are generated and added to show the equivalence of the design with the original format. The code simply adds the PP whereas, in actual designs, the PPs should be compressed using Wallace or Dadda reduction schemes. Finally, the module `FIRfilterCV` implements sign-extension elimination logic by computing a GCV. The vector is added in place of all the sign bits and 1s that are there to cater for two's complement in the PPs. The Verilog code of the three modules with stimulus is listed here.

```
// Module uses multipliers to implement an FIR filter
module FIRfilter(
input signed [15:0] x,
input clk,
output reg signed [31:0] yn);
reg signed [15:0] xn [0:4];
wire signed [31:0] yn_v;

// Coefficients of the filter
wire signed [15:0] h0 = 16'h0325;
wire signed [15:0] h1 = 16'h1e00;
wire signed [15:0] h2 = 16'h3DB6;
wire signed [15:0] h3 = 16'h1e00;
wire signed [15:0] h4 = 16'h0325;

// Implementing filters using multiplication and addition operators
assign yn_v = (h0*xn[0] + h1*xn[1] + h2*xn[2] + h3*xn[3] + h4*xn[4]);
always @(posedge clk)
begin
     // Tap delay line of the filter
     xn[0] <= x;
     xn[1] <= xn[0];
```

		1				1															1
1	1	x_{15}	x_{14}	x_{13}	x_{12}	x_{11}	x_{10}	x_9	x_8	x_7	x_6	x_5	x_4	x_3	x_2	x_1	x_0				
1	1	1	1	1	1	x_{15}	x_{14}	x_{13}	x_{12}	x_{11}	x_{10}	x_9	x_8	x_7	x_6	x_5	x_4	x_3	x_2	x_1	
1	1	0	1	1	1	1	0	0	0	0	0	0	0	0	0	0	0	0	0	0	1

Figure 6.7 Computation for CV_1 (see text)

```
    xn[2] <= xn[1];
    xn[3] <= xn[2];
    xn[4]<= xn[3];
    // Registering the output
    yn <= yn_v;
end
endmodule
// Module uses CSD coefficients for implementing the FIR filter
module FIRfilterCSD (
input signed [15:0] x,
input clk,
output reg signed [31:0] yncsd);
reg signed [31:0] yncsd_v;
reg signed [31:0] xn [0:4];
reg signed [31:0] pp[0:15];

always @(posedge clk)
begin
    // Tap delay line of FIR filter
    xn[0] <= {x, 16'h0};
    xn[1] <= xn[0];
    xn[2] <= xn[1];
    xn[3] <= xn[2];
    xn[4]<= xn[3];
    yncsd <= yncsd_v; // registering the output
end
always @ (*)
begin
    // Generating PPs using CSD representation of coefficients
    // PP using 4 significant digits in CSD value of coefficient h0
    pp[0] = xn[0]>>>5;
    pp[1] = -xn[0]>>>7;
    pp[2] = xn[0]>>>10;
    pp[3] = xn[0]>>>13;
    // PP using CSD value of coefficient h1
    pp[4] = xn[1]>>>2;
    pp[5] = - xn[1]>>>6;
    // PP using 4 significant digits in CSD value of coefficient h2
    pp[6] = xn[2]>>>1;
    pp[7] = -xn[2]>>>6;
    pp[8] = -xn[2]>>>9;
    pp[9] = -xn[2]>>>12;
    // PP using CSD value of coefficient h3
    pp[10] = xn[3]>>>2;
    pp[11] = -xn[3]>>>6;
    // PP using 4 significant digits in CSD value of coefficient h4
    pp[12] = xn[4]>>>5;
    pp[13] = -xn[4]>>>7;
    pp[14] = xn[4]>>>10;
    pp[15] = xn[4]>>> 13;
    // Adding all the PPs, the design to be implemented in a
                        16:2 compressor
```

```
    yncsd_v = pp[0]+pp[1]+pp[2]+pp[3]+pp[4]+pp[5]+
              pp[6]+pp[7]+pp[8]+pp[9]+pp[10]+pp[11]+
              pp[12]+pp[13]+pp[14]+pp[15];
end
endmodule

// Module uses a global correction vector by eliminating sign extension logic
module FIRfilterCV (
input signed [15:0] x,
input clk,
output reg signed [31:0] yn
);
reg signed [31:0] yn_v;
reg signed [15:0] xn_0, xn_1, xn_2, xn_3, xn_4;
reg signed [31:0] pp[0:15];
// The GCV is computed for sign extension elimination
reg signed [31:0] gcv = 32'b0110_1111_0111_0000_0001_0000_1001_0000;

always @ (posedge clk)
begin
    // Tap delay line of FIR filter
    xn_0 <= x;
    xn_1 <= xn_0;
    xn_2 <= xn_1;
    xn_3 <= xn_2;
    xn_4 <= xn_3;
    yn <= yn_v; // registering the output
end
always @ (*)
begin
    // Generating PPs for 5 coefficients
    // PPs for coefficient h0 with sign extension elimination
    pp[0]= {5'b0, ~xn_0[15], xn_0[14:0], 11'b0};
    pp[1] = {7'b0, xn_0[15], ~xn_0[14:0], 9'b0};
    pp[2] = {10'b0, ~xn_0[15], xn_0[14:0],6'b0};
    pp[3] = {13'b0, ~xn_0[15], xn_0[14:0],3'b0};
    // PPs for coefficient h1 with sign extension elimination
    pp[4] = {2'b0, ~xn_1[15], xn_1[14:0], 14'b0};
    pp[5] = {6'b0, xn_1[15], ~xn_1[14:0], 10'b0};
    // PPs for coefficient h2 with sign extension elimination
    pp[6] = {1'b0, ~xn_2[15], xn_2[14:0], 15'b0};
    pp[7] = {6'b0, xn_2[15], ~xn_2[14:0], 10'b0};
    pp[8] = {9'b0, xn_2[15], ~xn_2[14:0], 7'b0};
    pp[9] = {12'b0, xn_2[15], ~xn_2[14:0], 4'b0};
    // PPs for coefficient h3 with sign extension elimination
    pp[10] = {2'b0, ~xn_3[15], xn_3[14:0], 14'b0};
    pp[11] = {6'b0, xn_3[15], ~xn_3[14:0], 10'b0};
    // PPs for coefficient h4 with sign extension elimination
    pp[12]= {5'b0, ~xn_4[15], xn_4[14:0], 11'b0};
    pp[13] = {7'b0, xn_4[15], ~xn_4[14:0], 9'b0};
    pp[14] = {10'b0, ~xn_4[15], xn_4[14:0],6'b0};
    pp[15] = {13'b0, ~xn_4[15], xn_4[14:0],3'b0};
```

```
    // Adding all the PPs with GCV
    // The design to be implemented as Wallace or Dadda
                        reduction scheme
    yn_v = pp[0]+pp[1]+pp[2]+pp[3]+
    pp[4]+pp[5]+
    pp[6]+pp[7]+pp[8]+pp[9]+
    pp[10]+pp[11]+
    pp[12]+pp[13]+pp[14]+pp[15]+gcv;
end
endmodule

module stimulusFIRfilter;
reg signed [15:0] X;
reg CLK;
wire signed [31:0] YN, YNCV, YNCSD;
integer i;
// Instantiating all the three modules for equivalency checking
FIRfilterCV FIR_CV(X, CLK, YNCV);
FIRfilterCSD FIR_CSD(X, CLK, YNCSD);
FIRfilter FIR(X, CLK, YN);
initial
begin
    CLK = 0;
    X = 1;
    #1000 $finish;
end
// Generating clock signal
always
    #5 CLK = ~CLK;
// Generating a number of input samples
initial
begin
    for (i=0; i<256; i=i+1)
        #10 X = X+113;
end
initial
    $monitor ($time, " X=%h, YN=%h, YNCSD=%h, YNCV=%h\n",
                        X, YN<<1, YNCSD, YNCV);
endmodule
```

6.6.3 Transposed Direct-form FIR Filter

The direct-form FIR filter structure of Figure 5.5 results in a large combinational cloud of reduction tree and CPA. The cloud can be pipelined to reduce the critical path delay of the design.

Figure 6.8(a) shows a 5-coefficient FIR filter, pipelined to reduce the critical path delay of the design. Now the critical path consists of a multiplier and an adder. Pipelining causes latency and a large area overhead in implementing registers. This pipeline FIR filter can be best mapped on the Vertix™-4 and Vertix™-5 families of FPGAs with embedded DSP48 blocks. The effectiveness of this mapping is demonstrated in Chapter 5.

In many design instances, using general-purpose multipliers of DSP48 blocks may not be appropriate as they are a finite resource and should be used for parts of the algorithm that require

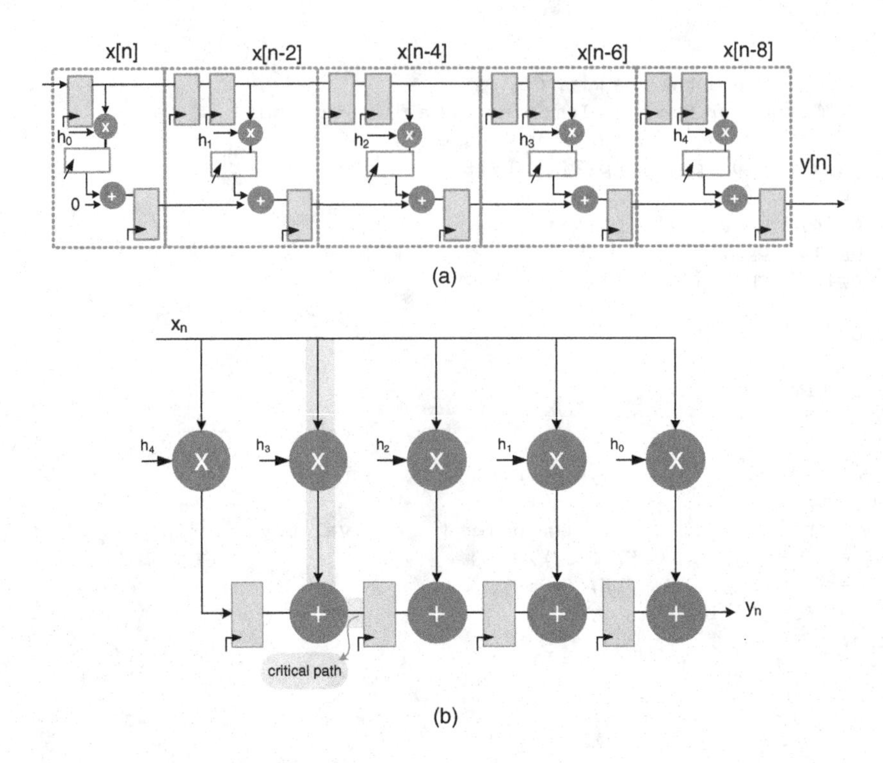

Figure 6.8 (a) Pipeline direct-form FIR filter best suited for implementation on FPGAs with DSP48-like building blocks. (b) TDF structure for optimal mapping of multipliers as CSD shift and add operation

multiplication by non-constant values. For many applications requiring multiplication by constants, CSD multiplication usually is the most optimal option.

Retiming is an effective technique to systematically move algorithmic delays in a design to reduce the critical path of the logic circuit. In Chapter 7, retiming is applied to transform the filter given in Figure 6.8(a) to get an equivalent filter shown in Figure 6.8(b). This new filter is the transposed direct form [6]. It is interesting to observe that in this form, without adding pipelining registers, the algorithm delays are systematically moved using retiming transformation from the upper edge of the DFG to the lower edge. This has resulted in reducing the critical path of an L-coefficient FIR filter from one multiplier and $L-1$ adders of direct form to one multiplier and an adder as shown in Figure 6.8(b).

In FDA designs of a TDF FIR filter, each constant in the multiplier is converted to CSD representation and then appropriate PPs are generated for non-zero digits in CSD representation of the coefficients. These PPs are reduced by carry save reduction to two layers. The outputs of these two layers are respectively stored in two registers. Although this doubles the number of registers in the design, it removes CPAs which otherwise are required to add the final two layers into one for each coefficient multiplication. Figure 6.9 shows the TDF FIR filter implemented as CSD multiplication and carry save addition. Each multiplier represents the generation of PPs for non-zero digits in CSD representation of the coefficients and their reduction to two layers using the carry save reduction scheme.

The critical path can be further reduced by adding pipeline layers in the design, as shown in Figure 6.10(a). Inserting appropriate levels of pipelining can reduce the critical path to just

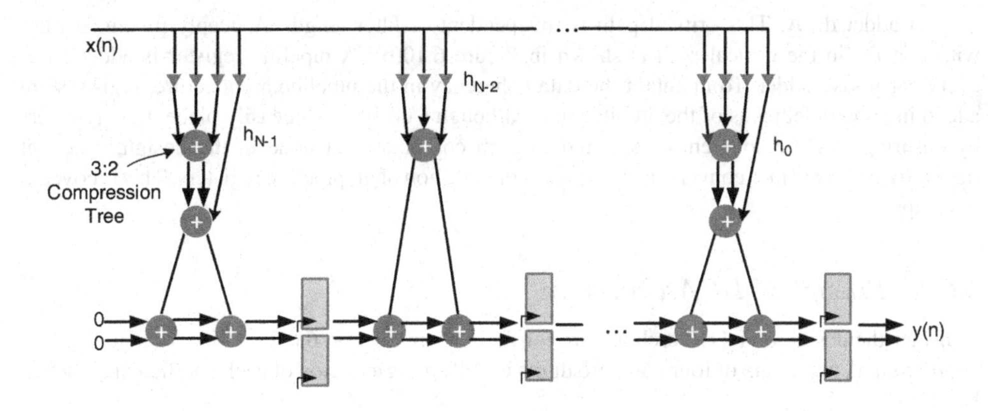

Figure 6.9 Transposed direct-form FIR filter with CSD multiplication and carry save addition

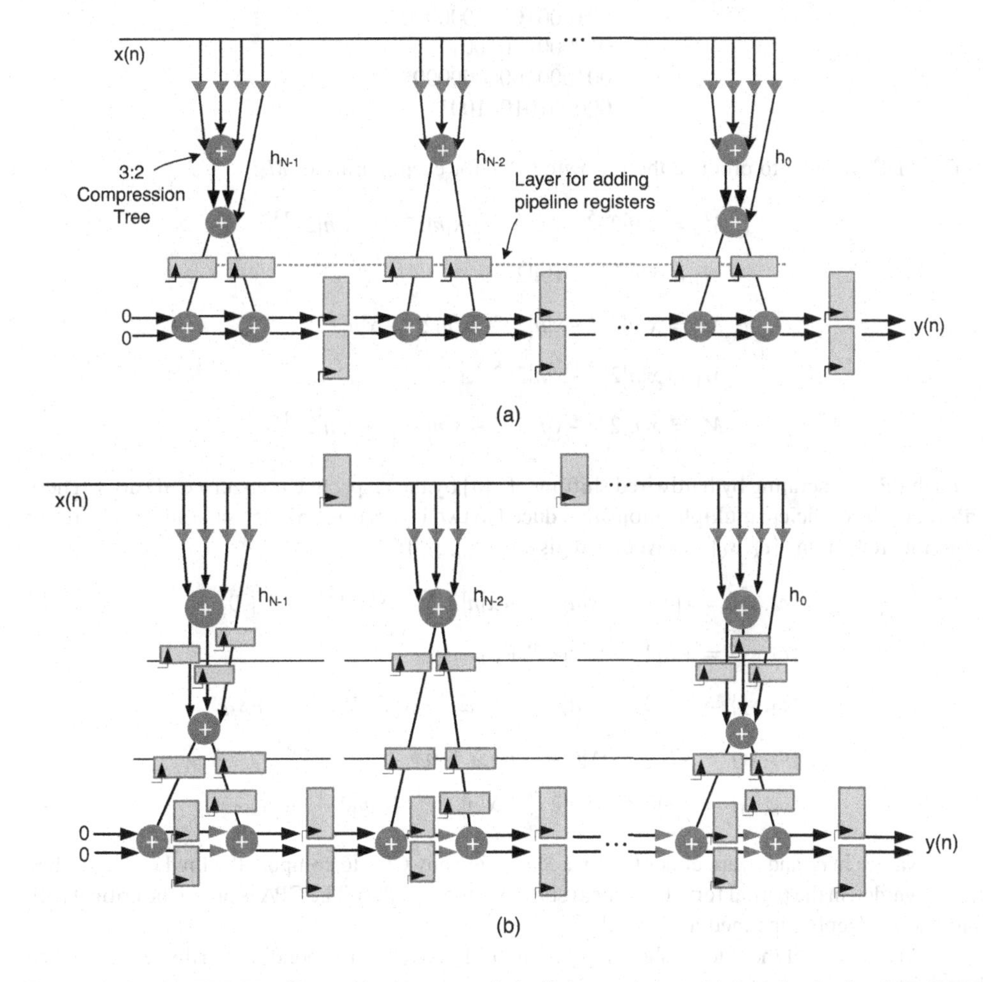

Figure 6.10 (a) TDF FIR filter with one stage of pipelining registers. (b) Deeply pipelined TDF FIR filter with critical path equal to one full adder delay

a fuller-adder delay. This critical path is independent of filter length. A deeply pipelined filter with one FA in the critical path is shown in Figure 6.10(b). A pipeline register is added after every carry save adder. To maintain the data coherency in the pipeline architecture, registers are added in all parallel paths of the architecture. Although a designer can easily place these registers by ensuring that the coherent data is fed to each computational node in the design, a cut-set transformation provides convenience in finding the location of all pipeline registers. This is covered in Chapter 7.

6.6.4 *Example: TDF Architecture*

Consider the design of a TDF architecture for the filter in Section 6.6.2.

Solution: A maximum of four non-zero digits in CSD representation of each coefficient are given here:

$$
\begin{array}{l}
0000010\bar{1}001001\\
001000\bar{1}0000000000\\
010000\bar{1}00\bar{1}00\bar{1}\\
001000\bar{1}000000000\\
0000010\bar{1}001001
\end{array}
$$

For TDF we need to produce the following PPs for each multiplier M_k:

$$M_4 = x[n]2^{-5} - x[n]2^{-7} + x[n]2^{-10} + x[n]2^{-13}$$

$$M_3 = x[n]2^{-2} - x[n]2^{-6}$$

$$M_2 = x[n]2^{-1} - x[n]2^{-6} - x[n]2^{-9} - x[n]2^{-12}$$

$$M_1 = x[n]2^{-2} - x[n]2^{-6}$$

$$M_0 = x[n]2^{-5} - x[n]2^{-7} + x[n]2^{-10} + x[n]2^{-13}$$

Each PP is generating by hardwired shifting of $x[n]$ by the respective non-zero CSD digit. These PPs for each coefficient multiplication are reduced to two layers using a carry save adder. The result from this reduction $\{c_k, s_k\}$ is saved in registers $\{c_{kd}, s_{kd}\}$:

$$\{c_4, s_4\} = x[n]2^{-5} - x[n]2^{-7} + x[n]2^{-10} + x[n]2^{-13} + 0 + 0$$

$$\{c_3, s_3\} = x[n]2^{-2} - x[n]2^{-6} + c_{4d} + s_{4d}$$

$$\{c_2, s_2\} = x[n]2^{-1} - x[n]2^{-6} - x[n]2^{-9} - x[n]2^{-12} + c_{3d} + s_{3d}$$

$$\{c_1, s_1\} = x[n]2^{-2} - x[n]2^{-6} + c_{2d} + s_{2d}$$

$$\{c_0, s_0\} = x[n]2^{-5} - x[n]2^{-7} + x[n]2^{-10} + x[n]2^{-13} + c_{1d} + s_{1d}$$

The values in c_0 and s_0 are either finally added using any CPA to compute the final result, or they are forwarded in the partial form to the next stage of the algorithm. The CPA is not in the critical path and can be deeply pipelined as desired.

An RTL design of the filter is shown in Figure 6.11 and the corresponding Verilog code is given here.

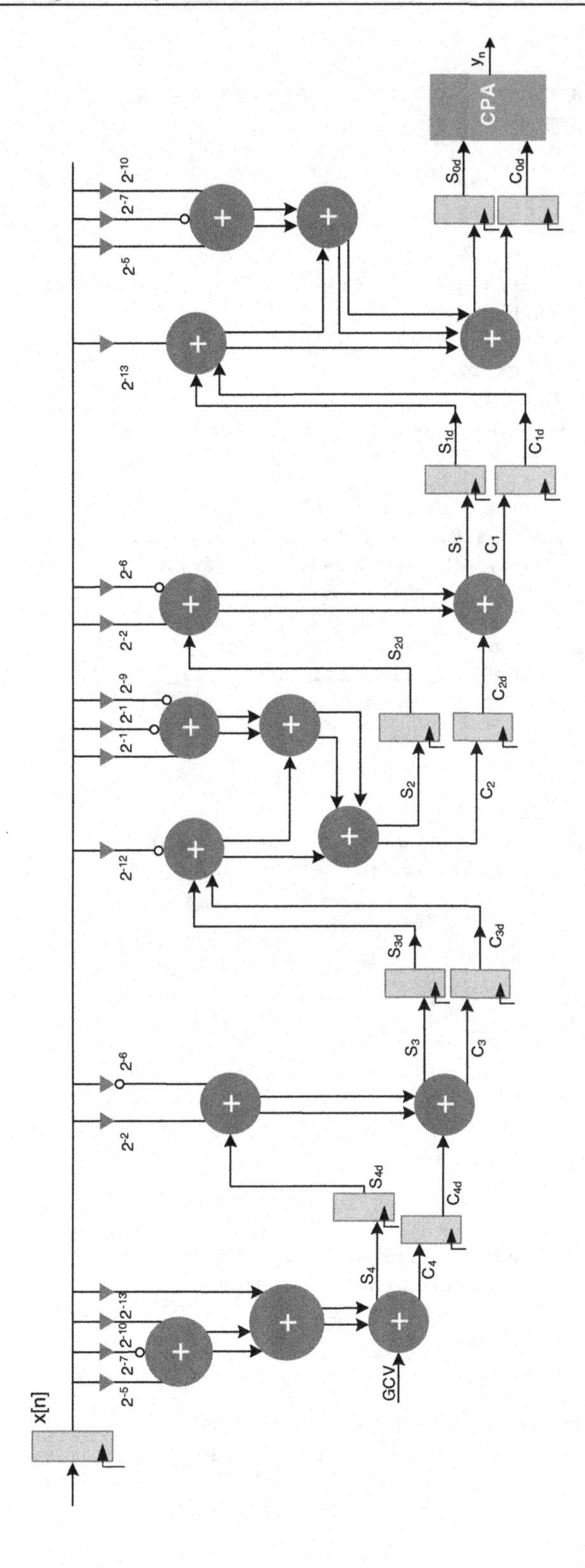

Figure 6.11 RTL design of the 5-coefficient TDF filter of the text example

```
// Module using TDF structure and 3:2 compressors
module FIRfilterTDFComp (
input signed [15:0] x,
input clk,
output signed [31:0] yn);
integer i;
reg signed [15:0] xn;
// Registers for partial sums, as filter is symmetric, the coefficients are numbered
// for 0, 1, 2, ... , 4 rather 4, 3, ... , 0
reg signed [31:0] sn_0, sn_1, sn_2, sn_3, sn_4;
// Registers for partial carries
reg signed [32:0] cn_0, cn_1, cn_2, cn_3, cn_4;
reg signed [31:0] pp_0, pp_1, pp_2, pp_3, pp_4, pp_5,
     pp_6, pp_7, pp_8, pp_9, pp_10, pp_11,
     pp_12, pp_13, pp_14, pp_15;
// Wires for compression tree for intermediate sums and carries
reg signed [31:0] s00, s01, s10, s11, s20, s200, s21, s30,
                    s31, s40, s400, s41, s02, s22, s42;
reg signed [32:0] c00, c01, c10, c11, c20, c200, c21, c30,
                    c31, c40, c400, c41, c02, c22, c42;
// The GCV computed for sign extension elimination
reg signed [31:0] gcv = 32'b0110_1111_0111_0000_0001_0000_1001_0000;
// Add final partial sum and carry using CPA
assign yn = cn_4+sn_4;
always @ (posedge clk)
begin
     xn <= x; // register the input sample
     // Register partial sums and carries in two sets of
     registers for every coeff multiplication
     cn_0 <= c02;
     sn_0 <= s02;
     cn_1 <= c11;
     sn_1 <= s11;
     cn_2 <= c22;
     sn_2 <= s22;
     cn_3 <= c31;
     sn_3 <= s31;
     cn_4 <= c42;
     sn_4 <= s42;
end
always @ (*)
begin
     // First level of 3:2 compressors, initialize 0 bit of
                    all carries that are not used
        c00[0]=0; c10[0]=0; c20[0]=0; c200[0]=0; c30[0]=0; c40[0]=0;
                    400[0]=0;
     for (i=0; i<32; i=i+1)
     begin
          // 3:2 compressor at level 0 for coefficient 0
          {c00[i+1],s00[i]} = pp_0[i]+pp_1[i]+pp_2[i];
          // 3:2 compressor at level 0 for coefficient 1
          {c10[i+1],s10[i]} = pp_4[i]+pp_5[i]+sn_0[i];
```

```
        // 3:2 compressor at level 0 for coefficient 2
        {c20[i+1],s20[i]} = pp_6[i]+pp_7[i]+pp_8[i];
        {c200[i+1],s200[i]} = pp_9[i]+sn_1[i]+cn_1[i];
        // 3:2 compressor at level 0 for coefficient 3
        {c30[i+1],s30[i]} = pp_10[i]+pp_11[i]+sn_2[i];
        // 3:2 compressor at level 0 for coefficient 4
        c40[i+1],s40[i]} = pp_12[i]+pp_13[i]+pp_14[i];
        {c400[i+1],s400[i]} = pp_15[i]+sn_3[i]+cn_3[i];
    end
        c01[0]=0; c11[0]=0; c21[0]=0; c31[0]=0; c41[0]=0;
    // Second level of 3:2 compressors
    for (i=0; i<32; i=i+1)
    begin
        // For coefficient 0
        {c01[i+1],s01[i]} = c00[i]+s00[i]+pp_3[i];
        // For coefficient 1: complete
        {c11[i+1],s11[i]} = c10[i]+s10[i]+cn_0[i];
        // For coefficient 2
        {c21[i+1],s21[i]} = c20[i]+s20[i]+c200[i];
        // For coefficient 3: complete
        {c31[i+1],s31[i]} = c30[i]+s30[i]+cn_2[i];
        // For coefficient 4
        {c41[i+1],s41[i]} = c40[i]+s40[i]+c400[i];
    end
// Third level of 3:2 compressors
  c02[0]=0; c22[0]=0; c42[0]=0;
  for (i=0; i<32; i=i+1)
  begin
 // Add global correction vector
 {c02[i+1],s02[i]} = c01[i]+s01[i]+gcv[i];
    // For coefficient 2: complete
 {c22[i+1],s22[i]} = c21[i]+s21[i]+s200[i];
 // For coefficient 4: complete
 {c42[i+1],s42[i]}= c41[i]+s41[i]+s400[i];
  end
end
always @(*)
begin
  // Generating PPs for 5 coefficients
  // PPs for coefficient h0 with sign extension elimination
  pp_0 = {5b0, ~xn[15], xn[14:0], 11b0};
  pp_1 = {7b0, xn[15], ~xn[14:0], 9b0};
  pp_2 = {10b0, ~ xn[15], xn[14:0],6b0};
  pp_3 = {13b0, ~xn[15], xn[14:0],3b0};
  // PPs for coefficient h1 with sign extension elimination
  pp_4 = {2b0, ~xn[15], xn[14:0], 14b0};
  pp_5 = {6b0, xn[15], ~xn[14:0], 10b0};
  // PPs for coefficient h2 with sign extension elimination
  pp_6 = {1b0, xn[15], xn[14:0], 15b0};
  pp_7 = {6b0, xn[15], ~xn[14:0], 10b0};
  pp_8 = {9b0, xn[15], ~xn[14:0], 7b0};
  pp_9 = {12b0, xn[15], ~ xn[14:0], 4b0};
```

```
  // PPs for coefficient h3 with sign extension elimination
  pp_10 = {2b0, ~xn[15], xn[14:0], 14b0};
  pp_11 = {6b0, xn[15], ~xn[14:0], 10b0};
  // PPs for coefficient h4 with sign extension elimination
  pp_12 = {5b0, ~xn[15], xn[14:0], 11b0};
  pp_13 = {7b0, xn[15], ~xn[14:0], 9b0};
  pp_14 = {10b0, ~xn[15], xn[14:0],6b0};
  pp_15 = {13b0, ~xn[15], xn[14:0],3b0};
end
endmodule
```

6.6.5 *Hybrid FIR Filter Structure*

Earlier sections have elaborated on architecture for a direct and transposed form FIR filter. These architectures are shown in Figures 6.12(a) and (b). Both the architectures have their respective

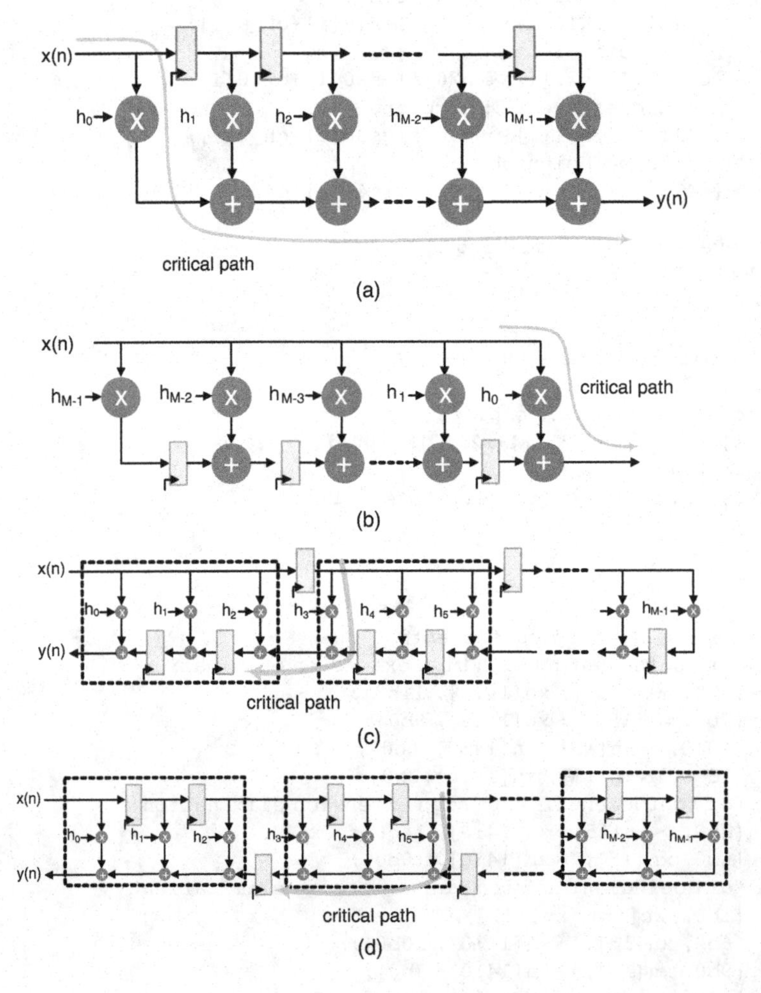

Figure 6.12 FIR filter structures. (a) Direct-form structure. (b) Transposed-form structure. (c) Hybrid. (d) Direct–transposed hybrid

benefits and tradeoffs in HW mapping. The direct form requires only one set of registers and a unified compression tree to optimize the implementation and provide area-efficient designs, whereas the transposed form works on individual multiplications and keeps the results of these multiplications in sum and carry forms that require two sets of registers for holding the intermediate results. This form offers time-efficient designs. A mix of the two forms can also be used for achieving the best time–area tradeoffs. These forms are shown in Figures 6.12 (c) and (d).

6.7 Complexity Reduction

The TDF structure of an FIR filter implements multiplication of all coefficients with $x[n]$, as shown in Figure 6.12(b). These multiplications can be implemented as one unit in a block. This consideration of one unit helps in devising techniques to further simplify the hardware of the block. All the multiplications in the block are searched to share common sub-expressions. This simplification targets reducing either the number of adders or the number of adder levels. The two approaches used for the HW reduction are *sub-graph sharing* and *common sub-expression elimination* [9].

6.7.1 Sub-graph Sharing

The algorithms that can be represented as a dependency graph can be optimized by searching constituent sub-graphs that are shared in the original graph. An algorithm that involves multiplication by constants of the same variable has great potential of sub-graph sharing. The sub-graphs, in these algorithms, are formed by generating all possible MSD (minimum signed digit) factors of each multiplication. An optimization algorithm minimizes the number of adders by selecting those options of the sub-graphs that are maximally shared among multiplications. The problem of generating all possible graphs and finding the ones that minimize the number of adders is a non-deterministic polynomial-time (NP)-complete problem. Many researchers have presented heuristic solutions for finding near optimal solutions of this optimization problem. A detailed description of these solutions is outside the scope of this book, so interested readers will find the references listed in this section very relevant. An n-dimension reduced adder graph (n-RAG) can be sub-optimally computed using efficient heuristics. Such a heuristic is listed in [10].

Example: This example is derived from [11]. An optimal algorithm in the graphical technique first generates multiple options of adder graphs for each multiplication. The algorithm then selects the graph out of all options for each multiplication that shares maximum nodes among all the graphs implementing multiplications by coefficients.

Consider a 3-coefficient FIR filter with the following 12-bit fixed-point values:

$$h_0 = 12'b0000\_0000\_0011 = 3$$

$$h_1 = 12'b0000\_0011\_0101 = 53$$

$$h_2 = 12'b0010\_0010\_1001 = 583$$

These coefficients are decomposed in multiple options. The decomposition is not unique and generates a search space. For the design in the example, some of the MSD decomposition options for each coefficient are:

$$
\begin{aligned}
h_0 &= 3 = 1 + 2^1 \\
h_0 &= 3 = 2^2 - 1 \\
h_1 &= 53 = 1 + 13 \times 2^2 \\
13 &= 2^4 - 3 \\
h_1 &= 53 = 65 - 3 \times 2^2 \\
65 &= 2^6 + 1 \\
h_1 &= 53 = 56 - 3 \\
56 &= 2^6 - 2^3 \\
h_2 &= 585 = 293 \times 2^1 - 1 \\
293 &= 2^8 + 37 \\
37 &= 3 \times 2^3 + 13 \\
h_2 &= 585 = 65 \times 2^3 + 65
\end{aligned}
$$

The graphical representations of these options are shown in Figure 6.13. Although there are several other options for each coefficient, only a few are shown to demonstrate that generating all these options and then finding the ones that maximize sharing of intermediate and final results is complex. For this example, the best solution picks the first sub-graph for h_0 and second sub-graphs for h_1 and h_2. An optimized architecture that uses these options is shown in Figure 6.14.

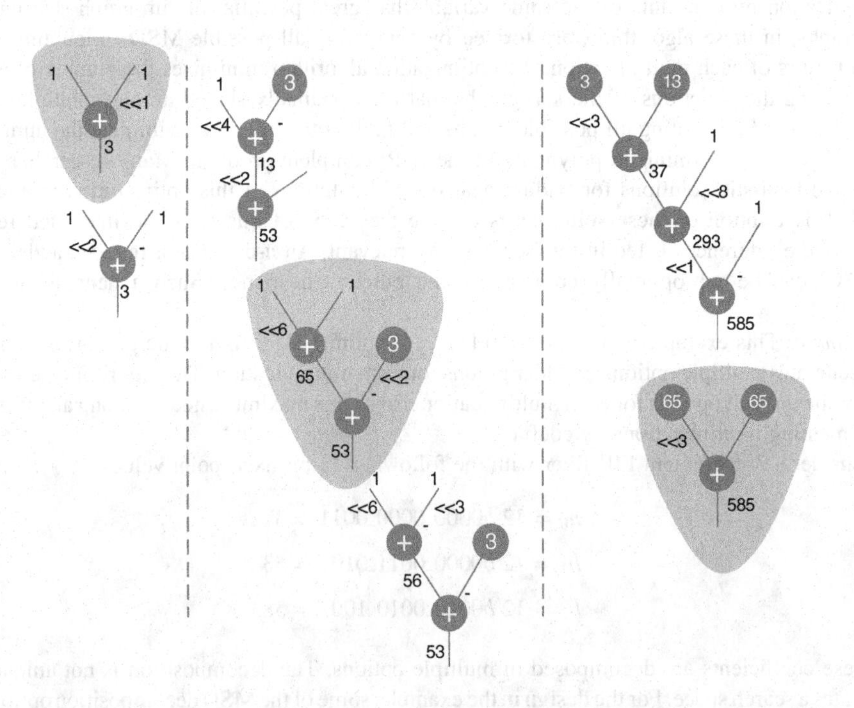

Figure 6.13 Decomposition of {3,53,585} in sub-graphs for maximum sharing

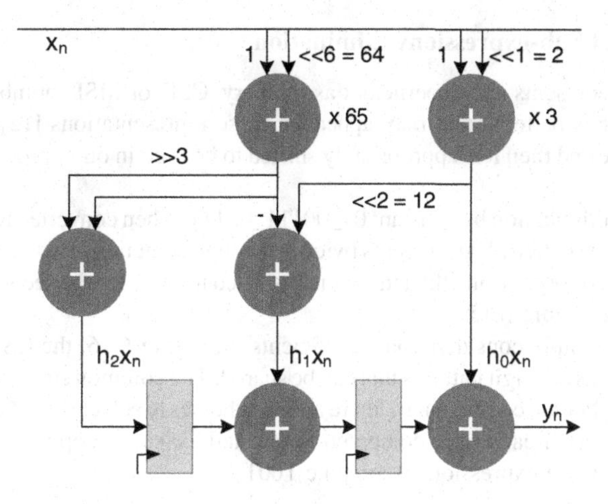

Figure 6.14 Selecting the optimal out of multiple decomposition options for optimized implementation of the FIR filter

6.7.2 Common Sub-expression Elimination

In contrast to graphical methods that decompose each coefficient into possible MSD factors and then find the factors for each multiplication that maixmizes sharing, common sub-expression elimination (CSE) exploites the repetition of bits or digit patterns in binary, CSD or MSD representations of coefficents.

CSE works equally well for binary representation of coefficients, but to reduce the hardware cost the coefficients are first converted to CSD or MSD and are then searched for the common expressions. The technique is explained with the help of a simple example.

Example: Consider a two-tap FIR filter with $h_0 = 5'b01110$ and $h_1 = 5'b01011$ in Q1.4 format. Implementing this filter using TDF structure requires calculation of two products, h_0x_n and h_1x_n. These multiplications independently require the following shift and add operations:

$$\begin{aligned} h_0x_n &= (x_n \gg 1) + (x_n \gg 2) + (x_n \gg 3) \\ h_1x_n &= (x_n \gg 1) + (x_n \gg 3) + (x_n \gg 4) \end{aligned}$$

It is trivial to observe that two of the expressions in both the multiplications are common, and so needs to be computed once for computation of both the products. This reuse of common sub-expressions is shown here:

$$c_0 = (x_n \gg 1) + (x_n \gg 3)$$

This value is then reused in both the expressions:

$$\begin{aligned} h_0x_n &= c_0 + (x_n \gg 2) \\ h_1x_n &= c_0 + (x_n \gg 4) \end{aligned}$$

Several algorithms have been proposed in the literature to remove common sub-expressions [12, 13]. All these algorithms search for direct or indirect commonality that helps in reducing the HW complexity. Some of these methods are elucidated below for CSD and binary representations of coefficients.

6.7.2.1 Horizontal Sub-expressions Elimination

This method first represents each coefficient as a binary, CSD or MSD number and searches for common bits or digits patterns that may appear in these representations [12]. The expression is computed only once and then it is appropriately shifted to be used in other products that contain the pattern.

Example: For multiplication by constant 01_00111_00111, when converted to CSD format, gives 01_0100$\bar{1}$_0100$\bar{1}$. The pattern 100$\bar{1}$ appears twice in the representation. To optimize HW the pattern can be implemented only once and then the value is shifted to cater for the second appearance of the pattern, as shown in Figure 6.15.

Example: This example considers four coefficients. In Figure 6.16, the CSD representation of these constants reveals the digit patterns that can be shared. The common sub-expressions are shown as connected boxes. For the coefficient h_3, there are two choices to select from. One option shares the $\bar{1}0\bar{1}$ pattern, which has already been computed for h_0, and the second option computes 100$\bar{1}$ once and reuses it in the same expression to compute $\bar{1}$001.

6.7.2.2 Vertical Sub-expressions Elimination

This technique searches for bit or digit patterns in columns and those expression are computed once and reused across different multiplications [14]. The following example illustrates the methodology:

$$h_3 = 1\ 0\ 0\ \ 0\ 0$$
$$h_2 = \bar{1}\ 0\ \bar{1}\ \ 0\ 0$$
$$h_1 = 1\ 0\ 1\ \ 0\ \bar{1}$$
$$h_0 = \bar{1}\ 0\ 0\ \ 0\ 1$$

Here, $\left[\begin{smallmatrix}1\\ \bar{1}\end{smallmatrix}\right]$ apears four times. This expression can be computed once and then shared in the rest of the computation. To illustrate invertical bit locations methodology, the convolution summation is first written as:

$$y_n = x_n z^{-3} h_3 + x_n z^{-2} h_2 + x_n z^{-1} h_1 + x_n h_0 \tag{6.6}$$

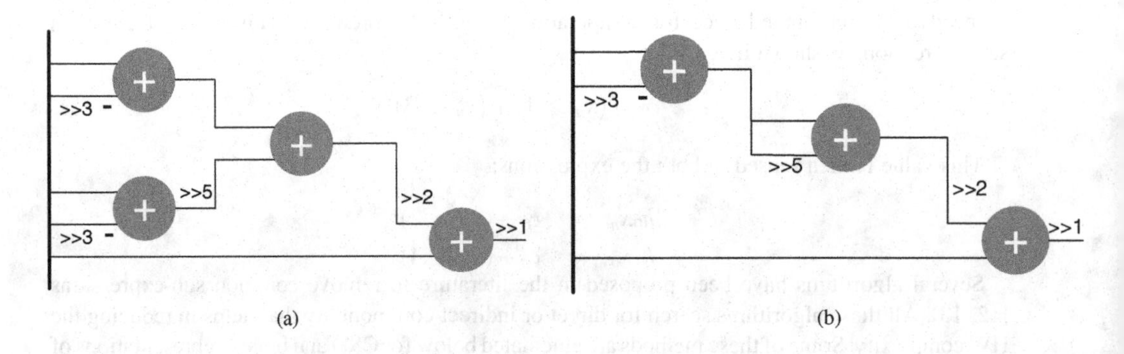

Figure 6.15 To simplify hardware, sub-expression 01001 which is repeated twice in (a) is only computed once in (b)

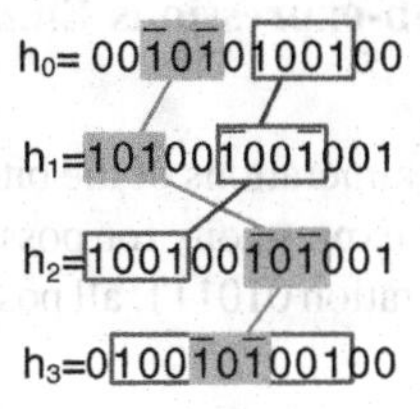

Figure 6.16 Common Sub-expressions for the example in the text

where $x_n z^{-k}$ represents $x[n-k]$ for $k = 0 \ldots 3$. If the multiplications for the given coefficients are implemented as shift and add operations, then convolution summation of (6.6) can be written as:

$$\begin{aligned} y_n &= x_n z^{-3} \\ &\quad - x_n z^{-2} - x_n z^{-2} 2^{-2} \\ &\quad + x_n z^{-1} + x_n z^{-1} 2^{-2} - x_n z^{-1} 2^{-4} \\ &\quad - x_n + x_n z^{-1} 2^{-4} \end{aligned} \tag{6.7}$$

as z^{-1} represents a delay that makes $x_n z^{-1} = x_{n-1}$, using it to extract common sub-expressions:

$$\begin{aligned} y_n &= x_{n-1} z^{-2} \\ &\quad - x_n z^{-2} - x_{n-1} z^{-1} 2^{-2} \\ &\quad + x_{n-1} + x_n z^{-1} 2^{-2} - x_{n-1} 2^{-4} \\ &\quad - x_n + X_{n-1} 2^{-4} \end{aligned} \tag{6.8}$$

Let $x_{n-1} - x_n = w_n$. Then the convolution summation of (6.8) can be rewritten by using the common sub-expression of w_n as:

$$y_n = w_n z^{-2} - w_n z^{-1} 2^{-2} + w_n - w_n 2^{-4} = w_{n-2} - w_{n-1} 2^{-2} + w_n - w_n 2^{-4}$$

The number of additions/subtractions is now reduced from eight to four. Figure 6.17 shows the optimized implementation.

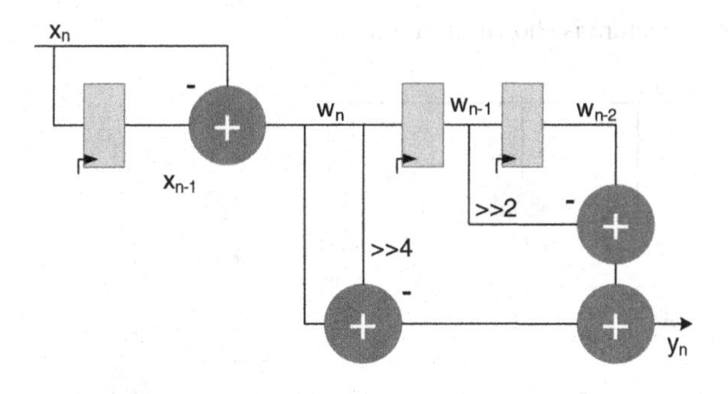

Figure 6.17 Optimized implementation exploiting vertical common sub-expressions

6.7.2.3 Horizontal and Vertical Sub-expressions Elimination with Exhaustive Enumeration

This method searches for all possible enumerations of the bit or digit patterns with at least two non-zero bits in the horizontal and vertical expressions for possible HW reduction [13, 15].

For a coefficient with binary representation 010111, all possible enumerations with a minimum of two non-zero bits are as follows:

010100, 010010, 010001, 000110, 000101, 000011, 010110, 010101, 000111

Similarly, after writing all coefficients in binary representation, each of the columns is also enumerated with all patterns with more than one non-zero entries. Any of these expressions appearing in other columns can be computed once and used multiple times. Exhaustive enumeration, though, gives minimum hardware for the CSE problem but grows exponentially and becomes intractable for large filter lengths.

Example:

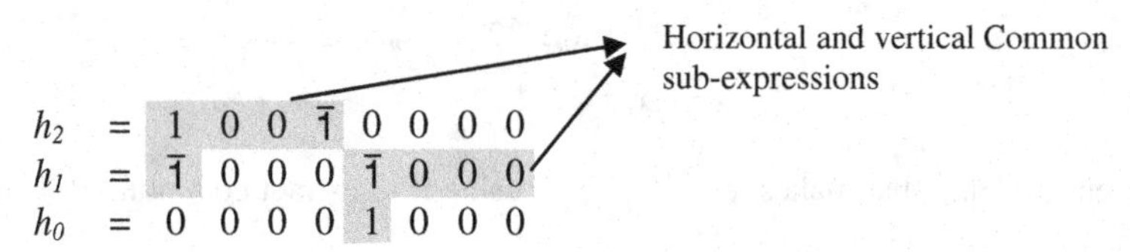

Combining the horizontal and vertical sharing of expressions, the convolution summation can be written as:

$$\begin{aligned} y_n = {} & x_{n-1}z^{-1} - x_{n-1}z^{-1}2^{-3} \\ & - x_n z^{-1} - x_{n-1}2^{-4} + x_{n-1}2^{-7} \\ & + x_n 2^{-4} \end{aligned}$$

A common expression that shares vertical and horizontal sub-expressions can result in optimized logic. Let the expression be:

$$w_n = x_{n-1} - x_{n-1}2^{-3} - x_n$$

Then the convolution summation becomes:

$$y_n = w_{n-1} - w_n 2^{-4}$$

The optimized architecture is shown in Figure 6.18.

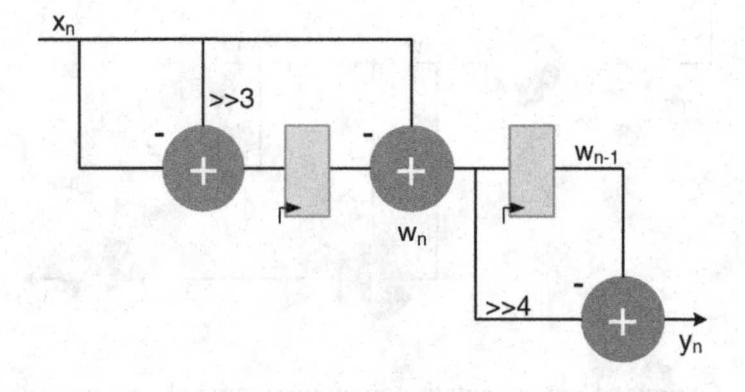

Figure 6.18 Example of horizontal and vertical sub-expressions elimination

6.7.3 Common Sub-expressions with Multiple Operands

In many signal processing applications, the multiplication is with multiple operands, as in DCT, DFT or peak-to-average power ratio (PAPR) reduction using a precoding matrix in an orthogonal frequency-division multiplexed (OFDM) transmitter [16]. In these algorithms an extension of the method for common sub-expression elimination can also be used while computing more than one output sample as a linear combination of input samples [17]. Such a system requires computing matrix multiplication of the form:

$$\begin{bmatrix} y_0[n] \\ y_1[n] \\ y_2[n] \end{bmatrix} = \begin{bmatrix} c_{00} & c_{01} & c_{02} \\ c_{10} & c_{11} & c_{12} \\ c_{20} & c_{21} & c_{22} \end{bmatrix} \begin{bmatrix} x_0[n] \\ x_1[n] \\ x_2[n] \end{bmatrix}$$

This multiplication results in computing a linear combination of input sample for each output sample. The first equation is:

$$y_0[n] = c_{00}x_0[n] + c_{01}x_1[n] + c_{02}x_2[n]$$

All the multiplications in the equation are implemented as multiply shift operations where the techniques discussed in this section for complexity reduction can be used for sub-graph sharing and common expression elimination. The example below highlights use of the common sub-expression elimination method on a set of equations for given constants.

Example:

In the following set of equations all the constants are represented in CSD format:

$$\begin{bmatrix} y_0[n] \\ y_1[n] \\ y_2[n] \end{bmatrix} = \begin{bmatrix} 010\bar{1}0 & \bar{1}0010 & 0100\bar{1} \\ 0010\bar{1} & 0\bar{1}010 & 0010\bar{1} \\ 01000 & 00100 & 000\bar{1}0 \end{bmatrix} \begin{bmatrix} x_0[n] \\ x_1[n] \\ x_2[n] \end{bmatrix}$$

The required matrix multiplication using shift and add operations can be optimized by using the following sub-expressions:

$$p_0[n] = (x_0[n] \gg 1) - (x_0[n] \gg 3) - x_1[n] + (x_2[n] \gg 1)$$
$$p_1[n] = x_2[n] \gg 2 - x_2[n] \gg 3$$

These sub-expressions simplify the computation of matrix multiplication as shown here:

$$y_0[n] = p_0[n] + p_1[n] \gg 1$$
$$y_1[n] = (p_0[n] \gg 1) + (p_1[n] \gg 1)$$
$$y_2[n] = (x_0[n] \gg 1) + p_1[n]$$

6.8 Distributed Arithmetic

6.8.1 Basics

Distributed arithmetic (DA) is another way of implementing a dot product where one of the arrays has constant elements. The DA can be effectively used to implement FIR, IIR and FFT type algorithms [18–24]. For example, in the case of an FIR filter, the coefficients constitute an array of constants in some signed Q-format where the tapped delay line forms the array of variables which changes every sample clock. The DA logic replaces the MAC operation of convolution summation of (6.5) into a bit-serial look-up table read and addition operation [18]. Keeping in perspective the architecture of FPGAs, time/area effective designs can be implemented using DA techniques [19].

The DA logic works by first expanding the array of variable numbers in the dot product as a binary number and then rearranging MAC terms with respect to weights of the bits. A mathematical explanation of this rearrangement and grouping is given here.

Let the different elements of arrays of constants and variables be A_k and x_k, respectively. The length of both the arrays is K. Then their dot product can be written as:

$$y = \sum_{k=0}^{K-1} A_k x_k \tag{6.9}$$

Without lost of generality, let us assume x_k is an N-bit Q1. $(N-1)$-format number:

$$x_k = -x_{k0}2^0 + \sum_{b=1}^{N-1} x_{kb}2^{-b} = -x_{k0}2^0 + x_{k1}2^{-1} + \cdots x_{k(N-1)}2^{N-1}$$

The dot product of (6.9) can be written as:

$$y = \sum_{k=0}^{K-1}\left(-x_{k0}2^0 + \sum_{b=1}^{N-1} x_{kb}2^{-b}\right)A_k$$

$$y = \sum_{k=0}^{K-1}(-x_{k0}2^0 + x_{k1}2^{-1} + \cdots x_{k(N-1)}2^{N-1})A_k$$

Rearranging the terms yields:

$$y = -\sum_{k=0}^{K-1} x_{k0}A_k2^0 + \sum_{b=1}^{N-1} 2^{-b}\sum_{k=0}^{K-1} x_{kb}A_k$$

For $K=3$ and $N=4$, the rearrangement forms the following entries in the ROM:

$$\begin{aligned}
&-(x_{00}A_0 + x_{10}A_1 + x_{20}A_2)2^0\\
&+(x_{01}A_0 + x_{11}A_1 + x_{21}A_2)2^{-1}\\
&+(x_{02}A_0 + x_{12}A_1 + x_{22}A_2)2^{-2}\\
&+(x_{03}A_0 + x_{13}A_1 + x_{23}A_2)2^{-3}
\end{aligned}$$

The DA technique pre-computes all possible values of

$$\sum_{k=0}^{K-1} x_{kb}A_k$$

For the example under consideration, the summations for all eight possible values of x_{kb} for a particular b and $k=0$, 1 and 2 are computed and stored in ROM. The ROM is P bits wide and 2^K deep and implements a look-up table. The value of P is:

$$P = \left\lfloor \log_2 \sum_{k=0}^{K-1} |A_k| \right\rfloor + 1$$

Table 6.1 ROM for distributed arithmetic

x_{2b}	x_{1b}	x_{0b}	Contents of ROM
0	0	0	0
0	0	1	A_0
0	1	0	A_1
0	1	1	$A_1 + A_0$
1	0	0	A_2
1	0	1	$A_2 + A_0$
1	1	0	$A_2 + A_1$
1	1	1	$A_2 + A_1 + A_0$

where $\lfloor . \rfloor$ is the floor operator that rounds a fraction value to its lower integer. The contents of the look-up table are given in Table 6.1.

All the elements of the vector $\mathbf{x} = [x_0\ x_1. \ldots x_{K\text{-}1}]$ are stored in shift registers. The architecture considers in each cycle the bth bit of all the elements and concatenates them to form the address to the ROM. For the most significant bits (MSBs) the value in the ROM is subtracted from a running accumulator, and for the rest of the bit locations values from ROM are added in the accumulator. To cater for weights of different bit locations, in each cycle the accumulator is shifted to the right. To keep the space for the shift, the size of the accumulator is set to $P + N$, where a (P-bit adder adds the current output of the ROM in the accumulator and N bits of the accumulator are kept to the right side to cater for the shift operation. The data is input to the shift registers from LSB. The dot product takes N cycles to compute the summation. The architecture implementing the dot product for $K = 3, P = 5$ and $N = 4$ is shown in Figure 6.19. FPGAs with look-up tables suit well DA-based filter design [20].

Example: Consider a ROM to compute the dot product of a 3-element vector with a vector of constants with the following elements: $A_0 = 3, A_1 = -1$ and $A_2 = 5$. Test the design for the following values in vector $\mathbf{x}$:

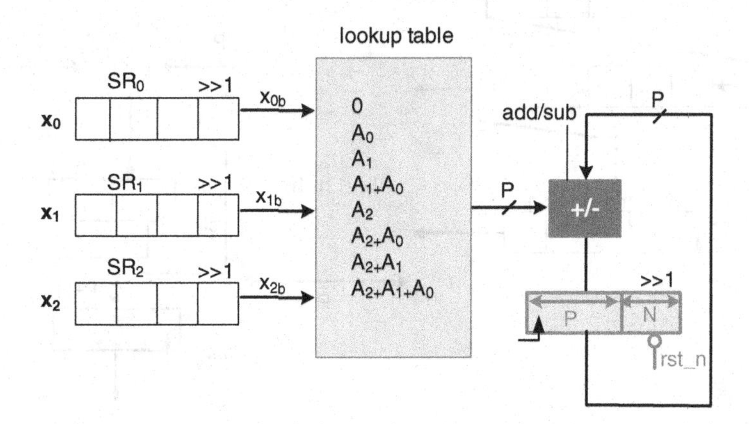

Figure 6.19 DA for computing the dot product of integer numbers for $N = 4$ and $K = 3$

Table 6.2 Look-up table (LUT) for the text example

x_{2b}	x_{1b}	x_{0b}	Contents of ROM	
0	0	0	0	0
0	0	1	A_0	3
0	1	0	A_1	−1
0	1	1	A_1+A_0	2
1	0	0	A_2	5
1	0	1	A_2+A_0	8
1	1	0	A_2+A_1	4
1	1	1	$A_2+A_1+A_0$	7

$$x_0 = -6 = 4'b1010$$

$$x_1 = 6 = 4'b0110$$

$$x_2 = -5 = 4'b1011$$

The contents of a 5-bit wide and 8-bit deep ROM are given in Table 6.2. The shift registers are assumed to contain elements of vector **x** with LSB in the right-most bit location.

The cycle-by-cycle working of the DA architecture of Figure 6.20 for the case in consideration is given in Table 6.3. After four cycles the accumulator contains the value of the dot product: $9'b111001\_111 = -49_{10}$.

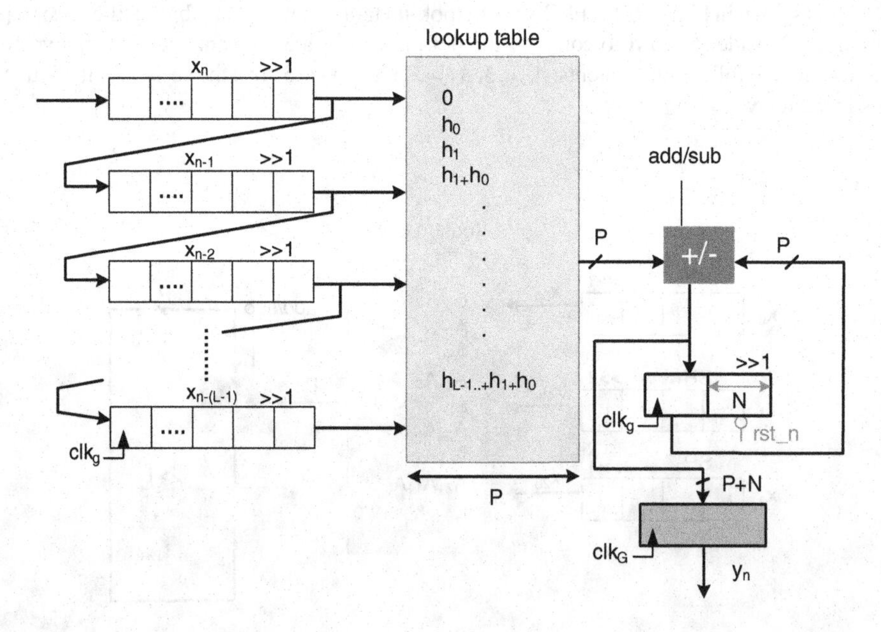

Figure 6.20 DA-based architecture for implementing an FIR filter of length L and N-bit data samples

Table 6.3 Cycle by cycle working of DA for the text example

Cycle	Address	ROM	Accumulator
0	$3'b100$	5	00010_1000
1	$3'b111$	7	00100_1100
2	$3'b010$	−1	00001_1110
3	$3'b101$	8	11100_1111

6.8.2 Example: FIR Filter Design

The DA architecture can be effectively used for implementing an FIR filter. The technique is of special interest in applications where the data is input to the system in bit-serial fashion. A DA-based design eliminates the use of a hardware multiplier and uses only a look-up table to provide high throughput execution at bit-rate irrespective of the filter length and width of the coefficients [21].

Figure 6.21 shows the architecture of an L-coefficient FIR filter. The data is input to the design in bit-serial fashion into shift register SR_0, and all the shift registers are connected in a daisy-chain to form a tap delay line. The design works at bit clock clk_g, and the design computes an output at every sample clk_G, where sample clk_G is N times slower than clk_g.

RTL Verilog code of a DA-based 4-coefficient FIR filter is given below. The code also implements an FIR filter using convolution summation for equivalence checking:

```
/* Distributed arithmetic FIR filter
module FIRDistributedArithmetics
(
input xn_b, clk_g, rst_n,
input [3:0] contr,
output reg signed [31:0] yn,
output valid);
reg signed [31:0] acc; // accumulator
reg [15:0] xn_0, xn_1, xn_2, xn_3; // tap delay line
reg [16:0] rom_out;
reg [3:0] address;
wire signed [31:0] sum;
wire msb;

// DA ROM storing all the pre-computed values
always @(*)
begin
  //lsb of all registers
  address={xn_3[0],xn_2[0],xn_1[0],xn_0[0]};
  case(address)
      4'd0: rom_out=17'b00000000000000000;
      4'd1: rom_out=17'b00000001000100100; // h0
      4'd2: rom_out=17'b00011110111011100; // h1
      4'd3: rom_out=17'b00100000000000000; // h0+h1
      4'd4: rom_out=17'b00011110111011100; // h2
      4'd5: rom_out=17'b00100000000000000; // h2+h0
```

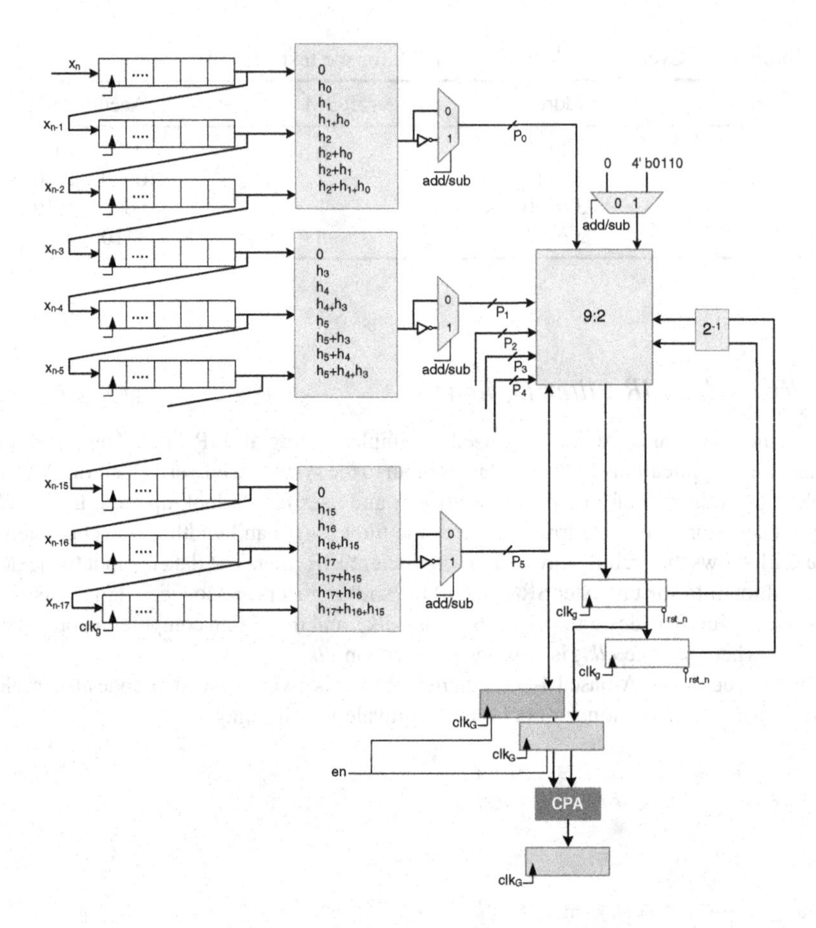

Figure 6.21 DA-based parallel implementation of an 18-coefficient FIR filter setting $L=3$ and $M=6$

```
        4'd6: rom_out=17'b00111101110111000; // h2+h1
        4'd7: rom_out=17'b00111110111011100; // h2+h1+h0
        4'd8: rom_out=17'b00000001000100100; // h3
        4'd9: rom_out=17'b00000010001001000; // h3+h0
        4'd10: rom_out=17'b00100000000000000; // h3+h1
        4'd11: rom_out=17'b00100001000100100; // h3+h1+h0
        4'd12: rom_out=17'b00100000000000000; // h3+h2
        4'd13: rom_out=17'b00100001000100100; // h3+h2+h0
        4'd14: rom_out=17'b00111110111011100; // h3+h2+h1
        4'd15: rom_out=17'b01000000000000000; // h3+h2+h1+h0
        default: rom_out= 17'bx;
    endcase
end
assign valid = ~ (|contr); // output data valid signal
assign msb = contr // msb = 1 for contr = ff
assign sum = (acc +
   {rom_out^{17{msb}}, 16'b0} + // takes 1's complement for msb
```

```
   {15'b0, msb, 16'b0}) > > >1; // add 1 at 16th bit location for 2's
complement logic
always @ (posedge clk_g or negedge rst_n)
begin
   if(!rst_n)
      begin
             // Initializing all the registers
                 xn_0 <= 0;
                 xn_1 <= 0;
                 xn_2 <= 0;
                 xn_3 <= 0;
                 acc <= 0;
      end
else
      begin
      // Implementing daisy-chain for DA computation
                xn_0 <= {xn_b, xn_0[15:1]};
                xn_1 <= {xn_0[0], xn_1[15:1]};
                xn_2 <= {xn_1[0], xn_2[15:1]};
                xn_3 <= {xn_2[0], xn_3[15:1]};
  // A single adder should be used instead, shift right to multiply by 2-i
  if(&contr)
 begin
yn <= sum;
acc <= 0;
 end
  else
 acc <= sum;
   end
end
endmodule

// Module uses multipliers to implement an FIR filter for verification of DA arch
module FIRfilter
(
input signed [15:0] x,
input clk_s, rst_n,
output reg signed [31:0] yn);
reg signed [15:0] xn_0, xn_1, xn_2, xn_3;
wire signed [31:0] yn_v, y1, y2, y3, y4;

// Coefficients of the filter
wire signed [15:0] h0 = 16'h0224;
wire signed [15:0] h1 = 16'h3DDC;
wire signed [15:0] h2 = 16'h3DDC;
wire signed [15:0] h3 = 16'h0224;

// Implementing filters using multiplication and addition operators
assign y1=h0*xn_0;
assign y2=h1*xn_1;
assign y3=h2*xn_2;
assign y4=h3*xn_3;
```

```
assign yn_v = y1+y2+y3+y4;
always @(posedge clk_s or negedge rst_n)
begin
   if (!rst_n)
  begin
 xn_0<=0;
 xn_1<=0;
 xn_2<=0;
 xn_3<=0;
  end
   else
  begin
  // Tap delay line of the filter
 xn_0 <= x;
 xn_1 <= xn_0;
 xn_2 <= xn_1;
 xn_3 <= xn_2;
  end
   // Registering the output
  yn <= yn_v;
end
endmodule

module testFIRfilter;
reg signed [15:0] X_reg, new_val;
reg CLKg, CLKs, RST_N;
reg [3:0] counter;
wire signed [31:0] YND;
wire signed [31:0] YN;
wire VALID, inbit;
integer i;

// Instantiating the two modules, FIRfilter is coded for equivalence checking
FIRDistributedArithmetics FIR_DA(X_reg[counter],
CLKg, RST_N, counter, YND, VALID);
FIRfilter FIR(new_val, CLKs, RST_N, YN);
initial
begin
   CLKg = 0; // bit clock
   CLKs = 0; // sample clock
   counter = 0;
   RST_N = 0;
   #1 RST_N = 1;
   new_val = 1;
   X_reg = 1;
   #10000 $finish;
end
// Generating clock signal
always
   #2 CLKg = ~ CLKg; // fast clock
always
   #32 CLKs = ~ CLKs; // 16 times slower clock
```

```
// Generating a number of input samples
always @ (counter)
begin
   // A new sample at every sample clock
   if (counter == 15)
  new_val = X_reg-1;
end

// Increment counter that controls the DA architecture to be placed in a con-
troller
always @ (posedge CLKg)
begin
   counter <= counter+1;
   X_reg <= new_val;
end
initial
   $monitor ($time, " X_reg=%d, YN=%d, YND=%d\n", X_reg, YN, YND);
endmodule
```

It is obvious from the configuration of a DA-based design that the size of ROM increases with an increase in the number of coefficients of the filter. For example, a 128-coefficients FIR filter requires a ROM of size 2^{128}. This size is prohibitively large and several techniques are used to reduce the ROM requirement [22–25].

6.8.3 *M*-parallel Sub-filter-based Design

This technique divides the filter into *M* sub-filters of length *L*, and each sub-filter is implemented as an independent DA-based module. For computing the output of the filter, the results of all *M* sub-filters are first compressed using any reduction tree, and the final sum and carry are added using a CPA. For a filter of length *K*, the length of each sub-filter is $L=K/M$. The filter is preferably designed to be of length *LM*, or one of the sub-filters may be of a little shorter length than the rest. For the parallel case, the convolution summation can be rewritten as:

$$y[n] = \sum_{i=0}^{M-1}\sum_{k=0}^{L-1} h[i*L+k]x[n-(i*L+k)]$$

The inner summation implements each individual sub-filter. These filters are designed using a DA-based MAC calculator. The outer summation then sums the outputs of all sub-filters. The summation is implemented as a compression tree and a CPA. The following example illustrates the design.

Example: The architecture designs an 18-coefficient FIR filter using six sub-filters. For the design $L=3$ and $M=6$, and the architecture is shown in Figure 6.21. Each sub-filter implements a $2^3=8$ deep ROM. The width of each ROM, P_i for $i=0\ldots5$, depends on the maximum absolute value of its contents. The output of each ROM is input to a compression tree. For the MSBs in the respective daisy-chain tap delay-line, the result needs to be subtracted. To implement this subtraction the architecture selects the one's complement of the output from the ROM and a cumulative correction term for all the six sub-filters is added as $4'b0110$ in the compression tree. The CPA is moved outside the accumulation module and the partial sum and partial carry from the compression tree is latched in the two sets of accumulator registers. The contents in the registers are also input to the compression tree. This makes the compression tree 9:2. If necessary the CPA adder needs to work on slower output sample-clock clk_G, whereas the compression tree operates on fast bit-clock clk_g. The final results from the compression trees are latched into two sets of registers clocked with clk_G for final addition using a CPA and the two accumulator registers are reset to perform next set of computation.

6.8.4 DA Implementation without Look-up Tables

LUT-less DA implementation uses multiplexers. If the parallel implementation is extended to use $M=K$, then each shift register is connected to a two-entry LUT that either selects a 0 or the corresponding coefficient. The LUT can be implemented as a 2:1 MUX.

Designs for a 4-coefficient FIR filters are shown in Figure 6.22, using compression- and adder tree-based implementation. For the adder tree design the architecture can be pipelined at each adder stage if required.

The architectures of LUT and LUT-less implementation can be mixed to get a hybrid design. The resultant design has a mix of MUX- and LUT-based implementation. The design requires reduced sized LUTs.

Example: This example implements a DA-based biquadrature IIR filter. The transfer function of the filter is:

$$H(z) = \frac{b_0 + b_1 z^{-1} + b_2 z^{-2}}{1 - a_1 z^{-1} - a_2 z^{-2}}$$

This transfer function translates into a difference equation given by:

$$y[n] = b_0 x[n] + b_1 x[n-1] + b_2 x[n-2] + a_1 y[n-1] + a_2 y[n-2]$$

The difference equation can be easily mapped on DA-based architecture. Either two ROMs can be designed for feed forward and feed back coefficients, or a unified ROM-based design can be realized. The two designs are shown in Figure 6.23. The value of the output, once computed, is loaded in parallel to a shift register for $y[n-1]$.

6.9 FFT Architecture using FIR Filter Structure

To fully exploit the potential optimization in mapping a DFT algorithm in hardware using techniques listed in this chapter, the DFT algorithm can be implemented as an FIR filter. This requires rewriting of the DFT expression as convolution summation. The Bluestein Chirp-z Transform (CZT) algorithm transforms the DFT computation problem into FIR filtering [25]. The CZT translates the nk term in the DFT summation in terms of $(k-n)$ for it to be written as a convolution summation.

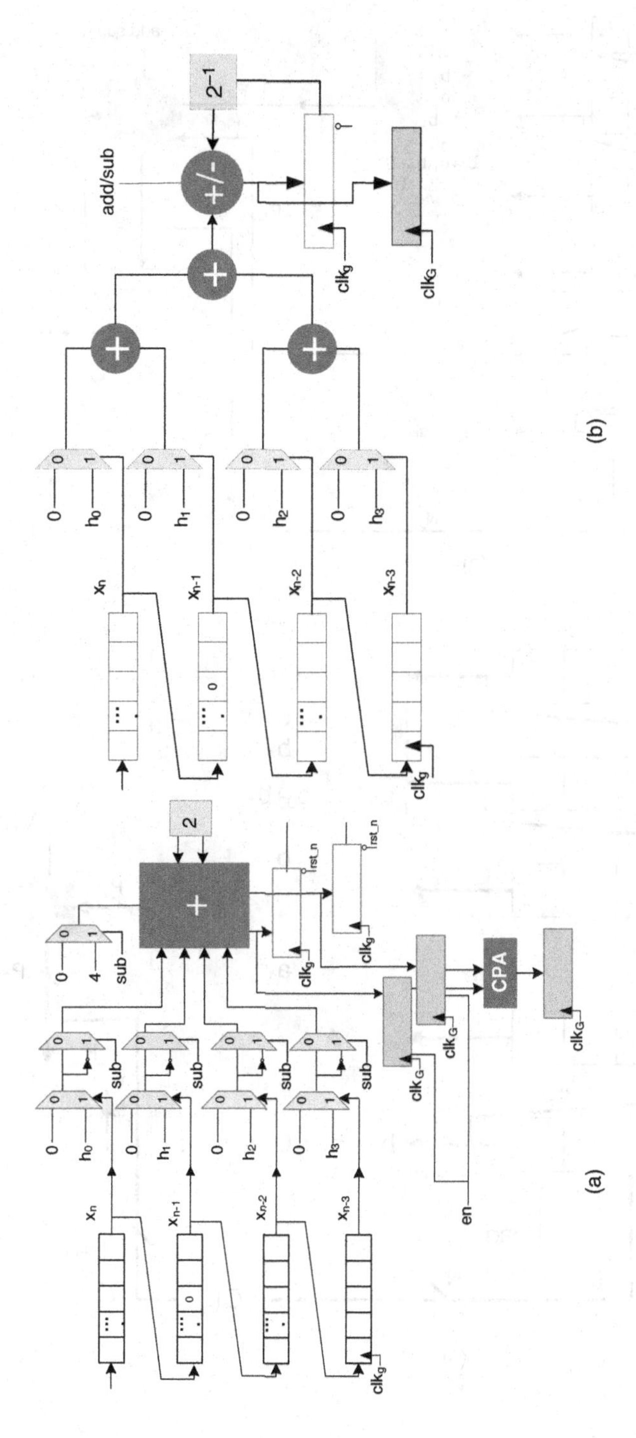

Figure 6.22 A LUT-less implementation of a DA-based FIR filter. (a) A parallel implementation for $M = K$ uses a 2:1 MUX, compression tree and a CPA. (b) Reducing the output of the multiplexers using a CPA-based adder tree and one accumulator

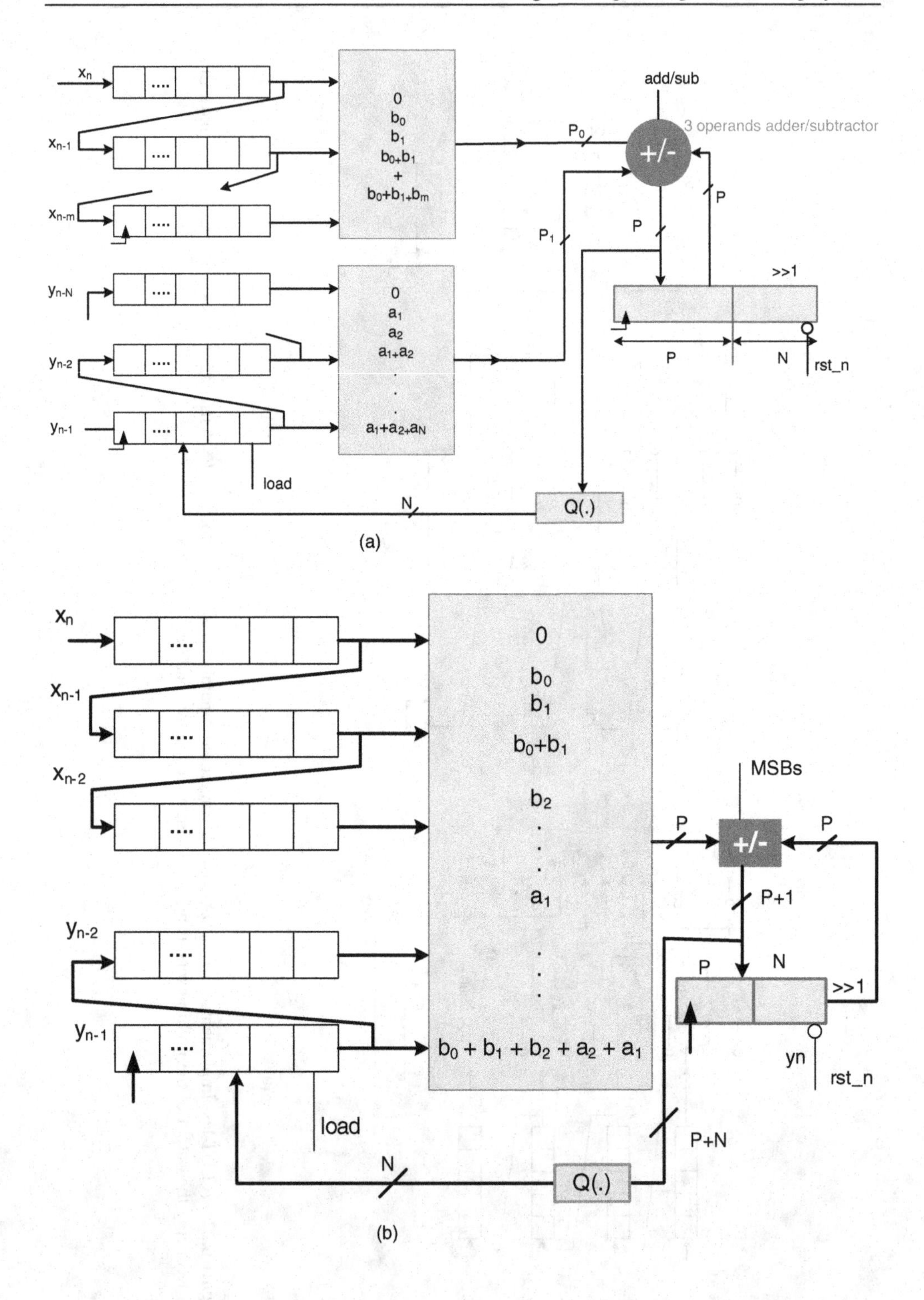

Figure 6.23 DA-based IIR filter design (a) Two ROM-based design. (b) One ROM-based design

The DFT summation is given as:

$$X[k] = \sum_{n=0}^{N-1} x[n]W_N^{nk} \text{ for } k = 0, 1, 2, \ldots, N-1; \text{ where } W_N = e^{-2\pi/N} \tag{6.10}$$

The nk term in (6.10) can be expressed as:

$$nk = \frac{-(k-n)^2 + n^2 + k^2}{2} \tag{6.11}$$

Substituting (6.10) in (6.11) produces the DFT summation as:

$$X[k] = W_N^{k^2/2} \sum_{n=0}^{N-1} \left(x[n]W_N^{n^2/2}\right) W_N^{-(k-n)^2/2} \text{ for } k = 0, 1, 2, \ldots, N-1 \tag{6.12}$$

Using the circular convolution notation $\otimes$, the expression in (6.12) can be written as:

$$X[k] = W_N^{\frac{k^2}{2}} \left(x[n]W_N^{\frac{n^2}{2}} \otimes W_N^{\frac{-k^2}{2}} \right)$$

To compute an N-point DFT, the signal is first multiplied by an array of constants $W_N^{n^2/2}$. Then an N-point circular convolution is performed with an impulse response $W_N^{-n^2}$. Finally the output of the convolution operation is again multiplied with an array of constants $W_N^{k^2/2}$. A representative serial architecture for the algorithm for an 8-point FFT computation is given in Figure 6.24.

The array of constants is stored in a ROM. The input data $x[n]$ is serially input to the design. The input data is multiplied by the corresponding value from the ROM to get $x[n]W_N^{n^2/2}$. The output of the multiplication is fed to an FIR filter with constant coefficients $W_N^{-n^2/2}$. The output of the FIR filter is passed through a tapped delay line to compute the circular convolution. The final output is again serially multiplied with the array of constants in the ROM. By exploiting the symmetry in the array of constants the size of ROM can be reduced. The multiplication by coefficients for the convolution implementation can also be reduced using a CSE elimination and sub-graph optimization techniques. The DFT implemented using this technique has been demonstrated to use less hardware and have better fixed-point performance [25].

The following MATLAB® code implements an 8-point FFT using this technique:

```
clear all
xn =[1+j 1 1 -1-2j 0 0 0 1]; % Generate test data
N=8;
n=0:N-1;
Wsqr = exp((-j*2*pi*n.^2)/(2*N));
xnW = Wsqr .* xn;
WsqrT = Wsqr';
yn = conv(xnW,WsqrT.');
yn_cir = yn(1:N) + [yn(N+1:end) 0];
Xk=Wsqr.*yn_cir;
% To compare with FFT
Xk_fft=fft(xn);
diff = sum(abs(Xk-Xk_fft))
```

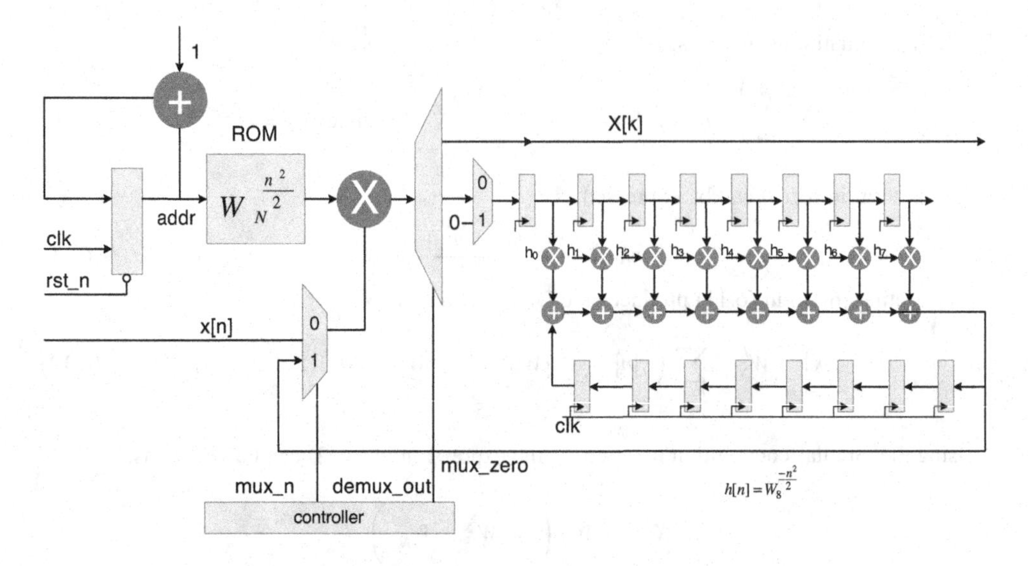

Figure 6.24 DFT implementation using circular convolution

The following example optimizes the implementation of DFT architecture.

Example: Redesign the architecture of Figure 6.24 using a TDF FIR filter structure. Optimize the multiplications using the CSE technique.

The filter coefficients for $N=8$ are computed by evaluating the expression $W_N^{-n^2/2}$ for $n=0\ldots7$. The values of the coefficients are:

$$h[n] = [1, 0.92+0.38j, j\text{–}0.92\text{–}0.38j,\ 1\text{–}0.92\text{–}0.38j,\ j\,0.92+0.38j]$$

These values of coefficients require just one multiplier and swapping of real and imaginary components of $x[n]$ for realizing multiplication by j. The FIR filter structure of Figure 6.24 is given in Figure 6.25.

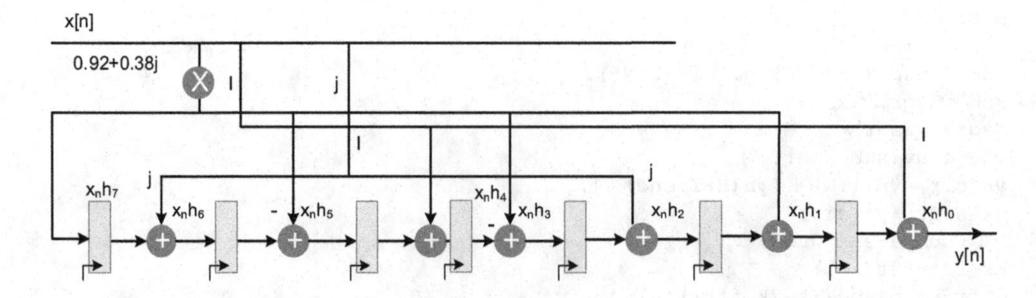

Figure 6.25 Optimized TDF implementation of the DF implementation in Figure 6.24

Exercises

Exercise 6.1

Convert the following expression into its equivalent 8-bit fixed-point representation:

$$y[n] = -0.231x[n] + 0.396x[n-1] + 0.1111x[n-5]$$

Further convert the fixed-point constants into their respective CSD representations. Consider $x[n]$ is an 8-bit input in Q1.7 format. Draw an RTL diagram to represent your design. Each multiplication should be implemented as a CSD multiplier. Consider only the four most significant non-zero bits in your CSD representation.

Exercise 6.2

Implement the following difference equation in hardware:

$$y[n] = -0.9821y[n-1] + x[n]$$

First convert the constant to appropriate 8-bit fixed-point format, and then convert fixed-point number in CSD representation. Implement CSD multipliers and code the design in RTL Verilog.

Exercise 6.3

Draw an optimal architecture that uses a CSD representation of each constant with four non-zero bits. The architecture should only use one CPA outside the filter structure of Figure 6.26.

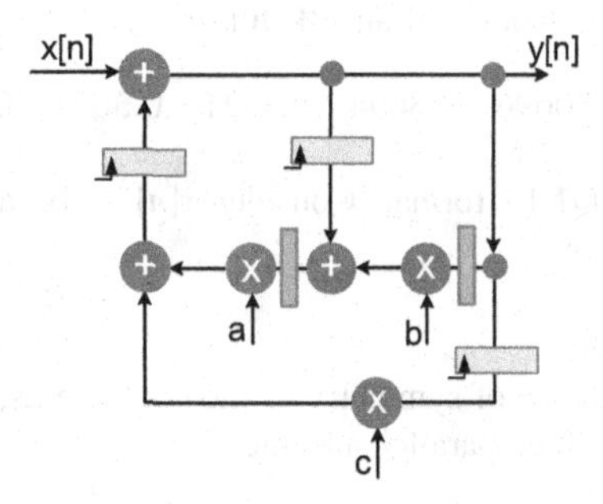

Figure 6.26 Design for exercise 6.3

Exercise 6.4

Optimize the hardware design of a TDF FIR filter with the following coefficients in fixed-point format:

$$h[n] = [3 \quad 13 \quad 219 \quad 221]$$

Minimize the number of adder levels using sub-graph sharing and CSE techniques.

Exercise 6.5

In the following matrix of constant multiplication by a vector, design optimized HW using a CSE technique:

$$\begin{bmatrix} y_0 \\ y_1 \\ y_2 \\ y_3 \end{bmatrix} = \begin{bmatrix} 7 & -3 & 2 & 7 \\ 14 & -6 & 4 & 3 \\ -1 & -12 & 2 & 15 \\ 2 & -7 & 8 & 7 \end{bmatrix} \begin{bmatrix} x_0 \\ x_1 \\ x_2 \\ x_3 \end{bmatrix}$$

Exercise 6.6

Compute the dot product of two vectors **A** and **x**, where **A** is a vector of constants and **x** is a vector of variable data with 4-bit elements. The coefficients of **A** are:

$$A_0 = 13,\ A_1 = -11,\ A_2 = -11,\ A_3 = 13$$

Use symmetry of the coefficients to optimize the DA-based architecture. Test the design for the following:

$$\mathbf{x} = [3\ 13 - 11\ 3]$$

Exercise 6.7

Consider the following nine coefficients of an FIR filter:

$$h[n] = [-0.0456 \quad -0.1703 \quad 0.0696 \quad 0.3094 \quad 0.4521 \quad 0.3094 \quad 0.0696 \quad -0.1703 \quad -0.0456]$$

Convert the coefficients into Q1.15 format. Consider $x[n]$ to be an 8-bit number. Design the following DA-based architecture:

1. a unified ROM-based design;
2. reduced size of ROM by the use of symmetry of the coefficients;
3. a DA architecture based on three parallel sub-filters;
4. a ROM-less DA-based architecture

Exercise 6.8

A second-order IIR filter has the following coefficients:

$$\begin{aligned} b &= [0.2483 \quad 0.4967 \quad 0.2483] \\ a &= [1 \quad -0.1842 \quad -0.1776] \end{aligned}$$

1. Convert the coefficients into 16-bit fixed-point numbers.
2. Design a DA-based architecture that uses a unified ROM as LUT. Consider $x[n]$ to be an 8-bit number. Use two LUTs, one for feedback and the other for feedforward coefficients.

Exercise 6.9

The architecture of Figure 6.24 computes the DFT of an 8-point sequence using convolution summation. Replace the DF FIR structure in the architecture with a TDF filter structure. Exploit the symmetry in the coefficients and design architecture by minimizing multipliers and ROM size. Draw an RTL diagram and write Verilog code to implement the design. Test the design using a 16-bit complex input vector of eight elements. Select an appropriate Q-format for the architecture, and display your result as 16-bit fixed-point numbers.

References

1. A. Avizienis, "Signed-digit number representation for fast parallel arithmetic," *IRE Transactions on Electronic Computers*, 1961, vol. 10, pp. 389–400.
2. R. Hashemian, "A new method for conversion of a 2's complement to canonic signed digit number system and its representation," in *Proceedings of 30th IEEE Asilomar Conference on Signals, Systems and Computers*, 1996, pp. 904–907.
3. H. H. Loomisand B. Sinha, "High-speed recursive digital filter realization," *Circuits, Systems and Signal Processing*, 1984, vol. 3, pp. 267–294.
4. K. K. Parhi and D. G. Messerschmitt, "Pipeline interleaving parallelism in recursive digital filters. Pt I: Pipelining using look-ahead and decomposition," *IEEE Transactions on Acoustics, Speech Signal Processing*, 1989, vol. 37, pp. 1099–1117.
5. K. K. Parhi and D. G. Messerschmitt, "Pipeline interleaving and parallelism in recursive digital filters. Pt II: Pipelining incremental block filtering," *IEEE Transactions on Acoustics, Speech Signal Processing*, 1989, vol. 37, pp. 1118–1134.
6. A. V. Oppenheim and R. W. Schafer, *Discrete-time Signal Processing*, 3rd, 2009, Prentice-Hall.
7. Y. C. Lim and S. R. Parker, "FIR filter design over a discrete powers-of-two coefficient space," *IEEE Transactions on Acoustics, Speech Signal Processing*, 1983, vol. 31, pp. 583–691.
8. H. Samueli, "An improved search algorithm for the design of multiplierless FIR filters with powers-of-two coefficients," *IEEE Transactions on Circuits and Systems*, 1989, vol. 36, pp. 1044–1047.
9. J.-H. Han and I.-C. Park, "FIR filter synthesis considering multiple adder graphs for a coefficient," *IEEE Transactions on Computer-Aided Design of Integrated Circuits and Systems*, 2008, vol. 27, pp. 958–962.
10. A. G. Dempster. and M. D. Macleod, "Use of minimum-adder multiplier blocks in FIR digital filters," *IEEE Transactions on Circuits and Systems II*, 1995, vol. 42, pp. 569–577.
11. J.-H. Han and I.-C. Park, "FIR filter synthesis considering multiple adder graphs for a coefficient," *IEEE Transactions on Computer-Aided Design of Integrated Circuits and Systems*, 2008, vol. 27, pp. 958–962.
12. R. I. Hartley, "Subexpression sharing in filters using canonic signed digit multipliers," *IEEE Transactions on Circuits and Systems II*, 1996, vol. 43, pp. 677–688.
13. S.-H. Yoon, J.-W. Chong and C.-H. Lin, "An area optimization method for digital filter design," *ETRI Journal*, 2004, vol. 26, pp. 545–554.
14. Y. Jang. and S. Yang, "Low-power CSD linear phase FIR filter structure using vertical common sub-expression," *Electronics Letters*, 2002, vol. 38, pp. 777–779.
15. A. P. Vinod, E. M.-K. Lai, A. B. Premkuntar and C. T. Lau, "FIR filter implementation by efficient sharing of horizontal and vertical sub-expressions," *Electronics Letters*, 2003, vol. 39, pp. 251–253.
16. S. B. Slimane, "Reducing the peak-to-average power ratio of OFDM signals through precoding," *IEEE Transactions on Vehicular Technology*, 2007, vol. 56, pp. 686–695.
17. A. Hosnagadi, F. Fallah and R. Kastner, "Common subexpression elimination involving multiple variables for linear DSP synthesis," in *Proceedings of 15th IEEE International Conference on Application-specific Systems, Architectures and Processors*, Washington 2004, pp. 202–212.
18. D. J. Allred, H. Yoo, V. Krishnan, W. Huang and D. V. Anderson, "LMS adaptive filters using distributed arithmetic for high throughput," *IEEE Transactions on Circuits and Systems*, 2005, vol. 52, pp. 1327–1337.
19. P. K. Meher, S. Chandrasekaran and A. Amira, "FPGA realization of FIR filters by efficient and flexible systolization using distributed arithmetic," *IEEE Transactions on Signal Processing*, 2008, vol. 56, pp. 3009–3017.

20. W. Sen, T. Bin and Z. Jim, "Distributed arithmetic for FIR filter design on FPGA," in *Proceedings of IEEE international conference on Communications, Circuits and Systems*, Japan, 2007, vol. 1, pp. 620–623.
21. S. A. White, "Applications of distributed arithmetic to digital signal processing: a tutorial review," *IEEE ASSP Magazine*, 1989, vol. 6, pp. 4–19.
22. P. Longa and A. Miri, "Area-efficient FIR filter design on FPGAs using distributed arithmetic," *Proceedings of IEEE International Symposium on Signal Processing and Information Technology*, 2006, pp. 248–252.
23. S. Hwang, G. Han, S. Kang and J. Kim, "New distributed arithmetic algorithm for low-power FIR filter implementation," *IEEE Signal Processing Letters*, 2004, vol. 11, pp. 463–466.
24. H. Yoo and D. V. Anderson, "Hardware-efficient distributed arithmetic architecture for high-order digital filters," *Proceedings of IEEE International Conference on Acoustics, Speech and Signal Processing*, 2005, vol. 5, pp. 125–128.
25. H. Natarajan, A. G. Dempster and U. Meyer-Bäse, "Fast discrete Fourier transform computations using the reduced adder graph technique," *EURASIP Journal on Advances in Signal Processing*, 2007, December, pp. 1–10.

7

Pipelining, Retiming, Look-ahead Transformation and Polyphase Decomposition

7.1 Introduction

The chapter discusses pipelining, retiming and look-ahead techniques for transforming digital designs to meet the desired objectives. Broadly, signal processing systems can be classified as *feedforward* or *feedback* systems. In feedforward systems the data flows from input to output and no value in the system is fed back in a recursive loop. Finite impulse response (FIR) filters are feedforward systems and are fundamental to signal processing. Most of the signal processing algorithms such as fast Fourier transform (FFT) and discrete cosine transform (DCT) are feedforward. The timing can be improved by simply adding multiple stages of pipelining in the hardware design.

Recursive systems such as infinite impulse response (IIR) filters are also widely used in DSP. The feedback recursive algorithms are used for timing, symbol and frequency recovery in digital communication receivers. In speech processing and signal modeling, autoregressive moving average (ARMA) and autoregressive (AR) processes involve IIR systems. These systems are characterized by a *difference equation*. To compute an output sample, the equation directly or indirectly involves previous values of the output samples along with the current and previous values of input samples. As the previous output samples are required in the computation, adding pipeline registers to improve timing is not directly available to the designer. The chapter discusses cut-set retiming and node transfer theorem techniques for systematically adding pipelining registers in feedforward systems. These techniques are explained with examples. This chapter also discusses techniques that help in meeting timings in implementing fully dedicated architectures (FDAs) for feedback systems.

The chapter also defines some of the terms relating to digital design of feedback systems. Iteration, the iteration period, loop, loop bound and iteration period bound are explained with examples. While designing a feedback system, the designer applies mathematical transformations to bring the critical path of the design equal to the iteration period bound (IPB), which is defined as the loop bound of the

Digital Design of Signal Processing Systems: A Practical Approach, First Edition. Shoab Ahmed Khan.

critical loop of the system. The chapter then discusses the node transfer theorem and cut-set retiming techniques in reference to feedback systems. The techniques help the designer to systematically move the registers to retime the design in such a way that the transformed design has reduced critical path and ideally it is equal to the IPB.

Feedback loops pose a great challenge. It is not trivial to add pipelining and extract parallelism in these systems. The IPB limits any potential improvement in the design. The chapter highlights that if the IPB does not satisfy the sampling rate requirement even after using the fastest computational units in the design, the designer should then resort to look-ahead transformation. This transformation actually reduces the IPB at the cost of additional logic. Look-ahead transformation, by replacing the previous value of the output by its equivalent expression, adds additional registers in the loop. These additional delays reduce the IPB of the design. These delays then can be retimed to reduce the critical path delay. The chapter illustrates the effectiveness of the methodology by examples. The chapter also introduces the concept of 'C-slow retiming', where each register in the design is first replaced by C number of registers and these registers are then retimed for improved throughput. The design can then handle C independent streams of inputs.

The chapter also describes an application of the methodology in designing effective decimation and interpolation filters for the front end of a digital communication receiver, where the required sampling rate is much higher and any transformation that can help in reducing the IPB is very valuable.

IIR filters are traditionally not used in multi-rate decimation and interpolation applications. The recursive nature of IIR filters requires computing all intermediate values, so it cannot directly use polyphase decomposition. The chapter presents a novel methodology that makes an IIR filter use Nobel identities and effectively run at a slower clock rate in multi-rate decimation and interpolation applications.

7.2 Pipelining and Retiming

7.2.1 Basics

The advent of very large scale integration (VLSI) has reduced the cost of hardware devices. This has given more flexibility to system designers to implement computationally intensive applications in small form factors. Designers are now focusing more on performance and less on densities, as the technology is allowing more and more gates on a single piece of silicon. Higher throughout and data rates of applications are becoming the driving objectives. From the HW perspective, pipelining and parallel processing are two methods that help in achieving high throughput.

A critical path running through a *combinational cloud* in a feedforward system can be broken by the addition of pipeline registers. A feedforward system is one in which the current output depends only on current and previous input samples and no previous output is fed back in the system for computation of the current output. Pipeline registers only add latency. If L pipeline registers are added, the transfer function of the system is multiplied by z^{-L}. In a feedback combinational cloud, pipeline registers cannot be simply added because insertion of delays modifies the transfer function of the system and results in change of the order of the difference equation.

Figure 7.1(a) shows a combinational cloud that is broken into three clouds by placement of two pipeline registers. This results in one-third potential reduction in the path delay of the combinational cloud. In general, in a system with L levels of pipeline stages, the number of delay elements in any

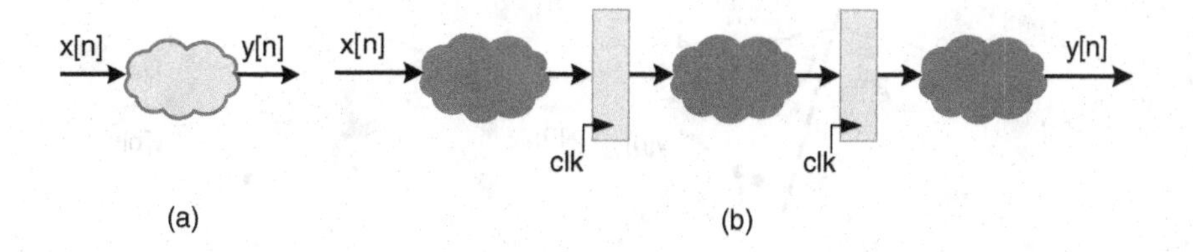

Figure 7.1 Pipeline registers are added to reduce the critical path delay of a combinational cloud. (a) Original combinational logic. (b) Design with three levels of pipelining stages with two sets of pipeline registers

path from input to output is $L-1$ greater than that in the same path in the original logic. Pipelining reduces the critical path, but it increases the system latency and the output $y[n]$ corresponds to the previous input sample $x[n-2]$. Pipelining also adds registers in all the paths, including the paths running parallel to the critical path. These extra registers add area as well as load the clock distribution. Therefore pipelining should only be added when required.

Pipelining and retiming are two different aspects of digital design. In pipelining, additional registers are added that change the transfer function of the system, whereas in retiming, registers are relocated in a design to maximize the desired objective. Usually the retiming has several objectives: to reduce the critical path, reduce the number of registers, minimize power use [1, 2], increase testability [3, 4], and so on. Retiming can either move the existing registers in the design or can be followed by pipelining, where the designer places a number of registers on a cut-set line and then applies a retiming transformation to place these registers at appropriate edges to minimize the critical path while keeping all other objectives in mind.

Retiming, then, is employed to optimize a set of design objectives. These usually involve conflicting design solutions, such as maximizing performance may result in an increase in area, for example. The objectives must therefore be weighted to get the best tradeoff.

Retiming also helps in maximizing the testability of the design [4, 5]. With designs based on field-programmable gate arrays (FPGAs), where registers are in abundance, retiming and pipelining transformation can optimally use this available pool of resources to maximize performance [6].

A given dataflow graph (DFG) can be retimed using cut-set or delay transfer approaches. This chapter interchangeably represents delays in the DFG with dots or registers.

7.2.2 Cut-set Retiming

Cut-set retiming is a technique to retime a dataflow graph by applying a valid cut-set line. A valid cut-set is a set of forward and backward edges in a DFG intersected by a cut-set line such that if these edges are removed from the graph, the graph becomes disjoint. The cut-set retiming involves transferring a number of delays from edges of the same direction across a cut-set line of a DFG to all edges of opposing direction across the same line. These transfers of delays do not alter the transfer function of the DFG.

Figure 7.2(a) shows a DFG with a valid cut-set line. The line breaks the DFG into two disjoint graphs, one consisting of nodes N_0 and N_1 and the other having node N_2. The edge $N_1 \rightarrow N_2$ is a forward cut-set edge, while $N_2 \rightarrow N_0$ and $N_2 \rightarrow N_1$ are backward cut-set edges. There are two delays on $N_2 \rightarrow N_1$ and one delay on $N_2 \rightarrow N_0$; one delay each is moved from these backward edges to the forward edge. The retimed DFG minimizing the number of registers is shown in Figure 7.2(b).

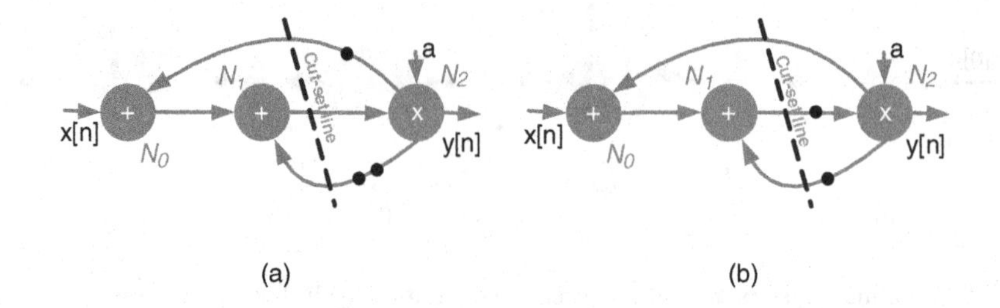

Figure 7.2 Cut-set retiming (a) DFG with cut-set line. (b) DFG after cut-set retiming

7.2.3 *Retiming using the Delay Transfer Theorem*

The delay transfer theorem helps in systematic shifting of registers across computational nodes. This shifting does not change the transfer function of the original DFG.

The theorem states that, without affecting the transfer function of the system, N registers can be transferred from each incoming edge of a node of a DFG to all outgoing edges of the same node, or vice versa. Figure 7.3 shows this theorem applied across a node N_0, where one register is moved from each of the outgoing edges of the DFG to all incoming edges.

Delay transfer is a special case of cut-set retiming where the cut line is placed to separate a node in a DFG. The cut-set retiming on the DFG then implements the delay transfer theorem.

7.2.4 *Pipelining and Retiming in a Feedforward System*

Pipelining of a feedforward system adds the appropriate number of registers in such a way that it transforms a given dataflow graph G to a pipelined one G_r, such that the corresponding transfer functions $H(z)$ of G and $H_r(z)$ of G_r differ only by a pure delay z^{-L}, where L is the number of pipeline stages added in the DFG. Adding registers followed by retiming facilitates the moving of pipeline registers in a feedforward DFG from a set of edges to others to reduce the critical path of the design.

7.2.5 *Re-pipelining: Pipelining using Feedforward Cut-set*

Feedforward cut-set is a technique used to add pipeline registers in a feedforward DFG. The longest (critical) path can be reduced by suitably placing pipelining registers in the architecture.

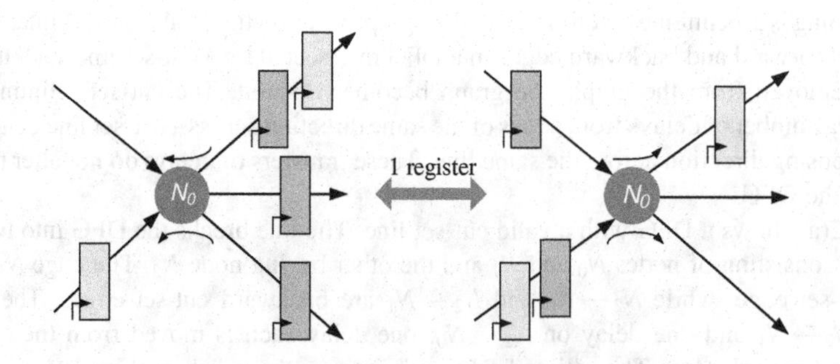

Figure 7.3 Delay transfer theorem moves one register across the node N_0

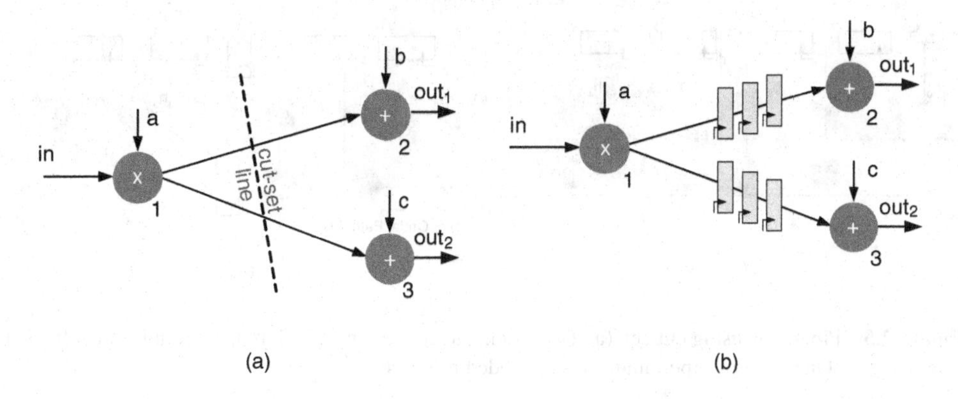

Figure 7.4 (a) Feedforward cut-set. (b) Pipeline registers added on each edge of the cut-set

Maintaining data coherency is a critical issue, but a cut-set eases the task of the designer in figuring out the coherency issues in a complex design.

Figure 7.4 shows an example of a feedforward cut-set. If the two feedforward edges $1 \rightarrow 2$ and $1 \rightarrow 3$ are removed, the graph becomes disjoint consisting of node 1 in one graph and nodes 2 and 3 in the other graph, and two parallel paths from input to output, in $\rightarrow 1 \rightarrow 2 \rightarrow$ out_1 and in $\rightarrow 1 \rightarrow 3 \rightarrow$ out_2, intersect this cut-set line once.

Adding N registers to each edge of a feedforward cut-set of a DFG maintains data coherency, but the respective output is delayed by N cycles. Figure 7.4(b) shows three pipeline registers that are added on each edge of the cut-set of Figure 7.4(a). The respective outputs are delayed by three cycles.

Example: The difference equation and corresponding transfer function $H(z)$ of a 5-coefficient FIR filter are:

$$y_n = h_0 x_n + h_1 x_{n-1} + h_2 x_{n-2} + h_3 x_{n-3} + h_4 x_{n-4} \tag{7.1a}$$

$$H(z) = h_0 + h_1 z^{-1} + h_2 z^{-2} + h_3 z^{-3} + h_4 z^{-4} \tag{7.1b}$$

Assuming that general-purpose multipliers and adders are used, the critical path delay of the DFG consists of accumulated delay of the combinational cloud of one multiplier T_{mult}, and four adders $4 \times T_{\text{adder}}$. Assuming T_{mult} is 2 time units (tu) and $T_{\text{adder}} = 1$ tu, then the critical path of the design is 6 tu. It is desired to reduce this critical path by partitioning the design into two levels of pipeline stages. One feedforward cut-set can be used for appropriately adding one pipeline register in the design. Two possible cut-set lines that reduce the critical path of the design to 4 tu are shown in Figure 7.5(a). Cut-set line 2 is selected for adding pipelining as it requires the addition of only two registers. The registers are added and the pipeline design is shown in Figure 7.5(b). This reflects that the pipeline improvement is limited by the slowest pipeline stage. In this example the slowest pipeline stage consists of a multiplier and two adders.

Although the potential speed-up of two-level pipelining is two times the original design, the potential speed-up is not achieved owing to the unbalanced length of pipeline stages. The optimal speed-up requires breaking the critical path into two exact lengths of 3 tu. This requires pipelining inside the adder, which may not be very convenient to implement or may require

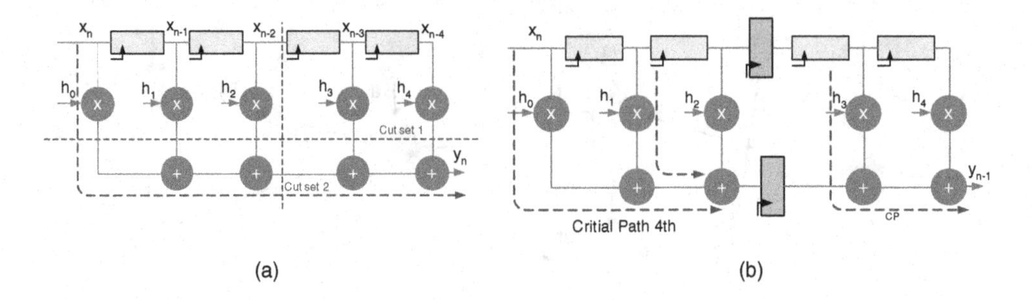

Figure 7.5 Pipelining using cut-set (a) Two candidate cut-sets in a DFG implementing a 5-coefficient FIR filter. (b) One level of pipelining registers added using cut-set 2

more registers. It is also worth mentioning that in many designs convenience is preferred over absolute optimality.

The pipeline design implements the following transfer function:

$$H_r(z) = z^{-1}H(z) \tag{7.2}$$

7.2.6 *Cut-set Retiming of a Direct-form FIR Filter*

Cut-set retiming can be applied to a direct-form FIR filter to get a transposed direct-form (TDF) version. The TDF is formed by first reversing the direction of addition, successively placing cut-set lines on every delay edge. Each cut-set line cuts two edges of the DFG, one in the forward and the other in the backward direction. Then retiming moves the registers from the forward edge to the backward edge. This transforms the filter from DF to TDF. This form breaks the critical path by placing a register before every addition operation. Figure 7.6 shows the process.

Example: Consider an example of fine-grain pipelining where the registers are added inside a computational unit. Figure 7.7(a) shows a 4-bit ripple carry adder (RCA). We need to reduce the critical path of the logic by half. This requires exactly dividing the combinational cloud into two stages. Figure 7.7(a) shows a cut-set line that divides the critical path into equal-delay logic, where

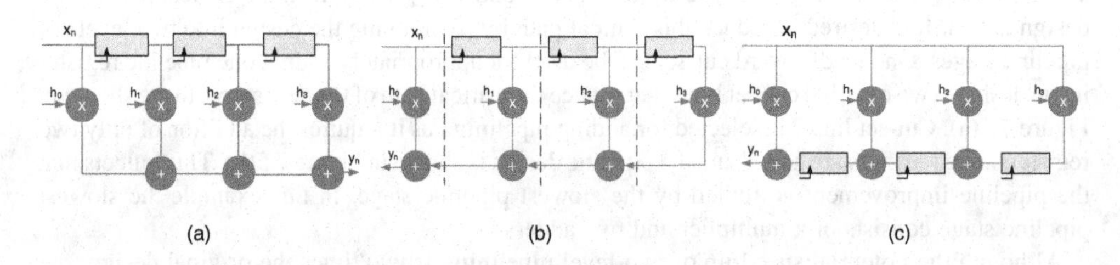

Figure 7.6 FIR filter in direct-form transformation to a TDF structure using cut-set retiming. (a) A 4-coefficient FIR filter in DF. (b) Reversing the direction of additions in the DF and applying cut-set retiming. (c) The retimed filter in TDF

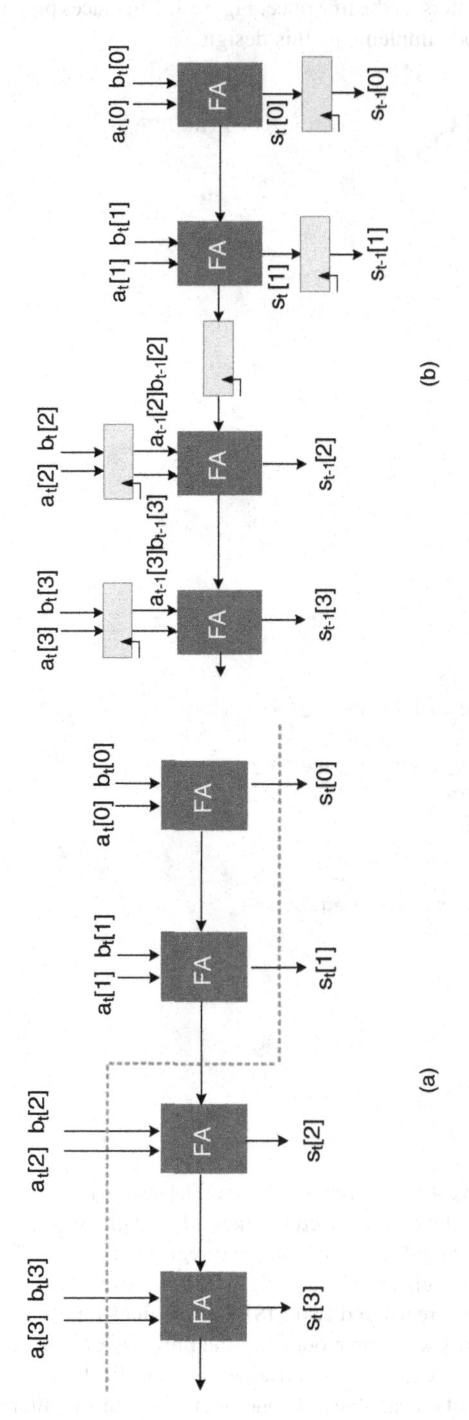

Figure 7.7 Pipelining using cut-set (a) A cut-set to pipeline a 4-bit RCA. (b) Placing pipeline registers along the cut-set line

all paths from input to output intersect the line once. Figure 7.7(b) places pipeline registers along the cut-set line. The following code implements this design:

```
// Module to implement a 4-bit 2-stage pipeline RCA
module pipeline_adder
(
input clk,
input [3:0] a, b,
input cin,
output reg [3:0] sum_p,
output reg cout_p);

// Pipeline registers
reg [3:2] a_preg, b_preg;
reg [1:0] s_preg;
reg c2_preg;
// Internal wires
reg [3:0] s;
reg c2;

// Combinational cloud
always @*
begin
	// Combinatinal cloud 1
	{c2, s[1:0]} = a[1:0] + b[1:0] + cin;
	// Combinational cloud 2
	{cout_p, s[3:2]} = a_preg + b_preg + c2_preg;
	// Put the output together
	sum_p = {s[3:2], s_preg};
end

// Sequential circuit: pipeline registers
always @(posedge clk)
begin
	s_preg <= s[1:0];
	a_preg <= a[3:2];
	b_preg <= b[3:2];
	c2_preg <= c2;
end
endmodule
```

The implementation adds two 4-bit operands a and b. The design has one set of pipeline registers that divides the combinational cloud into two equal stages. In the first stage, two LSBs of a and b are added with `cin`. The sum is stored in a 2-bit pipeline register `s_preg`. The carry out from the addition is saved in pipeline register `c2_preg`. In the same cycle, the combinational cloud simultaneously adds the already registered two MSBs of previous inputs a and b in 2-bit pipeline registers `a_preg` and `b_prag` with carry out from the previous cycle stored in pipeline register `c2_preg`. This 2-stage pipeline RCA has two full adders in the critical path, and for a set of inputs the corresponding sum and carry out is available after one clock cycle; that is, after a latency of one cycle.

Example: Let us extend the previous example. Consider that we need to reduce the critical path to one full adder delay. This requires adding a register after the first, second and third FAs. We need to

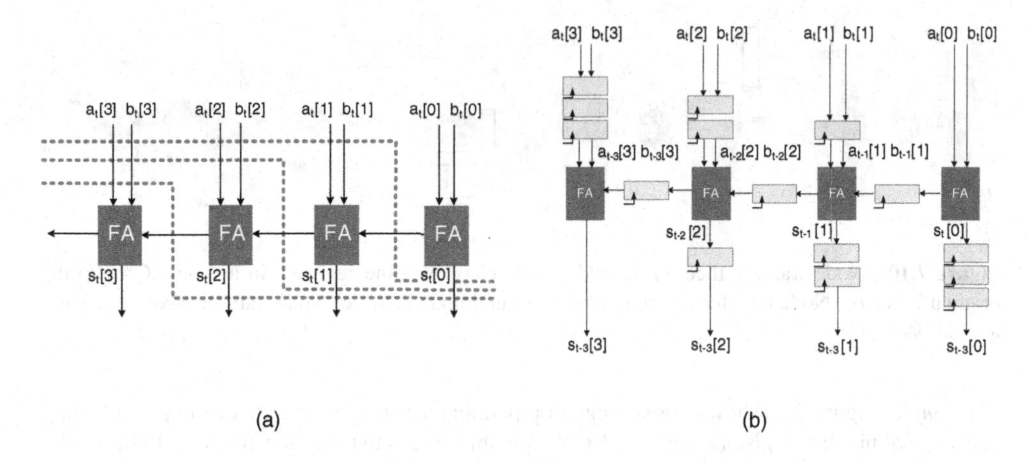

Figure 7.8 (a) Three cut-sets for adding four pipeline stages in a 4-bit RCA. (b) Four-stage pipelined 4-bit RCA

apply three cut-sets in the DFG, as shown in Figure 7.8(a). To maintain data coherency, the cut-sets ensure that all the paths from input to output have three registers. The resultant pipeline digital design is given in Figure 7.8(b). It is good practice to line up the pipeline stages or to straighten out the cut-sets lines for better visualization, as depicted in Figure 7.9.

7.2.7 *Pipelining using the Delay Transfer Theorem*

A convenient way to implement pipelining is to add the desired number of registers to all input edges and then, by repeated application of the node transfer theorem, systematically move the registers to break the delay of the critical path.

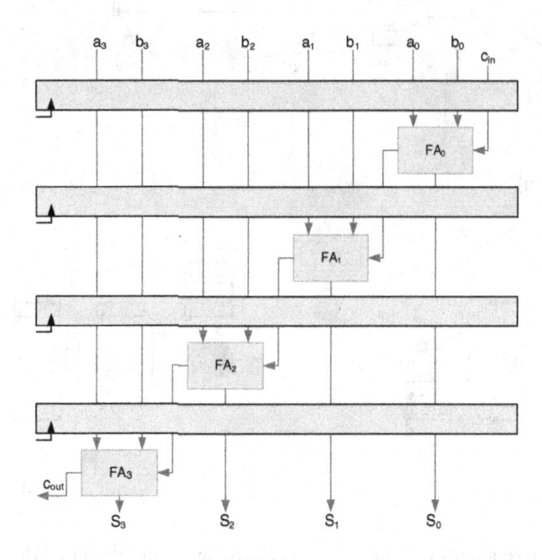

Figure 7.9 Following good design practice by lining up of different pipeline stages

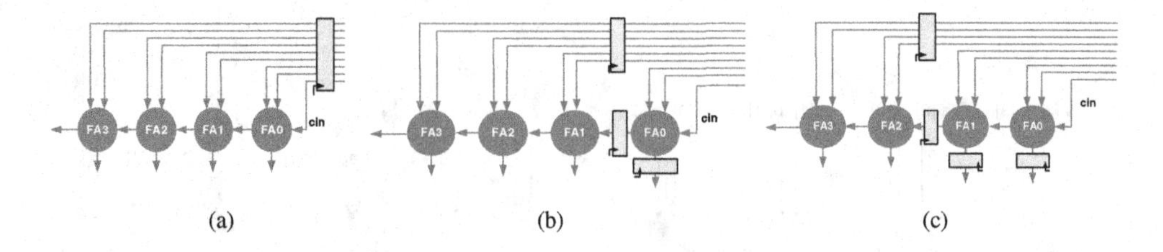

Figure 7.10 Node transfer theorem to add one level of pipeline registers in a 4-bit RCA. (a) the original DFG. (b) Node transfer theorem applied around node FA0. (c) Node transfer theorem applied around FA1

Example: Figure 7.10 shows this strategy of pipelining applied to the earlier example of adding one stage of pipeline registers in a 4-bit RCA. The objective is to break the critical path that is the carry out path from node FA1 to FA2 of the DFG. To start with, one register is added to all the input edges of the DFG, as shown in Figure 7.10(a). The node transfer theorem is applied on the FA0 node. One register each is transferred from all incoming edges of node FA0 to all outgoing edges. Figure 7.10(b) shows the resultant DFG. Now the node transfer theorem is applied on node FA1. One delay each is again transferred from all incoming edges of node FA1 to all outgoing edges. This has moved the pipeline register in the critical path of the DFG while keeping the data coherency intact. The final pipelined DFG is shown in Figure 7.10(c).

Example: This example adds three pipeline registers by applying the delay transfer theorem to a 4-bit RCA. Figure 7.11 shows the node transfer theorem applied repeatedly to place three pipeline

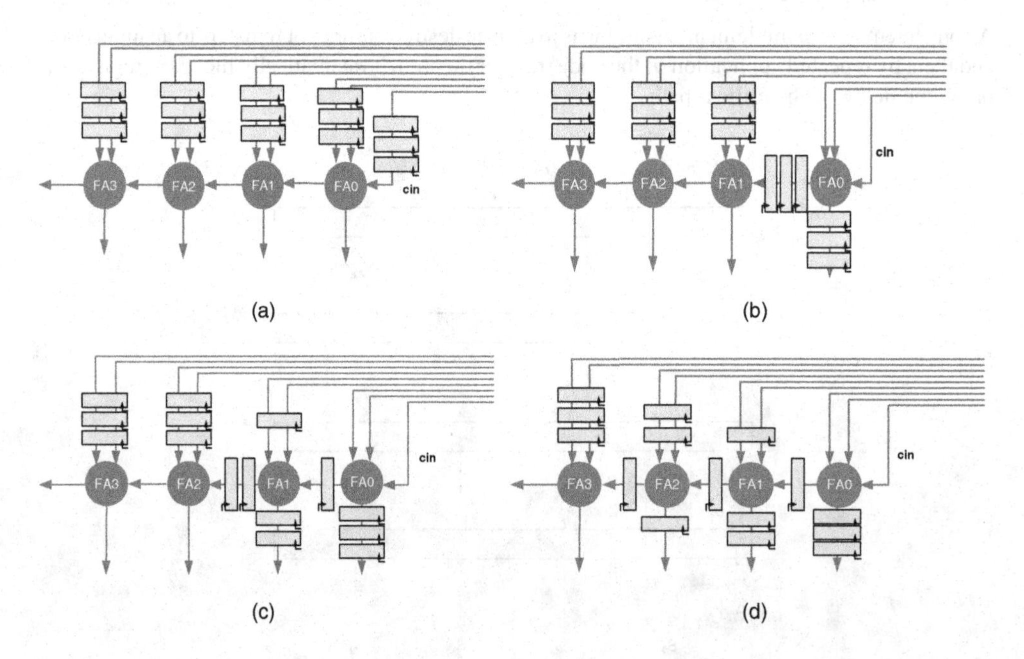

Figure 7.11 Adding three stages of pipeline registers by applying the node delay transfer theorem on (a) the original DFG, around (b) FA0, (c) FA1, and (d) FA2

registers at different edges. To start with, three pipeline registers are added to all input edges of the DFG, as shown in Figure 7.11(a). In the first step, three registers of first node FA0 are moved across the node to output edges, and registers at other input edges are intact. The resultant DFG is shown in Figure 7.11(b). Two registers from all input edges of node FA1 are moved to the output, as depicted in Figure 7.11(c). And finally one register from each input edge of node FA2 is moved to all output edges. This leaves three registers at input node FA3 and systematically places one register each between two consecutive FA nodes, thus breaking the critical path to one FA delay while maintaining data coherency. The final DFG is shown in Figure 7.11(d).

7.2.8 Pipelining Optimized DFG

Chapter 6 explained that a general-purpose multiplier is not used for multiplication by a constant and a carry propagate adder (CPA) is avoided in the datapath. An example of a direct-form FIR filter is elaborated in that chapter, where each coefficient is changed to its corresponding CSD representation. For 60–80 dB stop-band attenuation, up to four significant non-zero digits in the CSD representation of the coefficients are considered. Multiplication by a constant is then implemented by four hardwired shift operations. Each multiplier generates a maximum of four partial products (PPs). A global CV is computed which takes care of sign extension elimination logic and two's complement logic of adding 1 to the LSB position of negative PPs. This CV is also added in the PPs reduction tree. A Wallace reduction tree then reduces the PPs and CV into two layers, which are added using any CPA to compute the final output.

A direct-form FIR filter with three coefficients is shown in Figure 7.12. An optimized design with CSD multiplications, PPs and CV reduction into two layers and final addition of the two layers using CPA is shown in Figure 7.13(a).

The carry propagate adder is a basic building block to finally add the last two terms that are at the output of a compression tree. The trees are extensively used in fully dedicated architecture. Adding pipeline registers in compression trees using cut-set retiming is straightforward as all the logic levels are aligned and all the paths run in parallel. A cut-set can be placed after any number of logic levels and pipeline registers can be placed at the location of the cut-set line.

The pipelined optimized design of the DF FIR filter of Figure 7.13(a) with five sets of pipeline registers in the compression tree is shown in Figure 7.13(b).

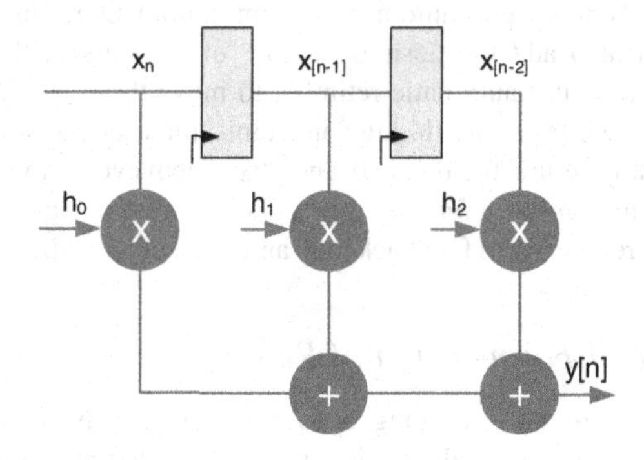

Figure 7.12 Direct-form FIR filter with three coefficients

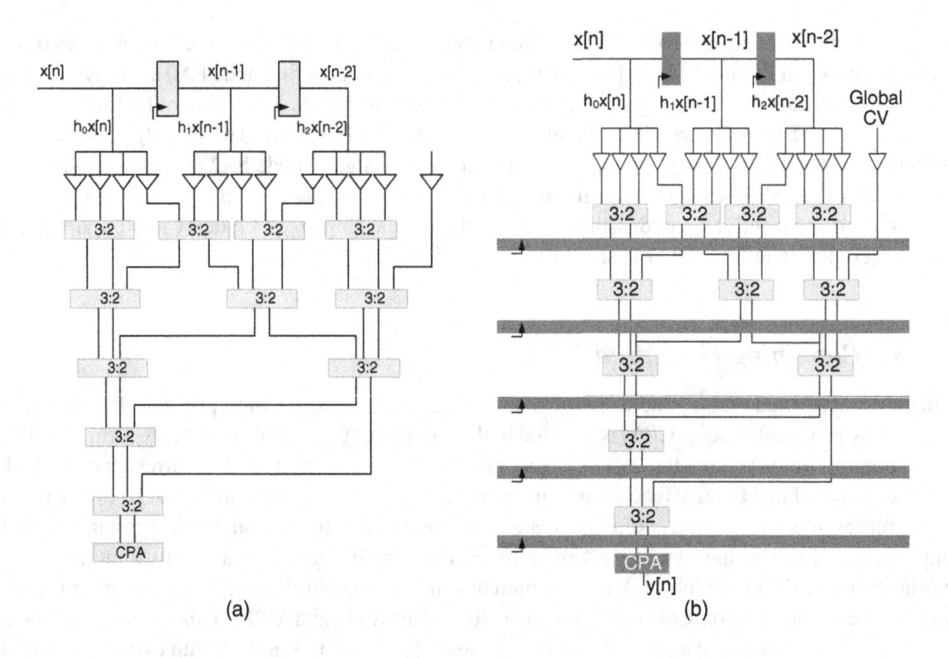

Figure 7.13 (a) Optimized implementation of 3-coefficient direct-form FIR filter. (b) Possible locations to apply cut-set pipelining to a compression tree

7.2.9 Pipelining Carry Propagate Adder

Some of the CPAs are very simple to pipeline as the logic levels in their architectures are aligned and all paths from input to output run in parallel. The conditional sum adder (CSA) is a good example where cut-set lines can be easily placed for adding pipeline registers, as shown in Figure 7.14. For other CPAs, it is not trivial to augment pipelining. An RCA, as discussed in earlier examples, requires careful handling to add pipeline registers, as does a carry generate adder.

7.2.10 Retiming Support in Synthesis Tools

Most of the synthesis tools do support automatic or semi-automatic retiming. To get assistance in pipelining, the user can also add the desired number of registers at the input or output of a feedforward design and then use automatic retiming to move these pipeline registers to optimal places. In synthesis tools, while setting timing constraints, the user can specify a synthesis option for the tool to automatically retime the registers and place them evenly in the combinational logic. The Synopsys design compiler has a *balance_registers* option to achieve automatic retiming [7]. This option cannot move registers in a feedback loop and some other restrictions are also applicable.

7.2.11 Mathematical Formulation of Retiming

One way of performing automatic retiming is to first translate the logic into a mathematical model [2] and then apply optimization algorithms and heuristics to optimize the objectives set in the model [4, 5]. These techniques are mainly used by synthesis tools and usually a digital designer does

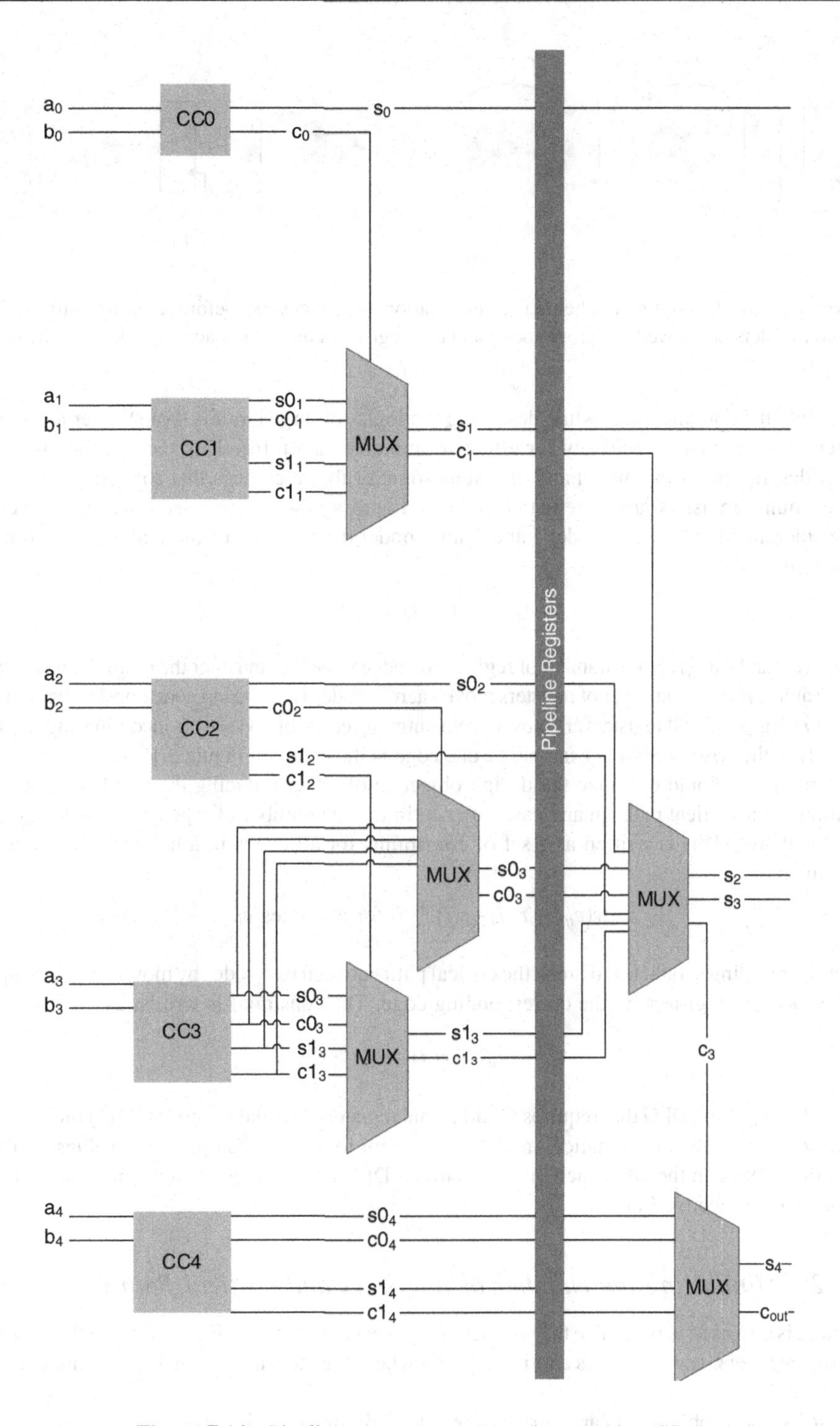

Figure 7.14 Pipelining a 5-bit conditional sum adder (CSA)

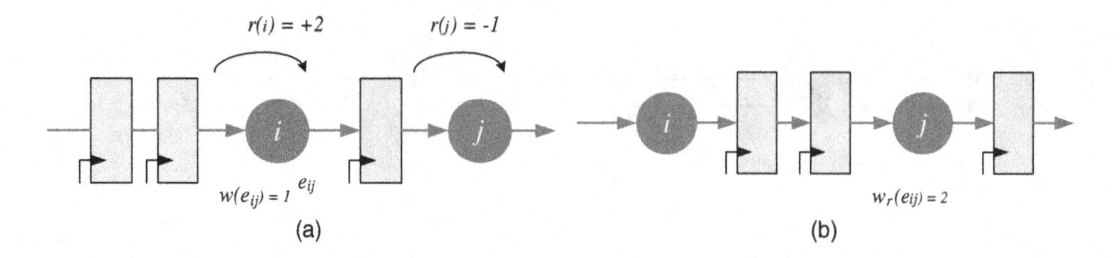

Figure 7.15 Node retiming mathematical formulation. (a) An edge e_{ij} before retiming with $w(e_{ij}) = 1$. (b) Two registers are moved-in across node i and one register is moved out across node j, for the retimed edge $w_r(e_{ij}) = 2$

not get into the complications while designing the logic for large circuits. For smaller designs, the problem can be solved intuitively for effective implementation [6]. This section, therefore, only gives a description of the model and mentions some of the references that solve the models.

In retiming, registers are systematically moved across a node from one edge to another. Let us consider an edge between nodes i and j, and model the retiming of the edge e_{ij}. The retiming equation is:

$$w_r(e_{ij}) = w(e_{ij}) + r(j) - r(i)$$

where $w(e_{ij})$ and $w_r(e_{ij})$ are the number of registers on edge e_{ij} before and after the retiming, respectively, and $r(i)$ and $r(j)$ are the number of registers moved across nodes i and j using node transfer theorem. The value of $r(j)$ is positive if registers are moved from outgoing edges of node j to its incoming edge e_{ij}, and it is negative otherwise. A simple example of one edge is illustrated in Figure 7.15.

The retiming should optimize the design objective of either reducing the number of delays or minimizing the critical path. In any case, after retiming the number of registers on any edge must not be negative. This is written as a set of constraints for all edges in a mathematical modeling problem as:

$$w(e_{ij}) + r(j) - r(i) \geq 0 \text{ for all edges } e_{ij}$$

Secondly, retiming should try to break the critical path between two nodes by moving the appropriate number of extra registers on the corresponding edge. The constraint is written as:

$$w(e_{ij}) + r(j) - r(i) \geq C$$

for all edges e_{ij} in the DFG that requires C additional registers to make the nodes i and j meet timings.

The solution of the mathematical modeling problem for a DFG computes the values of all $r(v)$ for all the nodes v in the DFG such that the retimed DFG does not violate any constraints and still optimizes the objective function.

7.2.12 Minimizing the Number of Registers and Critical Path Delay

Retiming is also used to minimize the number of registers in the design. Figure 7.16(a) shows a graph with five registers, and (b) shows a retimed graph where registers are retimed to minimize the total number to three.

To solve the problem for large graphs, register minimization is modeled as an optimization problem. The problem is also solved using incremental algorithms where registers are recursively

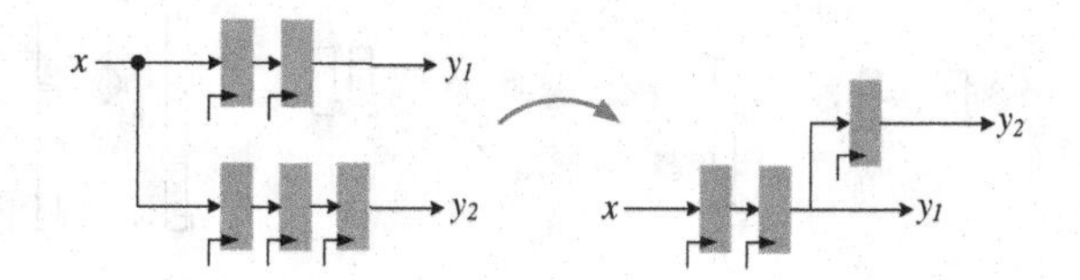

Figure 7.16 Retiming to minimize the number of registers

moved to vertices, where the number of registers is minimized and the retimed graph still satisfies the minimum clock constraint [7].

As the technology is shrinking, wire delays are becoming significant, so it is important for the retiming formulation to consider these too [8].

7.2.13 Retiming with Shannon Decomposition

Shannon decomposition is one transformation that can extend the scope of retiming [8, 14]. It breaks down a multivariable Boolean function into a combination of two equivalent Boolean functions:

$$f(x_0, x_1, \ldots, x_{N-1}) = \bar{x}_0 . f(0, x_1, \ldots, x_{N-1}) + x_0 . f(1, x_1, \ldots, x_{N-1})$$

The technique identifies a late-arriving signal to a block and duplicates the logic in the block with x_0 assigning values of 0 and 1. A 2:1 multiplexer then selects the correct output from the duplicated logic. A carry select adder is a good example of Shannon decomposition. The carry path is the slowest path in a ripple carry adder. The logic in each block is duplicated with fixed carry 0 and 1 and finally the correct value of the output is selected using a 2:1 multiplexer. A generalized Shannon decomposition can also hierarchically work on a multiple-input logic, as in a hierarchical carry select adder or a conditional sum adder.

Figure 7.17(a) shows a design with a slowest input x_0 to node f_0. The dependency of f_0 on x_0 is removed by Shannon decomposition that duplicates the logic in f_0 and computes it for both possible

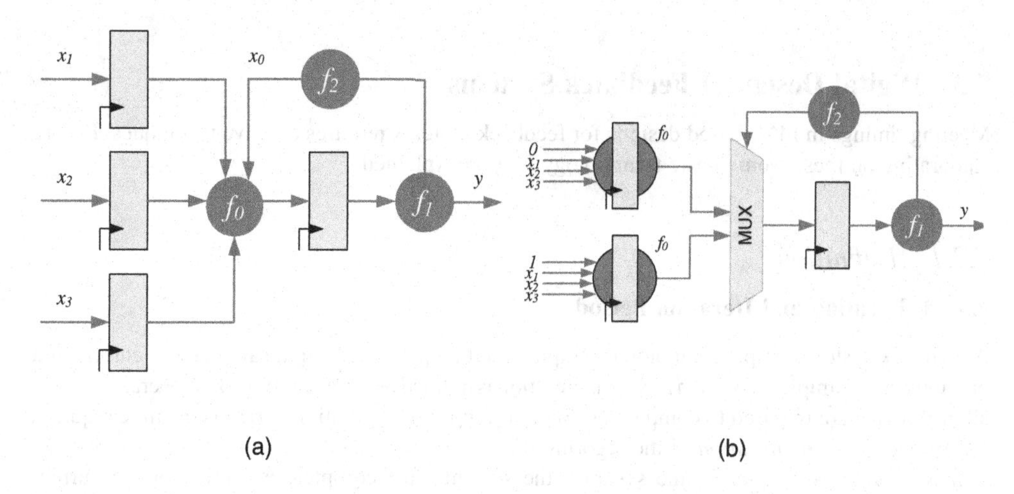

Figure 7.17 (a) Shannon decomposition removing the slowest input x_0 to f_0 and duplicating the logic in f_0 with 0 and 1 fixed input values for x_0. (b) The design is then retimed for effective timing

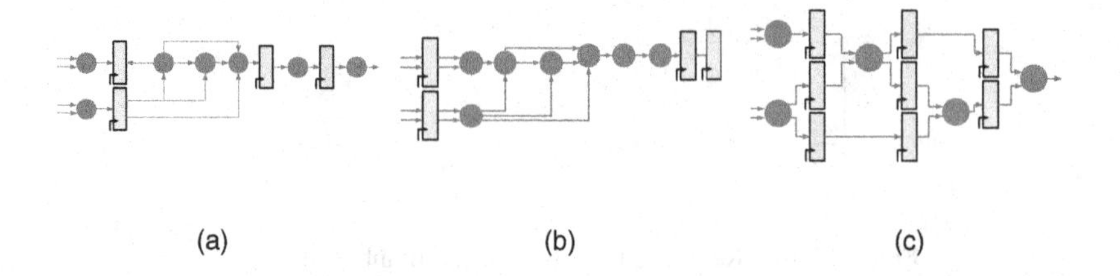

Figure 7.18 Peripheral retiming. (a) The original design. (b) Moving registers to the periphery. (c) Retiming and optimization

input values of x_0. A 2:1 multiplexer then selects the correct output. The registers in this Shannon decomposed design can be effectively retimed, as shown in Figure 7.17(b).

7.2.14 Peripheral Retiming

In peripheral retiming all the registers are moved to the periphery of the design either at the input or output of the logic. The combinational logic is then globally optimized and the registers are retimed into the optimized logic for best timing. This movement of registers on the periphery requires successive application of the node transfer theorem. In cases where some of the edges connected to the node do not have requisite registers, negative registers are added to make the movement of registers on the other edges possible. These negative registers are then removed by moving them to edges with positive retimed registers.

Figure 7.18 shows a simple case of peripheral retiming where the registers in the logic are moved to the input or output of the design. The gray circles represent combinational logic. The combinational logic is then optimized, as shown in Figure 7.18(c) where the registers are retimed for optimal placement.

7.3 Digital Design of Feedback Systems

Meeting timings in FDA-based designs for feedback systems requires creative techniques. Before elaborating on these, some basic terminology will be explained.

7.3.1 Definitions

7.3.1.1 Iteration and Iteration Period

A feedback system computes an output sample based on previous output samples and current and previous input samples. For such a system, iteration is defined as the execution of all operations in an algorithm that are required to compute one output sample. The iteration period is the time required for execution of *one iteration* of the algorithm.

In synchronous hard–real time systems, the system must complete execution of the current iteration before the next input sample is acquired. This imposes an upper bound on the iteration period to be less than or equal to the sampling rate of the input data.

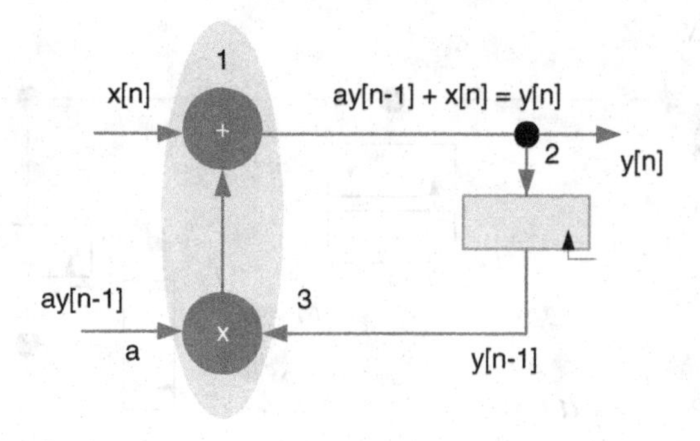

Figure 7.19 Iteration period of first-order IIR system is equal to $T_m + T_a$

Example: Figure 7.19 shows an IIR system implementing the difference equation:

$$y[n] = ay[n-1] + x[n] \tag{7.3}$$

The algorithm needs to perform one multiplication and one addition to compute an output sample in one iteration. Assume the execution times of multiplier and adder are T_m and T_a, respectively. Then the iteration period for this example is $T_m + T_a$. With respect to the sample period T_s of input data, these timings must satisfy the constraint:

$$T_m + T_a \leq T_s \tag{7.4}$$

7.3.1.2 Loop and Loop Bound

A loop is defined as a directed path that begins and ends at the same node. In Figure 7.19 the directed path consisting of nodes $1 \rightarrow 2 \rightarrow 3 \rightarrow 1$ is a loop. The loop bound of the *i*th loop is defined as T_i/D_i, where T_i is the loop computation time and D_i is the number of delays in the loop. For the loop in Figure 7.19 the loop bound is $(T_m + T_a)/1$.

7.3.1.3 Critical Loop and Iteration Bound

A critical loop of a DFG is defined as the loop with maximum loop bound. The iteration period of a critical loop is called the iteration period bound (IPB). Mathematically it is written as:

$$IPB = \max_{all\, L_i} \left\{ \frac{T_i}{D_i} \right\}$$

where T_i and D_i are the cumulative computational time of all the nodes and the number of registers in the loop L_i, respectively.

7.3.1.4 Critical Path and Critical Path Delay

The critical path of a DFG is defined as the path with the longest computation time delay among all the paths that contain zero registers. The computation time delay in the path is called the critical path delay of the DFG.

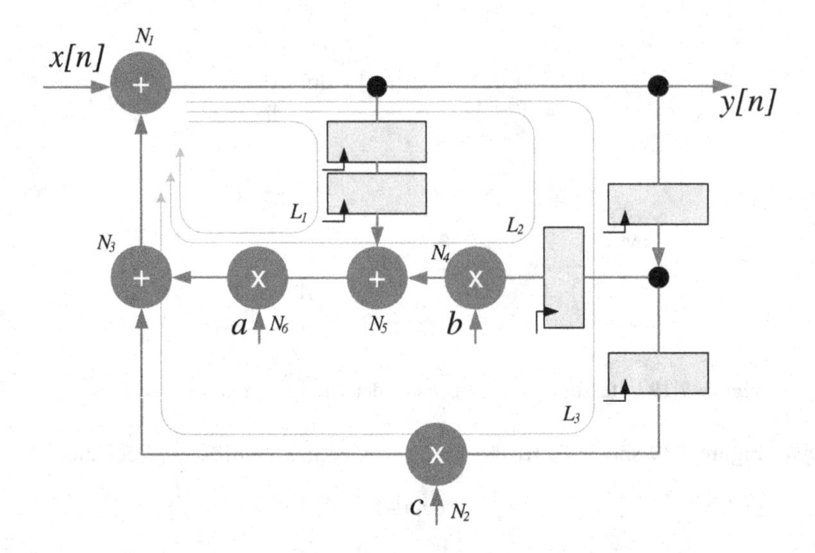

Figure 7.20 Dataflow graph with three loops. The critical loop L_2 has maximum loop bound of 3.5 timing units

The critical path has great significance in digital design as the clock period is lower bounded by the delay of the critical path. Reduction in the critical path delay is usually the main objective in many designs. Techniques such as selection of high-performance computational building blocks of adders, multipliers and barrel shifters and application of retiming transformations help in reducing the critical path delay.

As mentioned earlier, for feedforward designs additional pipeline registers can further reduce the critical path. Although in feedback designs pipelining cannot be directly used, retiming can still help in reducing the critical path delay. In feedback designs, the IPB is the best time the designer can achieve using retiming without using more complex pipelining transformations. The transformations discussed in this chapter are very handy to get to the target delay specified by the IPB.

Example: The DFG shown in Figure 7.20 has three loops, L_1, L_2 and L_3. Assume that multiplication and addition respectively take 2 and 1 time units. Then the loop bounds (LBs) of these loops are:

$$LB1 = \frac{T1}{D1} = \frac{(1+1+2+1)}{2} = 2.5$$

$$LB2 = \frac{T2}{D2} = \frac{(1+2+1+2+1)}{2} = 3.5$$

$$LB3 = \frac{T3}{D3} = \frac{(1+2+1)}{2} = 2$$

L_2 is the critical loop as it has the maximum loop bound; that is, IPB $= \max\{2.5, 3.5, 2\} = 3.5$ time units The critical path or the longest path of the DFG is $N_4 \rightarrow N_5 \rightarrow N_6 \rightarrow N_3 \rightarrow N_1$, and the delay on the path is 7 time units. The IPB of 3.5 time units indicates that there is still significant potential

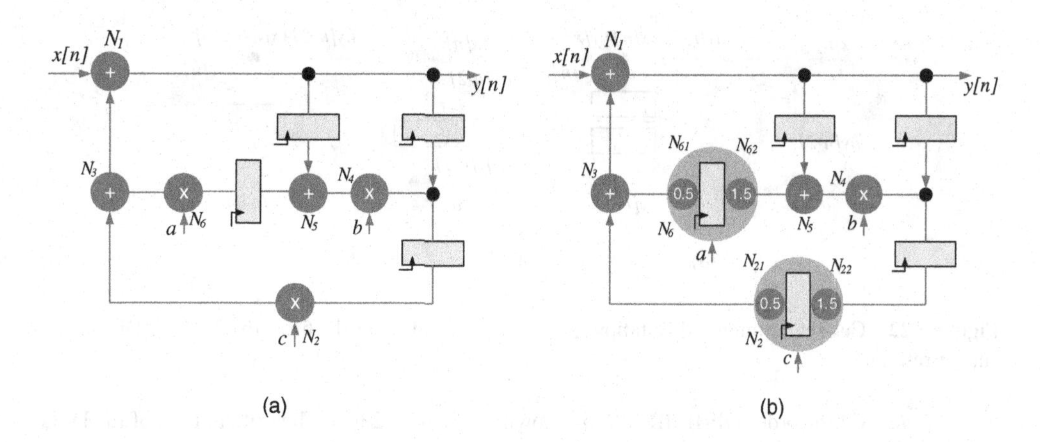

Figure 7.21 (a) Retimed DFG with critical path of 4 timing units and IPB of 3.5 timing units. (b) Fine-grain retimed DFG with critical path delay = IPB = 3.5 timing units

to reduce the critical path using retiming, although this requires fine-grain insertion of algorithmic registers inside the computational units.

The designer may choose not to get to this complexity and may settle for a sub-optimal solution where the critical path is reduced but not brought down to the IPB. In the dataflow graph, recursively applying the delay transfer theorem around node N_4 and then around N_5 modifies the critical paths to $N_6 \rightarrow N_3 \rightarrow N_1$ or $N_2 \rightarrow N_3 \rightarrow N_1$ with critical path delay of both the paths equal to 4 time units. The DFG is shown in Figure 7.21(a).

Further reducing the critical path delay to the IPB requires a fine-grain pipelined multiplier. Each multiplier node N_6 and N_2 are further partitioned into two sub-nodes, N_{61},N_{62} and N_{21},N_{22}, respectively, such that the computational delay of the multiplier is split as 1.5 tu and 0.5 tu, respectively, and the critical path is reduced to 3.5 tu, which is equal to the IPB. The final design is shown in Figure 7.21(b).

7.3.2 *Cut-set Retiming for a Feedback System*

In a feedback system, retiming described in section 7.2.2 can be employed for systematic shifting of algorithmic delays of a dataflow graph G from one set of edges to others to maximize defined design objectives such that the retimed DFG, G_r, has same transfer function. The defined objectives may be to reduce the critical path delay, the number of registers, or the power dissipation. The designer may desire to reduce a combination of these.

Example: A second-order IIR filter is shown in Figure 7.22(a). The DFG implements the difference equation:

$$y[n] = ay[n-2] + x[n] \tag{7.5}$$

The critical path of the system is $T_m + T_a$, where T_m and T_a are the computational delays of a multiplier and adder.

Figure 7.22(a) also shows the cut-set line to move a delay for breaking the critical path of the DFG. The retimed DFG is shown in (b) with a reduced critical path equal to $\max\{T_m, T_a\}$. The same retiming can also be achieved by applying the delay transfer theorem around the multiplier node.

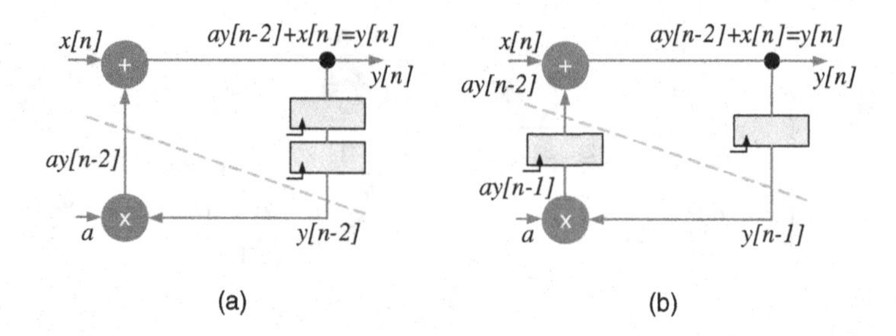

Figure 7.22 Cut-set retiming (a) Dataflow graph for a second-order IIR filter. (b) Retimed DFG using cut-set retiming

Example: A third-order DF-II IIR filter is shown in Figure 7.23(a). The critical path of the DFG consists of one multiplier and three adders. The path is shown in the figure as a shaded region. Two cut-sets are applied on the DFG to implement feedback cut-set retiming. On each cut-set, one register from the cut-set line up-to-down is moved to two cut-set lines from down-to-up. The retimed DFG is shown in Figure 7.23(b) with reduced critical path of one multiplier and two adders.

7.3.3 Shannon Decomposition to Reduce the IPB

For recursive systems, the IPB defines the best timing that can be achieved by retiming. Shannon decomposition can reduce the IPB by replicating some of the critical nodes in the loop and computing them for both possible input bits and then selecting the correct answer when the output of the preceding node of the selected path is available [14].

Figure 7.24(a) shows a critical loop with one register and four combinational nodes N_0, N_1, N_2 and N_3. Assuming each node takes 2 time units for execution, the IPB of the loop is 10 tu. By taking the nodes N_2 and N_3 outside the loop and computing values for both the input 0 and 1, the IPB of the loop is now reduced to around 4 tu because only two nodes and a MUX are left in the loop, as shown in Figure 7.24(b).

7.4 C-slow Retiming

7.4.1 Basics

The C-slow retiming technique replaces every register in a dataflow graph with C registers. These can then be retimed to reduce the critical path delay. The resultant design now can operate on C distinct streams of data. An optimal use of C-slow design requires multiplexing of C streams of data at the input and demultiplexing of respective streams at the output.

The critical path of a DFG implementing a first-order difference equation is shown in Figure 7.25(a). It is evident from Figure 7.25(b) that the critical path delay of the DFG cannot be further reduced using retiming.

C-slow is very effective in further reducing the critical path of a feedback DFG for multiple streams of inputs. This technique works by replicating each register in the DFG with C registers and then retiming it for critical path reduction. The C-slow technique requires C streams of input data or periodically feeding $C-1$ nulls after every valid input to the DFG. Thus effective throughput of the design for one stream will be unaffected by the C-slow technique, but it provides an effective way to implement coarse-grain parallelism without adding any redundant combinational logic as only

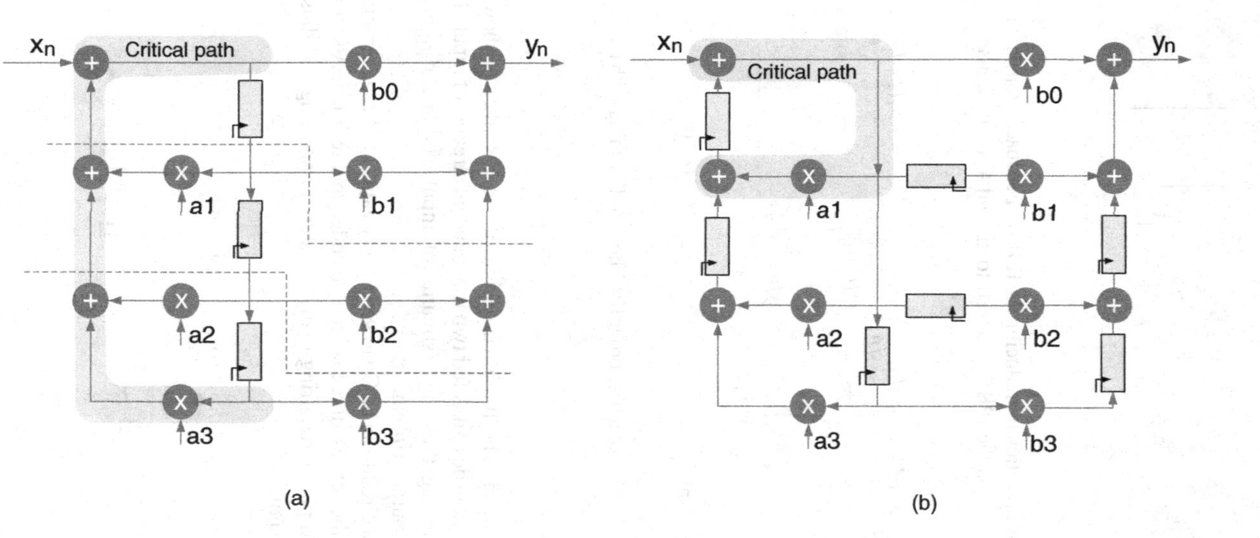

Figure 7.23 Feedback cut-set retiming on a third-order direct-form IIR filter. (a) Two cut-sets are applied on the DFG. (b) The registers are moved by applying feedback cut-set retiming

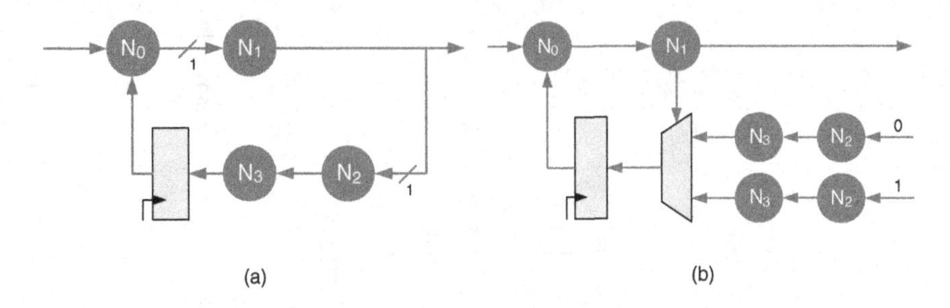

Figure 7.24 Shannon decomposition for reducing the IPB (a) Feedback loop with IPB of 8 timing units. (b) Using Shannon decomposition the IPB is reduced to around 4 timing units

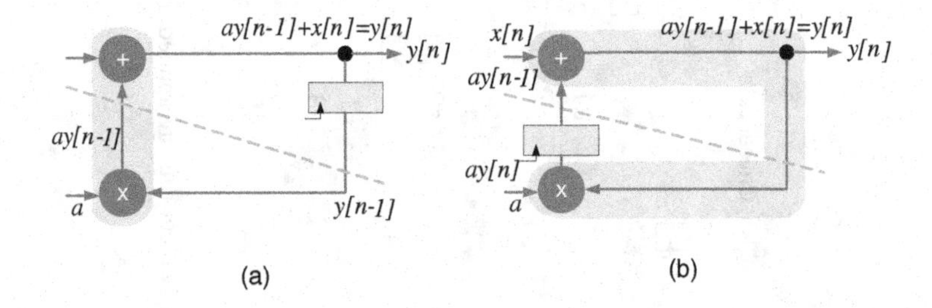

Figure 7.25 (a) Critical path delay of a first-order IIR filter. (b) Retiming with one register in a loop will result in the same critical path

registers need to be replicated. The modified design of the IIR filter is shown in Figure 7.26, implementing 2-slow. The new design is fed two independent streams of input. The system runs in complete lock step processing these two streams of input $x[n]$ and $x'[n]$ and producing two corresponding streams of output $y[n]$ and $y'[n]$.

Although theoretically a circuit can be C-slowed by any value of C, register setup time and clock-to-Q delay bound the number of resisters that can be added and retimed. A second limitation is in designs that may have odd places requiring many registers in retiming and thus results in a huge increase in area of the design.

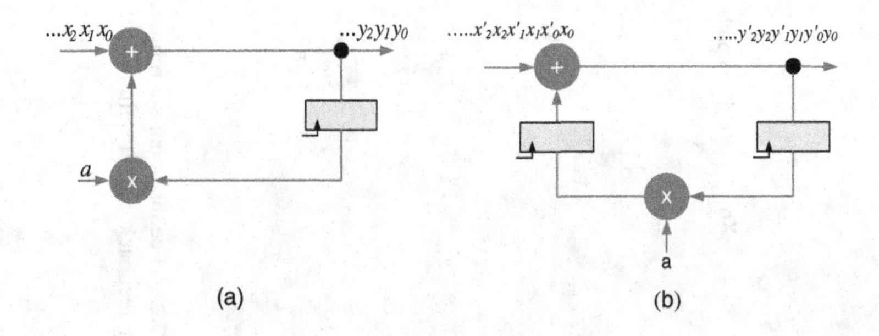

Figure 7.26 C-slow retiming (a) Dataflow graph implementing a first-order difference equation. (b) A 2-slow DFG with two delays to be used in retiming

7.4.2 C-slow for Block Processing

The C-slow technique works well for block processing algorithms. *C* blocks from a single stream of data can be simultaneously processed by the C-slow architecture. A good example is AES encryption where a block of 128 bits is encrypted. By replicating every register in an AES architecture, with *C* registers the design can simultaneously encrypt *C* blocks of data.

7.4.3 C-slow for FPGAs and Time-multiplexed Reconfigurable Design

As FPGAs are rich in registers, the C-slow technique can effectively use these registers to process multiple streams of data. In many designs the clock frequency is fixed to a predefined value, and the C-slow technique is then very useful to meet the fixed frequency of the clock.

C-slow and retiming can also be used to convert a fully parallel architecture to a time-multiplexed design where *C* partitions can be mapped on run-time reconfigurable FPGAs. In a C-slow design, as the input is only valid at every *C*th clock, retiming can move the registers to divide the design into *C* partitions where the partitions can be optimally created to equally divide the logic into *C* parts. The time-multiplexed design requires less area and is also ideal for run-time reconfigurable logic.

Figure 7.27 shows the basic concept. The DFG of (a) is 2-slowed as shown in (b). Assuming all the three nodes implement some combinational logic, that is repeated twice in the design. The design is retimed to optimally place the registers while creating two equal parts. The C-slowed and retimed design is given in Figure 7.27(c). This DFG can now be mapped on time-multiplexed logic, where the data is fed into the design at one clock cycle and the results are saved in the registers. These values are input to the design that reuses the same computational nodes so C-slow reduces the area.

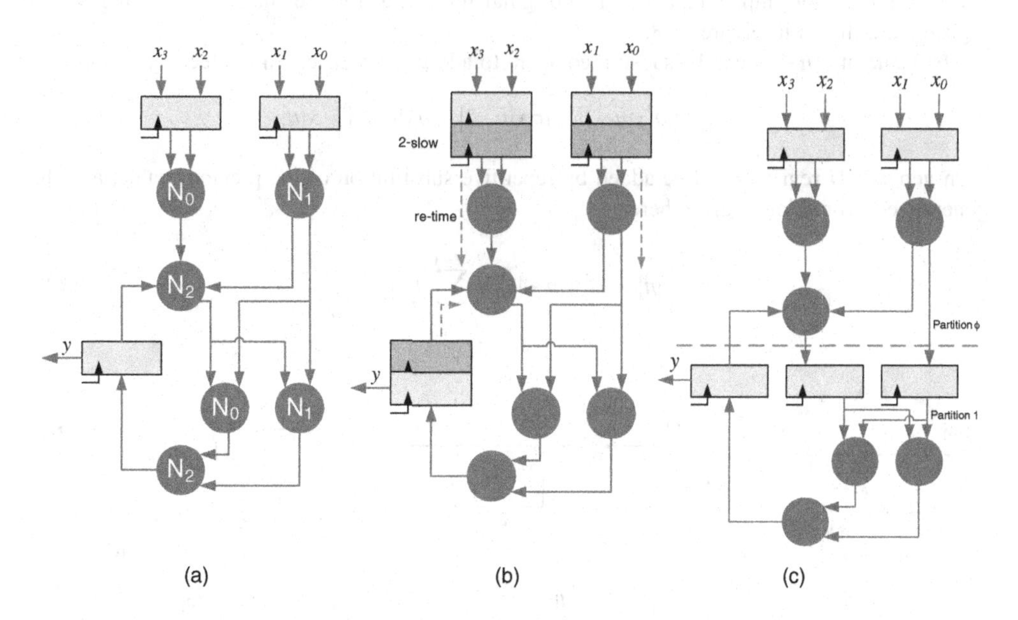

Figure 7.27 C-slow technique for time-multiplexed designs. (a) Original DFG with multiple nodes and few registers and large critical path. (b) Retimed DFG. (c) The DFG partitioned into two equal sets of logic where all the three nodes can be reused in a time-multiplexed design

7.4.4 C-slow for an Instruction Set Processor

C-slow can be applied to microprocessor architecture. The architecture can then run a multi-threaded application where C threads are run in parallel. This arrangement does require a careful design of memories, register files, caches and other associated units of the processor.

7.5 Look-ahead Transformation for IIR filters

Look-ahead transformation (LAT) can be used to add pipeline registers in an IIR filter. In a simple configuration, the technique looks ahead and substitutes the expressions for previous output values in the IIR difference equation. The technique will be explained using a simple example of a difference equation implementing a first-order IIR filter:

$$y[n] = ay[n-1] + x[n] \tag{7.6}$$

Using this expression, $y[n-1]$ can be written as:

$$y[n-1] = ay[n-2] + x[n-1] \tag{7.7}$$

On substituting (7.7) in (7.6) we obtain:

$$y[n] = a^2y[n-2] + ax[n-1] + x[n]$$

This difference equation is implemented with two registers in the feedback loop as compared to one in the original DFG. Thus the transformation improves the IPB. These registers can be retimed for better timing in the implementation. The original filter, transformed filter and retiming of the registers are shown in Figure 7.28.

The value of $y[n-2]$ can be substituted again to add three registers in the feedback loop:

$$y[n] = a^3y[n-3] + a^2x[n-2] + ax[n-1] + x[n]$$

In general, M registers can be added by repetitive substitution of the previous value, and the generalized expression is given here:

$$y[n] = a^M y[n-M] + \sum_{i=0}^{M-1} a^i x[n-i] \tag{7.8}$$

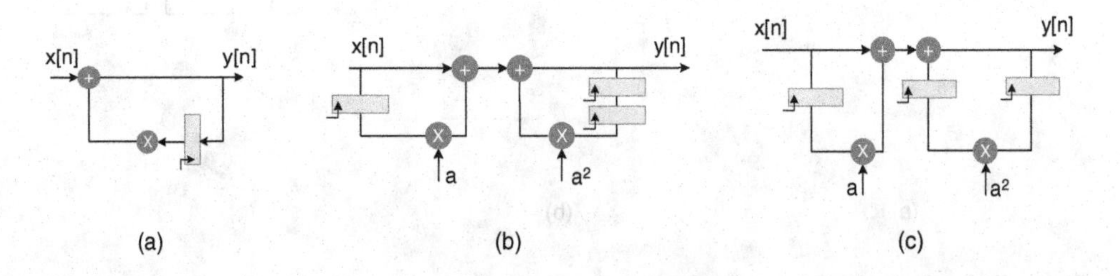

Figure 7.28 Look-ahead transformation. (a) Original first-order IIR filter. (b) Adding one register in the feedback path using look-ahead transformation. (c) Retiming the register to get better timing

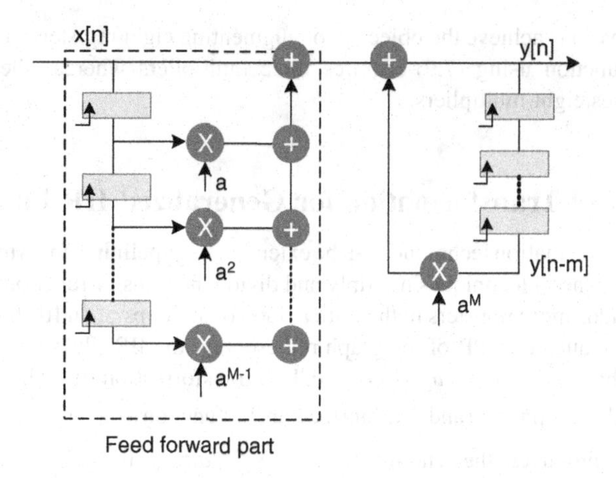

Figure 7.29 Look-ahead transformation to add M registers in the feedback path of a first-order IIR filter

The generalized IIR filter implementing this difference equation with M registers in the feedback loop is given in Figure 7.29. This reduces the IPB of the orginal DFG by a factor of M. The difference equation implements the same transfer function $H(z)$:

$$H(z) = \frac{1}{1-az^{-1}} = \frac{\sum_{i=0}^{M-1} a^i z^{-i}}{1-(az^{-1})^M} \tag{7.9}$$

The expression in (7.8) adds an extra $M-1$ coefficients in the numerator of the transfer function of the system given in (7.9), requiring $M-1$ multiplications and $M-1$ additions in its implementation. These extra computations can be reduced by constraining M to be a power of 2, so that $M=2^m$. This makes the transfer function of (7.9) as:

$$H(z) = \frac{1}{1-az^{-1}} = \frac{\prod_{i=1}^{m}\left(1+(az^{-1})^{2^i}\right)}{1-(az^{-1})^M} \tag{7.10}$$

The numerator of the transfer function now can be implemented as a cascade of m FIR filters each requiring only one multiplier, where $m=\log_2 M$.

Example: This example increases the IPB of a first-order IIR filter by $M = 8$. Applying the look-ahead transformation of (7.9), the transfer function of a first-order system with a pole at 0.9 is given as:

$$H(z) = \frac{1}{1-0.9z^{-1}} = \frac{1+0.9z^{-1}+0.81z^{-2}+0.729z^{-3}+0.6561z^{-4}+0.5905z^{-5}+0.5314z^{-6}+0.4783z^{-7}}{1-0.4305z^{-8}}$$

Using (7.9), this transfer function can be equivalently written as:

$$H(z) = \frac{1}{1-0.9z^{-1}} = \frac{(1+0.9z^{-1})(1+0.81z^{-2})(1+0.6561z^{-4})}{1-0.4305z^{-8}}$$

Both these expressions achieve the objective of augmenting eight registers in the feedback loop, but the transfer function using (7.9) requires three multipliers whereas the transfer function using (7.10) requires eight multipliers.

7.6 Look-ahead Transformation for Generalized IIR Filters

The look-ahead transformation techniques can be extended for pipelining any Nth-order generalized IIR filter. All these general techniques multiply and divide the transfer function of the IIR filter by a polynomial that adds more registers in the critical loop or all loops of an IIR filter. These registers are then moved to reduce the IPB of the graph representing the IIR filter.

An effective technique is *cluster look-ahead* (CLA) transformation [9–11]. This technique adds additional registers by multiplying and dividing the transfer function by a polynomial $\left(1+\sum_{i=1}^{M} c_i z^{-i}\right)$ and then solving the product in the denominator $\left(1+\sum_{i=1}^{N} a_i z^{-i}\right)\left(1+\sum_{i=1}^{M} c_i z^{-i}\right)$ to force the first M coefficients to zero:

$$H(z) = \frac{1+\sum_{i=1}^{M} c_i z^{-i}}{\left(1+\sum_{i=1}^{N} a_i z^{-i}\right)\left(1+\sum_{i=1}^{M} c_i z^{-i}\right)} = \frac{1+\sum_{i=1}^{M} c_i z^{-i}}{1+\sum_{i=M}^{N+M} d_i z^{-i}} \tag{7.11}$$

This technique does not guarantee stability of the resultant IIR filter. To solve this, *scatteredcluster look-ahead* (SCA) is proposed. This adds, for every pole, an additional $M-1$ poles and zeros such that the resultant transfer function is stable. The transfer function $H(z)$ is written as a function of z^M that adds M registers for every register in the design. These registers are then moved to reduce the IPB of the dataflow graph representing the IIR filter. For conjugate poles at $re^{\pm j\theta}$, that represents a second-order section $(1-2r\cos\theta z^{-1}+r^2z^{-2})$. Constraining M to be a power of 2, and applying SCA using expression (7.10), the transfer function corresponding to this pole pair changes to:

$$H(z) = \frac{\prod_{i=1}^{\log_2 M}\left(1+\left(re^{j\theta}z^{-1}\right)^{2^i}\right)\left(1+\left(re^{-j\theta}z^{-1}\right)^{2^i}\right)}{\left(1-\left(re^{j\theta}z^{-1}\right)^{M}\right)\left(1-\left(re^{-j\theta}z^{-1}\right)^{M}\right)} = \frac{\prod_{i=1}^{\log_2 M}\left(1+2r^{2^i}\cos 2^i\theta z^{-2^i}+r^{2i+1}z^{-2^{i+1}}\right)}{(1-2r^M\cos(M\theta)z^{-M}+r^{2M}z^{2M})} \tag{7.12}$$

which is always stable.

There are other transformations. For example, a generalized cluster look-ahead (GCLA) does not force the first M coefficients in the expression to zero; rather it makes them equal to ±1 or some signed power of 2 [11]. This helps in achieving a stable pipelined digital IIR filter.

7.7 Polyphase Structure for Decimation and Interpolation Applications

The sampling rate change is very critical in many signal processing applications. The front end of a digital communication receiver is a good example. At the receiver the samples are acquired at a very high rate, and after digital mixing they go through a series of decimation filters. At the transmitter the samples are brought to a higher rate using interpolation filters and then are digitally mixed before passing the samples to a D/A converter. A polyphase decomposition of an FIR filter conveniently architects the decimation and interpolation filters to operate at slower clock rate [12].

For example, a decimation by D filters computes only the Dth sample at an output clock that is slower by a factor of D than the input data rate. Similarly, for interpolation by a factor of D, theoretically the first $D-1$ zeros are inserted after every sample and then the signal is filtered using an interpolation FIR filter. The polyphase decomposition skips multiplication by zeros while implementing the interpolation filter.

The polyphase decomposition achieves this saving by first splitting a filter with L coefficients into D sub-filters denoted by $e_k[n]$ for $k=0 \ldots D-1$, where:

$$h[n] = \sum_{k=0}^{D-1} e_k[n] \text{ where } e_k[n] = h[nD+k] \tag{7.13}$$

By simple substitution the desired decomposition is formulated as [13]:

$$\begin{aligned} H(z) &= h_0 + h_D z^{-D} + h_D z^{-2D} + \ldots \\ &+ h_1 z^{-1} + h_{D+1} z^{-(D+1)} + h_{2D+1} z^{-(2D+1)} + \cdots \\ &+ h_2 z^{-2} + h_{D+2} z^{-(D+2)} + h_{2D+2} z^{-(2D+2)} + \cdots \\ &\vdots \quad \vdots \quad \vdots \quad \vdots \quad \vdots \\ &+ h_{D-1} z^{-(D-1)} + h_{2D-1} z^{-(2D-1)} + h_{3D-1} z^{-(3D-1)} + \cdots \end{aligned}$$

Regrouping of the expressions results in:

$$\begin{aligned} H(z) &= h_0 + h_D z^{-D} + h_{2D} z^{-2D} + \cdots \\ &+ z^{-1}(h_1 + h_{D+1} z^{-D} + h_{2D+1} z^{-2D} + \cdots) \\ &+ z^{-2}(h_2 + h_{D+2} z^{-D} + h_{2D+2} z^{-2D} + \cdots) \\ &\vdots \quad \vdots \quad \vdots \quad \vdots \quad \vdots \\ &+ z^{-(D-1)}(h_{D-1} + h_{2D-1} z^{-D} + h_{3D-1} z^{-2D} + \cdots) \\ &= E_0(z^D) + z^{-1}E_1(z^D) + z^{-2}E_2(z^D) + \cdots + z^{-(D-1)}E_{D-1}(z^D) \end{aligned}$$

In closed form $H(z)$ is written as:

$$H(z) = \sum_{k=0}^{D-1} z^{-k} E_k(z^D) \tag{7.14}$$

This expression is the polyphase decomposition. For decimation and interpolation applications, $H(z)$ is implemented as D parallel sub-systems $z^{-k}E_k(z^D)$ realizing (7.14), as shown in Figure 7.30.

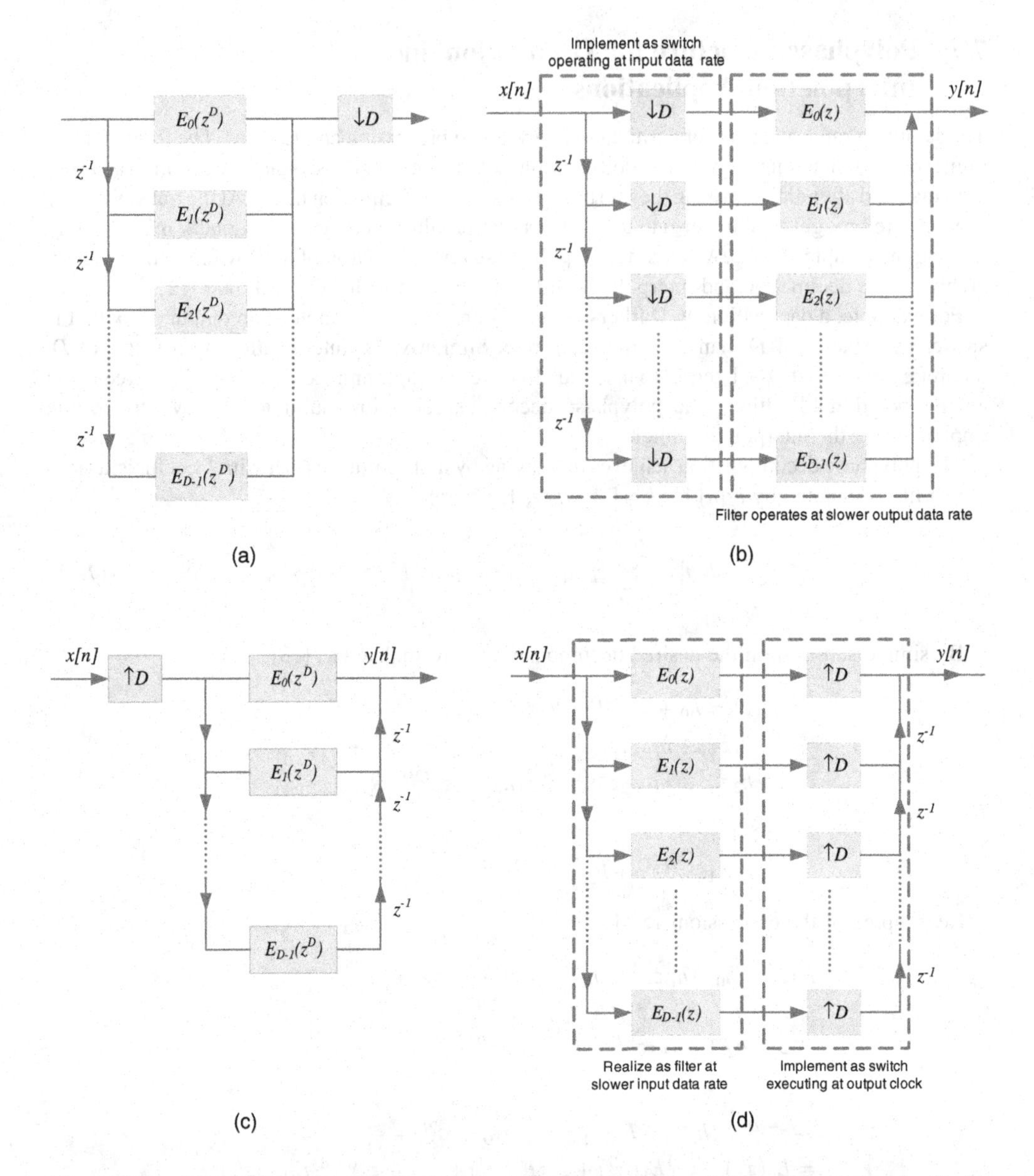

Figure 7.30 (a) Polyphase decomposition for decimation by *D*. (b) Application of the Noble identity for effective realization of a polyphase decomposed decimation filter. (c) Polyphase decomposition for interpolation by *D*. (d) Application of the Noble identity for effective realization of a polyphase decomposition interpolation filter

The Noble identities can be used for optimizing multi-rate structures to execute the filtering logic at a slower clock. The Noble identities for decimation and interpolation are shown in Figure 7.31.

The polyphase decomposition of a filter followed by application of the corresponding Noble identity for decimation and interpolation are shown in Figures 7.30(c) and (d).

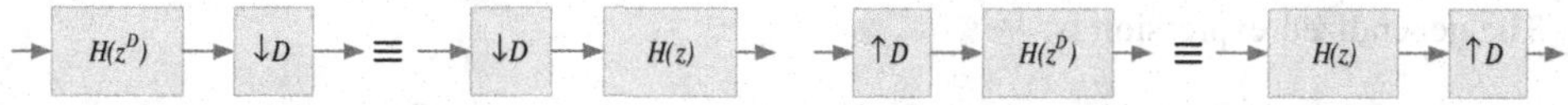

$\rightarrow H(z^D) \rightarrow \downarrow D \rightarrow \; \equiv \; \rightarrow \downarrow D \rightarrow H(z) \rightarrow$

(a)

$\rightarrow \uparrow D \rightarrow H(z^D) \rightarrow \; \equiv \; \rightarrow H(z) \rightarrow \uparrow D \rightarrow$

(b)

Figure 7.31 Noble identities for (a) decimation and (b) interpolation

7.8 IIR Filter for Decimation and Interpolation

IIR filters are traditionally not used in multi-rate signal processing because they do not exploit the polyphase structure that helps in performing only needed computations at a slower clock rate. For example, while decimating a discrete signal by a factor of D, the filter only needs to compute every Dth sample while skipping the intermediate computations. For a recursive filter, $H(z)$ cannot be simply written as a function of z^D.

This section presents a novel technique for using an IIR filter in decimation and interpolation applications [15]. The technique first represents the denominator of $H(z)$ as a function of z^D. For this the denominator is first written in product form as:

$$H(z) = \frac{B(z)}{A(z)} = \frac{\sum_{k=0}^{M-1} b_k z^{-k}}{\prod_{k=0}^{N_1-1}(1-p_k z^{-1}) \prod_{k=0}^{N_2-1}(1-r_k e^{j\theta_k} z^{-1})(1-r_k e^{-j\theta_k} z^{-1})}$$

For each term in the product, the transformation converts the expression of $A(z)$ with N_1 real and N_2 complex conjugate pairs into a function of z^D by multiplying and dividing it with requisite polynomials. For a real pole p_0, the requisite polynomial and the resultant polynomial for different values of D are given here:

$$\begin{array}{llll}
D=2: & (1-p_0 z^{-1}) & (1+p_0 z^{-1}) & = (1-p_0^2 z^{-2}) \\
D=3: & (1-p_0 z^{-1}) & (1+p_0 z^{-1}+p_0^2 z^{-2}) & = (1-p_0^3 z^{-3}) \\
 & \vdots & \vdots & \vdots \\
 & (1-p_0 z^{-1}) & (1+\sum_{k=1}^{D} p_0^K Z^{-K}) & = (1-p_0^D z^{-D})
\end{array}$$

Requisite polynomial

Similarly for a complex conjugate pole pair $re^{j\theta}$ and $re^{-j\theta}$, the use of requisite polynomial $P(z)$ to transform the denominator $A(z)$ as a function of z^{-D} is given here:

$$\begin{aligned}
D &= 2: \left(1-re^{j\theta}z^{-1}\right)\left(1-re^{-j\theta}z^{-1}\right)\left(1+re^{j\theta}z^{-1}\right)\left(1+re^{-j\theta}z^{-1}\right) \\
&= \left(1-r^2 e^{2j\theta}z^{-2}\right)\left(1-r^2 e^{-2j\theta}z^{-2}\right) \\
&= 1-2r^2\cos 2\theta\, z^{-2}+r^4 z^{-4} \\
D &= 3: \left(1-re^{j\theta}z^{-1}\right)\left(1-re^{-j\theta}z^{-1}\right)\left(1+re^{j\theta}z^{-1}+r^2 e^{2j\theta}z^{-2}\right) \\
&\quad \left(1+re^{-j\theta}z^{-1}+r^2 e^{-2j\theta}z^{-2}\right) \\
&= 1-2r^3\cos 3\theta\, z^{-3}+r^6 z^{-6}
\end{aligned}$$

The generalized expression is:

$$\left(1-re^{j\theta}z^{-1}\right)\left(1+\sum_{k=1}^{D} r^k e^{jk\theta}z^{-k}\right)\left(1-re^{-j\theta}z^{-1}\right)\left(1+\sum_{k=1}^{D} r^k e^{-jk\theta}z^{-k}\right)$$
$$= 1-2r^D\cos D\theta z^{-D}+r^{2D}z^{-2D}$$

This manipulation amounts to adding $D-1$ complex conjugate poles and their corresponding zeros. A complex conjugate zero pair contributes a $2(D-1)$th-order polynomial $P(z)$ with real coefficients in the numerator. A generalized close form expression of $P_i(z)$ for a pole pair i can be computed as:

$$P_i(z) = \left(\sum_{k=0}^{D-1} r^k e^{jk\theta}z^{-k}\right)\left(\sum_{l=0}^{D-1} r^l e^{-jl\theta}z^{-l}\right) = \sum_{k=0}^{D-1}\sum_{l=0}^{D-1} r^{k+l}e^{j(k-l)\theta}z^{-(k+l)}$$

The complex exponentials in the expression can be paired for substitution as cosines using Euler's identity. The resultant polynomial will have real coefficients.

All these expressions of $P_i(z)$ are multiplied with $B(z)$ to formulate a single polynomial $B'(z)$ in the numerator. This polynomial is decomposed in FIR polyphase filters for decimation and interpolation applications. The modified expression $A'(z)$ which is now a function of z^{-D} is implemented as an all-pole IIR system. This system can be broken down into parallel form for effective realization. All the factors of the denominator can also be multiplied to write the denominator as a single polynomial where all powers of z are integer multiples of D.

The respective Noble identities can be applied for decimation and interpolation applications. The IIR filter decomposition for decimation by a factor of D is shown in Figure 7.32. An example below illustrates the decomposition for a fifth-order IIR filter.

The decomposition can also be used for interpolation applications. For interpolation, the all-pole IIR component is implemented first and then its output is passed through a polyphase decomposed numerator. Both the components execute at input sample rate and output is generated at a fast running switch at output sampling rate. The design is shown in Figure 7.33.

Example: Use the given fifth-order IIR filter with cut-off frequency $\pi/3$ in decimation and interpolation by three structures. The MATLAB® code for designing the filter with 60 dB stop-band attenuation is given here:

```
Fpass = 1/3; % Passband Frequency
Apass = 1; % Passband Ripple (dB)
Astop = 60; % Stopband Attenuation (dB)

% Construct an FDESIGN object and call its ELLIP method.
h = fdesign.lowpass('N,Fp,Ap,Ast', N, Fpass, Apass,Astop);
Hd = design(h, 'ellip');
% Get the transfer function values.
[b, a] = tf(Hd);
```

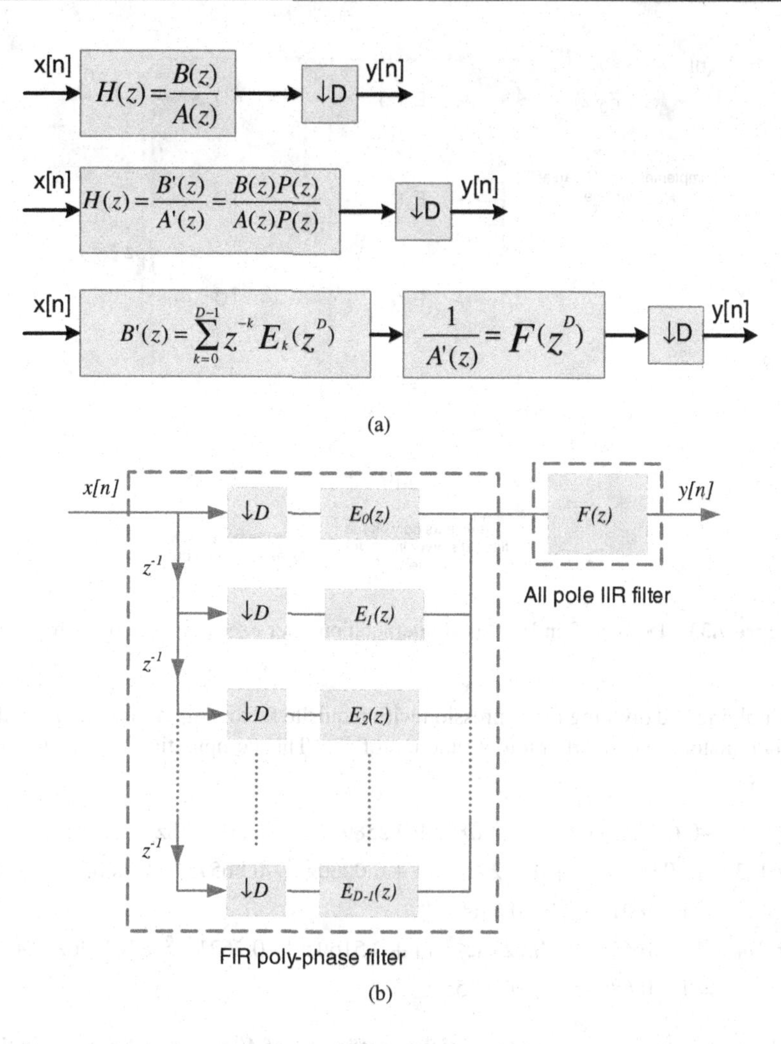

Figure 7.32 (a) and (b) Implementation of an IIR filter for decimation applications

The transfer function is:

$$H(z)=\frac{B(z)}{A(z)}=\frac{0.0121+0.0220z^{-1}+0.0342z^{-2}+0.0342z^{-3}+0.0220z^{-4}+0.0121z^{-5}}{1-2.7961z^{-1}+4.0582z^{-2}-3.4107z^{-3}+1.6642z^{-4}-0.3789z^{-5}}$$

The denominator $A(z)$ is written in product form having first- and second-order polynomials with real coefficients:

$$H(z)=\frac{0.0121+0.0220z^{-1}+0.0342z^{-2}+0.0342z^{-3}+0.0220z^{-4}+0.0121z^{-5}}{(1-0.6836z^{-1})(1-0.9460z^{-1}+0.8828z^{-2})(1-1.1665z^{-1}+0.6278z^{-2})}$$

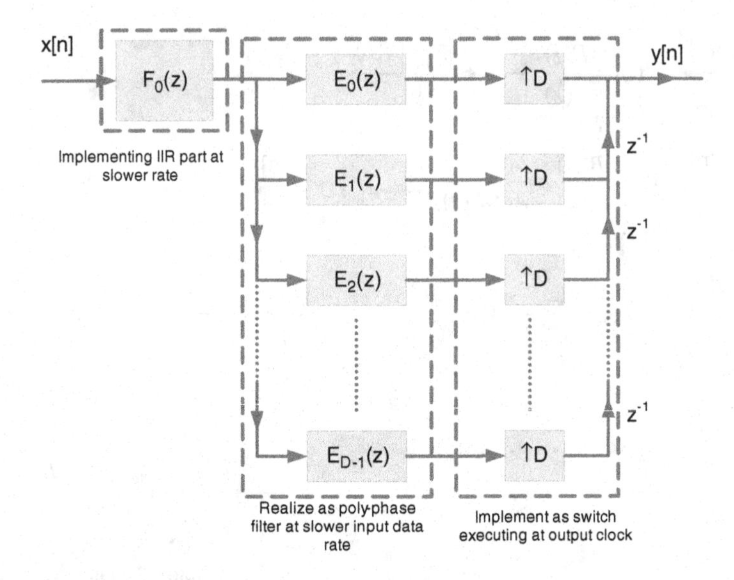

Figure 7.33 Design of an IIR-based interpolation filter executing at input sample rate

By multiplying and dividing the expression with requisite factors of polynomial $P(z)$, the factors in the denominator are converted into a function of z^{-3}. This computation for the denominator is given below:

$$\text{real pole } 0 : (1-0.6836z^{-1})(1+0.5088z^{-1}+0.2589z^{-2}) = 1-0.1318z^{-3}$$

$$\begin{aligned}\text{pole pair } 1,2 : &(1-0.9460z^{-1}+0.8828z^{-2})(1+0.0060z^{-1}-0.8657z^{-2}+0.0052z^{-3}+0.7495z^{-4})\\ &= 1+0.0156z^{-3}+0.6489z^{-6}\end{aligned}$$

$$\begin{aligned}\text{pole pair } 3,4 : &(1-1.1665z^{-1}+0.6278z^{-2})(1+0.5160z^{-1}-0.2621z^{-2}+0.2726z^{-3}+0.2791z^{-4})\\ &= 1+0.6804z^{-3}+0.1475z^{-6}\end{aligned}$$

By multiplying all the above terms we get the coefficients of $A'(z) = A(z)P(z)$ as a function of z^{-3}:

$$A'(z) = 1+0.5643z^{-3}+0.7153z^{-6}+0.3375z^{-9}+0.0372z^{-12}-0.0126z^{-15}$$

The coefficients of $B'(z) = B(z)P(z)$ are:

$$\begin{aligned}B'(z) = &0.0121+0.0345z^{-1}+0.0496z^{-2}+0.0489z^{-3}+0.0341z^{-4}+0.0220z^{-5}\\ &+0.0247z^{-6}+0.0347z^{-7}+0.0394z^{-8}+0.0331z^{-9}+0.0202z^{-10}+0.0136z^{-11}\\ &+0.0120z^{-12}+0.0078z^{-13}+0.0031z^{-14}+0.0007z^{-15}\end{aligned}$$

The coefficients are computed using the following MATLAB® code:

```
p=roots(a);
% For real coefficient, generate the requisite polynomial
rP0 = [1 p(5) p(5)^2];
p0 = [1 -p(5)]; % the real coefficient
```

```
% Compute the new polynomial as power of z^-3
Adash0=conv(p0, rP0);% conv performs poly multiplication
% Combine pole pair to form poly with real coefficients
p1 = conv([1 -p(1)], [1 -p(2)]);
% Generate requisite poly for first conjugate pole pair
rP1 = conv([1 p(1) p(1)^2], [1 p(2) p(2)^2]);
% Generate polynomial in z^-3 for first conjugate pole pair
Adash1 = conv(p1, rP1);
% Repeat for second pole pair
p2 = conv([1 -p(3)], [1 -p(4)]); %
rP2 = conv([1 p(3) p(3)^2], [1 p(4) p(4)^2]);
Adash2 = conv(p2, rP2);
% The modified denominator in power of 3
Adash = conv(conv(Adash0, Adash1), Adash2);
% Divide numerator with the composite requisite polynomial
Bdash = conv(conv(conv(b, rP0), rP1), rP2);
% Frequency and Phase plots
figure (1)
freqz(Bdash, Adash, 1024)
figure (2)
freqz (b,a,1024);
hold off
```

The $B'(z)$ is implemented as a polyphase decimation filter, whereas $1/A'(z)$ is either implemented as single-order all-pole IIR filter or converted into serial form for effective realization in fixed-point arithmetic.

Figure 7.34 shows the optimized realization of decimation by 3 using the decomposition and Noble identity where the denominator is implemented as a cascade of three serial filters. The multiplexer runs at the input data rate, whereas the FIR and IIR part of the design works at a slower clock.

The MATLAB® code below implements the decimation using the structure given in Figure 7.34 and the result of decimation on a randomly generated data sample is shown in Figure 7.35(a). The figure shows the equivalence of this technique with the standard procedure of filtering then decimation.

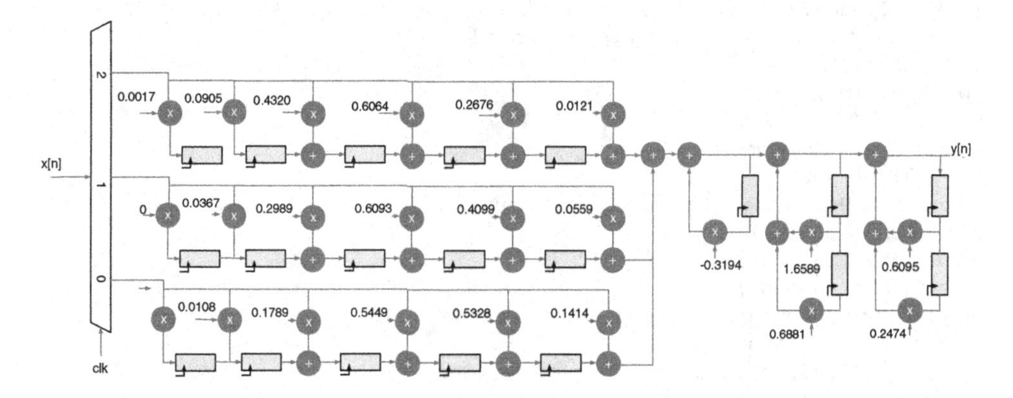

Figure 7.34 Optimized implementation of a polyphase fifth-order IIR filter in decimation by a factor of 3

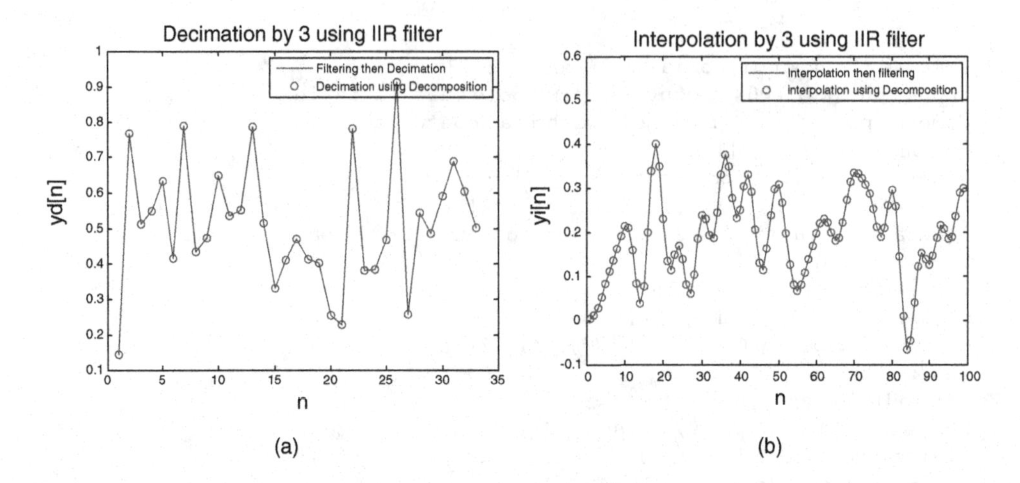

Figure 7.35 Results of equivalence for decimation and interpolation. (a) Decimation performing filtering then decimation and using the decomposed structure of Figure 7.34. (b) Interpolation by first inserting zeros then filtering and using efficient decomposition of Figure 7.34

```
L=99;
x=rand(1,L); % generating input samples
% Filter then decimate
y=filter(b,a,x);
yd = y(3:3:end);
% Performing decimation using decomposition technique
% Polyphase decomposition of numerator
b0=Bdash(3:3:end);
b1=Bdash(2:3:end);
b2=Bdash(1:3:end);
% Dividing input in three steams
x0=x(1:3:end);
x1=x(2:3:end);
x2=x(3:3:end);
% Filtering the streams using polyphase filters
yp0=filter(b0,1,x0);
yp1=filter(b1,1,x1);
yp2=filter(b2,1,x2);
% Adding all the samples
yp=yp0+yp1+yp2;
% Applying Nobel identity on denominator
a0=Adash0(1:3:end);
a1=Adash1(1:3:end);
a2=Adash2(1:3:end);
% Filtering output of polyphase filters through IIR
% Cascaded sections
yc0=filter(1,a0,yp); % first 1st order section
yc1=filter(1,a1,yc0);% Second 2nd order section
yc = filter(1,a2,yc1); % Third 2nd order section
% Plotting the two outputs
```

```
plot(yd);
hold on
plot(yc, 'or');
xlabel('n')
ylabel('yd[n]')
title('Decimation by 3 using IIR filter');
legend('Filtering then Decimation',' Decimation using Decomposition');
hold off
```

The novelty of the implementation is that it uses an IIR filter for decimation and only computes every third sample and runs at one-third of the sampling clock.

Example: The same technique can be used for interpolation applications. Using an IIR filter for interpolation by 3 requires a low-pass IIR filter with cut-off frequency $\pi/3$. The interpolation requires first inserting two zeros after every other sample and then passing the signal through the filter. Using a decomposition technique, the input data is first passed through the all-pole cascaded IIR filters and then the output is passed to three polyphase numerator filters. The three outputs are picked at the output of a fast running switch.

The design of an interpolator running at slower input sampling rate is given in Figure 7.36. The MATLAB® code that implements the design of this is given here. Figure 7.35(b) shows the equivalence of the efficient technique that filters a slower input sample rate with standard interpolation procedure that inserts zeros first and then filters the signal at fast output data rate.

```
L = 33;
x=rand(1,L); % generating input samples
% Standard technique of inserting zeros and then filtering
% Requires filter to execute at output sample rate
xi=zeros(1,3*L);
xi(1:3:end)=x;
y=filter(b,a,xi);
% Performing interpolation using decomposition technique
% Applying Nobel identity on interpolator
a0=Adash0(1:3:end);
a1=Adash1(1:3:end);
```

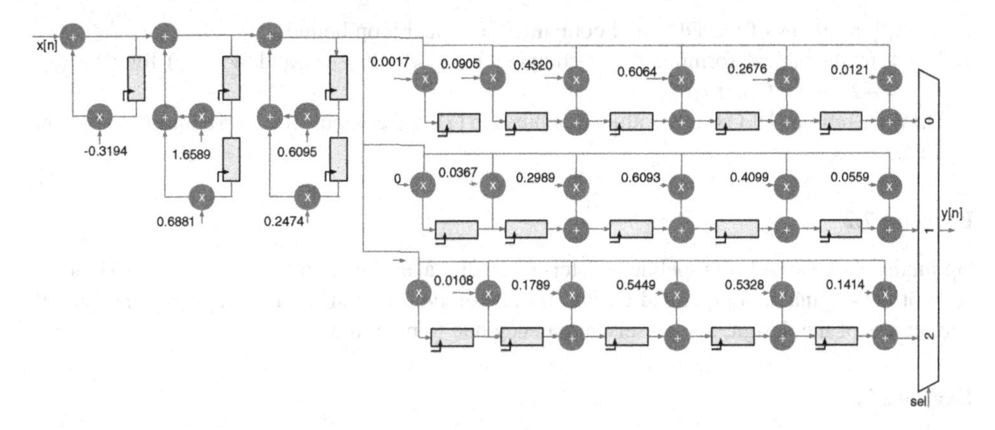

Figure 7.36 IIR filter decomposition interpolating a signal by a factor of 3

```
a2=Adash2(1:3:end);
% Filtering input data using three cascaded IIR sections
yc0=filter(1,a0,x); % first 1st order section
yc1=filter(1,a1,yc0);% Second 2nd order section
yc = filter(1,a2,yc1); % Third 2nd order section
% Polyphase decomposition of numerator
b0=Bdash(1:3:end);
b1=Bdash(2:3:end);
b2=Bdash(3:3:end);
% Filtering the output using polyphase filters
yp0=filter(b0,1,yc);
yp1=filter(b1,1,yc);
yp2=filter(b2,1,yc);
% Switch/multiplexer working at output sampling frequency
% Generates interpolated signal at output sampling rate
y_int=zeros(1,3*L);
y_int(1:3:end)=yp0;
y_int(2:3:end)=yp1;
y_int(3:3:end)=yp2;
% Plotting the two outputs
plot(y);
hold on
plot(y_int, 'or');
xlabel('n')
ylabel('yi[n]')
title('Interpolation by 3 using IIR filter');
legend('Interpolation then filtering','interpolation using Decomposition');
hold off
```

Exercises

Exercise 7.1

For the DFG of Figure 7.37, assume multipliers and adders take 1 time unit, perform the following:

1. Identify all loops of the DFG and compute the critical loop bound.
2. Use a mathematical formulation to compute $W_r(e_{2\_5})$, $W_r(e_{4\_5})$ and $W_r(e_{5\_6})$ for $r(5)=-1$, $r(2)=-2$, $r(4)=0$ and $r(6)=0$.
3. Draw the retimed DFG for the values computed in (2), and compute the loop bound of the retimed DFG.

Exercise 7.2

Optimally place two sets of pipeline registers in the digital design of Figure 7.38. Write RTL Verilog code of the original and pipelined design. Instantiate both designs in a stimulus for checking the correctness of the design, also observing latency due to pipelining.

Exercise 7.3

Retime the DFG of Figure 7.39. Move the two set of registers at the input to break the critical path of the digital logic. Each computational node also depicts the combinational time delay of the node in the logic.

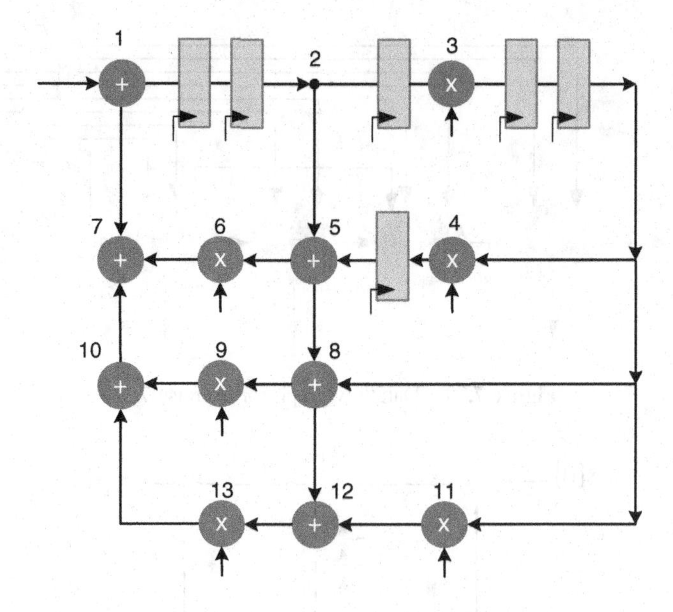

Figure 7.37 Dataflow graph implementing an IIR filter for exercise 7.1

Exercise 7.4

Identify all the loops in the DFG of Figure 7.40, and compute the critical path delay assuming the combinational delays of the adder and the multiplier are 4 and 8 tu, respectively. Compute the IPB of the graph. Apply the node transfer theorem to move the algorithmic registers for reducing the critical path and achieving the IPB.

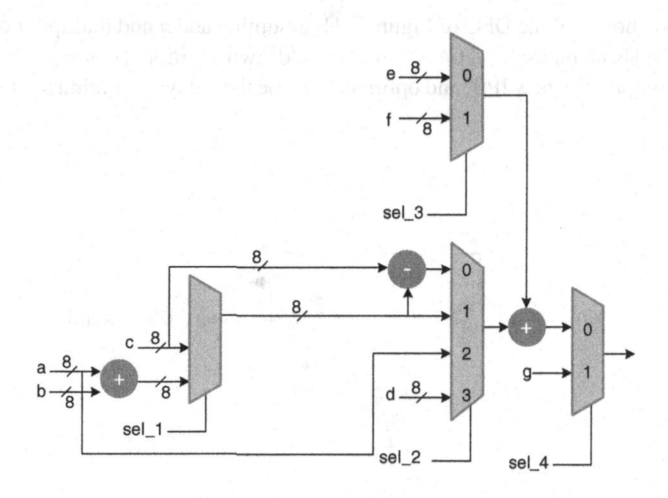

Figure 7.38 Digital logic design for exercise 7.2

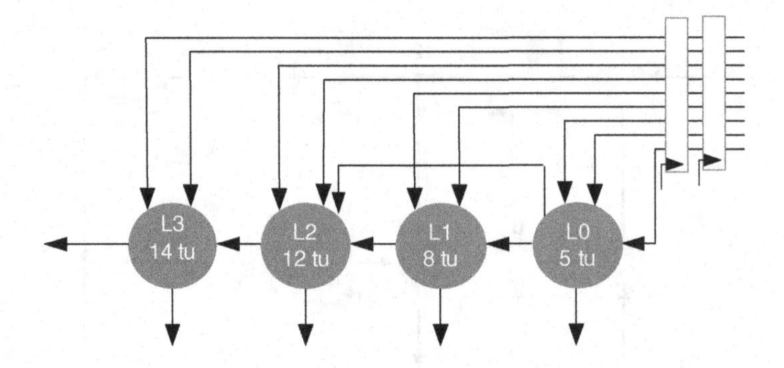

Figure 7.39 Dataflow graph for exercise 7.3

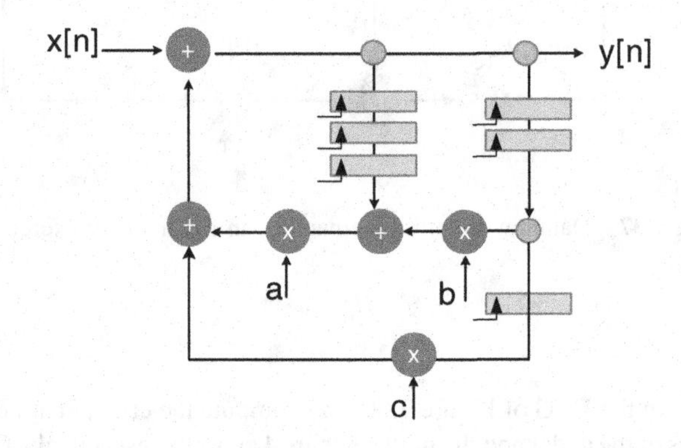

Figure 7.40 An IIR filter for exercise 7.4

Exercise 7.5

Compute the loop bound of the DFG of Figure 7.41, assuming adder and multiplier delays are 4 and 6 tu, respectively. Using look-ahead transformation adds two additional delays in the feedback loop of the design. Compute the new IPB, and optimally retime the delays to minimize the critical path of the system.

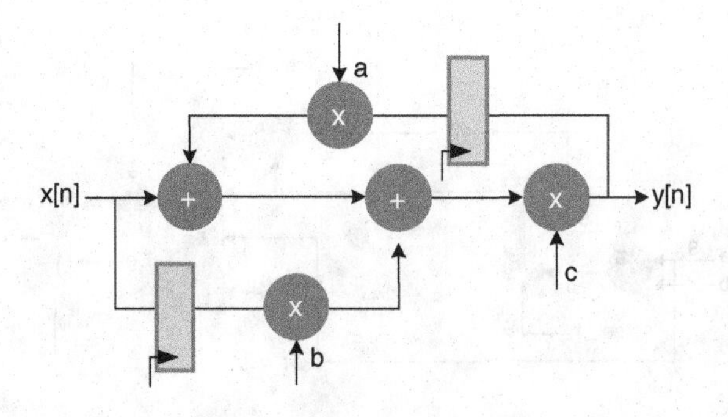

Figure 7.41 Dataflow graph for exercise 7.5

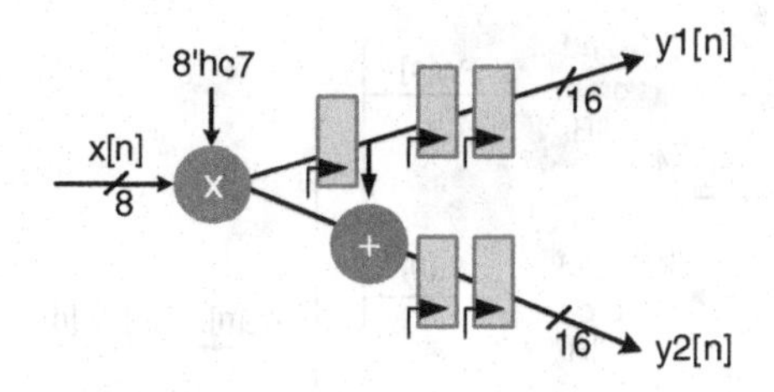

Figure 7.42 Digital design for exercise 7.6

Exercise 7.6

Design an optimal implementation of Figure 7.42. The DFG multiplies an 8-bit signed input $x[n]$ with a constant and adds the product in the previous product. Convert the constant to its equivalent CSD number. Generate PPs and append the adder and the CV as additional layers in the compression tree, where $x[n]$ is an 8-bit signed number. Appropriately retime the algorithmic registers to reduce the critical path of the compression tree. Write RTL Verilog code of the design.

Exercise 7.7

Design DF-II architecture to implement a second-order IIR filter given by the following difference equation:

$$y[n] = 0.3513y[n-1] + 0.3297y[n-1] + 0.2180x[n] - 0.0766x[n-1] + 0.0719x[n-2]$$

Now apply C-slow retiming for $C = 3$, and retime the registers for reducing the critical path of the design.

1. Write RTL Verilog code of the original and 3-slow design. Convert the constants into Q1.15 format.
2. In the top-level module, make three instantiations of the module for the original filter and one instantiation of the module for the 3-slow filter. Generate two synchronous clocks, `clk1` and `clk3`, where the latter is three times faster than the former. Use `clk1` for the original design and `clk3` for the 3-slow design. Generate three input steams of data. Input the individual streams to three instances of the original filter on `clk1` and time multiplex the stream and input to C-slow design using `clk3`. Compare the results from all the modules for correctness of the 3-slow design.This description is shown in Figure 7.43.

Exercise 7.8

Design a time multiplex-based architecture for the C-slowed and retimed design of Figure 7.27. The design should reuse the three computational nodes by appropriately selecting inputs from multiplexers.

Exercise 7.9

Draw a block diagram of a three-stage pipeline 12-bit carry skip adder with 4-bit skip blocks.

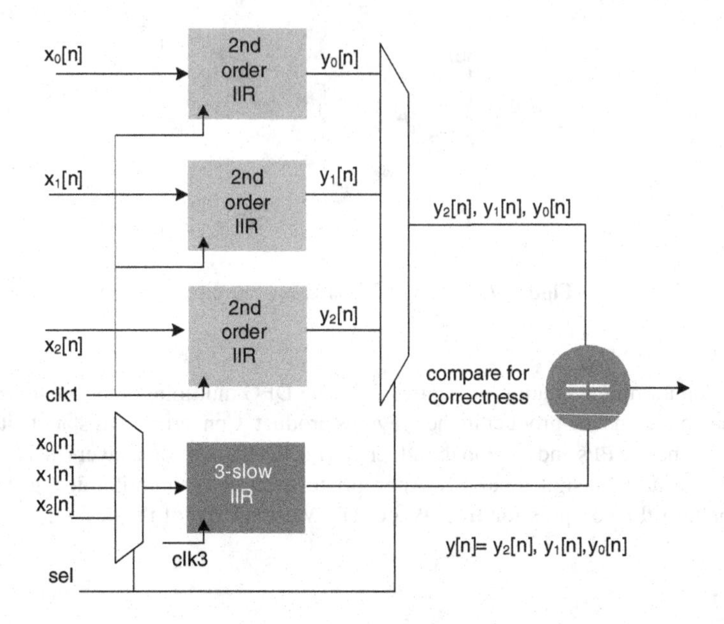

Figure 7.43 Second-order IIR filter and 3-slow design for exercise 7.7

Exercise 7.10

Apply a look-ahead transformation to represent $y[n]$ as $y[n-4]$. Give the IPBs for the original and transformed DFG for the following equation:

$$y[n] = 0.13y[n-1] + 0.2y[n-2] - x[n]$$

Assume both multiplier and adder take 1 time unit for execution. Also design a scattered-cluster look-ahead design for $M = 4$.

Exercise 7.11

Design a seventh-order IIR filter with cutoff frequency $\pi/7$ using the filter design and analysis toolbox of MATLAB®. Decompose the filter for decimation and interpolation application using the technique of Section 7.8. Code the design of interpolator and decimator in Verilog using 16-bit fixed-point signed arithmetic.

References

1. C. V. Schimpfle, S. Simon and J. A. Nossek, "Optimal placement of registers in data paths for low power design," in *Proceedings of IEEE International Symposium on Circuits and Systems*, 1997, pp. 2160–2163.
2. J. Monteiro, S. Devadas and A. Ghosh, "Retiming sequential circuits for low power," in *Proceedings of IEEE International Conference on Computer-aided Design*, 1993, pp. 398–402.
3. A. El-Maleh, T. E. Marchok, J. Rajski and W. Maly, "Behavior and testability preservation under the retiming transformation," *IEEE Transactions on Computer-Aided Design of Integrated Circuits and Systems*, 1997, vol. 16, pp. 528–542.

4. S. Dey and S. Chakradhar, "Retiming sequential circuits to enhance testability," in *Proceedings of IEEE VLSI Test Symposium*, 1994, pp. 28–33.
5. S. Krishnawamy, I. L. Markov and J. P. Hayes, "Improving testability and soft-error resilience through retiming," in *Proceedings of DAC*, ACM/IEEE 2009, pp. 508–513.
6. N. Weaver, Y. Markovshiy, Y. Patel and J. Wawrzynek, "Post-placement C-slow retiming for the Xilinx Virtex FPGA," in *Proceedings of SIGDA*, ACM 2003, pp. 185–194.
7. Synopsys/Xilinx. "High-density design methodology using FPGA complier," application note, 1998.
8. D. K. Y. Tong, E. F. Y. Young, C. Chu and S. Dechu, "Wire retiming problem with net topology optimization," *IEEE Transactions on Computer-Aided Design of Integrated Circuits and Systems*, 2007, vol. 26, pp. 1648–1660.
9. H. H. Loomis and B. Sinha, "High-speed recursive digital filter realization," *Circuits, Systems Signal Processing*, 1984, vol. 3, pp. 267–294.
10. K. K. Parhi and D. G. Messerschmitt, "Pipeline interleaving and parallelism in recursive digital filters.-Pts I and II," *IEEE Transactions on Acoustics, Speech and Signal Processing*, 1989, vol. 37, pp. 1099–1135.
11. Z. Jiang and A. N. Willson, "Design and implementation of efficient pipelined IIR digital filters," *IEEE Transactions on Signal Processing*, 1995, vol. 43, pp. 579–590.
12. F. J. Harris, C. Dick and M. Rice, "Digital receivers and transmitters using polyphase filter banks for wireless communications," *IEEE Transactions on Microwave Theory and Technology*, 2003, vol. 51, pp. 1395–1412.
13. A. V. Oppenheim and R. W. Schafer, *Discrete-time Signal Processing*, 3rd edn, 2009, Prentice-Hall.
14. C. Soviani, O. Tardieu, and S. A. Edwards, "Optimizing Sequential Cycles Through Shannon Decomposition and Retiming", *IEEE Transactions on Computer-Aided Design of Integrated Circuits and Systems*, vol. 26, no. 3, March 2007.
15. S. A. Khan, "Effective IIR filter architectures for decimation and interpolation applications", Technical Report CASE May 2010.

8

Unfolding and Folding of Architectures

8.1 Introduction

Major decisions in digital design are based on the ratio of sampling clock to circuit clock. The sampling clock is specific to an application and is derived from the Nyquist sampling criteria or band-pass sampling constraint. The circuit clock, on the other hand, primarily depends on the design and the technology used for implementation. In many high-end applications, the main focus of design is to run the circuit at the highest possible clock rate to get the desired throughput. If a simple mapping of the dataflow graph (DFG) on hardware cannot be synthesized at the required clock rate, the designer opts to use several techniques.

In feedforward designs, an unfolding transformation makes parallel processing possible. This results in an increase in throughput. Pipelining is another option in feedforward design for better timing. Pipelining is usually the option of choice as it results in a smaller area than with an unfolded design. In feedback DFGs, the unfolding transformation does not result in true parallel processing as the circuit clock rate needs to be reduced and hence does not give any potential iteration period bound (IPB) improvement. The only benefit of the unfolding transformation is that the circuit can be run at slower clock as each register is slower by the unfolding factor. In FPGA-based design, with a fixed number of registers and embedded computational units, unfolding helps in optimizing designs that require too many algorithmic registers. The design is first unfolded and then the excessive registers are retimed to give better timing. The chapter presents designs of FIR and IIR filters where unfolding and then retiming achieves better performance.

In contrast to dedicated or parallel architectures, time-shared architectures are designed in instances where the circuit clock is at least twice as fast as the sampling clock. The design running at circuit clock speed can reuse its hardware resources, as the input data remains valid for multiple circuit clocks. For many applications the designer can easily come up with a time-shared architecture. The chapter describes several examples to highlight these design issues.

For many applications, designing an optimal time-shared architecture may not be simple. This chapter covers mathematical transformation techniques for folding in time-multiplexed architectures. These transformations take the DFG representation of a synchronous digital signal processing (DSP) algorithm, a folding factor along with schedule of folding and then they systematically

Digital Design of Signal Processing Systems: A Practical Approach, First Edition. Shoab Ahmed Khan.

generate a folded architecture that maps the DFG on fewer hardware computational units. The folded architecture with the schedule can then be easily implemented using, respectivly, a datapath consisting of computational nodes and a controller based on a finite state machine. The chapter gives examples to illustrate the methodology. These examples are linked with their implementation as state machine-based architecture.

8.2 Unfolding

Design decisions are based on the ratio of sampling and circuit clock speeds. The sampling clock is specific to an application and is based on Nyquist sampling criteria or band-pass sampling techniques [1–3]. These sampling rates lower-bound the sampling clock. The sampling clock also imposes a throughput constraint on the design. In many applications the samples are also passed to a digital-to-analog (D/A) converter. On the other hand, the circuit clock depends on several factors, such as the target technology and power considerations. In many high-throughput applications, the main focus of the design is to run the circuit at the highest achievable clock rate. If fully dedicated architecture (FDA) mapping of a dataflow graph on to hardware cannot be synthesized at the required clock rate, the designer must look for other options.

Unfolding increases the area of the design without affecting the throughput. A slower clock does effect the power dissipation but increases the area of the design. The area–power tradeoff must be carefully studied if the unfolding is performed with the objective of power reduction. The unfolding transformation is very effective if there are more registers in the original DFG that can be effectively retimed for reducing the critical path delay of the design. This is especially true if the design is mapped on an FPGA with embedded computational units. These units have fixed number of registers that can be effectively mapped in an unfolded design. Similarly for feedback designs, the timing performance of the design can be improved by first adding and retiming pipeline registers and then applying the unfolding transformation to evenly distribute the registers with replicated functional units. It is important to point out that in many design instances pipelining and retiming is usually the option of choice because it results in less area than an unfolded design [4].

8.3 Sampling Rate Considerations

As the sampling and circuit clocks dictate the use of unfolding and folding transformations, it is important to understand the requirements placed on the sampling clock. This section explains Nyquist and band-pass sampling criteria.

8.3.1 *Nyquist Sampling Theorem and Design Options*

For digitization of an analog signal, the Nyquist sampling theorem defines the minimum constraint on the sampling frequency of the analog signal. The sampling frequency defines the number of samples the system needs to process every second. For perfect reconstruction, the Nyquist sampling criterion constrains the minimum sampling rate to be greater than or equal to twice the maximum frequency content in the signal:

$$f_s \geq 2f_N$$

where f_s and f_N represent sampling frequency and maximum frequency content in the analog signal, respectively.

The sampling frequency is the most critical constraint on a digital design. Usually the samples from an A/D converter are placed in a FIFO (first-in first-out) for processing. In a synchronous design a ping-pong buffer may also be used. The required number of samples in the buffer generates an activation signal for the digital component to start processing the buffer. For a design that processes the discrete signal on a sample-by-sample basis, the sampling clock is also used as circuit clock for the hardware design. Each newly acquired sample is stored in a register to be then processed by the digital system.

The sampling frequency imposes strict constraints even if the sampled data is first stored in a buffer. The designer, knowing the algorithmic constraint of processing a buffer of data of predefined size, allocates required memory to FIFO or ping-pong buffers. The designer also architects the HW to process this buffer of data before the next buffer is ready for processing. If the sampling rate is high enough and is approximately equal to the best achievable circuit clock, mapping the algorithm as FDA and synthesizing the design to achieve sampling clock timing is the most viable design option. If the sampling clock is faster than an achievable circuit clock, then the designer needs to explore parallel processing or pipelining options.

A software-defined radio (SDR) is an interesting application to explain different aspects of digital design. The following section presents SDR architecture to illustrate how band-pass sampling effectively reduces the requirement of sampling rate in digital communication applications. If the designer decides to sample the signal following Nyquist sampling criteria, the design requires the processing of more samples.

8.3.2 Software-defined Radio Architecture and Band-pass Sampling

In many designs, although strict compliance to the Nyquist sampling criterion requires the processing of a large number of samples, the sampling frequency of the A/D converter can be set to lower values without effecting algorithmic performance. This technique is called *band-pass sampling*.

An ideal SDR receiver requires direct digital processing of the analog signal after the signal passes through low-pass filtering and amplification. However, this implies a very high-speed A/D converter and processing requirement as per the Nyquist sampling criterion. For many applications this requires sampling and then processing of the signal at a few gigahertz range. In many designs the received analog signal is first brought down to an intermediate frequency (IF) by a radio frequency (RF) section and the signal is then digitized for processing by a mix of computing platforms such as FPGAs, DSPs and GPPs. The design of a typical SDR is given in Figure 8.1.

In a digital receiver a baseband signal primarily consists of digitized compressed voice, video, images or data that is modulated to take a limited bandwidth $B = 2f_N$, where f_N is the highest frequency content of the baseband signal. According to the frequency allocation assignment to a particular band in the spectrum, this baseband signal at the transmitter is multiplied with a carrier of high frequency f_0. This process is shown in Figure 8.2. This carrier-ridden signal occupies the same bandwidth $B = f_H - f_L$ around f_0, where f_H and f_L are the highest and lowest frequency content of this signal. The transmitted signal spectrum is usually populated with adjacent channels from other transmitters in the area. These channels have to be filtered out before the signal is digitized and demodulated in the baseband.

At the receiver, the signal from the antenna is first passed through a band-pass filter and then to a low-noise amplifier. This signal usually has some of the adjacent channels still left to be removed in the digital domain with better and sharper filters. Usually the RF signal of interest has a very narrow bandwidth and is centered at a very high frequency. Designing a narrow RF band-pass filter operating

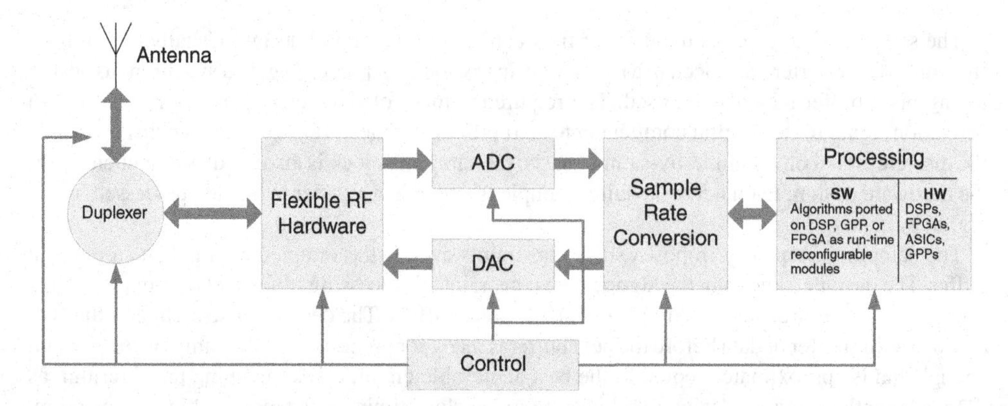

Figure 8.1 Typical software-defined radio architecture

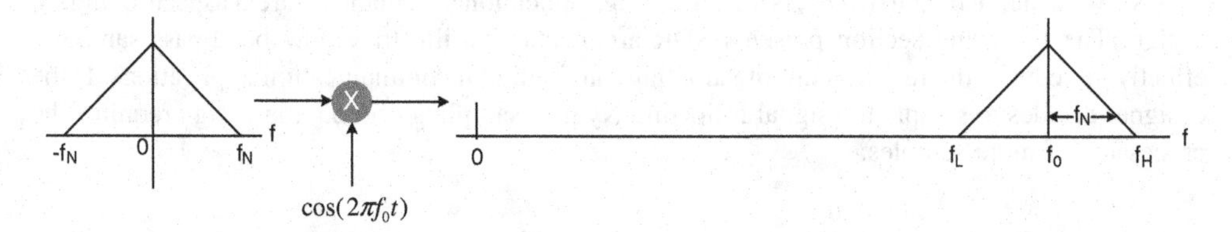

Figure 8.2 Baseband signal mixed with a carrier for signal translation in the frequency band

at very high frequency is a difficult engineering problem, and a relaxation in the band results in cost-effective designs. Although this relaxation results in adjacent channels in the filtered signals and a wider bandwidth of the signal to be digitized, the adjacent channels are then easily filtered out in the digital domain as shown in Figure 8.3(a).

In many designs the analog signal is first translated to a common intermediate frequency f_{IF}. The signal is also filtered to remove images and harmonics. The highest frequency content of this signal is $f_{IF} + B'/2$, where B' is the bandwidth of the filtered signal with components of the adjacent channels still part of the signal. Applying Nyquist, the signal can be digitized at twice this frequency. The sampling frequency in this case needs to be greater than $2(f_{IF} + B'/2)$. The digital signal is then digitally mixed using a quadrature mixer, and low-pass filters then remove adjacent channels from the quadrature I and Q signals. The signal is then decimated to only keep the required number of samples at base band. A typical receiver that digitizes a signal centered at f_{IF} for further processing is shown in Figure 8.3(b).

As the sampling frequency is so critical, further reduction in this frequency should be explored. In this context, for cases where the rest of the spectrum of the band-pass received signal or the IF signal is cleaned of noise, the signal can be directly sampled at a lower sampling frequency. This technique sub-samples an analog signal for intentional aliasing and down-conversion to an unused lower spectrum. The minimum sampling rate for band-pass sampling that avoides any possible aliasing of a band-pass signal is [1]:

$$\frac{2f_L}{K} \leq f_s \leq \frac{2f_H}{K-1} \tag{8.1}$$

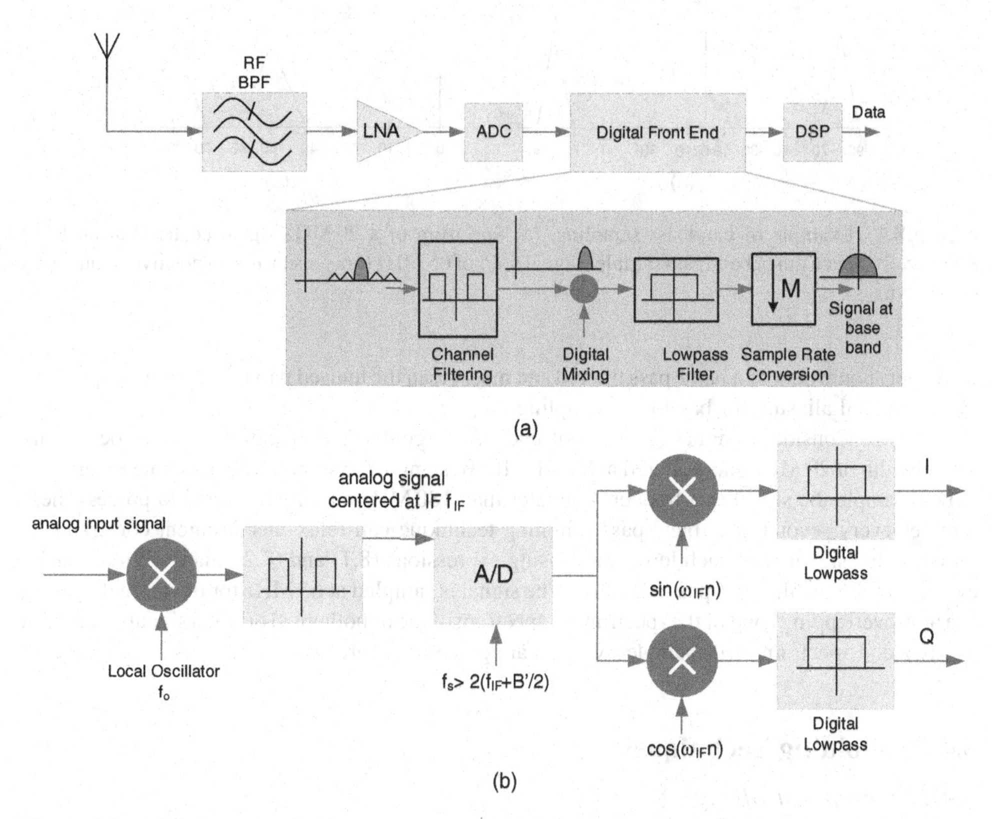

Figure 8.3 Digital communication receiver. (a) Direct conversion receiver. (b) IF conversion receiver

where, for minimum sampling frequency that avoids any aliasing, the maximum value of *K* is lower-bounded by:

$$K_{MAX} \leq \frac{f_H}{B} \tag{8.2}$$

This relationship usually makes the system operate 'on the edge', so care must be exercised to relax the condition of sampling frequency such that the copies of the aliased spectrum must fall with some guard bands between two copies. The sampling frequency options for different values of *K* are given in [1].

8.3.3 A/D Converter Bandwidth and Band-pass Sampling

It is important to point out that, athough the sampling rate is the fastest frequency at which an A/D converter can sample an analog signal, the converter's bandwidth (BW) is also an important consideration. This corresponds to the highest frequency at which the internal electronics of the converter can pass the signal without any attenuation. The Nyquist filter at the front of the A/D

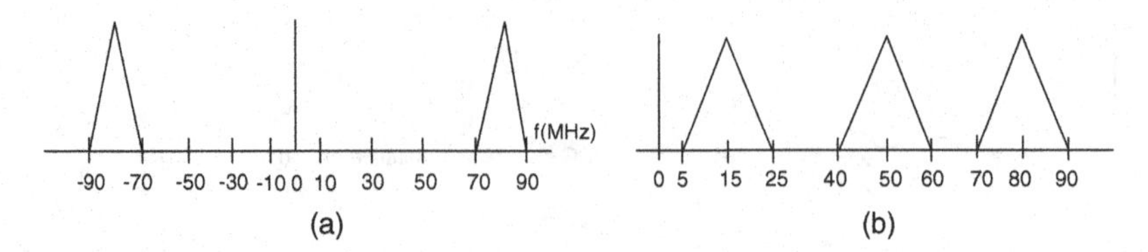

Figure 8.4 Example of bandpass sampling. (a) Spectrum of a 20 MHz signal centered at an IF of 80 MHz. (b) Spectrum of bandpass sampled signal for f_s of 65 MHz (only spectrum on positive frequencies are shown)

converter should also be a band-pass filtered and must clean the unused part of the original spectrum for intentional aliasing for band-pass sampling.

Example: Consider a signal at the front end of a digital communication receiver occupying a bandwidth of 20 MHz and centered at 80 MHz IF. To sample the signal at Nyquist rate requires an A/D to sample the signal at a frequency greater than 180 MHz and the front end to process these samples every second. The band-pass sampling technique can relax this stringent constraint by sampling the signal at a much lower rate. Using expressions (8.1) and (8.2), many options can be evaluated for feasible band-pass sampling. The signal is sampled at 65 MHz for intentional aliasing in a non-overlapping band of the spectrum. A spectrum of the hypothetical original signal centered at 80 MHz and spectrum of its sampled version are given in Figure 8.4.

8.4 Unfolding Techniques

8.4.1 Loop Unrolling

In the software context, unfolding transformation is the process of 'unrolling' a loop so that several iterations of the loop are executed in one unrolled iteration. For this reason, unfolding is also called 'loop unrolling'. It is a technique widely used by compilers for reducing the loop overhead in code written in high-level languages. For hardware design, unfolding corresponds to applying a mathematical transformation on a dataflow graph to replicate its functionality for computing multiple output samples for given relevant multiple input samples. The concept can also be exploited while performing SW to HW mapping of an application written in a high-level language. Loop unrolling gives more flexibility to the HW designer to optimize the mapping. Often the loop is not completely unrolled, but unfolding is performed to best use the HW resources, retiming and pipelining.

Loop unrolling can be explained using a dot-product example. The following code depicts a dot-product computation of two linear arrays of size N:

```
sum=0;
for (i=0; i<N; i++)
sum += a[i]*b[i];
```

For computing each MAC (multiplication and accumulation) operation, the program executes a number of cycles to maintain the loop; this is known as the *loop overhead.* This overhead includes incrementing the counter variable i and comparing the incremented value with N, and then taking a branch decision to execute or not to execute the loop iteration.

In addition to this, many DSPs have more than one MAC unit. To reduce the loop overhead and to utilize the multiple MAC units in the DSP architecture, the loop can be unrolled by a factor of J. For example, while mapping this code on a DSP with four MAC units, the loop should first be unrolled for $J=4$. The unrolled loop is given here:

```
J=4
sum1=0;
sum2=0;
sum3=0;
sum4=0;
for(i=0; i<N; i=i+J)
{
    sum1 += a[i]*b[i];
    sum2 += a[i+1]*b[i+1];
    sum3 += a[i+2]*b[i+2];
    sum4 += a[i+3]*b[i+3];
}
sum = sum0+sum1+sum2+sum3;
```

Similarly, the same loop can be unrolled by any factor J depending on the value of loop counter N and the number of MAC units on a DSP. In software loop unrolling, a designer is trading off code size with performance.

8.4.2 Unfolding Transformation

Algorithms for unfolding a DFG are available [5–8]. Any DFG can be unfolded by an unfolding factor J using the following two steps:

S0 To unfold the graph, each node U of the original DFG is replicated J times as $U_0, \ldots, U_{J-1}$ in the unfolded DFG.

S1 For two connected nodes U and V in the original DFG with w delays, draw J edges such that each edge j $(=0\ldots J-1)$ connects node U_j to node $V(j+w)\%J$ with $\lfloor (j+w)/J \rfloor$ delays, where % and $\lfloor . \rfloor$ are, respectively, remainder and floor operators.

Making J copies of each node increases the area of the design many-fold. but the number of delays in the unfolded transformation remains the same as in the original DFG. This obviously increases the critical path delay and the IPB for a recursive DFG by a factor of J.

Example: This example applies the unfolding transformation for an unfolding factor of 2 to a second-order IIR filter in TDF structure, shown in Figure 8.5(a). Each node of the DFG is replicated twice and the edges without delays are simply connected as they are. For edges with delays, the step S1 of unfolding transformation is applied. The nodes (U, V) on an edge with one delay in the original DFG are first replicated twice as (U_0, U_1) and (V_0, V_1). Now for $j=0$ and $w=1$, node U_0 is connected to node $V_{(0+1)\%2=1}$ with delays $\lfloor (0+1)/2 \rfloor = 0$; and for $j=1$ and $w=1$, node U_1 is connected to node $V_{(1+1)\%2=0}$ with delays $\lfloor (1+1)/2 \rfloor = 1$. This is drawn by the joining of node U_0 to corresponding node V_1 with 0 delay and node U_1 to corresponding node V_0 with 1 delay. Similarly, unfolding is performed on the second edge with a delay and the resultant unfolded DFG is shown in Figure 8.5(b).

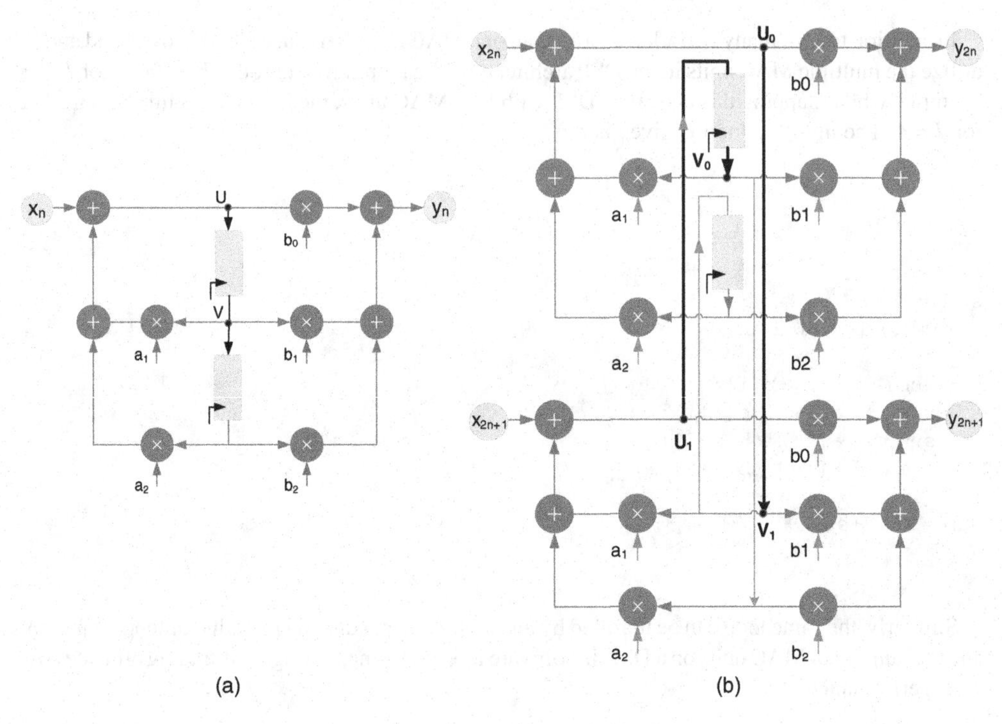

Figure 8.5 Unfolding transformations (a) Second-order TDF structure. (b) Unfolding with a factor of 2

8.4.3 Loop Unrolling for Mapping SW to HW

In many applications, algorithms are developed in high-level languages. Their implementation usually involves nested loops. The algorithm processes a defined number of input samples every second. The designer needs to map the algorithm in HW to meet the computational requirements. Usually these requirements are such that the entire algorithm need not to be unrolled, but a few iterations are unrolled for effective mapping. The unfolding should be carefully designed as it may lead to more memory accesses.

Loop unrolling for SW to HW mapping is usually more involved than application of an unfolding transformation on DFGs. The code should be carefully analyzed because, in instances with several nested loops, unrolling the innermost loop may not generate an optimal architecture. The architect should explore the design space by unrolling different loops in the nesting and also try merging a few nested loops together to find an effective design.

For example, in the case of a code that filters a block of data using an FIR filter, unrolling the outer loop that computes multiple output samples, rather than the inner loop that computes one output sample, offers a better design option. In this example the same data values are used for computing multiple output samples, thus minimizing the memory accesses. This type of unrolling is difficult to achieve using automatic loop unrolling techniques. The following example shows an FIR filter implementation that is then unrolled to compute four output samples in parallel:

```
#include <stdio.h>
#define N 12
#define L 8
// Test data and filter coefficients
short xbf[N+L-1]={1, 1, 2, 3, 4, 1, 2, 3, 1, 2, 1, 4, 5, -1, 2, 0, 1, -2, 3};
short hbf[L]={5, 9, -22, 11, 8, 21, 64, 18};
short ybf[N];
// The function performs block filtering operation
void BlkFilter(short *xptr, short *hptr, int len_x, int len_h, short *ybf)
{
  int n, k, m;
  for (n=0; n<len_x; n++)
  {
    sum = 0;
    for(k=0, m=n; k<len_h; k++, m-)
      sum += xptr[m]*hptr[k]; // MAC operation
    ybf[n] = sum;
  }
}
// Program to test BlkFilter and BlkFilterUnroll functions
void main(void)
{
   short *xptr, *hptr;
   xptr = &[L-1]; // xbf has L-1 old samples
   hptr = hbf;
   BlkFilterUnroll (xptr, hptr, N, L, ybf);
  BlkFilter (xptr, hptr, N, L, ybf);
}
```

The unrolled loop for effective HW mapping is given here:

```
// Function unrolls the inner loop to perform 4 MAC operations
// while performing block filtering function
void BlkFilterUnroll(short *xptr, short *hptr,
                     int len_x, int len_h, short *ybf)
{
  short sum_n_0, sum_n_1, sum_n_2, sum_n_3;
  int n, k, m;
  // Unrolling outer loop by a factor of 4
  for (n=4-1; n<len_x; n+=4)
  {
    sum_n_0 = 0;
    sum_n_1 = 0;
    sum_n_2 = 0;
    sum_n_3 = 0;
    for (k=0, m=n; k<len_h; k++, m-)
    {
      sum_n_0 += hptr[k] * xptr[m];
      sum_n_1 += hptr[k] * xptr[m-1];
      sum_n_2 += hptr[k] * xptr[m-2];
      sum_n_3 += hptr[k] * xptr[m-3];
    }
```

```
        ybf[n] = sum_n_0;
        ybf[n-1] = sum_n_1;
        ybf[n-2] = sum_n_2;
        ybf[n-3] = sum_n_3;
    }
}
```

The code processes N input samples in every iteration to compute N output samples in L iterations of the innermost loop. The processing of the innermost loop requires the previous $L-1$ input samples for computation of convolution summation. All the required input samples are stored in a buffer `xbf` of size $N+L-1$. The management of the buffer is conveniently performed using circular addressing, where every time N new input data values are written in the buffer replacing the oldest N values. This arrangement always keeps the $L-1$ most recent previous values in the buffer.

Figure 8.6 shows mapping of the code listed above in hardware with four MAC and three delay registers. This architecture only requires reading just one operand each from `xbf` and `hbf` memories. These values are designated as $x[m]$ and $y[k]$. At the start of every iteration the address registers for `xbf` and `hbf` are initialized to $m = L+4-1$ and $k = 0$, respectively. In every cycle address, register m is decremented and k is incremented by 1. In L cycles the architecture computes four output samples. The address m is then incremented for the next iteration of the algorithm. Three cycles are initially required to fill the tap delay line. The architecture can be unfolded to compute as many output samples as required and still needs to read only one value from each buffer.

There is no trivial transformation or loop unrolling technique that automatically generates such an optimized architecture. The designer's intellect and skills are usually required to develop an optimal architecture.

8.4.4 Unfolding to Maximize Use of a Compression Tree

For hardware design of feedforward DFGs, effective architectural options can be explored using unfolding transformation. An unrolled design processes multiple input samples by using

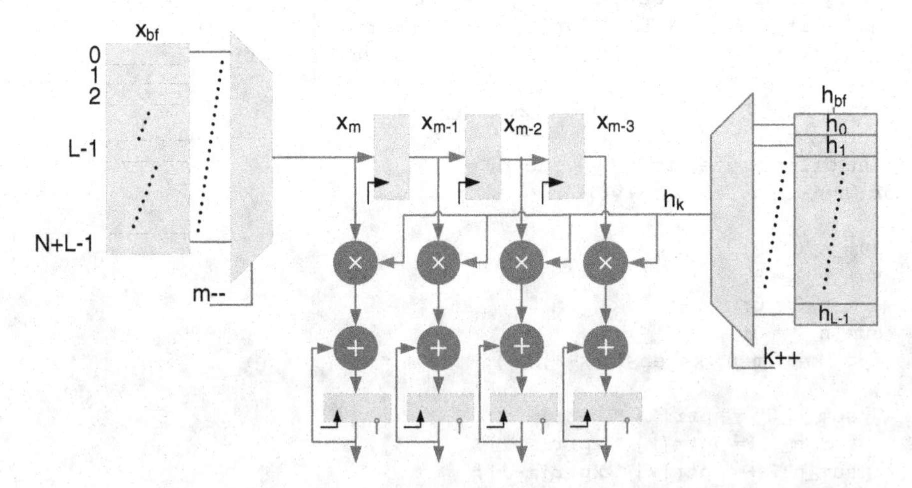

Figure 8.6 Hardware mapping of the code listed in the text

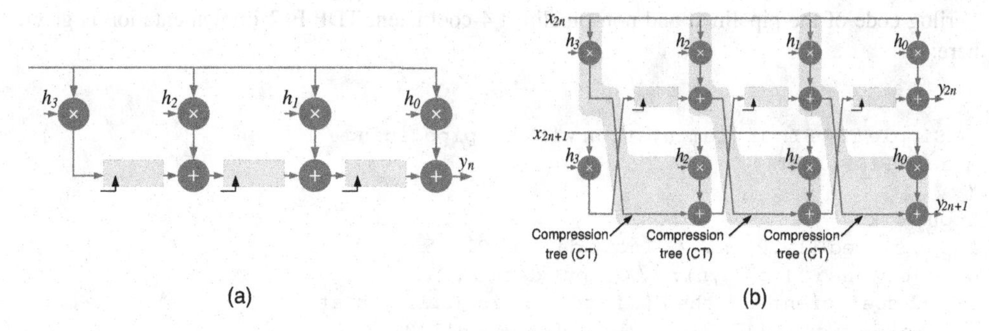

Figure 8.7 Unrolling an FIR filter. (a) Four-coefficient FIR filter. (b) The filter is unrolled by a factor of 2

additional resources. The unfolded architecture can now be explored for further optimization. Figure 8.7 shows a 4-coefficient FIR filter and a design after unfolding by a factor of 2. The designer can now design a computational unit consisting of two CSD multipliers and two adders as one computational unit. This unit can be implemented as a compression tree producing a sum and a carry. The architecture can also further exploit common sub-expression elimination (CSE) techniques (see Chapter 6).

It is important to point out that the design can also be pipelined for effective throughput increase. In many designs, simple pipelining without any folding may cost less in terms of HW than unfolding, because unfolding creates a number of copies of the entire design.

8.4.5 *Unfolding for Effective Use of FPGA Resources*

Consider a design instance where the throughput is required to be increased by a factor of 2. Assume the designer is using an FPGA with embedded DSP48 blocks. The designer can easily add additional pipeline registers and retime them between a multiplier and an adder, as shown in Figure 8.8. The

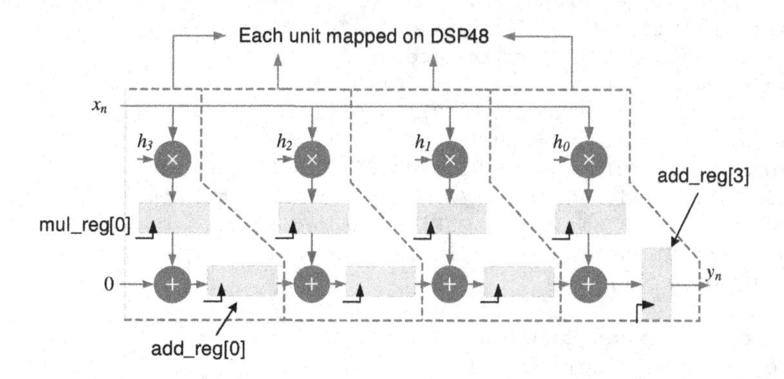

Figure 8.8 Pipelined FIR filter for effective mapping on FPGAs with DSP48 blocks

Verilog code of the pipelined and non-pipelined 4-coefficient TDF FIR implementation is given here:

```
/* Simple TDF FIR implementation without pipelining /
module FIRFilter
(
input clk,
input signed [15:0] xn, // Input data in Q1.15
output signed [15:0] yn); // Output data in Q1.15
// All coefficients of the FIR filter are in Q1.15 format
parameter signed [15:0] h0 = 16b1001100110111011;
parameter signed [15:0] h1 = 16b1010111010101110;
parameter signed [15:0] h2 = 16b0101001111011011;
parameter signed [15:0] h3 = 16b0100100100101000;
reg signed [31:0] add_reg[0:2] ; // Transpose Direct Form delay line
wire signed [31:0] mul_out[0:4]; // Wires for intermediate results
always @(posedge clk)
begin
  // TDF delay line
  add_reg[0] <= mul_out[0];
  add_reg[1] <= mul_out[1]+add_reg[0];
  add_reg[2] <= mul_out[2]+add_reg[1];
end
  assign mul_out[0]= xn * h3;
  assign mul_out[1]= xn * h2;
  assign mul_out[2]= xn * h1;
  assign mul_out[3]= xn * h0;
  assign mul_out[4] = mul_out[3]+add_reg[2];
  // Quantizing it back to Q1.15 format
  assign yn = mul_out[4][31:16];
endmodule
/* Pipelined TDF implementation of a 4-coefficient
         FIR filter for effective mapping
         on embedded DSP48 based FPGAs /
module FIRFilterPipeline
(
input clk,input signed [15:0] xn, // Input data in Q1.15
output signed [15:0] yn); // Output data in Q1.15
// All coefficients of the FIR filter are in Q1.15 format
parameter signed [15:0] h0 = 16b1001100110111011;
parameter signed [15:0] h1 = 16b1010111010101110;
parameter signed [15:0] h2 = 16b0101001111011011;
parameter signed [15:0] h3 = 16b0100100100101000;
reg signed [31:0] add_reg[0:2] ; // Transposed direct-form delay line
reg signed [31:0] mul_reg[0:3]; // Pipeline registers
wire signed [31:0] yn_f; // Full-precision output
always @( posedge clk)
begin
  // TDF delay line and pipeline registers
  add_reg[0] <= mul_reg[0];
```

```
    add_reg[1] <= mul_reg[1]+add_reg[0];
    add_reg[2] <= mul_reg[2]+add_reg[1];
    add_reg[3] <= mul_reg[3]+add_reg[2];
    mul_reg[0] <= xn*h3;
    mul_reg[1] <= xn*h2;
    mul_reg[2] <= xn*h1;
    mul_reg[3] <= xn*h0;
end
    // Full-precision output
    assign yn_f = mul_reg[3];
    // Quantizing to Q1.15
    assign yn = yn_f[31:16];
endmodule
```

When the requirement on throughput is almost twice what is achieved through one stage of pipelining and mapping on DSP48, will obviously not improve the throughput any further. In these cases unfolding can become very handy. The designer can add pipeline registers as shown in Figure 8.9(a). The number of pipeline registers should be such that each computational unit must have two sets of registers. The DFG is unfolded and the registers are then retimed and appropriately placed for mapping on DSP48 blocks. In the example shown the pipeline DFG is unfolded by a factor of 2. Each pipelined MAC unit of the unfolded design can then be mapped on a DSP48 where the architecture processes two input samples at a time. The pipeline DFG and its unfolded design are shown in Figure 8.9(b). The mapping on DSP48 units is also shown with boxes with one box shaded in gray for easy identification. The RTL Verilog code of the design is given here:

```
/* Pipelining then unfolding for effective mapping on
                        DSP48-based FPGAs with twice speedup /
module FIRFilterUnfold
(
input clk,
input signed [15:0] xn1, xn2, // Two inputs in Q1.15
output signed [15:0] yn1, yn2); // Two outputs in Q1.15
// All coefficients of the FIR filter are in Q1.15 format
parameter signed [15:0] h0 = 16'b1001100110111011;
parameter signed [15:0] h1 = 16'b1010111010101110;
parameter signed [15:0] h2 = 16'b0101001111011011;
parameter signed [15:0] h3 = 16'b0100100100101000;
// Input sample tap delay line for unfolding design
reg signed [15:0] xn_reg[0:4];
// Pipeline registers for first layer of multipliers
reg signed [31:0] mul_reg1[0:3];
// Pipeline registers for first layer of adders
reg signed [31:0] add_reg1[0:3];
// Pipeline registers for second layer of multipliers
reg signed [31:0] mul_reg2[0:3];
// Pipeline registers for second layer of adders
reg signed [31:0] add_reg2[0:3];
// Temporary wires for first layer of multiplier results
wire signed [31:0] mul_out1[0:3];
// Temporary wires for second layer of multiplication results
wire signed [31:0] mul_out2[0:3];
```

```
always @ ( posedge clk)
begin
     // Delay line for the input samples
     xn_reg[0] <= xn1;
     xn_reg[4] <= xn2;
     xn_reg[1] <= xn_reg[4];
     xn_reg[2] <= xn_reg[0];
     xn_reg[3] <= xn_reg[1];
end
always @ ( posedge clk)
begin
     // Registering the results of multipliers
     mul_reg1[0] <= mul_out1[0];
     mul_reg1[1] <= mul_out1[1];
     mul_reg1[2] <= mul_out1[2];
     mul_reg1[3] <= mul_out1[3];
     mul_reg2[0] <= mul_out2[0];
     mul_reg2[1] <= mul_out2[1];
     mul_reg2[2] <= mul_out2[2];
     mul_reg2[3] <= mul_out2[3];
end
always @ ( posedge clk)
begin
     // Additions and registering the results
     add_reg1[0] <= mul_reg1[0];
     add_reg1[1] <= add_reg1[0]+mul_reg1[1];
     add_reg1[2] <= add_reg1[1]+mul_reg1[2];
     add_reg1[3] <= add_reg1[2]+mul_reg1[3];
     add_reg2[0] <= mul_reg2[0];
     add_reg2[1] <= add_reg2[0]+mul_reg2[1];
     add_reg2[2] <= add_reg2[1]+mul_reg2[2];
     add_reg2[3] <= add_reg2[2]+mul_reg2[3];
end
     // Multiplications
     assign mul_out1[0]= xn_reg[0] * h3;
     assign mul_out1[1]= xn_reg[1] * h2;
     assign mul_out1[2]= xn_reg[2] * h1;
     assign mul_out1[3]= xn_reg[3] * h0;
     assign mul_out2[0]= xn_reg[4] * h3;
     assign mul_out2[1]= xn_reg[0] * h2;
     assign mul_out2[2]= xn_reg[1] * h1;
     assign mul_out2[3]= xn_reg[2] * h0;
     // Assigning output in Q1.15 format
     assign yn1 = add_reg1[3][31:16];
     assign yn2 = add_reg2[3][31:16];
endmodule
```

8.4.6 Unfolding and Retiming in Feedback Designs

It has already been established that an unfolding transformation does not improve timing; rather, it results in an increase in critical path delay and, for feedback designs, an increase in IPB by the

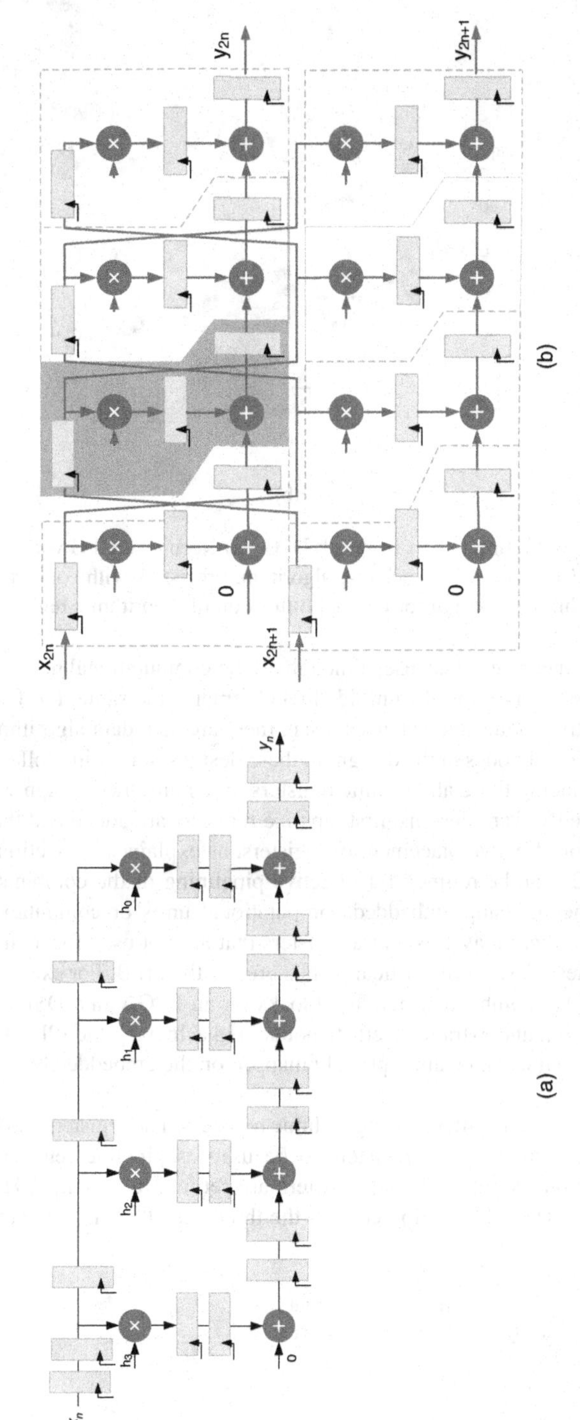

Figure 8.9 Unfolding for mapping on an embedded pipeline MAC unit. (a) The DFG is appropriately pipelined. (b) The pipeline DFG is unfolded for effective mapping on FGPAs with embedded pipelined MAC blocks

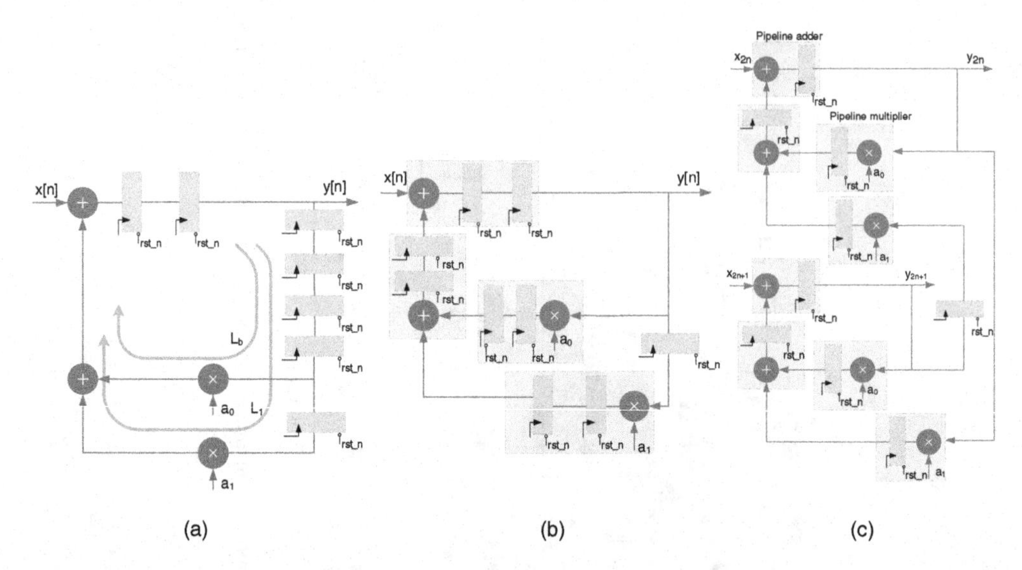

Figure 8.10 Unfolding and retiming of a feedback DFG. (a) Recursive DFG with seven algorithmic registers. (b) Retiming of resisters for associating algorithmic registers with computational nodes for effective unfolding. (c) Unfolded design for optimal utilization of algorithmic registers

unfolding factor J. This increase is because, although all the computational nodes are replicated J times, still the number of registers in the unfolded DFG remains the same. For feedback designs, unfolding may be effective for design instances where there are abundant algorithmic registers for pipelining the combinational nodes in the design. In these designs, unfolding followed by retiming provides flexibility of placing these algorithmic registers in the unfolded design while optimizing timing. Similarly for feedforward designs, first pipeline registers are added and the design is then unfolded and retimed for effective placement of registers, as explained in Section 8.4.5.

The registers in DFGs can be retimed for effective pipelining of the combinational cloud. In cases where the designer is using embedded computational units or computational units with limited pipeline support, there may exist extra registers that are not used for reducing the critical path of the design. In these designs the critical path is greater than IPB. For example, the designer might intend to use already embedded building blocks on an FPGA like DSP48. These blocks have a fixed pipeline option and extra registers do not help in achieving the IPB. By unfolding and retiming, the unfolded design can be appropriately mapped on the embedded blocks to effectively use all the registers.

Figure 8.10(a) shows a design with seven algorithmic registers. The registers can be retimed such that each computational unit has two registers to be used as pipeline registers, as shown in Figure 8.10(b). The design is unfolded and registers are retimed for optimal HW mapping, as shown in Figure 8.10(c). The RTL Verilog code of the three designs is listed here:

```
/* IIR filter of Fig. 8.10(a), having excessive
                        algorithmic registers /
module IIRFilter
(
input clk, rst_n,
input signed [15:0] xn, //Q1.15
```

```
output signed [15:0] yn); //Q1.15
parameter signed [15:0] a0=16'b0001_1101_0111_0000; //0.23 in Q1.15
parameter signed [15:0] a1=16'b1100_1000_1111_0110; //=0.43 in Q1.15
reg signed [15:0] add_reg[0:1] ;
reg signed [15:0] y_reg[0:3] ; // Input sample delay line
reg signed [15:0] delay;
wire signed [15:0] add_out[0:1];
wire signed [31:0] mul_out[0:1];
always @ ( posedge clk or negedge rst_n)
begin
    if(!rst_n) // Reset all the registers in the feedback loop
    begin
        add_reg[0] <= 0;
        add_reg[1] <= 0;
        y_reg[0] <= 0;
        y_reg[1] <= 0;
        y_reg[2] <= 0;
        y_reg[3] <= 0;
        delay <= 0;
    end
    else
    begin
        // Assign values to registers
        add_reg[0] <= add_out[0];
        add_reg[1] <= add_reg[0];
        y_reg[0] <= yn;
        y_reg[1] <= y_reg[0];
        y_reg[2] <= y_reg[1];
        y_reg[3] <= y_reg[2];
        delay <= y_reg[3];
    end
end

    // Implement combinational logic of two additions and
                    two multiplications
    assign add_out[0] = xn + add_out[1];
    assign mul_out[0]= y_reg[3] * a0;
    assign mul_out[1]= delay * a1;
    assign add_out[1] = mul_out[1][31:16]+mul_out[0][31:16];
    assign yn = add_reg[1];
endmodule

/* Retime the algorithmic registers to reduce the critical
                    path while mapping the design on FPGAs
                    with DSP48-like blocks /
module IIRFilterRetime
(
input clk, rst_n,
input signed [15:0] xn, //Q1.15
output signed [15:0] yn); //Q1.15
parameter signed [15:0] a0 = 16'b0001_1101_0111_0000;
                        //0.23 in Q1.15
```

```
parameter signed [15:0] a1 = 16'b1100_1000_1111_0110;
                        //=0.43 in Q1.15
reg signed [15:0] mul0_reg[0:1];
reg signed [15:0] mul1_reg[0:1]; // Input sample delay line
reg signed [15:0] add1_reg[0:1];
reg signed [15:0] add0_reg[0:1];
reg signed [15:0] delay;
wire signed [15:0] add_out[0:1];
wire signed [31:0] mul_out[0:1];
// Block Statements
always @(posedge clk or negedge rst_n)
if(!rst_n) // Reset registers in the feedback loop
    begin
        add0_reg[0] <= 0;
        add0_reg[1] <= 0;
        mul0_reg[0] <= 0;
        mul0_reg[1] <= 0;
        mul1_reg[0] <= 0;
        mul1_reg[1] <= 0;
        add1_reg[0] <= 0;
        add1_reg[1] <= 0;
        delay <= 0;
    end
else
    begin
        // Registers are retimed to reduce the critical path
        add0_reg[0] <= add_out[0];
        add0_reg[1] <= add0_reg[0];
        mul0_reg[0] <= mul_out[0][31:16];
        mul0_reg[1] <= mul0_reg[0];
        mul1_reg[0] <= mul_out[1][31:16];
        mul1_reg[1] <= mul1_reg[0];
        add1_reg[0] <= add_out[1];
        add1_reg[1] <= add1_reg[0];
        delay <= yn;
    end
    // Combinational logic implementing additions
                        and multiplications
    assign add_out[0]= xn + add1_reg[1];
    assign mul_out[0]= yn * a0;
    assign mul_out[1]= delay * a1;
    assign add_out[1]= mul0_reg[1]+mul1_reg[1];
    assign yn = add0_reg[1];
endmodule

/* Unfolding and retiming to fully utilize the algorithmic
                        registers for reducing the critical path
                        of the design for effective mapping on
                        DSP48-based FPGAs /
module IIRFilterUnfold
(
input clk, rst_n,
input signed [15:0] xn0, xn1, // Two inputs in Q1.15
```

```
output signed [15:0] yn0, yn1); //Two outputs in Q1.15
parameter signed [15:0] a0=16'b0001_1101_0111_0000;
                        //0.23 in Q1.15
parameter signed [15:0] a1=16'b1100_1000_1111_0110;
                        //=0.43 in Q1.15
reg signed [15:0] mul0_reg[0:1];
reg signed [15:0] mul1_reg[0:1]; // Input sample delay line
reg signed [15:0] add1_reg[0:1];
reg signed [15:0] add0_reg[0:1];
reg signed [15:0] delay;
wire signed [15:0] add0_out[0:1], add1_out[0:1];
wire signed [31:0] mul0_out[0:1], mul1_out[0:1];
// Block Statements
always @ ( posedge clk or negedge rst_n)
if(!rst_n)
    begin
        add0_reg[0] <= 0;
        add0_reg[1] <= 0;
        mul0_reg[0] <= 0;
        mul0_reg[1] <= 0;
        mul1_reg[0] <= 0;
        mul1_reg[1] <= 0;
        add1_reg[0] <= 0;
        add1_reg[1] <= 0;
        delay <= 0;
    end
else
    begin
        // Same number of algorithmic registers, retimed differently
        add0_reg[0] <= add0_out[0];
        add1_reg[0] <= add0_out[1];
        mul0_reg[0] <= mul0_out[0][31:16];
        mul1_reg[0] <= mul0_out[1][31:16];
        add0_reg[1] <= add1_out[0];
        add1_reg[1] <= add1_out[1];
        mul0_reg[1] <= mul1_out[0][31:16];
        mul1_reg[1] <= mul1_out[1][31:16];
        delay <= yn1;
    end

    /* Unfolding by a factor of 2 makes two copies of the
                      combinational nodes /
    assign add0_out[0]= xn0 + add1_reg[0];
    assign mul0_out[0]= yn0 * a0;
    assign mul0_out[1]= delay * a1;
    assign add0_out[1]= mul0_reg[0]+mul1_reg[0];
    assign yn0 = add0_reg[0];
    assign add1_out[0]= xn1 + add1_reg[1];
    assign mul1_out[0]= yn1 * a0;
    assign mul1_out[1]= yn0 * a1;
    assign add1_out[1]= mul0_reg[1]+mul1_reg[1];
    assign yn1 = add0_reg[1];
endmodule
```

```
module testIIRfilter;
reg signed [15:0] Xn, Xn0, Xn1;
reg RST_N, CLK, CLK2;
wire signed [15:0] Yn, Yn0, Yn1;
wire signed [15:0] Yn_p;
integer i;
// Instantiating the two modules for equivalence checking
IIRFilter IIR(CLK, RST_N, Xn, Yn);
IIRFilterRetime IIRRetime(CLK, RST_N, Xn, Yn_p);
IIRFilterUnfold IIRUnfold(CLK2, RST_N, Xn0, Xn1, Yn0, Yn1);
initial
begin
    CLK = 0; // Sample clock
    CLK2 = 0;
    Xn = 215;
    Xn0 = 215;
    Xn1 = 430;
    #1 RST_N = 1; // Generate reset
    #1 RST_N = 0;
    #3 RST_N = 1;
end
// Generating clock signal
always
    #4 CLK = ~CLK; // Sample clock
// Generating clock signal for unfolded module
always
    #8 CLK2 = ~CLK2; // Twice slower clock
always @(posedge CLK)
begin
    Xn <= Xn+215;
end
always @(posedge CLK2)
begin
    Xn0 <= Xn0+430;
    Xn1 <= Xn1+430;
end
initial
    $monitor ($time, " Yn=%d, Yn_p=%d, Yn0=%d, Yn1=%d\n",
                          Yn, Yn_p, Yn0, Yn1);
endmodule
```

8.5 Folding Techniques

Chapter 9 covers time-shared architectures. These are employed when the circuit clock is at least twice as fast as the sampling clock. The design running at circuit clock speed can reuse its hardware resources because the input data remains valid for multiple circuit clocks. In many design problems the designer can easily come up with time-shared architecture. Examples in Chapter 9 highlight a few of the design problems. For complex applications, designing an optimal time-shared architecture is a complex task. This section also covers a mathematical technique for designing time-shared architectures.

8.5.1 Definitions and the Folding Transformation

- *Folding* is a mathematical technique for finding a time-multiplexed architecture and a schedule of mapping multiple operations of a dataflow graph on fewer hardware computational units.
- The *folding factor* is defined as the maximum number of operations in a DFG mapped on a shared computational unit.
- A *folding set* or *folding scheduler* is the sequence of operations of a DFG mapped on a single computational unit.

The *folding transformation* has two parts. The first part deals with finding a folding factor and the schedule for mapping different operations of the DFG on computational units in the folded DFG. The *optimal folding factor* is computed as:

$$N = \left\lfloor \frac{f_c}{f_s} \right\rfloor$$

where f_c and f_s are circuit and sampling clock frquencies. Based on N and a schedule of mapping multiple operations in the DFG on shared resources, a folded architecture automatically saves intermediate results in registers and inputs them to appropriate units in the scheduled cycle for correct computation.

Folding transformation by a factor of N introduces latency into the system. The corresponding output for an input appears at the output after N clock cycles. The sharing of a computational unit by different operations also requires a controller that schedules these operations in time slots. The controller may simply be a counter, or it can be based on a finite state machine (FSM) that periodically generates a sequence of control signals for the datapath to select correct operands for the computational units.

Example: The DFG shown in Figure 8.11(a) is to be folded. The DFG has two nodes (A_1 and A_2) performing addition and two (M_1 nd M_2) performing multiplication. Assuming the data is valid for two circuit clock cycles, the DFG is folded by a factor of 2 using one multiplier and one adder. The folded architecture is shown in Figure 8.11(b). The mapping of the algorithm on the folded architecture requires a schedule, which for this example is very simple to work out. For the two cycles of the circuit clock, the adder and multiplier perform operations in the order $\{A_1, A_2\}$ and $\{M_1, M_2\}$. The folded architecture requires two additional registers to hold the values of intermediate results to be used in the next clock cycle. All the multiplexers select port 0 in the first cycles and port 1 in the second cycle for feeding correct inputs to the shared computational units.

8.5.2 Folding Regular Structured DFGs

Algorithms that exhibit a regular structure can be easily folded. Implementation of an FIR filter is a good example of this. The designer can fold the structure by any order that is computed as the ratio of the circuit and sampling clocks.

For any regularly structured algorithm the folding factor is simply computed as:

$$N = \frac{f_c}{f_s} \tag{8.3}$$

The DFG of the regular algorithm is partitioned into N equal parts. The regularity of the algorithm means implementing the datapath of a single partition as a shared resource. The delay line or temporary results are stored in registers and appropriately input to the shared data path.

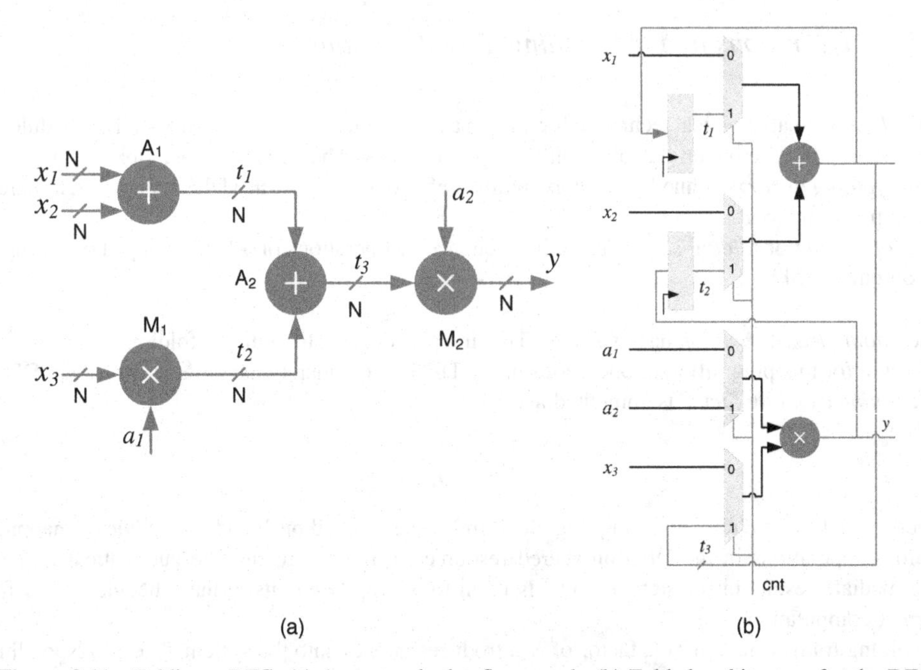

Figure 8.11 Folding a DFG (a) An example dataflow graph. (b) Folded architecture for the DFG

Example: An L-coefficient FIR filter implements the convolution equation:

$$y[n] = \sum_{k=0}^{L-1} h[k]x[n-k] \tag{8.4}$$

This equation translates into a regularly structured dataflow graph as DF or TDF realizations. The index of summation can be implemented as N summation for $k = mM + l$, where $M = L/N$, and the equation can be written as:

$$y[n] = \sum_{m=0}^{N-1} \sum_{l=0}^{M-1} h[mM + l]x[n - (mM + l)] \tag{8.5}$$

This double summation can be realized as folded architecture. For DF, the inner equation implements the shared resource and the outer equation executes the shared resource N times to compute each output sample $y[n]$. The realizations of the folded architecture for $L = 9$ and $N = 3$ as DF and TDF are given in Figures 8.12. In the DF case the tap delay line is implemented 'as is', whereas the folded architecture implements the inner summation computing 3-coefficient filtering in each cycle. This configuration implements three multipliers and adders as shared resources. The multiplexer appropriately sends the correct inputs to, respectively, multiplier and adder. The adder at the output adds the partial sums and implements the output summation of (8.5).

Figure 8.12(b) shows the TDF realization. Here again all the registers are realized as they are in the FDA implementation whereas the hardware resources are shared. The *cntr* signal is used to control the multiplexers and demultiplexers to input the correct data to the shared resources and to store the computational results in appropriate registers in each clock cycle.

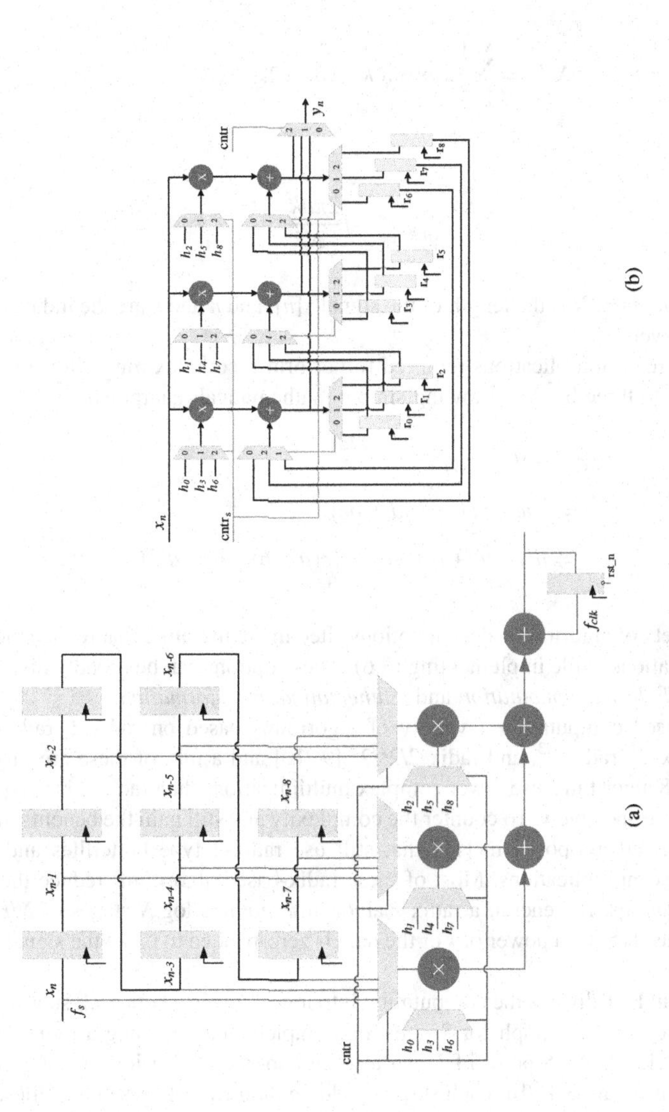

Figure 8.12 Folded-by-3 architecture for a 9-coefficient FIR filter. (a) Folded DF architecture. (b) Folded TDF architecture

8.5.3 Folded Architectures for FFT Computation

Most signal processing algorithms have a regular structure. The fast Fourier transform (FFT) is another example to demonstrate the folding technique. An FFT algorithm efficiently implements DFT summation:

$$X[k] = \sum_{n=0}^{N-1} x[n]W_N^{nk} k = 0,1,2,\ldots,N-1 \tag{8.6}$$

where

$$W_N^{nk} = e^{-j\frac{2\pi nk}{N}}$$

are called *twiddle factors*. N is the length of the signal $x[n]$, and n and k are the indices in time and frequency, respectively.

The number of real multiplications required to perform a complex multiplication can be first reduced from four to three by the following simple mathematical manipulation:

$$\begin{aligned}&(a+jb)(c+jd)\\&= (ac-bd)+j(ad+bc)\\&= d(a-b)+a(c-d)+(c(a+b)-a(c-d))j\end{aligned}$$

There are a variety of algorithmic design options cited in the literature that reduce the number of complex multiplications while implementing (8.6). These options can be broadly divided into two categories: *butterfly-based computation* and *mathematical transformation*.

For butterfly-based computation a variety of algorithms based on radix-2, radix-4, radix-8, hybrid radix, radix-2^2, radix-2^3, and radix-$2/2^2/2^3$ [9, 10] and a mix of these are proposed. The radix-4 and radix-8 algorithms use fewer complex multiplications than radix-2 but require N to be a power of 4 and 8, respectively. To counter the complexity and still gain the benefits, radix-2^2 and radix-2^3 algorithms are proposed in [11] that still use radix-2 type butterflies and reduce the number of complex multiplications. Most of these radix-based algorithms reduce the number of complex multiplications. In general, a radix-r algorithm requires $\log_r N$ stages of N/r butterflies, and preferably N needs to be a power of r or the data is zero-padded to make the signal of requisite length.

A radix-2 N-point FFT divides the computation into $\log_2 N$ stages where each stage executes $N/2$ two-point butterflies. A flow graph for 8-point FFT implementation using a radix-2 butterfly is shown in Figure 8.13(a). An 8-point FFT can also be computed by using two stages of radix-4 butterflies,shown in Figure 8.13(b); each stage would contain two of these butterflies.

A pipelined FDA requires the data to be placed at the input of every butterfly in every cycle, and then the design works in lock step to generate N output in every clock cycle with an initial latency of the number of pipeline stages in the design. This design requires placing of N samples at the input of the architecture in every clock cycle. This for large values of N is usually not possible, so then the designer needs to resort to folded architecture.

When there are more clock cycles available for the design, the architecture is appropriately folded. In general, a folded architecture can realize M butterflies in parallel to implement an N-point FFT,

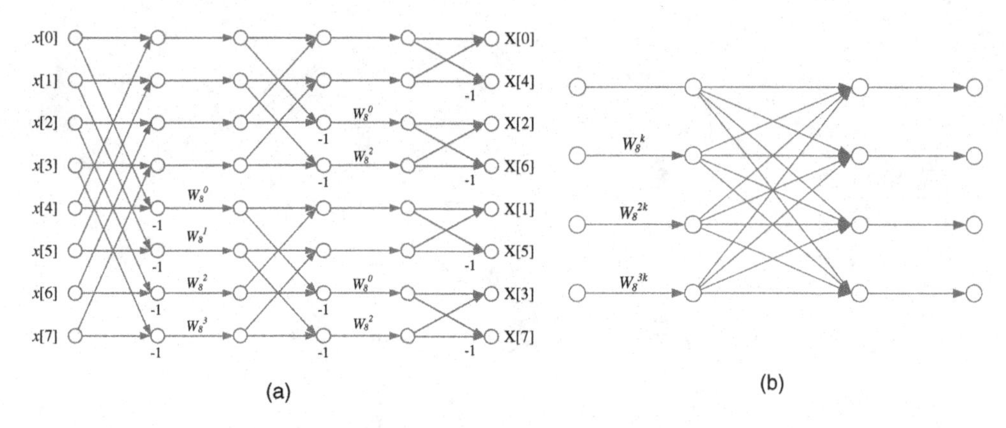

Figure 8.13 (a) Flow graph realizing an 8-point FFT algorithm using radix-2 butterflies, and (b) A radix-4 butterfly

where $N = JM$ and J is the folding factor. The design executes the algorithm in J clock cycles. Feeding the input data to the butterflies and storing the output for subsequent stages of the design is the most critical design consideration. The data movement and storage should be made such that data is always available to the butterflies without any contention. Usually the data is either stored in memory or input to the design using registers. The intermediate results are either stored back in memory or are systematically placed in registers such that they are conveniently available for the next stages of computation. Similarly the output can be stored either in memory or in temporary registers. For register-based architectures the two options are MDC (multi-delay commutator) [12–14] or SDF (single-path delay feedback) [15, 16]. In MDC, feedforward registers and multiplexers are used on r stream of input data to a folded radix-r butterfly.

Figure 8.14 shows a design for 8-point decimation in a frequency FFT algorithm that folds three stages of eight radix-2 butterflies on to three butterflies. The pipeline registers are appropriately placed to fully utilize all the butterflies in all clock cycles.

8.5.4 Memory-based Folded FFT Processor

The designer can also produce a *generic* FFT processor that uses a number of butterflies in the datapath and multiple dual-ported memories and respective address generation units for computing an N-point FFT [17–19]. Here a memory-based designed is discussed.

For memory-based HW implementation of an N-point radix-2 FFT algorithm, one or multiple butterfly units can be placed in the datapath. Each butterfly is connected to two dual-ported memories and a coefficient ROM for providing input data to the butterfly. The outputs of the butterflies are appropriately saved in memories to be used for the next stages of execution. The main objective of the design is to place input and output data in a way that two corresponding inputs to execute a particular butterfly are always stored in two different dual-port memories. The memory management should be done in a way that avoids access conflicts.

A design using a radix-2 butterfly processing element (PE) with two dual-ported RAMs and a coefficient ROM in the datapath is shown in Figure 8.15(a). While implementing decimation in frequency FFT computation, the design initially stores $N/2$ points of input data, $x_0, x_1, \ldots, x_{N/2-1}$ in

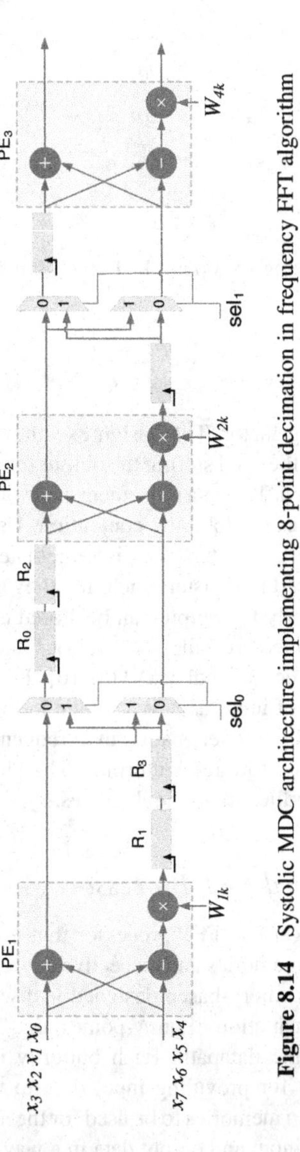

Figure 8.14 Systolic MDC architecture implementing 8-point decimation in frequency FFT algorithm

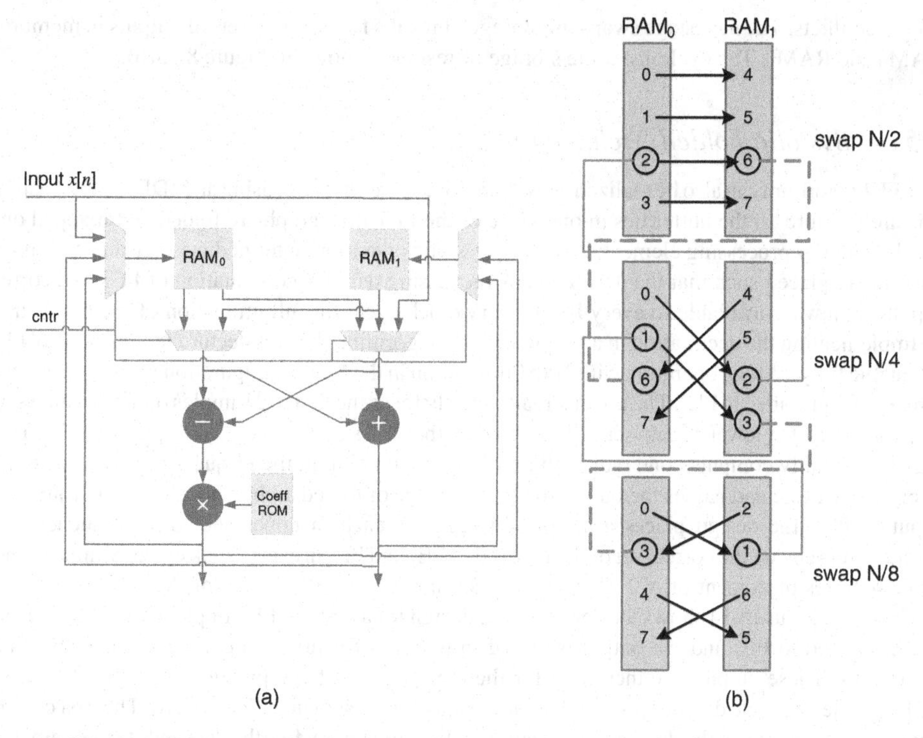

Figure 8.15 (a) FFT processor with two dual-port memories, with one radix-2 butterfly in the datapath. (b) Memory storage pattern for effective FFT implementation

RAM_0, and then with the arrival of every new sample, $x_{N/2}$, $x_{N/2+1}$, ..., $x_N - 1$, executes the following butterfly computations:

$$y_k = x_n + x_{n+\frac{N}{2}} \tag{8.7}$$

$$y_{k+1} = \left(x_n - x_{n+\frac{N}{2}}\right) W_N^{nk} \tag{8.8}$$

The output of the computations from (8.7) and (8.8) are stored in RAM_0 and RAM_1, respectively, for the first $N/2$ samples,and then reversing the storage arrangements whereby the outputs of (8.7) and (8.8) are stored in RAM_1 and RAM_0, respectively. This arrangement enables the next stage of computation without any memory conflicts. In every stage of FFT computation, the design swaps the storage pattern for every $N/2^s$ output samples, where s is the stage of butterfly computation for $s = 1, 2, \ldots, \log_2 N$.

Example: Figure 8.13(a) shows a flow graph realizing an 8-point decimation in frequency FFT algorithm. The first four samples of input, $x_0 \ldots x_3$, are stored in RAM_0. For the next input samples the architecture starts executing (8.7) and (8.8) and stores the first four results in RAM_0 and RAM_1, respectively, and the results of the next four computations are stored in RAM_1 and RAM_0, respectively. This arrangement makes the computation of the next stage of butterfly possible without memory access conflicts. The next stage swaps the storage of results for evry two sets of outputs in memories RAM_0 and RAM_1. The cycle-by-cycle storage of results is shown in Figure 8.13(b).

8.5.5 Systolic Folded Architecture

The FFT algorithm can also be realized as systolic folded architecture, using an MDF or SDF basis. In this architecture all the butterflies in one stage of the FFT flow graph are folded and mapped on a single butterfly processing element. To ensure systolic operation, intermediate registers and multiplexers are placed such that the design keeps processing data for computation of FFT and correct inputs are always available to every PE in every clock cycle for full utilization of the hardware.

Implementing the decimation in a frequency FFT algorithm, the architecture requires $\log_2 N$ PEs. To enable fully systolic operation with 100% utilization in the MDF configuration, the two input data streams are provided to PE_1. The upper stream serially feeds the first $N/2$ samples of input data, $x_0, x_1, \ldots, x_{N/2-1}$, and the lower stream serially inputs the other $N/2$ samples, $x_{N/2}, x_{N/2+1}, \ldots, x_N - 1$. Then PE_1 sequentially computes one radix-2 butterfly of decimation in the frequency FFT algorithm in every clock cycle and passes the output to the next stage of folded architecture. To synchronize the input to PE_2, the design places a set of $N/4$ registers each in upper and lower branches. The multiplexers ensure the correct set of data are fed to PE_1 and PE_2 in every clock cycle. Similarly each stage requires placement of $N/2^s$ registers at each input for $s = 2, 3, \ldots, \log_2 N$.

Example: Figure 8.14 shows the details of the design for a systolic FFT implementation. The data is serially fed to PE_1 and the output is stored in registers for first two computations of butterfly operations. These outputs are then used for the next two set of computation.

The cycle-by-cycle detailed working of the architecture is shown in Figure 8.16. The first column shows the cycles where the design repeats computation after every fourth cycle and starts computing the 8-point FFT of a new sequence of 8 data points. The second column shows the data being serially fed to the design. The first cycle takes x_0 and x_4, and then in each subsequent cycle two data values are fed as shown in this column. The columns labeled as PE_1, PE_2 and PE_3 show the butterfly computations. The intermediate values stored in registers R_0, R_1, R_2 and R_3 are also shown in the table as y_{ij}, where i is the level and j is the index..

The multiplexer selects signals given in Table 8.1. The controller is a simple 2-bit counter, where the most significant bit (MSB) of the counter is used for $\texttt{sel}_1$ and the least significant bit (LSB) is used for $\texttt{sel}_2$. The counter is used to read the values of twiddle factors from four-deep ROM. The RTL Verilog code is given here:

```
// Systolic FFT architecture for 8-point FFT computation
module FFTSystolic
(
input signed [15:0] xi_re, xi_im, xj_re, xj_im,
input clk, rst_n,
output signed [15:0] yi_re, yi_im, yj_re, yj_im);
reg signed [15:0] R_re[0:5], R_im[0:5];
reg [1:0] counter;
wire signed [15:0] W_re[0:3], W_im[0:3];
// Twiddle factors of three butterflies
reg signed [15:0] W0_re, W0_im, W1_re, W1_im, W2_re, W2_im;
wire sel1, sel2;
wire signed [15:0] xj_re1, xj_im1, xj_re2, xj_im2;
wire signed [15:0] yi_re1, yi_im1, yj_re1, yj_im1,
                  yi_re2, yi_im2, yj_re2, yj_im2;
```

```
wire signed [15:0] yi_out_re1, yi_out_im1, yi_out_re2,
                        yi_out_im2;
// The control signals
assign sel1 = ~counter[1];
assign sel2 = ~counter[0];
// Calling butterfly tasks
ButterFly B0(xi_re, xi_im, xj_re, xj_im, W0_re, W0_im,
                        yi_re1, yi_im1, yj_re1, yj_im1);
ButterFly B1(R_re[3], R_im[3], xj_re1, xj_im1, W1_re,
                        W1_im, yi_re2, yi_im2, yj_re2, yj_im2);
ButterFly B2(R_re[5], R_im[5], xj_re2, xj_im2, W2_re,
                        W2_im, yi_re, yi_im, yj_re, yj_im);
Mux2To2 MUX1(yi_re1, yi_im1, R_re[1], R_im[1], sel1,
                        i_out_re1, yi_out_im1, xj_re1, xj_im1);
Mux2To2 MUX2(yi_re2, yi_im2, R_re[4], R_im[4], sel2,
                        yi_out_re2, yi_out_im2, xj_re2, xj_im2);
always @(*)
begin
    // Reading values of twiddle factors for three butterflies
    W0_re = W_re[counter]; W0_im = W_im[counter];
    W1_re = W_re[{counter[0],1'b0}];W1_im = W_im[{counter[0],1'b0}];
    W2_re = W_re[0]; W2_im = W_im[0];
end
always @(posedge clk or negedge rst_n)
begin
    if(!rst_n)
    begin
        R_re[0] <= 0; R_im[0] <= 0;
        R_re[1] <= 0; R_im[1] <= 0;
        R_re[2] <= 0; R_im[2] <= 0;
        R_re[3] <= 0; R_im[3] <= 0;
        R_re[4] <= 0; R_im[4] <= 0;
        R_re[5] <= 0; R_im[5] <= 0;
    end
    else
    begin
        R_re[0] <= yj_re1; R_im[0] <= yj_im1;
        R_re[1] <= R_re[0]; R_im[1] <= R_im[0];
        R_re[2] <= yi_out_re1; R_im[2] <= yi_out_im1;
        R_re[3] <= R_re[2]; R_im[3] <= R_im[2];
        R_re[4] <= yj_re2; R_im[4] <= yj_im2;
        R_re[5] <= yi_out_re2; R_im[5] <= yi_out_im2;
    end
end
always @(posedge clk or negedge rst_n)
if(!rst_n)
    counter <= 0;
else
    counter <= counter+1;
    // 1, 0
    assign W_re[0] = 16'h4000; assign W_im[0] = 16'h0000;
    // 0.707, -0.707
```

```
    assign W_re[1] = 16'h2D41; assign W_im[1] = 16'hD2BF;
    // 0, -1
    assign W_re[2] = 16'h0000; assign W_im[2] = 16'hC000;
    // -0.707, -0.707
    assign W_re[3] = 16'hD2BF; assign W_im[3] = 16'hD2BF;
endmodule

module ButterFly
(
input signed [15:0] xi_re, xi_im, xj_re, xj_im, // Input data
input signed [15:0] W_re, W_im, // Twiddle factors
output reg signed [15:0] yi_re, yi_im, yj_re, yj_im);
// Extra bit to cater for overflow
reg signed [16:0] tempi_re, tempi_im;
reg signed [16:0] tempj_re, tempj_im;
reg signed [31:0] mpy_re, mpy_im;
always @(*)
begin
     // Q2.14
    tempi_re = xi_re + xj_re; tempi_im = xi_im + xj_im;
    // Q2.14
    tempj_re = xi_re - xj_re; tempj_im = xi_im - xj_im;
    mpy_re = tempj_re*W_re - tempj_im*W_im;
                         mpy_im = tempj_re*W_im + tempj_im*W_re;
    // Bring the output format to Q3.13 for first stage
    // and to Q4.12 and Q5.11 for the second and third stages
    yi_re = tempi_re>>>1; yi_im = tempi_im>>>1;
    // The output for Q2.14 x Q 2.14 is Q4.12
    yj_re = mpy_re[30:15]; yj_im = mpy_im[30:15];
end
endmodule

module Mux2To2
(
input [15:0] xi_re, xi_im, xj_re, xj_im,
input sel1,
output reg [15:0] yi_out_re, yi_out_im, yj_out_re, yj_out_im);
always @ (*)
begin
     if (sel1)
     begin yi_out_re = xj_re; yi_out_im = xj_im;
                    yj_out_re = xi_re; yj_out_im = xi_im; end
     else
     begin yi_out_re = xi_re; yi_out_im = xi_im;
                    yj_out_re = xj_re; yj_out_im = xj_im; end
end
endmodule
```

8.6 Mathematical Transformation for Folding

Many feedforward algorithms can be formulated to recursively compute output samples. DCT [20] and FFT are good examples that can be easily converted to use recursions. Many digital designs for

Clock cycle		PE_1	R0 / R1	R2 / R3	PE_3	R4 / R5	PE_3
0	$x_3x_2x_1x_0$	$y_{10} = x_0 + x_4$					
	$x_7x_6x_5x_4$	$y_{14} = (x_0 - x_4)\, W_8^0$					
1	$x_0x_3x_2x_1$	$y_{11} = x_1 + x_5$	y_{10}				
	$x_4x_7x_6x_5$	$y_{15} = (x_1 - x_5)\, W_8^1$	y_{14}				
2	$x_1x_0x_3x_2$	$y_{12} = x_2 + x_6$	y_{11}	y_{10}	$y_{20} = y_{10} + y_{12}$		
	$x_5x_4x_7x_6$	$y_{16} = (x_2 - x_6)\, W_8^2$	y_{15}	y_{14}	$y_{22} = (y_{10} - y_{12})\, W_8^0$		
3	$x_2x_1x_0x_3$	$y_{13} = x_3 + x_7$	y_{14}	y_{11}	$y_{21} = y_{11} + y_{13}$	y_{20}	$x(0) = y_{20} + y_{21}$
	$x_7x_6x_5x_7$	$y_{17} = (x_3 - x_7)\, W_8^3$	y_{16}	y_{15}	$y_{23} = (y_{11} - y_{13})\, W_8^2$	y_{22}	$x(4) = (y_{20} - y_{21})\, W_8^0$
4			y_{15}	y_{14}	$y_{24} = y_{14} + y_{16}$	y_{22}	$x(2) = y_{22} + y_{23}$
			y_{17}	y_{16}	$y_{26} = (y_{14} - y_{16})\, W_8^0$	y_{23}	$x(6) = (y_{22} - y_{23})\, W_8^0$
5				y_{15}	$y_{25} = y_{15} + y_{17}$	y_{24}	$x(1) = y_{24} + y_{25}$
				y_{17}	$y_{27} = (y_{15} - y_{17})\, W_8^2$	y_{26}	$x(5) = (y_{24} - y_{25})\, W_8^0$
6						y_{26}	$x(3) = y_{26} + y_{27}$
						y_{27}	$x(7) = (y_{26} - y_{27})\, W_8^0$

Figure 8.16 Cycle-by-cycle description of the architecture, depicting values in registers and the computation in different PEs

Table 8.1 Select signals for multiplexers of the text example

Clock cycle	sel_1	sel_2
0	0	0
1	0	1
2	1	0
3	1	1

signal processing applications may require folding of algorithms for effective hardware mapping that minimizes area. The folding can be accomplished by using a folding transformation. The mathematical formulation of folding transformations is introduced in [21, 22]. A brief description is given in this section.

For a given folding order and a folding set for a DFG, the folding transformation computes the number of delays on each edge in the folded graph. The folded architecture periodically executes operations of the DFG according to the folding set. Figure 8.17 shows an edge that connects nodes U_i and V_j with W_{ij} delays. The W_{ij} delays on the edge $U_i \rightarrow V_j$ signify that the output of node U_i is used by node V_j after W_{ij} cycles or iterations in the original DFG. If the DFG is folded by a folding factor N, then the folded architecture executes each iteration of the DFG in N cycles. All the nodes of type U and V in the DFG are scheduled on computational units H_u and H_v, respectively, in clock cycles u_i and v_j such that $0 \leq u_i, v_i \leq N-1$. If nodes U and V are of the same type, they may be scheduled on the same computational unit in different clock cycles.

For the folded architecture, node U_i is scheduled in clock cycle u_i in the current iteration and node V_j is scheduled in v_j clock cycle in the W_{ij} iteration. In the original DFG the output of node U_i is used after W_{ij} clock cycles, and now in the folded architecture, as the node U_i is scheduled in the u_i clock cycle of the current iteration and it is used in the W_{ij} iteration in the original DFG, as each iteration takes N clock cycle thus the folded architecture starts executing the W_{ij} iteration in the $N \times W_{ij}$ clock cycle, where node V_j is scheduled in the v_j clock cycle in this iteration. This implies that, with respect to the current iteration, node V_j is scheduled in the $N \times W_{ij} + v_j$ clock cycle, so in the folded architecture the result from node U_i needs to be stored for F_{ij} number of register clock cycles, where:

$$F_{ij} = N \times W_{ij} + v_j - u_i$$

If the node of type U is mapped on a computational unit H_u with P_u pipeline stages, these delays will also be incorporated in the above equation, and the new equation becomes:

$$F_{ij} = N \times W_{ij} + v_j - u_i - P_u$$

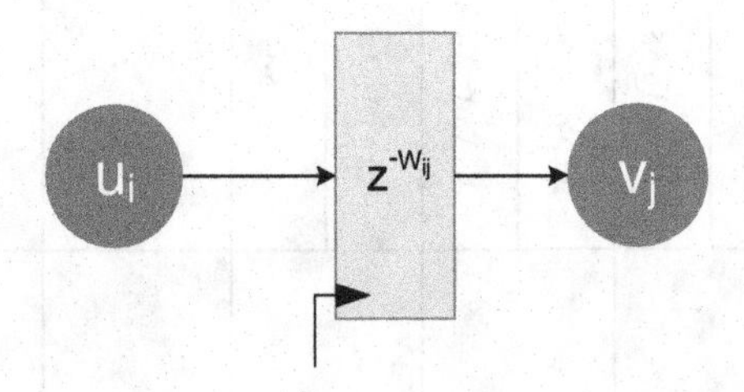

Figure 8.17 An edge in a DFG connecting nodes U_i and V_j with W_{ij} delays

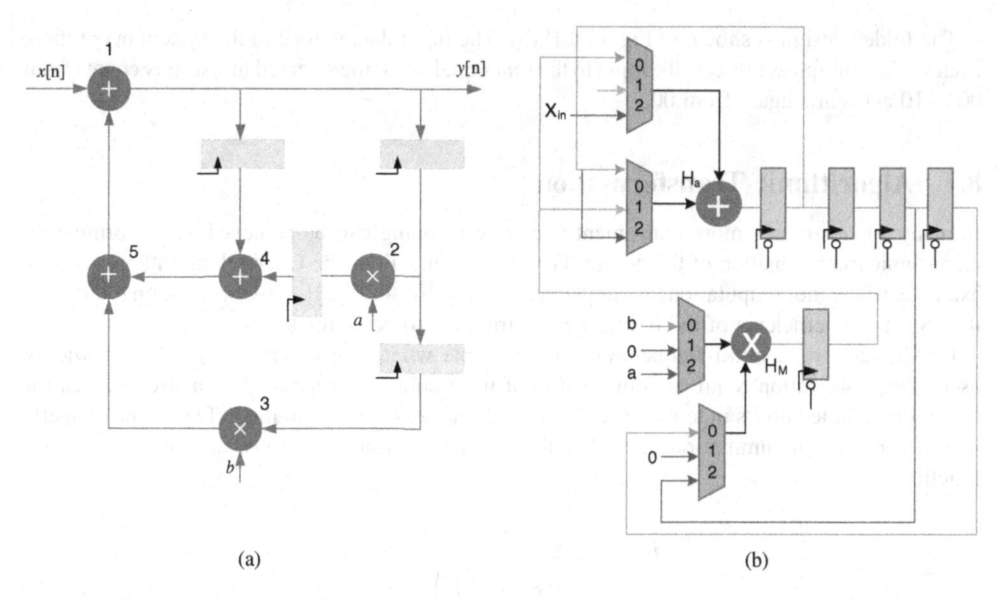

Figure 8.18 Second-order IIR system. (a) Dataflow graph of the system. (b) Folded design with folding factor 3

Example: A DFG of second order IIR filter is given in Figure 8.18(a) with three addition nodes 1, 4 and 5, and two multiplication nodes 2 and 3. Fold the architecture by a folding factor of 3 with the folding set for the adder $S_a = \{4, 5, 1\}$ and the folding set for the multiplier as $S_m = \{3,\_,2\}$. The number of registers for each edge in the folded architecture using the equation for the folding transformation assuming no pipeline stage in multiplier and adder, is as follows:

$$
\begin{aligned}
F_{ij} &= N \times W_{ij} + v_j - u_i \\
F_{12} &= 3 + 2 - 2 = 3 \\
F_{13} &= 3 \times 2 + 0 - 2 = 4 \\
F_{24} &= 3 + 0 - 2 = 1 \\
F_{14} &= 3 + 0 - 2 = 1 \\
F_{45} &= 0 + 1 - 0 = 1 \\
F_{35} &= 0 + 1 - 0 = 1 \\
F_{51} &= 0 + 2 - 1 = 1
\end{aligned}
$$

After figuring out the number of registers required for storing the intermediate results, the architecture can be easily drawn. One adder and one multiplier are placed with two sets of 3:1 multiplexers with each functional unit. Now observing the folding set, connections are made from the registers to the multiplier. For example, the adder first executes node 4. This node requires inputs from node 2 and node 1. The values of F_{24} and F_{14} are 1 and 1, where node 2 is the multiplier node. The connections to port 0 of the two multiplexers at the input of the adder are made by connecting the output of one register after multiplier and one register after adder. Similarly, connections for all the operations are made based on folding set and values of F_{ij}.

The folded design is shown in Figure 8.18(b). The input data is feed to the system every three cycles. The multiplexer selects the input to the functional units, the selected line simply counts from 00 to 10 and starts again from 00.

8.7 Algorithmic Transformation

In many applications a more convenient hardware mapping can be achieved by performing an algorithmic transformation of the design. For FFT computations the Goertzel algorithm is a good example, where the computations are implemented as an IIR filter [23]. This formulation is effective if only a few coefficients of the frequency spectrum are to be computed.

DTMF (dual-tone multi-frequency) is an example where the Goertzel algorithm is widely used. This application requires computation of the frequency content of eight frequencies for detection of dialed digits in telephony. The algorithm takes the formulation of (8.6) and converts it to a convolution summation of an IIR linear time-invariant (LTI) system with the transfer function:

$$H(z) = \frac{1 - W_N^k z^{-1}}{1 - 2\cos\left(\frac{2\pi}{N}k\right)z^{-1} + z^{-2}}$$

The Nth output sample of the system gives a DFT of N data samples at the kth frequency index. Figure 8.19(a) shows the IIR filter realization of the transfer function. The feedforward path involves a multiplication by $-W_N^k$ that only needs to be computed at every Nth sample, and the multiplication by -1 and two additions in the feedback loop can be implemented by a compression tree and an adder. The design can be effectively implemented by using just one adder and a multiplier and is shown in Figure 8.19(b). The multiplier is reused for complex multiplication after the Nth cycle.

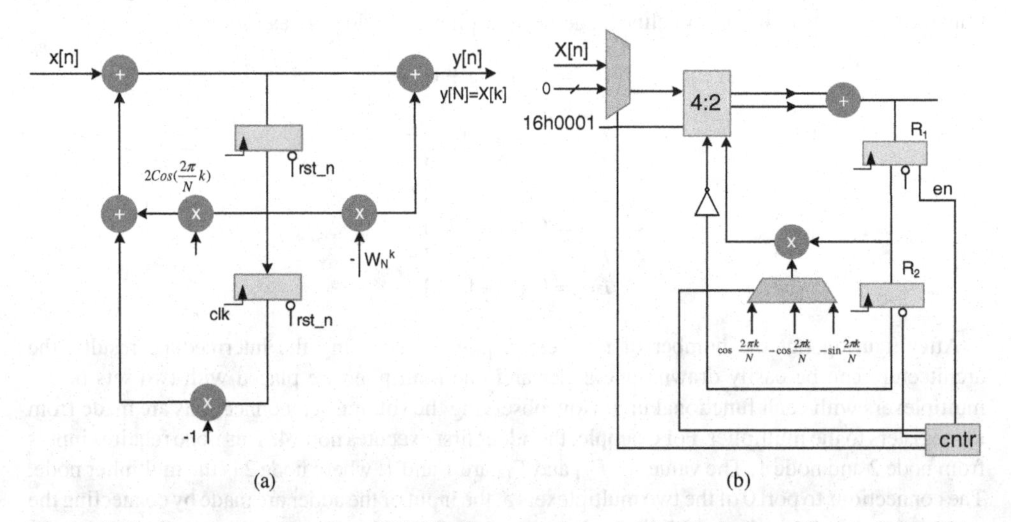

Figure 8.19 Iterative computation of DFT for frequency index k. (a) FDA design of Goertzel algorithm. (b) Design mapped on a multiplier and an adder

Assuming real data, 0, 1, ..., $N-1$ cycles compute the feedback loop. In the Nth cycle, the `en` of register R_1 is de-asserted and register R_2 is reset. The multiplexer selects multiplier for the multiplication by:

$$-W_N^k = -\cos\left(\frac{2\pi}{N}k\right) - j\sin\left(\frac{2\pi}{N}k\right)$$

in two consecutive cycles. In these cycles, a zero is fed in the compressor for $x[n]$.

Exercises

Exercise 8.1

A 4 MHz bandwidth signal is centered at an intermediate frequency of 70 MHz. Select the sampling frequency of an ADC that can bandpass sample the signal without any aliasing.

Exercise 8.2

The mathematical expression for computing correlation of a mask $m[n_1, n_2]$ of dimensions $L_1 \times L_2$ with an image $I[n_1, n_2]$ of size $N_1 \times N_2$ is:

$$c[n_1, n_2] = \sum_{k_2=0}^{L_2-1}\sum_{k_1=0}^{L_1-1} |m[k_1, k_2] - I[n_1+k_1, n_2+k_2]| \; for\, n_1 = 0, 1, \ldots, N_1 - L_1 \; and\, n_2 = 0, 1, \ldots, N_2 - L_2$$

Write C code to implement the nested loop that implements the correlation summation. Assuming $L_1 = L_2 = 3$ and $N_1 = N_2 = 256$, unroll the loops to speed up the computation by a factor of 9 such that the architecture needs to load the minimum number of new image samples and maximize data reuse across iterations.

Exercise 8.3

A 48th-order IIR filter is given by the following mathematical expression:

$$y[n] = \sum_{k=0}^{47} b_k x[n-k] + \sum_{k=1}^{47} a_k y[n-k]$$

Write C code to implement the equation. Design a folded architecture by appropriately unrolling loops in C implementation of the code. The design should perform four MAC operations of the feedforward and four MAC operations of the feedback summations. The design should require the minimum number of memory accesses.

Exercise 8.4

Unfold the second-order IIR filter in the TDF structure of Figure 8.5, first by a factor of 3 and then by a factor of 4. Identify loops in the unfolded structures and compute their IPBs assuming multiplication and addition take 3 and 2 time units, respectively.

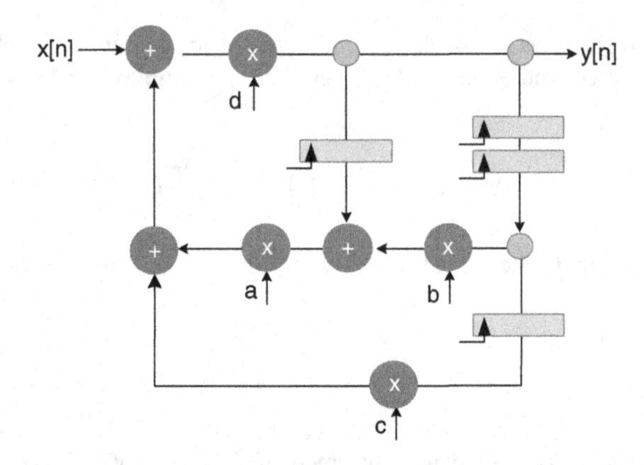

Figure 8.20 Design for folding by factor of 4 in exercise 8.7

Exercise 8.5

Figure 8.7(b) gives an unfolded architecture of Figure 8.7(a) for maximizing the use of compression trees for optimal HW design. Unfold the FIR filter in Figure 8.7(a) by a factor of 4, assuming each coefficient is represented by four non-zero digits in CSD format. Design an architecture with compression trees, keeping the internal datapath in partial sum and partial carry form.

Exercise 8.6

The objective is to design a high-performance 8-coefficient FIR filter for mapping on an FPGA with excessive DSP48 embedded blocks. Appropriately pipeline the DF structure of the filter to unfold the structure by a factor of 4 for effective mapping on the FPGA.

Exercise 8.7

Fold the architecture of Figure 8.20 by a factor of 4. Find an appropriate folding set. Draw the folded architecture and write RTL Verilog code of the original and unfolded designs.

References

1. R. G. Vaughan and N. L. Scott, "The theory of bandpass sampling," *IEEE Transactions on Signal Processing*, 1991, vol. 39, pp. 1973–1984.
2. A. J. Coulson, R. G. Vaughan and M. A. Poletti, "Frequency-shifting using bandpass sampling," *IEEE Transactions on Signal Processing*, 1994, vol. 42, pp. 1556–1559.
3. X. Wang and S. Yu, "A feasible RF bandpass sampling architecture of single-channel software-defined radio receiver," in *Proceedings of WRI International Conference on Communications and Mobile Computing*, 2009, pp. 74–77.
4. Q. Zhuge *et al.* "Design optimization and space minimization considering timing and code size via retiming and unfolding," *Microprocessors and Microsystems*, 2006, vol. 30, no. 4, pp. 173–183.
5. L. E. Lucke, A. P. Brown and K. K. Parhi, "Unfolding and retiming for high-level DSP synthesis," in *Proceedings of IEEE International Symposium on Circuits and systems*, 1991, pp. 2351–2354.

6. K. K. Parhi and D. G. Messerschmitt, "Static rate-optimal scheduling of iterative data-flow programs via optimum unfolding," *IEEE Transactions on Computing*, 1991, vol. 40, pp. 178–195.
7. K. K. Parhi, "High-level algorithm and architecture transformations for DSP synthesis," *IEEE Journal of VLSI Signal Processing*, 1995, vol. 9, pp. 121–143.
8. L.-G. Jeng and L.-G. Chen, "Rate-optimal DSP synthesis by pipeline and minimum unfolding," *IEEE Transactions on Very Large Scale Integration Systems*, 1994, vol. 2, pp. 81–88.
9. L. Jia, Y. Gao and H. Tenhunen, "Efficient VLSI implementation of radix-8 FFT algorithm," in *Proceedings of IEEE Pacific Rim Conference on Communications, Computers and Signal Processing*, 1999, pp. 468–471.
10. L. Jia, Y. Gao, J. Isoaho and H. Tenhunen, "A new VLSI-oriented FFT algorithm and implementation," in *Proceedings of 11th IEEE International ASIC Conference*, 1998, pp. 337–341.
11. G. L. Stuber *et al.*, "Broadband MIMO–OFDM wireless communications," *Proceedings of the IEEE*, 2004, vol. 92, pp. 271–294.
12. J. Valls, T. Sansaloni *et al.* "Efficient pipeline FFT processors for WLAN MIMO–OFDM systems," *Electronics Letters*, 2005, vol. 41, pp. 1043–1044.
13. B. Fu and P. Ampadu, "An area-efficient FFT/IFFT processor for MIMO–OFDM WLAN 802.11n," *Journal of Signal Processing Systems*, 2008, vol. 56, pp. 59–68.
14. C. Cheng and K. K. Parhi, "High-throughput VLSI architecture for FFT computation," *IEEE Transactions on Circuits and Systems II*, 2007, vol. 54, pp. 863–867.
15. S. He and M. Torkelson, "Designing pipeline FFT processor for OFDM (de)modulation," in *Proceedings of International Symposium on Signals, Systems and Electronics*, URSI, 1998, pp. 257–262.
16. C.-C. Wang, J.-M. Huang and H.-C. Cheng, "A 2K/8K mode small-area FFT processor for OFDM demodulation of DVB-T receivers," *IEEE Transactions on Consumer Electronics*, 2005, vol. 51, pp. 28–32.
17. Y.-W. Lin, H.-Y. Liu and C.-Y. Lee, "A dynamic scaling FFT processor for DVB-T applications," *IEEE Journal of Solid-State Circuits*, 2004, vol. 39, pp. 2005–2013.
18. C.-L. Wang and C.-H. Chang, "A new memory-based FFT processor for VDSL transceivers," in *Proceedings of IEEE International Symposium on Circuits and Systems*, 2001, pp. 670–673.
19. C.-L. Wey, W.-C. Tang and S.-Y. Lin, "Efficient VLSI implementation of memory-based FFT processors for DVB-T applications," in *Proceedings of IEEE Computer Society Annual Symposium on VLSI*, 2007, pp. 98–106.
20. C.-H. Chen, B.-D. Liu, J.-F. Yang and J.-L. Wang, "Efficient recursive structures for forward and inverse discrete cosine transform," *IEEE Transactions on Signal Processing*, 2004, vol. 52, pp. 2665–2669.
21. K. K. Parhi, *VLSI Digital Signal Processing Systems: Design and Implementation*, 1999, Wiley.
22. T. C. Denk and K. K. Parhi, "Synthesis of folded pipelined architectures for multirate DSP algorithms," *IEEE Transactions on Very Large Scale Integration Systems*, 1998, vol. 6, pp. 595–607.
23. G. Goertzel, "An algorithm for evaluation of finite trigonometric series," *American Mathematical Monthly*, 1958, vol. 65, pp. 34–35.

9

Designs based on Finite State Machines

9.1 Introduction

This chapter looks at digital designs in which hardware computational units are shared or time-multiplexed to execute different operations of the algorithm. To highlight the difference between time-shared and fully dedicated architecture (FDA), the chapter first examines examples while assuming that the circuit clock is at least twice as fast as the sampling clock. It is explained that, if instances of these applications are mapped on a dedicated fully parallel architecture, they will not utilize the HW in every clock cycle. Time sharing is the logical design decision for mapping these applications in HW. These designs use the minimum required HW computational resources and then share them for multiple computations of the algorithm in different clock cycles. The examples pave the way to generalize the discussion to time-shared architecture.

A synchronous digital design that shares HW building blocks for computations in different cycles requires a *controller*. The controller implements a *scheduler* that directs the use of resources in a time-multiplexed way. There are several options for the controller, but this chapter covers a hard-wired state machine-based controller that cannot be reprogrammed.

The chapter describes both Mealy and Moore state machines. With the Mealy machine the output and next state are functions of the input and current state, whereas with the Moore machine the input and current state only compute the next state and the output only depends on the current state. Moore machines provide stable control input to the datapath for one complete clock cycle. In designs using the Mealy machine, the output can change with the change of input and may not remain stable for one complete cycle. For digital design of signal processing systems, these output signals are used to select the logic in the datapath. Therefore these signals are time-critical and they should be stable for one complete cycle. Stable output signals can also be achieved by registering output from a Mealy machine.

The current state is latched in a state register in every clock cycle. There are different state encoding methods that affect the size of the state register. A *one-hot state machine* uses one flip-flop per state. This option is attractive because it results in simple timing analysis, and addition and deletion of newer states is also trivial. This machine is also of special interest to field-programmable

Digital Design of Signal Processing Systems: A Practical Approach, First Edition. Shoab Ahmed Khan.

gate arrays (FPGAs) that are rich in flip-flops. If the objective is to conserve the number of flip-flops of a state register, a binary-coded state machine should be used.

The chapter gives special emphasis to RTL coding guidelines for state machine design and lists RTL Verilog code for examples. The chapter then focuses on digital design for complex signal processing applications that need a finite state machine to generate control signals for the datapath and have algorithm-like functionality. The conventional bubble representation of a state machine is described, but it is argued that this is not flexible enough for describing complex behavior in many designs. Complex algorithms require gradual refinement and the bubble diagram representation is not appropriate. The bubble diagram is also not algorithm-like, whereas in many instances the digital design methodology requires a representation that is better suited for an algorithm-like structure.

The algorithmic state machine (ASM) notation is explained. This is a flowchart-like graphical notation to describe the cycle-by-cycle behavior of an algorithm. To demonstrate the differences, the chapter represents in ASM notation the same examples that are described using a bubble diagram. The methodology is illustrated by an example of a first-in first-out (FIFO).

9.2 Examples of Time-shared Architecture Design

To demonstrate the need for a scheduler or a controller in time-shared architectures, this section first describes some simple applications that require mapping on time-shared HW resources. These applications are represented by simple dataflow graphs (DFGs) and their mappings on time-shared HW require simple schedulers. For complex applications, finding an optimal hardware and its associated scheduler is an 'NP complete' problem, meaning that the computation of an optimal solution cannot be guaranteed in measurable time. Smaller problems can be optimally solved using integer programming (IP) techniques [1], but for larger problems near-optimal solutions are generated using heuristics [2].

9.2.1 Bit-serial and Digit-serial Architectures

Bit-serial architecture works on a bit-by-bit basis [3, 4]. This is of special interest where the data is input to the system on bit-by-bit basis on a serial interface. The interface gives the designer motivation to design a bit-serial architecture. Bit-by-bit processing of data, serially received on a serial interface, minimizes area and in many cases also reduces the complexity of the design [5], as in this case the arrangement of bits in the form of words is not required for processing.

An extension to bit-serial is a digit-serial architecture where the architecture divides an N-bit operand to $P = N/M$-bit digits that are serially fed, and the entire datapath is P-bit wide, where P should be an integer [6–8]. The choice of P depends on the throughput requirement on the architecture, and it could be 1 to N bits wide.

It is pertinent to point out that, as a consequence of the increase in device densities, the area usually is not a very stringent constraint, so the designer should not unnecessarily get into the complications of bit-serial designs. Only designs that naturally suit bit-serial processing should be mapped on these architectures. A good example of a bit-serial design is given in [5].

Example: Figure 9.1 shows the design of FDA, where we assume the sampling clock equals the circuit clock. The design is pipelined to increase the throughput performance of the architecture. A node-merging optimization technique is discussed in Chapter 5. The technique suggests the use of CSA and compression trees to minimize the use of CPA.

Now assume that for the DFG in Figure 9.1(a) the sampling clock frequency f_s is eight times slower than the circuit clock frequency f_c. This ratio implies the HW can be designed such that it

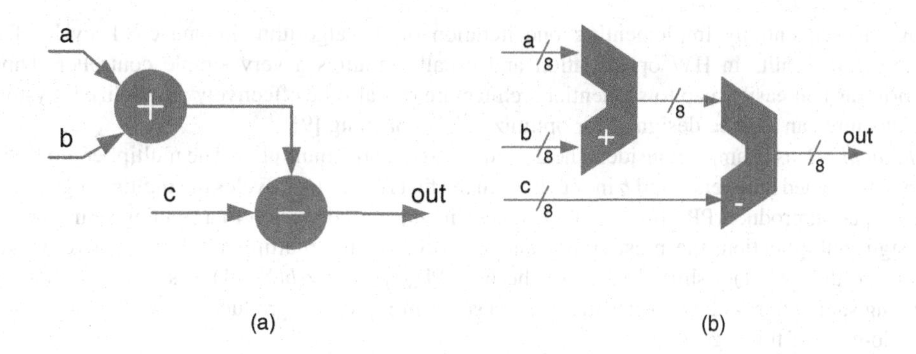

Figure 9.1 (a) Algorithm requiring equal sampling and clock frequencies. (b) Mapping on FDA

can be shared eight times. While mapping the DFG for $f_c = 8f_s$, the design is optimized and transformed to a bit-serial architecture where operations are performed on bit-by-bit basis, thus effectively using eight clocks for every input sample. A pipelined bit-serial HW implementation of the algorithm is given in Figure 9.2, where for bit-serial design $P = 1$. The design consists of one full adder (FA) and a full subtractor (FS) replacing the adder and subtractor of the FDA design. Associated with the FA and FS for 1-bit addition and subtraction are two flip-flops (FFs) that feed the previous carry and borrow back to the FA and FS for bit-serial addition and subtraction operations, respectively. The FFs are cleared at every eighth clock cycle when the new 8-bit data is input to the architecture. The sum from the FA is passed to the FS through the pipeline FF. The 1-bit wide datapath does not require packing of intermediate bits for any subsequent calculations. Many algorithms can be transformed and implemented on bit-serial architectures.

It is interesting to note that the architecture of Figure 9.2 can be easily extended to a digit-serial architecture by feeding digits instead of bits to the design. The flip-flop for carry and borrow will remain the same, while the adder, subtractor and other pipeline registers will change to digit size. For example, if the digit size is taken as 4-bit (i.e. $P = 4$), the architecture will take two cycles to process two sets of 4-bit inputs, producing out [3:0] in the first cycle and then [7:4] in the second cycle.

9.2.2 *Sequential Architecture*

In many design instances the algorithm is so regular that hardware can be easily designed to sequentially perform each iteration on shared HW. The HW executes the algorithm like a software

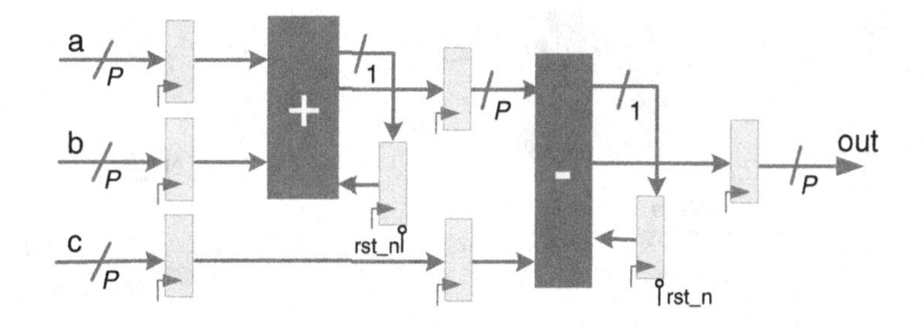

Figure 9.2 Bit-serial and digit-serial architectures for the DFG in Figure 9.1(a)

program, sequentially implementing one iteration of the algorithm in one clock cycle. This architecture results in HW optimization and usually requires a very simple controller. Those algorithms that easily map to sequential architecture can also be effectively parallelized. Systolic architecture can also be designed for optimized HW mapping [9].

Example: This example considers the design of a sequential multiplier. The multiplier multiplies two N-bit signed numbers a and b in N cycles. In the first $N-1$ clock cycles the multiplier generates the ith partial product (PP_i) for $a \times b(i)$ and adds its ith shifted version to a running sum. For sign by sign multiplication, the most significant bit (MSB) of the multiplier b has negative weight; therefore the $(N-1)$th shifted value of the final PP_{N-1} for $a \times b(N-1)$ is subtracted from the running sum. After N clock cycles the running sum stores the final product of $a \times b$ operation. The pseudo-code of this algorithm is:

$$\text{sum} = 0$$
$$\text{for}(i = 0; i < N-1; i++)$$
$$\text{sum}+ = a \times b[i] \times 2^i$$
$$\text{sum}- = a \times b[N-1] \times 2^{N-1}$$
$$\text{prod} = \text{sum}$$

Figure 9.3 shows a hardware implementation of the $(N \times N)$-bit sequential multiplier. The $2N$-bit shift register `prod_reg` is viewed as two concatenated registers: the register with N least significant bits (LBSs) is tagged as `prod_reg_L`, and the register with N MSBs is tagged as `prod_reg_H`, and for convenience of explanation this register is also named `reg_b`. Register `reg_a` in the design stores the multiplicand a. An N-bit adder/subtractor adds the intermediate PPs and subtracts the final PP from a running sum that is stored in `prod_reg_L`. The ith left shift of the PP_i is successively performed by shifting the running sum in the `prod_reg` to the right in every clock cycle. Before starting the execution of sequential multiplication, `reg_a` is loaded with

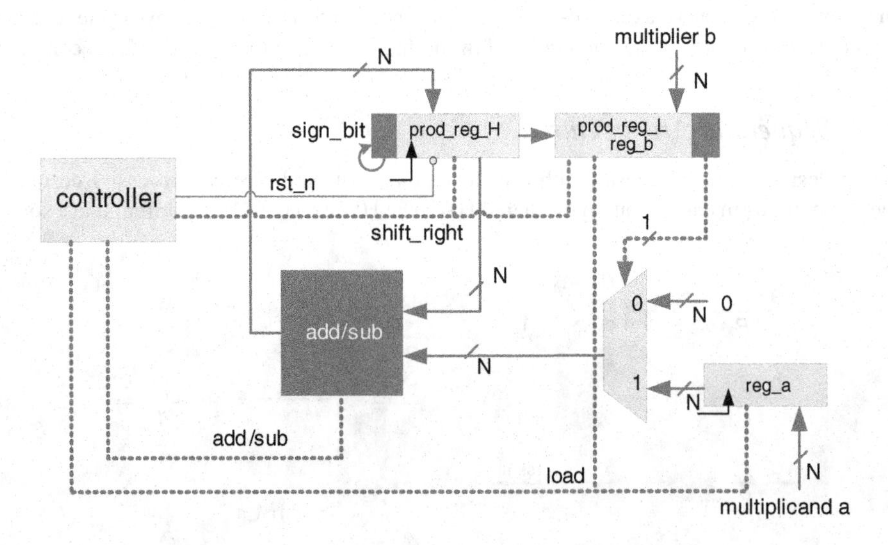

Figure 9.3 An $N \times N$-bit sequential multiplier

multiplicand a, prod_reg_H (reg_b) is loaded with multiplier b, and prod_reg_H is set to zero. Step-by-step working of the design is described below.

S1 Load the registers with appropriate values for multiplication:

```
prod_reg_H → 0,
reg_a → multiplicand a,
prod_reg_L (reg_b) → multiplier b
```

S2 Each cycle adds the multiplicand to the accumulator if the ith bit of the multiplier is 1 and adds 0 when this bit is a 0:

```
if reg_b[i]==1 then
    prod_reg_H += reg_a
else
    prod_reg_H+=0
```

In every clock cycle, prod_reg is also shifted right by 1. For sign by sign multiplication, right shift is performed by writing back the MSB of prod_reg to its position while all the bits are moved to the right. For unsigned multiplication, a logical right shift by 1 is performed in each step.

S3 For signed multiplication, in the Nth clock cycle, the adder/subtractor is set to perform subtraction, whereas for unsigned multiplication addition is performed in this cycle as well.

S4 After N cycles, prod_reg stores the final product.

A multiplexer is used for the selection of contents of reg_a or 0. The LSB of reg_b is used for this selection. This is a good example of time-shared architecture that uses an adder and associated logic to perform multiplication in N clock cycles.

The architecture can be optimized by moving the CPA out from the feedback loop, as in Figure 9.4(a). The results of accumulation are saved in a partial sum and carry form in prod_reg_Hs and prod_reg_Hc, respectively. The compression tree takes the values saved in these two registers and the new PP$_i$ and compresses them to two layers. This operation always produces one free product bit, as shown in Figure 9.4(b). The compressed partial sum and carry are saved one bit position to the right. In N clock cycles the design produces N product LSBs, whereas N product MSBs in sum and carry form are moved to a set of registers clocking at slower clock f_s and are then added using a CPA working at slower clock speed. The compression tree has only one full adder delay and works at faster clock speed, allowing the design to work in high-performance applications.

Example: A sequential implementation of a 5-coefficient FIR filter is explained in this example. The filter implements the difference equation:

$$y_n = h_0x_n + h_1x_{n-1} + h_2x_{n-2} + h_3x_{n-3} + h_4x_{n-4} \tag{9.1}$$

For the sequential implementation, let us assume that f_c is five times faster than f_s. This implies that after sampling an input value at f_s the design running at clock frequency f_c has five clock cycles to compute one output sample. An FDA implementation will require five multipliers and four adders. Using this architecture, the design running at f_c would compute the output in one clock cycle and would then wait four clock cycles for the next input sample. This obviously is a waste of resources, so a time-shared architecture is preferred.

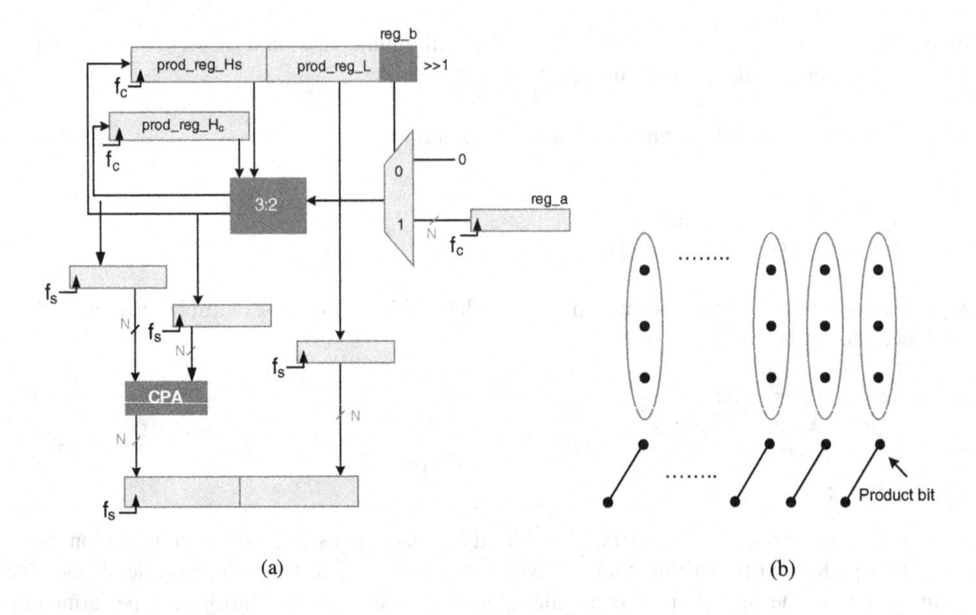

Figure 9.4 Optimized sequential multiplier. (a) Architecture using a 3:2 compression tree and CPA operating at f_c and f_s. (b) Compression tree producing one free product bit

A time-shared sequential architecture implements the same design using one MAC (multiplier and accumulator) block. The pseudo-code for the sequential design implementing (9.1) is:

$$\mathrm{acc} = 0$$
$$\mathrm{for}(i = 0; i < 4; i++)$$
$$\mathrm{acc} += h_i x_{n-i}$$
$$y_n = \mathrm{acc}$$

The design is further optimized by moving the CPA out of the running multiplication and accumulation operation. The results from the compression tree are kept in partial sum and carry form in two sets of accumulator registers, $\mathtt{acc}_s$ and $\mathtt{acc}_c$, and the same are fed back to the compression tree for MAC operation. The partial results are moved to two registers clocked at slower clock f_s. The CPA adds the two values at slower clock f_s. The pseudo-code for the optimized design is:

$$acc_s = 0$$
$$acc_c = 0$$
$$\mathrm{for}(i = 0; i < 4; i++)$$
$$\{acc_c, acc_s\} += h_i x_{n-i}$$
$$y_n = acc_s + acc_c$$

A generalized hardware design for an N-coefficient FIR filter is shown in Figure 9.5. The design assumes $f_c/f_s = N$. The design has a tap delay line that samples new input data at every clock f_s at the top of the line and shifts all the values down by one register. The coefficients are either saved in

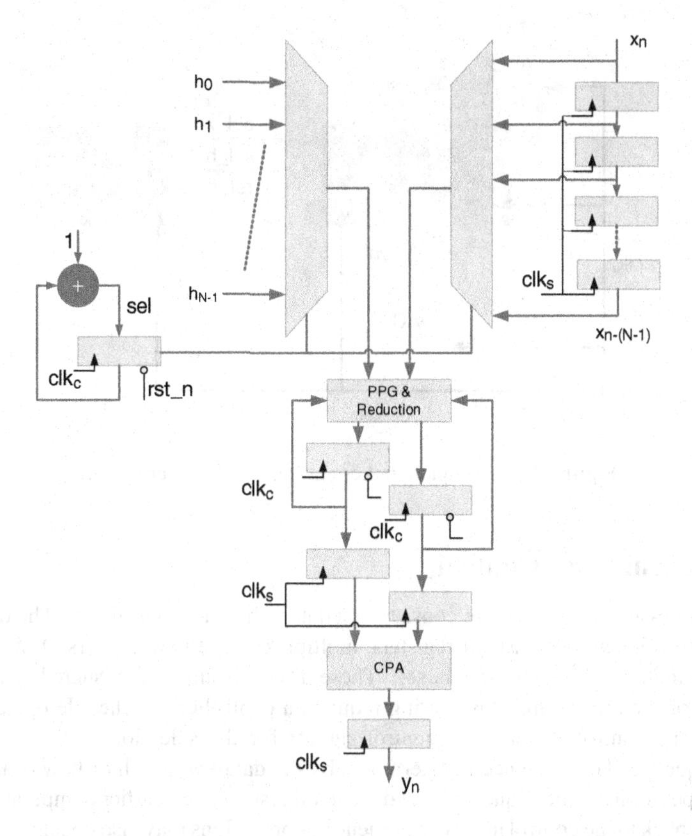

Figure 9.5 Time-shared *N*-coefficient FIR filter

ROM, or registers may also be used if the design is required to work for adaptive or changed coefficients. A new input is sampled in register `x_n` of the tap delay line of the design, and $\mathtt{acc}_\mathtt{s}$ and $\mathtt{acc}_\mathtt{c}$ are reset. The design computes one output sample in *N* clock cycles. After *N* clock cycles, a new input is sampled in the tap delay line and partial accumulator registers are saved in intermediate registers clocked at f_s. The registers are initialized in the same cycle for execution of the next iteration. A CPA working at slower clock f_s adds the sum of partial results in the intermediate registers and produces y_n.

The HW design assumes that the sampling clock is synchronously derived from the circuit clock. The shift registers storing the current and previous input samples are clocked using the sampling clock $\mathtt{clk}_\mathtt{S}$, and so is the final output register. At every positive edge of the sampling clock, each register stores a new value previously stored in the register located above it, with the register at the top storing the new input sample. The datapath executes computation on the circuit clock $\mathtt{clk}_\mathtt{C}$, which is *N* times faster than the sampling clock. The top-level design consisting of datapath and controller is depicted in Figure 9.6. The controller clears the partial sum and carry registers and starts generating the select signals `sel_h` and `sel_x` for the two multiplexers, starting from 0 and then incrementing by 1 at every positive edge of $\mathtt{clk}_\mathtt{C}$. At the *N*th clock cycle the final output is generated, and the controller restarts again to process a new input sample.

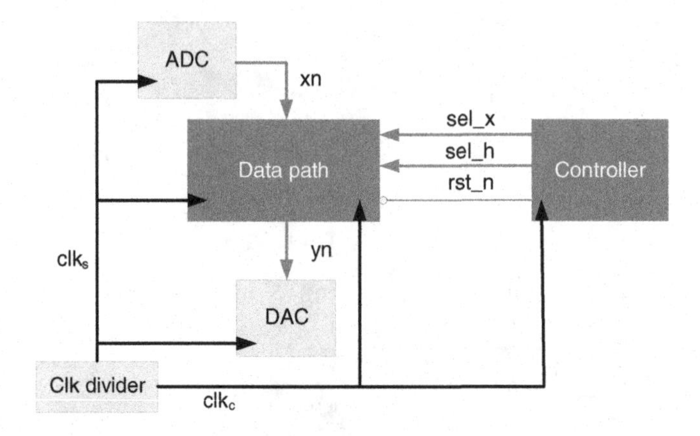

Figure 9.6 Datapath and controller of FIR filter design

9.3 Sequencing and Control

In general, a time-shared architecture consists of a datapath and a control unit. The datapath is the computational engine and consists of registers, multiplexers, de-multiplexers, ALUs, multipliers, shifters, combinational circuits and buses. These HW resources are shared across different computations of the algorithm. This sharing requires a controller to schedule operations on sets of operands. The controller generates control signals for the selection of these operands in a predefined sequence. The sequence is determined by the dataflow graph or flow of the algorithm. Some of the operations in the sequence may depend on results from earlier computations, so status signals are fed back to the control unit. The sequence of operations may also depend on input signals from other modules in the system.

Figure 9.7 shows these two basic building blocks. The control unit implements the schedule of operations using a finite state mMachine.

9.3.1 Finite State Machines

A synchronous digital design usually performs a sequence of operations on the data stored in registers and memory of its datapath under the direction of a finite state machine (FSM). The FSM is a sequential digital circuit with an N-bit state register for storing the current state of the sequence of operations. An N-bit state register can have 2^N possible values, so the number of possible states in the

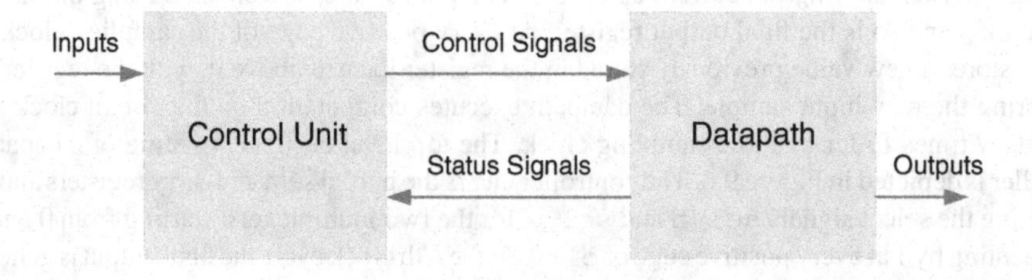

Figure 9.7 Time-shared architecture has two components, the datapath and a control unit

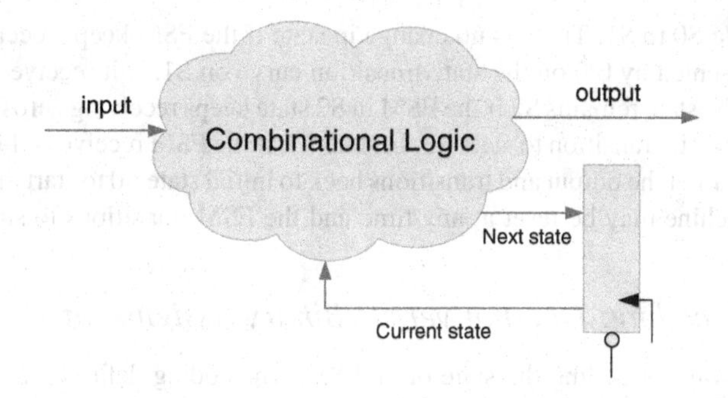

Figure 9.8 Combinational and sequential components of an FSM design

sequence is finite. The FSM can be described using a bubble (state) diagram or an algorithmic state machine (ASM) chart.

FSM implementation in hardware has two components, a combinational cloud and a sequential logic. The combination cloud computes the output and the next state based on the current state and input, whereas the sequential part has the resetable state register. A FSM with combinational and sequential components is shown in Figure 9.8.

Example: This example designs an FSM that counts four 1s on a serial interface and generates a 1 at the output. One bit is received on the serial interface at every clock cycle. Figure 9.9 shows a bubble diagram in which each circle represents a state, and lines and curves with arrowheads represent state transitions. There are a total of four states, S0 ... S3.

S0 represents the initial state where number of 1s received on the interface is zero. A 0/0 on the transition curve indicates that, in state S0, if the input on the serial interface is 0 then the state machine maintains its state and generates a 0 at the output. In the same state S0, if 1 is received at the input, the FSM changes its state to S1 and generates a 0 at the output. This transition and input-output relationship is represented by 1/0 on the line or curve with an arrowhead showing state

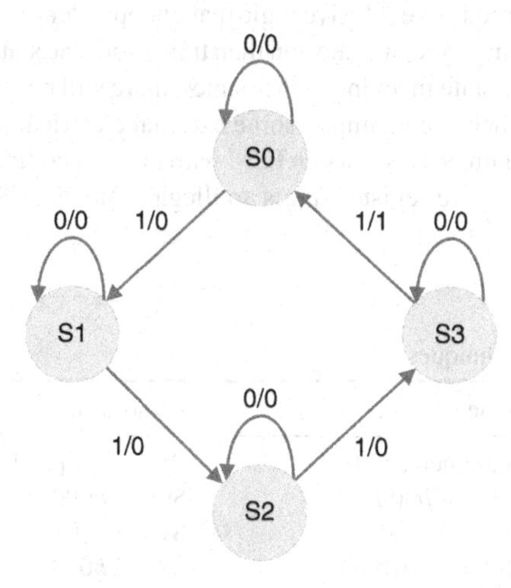

Figure 9.9 State diagram implementing counting of four 1s on a serial interface

transition from state S0 to S1. There is no change in state if the FSM keeps receiving 0s in the S1 state. This is represented by 0/0 on the state transition curve on S1. If it receives a 1 in state S1 it transitions to S2. The state remains S2 if the FSM in S2 state keeps receiving zeros, whereas another 1 at the input causes it to transition to state S3. In state S3, if the FSM receives a 1 (i.e. the fourth 1), the FSM generates a 1 at the output and transitions back to initial state S0 to start the counting again. Usually a state machine may be reset at any time and the FSM transitions to state S0.

9.3.2 State Encoding: One-hot versus Binary Assignment

There are various ways of coding the state of an FSM. The coding defines a unique sequence of numbers to represent each state. For example, in a 'one-hot' state machine with N states there are N bits to represent these states. An isolated 1 at a unique bit location represents a particular state of the machine. This obviously requires an N-bit state register. Although a one-hot state machine results in simple logic for state transitions, it requires N flip-flops as compared to $\log_2 N$ in a binary-coded design. The latter requires fewer flip-flops to encode states, but the logic that decodes the states and generates the next states is more complex in this style of coding.

The arrangement with a one-hot state machine, where each bit of the state register represents a state, works well for FPGA-based designs that are rich in flip-flops. This technique also avoids large fan-outs of binary-coded state registers, which does not synthesize well on FPGAs.

A variant, called 'almost one-hot', can be used for state encoding. This is similar, with the exception of the initial state that is coded as all zeros. This helps in easy resetting of the state register. In another variant, two bits instead of one are used for state coding.

For low-power designs the objective is to reduce the hamming distance among state transitions. This requires the designer to know the statistically most probable state transition pattern. The pattern is coded using the gray codes. These four types of encoding are given in Table 9.1.

If the states in the last example are coded as a one-hot state machine, for four states in the FSM, a state register of size 4 will be used where each flip-flop of the register will be assigned to one state of the FSM.

As a consequence of one-hot coding, only one bit of the state register can be 1 at any time. The designer needs to handle *illegal states* by checking whether more than one bit of the state register is 1. In many critical designs this requires exclusive logic that, independently of the state machine, checks whether the state machine is in an illegal state and then transitions the state machine to the reset state. Using N flip-flops to encode a state machine with N states, there will be a total of $2^N - N$ illegal states. These need to be detected when, for example, some external electrical phenomenon takes the state register to an illegal state. Figure 9.10 shows an FSM with one-hot coding and an exclusive logic for detecting illegal states. If the state register stores an illegal state, the FSM is transitioned to initial state *INIT*.

Table 9.1 State encoding techniques

Binary	One-hot	Almost one-hot	Gray
Parameter [1:0]	Parameter [3:0]	Parameter [2:0]	Parameter [1:0]
S0 = 2'*d*00	S0 = 4'*b*0001	S0 = 3'*b*000	S0 = 2'*b*00
S1 = 2'*d*01	S1 = 4'*b*0010	S1 = 3'*b*001	S1 = 2'*b*01
S2 = 2'*d*10	S2 = 4'*b*0100	S2 = 3'*b*010	S2 = 2'*b*11
S3 = 2'*d*11	S3 = 4'*b*1000	S3 = 3'*b*100	S3 = 2'*b*10

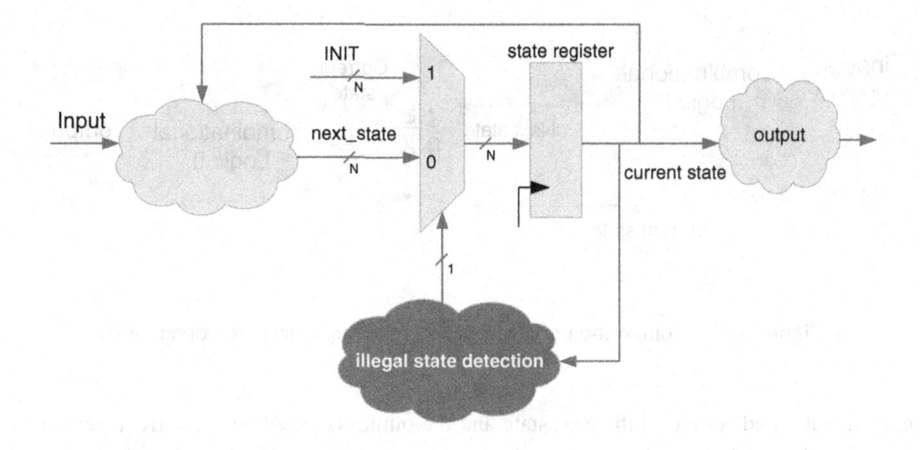

Figure 9.10 One-hot state machine with logic that detects illegal states

FPGA synthesis tools facilitate the coding of states to one-hot or binary encoding. This is achieved by selecting the appropriate encoding option in the FPGA synthesis tool.

9.3.3 Mealy and Moore State Machine Designs

In a Mealy machine implementation, the output of the FSM is a function of both the current state and the input. This is preferred in applications where all the inputs to the FSM are registered outputs from some other blocks and the generation of the next state and FSM output are not very complex. It results in simple combinational logic with a short critical path.

In cases where the input to the FSM is asynchronous (it may change within one clock cycle), the FSM output will also be asynchronous. This is not desirable especially in designs where outputs from the FSM are the control signals to the datapath. The datapath computational units are also combinational circuits, so an asynchronous control signal means that the control may not be stable for one complete clock cycle. This is undesirable as the datapath expects all control signals to be valid for one complete clock cycle.

In designs where the stability of the output signal is not a concern, the Mealy machine is preferred as it results in fewer states. A Mealy machine implementation is shown in Figure 9.11. The

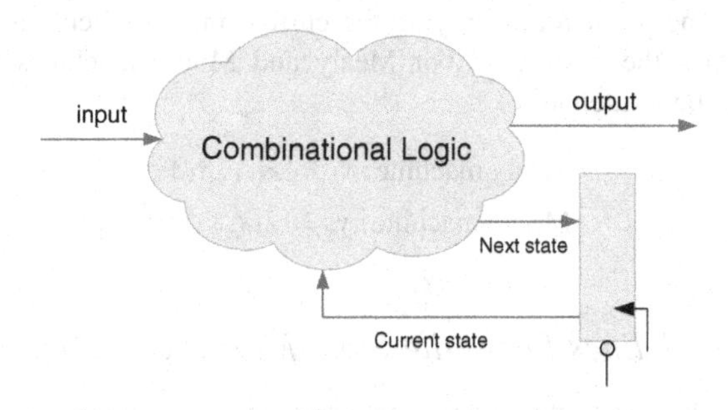

Figure 9.11 Composition of a Mealy machine implementation of an FSM

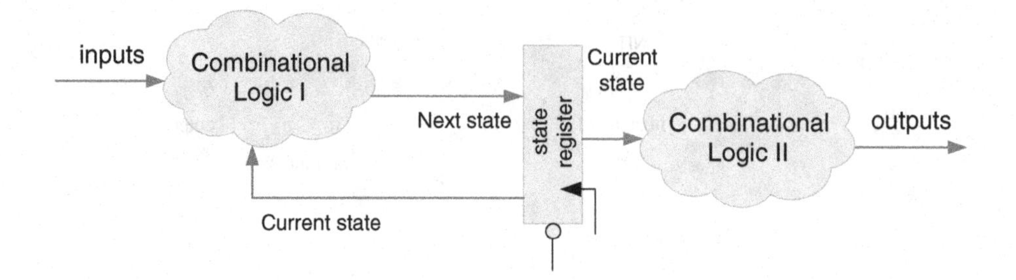

Figure 9.12 Composition of a Moore machine implementation of an FSM

combinational cloud computes the next state and the output is based on the current state and the input. The state register has an asynchronous reset to initialize the FSM at any time if desired.

In a Moore machine implementation, outputs are only a function of the current state. As the current sate is registered, output signals are stable for one complete clock cycle without any glitches. In contrast to the Mealy machine implementation, the output will be delayed by one clock cycle. The Moore machine may also result in more states as compared to the Mealy machine.

The choice between Mealy and Moore machine implementations is the designer's and is independent of the design problem. In many design instances it is a fairly simple decision to select one of the two options. In instances where some of the inputs are expected to glitch and outputs are required to be stable for one complete cycle, the designer should always use the Moore machine.

A Moore machine implementation is shown in Figure 9.12. The combinational logic-I of the FSM computes the next state based on inputs and the current state, and the combinational logic-II of the FSM computes the outputs based on the current state.

9.3.4 Mathematical Formulations

Mathematically an FSM can be formulated as a sextuple, $(X, Y, S, s_0, \delta, \lambda)$, where X and Y are sets of inputs and outputs, S is set of states, s_0 is the initial state, and δ and λ are the functions for computing the next state and output, respectively. The expression for next state computation can be written as:

$$s_{k+1} = \delta(x_k, s_k), \text{where } x_k \in X \text{ and } s_k \in S$$

Subscript k is the time index for specifying the current input and current state, so using this notation sk_{+1} identifies the next state. For Mealy and Moore machines the expressions for computing output can be written as:

$$\text{Mealy machine}: y_k = \lambda(s_k, x_k)$$

$$\text{Moore machine}: y_k = \lambda(s_k)$$

9.3.5 Coding Guidelines for Finite State Machines

Efficient hardware implementation of a finite state machine necessitates adhering to some coding guidelines [10, 11].

9.3.5.1 Design Partitioning in Datapath and Controller

The complete HW design usually consists of a datapath and a controller. The controller is implemented as an FSM. From the synthesis perspective, the datapath and control parts have different design objects. The datapath is usually synthesized for better timing whereas the controller is synthesized to take minimum area. The designer should keep the FSM logic and datapath logic in separate modules and then synthesize respective parts selecting appropriate design objectives.

9.3.5.2 FSM Coding in Procedural Blocks

The logic in an FSM module is coded using one or two `always` blocks. Two `always` blocks are preferred, where one implements the sequential part that assign the next state to the state register, and the second block implements the combinational logic that computes the next state $s_{k+1} = \delta(x_k, s_k)$.

The designer can include the output computations $y_k = \lambda(s_k, x_k)$ or $y_k = \lambda(s_k)$ for Mealy or Moore machines, respectively, in the same combinational block. Alternatively, if the output is easy to compute, they can be computed separately in a continuous assignment outside the combinational procedural block.

9.3.5.3 State Encoding

Each state in an FSM is assigned a code. From the readability perspective the designer should use meaningful tags using `'define` or `parameter` statements for all possible states. Use of `parameter` is preferred because the scope of `parameter` is limited to the module in which it is defined whereas `'define` is global in scope. The use of `parameter` enables the designer to use the same state names with different encoding in other modules.

Based on the design under consideration, the designer should select the best encoding options out of one-hot, almost one-hot, gray or binary. The developer may also let the synthesis tool encode the states by selecting appropriate directives. The Synopsis tool lets the user select binary, one-hot or almost one-hot by specifying it with the `parameter` declaration. Below, binary coding is invoked using the `enum` synopsis directive:

```
parameter [3:0] // Synopsys enum code
```

9.3.5.4 Synthesis Directives

In many designs where only a few of the possible states of the state register are used, the designer can direct the synthesis tool to ignore unused states while optimizing the logic. This directive for Synopsys is written as:

```
case(state) // Synopsys full_case
```

Adding `//Synopsis full_case` to the `case` statement indicates to the synthesis tool to treat all the cases that are explicitly not defined as 'don't care' for optimization. The designer knows that undefined cases will never occur in the design. Consider this example:

```
always @* begin
case (cntr) // Synopsys full_case
    2'b00: out = in1;
```

```
  2'b01: out = in2;
  2'b10: out = in3;
endcase
```

The user controls the synthesis tool by using the directive that the `cntr` signal will never take the unused value $2'b11$. The synthesis tool optimizes the logic by considering this case as 'don't care'.

Similarly, `//Synopsis parallel_case` is used where all the cases in a `case`, `casex` or `casez` statement are mutually exclusive and the designer would like them to be evaluated in parallel or the order in which they are evaluated does not matter. The directive also indicates that all cases must be individually evaluated in parallel:

```
always @* begin
// Code for setting the output to default comes here
casez (intr_req) // Synopsys parallel_case
3'b??1:
begin // Check bit 0 while ignoring rest
     // Code for interrupt 0 comes here
end
3'b?1?:
begin // Check bit 1 while ignoring rest
     // Code for interrupt 1 comes here
end
3'b??1:
begin // Check bit 2 while ignoring rest
     // Code for interrupt 2 comes here
end
endcase
```

The onus is on the designer to make sure that no two interrupts can happen at the same time. On this directive, the synthesis tool optimizes the logic assuming non-overlapping cases.

While using one-hot or almost one-hot encoding, the use of `//Synopsys full_case_parallel_case` signifies that all the cases are non-overlapping and only one bit of the state register will be set and the tool should consider all other bit patterns of the state register as 'don't care'. This directive generates the most optimal logic for the FSM.

Instead of using the `default` statement, it is preferred to use `parallel_case` and `full_case` directives for efficient synthesis. The `default` statement should be used only in simulation and then should be turned off for synthesis using the `compiler` directive. It is also important to know that these directives have their own consequences and should be cautiously use in the implementation.

Example: Using the guidelines, RTL Verilog code to implement the FSM of Figure 9.9 is given below. Listed first is the design using *binary encoding*, where the output is computed inside the combinational block:

```
// This module implements FSM for the detection of
// four ones in a serial input stream of data
module fsm_mealy(
  input   clk,            //system clock
  input   reset,          //system reset
  input   data_in,        //1-bit input stream
 output reg four_ones_det //1-bit output to indicate 4 ones are detected or not
  );
```

```
// Internal Variables
 reg [1:0] current_state,   //4-bit current state register
           next_state;      //4-bit next state register

// State tags assigned using binary encoding
 parameter STATE_0 = 2'b00,
           STATE_1 = 2'b01,
           STATE_2 = 2'b10,
           STATE_3 = 2'b11;

 // Next State Assignment Block
 // This block implements the combination cloud of next state assignment logic

always @(*)
begin : next_state_bl
  case(current_state)

  STATE_0 :
  begin
   if(data_in)
   begin
    //transition to next state
    next_state   = STATE_1;
    four_ones_det = 1'b0;
   end
   else
   begin
   //retain same state
    next_state   = STATE_0;
    four_ones_det = 1'b0;
   end
 end

  STATE_1:
  begin
   if(data_in)
   begin
    //transition to next state
    next_state = STATE_2;
    four_ones_det = 1'b0;
   end
   else
   begin
    //retain same state
    next_state = STATE_1;
    four_ones_det = 1'b0;
   end
  end

  STATE_2 :
  begin
   if(data_in)
   begin
    //transition to next state
```

```
    next_state = STATE_3;
    four_ones_det = 1'b0;
   end
   else
   begin
    //retain same state
    next_state = STATE_2;
    four_ones_det = 1'b0;
   end
  end

  STATE_3 :
  begin
   if(data_in)
   begin
    //transition to next state
    next_state = STATE_0;
    four_ones_det = 1'b1;
   end
   else
   begin
    //retain same state
    next_state = STATE_3;
    four_ones_det = 1'b0;
   end
  end

    endcase
 end

  // Current Register Block
  always @(posedge clk)
  begin : current_state_bl

    if(reset)
      current_state <= #1 STATE_0;
  else
   current_state <= #1 next_state;

     end
 endmodule
```

The following codes the same design using *one-hot coding* and computes the output in a separate continuous assignment statement. Only few states are coded for demonstration. For effective HW mapping, an inverse `case` statement is used where each case is only evaluated to be TRUE or FALSE:

```
always@(*)
begin
next_state = 4'b0 ;
case (1'b1) // Synopsys parallel_case full_case
    current_state[S0]:
    if (in)
        next_state[S1] = 1'b1 ;
```

```
        else
            next_state[S0] = 1'b1 ;
        current_state[S1]:
        if (in)
            next_state[S2] = 1'b1 ;
        else
            next_state[S1] = 1'b1 ;
    endcase
    end

    // Separately coded output function
    assign output = state[S3];
```

9.3.6 SystemVerilog Support for FSM Coding

Chapter 3 gives an account of SystemVerilog. The language supports multiple procedural blocks such as `always_comb`, `always_ff` and `always_latch`, along with `unique` and `priority` keywords. The directives of `full_case`, `parallel_case` and `full_case_parallel_case` are synthesis directives and are ignored by simulation and verification tools. There may be instances when the wrong designer perception can go unchecked. To cover these short comings, SystemVerilog supports these statements and keywords to provide unified behavior in simulation and synthesis. The statements cover all the synthesis directives and the user can appropriately select the statements that are meaningful for the design. These statements ensure coherent behavior of the simulation and post-synthesis code [27].

Example: The compiler directives are ignored by simulation tools. Use of the directives may create mismatches between pre- and post-synthesis results. The following example of a Verilog implementation explains how SystemVerilog covers these short comings:

```
always@(cntr, in1, in2, in3, in4)
casex (cntr) /* Synopsys full_case parallel_case */
    4'b1xxx: out = in1;
    4'bx1xx: out = in2;
    4'bxx1x: out = in3;
    4'bxxx1: out = in4;
endcase
```

As the simulator ignores the compiler directives, a mismatch is possible when more than two bits of `cntr` are 1 at any time. The code can be rewritten using SystemVerilog for better interpretation:

```
always_comb
unique casex (cntr)
    4'b1xxx: out = in1;
    4'bx1xx: out = in2;
    4'bxx1x: out = in3;
    4'bxxx1: out = in4;
endcase
```

In this case the simulator will generate a warning if the `cntr` signal takes values that are not one-hot. This enables the designer to identify problems in the simulation and functional verification stages of the design cycle.

9.4 Algorithmic State Machine Representation

9.4.1 Basics

Hardware mapping of signal processing and communication algorithms is the main focus of this book. These algorithms are mostly mapped on time-shared architectures. The architectures require the design of FSMs to generate control signals for the datapath. In many designs the controllers have algorithm-like functionality. The functionality also encompasses decision support logic. Bubble diagrams are not flexible enough to describe the complex behavior of these finite state machines. Furthermore, many design problems require gradual refinement and the bubble diagram representation is not appropriate for this incremental methodology. The bubble diagram is also not algorithm-like. The algorithmic state machine (ASM) notation is the representation of choice for these design problems.

ASM is a flowchart-like graphical notation that describes the cycle-by-cycle behavior of an algorithm. Each step transitioning from one state to another or to the same state takes one clock cycle. The ASM is composed of three basic building blocks: rectangles, diamonds and ovals. Arrows are used to interconnect these building blocks. Each *rectangle* represents a state and the state output is written inside the rectangle. The state output is always an unconditional output, which is asserted when the FSM transitions to a state represented by the respective rectangle. A *diamond* is used for specifying a condition. Based on whether the condition is TRUE or FALSE, the next state or conditional output is decided. An *oval* is used to represent a conditional output. As Moore machines only have state outputs, Moore FSM implementations do not have ovals, but Mealy machines may contain ovals in their ASM representations.

Figure 9.13 shows the relationship of three basic components in an ASM representation. For TRUE or FALSE, T and F are written on respective branches. The condition may terminate in an oval, which lists conditional output.

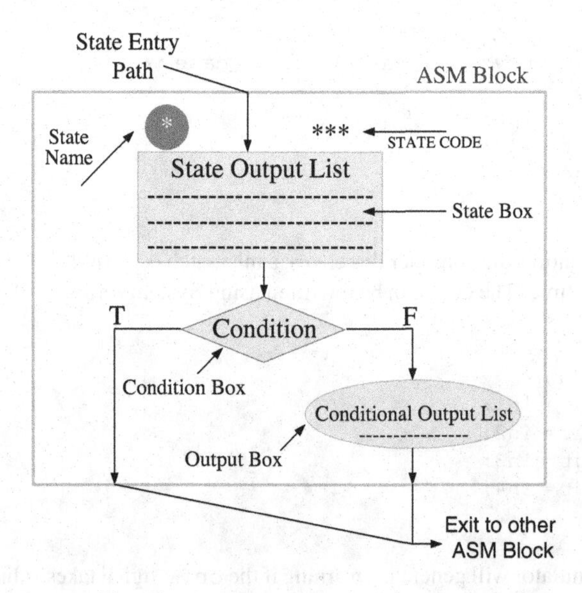

Figure 9.13 Rectangle, diamond and oval blocks in an ASM representation

The ASM is very descriptive and provides a mechanism for a systematic and step-by-step design of synchronous circuits. Once defined it can be easily translated into RTL Verilog code. Many off-the-shelf tools are available that directly convert an ASM representation into RTL Verilog code [11].

Example: Consider the design of Figure 9.9. The FSM implements counting of four 1s on a serial interface using Mealy or Moore machines. This example describes the FSM using ASM representation. Figure 9.14 shows the representations. The Moore machine requires an additional state.

9.4.2 Example: Design of a Four-entry FIFO

Figure 9.15 shows the datapath and control unit for a 4-entry FIFO queue. The inputs to the controller are two operations, `Write` and `Del`. The former moves data from the `fifo_in` to the tail of the FIFO queue, and the latter deletes the head of the queue. The head of the queue is always available at `fifo_out`. A write into a full queue or deletion from an empty queue causes an error condition. Assertion of `Write` and `Del` at the same time also causes an error condition.

The datapath consists of four registers, R_0, R_1, R_2 and R_3, and a multiplexer to select the head of the queue. The input to the FIFO is stored in R_0 when the `write_en` is asserted. The `write_en` also moves the other entries in the queue down by one position. With every new write and delete in the queue, the controller needs to select the head of the queue from its new location using an `out_sel` signal to the multiplexer. The FSM controller for the datapath is described using an ASM chart in Figure 9.16.

The initial state of FSM is S_0 and it identifies the status of the queue as empty. Any `Del` request to an empty queue will cause an error condition, as shown in the ASM chart. On a `Write` request the controller asserts a `write_en` signal and the value at `fifo_in` is latched in register R_0. The controller also selects R_0 at the output by assigning value $3'b00$ to `out_sel`. On a `Write` request the controller also transitions to state S_1. When the FSM is in S_1 a `Del` takes the FSM back to S_0 and a `Write` takes it to S_2. Similarly, another `Write` in state S_2 transitions the FSM to S_3. In this state, another `Write` generates an error as the FIFO is now completely filled. In every state, `Write` and `Del` at the same time generates an error.

Partial RTL Verilog code of the datapath and the FSM is given below. The code only implements state S_0. All the other states can be easily coded by following the ASM chart of Figure 9.16.

```
// Combinational part only for S0 and default state is given
always @(*)
begin
	next_state=0;
	case(current_state)

	`S0:
	begin
		if(!Del&& Write)
		begin
			next_state = `S1;
			write_en = 1'b1;
			Error= 1'b0;
			out_sel = 0;
		end
		else if(Del)
		begin
```

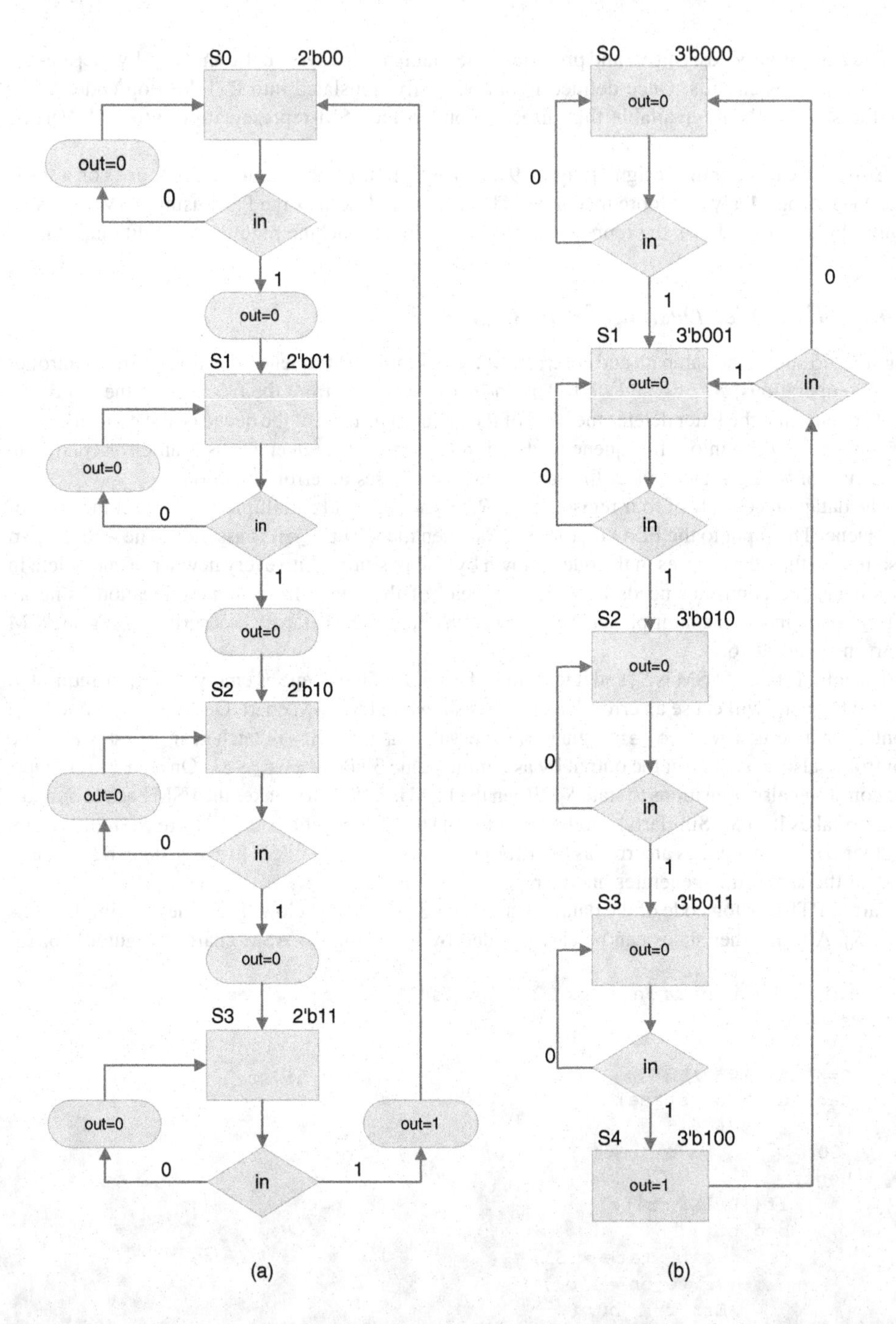

Figure 9.14 ASM representations of four 1s detected problem of Figure 9.9: (a) Mealy machine; (b) Moore machine

```
                next_state=`S0;
                write_en =1'b0;
                Error = 1'b1;
                out_sel=0;
            end
            else
            begin
                next_state=`S0;
                write_en=1'b0;
                out_sel = 1'b0;
            end

        // Similarly, rest of the states are coded //

            default:
            begin
                next_state=`S0;
                write_en = 1'b0;
                Error = 1'b0;
                out_sel =0;
            end
        endcase
end

// Sequential part
always @ (posedge clk or negedge rst_n)
if(!rst_n)
        current_state <= #1 'S0;
else
        current_state <= #1 next_state;
```

9.4.3 Example: Design of an Instruction Dispatcher

This section elaborates the design of an FSM-based instruction dispatcher for a programmable processor that can read 32-bit words from program memory (PM). The instruction words are written into two 32-bit instruction registers, IR_0 and IR_1. The processor supports short and long instructions of lengths 16 and 32-bit, respectively. The LSB of the instruction is coded to specify the instruction type. A 0 in the LSB indicates a 16-bit instruction and a 1 depicts the instruction is 32-bit wide. The dispatcher identifies the instruction type and dispatches the instruction to the respective Instruction Decoder (ID).

Figure 9.17(a) shows the bubble diagram that implements the instruction dispatcher. Following is the description of the working of the design.

1. The dispatcher's reset state is S_0.
2. The FSM reads the first 32-bit word from PM, transitions into state S_1 and stores it in instruction register IR_0.
3. The FSM then reads another 32-bit word, which is latched in IR_0 while it shifts the contents of IR_0 to IR_1 in the same cycle and makes transitions from S_1 to S_2. In state S_2 the FSM reads the LSB of IR_1 to check the instruction type. If the instruction is of type long (IL) i.e. 32-bit instruction it is dispatched to the ID for long instruction decoding and another 32-bit word is read in IR_0 where the

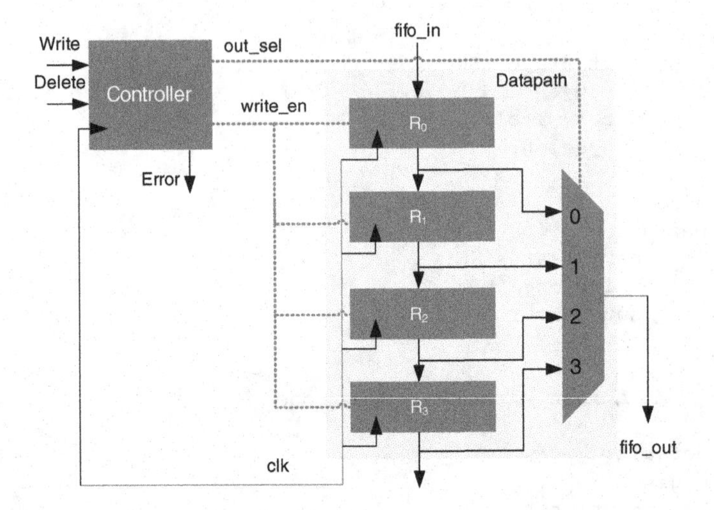

Figure 9.15 Four-entry FIFO queue with four registers and a multiplexer in the datapath

contents of IR_0 are shifted to IR_1. The FSM remains in state S_2. On the other hand, if the FSM is in state S_2 and the current instruction is of type short (IS), then the instruction is dispatched to the Instruction Decoder Short (IDS) and the FSM transition into S_3 without performing any read operation.

4. In state S_3, the FSM readjusts data. IR_{1H} representing 16 MSBs of register IR_1 is moved to IR_{1L} that depicts 16 LSBs of the instruction register IR_1. In the same sequence of operations, IR_{0H} and IR_{0L} are moved to IR_{0L} and IR_{1L}, respectively. In state S_3, the FSM checks the instruction type. If the instruction is of type long the FSM transitions to state S_4; otherwise if the instruction is of type short the FSM transitions to S_5.In both cases FSM initiates a new read from PM.
5. The behavior of the FSM in states S_4 and S_5 is given in Figure 9.17(a). To summarize, the FSM adjusts the data in the instruction registers such that the LSB of the current instruction is in the LSB of IR_1 and the dispatcher brings the new 32-bit data from PM in IR_0 if required. Figure 9.17(b) shows the associated architecture for readjusting the contents of instruction registers. The control signals to multiplexers are generated to bring the right inputs to the four registers IR_{0H}, IR_{0L}, IR_{1H} and IR_{1L} for latching in the next clock edge.

The RTL Verilog code of the dispatcher is given here:

```
// Variable length instruction dispatcher
moduleVariableInstructionLengthDispatcher(inputclk,rst_n,output [31:0]IR1);

  reg  [31:0] program_mem [0:255];
  reg   [7:0] PC;                  // program counter for memory (physical reg)
  wire [15:0] data_H;             // contains 16 MSBs of data from program memory
  wire [15:0] data_L;             // contains 16 LSBs of data from program_mem
  reg   [5:0] next_state, current_state;
  reg         read;
  reg         IR0_L_en, IR0_H_en;         // write enable for IR0 regsiter
  reg   [1:0] mux_sel_IR0_L;
  reg         IR1_L_en, IR1_H_en;       // write enable for IR1 register
```

```
 reg   [1:0] mux_sel_IR1_H;
 reg         mux_sel_IR1_L;
 reg  [15:0] mux_out_IR0_L, mux_out_IR1_H, mux_out_IR1_L;
 reg  [15:0] IR0_H, IR0_L, IR1_H, IR1_L;

 // state assignments
 parameter [5:0] S0 = 1;
 parameter [5:0] S1 = 2;
 parameter [5:0] S2 = 4;
 parameter [5:0] S3 = 8;
 parameter [5:0] S4 = 16;
 parameter [5:0] S5 = 32;

// loading memory "program_memory.txt"
 initial
 begin
    $readmemh("program_memory.txt",program_mem);
end

 assign {data_H, data_L} = program_mem[PC];
 assign IR1 = {IR1_H, IR1_L};              //instructions in IR1 get dispatched

always @ (*)
begin
   // default settings
  next_state = 6'd0;
  read = 0;
       IR0_L_en = 1'b0; IR0_H_en = 1'b0;                // disable registers
       mux_sel_IRO_L = 2'b0;
       IR1_L_en = 1'b0; IR1_H_en = 1'b0;
       mux_sel_IR1_H = 2'b0;
       mux_sel_IR1_L = 1'b0;

  case (current_state)
  S0:
  begin
    read=1'b1;
    next_state = S1;
    IR0_L_en = 1'b1; IR0_H_en = 1'b1;       // load 32 bit data from PM into
the IR0 register
        mux_sel_IRO_L = 2'd2;
  end

  S1:
    begin
       read = 1'b1;
       next_state = S2;
    IR0_L_en = 1'b1; IR0_H_en = 1'b1;       // load 32 bit data from PM into
the IR0 register
       mux_sel_IRO_L = 2'd2;
       IR1_L_en = 1'b1; IR1_H_en = 1'b1;       // load the contents of IR0 in
IR1 (full 32bits)
       mux_sel_IR1_H = 2'd0;
```

```
        mux_sel_IR1_L = 1'b1;
    end

  S2:
    begin
      if (IR1_L[0])                              // instruction type: long
      begin
        read=1'b1;
        next_state = S2;
        IR0_L_en = 1'b1; IR0_H_en = 1'b1;        // load 32 bit data from PM into
the IR0 register
        mux_sel_IR0_L = 2'd2;
        IR1_L_en = 1'b1; IR1_H_en = 1'b1;        // load the contents of IR0 in
IR1 (full 32bits)
        mux_sel_IR1_H = 2'd0;
        mux_sel_IR1_L = 1'b1;
      end
      else
      begin
        read=1'b0;
        next_state = S3;
      IR0_L_en = 1'b1; IR0_H_en = 1'b0;         // move IR0_H -> IR0_L ;
        mux_sel_IR0_L = 2'd0;
        IR1_L_en = 1'b1; IR1_H_en = 1'b1;       // move IR0_L -> IR1_H and
IR1_H -> IR1_L
        mux_sel_IR1_H = 2'd2;
        mux_sel_IR1_L = 1'b0;
      end
    end

  S3:
    begin
     if (IR1_L[0])                               // instruction type: long
     begin
     read=1'b1;
     next_state = S4;
      IR0_L_en = 1'b1; IR0_H_en = 1'b0;
     mux_sel_IR0_L = 2'd1;                      // move dataH -> IR0_L
      IR1_L_en = 1'b1; IR1_H_en = 1'b1;
     mux_sel_IR1_H = 2'd1;                          // move dataL -> IR1_H
     mux_sel_IR1_L = 1'b1;                          // move IR0_L -> IR1_L
     end
     else                                           // instruction type: short
     begin
       read = 1'b1;                                 // no read
       next_state = S5;
     IR0_L_en = 1'b1; IR0_H_en = 1'b1;          // load 32 bit data from PM into
the IR0 register
       mux_sel_IR0_L = 2'd2;
       IR1_L_en = 1'b1; IR1_H_en = 1'b1;        // move IR0_L -> IR1_H and
IR1_H -> IR1_L
       mux_sel_IR1_H = 2'd2;
       mux_sel_IR1_L = 1'b0;
```

```
  end
 end

 S4:
    begin
      if (IR1_L[0])                    // instruction type: long
      begin
      read=1'b1;
     next_state = S4;
      IR0_L_en = 1'b1; IR0_H_en = 1'b0;
      mux_sel_IRO_L = 2'd1;                        // move dataH -> IR0_L
      IR1_L_en = 1'b1; IR1_H_en = 1'b1;
      mux_sel_IR1_H = 2'd1;                        // move dataL -> IR1_H
      mux_sel_IR1_L = 1'b1;                        // move IR0_L -> IR1_L
      end
      else                                         // instruction type: short
      begin
          read = 1'b1;                             // no read
          next_state = S5;
     IR0_L_en = 1'b1; IR0_H_en = 1'b1;            // load 32 bit data from PM into
the IR0 register
          mux_sel_IRO_L = 2'd2;
          IR1_L_en = 1'b1; IR1_H_en = 1'b1;        // move IR0_L -> IR1_H and
IR1_H -> IR1_L
          mux_sel_IR1_H = 2'd2;
          mux_sel_IR1_L = 1'b0;
     end
    end

    S5:
    begin
    if (IR1_L[0])                                      // instruction type: long
     begin
     read=1'b1;
     next_state = S2;
      IR0_L_en = 1'b1; IR0_H_en = 1'b1;           // load 32 bit data from PM into
the IR0 register
     mux_sel_IRO_L = 2'd2;
     IR1_L_en = 1'b1; IR1_H_en = 1'b1;            // load the contents of IR0
 in IR1 (full 32bits)
     mux_sel_IR1_H = 2'd0;
     mux_sel_IR1_L = 1'b1;
    end

    else                                               // instruction type: short
    begin
      read=1'b0;
      next_state = S3;
   IR0_L_en = 1'b1; IR0_H_en = 1'b0;                 // move IR0_H -> IR0_L ;
      mux_sel_IRO_L = 2'd0;
      IR1_L_en = 1'b1; IR1_H_en = 1'b1;              // move IR0_L -> IR1_H and
```

```
IR1_H -> IR1_L
          mux_sel_IR1_H = 2'd2;
          mux_sel_IR1_L = 1'b0;
        end
      end
  endcase
 end

always @ (*)
begin
   mux_out_IR0_L = 16'bx;
   case (mux_sel_IRO_L)
      2'b00: mux_out_IR0_L = IR0_H;
      2'b01: mux_out_IR0_L = data_H;
      2'b10: mux_out_IR0_L = data_L;
  endcase
 end

always @ (*)
begin
   mux_out_IR1_H = 16'bx;
   case (mux_sel_IR1_H)
      2'b00: mux_out_IR1_H = IR0_H;
      2'b01: mux_out_IR1_H = data_L;
      2'b10: mux_out_IR1_H = IR0_L;
  endcase
end

always @ (*)
 if(mux_sel_IR1_L)
   mux_out_IR1_L = IR0_L;
 else
   mux_out_IR1_L = IR1_H;

always @ (posedge clk or negedge rst_n)
begin
   if (!rst_n)
   begin
        PC <= 8'b0;
        current_state <= 6'b000_001;
   end
   else
   begin
      current_state <= next_state;
         if (read)
            PC <= PC+1;
   end

     // if enable load Instruction registers with valid values
  if (IR0_H_en)
    IR0_H <= data_H;
  if (IR0_L_en)
    IR0_L <= mux_out_IR0_L;
```

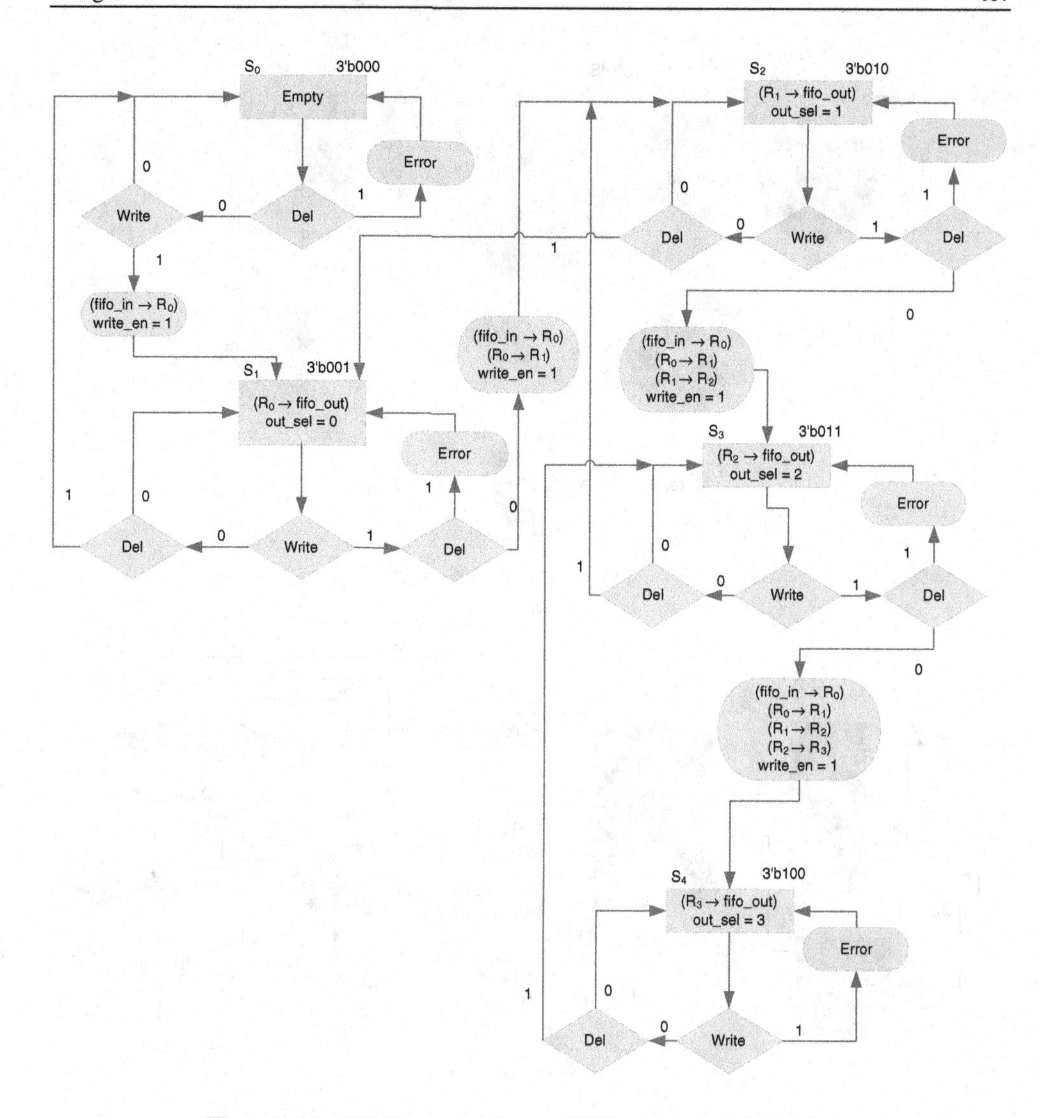

Figure 9.16 FSM design of 4-entry FIFO queue using ASM chart

```
  if (IR1_H_en)
    IR1_H <= mux_out_IR1_H;
  if (IR1_L_en)
    IR1_L <= mux_out_IR1_L;

  end
endmodule
```

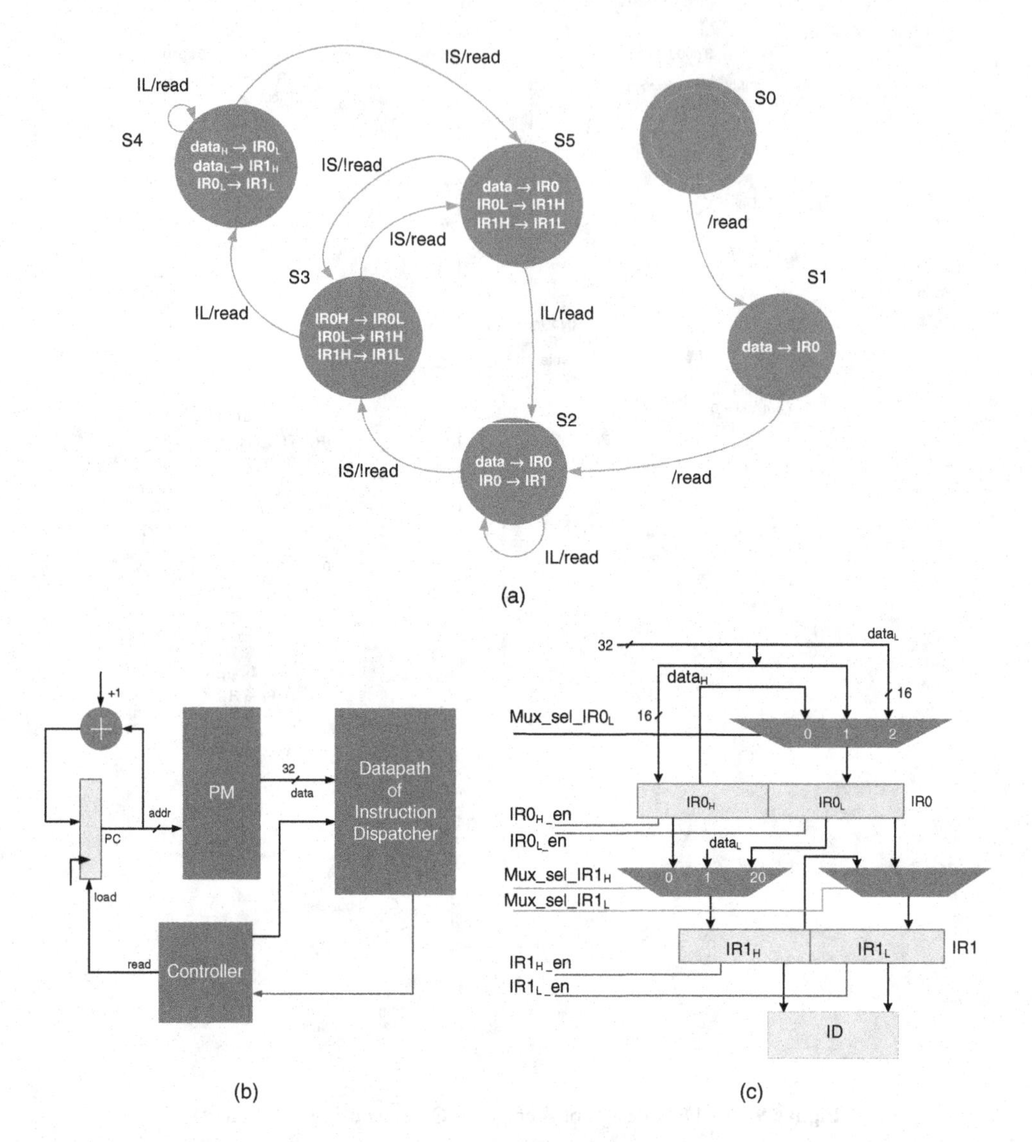

Figure 9.17 FSM implementing a variable-length instruction dispatcher. (a) Bubble-diagram representation of the FSM. (b) Architecture consisting of PM, PC, datapath and controller. (c) Datapath of the instruction dispatcher

9.5 FSM Optimization for Low Power and Area

There are a number of ways to implement an FSM. The designer, based on some set design objectives, selects one out of many options and then the synthesis tool further optimizes an FSM implementation [12, 13]. The tool, based on a selected optimization criterion, transforms a given FSM to functionally equivalent but topologically different FSMs. Usually FSMs are optimized for power and area.

The dynamic power dissipation in digital logic depends on switching activity. The tools, while performing state minimization for area reduction, also aim to minimize total switching activity. Out of

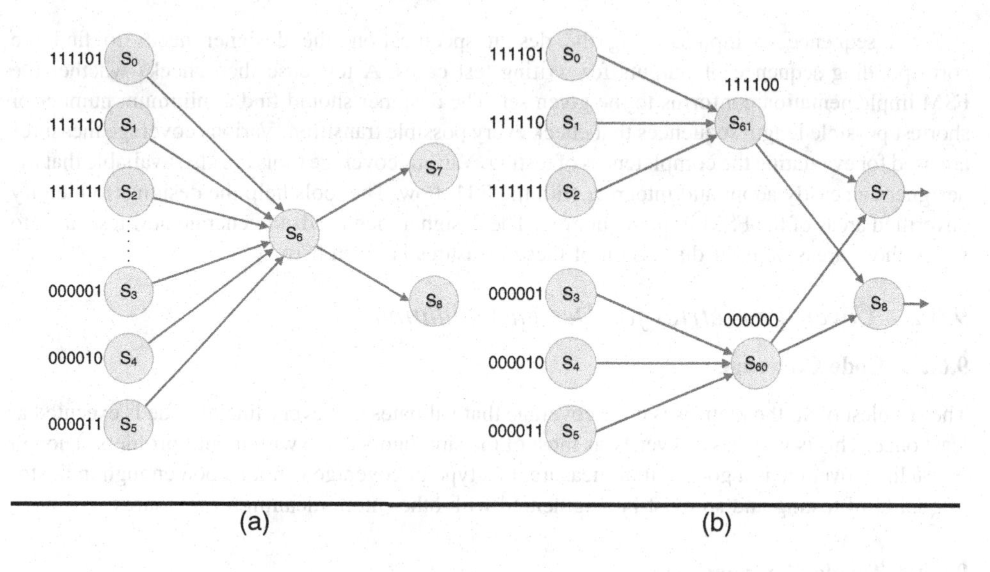

Figure 9.18 FSM optimization by splitting a critical state to reduce the hamming distance of state transitions. (a) Portion of original FSM with critical state. (b) Splitting of critical state into two

the many available optimization algorithms, one is to find a critical state that is connected to many states where the transitions into the critical state cause large hamming distances, thus generating significant switching activity. The algorithm, after identifying the critical state, splits it into two to reduce the hamming distance among the connected states and then analyzes the FSM to evaluate whether the splitting results in a power reduction without adding too much area in the design.

This is illustrated in Figure 9.18. Part (a) shows the original FSM, and (b) shows the transformed FSM with reduced switching activity as a result of any transition into the states formed by splitting the critical state. In an FSM with a large number of states, modeling the problem for an optimal solution is an 'NP complete' problem. Several heuristics are proposed in the literature, and an effective heuristic for the solution is given in [16].

9.6 Designing for Testability

From the testability perspective it is recommended to add reset and status reporting capabilities in an FSM. To provide an FSM with a reset capability, there should be an input that makes the FSM transition from any state to some defined initial state. An FSM with *status message capability* also increases the testability of the design. An FSM with this capability always returns its current state for a defined query input.

9.6.1 Methodology

The general methodology of testing digital designs discussed in Chapter 2 should be observed. Additionally, the methodology requires a specification or design document of the FSM that completely defines the FSM behavior for all possible sequences of inputs. This requires the developer to generate minimum length sequences of operations that must test all transitions of the FSM.

For initial white-box module-level testing, the stimulus should keep a record of all the transitions, corresponding internal states of the FSM, and inputs and output before a fault appears. This helps the designer to trace back and localize a bug.

For a sequence of inputs, using the design specification, the designer needs to find the corresponding sequence of outputs for writing test cases. A test case then checks whether the FSM implementation conforms to the given set. The designer should find a minimum number of shortest possible length sequences that check every possible transition. Various coverage measures are used for evaluating the completeness of testing. Various coverage tools are also available that the designer can easily adopt and integrate with the RTL flow. The tools help the designer to identify unverified areas of the FSM implementation. The designer then needs to generate new test cases to verify those areas. A brief discussion of these measures is given here.

9.6.2 Coverage Metrics for Design Validation

9.6.2.1 Code Coverage

The simplest of all the metrics is code coverage that validates that every line of code is executed at least once. This is very easy to verify as most of the simulators come with inbuilt profilers. Though 100% line coverage is a good initial measure, this type of coverage is not a good enough indicator of quality of testing and so must be augmented with other more meaningful measures.

9.6.2.2 Toggle Coverage

This measure checks whether every bit in all the registers has at least toggled once from 0 to 1 and 1 to 0. This is difficult to cover, as the content of registers are the outcome of some complex functions. The tester usually does not have deep enough understanding of the function to make its outcome generate the required bit patterns. Therefore the designer relies on random testing to verify this test metric. The toggling of all bits does not cover all sequences of state transitions so it does not completely test the design.

9.6.2.3 State Coverage

This measure keeps a record of the frequency of transitions into different states. Its objective is to ensure that the testing must make the FSM transit in all the states at least once. In many FSM designs a state can be reached from different states, so state coverage still does not guarantee complete testing of the FSM.

9.6.2.4 Transition or Arc Coverage

This measure requires the FSM to use all the transitions in the design at least once. Although it is a good measure, it is still not enough to ensure the correctness of the implementation.

9.6.2.5 Path Coverage

Path coverage is the most intensive measure of the correctness of a design. Starting from the initial state, traversing all paths in the design gives completeness to testing and verification. In many design instances with moderately large number of states, exhaustive generation of test vectors to transverse all paths may not be possible, so smart techniques to generate non-trivial paths for testing are critical. Commercially available and freeware tools [17–19] provide code coverage measurements and give a good indication of the thoroughness of testing.

One of these techniques is 'state space reduction'. This reduces the FSM to a smaller FSM by merging different nodes into one. The reduced FSM is then tested for path coverage.

There are various techniques for generating test vectors. *Specification-based testing* works on building a model from the specifications. The test cases are then derived from the model based on a

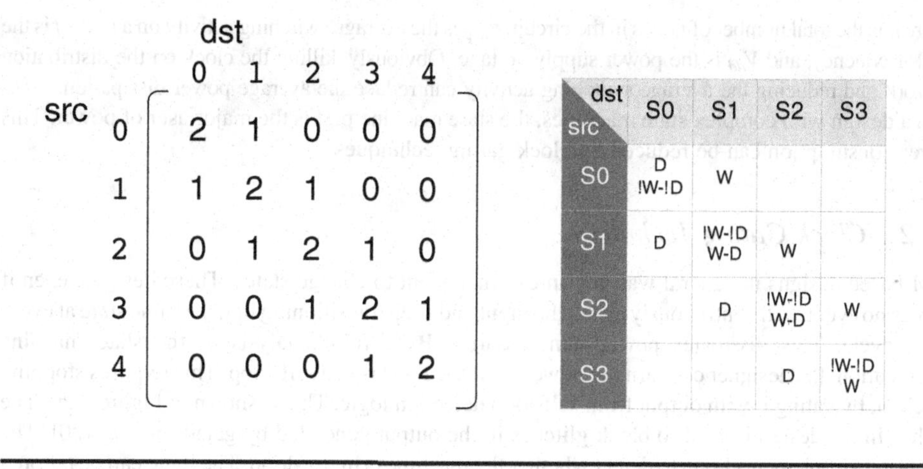

Figure 9.19 Adjacency matrix for automatic test vector generation. (a) Matrix showing possible entry paths from source to destination state. (b) Inputs for transitioning into different states

maximum coverage criterion. The test cases give a complete description of input and expected output of the test vectors. An *extended FSM model for exhaustive testing* works well for small FSMs but for large models it is computationally intractable. The ASM representation can be written as an 'adjacency matrix', where rows and columns represent source and destination nodes, respectively, and an entry in the matrix represents the number of paths from source node to destination node. An automatic test generation algorithm generates input that traverses the adjacency matrix by enumerating all the possible paths in the design.

Example: Write an adjacency matrix for the ASM of the 4-entry FIFO of Section 9.4.2 that traverses all possible paths from the matrix for automatic test vector generation.

Solution: All the paths start from the first node S_0 and terminate on other nodes in the FSM; the nodes are as shown in Figure 9.19. There are two paths that start and end at S_0. Corresponding to each path, a sequence of inputs is generated. For example, for a direct path originating from S_0 and ending at S_3, the following sequence of inputs generates the path:

$$\text{Write} \rightarrow \text{Write} \rightarrow \text{Write} \rightarrow \text{Write}$$

Similarly, the same path ending at S_3 with `error` output is:

$$\text{Write} \rightarrow \text{Write} \rightarrow \text{Write} \rightarrow \text{Write} \rightarrow \text{Write}$$

9.7 Methods for Reducing Power Dissipation

9.7.1 Switching Power

The distribution network of the clock and clock switching are the two major sources of power consumption. The switching power is given by the expression:

$$P_{\text{avg}} = n\alpha_{\text{avg}} f \frac{1}{2} c_{\text{avg}} V_{dd}^2$$

where n is the total number of nodes in the circuit, α_{avg} is the average switching activity on a node, f is the clock frequency, and V_{dd} is the power supply voltage. Obviously, killing the clock on the distribution network and reducing the average switching activity can reduce the average power dissipation.

In a design with complex state machines, the state machine part is the major user of power. This power consumption can be reduced by clock gating techniques.

9.7.2 Clock Gating Technique

FSM-based designs in general wait for an external event to change states. These designs, even if there is no event, still continuously check the input and keep transitioning into the same state at every clock cycle. This dissipates power unnecessarily. By careful analysis of the state machine specification, the designer can turn the power off if the FSM is in a self-loop. This requires stopping the clock by gating it with output from self-loop detection logic. This is shown in Figure 9.20. The latch L in the design is used to block glitches in the output generated by gated logic f_G [20]. The function f_G implements logic to detect whether the machine is in self-loop. The logic can be tailored to cater for only highly probable self-loop cases. This does require an *a priori* knowledge of the transition statistics of the implemented FSM.

Clock gating techniques are well known to ASIC designers [17–19], and now even for FPGAs effective clock gating techniques have been developed. FPGAs have dedicated clock networks to take the clock to all logic blocks with low skew and low jitter.

To illustrate the technique on a modern FPGA, the example of the Xilinx Virtex-5 FPGA is considered in [24]. An FPGA in this family has several embedded blocks such as block RAM, DSP48 and IOs, but for hardware configuration the basic building blocks of the FPGA are the configurable logic blocks (CLBs). Each CLB consists of two slices, and each slice contains four 6-input look-up tables (LUTs) and four flip-flops. The FPGA has 32 global clock buffers (BUFGCEs) to support these many internal or externally generated clocks. The FPGA is divided into multiple clock regions. Out of these 32 global clock signals, any 10 signals can be selected to be simultaneously fed to a clock region. Figure 9.21 shows a Virtex-5 device with 12 clock regions and associated logic of selecting 10 clocks out of 32 for these regions.

All the flip-flops in a slice also come with a clock enabling signal, as shown in Figure 9.22(a). Although disabling this signal will save some power, the dynamic power dissipated in the clock

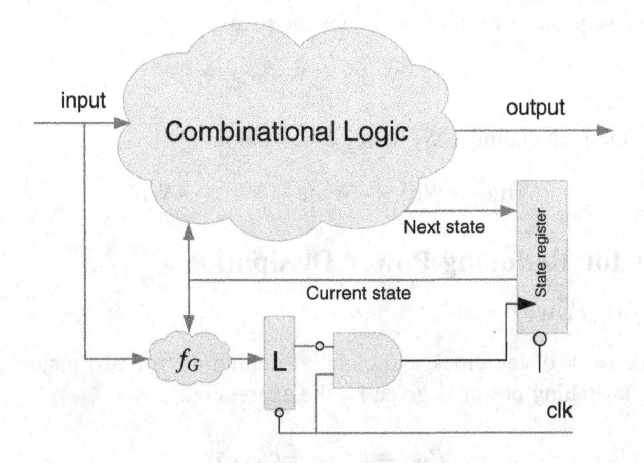

Figure 9.20 Low-power FSM design with gated clock logic

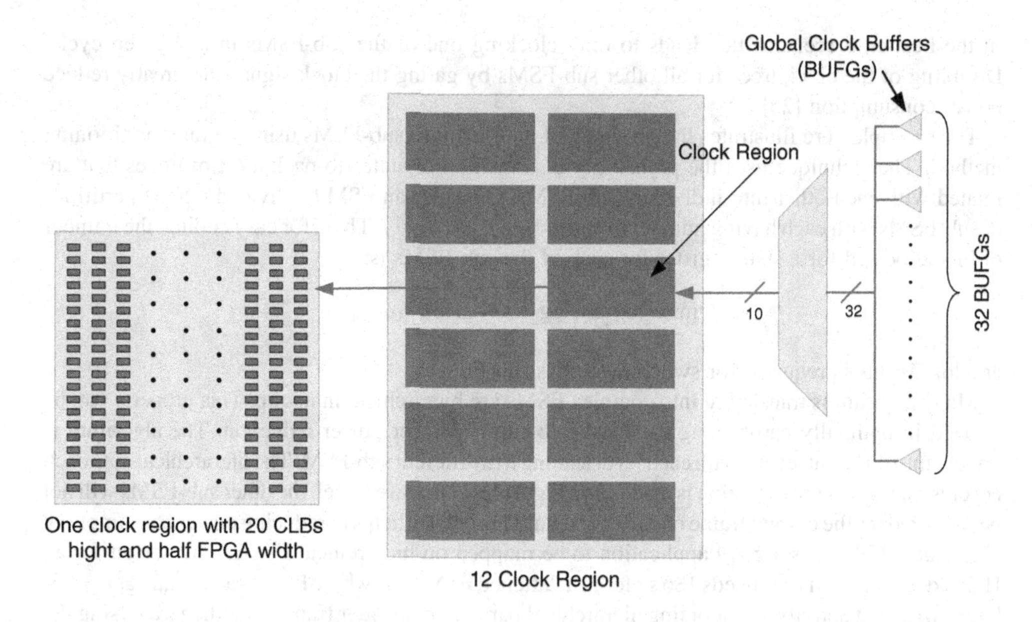

Figure 9.21 Virtex-5 with 32 global cock buffers with 10 clock regions and several CLBs in each region

distribution network will remain. To save this power, each clock buffer BUFGCE also comes with an enabling signal, as shown in Figure 9.22(b). This signal can also be disabled internally to kill the clock being fed to the clock distribution network. The methodology for using the clock enabling signal of flip-flops in a slice for clock gating is given in [24].

9.7.3 FSM Decomposition

FSM decomposition is a technique whereby an FSM is optimally partitioned into multiple sub-FSMs in such a way that the state machine remains OFF in most of the sub-FSMs for a considerable time.

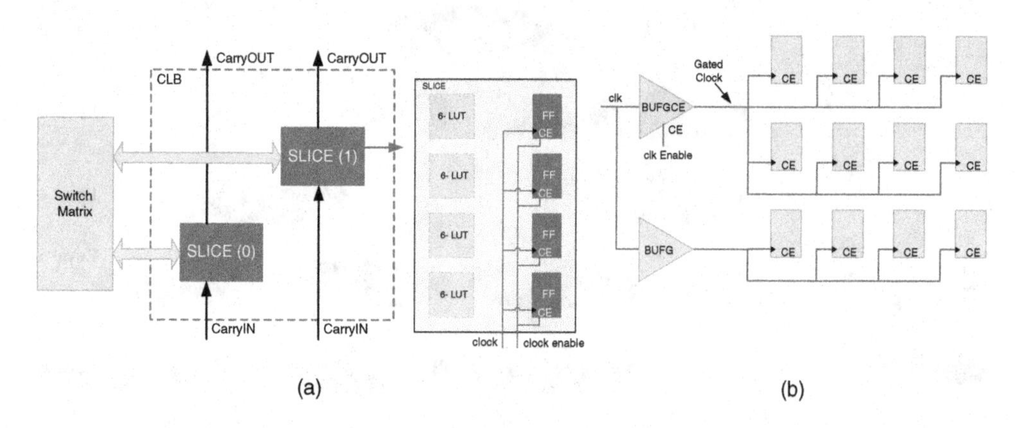

Figure 9.22 Clock gating in Virtex-5 FPGA. (a) CLB with two slices and a slice with four LUTs and four FFs with clock enable signal. (b) Gating the clocks to FFs

In the best-case scenario this leads to only clocking one of the sub-FSMs in any given cycle. Disabling of the clock trees for all other sub-FSMs by gating the clock signal can greatly reduce power consumption [25].

The example here illustrates the grouping of an FSM into sub-FSMs using a graph partitioning method. The technique uses the probability of transition of states to pack a set of nodes that are related with each other into individual sub-FSMs. Consider an FSM be divided into M partitions of sub-FSMs with each having number of states $S_0, S_1, \ldots, S_{M-1}$. Then, for easy coding, the number of bits allocated for a state register for each of the sub-FSMs is:

$$\max\{\log_2 S_0, \log_2 S_1, \ldots, \log_2 S_{M-1}\}$$

and $\log_2 M$ bits is required for switching across sub-FSMs.

Many algorithms translating into complex FSMs are hierarchical in nature. This property can be utilized in optimally partitioning the FSM into sub-FSMs for power reduction. The algorithm is broken into different levels, with each level leading to a different sub-FSM. The hierarchical approach ensures that, when the transition is made from higher level to lower level, the other sub-FSMs will not be triggered for the current frame or buffer of data. This greatly helps in reducing power consumption.

A video decoder is a good application to be mapped on hierarchical FSMs. The design for an H.264 decoder otherwise needs 186 states in a flattened FSM. The whole FSM is active and generates huge switching activity. By adopting hierarchical parsing of decoder frames and then traversing the FSM, the design can be decomposed into six levels with 13 sub-FSMs [25]. The designer, knowing the state transition patterns, can also assign state code that has minimum hamming distance among closely transitioned states to reduce switching activities.

Many algorithms in signal processing can be hierarchically decomposed. A representative decomposition is given in Figure 9.23. Each sub-FSM should preferably have a single point of entry to make it convenient to design and test. Each sub-FSM should also be self-contained to maximize transitions within the sub-FSM and to minimize transitions across sub-FSMs. The design is also best for power gating for power optimization.

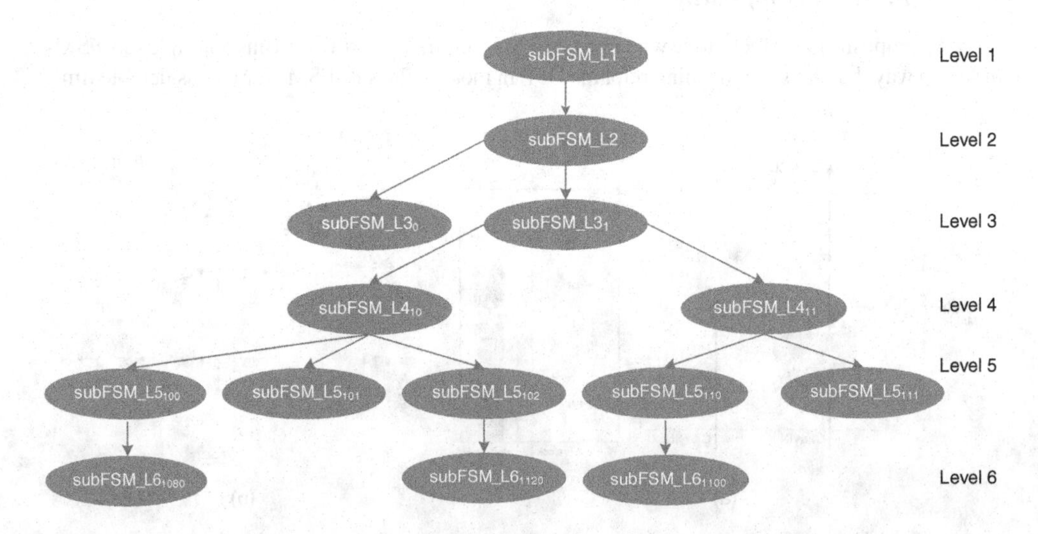

Figure 9.23 An FSM hierarchically partitioned into sub-FSMs for effective design, implementation, testing and power optimization

Exercises

Exercise 9.1

Design a 4 × 4-bit-serial unsigned multiplier that assumes bit-serial inputs and output [26]. Modify the multiplier to perform Q1.3 fractional signed × signed multiplication. Truncate the result to 4-bit Q1.3 format by ignoring least significant bits of the product.

Exercise 9.2

Using the fractional bit-serial multiplier of exercise 9.1, design a bit-serial 3-coefficient FIR filter. Assume both the coefficients and data are in Q1.3 format. Ignore least significant bits to keep the output in Q1.3 format.

Exercise 9.3

Design a sequential architecture for an IIR filter implementing the following difference equation:

$$y[n] = \sum_{k=0}^{10} b_k x[n-k] - \sum_{k=1}^{10} a_k y[n-k]$$

First write the sequential pseudo-code for the implementation, then design a single MAC-based architecture with associated logic. Assume the circuit clock is 21 times the sampling clock. All the coefficients are in Q1.15 format. Also truncate the output to fit in Q1.15 format.

Exercise 9.4

Design the sequential architectures of 4-coefficient FIR filters that use: (a) one sequential multiplier; and (b) four sequential multipliers.

Exercise 9.5

Design an FSM that detects an input sequence of 1101. Implement the FSM using Mealy and Moore techniques, and indentify whether Moore machine implementation takes more states. For encoding the states use a one-hot technique. Write RTL Verilog code of the design. Write test vectors to test all the possible transitions in the state machine.

Exercise 9.6

Design a smart traffic controller for the roads layout given in Figure 9.24. The signals on Main Road should cater for all the following:

1. The traffic on the minor roads, VIP and ambulance movements on the main roads (assume a sensing system is in place that informs about VIP, ambulance movements, and traffic on the minor roads). The system, when it turns the Main Road light to red, automatically switches lights on the minor roads to green.
2. The system keeps all traffic lights on Main Road green if it detects the movement of a VIP or an ambulance on the main road, while keeping the minor road lights red. The lights remain in this state for an additional two minutes after the departure of the designated vehicles from Main Road.

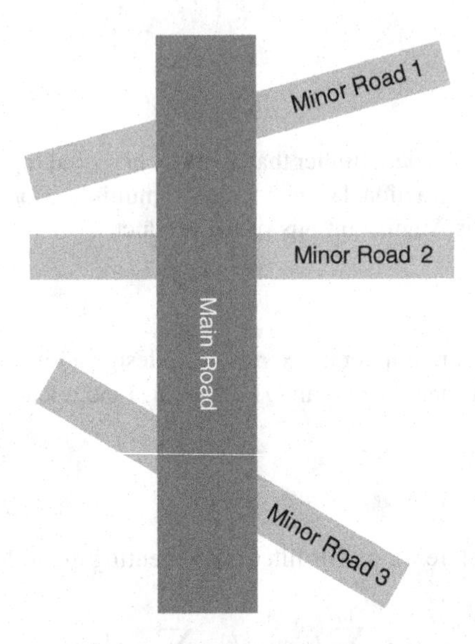

Figure 9.24 Road layout for exercise 9.6

3. If sensors on minor roads 1, 2 or 3 sense traffic, the lights for the traffic on Main Road are turned red for 80 seconds.
4. If the sensor on any one of the minor roads senses traffic, the light for Main Road traffic on that crossing is turned red for 40 seconds, and on the rest of the crossing for 20 seconds.
5. If the sensors on any two of the minor roads senses traffic, the lights on those crossings are turned red for 50 seconds while light on the left over third crossing is turned red for 30 seconds.
6. Once switched from red to green, the lights on Main Road remains green at least for the next 200 seconds.

Draw a bubble diagram of the state machine, clearly marking input and outputs, and write RTL Verilog code of the design. Design a comprehensive test-bench that covers all transitions of the FSM.

Exercise 9.7

Design a state machine that outputs 1 at every fifth 1 input as a Mealy and Moore machine. Check your design for the following bit stream of input:

```
1111110011111111111110011010111111011100011111111
```

Exercise 9.8

Design a Moore machine system to detect three transitions from 0 to 1 or 1 to 0. Describe your design using an ASM chart. Assume a serial interface where bits are serially received at every positive edge of the clock. Implement the design in RTL Verilog. Write a stimulus to check all transitions in the FSM.

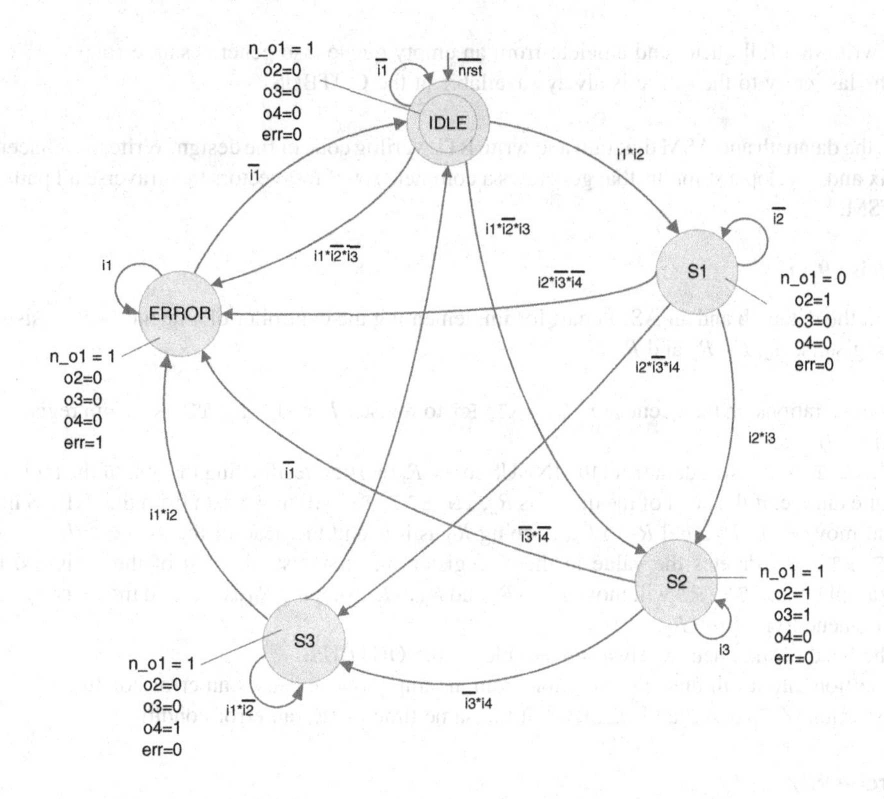

Figure 9.25 FSM from bench-marking database of exercise 9.9

Exercise 9.9

The state machine of Figure 9.25 is represented using a bubble diagram. The FSM is taken from a database that is used for bench-marking different testing and performance issues relating to FSMs.

1. Describe the state machine using a mathematical model. List the complete set of inputs X, states S and output Y.
2. Write the adjacency matrix for the state machine.
3. Write RTL Verilog code of the design that adheres to RTL coding guidelines for FSM design.
4. Write test vectors that traverse all the arcs at least once.

Exercise 9.10

Design an FSM-based 4-deep LIFO to implement the following:

1. A write into the LIFO writes data on INBUS at the end of the queue.
2. A delete from LIFO deletes the last entry in the queue.
3. A write and delete at the same time generates an error.

4. A write in a full queue and a delete from an empty queue also generates an error.
5. The last entry to the queue is always available at the OUTBUS.

Draw the datapath and ASM diagram and write RTL Verilog code of the design. Write the adjacency matrix and develop a stimulus that generates a complete set of test vectors that traverse all paths of the FSM.

Exercise 9.11

Design the datapath and an ASM chart for implementing the controller of a queue that consists of four registers, R_0, R_1, R_2 and R_3.

1. The operations on the queue are `INSERT_Ri` to register R_i and `DELETE_Ri` from register R_i, for $i = 0 \ldots 3$.
2. `INSERT_Ri` moves data from the INBUS to the R_i register, readjusting the rest of the registers. For example, if the head of the queue is R_2, `INSERT_R1` will move data from the INBUS in R_1 and move R_1 to R_2., and R_2 to R_3, keeping R_0 as it is and the head of the queue to R_3.
3. `DELETE_Ri` deletes the value in the R_i register, and readjusts the rest of the registers. For example, `DELETE_R1` will move R_2 to R_1 and R_3 to R_2, keeping R_0 as it is and move the head of the queue to R_2 from R_3.
4. The head of the queue is always available on the OUTBUS.
5. Insertion into a full queue or deletion from an empty queue causes an error condition.
6. Assertion of `INSERT` and `DELETE` at the same time causes an error condition

Exercise 9.12

Design a 3-entry FIFO, which supports `WRITE`, `DEL0`, `DEL1` and `DEL2`, where `WRITE` writes to the tail of the queue, and `DEL0`, `DEL1` and `DEL2` delete last, last two or all three entries from the queue, respectively. When insufficient entries are in the queue for a DELi operation, the FIFO controller generates an error, ERRORd. Similarly, `WRITE` in a completely filled FIFO also generates an error, ERRORw.

1. Design the datapath.
2. Design the state machine-based controller of the FIFO. Design the FSM implementing the controller and describe the design using an ASM diagram.
3. Write RTL Verilog code of the design.
4. Develop a stimulus that traverses all the states.

Exercise 9.13

Design a time-shared FIR filter that implements a 12-coefficient FIR filter using two MAC units. Clearly show all the registers and associated clocks. Use one CPA in the design. Write RTL code for the design and test the design for correctness.

Exercise 9.14

Develop a state machine-based instruction dispatcher that reads 64-bit words from the instruction memory into two 64-bit registers IR_0 and IR_1. The least significant two bits of the instruction depicts

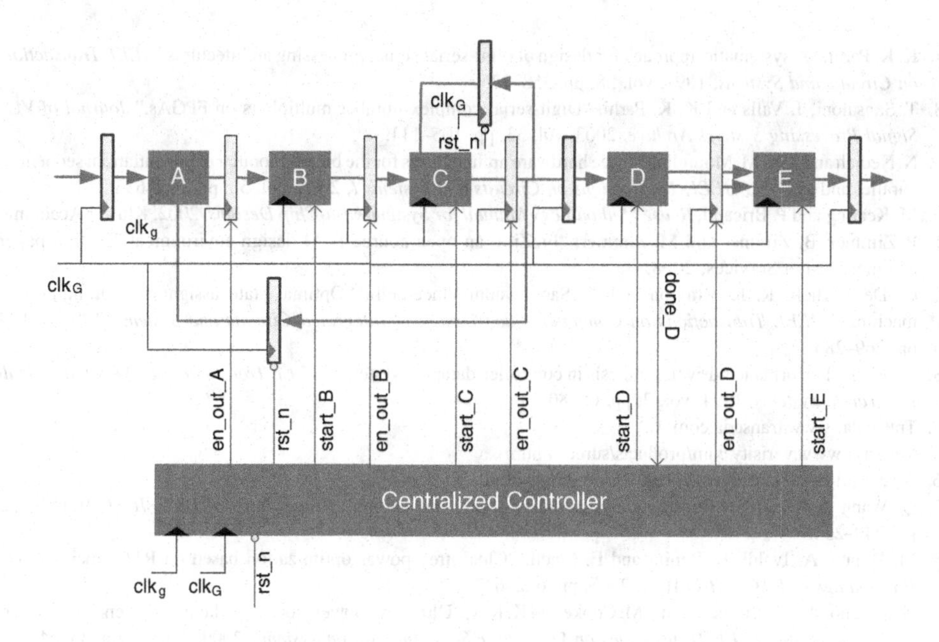

Figure 9.26 Centralized controller for exercise 9.15

whether the instruction is 16-bit, 32-bit, 48-bit or 64-bit wide. Use the two instruction registers such that IR_1 always keep a complete valid instruction.

Exercise 9.15

Write RTL Verilog code for the design given in Figure 9.26. Node A is a combinational logic, and nodes B, C and E take 7, 8 and 9 predefined number of circuit clocks, clk_g. Node D dynamically executes and takes a variable number of cycles between 3 and 8.7. Assume simple counters inside the nodes. Write a top-level module with two input clocks, clk_g and clk_G. All control signals are 1-bit and the data width is 8 bits. For each block A, B, C, D and E, only write module instances. Design a controller for generating all the control signals in the design.

References

1. B. A. Curtis and V. K. Madisetti, "A quantitative methodology for rapid prototyping and high-level synthesis of signal processing algorithms," *IEEE Transactions on Signal Processing*, 1994, vol. 42, pp. 3188–3208.
2. G. A. Constantinides, P. Cheung and W. Luk, "Optimum and heuristic synthesis of multiple word-length architectures," *IEEE Transactions on VLSI Systems*, 2005, vol. 13, pp. 39–57.
3. P. T. Balsara and D. T. Harper "Understanding VLSI bit-serial multipliers," *IEEE. Transactions on Education*, 1996, vol. 39, pp. 19–28.
4. S. A. Khan and A. Perwaiz, "Bit-serial CORDIC DDFS design for serial digital down-converter," in *Proceedings of Australasian Telecommunication Networks and Applications Conference*, 2007, pp. 298–302.
5. S. Matsuo *et al.* "8.9-megapixel video image sensor with 14-b column-parallel SA-ADC," *IEEE Transactions on Electronic Devices*, 2009, vol. 56, pp. 2380–2389.
6. R. Hartley and P. Corbett, "Digit-serial processing techniques," *IEEE Transactions on Circuits and Systems*, 1990, vol. 37, pp. 707–719.

7. K. K. Parhi, "A systematic approach for design of digit-serial signal processing architectures," *IEEE Transactions on Circuits and Systems*, 1991, vol. 38, pp. 358–375.
8. T. Sansaloni, J. Valls and K. K. Parhi, "Digit-serial complex-number multipliers on FPGAs," *Journal of VLSI Signal Processing Systems Archive*, 2003, vol. 33, pp. 105–111.
9. N. Nedjah and L. d. M. Mourelle, "Three hardware architectures for the binary modular exponentiation: sequential, parallel and systolic," *IEEE Transactions on Circuits and Systems I*, 2006, vol. 53, pp. 627–633.
10. M. Keating and P. Bricaud. *Reuse Methodology Manual for System-on-a-Chip Designs*, 2002, Kluwer Academic.
11. P. Zimmer, B. Zimmer and M. Zimmer, "FizZim: an open-source FSM design environment," design paper, Zimmer Design Services, 2008.
12. G. De Micheli, R. K. Brayton and A. Sangiovanni-Vincentelli, "Optimal state assignment for finite state machines," *IEEE Transactions on Computer-Aided Design of Integrated Circuits and Systems*, 1985, vol. 4, pp. 269–285.
13. W. Wolf, "Performance-driven synthesis in controller-datapath systems," *IEEE Transactions on Very Large Scale Integration Systems*, 1994, vol. 2, pp. 68–80.
14. Transeda: www.transeda.com
15. Verisity: www.verisity.com/products/surecov.html
16. Asic-world: www.asic-world.com/verilog/tools.html
17. Q. Wang and S. Roy, "Power minimization by clock root gating," *Proceedings of DAC, IEEE/ACM*, 2003, pp. 249–254.
18. M. Donno, A. Ivaldi, L. Benini and E. Macii, "Clock tree power optimization based on RTL clock-gating," *Proceedings of DAC, IEEE/ACM*, 2003, pp. 622–627.
19. H. Mahmoodi, V. Tirumalashetty, M. Cooke and K. Roy, "Ultra-low-power clocking scheme using energy recovery and clock gating," *IEEE Transactions on Very Large Scale Integration Systems*, 2009, vol. 17, pp. 33–44.
20. L. Benini, P. Siegel and G. De Micheli, "Saving power by synthesizing gated clocks for sequential circuits," *IEEE Design and Test of Computers*, 1994, vol. 11, pp. 32–41.
21. Q. Wang and S. Roy, "Power minimization by clock root gating," *Proceedings of DAC, IEEE/ACM*, 2003, pp. 249–254.
22. M. Donno, A. Ivaldi, L. Benini and E. Macii, "Clocktree power optimization based on RTL clock-gating," *Proceedings of DAC, IEEE/ACM*, 2003, pp. 622–627.
23. V. K. Madisetti and B. A. Curtis, "A quantitative methodology for rapid prototyping and high-level synthesis of signal processing algorithms," *IEEE Transactions on Signal Processing*, 1994, vol. 42, pp. 3188–3208.
24. Q. Wang, S. Gupta and J. H. Anderson, "Clock power reduction for Virtex-5 FPGAs," in *Proceedings of International Symposium on Field Programmable Gate Arrays, ACM/SIGDA*, 2009, pp. 13–22.
25. K. Xu, O. Chiu-sing Choy, C.-F. Chan and K.-P. Pun, "Power-efficient VLSI realization of a complex FSM for H.264/AVC bitstream parsing," *IEEE Transactions on Circuits and Systems II*, 2007, vol. 54, pp. 984–988.
26. R. Gnanasekaran, "On a bit-serial input and bit-serial output multiplier," *IEEE Transactions on Computers*, 1983, vol. 32, pp. 878–880.
27. The Benefits of SystemVerilog for ASIC Design and Verification, Ver 2.5 Synopsys jan 2007 (http://www.synopsys.com/Tools/Implementation/RTLSynthesis/CapsuleModule/sv_asic_wp.pdf)

10

Micro-programmed State Machines

10.1 Introduction

In time-shared architecture, computational units are shared to execute different operations of the algorithm in different cycles. As described in Chapter 9, time-shared architecture consists of a datapath and a control unit. In each clock cycle the control unit generates appropriate control signals for the datapath. In hardwired state machine-based designs, the controller is implemented as a Mealy or Moore finite state machine (FSM). However, in many applications the algorithms are so complex that a hardwired FSM design is not feasible. In other applications, either several algorithms need to be implemented on the same datapath or the designer wants to keep the flexibility of modifying the controller without repeating the entire design cycle, so a flexible controller is implemented to use the same datapath for different algorithms. The controller is made programmable [1].

This chapter describes the design of micro-programmed state machines with various capabilities and options. A methodology for converting a hardwired state machine to a micro-programmed implementation is introduced. The chapter gives an equivalent micro-programmed implementations of the examples already covered in previous chapters. The chapter then extends the implementation by describing a design of FIFO and LIFO queues on the same datapath running micro-programs.

The chapter then switches to implementation of DSP algorithms on time-shared architecture. Analysis of a few algorithms shows that in many cases a simple counter-based micro-programmed state machine can be used to implement a controller. The memory contains all the control signals to be generated in a sequence, whereas the counter keeps computing the address of the next micro-code stored in memory. In applications where decision-based execution of operations is required, this elementary counter-based state machine is augmented with decision support capability. The execution is based on a few condition lines coming either from the datapath or externally from some other blocks. These conditions are used in conditional jump instructions. Further to this, the counter in the state machine can be replaced by a program counter (PC) register, which in normal

---

Digital Design of Signal Processing Systems: A Practical Approach, First Edition. Shoab Ahmed Khan.

execution adds a fixed offset to its content to get the address of the next instruction. The chapter then describes cases where a program, instead of executing the next micro-code, needs to take a jump to execute a sequence of instructions stored at a different location in the program memory. This requires subroutine support in the state machine.

This chapter also covers the design of nested subroutine support whereby a subroutine can be called inside another subroutine. This is of special interest in image and signal processing applications, giving the capability to repeatedly execute a block of instructions a number of times without any loop maintenance overhead.

The chapter finishes with a complete example of a design that uses all the features of a micro-programmed state machine.

10.2 Micro-programmed Controller

10.2.1 Basics

The first step in designing a programmable architecture is to define an instruction set. This is defined such that a micro-program based on the instructions implements the target application effectively. The datapath and a controller are then designed to execute the instruction set. The controller in this case is micro-program memory-based. This design is required to have flexibility not only to implement the application but also to support any likely modifications or upgrading of the algorithm [2].

An FSM has two components, combinational and sequential. The sequential component consists of a state register that stores the current state of the FSM. In the Mealy machine implementation, combinational logic computes outputs and the next state from inputs and the current state. In a hardwired FSM this combinational logic cannot be programmed to change its functionality. In a micro-programmable state machine the combinational logic is replaced by a sequence of control signals that are stored in *program memory* (PM), as shown in Figure 10.1. The PM may be a read-only (ROM) or random-access (RAM).

A micro-programmed state machine stores the entire output sequence of control signals along with the associated next state in program memory. The address of the contents in the memory is determined by the current state and input to the FSM.

Assume a hypothetical example of a hardwired FSM design that has two 1-bit inputs, six 1-bit outputs and a 6-bit state register. Now, to add flexibility in the controller, the designer wishes to

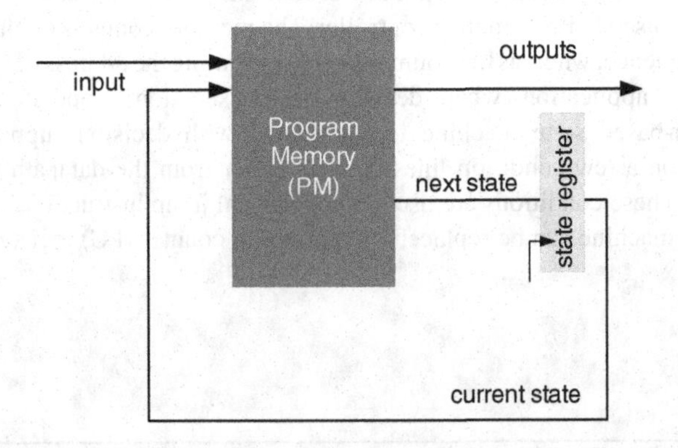

Figure 10.1 Micro-programmed state machine design showing program memory and state register

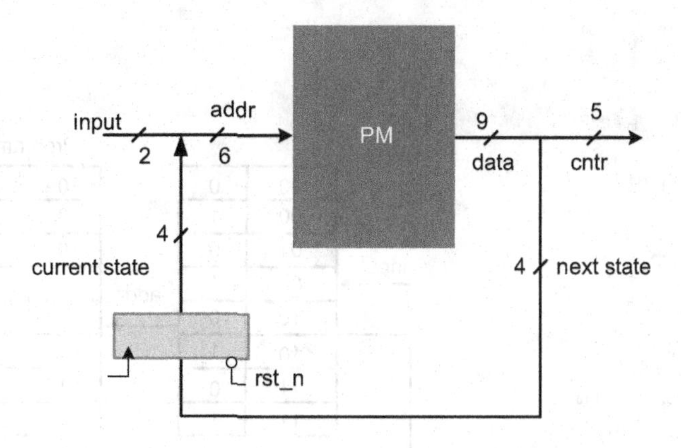

Figure 10.2 Hypothetical micro-programmed state machine design

replace the hardwired FSM. The designer evaluates all possible state transitions based on inputs and the current state and tabulates the outputs and next states as micro-coding for PM. These values are placed in the PM such that the inputs and the current state provide the index or address to the PM. Figure 10.2 shows the design of this micro-programmed state machine.

Mapping this configuration on the example under consideration makes the micro-program PM address bus 6 bits wide and its data bus 9 bits wide. The 2-bit input constitutes the two least significant bits (LSBs) of the address bus, `addr[1:0]`, whereas the current state forms the rest of the four most significant bits (MSBs) of the address bus, `addr[5:2]`. The contents of the memory are worked out from the ASM chart to produce desired output control signals and the next state for the state machine. The ASM chart for this example is not given here as the focus is more on discussing the main components of the design.

The five LSBs of the data bus, `data[4:0]`, are allocated for the outputs `cntr[4:0]`, and the four MSBs of the data bus, `data[8:5]`, are used for the next state. At every positive edge of the clock the state register latches the value of the next state to the state register. This value becomes the current state in the next clock cycle.

Example: This example designs a state machine that implements the four 1s detection problem of Chapter 9, where the same was realized as hardwired Mealy and Moore state machines.

The ASM chart for the Mealy machine implementation is given in Figure 10.3(a). This design has 1-bit input and 1-bit output and its state register is 2 bits wide. The PM-based micro-coded design requires a 3-bit address bus and a 3-bit data bus. The address bus requires the PM to be $2^3 = 8$ bits deep. The contents of the memory can be easily filled following the ASM chart of Figure 10.3(a). The address $3'b000$ defines the current state of S0 and the input $1'b0$. The next state and output for this current state and input can be read from the ASM chart. The chart shows that in state S0, if the input is 0 then the output is also 0 and the next state remains S0. Therefore the content of memory at address $3'b000$ is filled with $3'b000$. The LSB of the data bus is output and the rest of the 2 bits defines the next state. Similarly the next address of $3'b001$ defines the current state to be S0 and the input as 1. The ASM chart shows that in state S0, if input is 1 the next state is S1 and output is 0. The contents of the PM at address $3'b001$ is filled with $3'b010$. Similarly for each address, the current state and the input is parsed and the values of next state and output are read from the ASM chart. The contents of PM are accordingly filled. The PM for the four 1s detected problem is given in Figure 10.3(b) and the RTL Verilog of the design is listed here:

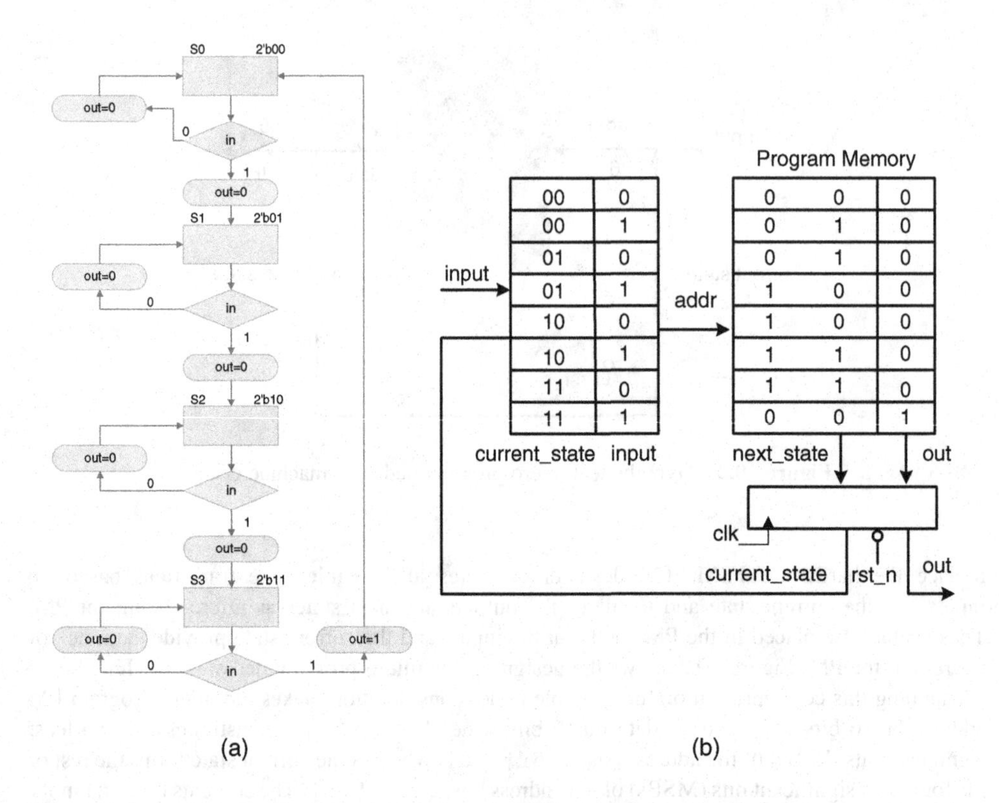

Figure 10.3 Moving from hardwired to micro-coded design. (a) ASM chart for four 1s detected problem of Figure 9.9 in Chapter 9. (b) Equivalent micro-programmed state machine design

```
// Module to implement micro-coded Mealy machine
module fsm_mic_mealy
(
input clk, // System clock
input rst_n, // System reset
input data_in, // 1-bit input stream
output four_ones_det // To indicate four 1s are detected
);

// Internal variables
reg [1:0] current_state; // 2-bit current state register
wire [1:0] next_state; // Next state output of the ROM
wire [2:0] addr_in; // 3-bit address for the ROM

// This micro-programmed ROM contains information of state transitions
mic_prog_rom mic_prog_rom_inst
(
.addr_in(addr_in),
.next_state(next_state),
.four_ones_det(four_ones_det)
);
```

```
// ROM address
assign addr_in = {current_state , data_in};
// Current state register
always @ (posedge clk or negedge rst_n)
begin : current_state_bl
 if(!rst_n)
  current_state <= #1 2b0; // One-hot encoded
 else
  current_state <= #1 next_state; // Next state from the ROM
end
endmodule

// Module to implement PM for state transition information
module mic_prog_rom
(
input [2:0] addr_in, // Address of the ROM
output reg [1:0] next_state, // Next state output
output reg four_ones_det // Detection of signal output
);

always @ (addr_in)
begin : ROM_bl
  case(addr_in)
       3b00_0 :
  {next_state, four_ones_det} = {4b00, 1b0};
       3b00_1 :
  {next_state, four_ones_det} = {4b01, 1b0};
       3b01_0 :
  {next_state, four_ones_det} = {4b01, 1b0};
       3b01_1 :
  {next_state, four_ones_det} = {4b10, 1b0};
       3b10_0 :
  {next_state, four_ones_det} = {4b10, 1b0};
       3b10_1 :
  {next_state, four_ones_det} = {4b11, 1b0};
       3b11_0 :
  {next_state, four_ones_det} = {4b11, 1b0};
       3b11_1 :
  {next_state, four_ones_det} = {4b00, 1b1};
  endcase
end
endmodule
```

Micro-programmed state machine implementation provides flexibility as changes in the behavior of the state machine can be simply implemented by loading a new micro-code in the memory. Assume the designer intends to generate a 1 and transition to state S3 when the current state is S2 and the input is 1. This can be easily accomplished by changing the memory contents of PM at location $3'b101$ to $3'b111$.

10.2.2 Moore Micro-programmed State Machines

For Moore machine implementation the micro-program memory is split into two parts, so the combinational logic-I and logic-II are replaced by PM-I and PM-II. The input and the current state

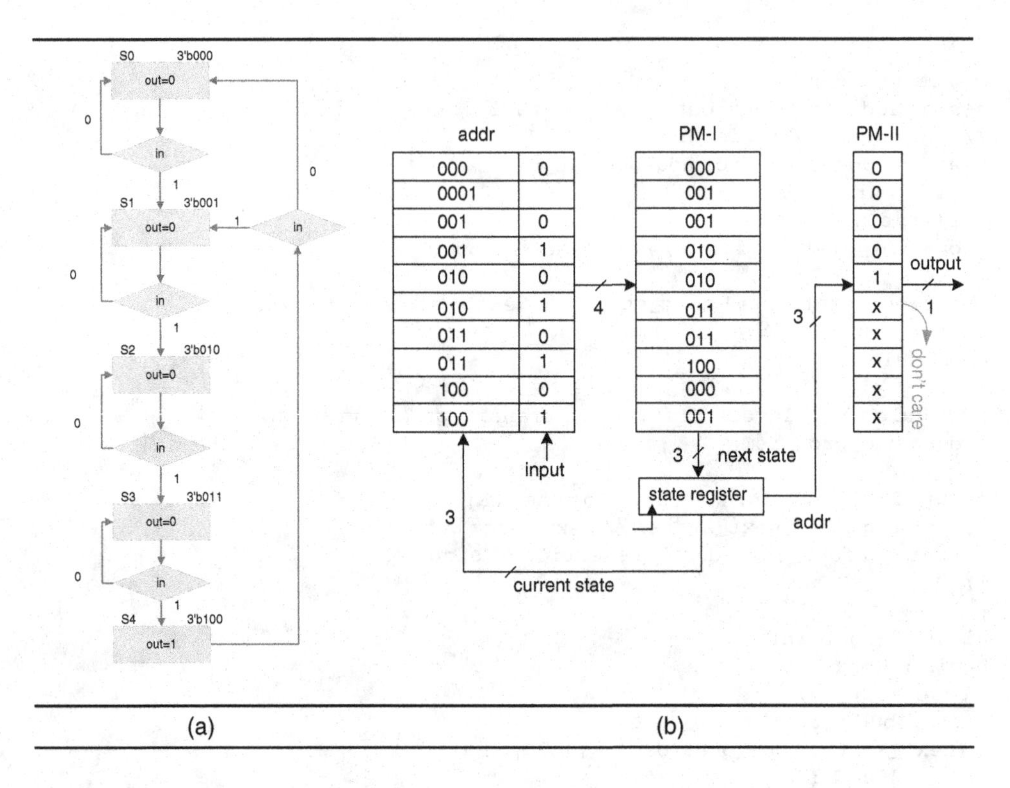

Figure 10.4 (a) Moore machine ASM chart for the four 1s detected problem. (b) Micro-program state machine design for the Moore machine

constitute the address for PM-I. The memory contents of PM-I are filled to appropriately generate the next state according to the ASM chart. The width of PM-I is equal to the size of the current state register, whereas its depth is $2^3 = 8$ (3 is the size of input plus current state). Only the current state acts as the address for PM-II. The contents of PM-II generate output signals for the datapath.

A micro-programmed state machine design for the four 1s detected problem is shown in Figure 10.4. One bit of the input and three bits of the current state constitute the 4-bit address bus for PM-I. The contents of the memory are micro-programmed to generate the next state following the ASM chart. Only three bits of the current state form the address bus of PM-II. The contents of PM-II are filled to generate a control signal in accordance with the ASM chart.

10.2.3 Example: LIFO and FIFO

This example illustrates a datapath that consists of four registers and associated logic to be used as LIFO (last-in first-out) or FIFO (first-in first-out). In both cases the user invokes different micro-codes consisting of a sequence of control signals to get the desired functionality. The top-level design consisting of datapath and controller is shown in Figure 10.5(a).

Out of a number of design options, a representative datapath is shown in Figure 10.5(b). The write address register `wr_addr_reg` is used for selecting the register for the write operation. A `wr_en` increments this register and the value on the `INBUS` is latched into the register selected by the `wr_addr` value. Similarly the value pointed by read address register `rd_addr_reg` is always

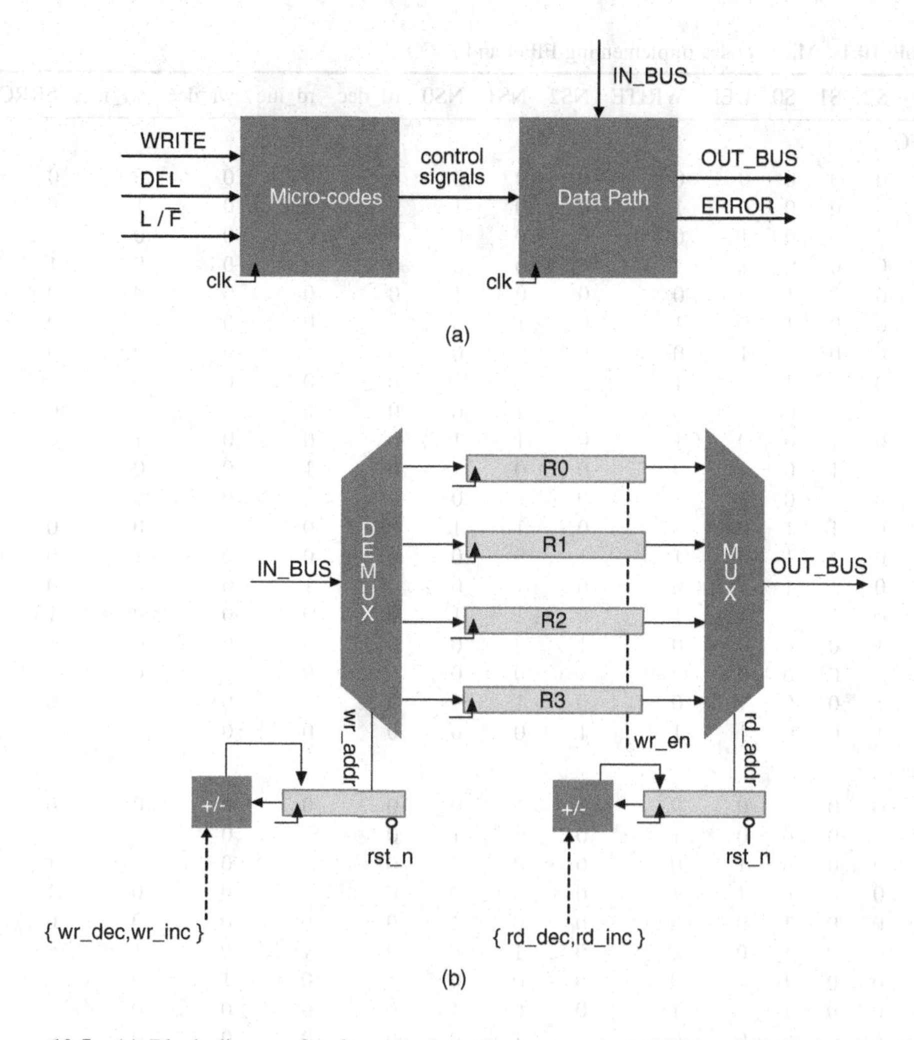

Figure 10.5 (a) Block diagram for datapath design to support both LIFO and FIFO functionality. (b) Representative datapath to implement both LIFO and FIFO functionality for micro-coded design

available on the `OUTBUS`. This value is the last or the first value input to a LIFO or a FIFO mode, respectively. A `DEL` operation increments the register `rd_add_reg`. The control signal `rd_inc` is asserted to increment the value on a valid `DEL` request. The micro-codes for the controller are listed in Table 10.1. The current states are S0, S1 and S2, and next states are depicted by NS0, NS1 and NS2.

10.3 Counter-based State Machines

10.3.1 Basics

Many controller designs do not depend on the external inputs. The FSMs of these designs generate a sequence of control signals without any consideration of inputs. The sequence is stored serially in the program memory. To read a value, the design only needs to generate addresses to the PM in a

Table 10.1 Micro-codes implementing FIFO and LIFO

L/F	S2	S1	S0	DEL	WRITE	NS2	NS1	NS0	rd_dec	rd_inc	wr_dec	wr_inc	ERROR
FIFO													
0	0	0	0	0	0	0	0	0	0	0	0	0	0
0	0	0	0	0	1	0	0	1	0	0	0	1	0
0	0	0	0	1	0	0	0	0	0	0	0	0	1
0	0	0	0	1	1	0	0	0	0	0	0	0	1
0	0	0	1	0	0	0	0	1	0	0	0	0	0
0	0	0	1	0	1	0	1	0	0	0	0	1	0
0	0	0	1	1	0	0	0	0	0	1	0	0	0
0	0	0	1	1	1	0	0	1	0	0	0	0	1
0	0	1	0	0	0	0	1	0	0	0	0	0	0
0	0	1	0	0	1	0	1	1	0	0	0	1	0
0	0	1	0	1	0	0	0	1	0	1	0	0	0
0	0	1	0	1	1	0	1	0	0	0	0	0	1
0	0	1	1	0	0	0	1	1	0	0	0	0	0
0	0	1	1	0	1	1	0	0	0	0	0	1	0
0	0	1	1	1	0	0	1	0	0	1	0	0	0
0	0	1	1	1	1	0	1	1	0	0	0	0	1
0	1	0	0	0	0	1	0	0	0	0	0	0	0
0	1	0	0	0	1	1	0	0	0	0	0	0	1
0	1	0	0	1	0	0	1	1	0	1	0	0	0
0	1	0	0	1	1	1	0	0	0	0	0	0	1
LIFO													
1	0	0	0	0	0	0	0	0	0	0	0	0	0
1	0	0	0	0	1	0	0	1	0	1	0	1	0
1	0	0	0	1	0	0	0	0	0	0	0	0	1
1	0	0	0	1	1	0	0	0	0	0	0	0	1
1	0	0	1	0	0	0	0	1	0	0	0	0	0
1	0	0	1	0	1	0	1	0	0	1	0	1	0
1	0	0	1	1	0	0	0	0	1	0	1	0	0
1	0	0	1	1	1	0	0	1	0	0	0	0	1
1	0	1	0	0	0	0	1	0	0	0	0	0	0
1	0	1	0	0	1	0	1	1	0	1	0	1	0
1	0	1	0	1	0	0	0	1	1	0	1	0	0
1	0	1	0	1	1	0	1	0	0	0	0	0	1
1	0	1	1	0	0	0	1	1	0	0	0	0	0
1	0	1	1	0	1	1	0	0	0	1	0	1	0
1	0	1	1	1	0	0	1	1	1	0	1	0	0
1	0	1	1	1	1	0	1	1	0	0	0	0	1
1	1	0	0	0	0	1	0	0	0	0	0	0	0
1	1	0	0	0	1	1	0	0	0	0	0	0	1
1	1	0	0	1	0	0	1	1	1	0	1	0	0
1	1	0	0	1	1	1	0	0	0	0	0	0	1

sequence starting from 0 and ending at the address in the PM that stores the last set of control signals. The addresses can be easily generated with a counter. The counter thus acts as the state register where the next state is automatically generated with an increment. The architecture results in a reduction in PM size as the memory does not store the next state and only stores the control signals. The machine

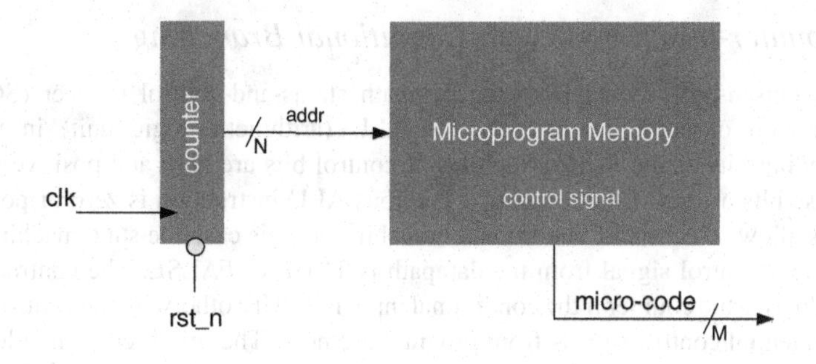

Figure 10.6 Counter-based micro-program state machine implementation

is reset to start from address 0 and then in every clock cycle it reads a micro-code from the memory. The counter is incremented in every cycle, generating the next state that is the address of the next micro-code in PM.

Figure 10.6 shows an example. An N-bit resetable counter is used to generate addresses for the program memory in a sequence from 0 to 2^N-1. The PM is M bits wide and 2^N-1 deep, and stores the sequence of signals for controlling the datapath. The counter increments the address to memory in every clock cycle, and the control signals in the sequence are output on the data bus. For executing desired operations, the output signals are appropriately connected to different blocks in the datapath.

10.3.2 *Loadable Counter-based State Machine*

As described above, a simple counter-based micro-programmed state machine can only generate control signals in a sequence. However, many algorithms once mapped on time-shared architecture may also require out-of-sequence execution, whereby the controller is capable of jumping to start generating control signals from a new address in the PM. This flexibility is achieved by incorporating the address to be branched as part of the micro-code and a loadable counter latches this value when the load signal is asserted. This is called 'unconditional branching'. Figure 10.7 shows the design of a micro-programmed state machine with a loadable counter.

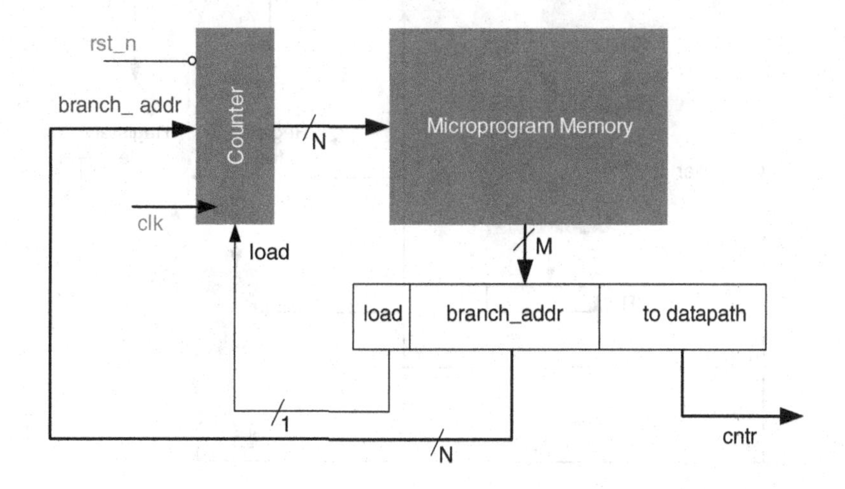

Figure 10.7 Counter-based controller with unconditional branching

10.3.3 Counter-based FSM with Conditional Branching

The control inputs usually come from the datapath status-and-control register (SCR). On the basis of execution of some micro-code, the ALU (arithmetic logic unit) in the datapath sets corresponding bits of the SCR. Examples of control bits are zero and positive status bits in the SCR. These bits are set if the result of a previous ALU instruction is zero or positive. These two status bits allow selection of conditional branching. In this case the state machine will check whether the input control signal from the datapath is TRUE or FALSE. The controller will load the branch address in the counter if the conditional input is TRUE; otherwise the controller will keep generating sequential control signals from program memory. The micro-code may look like this:

```
if(zero_flag) jump to label0
or
if(positive_flag) jump to label1
```

where `zero_flag` and `postivie_flag` are the zero and positive status bits of the SCR, and `label0` and `label1` are branch addresses. The controller jumps to this new location and reads the micro-code starting from this address if the conditional input is TRUE.

The datapath usually has one or more register files. The data from memory is loaded in the register files. The ALU operates on the data from the register file and stores the result in memory or in the register file. A representative block diagram of a datapath and controller is shown in Figure 10.8.

The conditional bits can also directly come from the datapath and be used in the same cycle by the controller. In this case the conditional micro-code may look like this:

```
if(r1==r2) jump to label1
```

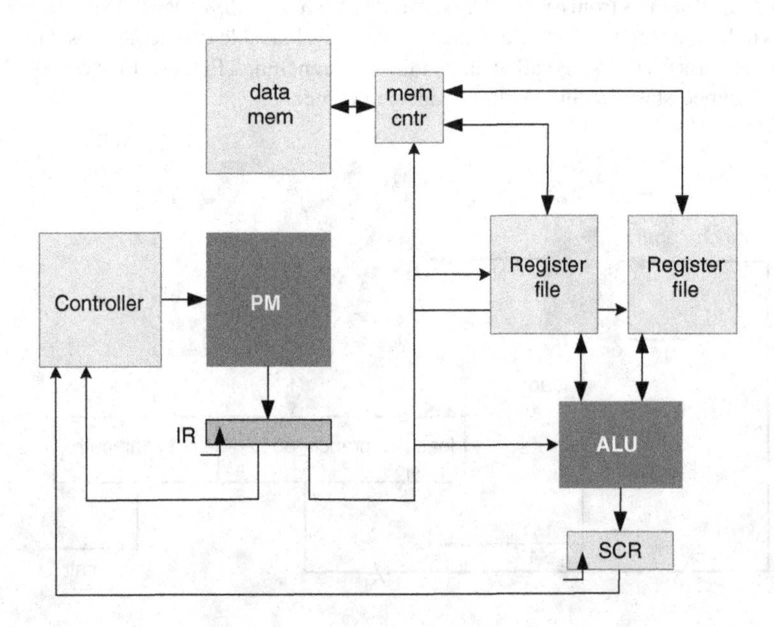

Figure 10.8 Representative design of controller, datapath with register file, data memory and SCR

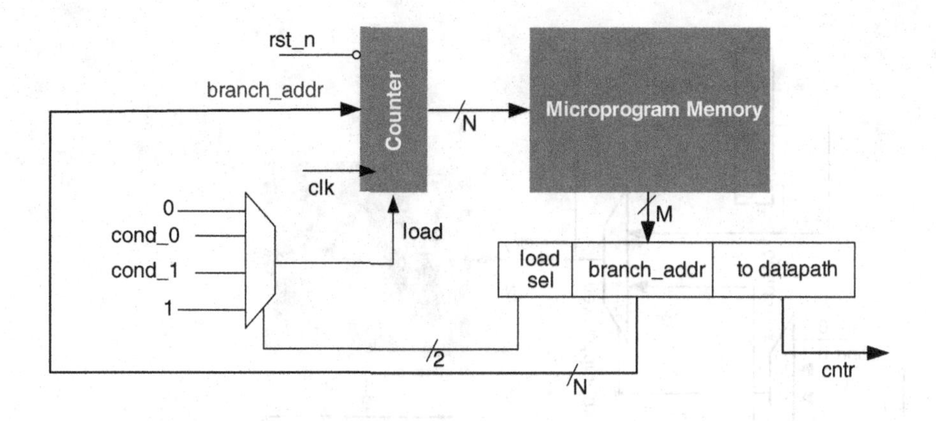

Figure 10.9 Micro-programmed state machine with conditional branch support

The datapath subtracts the value in register `r1` from `r2`, and if the result of this subtraction is 0 the state machine executes the micro-code starting from `label1`.

To provision the branching capabilities, a conditional multiplexer is added in the controller as shown in Figure 10.9. Two `load_sel` bits in the micro-code select one of the four inputs to the conditional MUX. If `load_sel` is set to $2'b00$, the conditional MUX selects 0 as the output to the load signal to the counter. With the load signal set to FALSE, the counter retains its normal execution and sequentially generates the next address for the PM. If, however, the programmer sets `load_sel` to $2'b11$, the conditional MUX selects 1 as the output to the load signal to the counter. The counter then loads the branch address and starts reading micro-code from this address. This is equivalent to unconditional branch support earlier. Similarly, if `load_sel` is set to $2'b01$ or $2'b10$, the MUX selects, respectively, one of the condition bits `cond_0` or `cond_1` for the load signal of the counter. If the selected condition bit is set to TRUE, the counter loads the `branch_addr`; otherwise it keeps its sequential execution.

Table 10.2 summarizes the options added to the counter-based state machine by incorporating a conditional multiplexer in the design.

10.3.4 Register-based Controllers

In many micro-programmed FSM designs an adder and a register replace the counter in the architecture. This register is called a *micro-program counter* (PC) register. The adder adds 1 to the content of the PC register to generate the address of the next micro-code in the micro-program memory.

Table 10.2 Summary of operations a micro-coded state machine performs based on `load_sel` and conditional inputs

load_sel	Load
$2'b00$	FALSE (normal execution: never load branch address)
$2'b01$	cond_0 (load branch address if cond_0 is TRUE)
$2'b10$	cond_1 (load branch address if cond_1 is TRUE)
$2'b11$	TRUE (unconditional jump: always load branch address)

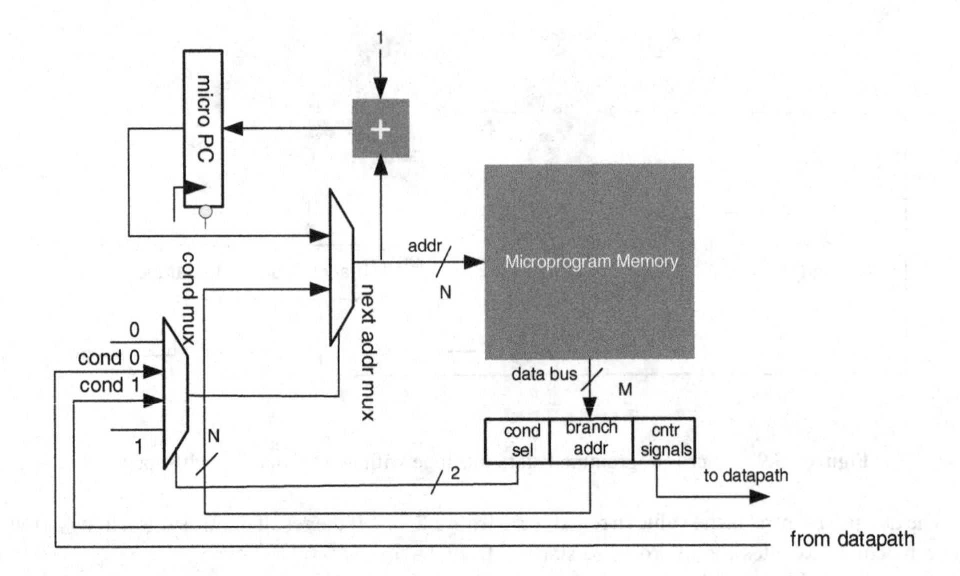

Figure 10.10 Micro-PC-based design

Figure 10.10 shows an example. The `next_addr_mux` selects the address to the micro-program memory. For normal execution, `cond_mux` selects 0. The output from `cond_mux` is used as the selected line to `next_addr_mux`. Then, `mirco_PC` and `branch_addr` are two inputs to `next_addr_mux`. A zero at the selected line to this multiplexer selects the value of `micro_PC` and a 1 selects the `branch_addr` to `addr_bus`. The `addr_bus` is also input to the adder that adds 1 to the content of `addr_bus`. This value is latched in `micro_PC` in the next clock cycle.

10.3.5 Register-based Machine with Parity Field

A parity bit can be added in the micro-code to expand the conditional branching capability of the state machine of Figure 10.10. The parity bit enables the controller to branch on both TRUE and FALSE states of conditional inputs. If the parity bit is set, the EXOR gate inverts the selected condition of `cond_mux`, whereas if the parity bit is not set the selected condition then selects the `next_addr` to the micro-program memory.

A controller with a parity field in the micro-code is depicted in Figure 10.11. The parity bit is set to 0 with the following micro-code:

```
if (cond_0) jump to label
```

The parity bit is set to 1 with the following micro-code:

```
if(!cond_0) jump to label
```

10.3.6 Example to Illustrate Complete Functionality

This example (see Figure 10.12) illustrates the complete functionality of a design based on a micro-programmed state machine. The datapath executes operations based on the control signals generated

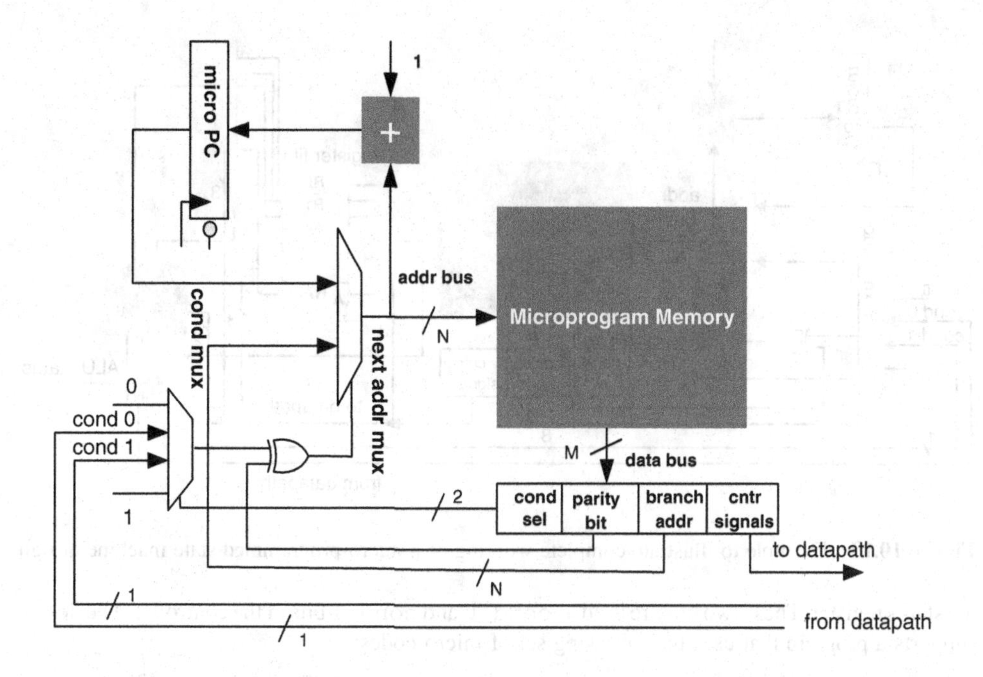

Figure 10.11 Register-based controller with a parity field in the micro-code

by the controller. These control signals are stored in the PM of the controller. The datapath has one register file. The register file stores operands. (In most designs these operands are loaded from memory. For simplicity this functionality is not included in the example.)

The register file has two read ports and one write port. For eight registers in the register file, 3 bits of the control signal are used to select a register to be read from each read port, and 3 bits of the control signal are used to select a register to be written by an ALU micro-code. The datapath based on the result of the executed micro-code updates the Z and N bits of the ALU status register. The Z bit is set if the result of an ALU micro-code is 0, while a negative value of the ALU result sets the N bit of

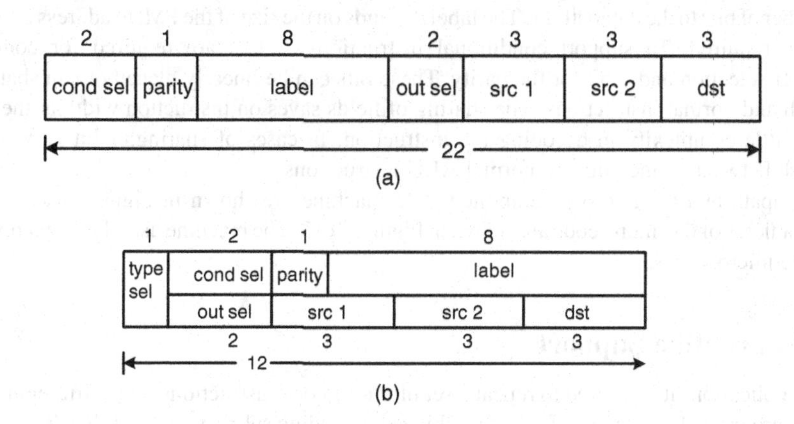

Figure 10.12 Different fields in the micro-code of the programmable state machine design of the text example. (a) Non-overlapping fields. (b) Overlapping fields to reduce the width of the PM

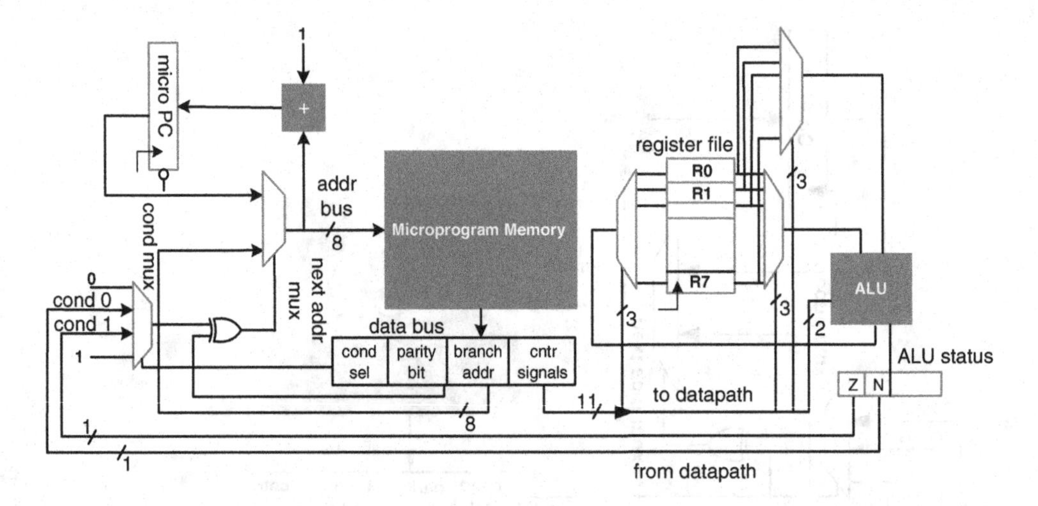

Figure 10.13 Example to illustrate complete working of a micro-programmed state machine design

the status register. These two bits are used as `cond_0` and `cond_1` bits of the controller. The design supports a program that uses the following set of micro-codes:

```
Ri = Rj op Rk, for i, j, k = 0, 1, 2, ..., 7 and op is any one of the four supported ALU
operations of +, -, AND, OR, an example micro code is R2 = R3 + R5
if(N) jump label
if(!N) jump label
if(Z) jump label
if(!Z) jump label
jump label
```

These micro-codes control the datapath and also support the conditional and unconditional branch instruction. For accessing three of eight registers R_i, R_j and R_k, three sets of 3 bits are required. Similarly, to select one of the outputs from four ALU operations, 2 bits are needed. This makes the total number of bits to the datapath 11. The label depends on the size of the PM: to address 256 words, 8 bits are required. To support conditional instructions, 2 bits are required for conditional multiplexer selection and 1 bit for the parity. These bits can be linearly placed or can share fields for branch and normal instructions. The sharing of fields saves on instruction width at the cost of adding a little complexity in decoding the instruction. In cases of sharing, 1 bit is required to distinguish between branching and normal ALU instructions.

The datapath and the micro-programmed state machine are shown in Figure 10.12. The two options for fields of the micro-code are shown in Figure 10.13. The machine uses the linear option for coding the micro-codes.

10.4 Subroutine Support

In many applications it is desired to repeat a set of micro-code instructions from different places in the micro-program. This can be effectively achieved by adding subroutine capability. Those micro-code instructions that are to be repeated can be placed in a subroutine that is called from any location in the program memory.

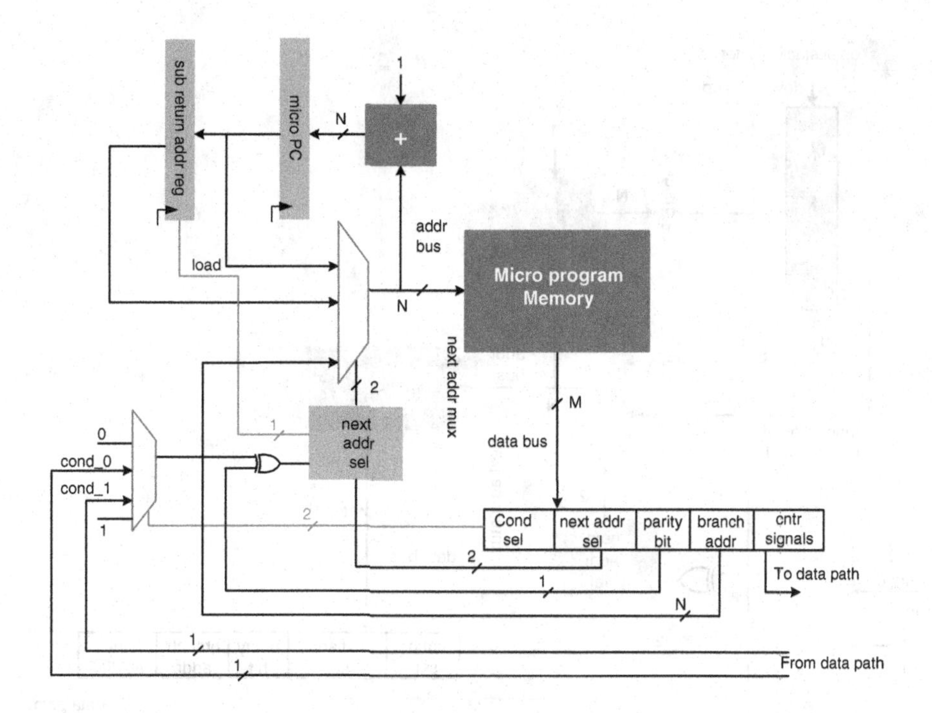

Figure 10.14 Micro-programmed state machine with subroutine support

The state machine, after executing the sequence of operations in the subroutine, needs to return to the next micro-code instruction. This requires storing of a return address in a register. This is easily accomplished in the cycle in which the call to the subroutine is made; `micro_PC` has the address of the next instruction, so this address is the subroutine return address. The state machine saves the contents of `micro_PC` in a special register called the *subroutine return address* (SRA) register. Figure 10.14 shows an example.

While executing the micro-code for the subroutine call, the next-address select logic generates a write enable signal to the SRA register. For returning from the subroutine, a return micro-code is added in the instruction-set of the state machine. While executing the return micro-code, the next-address select logic directs the next address multiplexer to select the address stored in the SRA register. The micro-code at the return address is read for execution. In the next clock cycle the micro-PC stores an incremented value of the next address, and a sequence of micro-codes is executed from there on.

10.5 Nested Subroutine Support

There are design instances where a call to a subroutine is made from an already called subroutine, as shown in Figure 10.15. This necessitates saving more than one return address in a last-in first-out (LIFO) stack. The depth of the stack depends on the level of nesting to be supported. A stack management controller maintains the top and bottom of the stack. To support J levels of nesting the LIFO stack needs J registers.

There are several options for designing the subroutine return address stack. A simple design of a LIFO is shown in Figure 10.16. Four registers are placed in the LIFO to support four nested subroutine calls. This LIFO is designed for a micro-program memory with a 10-bit address bus.

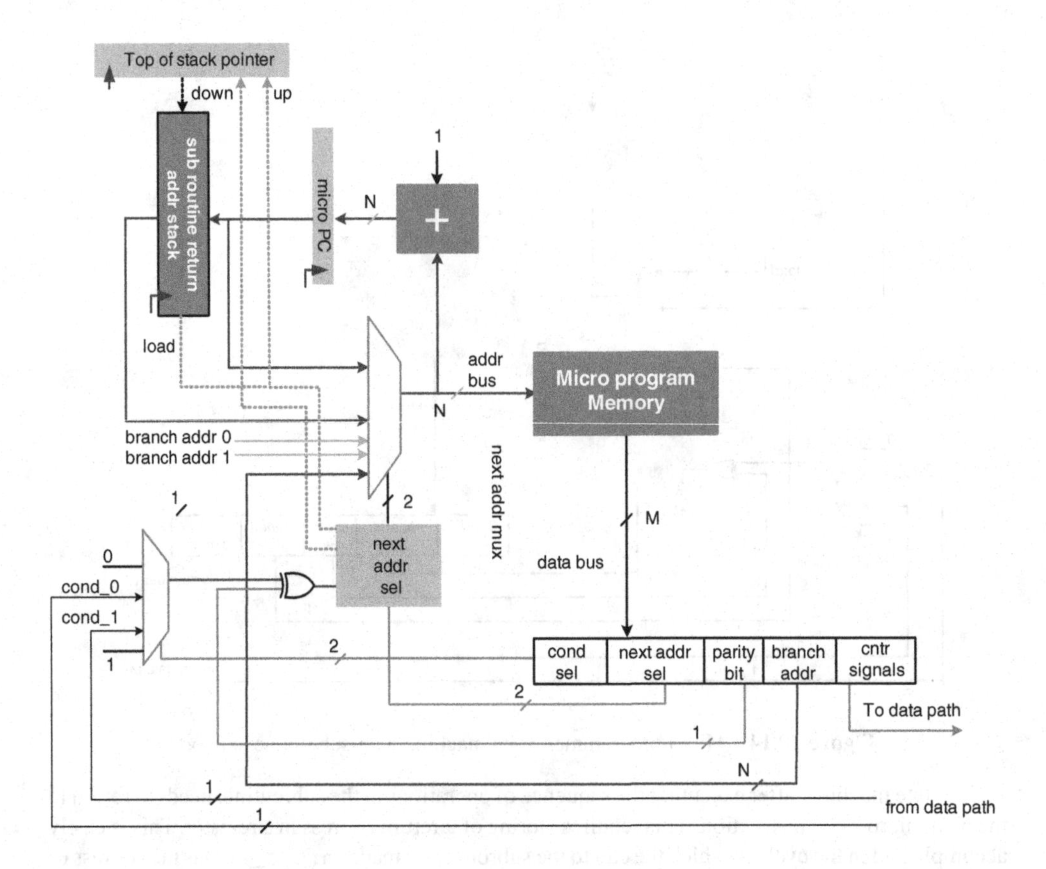

Figure 10.15 Micro-program state machine with nested subroutine support

A stack pointer register points to the last address in the LIFO with `read_lifo_addr` control signal. The signal is input to a 4:1 MUX and selects the last address stored in the LIFO and feeds it as one of the inputs to `next_address_mux`. To store the next return address for a nested subroutine call, `write_lifo_addr` control signal points to the next location in the LIFO. On a subroutine call, the `push` signal is asserted which enables the decoder, and the value in the PC is stored in the register pointed by `write_lifo_addr` and the value stored in the stack pointer is incremented by 1. On a return to subroutine call, the value stored in the stack pointer register is decremented by 1.

The design can be modified to check error conditions and to assert an error flag if a return from a subroutine without any call, or more than four nested subroutine calls, are made.

10.6 Nested Loop Support

Signal processing applications in general, and image and video processing applications in particular, require nested loop support. It is critical for the designer to minimize the loop maintenance overhead. This is associated with decrementing the loop counter, and checking where the loop does not expire and the program flow needs to be branched back to the loop start address. A micro-programmed state machine can be equipped to support the loop overhead in hardware. This zero-overhead loop support is especially of interest to multimedia controller designs that requires execution of regular repetitive loops [3].

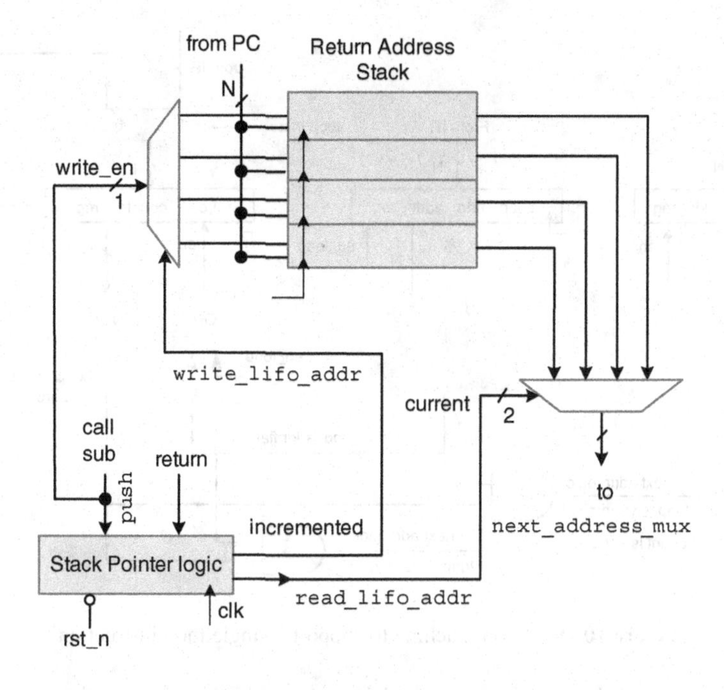

Figure 10.16 Subroutine return address stack

A representative instruction for a loop can be written as in Figure 10.17. The instruction provides the loop end-address and the count for the number of times the loop needs to be repeated. As counter runs down to zero therefore while specifying the loop count one must be subtracted from the actual count the loop needs to run. In the cycle in which the state machine executes the loop instruction, the PC stores the address of the next instruction that also is the start-address of the loop. In the example, the loop instruction is at address 80, PC contains the next address 81, and the loop instruction specifies the loop end-address and the loop counter as 100 and 20, respectively. The loop machine stores these three values in `loop_start_addr_reg`, `loop_end_addr_reg` and `loop_count_reg`, respectively. The value in the PC is continuously compared with the loop end-address. If the PC is equal to the loop end-address, the value in `loop_count_reg` is decremented by 1 and is checked whether the down count is equal to zero. If it is not, the next-address logic selects

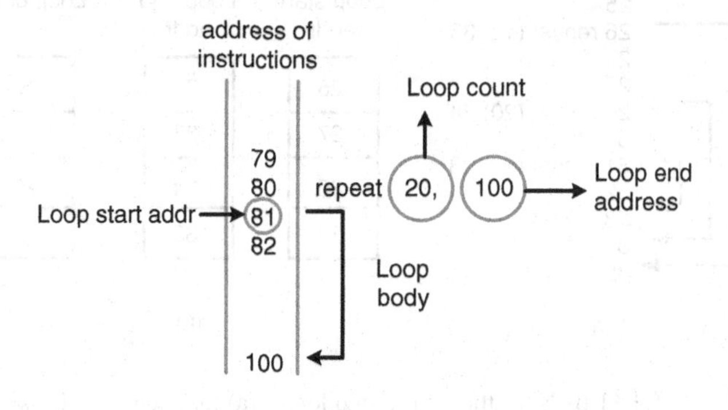

Figure 10.17 Representative instruction for a loop

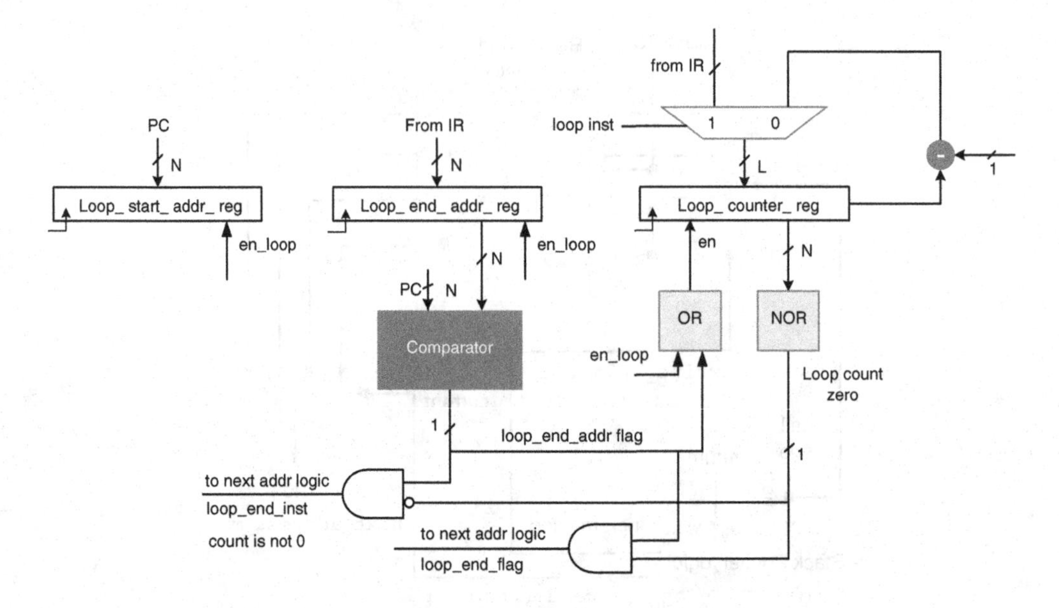

Figure 10.18 Loop machine to support a single loop instruction

the loop start-address on the PM address bus and the state machine starts executing the next iteration of the loop. When the PC gets to the last micro-code in the loop and the down count is zero, the next-address select logic starts executing the next instruction in the program. Figure 10.18 shows the loop machine for supporting a single loop instruction.

To provide nested loop support, each register in the loop machine is replaced with a LIFO stack with unified stack pointer logic. The stacks store the loops' start-addresses, end-addresses and counters for the nested loops and output the values for the currently executing loop to the loop control logic. When the current loop finishes, the logic selects the outer loop and in this way it iteratively executes nested loops.

An example of nested loops that correspondingly fill the LIFO registers is given in Figure 10.19. The loop machine supports four nested loops with three sets of LIFOs for managing loops start-

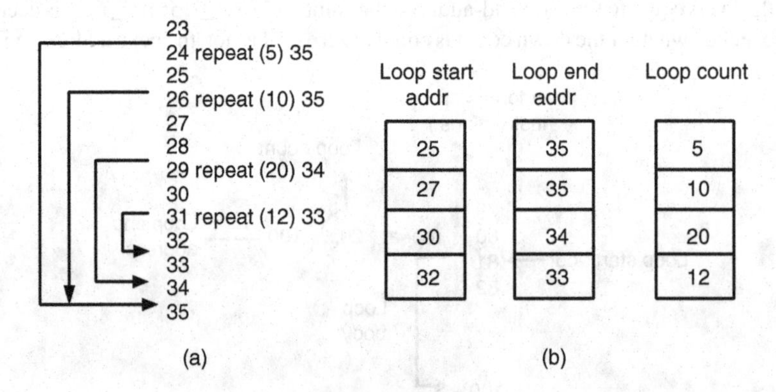

Figure 10.19 Filling of the LIFOs for the four nested loops. (a) Instruction addresses and nested loop instructions. (b) Corresponding filling of the LIFO registers

addresses, loops end-addresses and loops counters. The design of the machine is given in Figure 10.20.

All the three LIFOs in the loop machine work in lock step. The counter, when loaded with the loop count of the current loop instruction, decrements its value after executing the end instruction of the loop. The comparator sets the flag when the end of the loop is reached. Now based on the current value of the counter the two flags are assigned values. The `loop-end-flag` is set if the value of the counter is 0 and the machine is executing the last instruction of the loop. Similarly the `loop-end-instr-flag` is set if the processor is executing the end instruction but the counter value is still not 0. This signals the next-address logic to load the loop start-address as the next address to be executed. The state machine branches back to execute the loop again.

10.7 Examples

10.7.1 Design for Motion Estimation

Motion estimation is the most computationally intensive component of video compression algorithms and so is a good candidate for hardware mapping [4–6]. Although there are techniques for searching the target block in a pre-stored reference frame, a full-search algorithm gives the best result. This performs an exhaustive search for a target macro block in the entire reference frame to find the best match. The architecture designed for motion estimation using a full motion search algorithm can also be used for template matching in a machine vision application.

The algorithm works by dividing a video frame of size $N_h \times N_v$ into macro blocks of size $N \times M$. The motion estimation (ME) algorithm takes a target macro block in the current frame and searchers for its closest match in the previous reference frame. A full-search ME algorithm searches the macro block in the current frame with the macro block taking all candidate pixels of the previous frame. The algorithm computes the sum of absolute differences (SAD) as the matching criterion by implementing the expression:

$$SAD(i,j) = \sum_{k=0}^{N-1}\sum_{l=0}^{N-1} |S(k+i, l+j) - R(k,l)|$$

In this expression, R is the reference block in the current frame and S is the search area, which for the full-search algorithm is the entire previous frame while ignoring the boarder pixels. The ME algorithm computes the minimum value of SAD and stores the (i,j) as the point where the best match is found. The algorithm requires four summations over i, j, k and l.

Figure 10.21 shows an example. This design consists of a data movement block, two 2-D register files for storing the reference and the target blocks, and a controller that embeds a loop machine. The controller synchronizes the working of all blocks. The 2-D register file for the target block is extended in rows and columns making its size equal to $(N+1) \times (N+1) - 1$ registers, whereas the size of the register file for the reference block is $N \times N$.

A full-search algorithm executes by rows and, after completing calculation for the current macro block, the new target block is moved one pixel in a row. While working in a row, for each iteration of the algorithm, $N-1$ previous columns are reused. The data movement block brings the new column to the `extended-register-column` in the `ref-blk-reg-file`. The search algorithm works in a daisy-chain and, after completing one row of operation, the data movement block starts in the opposite direction and fetches a new row in the `extended-register-row` in the register file. The daisy-chain working of the algorithm for maximum reuse of data in a target block is illustrated in Figure 10.22.

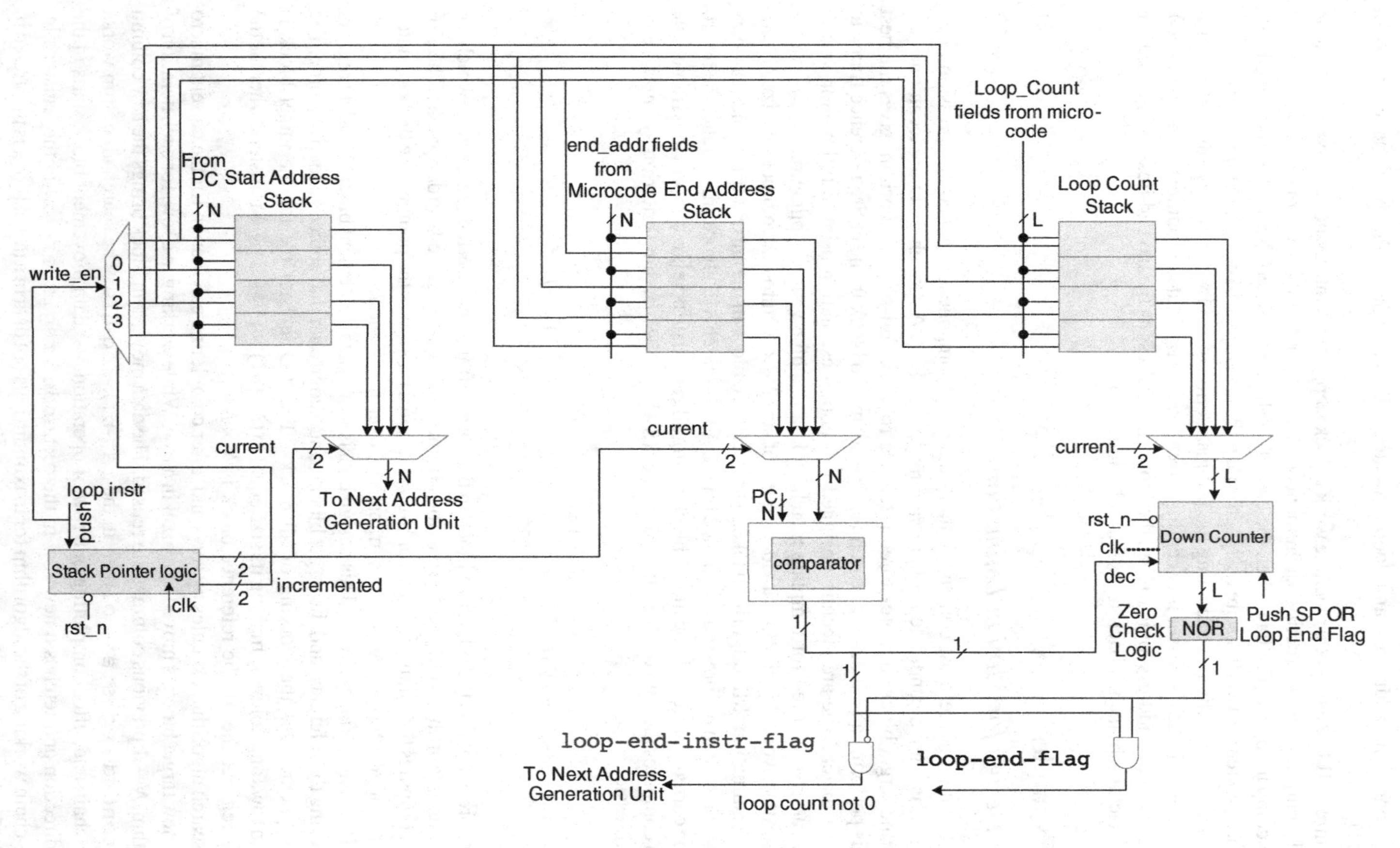

Figure 10.20 Loop machine with four-nested-loop support

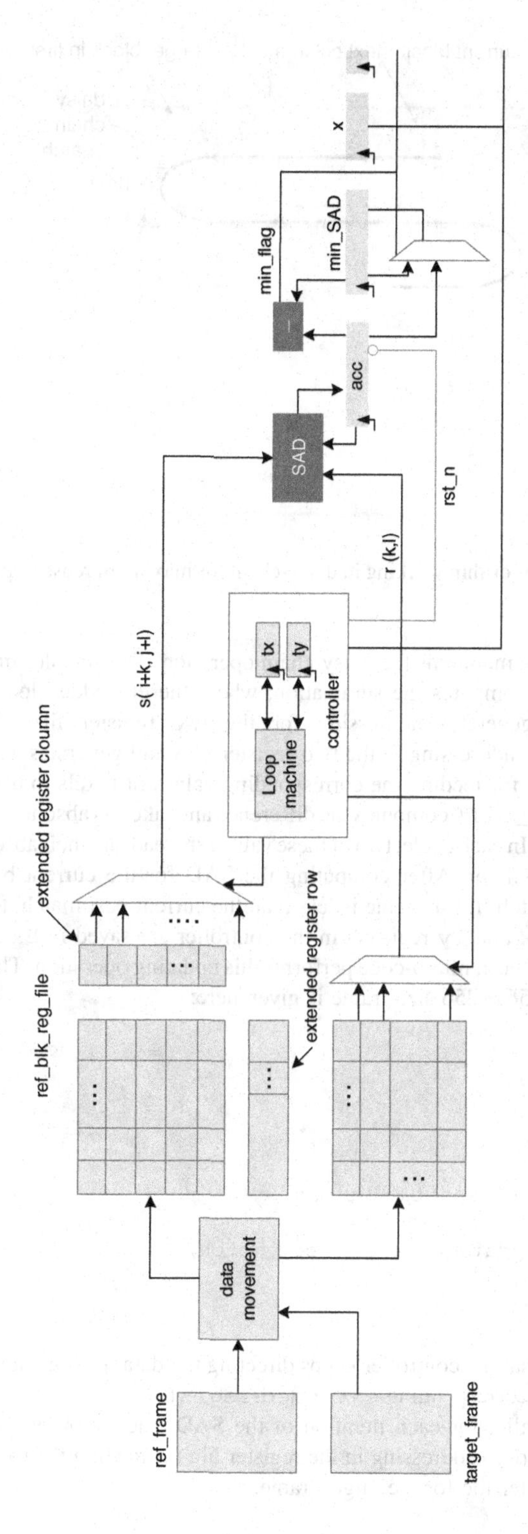

Figure 10.21 Micro-coded state machine design for full-search motion estimation algorithm

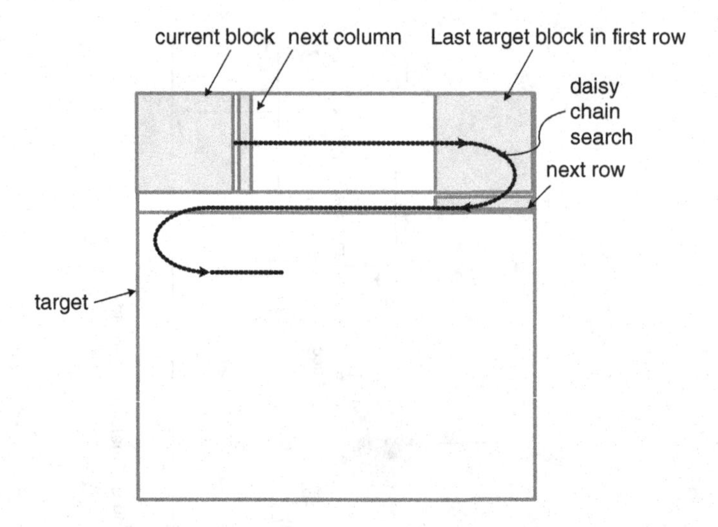

Figure 10.22 Full-search algorithm working in daisy-chain for maximum reuse of pixels for the target block

The `toggle` micro-code maintains the daisy-chain operation. This enables maximum reuse of data. The SAD calculator computes the summation, where the embedded loop machine in the controller automatically generates addressing for the two register files. The micro-code `repeat8x8` maintains the addressing in the two register files and generates appropriate control signals to two multiplexers for feeding the corresponding values of pixels to the SAD calculator block. The micro-code `abs_diff` computes the difference and takes its absolute value for addition into a running accumulator. In each cycle, two of these values are read and their absolute difference is added in a running accumulator. After computing the SAD for the current block, its value is compared with the best match. If this value is less than the current best match, the value with the corresponding values in `Tx` and `Ty` registers in the controller are saved in `Bx` and `By` registers, respectively. The `update_min` micro-code performs this updating operation. The micro-program for ME application for a 256 × 256 size frame is given here:

```
Tx = 0;
Ty = 0;
repeat 256-8, LABELy
  repeat 256-8, LABELx
  acc = 0;
  repeat_8x8 //embedded repeat 8x8 times
    acc+=abs_diff;
    (Bx, By, minSAD)=updata_min(acc, best_match, Tx, Ty)
LABELx toggle(Tx++/-);
LABELy Ty++;
```

This program assumes that the controller keeps directing the data movement block to bring the new row or column to the corresponding `extended-register-row` or `extended-register-column` files in parallel for each iteration of the SAD calculation and the `repeat8x8` appropriately performs modulo addressing in the register file for reading the pixels of the current target block from the register file for the target frame.

The architecture is designed in a way such that SAD calculation can be easily parallelized and the absolute difference for more than one value can be computed in one cycle.

10.7.2 Design of a Wavelet Processor

Another good example is the architecture of a wavelet processor. The processor performs wavelet transformations for image and video processing applications. There are various types of wavelet used [7–11], so efficient micro-program architectures to accelerate these applications are of special interest [12, 13]. A representative design is presented here for illustration.

A 2-D wavelet transform is implemented by two passes of a 1-D wavelet transform. The generic form of the 1-D wavelet transform passes a signal through a low-pass and a high-pass filter, $h[n]$ and $g[n]$, and then their respective outputs are down-sampled by a factor of 2. Figure 10.23(a) gives the 1-D decomposition of a 1-D signal $x[n]$ into low- and high-frequency components. For the 2-D wavelet transform, the first each column of the image is passed through a set of filters and the output is down-sampled; then in the second pass the rows of the resultant image are passed though the same set of filters. The pixel values of the output image are called the 'wavelet coefficients'. In a pyramid algorithm for computing wavelet coefficients for an image analysis application, an image is passed through multiple levels of 2-D wavelet transforms. Each level, after passing the image through two 1-D wavelet transforms, decimates the image by 2 in each dimension. Figure 10.23(b) shows 2-level 2-D wavelet decomposition in a pyramid algorithm. Similarly, Figures 2 10.23(c) and (d) shows the results of performing one-level and three-level 2-D wavelet transforms on an image.

There are a host of FIR filters designed for wavelet applications. The lifting scheme is an efficient tool for designing wavelets [14]. As a matter of fact, every wavelet transform with FIR filters can be

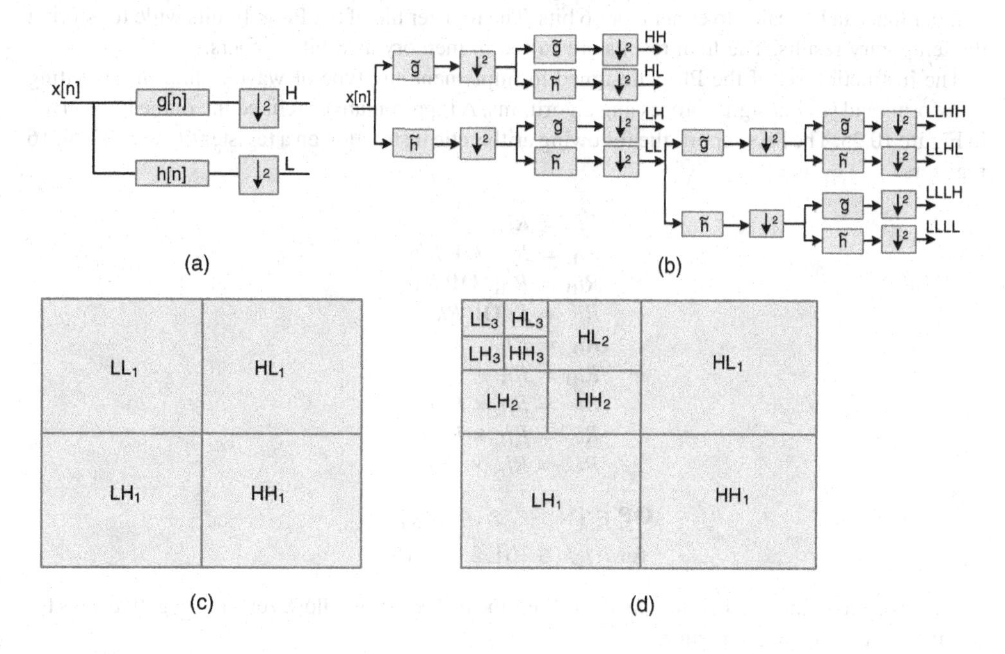

Figure 10.23 (a) One-dimensional wavelet decomposition. (b) Two-level 2-D pyramid decomposition. (c) One-level 2-D wavelet decomposition. (d) Three-level 2-D wavelet decomposition

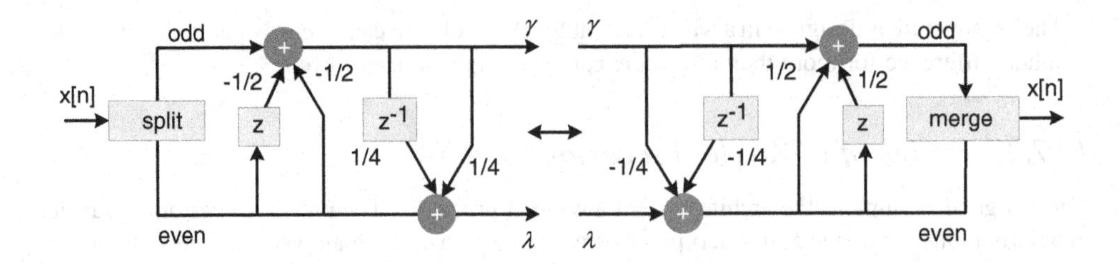

Figure 10.24 5/3 wavelet transform based on the lifting scheme

decomposed under the lifting scheme. Figure 10.24 shows an example of a 5/3 filter transformed as a lifting wavelet; this wavelet is good for image compression applications. Similarly, there is a family of lifting wavelets [12] that are more suitable for other applications, such as the 9/7 wavelet transform which is effective for image fusion applications (images of the same scene are taken from multiple cameras and then fused to one image for more accurate and reliable description) [14].

It is desired to design a programmable processor that can implement any such wavelet along with processing other signal processing functions. This requires the programmable processor to implement filters with different coefficients. For multimedia applications, multiple frames of images need to be processed every second. This requires several programmable processing elements (PEs) in architecture with shared and local memories placed in a configuration that enables scalability in inter-PE communications. This is especially so in video where multi-level pyramid-based wavelet transformations are performed on each frame. The coefficients of each lifting filter can be easily represented in 11 bits and an image pixel is represented in 8 bits. The multiplication results in a 19-bit number that can be scaled to either 8 or 16 bits. The register file of the PE is 16 bits wide for storing the temporary results. The final results are stored in memory as 8-bit numbers.

The instruction set of the PE is designed to implement any type of wavelet filtering as lifting transforms and similar signal processing algorithms. A representative architecture of the PE is given in Figure 10.25. The PE supports the following arithmetic instruction on a register file with 32-bit 16 registers:

$$
\begin{aligned}
Ri\ \ &= Rj \\
Ri_{\mathrm{L}} &= Rj_{\mathrm{L}}\ \ \mathbf{OP}\, Rk_{\mathrm{L}} \\
Ri_{\mathrm{H}} &= Rj_{\mathrm{H}}\ \ \mathbf{OP}\, Rk_{\mathrm{H}} \\
Ri\ \ &= Rj\, \mathbf{OP}\, Rk \\
Ri_{\mathrm{L}} &= Ri_{\mathrm{H}} \\
Ri_{\mathrm{H}} &= Ri_{\mathrm{L}} \\
Ri\ \ &= Rj_{\mathrm{L}} \times Rk_{\mathrm{L}} \\
Ri\ \ &= Rj_{\mathrm{L}} \times Rk_{\mathrm{H}} \\
Ri\ \ &= Rj_{\mathrm{H}} \times Rk_{\mathrm{H}}
\end{aligned}
$$

$$
\mathbf{OP} \in \{\times, +, \gg, \ll, \&, |, \wedge\}
$$
$$
\text{for } i,j,k \in \{01, 2, \ldots, 15\}
$$

Based on the arithmetic instruction, the PE sets the following overflow, zero and negative flags for use in the conditional instructions:

```
Flags: OV, Z,N
```

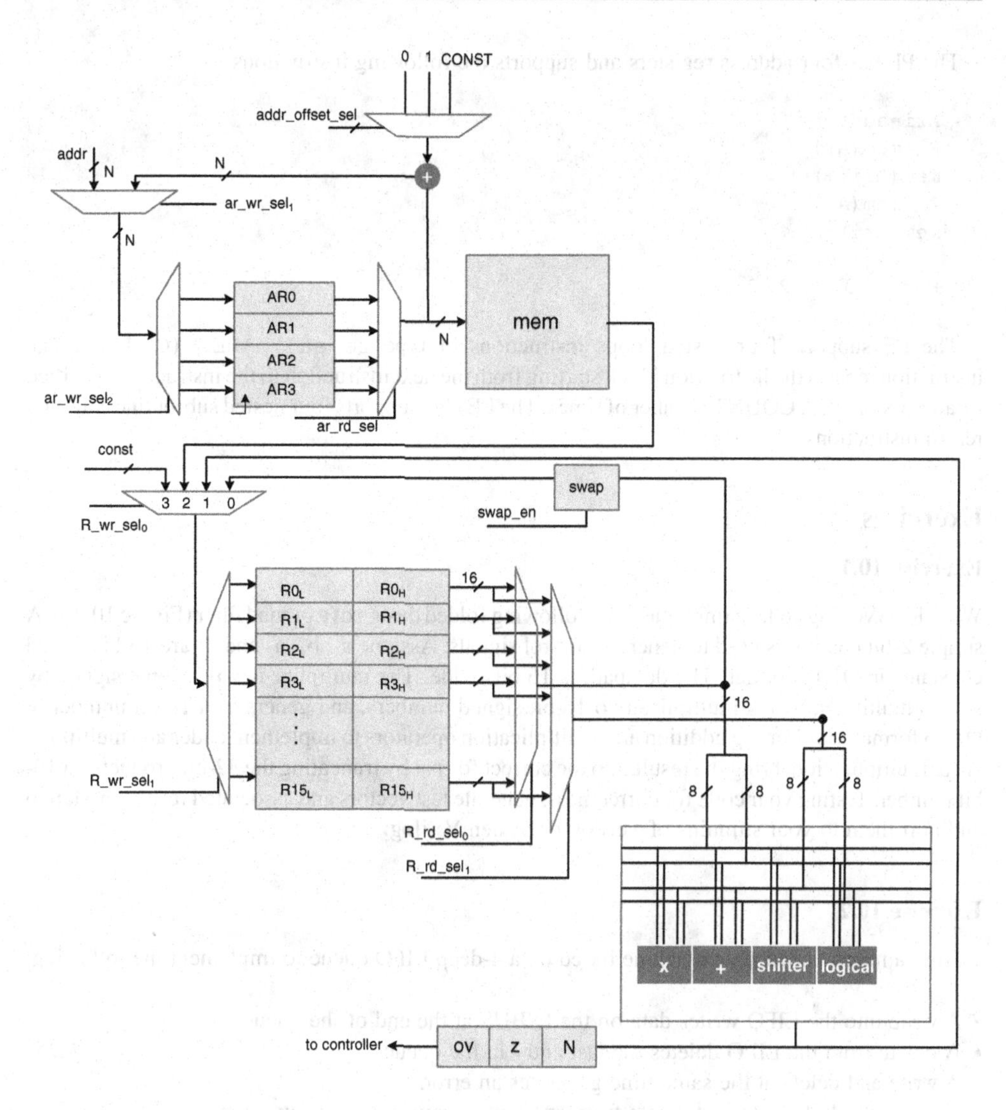

Figure 10.25 Datapath design for processing elements to compute a wavelet transform

The PE supports the following conditional and unconditional branch instructions:

```
if(Condi) jump Label,
else if(Condi) jump Label,
if(!Condi) jump Label,
else if (!Condi) jump Label,
else jump Label,
jump Label.
```

where `Condi` is one of the conditions 0V, Z and N, and `Label` is the address of the branch instruction.

The PE has four address registers and supports the following instructions:

```
Ari=addr
Ri=mem(Ari++)
mem(Ari++)=Rj
Ri=mem(Ari)
mem(Ari)=Rj
Ari=Ari+Const
For i = 0, 1, 2, 3
```

The PE supports four nested loops instructions of type `repeat LABEL (COUNT)`. The instruction repeats the instruction block starting from the next instruction to the instruction specified by address `LABEL`, COUNT number of times. The PE also supports four nested subroutine calls and return instructions.

Exercises

Exercise 10.1

Write RTL Verilog code to implement the following folded design of a biquad filter (Figure 10.26). A simple 2-bit counter is used to generate control signals. Assume a_1, b_1, a_2 and b_2 are 16-bit signed constants in Q1.15 format. The datapath is 16 bits wide. The multiplier is a fractional signed by signed multiplier, which multiplies two 16-bit signed numbers, and generates a 16-bit number in Q1.15 format. Use Verilog addition and multiplication operators to implement adder and multiplier. After multiplication, bring the result into the correct format by truncating the 32-bit product to a 16-bit number. Testing your code for correctness, generate test vectors and associated results in Matlab and port them to your stimulus of Verilog or SystemVerilog.

Exercise 10.2

Design a micro-coded state machine based on a 4-deep LIFO queue to implement the following:

- A write into the LIFO writes data on the INBUS at the end of the queue.
- A delete from the LIFO deletes the last entry in the queue.
- A write and delete at the same time generates an error.
- A write in a full queue and delete from an empty queue generates an error.
- The last entry to the queue is always available at the OUTBUS.

Draw the algorithmic state machine (ASM) diagram. Draw the datapath, and give the micro-coded memory contents implementing the state machine design.

Exercise 10.3

Design a micro-coded state machine-based FIFO queue with six 16-bit registers, R_0 to R_5. The FIFO has provision to write one or two values to the tail of the queue, and similarly it can delete one or two values from the head of the queue.

All the input and output signals of the FIFO are shown in Figure 10.27. A `WRITE_1` writes the values on `INBUS_1` at the tail of the queue in the R_0 register, and a `WRITE_2` writes values on

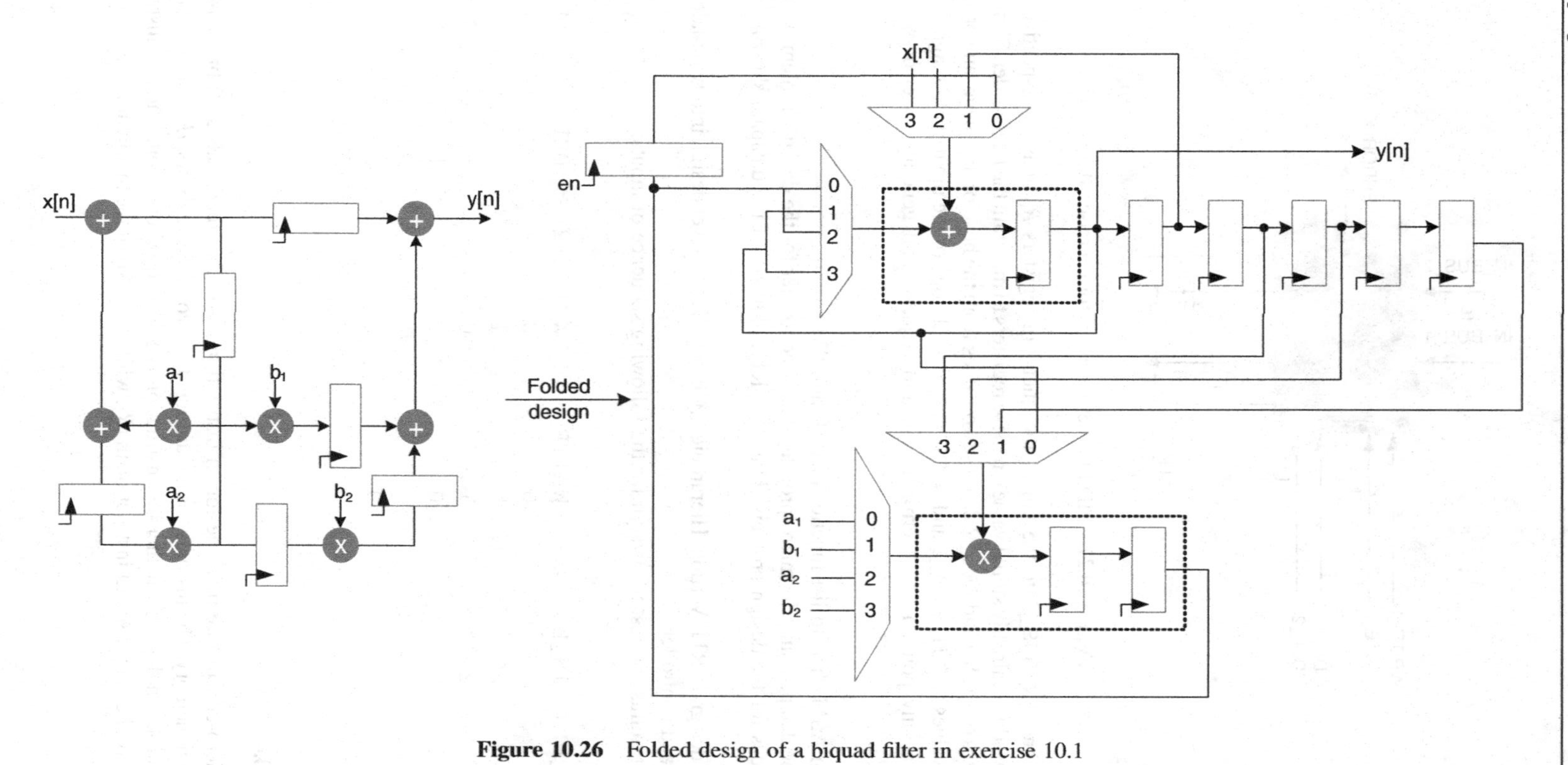

Figure 10.26 Folded design of a biquad filter in exercise 10.1

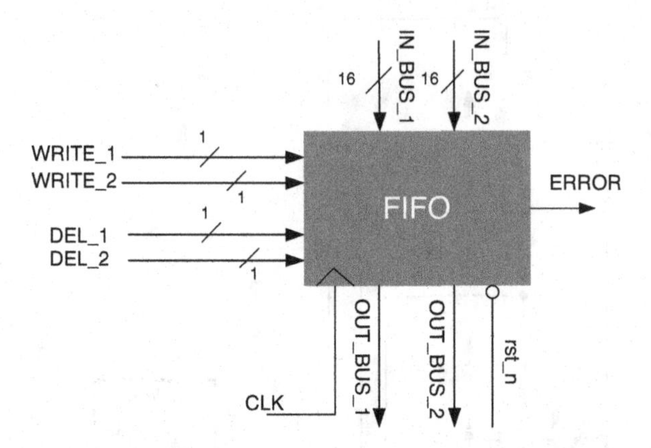

Figure 10.27 FIFO top-level design for exercise 10.3

IN_BUS_1 and IN_BUS_2 to the tail of the queue in registers R_0 and R_1, respectively, and appropriately adjusts already stored values in these registers to maintain the FIFO order. Similarly, a DEL_1 deletes one value and DEL_2 deletes two values from the head of the queue, and brings the next first-in values to OUT_BUS_1 and OUT_BUS_2. Only one of the input signals signal should be asserted at any point. List all erroneous/invalid requests and generate an error signal for these requests.

1. Design an ASM chart to list functionality of the FSM.
2. Design the datapath and the micro-program-based controller for the design problem. Write the micro-codes for the design problem. Draw the RTL diagram of the datapath showing all the signals.
3. Code the design in RTL Verilog. Hierarchically break the top-level design into two main parts, datapath and controller.
4. Write a stimulus and test the design for the following sequence of inputs:

IN_BUS_1	IN_BUS_2	WRITE_1	WRITE_2	DEL_1	DEL_2
0	0	0	0	1	0
10	20	0	1	0	0
30	29	1	0	0	0
-	-	0	0	1	0
10		-	1	00	0
-	-	0	0	0	1
-	-	0	1	1	0

Exercise 10.4

Design a micro-coded state machine for an instruction dispatcher that reads a 32-bit word from the instruction memory. The instructions are arranged in two 32-bit registers, IR_0 and IR_1. The first bit of the instruction depicts whether the instruction is 8-bit, 16-bit or 32-bit. The two instruction registers always keep the current instruction in IR_1, while the LSB of the instruction is moved to the LSB of IR_1.

Exercise 10.5

Design a micro-coded state machine to implement a bit-serial processor with the following micro-codes:

$$\{C_{out}, R_3[\mathrm{k}]\} = R_1[i] + R_2[j] + C_{in}$$
$$R_3[\mathrm{k}] = R_1[i]\wedge R_2[j]$$
$$R_3[\mathrm{k}] = R_1[i] \| R_2[j]$$
$$R_3[\mathrm{k}] = R_1[i]\&R_2[j]$$

```
Repeat (count) END_ADDRESS
Call Subroutine at ADDERSS
Return
```

where R_1, R_2 and R_3 are 8-bit registers, and indexes $i, j, k = 0, \ldots, 7$ specify a particular bit of these registers.

Exercise 10.6

Design a micro-coded state machine that operates on a register file with eight 8-bit registers. The machine executes the following instructions:

$$R_m = R_i \pm R_j \mp R_k$$
$$R_m = R_i \pm R_j$$
$$R_m = \pm R_i$$
$$R_m = R_i \times R_j \mp R_k$$
$$R_m = R_i \times R_j$$

```
Call Subroutine ADDRESS
Return
```

All multiplications are fractional multiplications that multiply two Q-format 8-bit signed numbers and produce an 8-bit signed number. In the register file, $i, j, k = 0, \ldots, 7$, and correspondingly R_i, R_j, R_k and R_m are any one of the eight registers.

Exercise 10.7

Design a micro-program state machine that supports the following instruction set:

$$R_4 = R_1 + R_2 + R_3 + R_4$$
$$R_4 = R_1 + R_2 + R_3$$
$$R_4 = \mathrm{R}_1 + R_2$$
$$R_4 = R_1 \,\&\, R_2 \,\&\, R_3 \,\&\, R_4$$
$$R_4 = R_1 \,\&\, R_2 \,\&\, R_3$$
$$R_4 = R_1 \,\&\, R_2$$
$$R_4 = R_1 | R_2$$
$$R_4 = R_1 \sim R_2$$

```
Load R5=ADDRESS
Jump R5
```

```
If (R4 == 0 ) jump R5
If (R4 > 0 ) jump R5
Call subroutine R5 and return support
```

R_1 to R_5 are 8-bit registers. A branch address or subroutine address can be written in R_5 using a load instruction. As given in the instruction set, jump and subroutine instructions are developed to use the contents of R_5 as address of the next instruction. The conditions are based on the content of R_4. Draw the RTL diagram of the datapath and micro-programmed state machine. Show all the control signals. Specify the instruction format and size. The datapath should be optimal using only one carry propagate adder (CPA) by appropriately using a compression block to compress the number of operands to two.

Exercise 10.8

Design a micro-coded state-machine based design, with the following features.

1. There are three register files, *P*, *Q* and *R*. Each register file has four 8-bit registers.
2. There is an adder and logic unit to perform addition, subtraction, AND, OR and XOR operations.
3. There is 264 deep micro-code memory.
4. The state machine supports the following instructions:
   (a) Load constant into *i*th register of P or Q:
   $P_i = \text{const}$
   $Q_i = \text{const}$
   (b) Perform any arithmetic or logic operation given in (2) on operands in register files *P* and *Q*, and store the result in register file R:
   $R_i = P_j \text{ op } Q_k$
   (c) Unconditional branch to any address LABEL in microcode memory:
   goto LABEL
   (d) Subroutine call support, where subroutine is stored at address LABEL:
   call subroutine LABEL
   return
   (e) Support repeat instruction, for repeating a block of code COUNT times:
   repeat (COUNT) LABEL
   LABEL is the address of the last instruction in the loop.

Exercise 10.9

Design a micro-programmed state machine to support 4-deep nested loops and conditional jump instructions with the requirements given in (1) and (2), respectively. Draw an RTL diagram clearly specifying sizes of instruction register, multiplexers, comparators, micro-PC register, and all other registers in your design.

1. The loop machine should support conditional loop instructions of the format:
   `if (COND`*i*`)` repeat END_LABEL COUNT
   The micro-code checks COND $i = 0,1,2,3$ and, if it is TRUE, the state machine repeats COUNT times a block of instructions starting from the next instruction to the instruction labeled with END_LABEL. Assume the size of PM is 256 words and a 6-bit counter that stores the

COUNT. Also assume that, in the instruction, 10 bits are kept for the datapath-related instruction. If CONDi is FALSE, the code skips the loop and executes the instruction after the instruction addressed by END_LABEL.

2. Add the following conditional support instruction in the instruction set:

```
if (CONDi ) Jump LABEL
if (!CONDi) Jump LABEL
jump LABEL
```

Exercise 10.10

Design a micro-program architecture with two 8-bit input registers R_1 and R_2, an 8-bit output register R_3, and a 256-word deep 8-bit memory *MEM*. Design the datapath and controller to support the following micro instructions. Give op-code of each instruction, draw the datapath, and draw the micro-coded state machine clearly specifying all control signals and their widths.

- `Load R1 Addr` loads the content of memory at address `Addr` into register R_1.
- `Load R2 Addr` loads the content of memory at address `Addr` into register R_2.
- `Store R3 Addr` stores the content of register R_3 into memory at address `Addr`.
- Arithmetic and logic instructions Add, Mult, Sub, OR and AND take operands from registers R_1 and R_2 and store the result in register R_3.
- The following conditional and unconditional branch instructions: `Jump Label`, `Jump Label if (R1==0)`, `Jump Label if (R2==0)`, `Jump Label if (R3==0)`.
- The following conditional and unconditional subroutine and return instructions: `Call Label`, `Call Label if (R1 > 0)`, `Call Label if (R2 > 0)`, `Call Label if (R3 > 0)`, `Return`, `Return if (R3<0)`, `Return if (R3==0)`.

Exercise 10.11

Using an instruction set of the wavelet transform processor, write a subroutine to implement 5/3 wavelet transform of Figure 10.24. Use this subroutine to implement a pyramid coding algorithm to decompose an image in three levels.

References

1. M. Froidevaux and J. M. Gentit, "MPEG1 and MPEG2 system layer implementation trade-off between micro-coded and FSM architecture," in *Proceedings of IEEE International Conference on Consumer Electronics*, 1995, pp. 250–251.
2. M. A. Lynch, *Microprogrammed State Machine Design*, 1993, CRC Press.
3. N. Kavvadias and S. Nikolaidis, "Zero-overhead loop controller that implements multimedia algorithms," *IEEE Proceedings on Computers and Digital Technology*, 2005, vol. 152, pp 517–526.
4. M. A. Daigneault, J. M. P. Langlois and J. P. David, "Application-specific instruction set processor specialized for block motion estimation," in *Proceedings of IEEE International Conference on Computer Design*, 2008, pp. 266–271.
5. T.-C. Chen, S.-Y. Chien, Y.-W. Huang, C.-H. Tsai and C.-Y. Chen, "Fast algorithm and architecture design of low-power integer motion estimation for H.264/AVC," *IEEE Transactions on Circuits and Systems for Video Technology*, 2007, pp. 568–577.
6. C.-C. Cheng, Y. J. Wang and T.-S. Chang, "A fast algorithm and its VLSI architecture for fractional motion estimation for H.264/MPEG-4 AVC video coding," *IEEE Transactions on Circuits and Systems for Video Technology*, 2007, pp. 578–583.

7. O. Rioul and M. Vetterli, "Wavelets and signal processing," *IEEE Signal Processing Magazine*, 1991, vol. 8, pp. 14–38.
8. I. Daubechies, *Ten Lectures on Wavelets,* 1992, Society of Industrial and Applied Mathematics (SIAM).
9. S. Mallat, *A Wavelet Tour of Signal Processing*, 1998, Elsevier Publishers, UK.
10. S. Burrus, R. Gopinath and H. Guo, *Introduction to Wavelets and Wavelet Transforms: a Primer,* 1997, Prentice-Hall.
11. M. Vetterli and J. Kovacevic, *Wavelets and Sub-band Coding*, 1995, Prentice-Hall.
12. S.-W. Lee and S.-C. Lim, "VLSI design of a wavelet processing core," *IEEE Transactions on Circuits and Systems for Video Technology*, 2006, vol. 16, pp. 1350–1361.
13. H. Olkkonen, J. T. Olkkonen and P. Pesola, "Efficient lifting wavelet transform for microprocessor and VLSI applications," *IEEE Signal Processing Letters*, 2005, vol. 12, pp. 120–122.
14. W. Sweldens, "The lifting scheme: a custom-design construction of biorthogonal wavelets," *Applied and Computational Harmonic Analysis Journal*, 1996, vol. 3, pp. 186–200.

11

Micro-programmed Adaptive Filtering Applications

11.1 Introduction

Deterministic filters are implemented in linear and time-invariant systems to remove out-of-band noise or unwanted signals. When the unwanted signals are known in terms of their frequency band, a system can be designed that does not require any adaptation in real time. In contrast, there are many scenarios where the system cannot be deterministically determined and is also time-variant. Then the system is designed as a linear time-invariant (LTI) filter in real time and, to cater for time variance, the filter has to update the coefficients periodically using an adaptive algorithm. Such an algorithm uses some error-minimization criterion whereby the output is compared with the desired result to compute an error signal. The algorithm adapts or modifies the coefficients of the filter such that the adapted system generates an output signal that *converges* to the desired signal [1, 2].

There are many techniques for adaptation. The selection of an algorithm for a particular application is based on many factors, such as complexity, convergence time, robustness to noise, ability to track rapid variations, and so on. The algorithm structure for effective implementation in hardware is another major design consideration.

11.2 Adaptive Filter Configurations

Adaptive filters are used in many settings, some of which are outlined in this section.

11.2.1 System Identification

The same input is applied to an unknown system $U(z)$ and to an adaptive filter $H(z)$. If perfectly identified, the output $y[n]$ of the adaptive filter should be the same as the output $d[n]$ of the unknown system. The adaptive algorithm first computes the error in the identification as $e[n] = d[n] - y[n]$. The algorithm then updates the filter coefficients such that the error is minimized and is zero in the ideal situation. Figure 11.1 shows an adaptive filter in system identification configuration.

Digital Design of Signal Processing Systems: A Practical Approach, First Edition. Shoab Ahmed Khan.

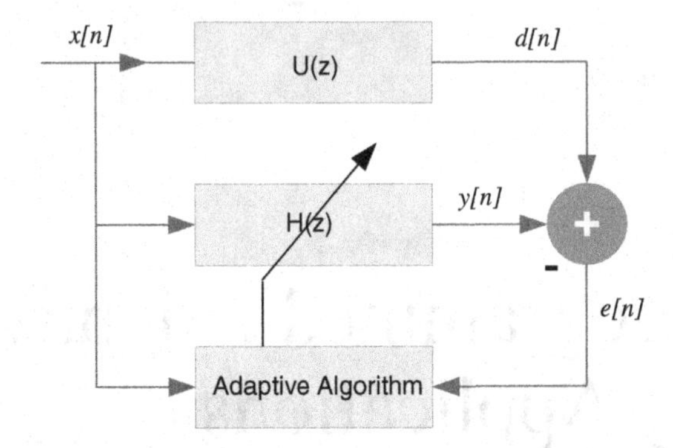

Figure 11.1 Adaptive filter in system identification configuration

11.2.2 Inverse System Modeling

The adaptive filter $H(z)$ is placed in cascade with the unknown system $U(z)$. A known signal $d[n]$ is periodically input to $U(z)$ and the output of the unknown system is fed as input to the adaptive filter. The adaptive filter generates $y[n]$ as output. The filter, in the ideal situation, should cancel the effect of $U(z)$ such that $H(z)U(z) = 1$. In this case the output perfectly matches with a delayed version of the input signal. Any mismatch between the two signals is considered as error $e[n]$. An adaptive algorithm computes the coefficients such that this error is minimized. This configuration of an adaptive filter is shown in Figure 11.2.

This configuration is also used to provide equalization in a digital communication receiver. For example, in mobile communication the signal transmitted from an antenna undergoes fading and multi-path effects on its way to a receiver. This is compensated by placing an equalizer at the receiver. A know input is periodically sent by the transmitter. The receiver implements an adaptive algorithm that updates the coefficients of the adaptive filter to cancel the channel effects.

11.2.3 Acoustic Noise Cancellation

An acoustic noise canceller removes noise from a speech signal. A reference microphone also picks up the same noise which is correlated with the noise that gets added in the speech signal. An adaptive filter cancels the effects of the noise from noisy speech. The signal $v_0[n]$ captures the noise from the noise source. The noise in the signal $v_1[n]$ is correlated with $v_0[n]$, and the filter $H(z)$ is adapted to

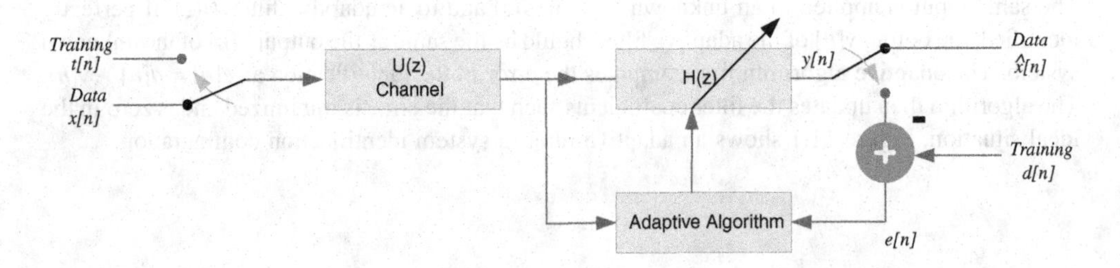

Figure 11.2 Adaptive filter in inverse system modeling

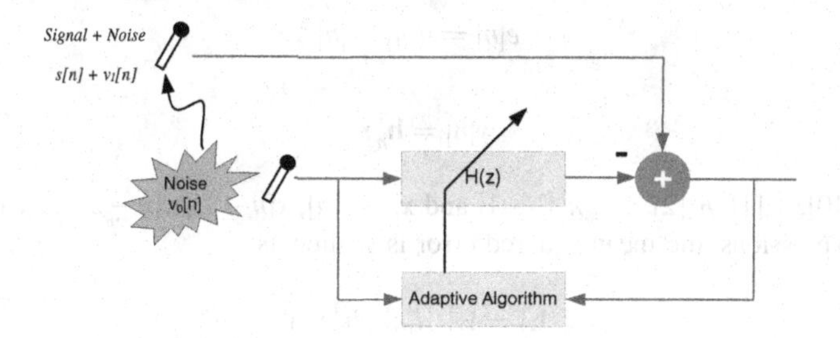

Figure 11.3 Acoustic noise canceller

produce an estimate of $v_1[n]$, which is then cancelled from the signal $x[n]$, where $x[n] = s[n] + v_1[n]$ and $s[n]$ is the signal of interest [3, 4]. This arrangement is shown in Figure 11.3.

11.2.4 Linear Prediction

Here an adaptive filter is used to predict the next input sample. The adaptive algorithm minimizes the error between the predicted and the next sample values by adapting the filter to the process that produces the input samples.

An adaptive filter in linear prediction configuration is shown in Figure 11.4. Based on N previous values of the input samples, $x[n-1], x[n-2], \ldots, x[n-N]$, the filter predicts the value of the next sample $x[n]$ as $\hat{x}[n]$. The error in prediction $e[n] = x[n] - \hat{x}[n]$ is fed to an adaptive algorithm to modify the coefficients such that this error is minimized.

11.3 Adaptive Algorithms

11.3.1 Basics

An ideal adaptive algorithm computes coefficients of the filter by minimizing an error criterion $\xi(\mathbf{h})$. The criterion used in many applications is the expectation of the square of the error or mean squared error:

$$\xi(\mathbf{h}) = \mathrm{E}[e[n]^2]$$

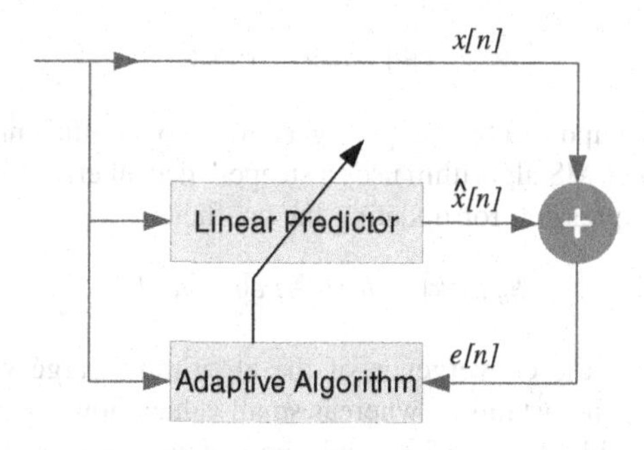

Figure 11.4 Adaptive filter as linear predictor

$$e[n] = d[n] - y[n]$$

$$y[n] = \mathbf{h}_n^{\mathrm{T}} \mathbf{x}_n$$

where $\mathbf{h}_n = h_n[0], h_n[1], h_n[2], \ldots, h_n[N-1]$ and $\mathbf{x}_n = x[n], x[n-1], x[n-2], \ldots, x[n-(N-1)]$.
Using these expressions, the mean squared error is written as:

$$\xi(\mathbf{h}) = \mathrm{E}\left[(d[n] - \mathbf{h}_n^{\mathrm{T}} x_n)^2\right]$$

Expanding this expression results in:

$$\xi(\mathbf{h}) = \mathrm{E}[d[n]^2] + \mathbf{h}_n^{\mathrm{T}} \mathrm{E}[\mathbf{x}_n \mathbf{x}_n^{\mathrm{T}}] \mathbf{h}_n - 2\mathbf{h}_n^{\mathrm{T}} \mathrm{E}[d[n] \mathbf{x}_n]$$

The error is a quadratic function of the values of the coefficients of the adaptive filter and results in a hyperboloid [1]. The minimum of the hyperboloid can be found by taking the derivative ofţ $\xi(\mathbf{h})$ with respect to the values of the coefficients. This results in:

$$\mathbf{h}_{\mathrm{opt}} = \arg\left(\min_h \xi(\mathbf{h})\right) = \mathbf{R}_x^{-1} \mathbf{p}$$

where $\mathbf{R}_x = \mathrm{E}[\mathbf{x}_n \mathbf{x}_n^{\mathrm{T}}]$ and $\mathbf{p} = \mathrm{E}[d[n] \mathbf{x}_n]$.

Computing the inverse of a matrix is computationally very expensive and is usually avoided. There are a host of adaptive algorithms that recursively minimize the error signal. Some of these algorithms are LMS, NLMS, RLS and the Kalman filter [1, 2].

The next sections present Least Mean Square (LMS) algorithms and a micro-programmed processor designed around algorithms.

11.3.2 Least Mean Square (LMS) Algorithm

This is one of the most widely used algorithms for adaptive filtering. The algorithm first computes the output from the adaptive filter using the current coefficients from convolution summation of:

$$y[n] = \sum_{k=0}^{L-1} \boldsymbol{h}_n[k] x[n-k] \tag{11.1}$$

The algorithm then computes the error using:

$$e[n] = d[n] - y[n] \tag{11.2}$$

The coefficients are then updated by computing a new set of coefficients that are used to process the next input sample. The LMS algorithm uses a steepest-decent error minimization criterion that results in the following expression for updating the coefficients [1]:

$$\boldsymbol{h}_{n+1}[k] = \boldsymbol{h}_n[k] - \mu e[n] x[n-k] \tag{11.3}$$

The factor μ determines the convergence of the algorithm. Large values of μ result in fast convergence but may result in instability, whereas small values slow down the convergence. There are algorithms that use variable step size for μ and compute its value based on some criterion for fast conversion [5, 6].

The algorithm for every input sample first computes the output $y[n]$ using the current set of coefficients $h_n[k]$, and then updates all the coefficients using (11.3).

11.3.3 Normalized LMS Algorithm

The main issue with the LMS algorithm is its dependency on scaling of $x[n]$ by a constant factor μ. An NLMS algorithm solves this problem by normalizing this factor with the power of signal to noise ratio (SNR) of the input signal. The equation for NLMS is written as:

$$\boldsymbol{h}_{n+1}[k] = \boldsymbol{h}_n[k] - \frac{\mu e[n]x[n-k]}{\boldsymbol{x}_n\boldsymbol{x}_n^T} \tag{11.4}$$

The step size is controlled by the power factor $\mathbf{x}_n\mathbf{x}_n^T$. If this factor is small, the step size is set large for fast convergence; if it is large, the step size is set to a small value [5, 6].

11.3.4 Block LMS

Here, the coefficients of the adaptive filter are changed only once for every block of input data, compared with the simple LMS that updates on a sample-by-sample basis. A block or frame of data is used for the adaptation of coefficients [7]. Let N be the block size, and $\mathbf{h}_{n-N}$ is the array of coefficients computed for the previous block of input data. The mathematical representation of the block LMS is presented here.

First the output samples $y[n]$ and the error signal $e[n]$ are computed using $\mathbf{h}_{n-N}$ for the current block i of data for $l=0, \ldots, N-1$:

$$y[n] = \sum_{m=0}^{L-1} h_{n-N}[m]x[n-m], \quad \text{for } n = iN + l$$

$$e[n] = d[n] - y[n]$$

Using the error signal and the block of data, the array of coefficients for the current block of input data is updated as:

$$\mathbf{h}_n = \mathbf{h}_{n-N} + \mu \sum_{m=iN}^{iN+N-1} e[m]x[m]$$

In the presence of noise, a leaky LMS algorithm is sometimes preferred for updating the coefficients. For $0<\alpha<1$, the expression for the algorithm is:

$$\mathbf{h}_n = (1-\alpha)\mathbf{h}_{n-N} + \mu \sum_{m=iN}^{iN+N-1} e[m]x[m]$$

11.4 Channel Equalizer using NLMS

11.4.1 Theory

A digital communication receiver needs to cater for the impairments introduced by the channel. The channel acts as $U(z)$ and to neutralize the effect of the channel an adaptive filter $H(z)$ is placed that works in cascade with the unknown channel and the coefficients of the filter are adapted to cancel the channel effects. One of the most challenging problems in wireless communication is to device algorithms and techniques that mitigate the impairments caused by multi-path fading effects of a wireless channel. These paths result in multiple and time-delayed echoes of the transmitted signal at the receiver. These delay spreads are up to tens of microseconds and result in inter-symbol interference spreading over many data symbols. To cater for the multi-path effects, either multi-carrier techniques like OFDM are used, or single sarrier (SC) techniques with more sophisticated time and frequency equalizations are performed.

A transmitted signal in a typical wireless communication system can be represented as:

$$s(t) = \mathrm{Re}\{s_b(t)e^{j2\pi f_c t}$$

where $s_b(t)$ and $s(t)$ are baseband and passband signals, respectively, and f_c is the carrier frequency. This transmitted signal goes through multiple propagation paths and experiences different delays $\tau_n(t)$ and attenuation $\alpha_n(t)$ while propagating to a receiver antenna. The received signal at the antenna is modeled as:

$$r(t) = \sum_n \alpha_n(t)s(t-\tau_n(t))$$

There are different channel models that statistically study the variability of the multi-path effects in a mobile environment. Rayleigh and Ricean fading distributions [10–13] are generally used for channel modeling. A time-varying tap delay-line system can model most of the propagation environment. The model assumes L multiple paths and expresses the channel as:

$$h_i[n] = \sum_{l=0}^{L-1} h_l\delta[n-\tau_l]$$

where h_l and τ_l are the complex-valued path gain and the path delay of the lth path, respectively. There are a total of L paths that are modeled for a particular channel environment.

The time-domain equalizer implements a transversal filter. The length of the filter should be greater than the maximum delay that spreads over many symbols. In many applications this amounts to hundreds of coefficients.

11.4.2 Example: NLMS Algorithm to Update Coefficients

This example demonstrates the use of an NLMS algorithm to update the coefficients of an equalizer. Each frame of communication appends a training sequence. The sequence is designed such that it helps in determining the start of a burst at the receiver. The sequence can also help in computing frequency offset and an initial channel estimation. This channel estimation can also be used as the initial guess for the equalizer. The receiver updates the equalizer coefficients and the rest of the data is passed through the updated equalizer filter. In many applications a blind equalizer is also used for generating the error signal from the transmitted data. This equalizer type compares the received symbol with closest ideal symbol in the transmitted constellation. This error is then used for updating the coefficients of the equalizer.

In this example, a transmitter periodically sends a training sequence $t[n]$, with $n = 0 \ldots N_{t-1}$, for the receiver to adapt the equalizer coefficients using the NLMS algorithm. The transmitter then appends the data $d[n]$ to form a frame. The frame consisting of data and the training is modulated using QAM modulation technique. The modulated signal is transmitted. The signal goes through a channel and is received at the receiver. The algorithm first computes the start of the frame. A training sequence is selected that helps in establishing the start of the burst [12]. After marking the start of the frame, the received training sequence helps in adapting the filter coefficients. The adaption is continued in blind mode on transmitted data. In this mode, the algorithm keeps updating the equalizer coefficients using knowledge of the transmitted signal. For example, in QAM the received symbol is mapped to the closest point of the constellation and the Euclidian distance of the received point from the closest point in the constellation is used to derive the adaptive algorithm.

The error for the training part of the signal is computed as:

$$e[n] = ||t[n]-t_r[n]||_2$$

where $t[n]$ and $t_r[n]$ are the transmitted and received training symbol, respectively. For the rest of the data the equalizer works in blind mode and the error in this case is the distance between the received and the closest symbols $d[n]$ and $d_r[n]$, resepectively:

$$e[n] = ||d[n]-d_r[n]||_2$$

The MATLAB® code implementing an equalizer using an NLMS algorithm is given here:

```
clc
close all
clear all

% Setting the simulation pParameters
NO_OF_FRAMES = 20;   %Blocks of data sequence to be sent
CONST = 3;   % Modulation type
SNR = 30;   % Signal to noise ratio
L = 2;   % Selection of a training sequence
DATA_SIZE = 256;   % Size of data to be sent in every frame

% Setting the channel parameters
MAX_SPREAD = 10;   % Maximum intersymbol interference channel can cause
tau_l = [1 2 5 7 ];   % Multipath delay locations
gain_l =[1 0.1 0.2 0.02];   % Corresponding gain

% Equalizer parameters
mue = 0.4;

% Channel %
% Creating channel impulse response
hl = zeros(1, MAX_SPREAD-1);
for ch = 1:length(tau_l)
    hl(tau_l(ch)) = gain_l(ch);
end

% Constellation table %
if CONST == 1
   Const_Table = [(-1 + 0*j) (1 + 0*j)];
elseif CONST == 2
```

```
   Const_Table = [(-.7071 - .7071*j) (-.7071 + .7071*j)
                  (.7071 - .7071*j) (.7071 + .7071*j)];
elseif ( CONST == 3)
   Const_Table = [(1 + 0j) (.7071 + .7071i) (-.7071 + .7071i)
                  (0 + i) (-1 + 0i) (-.7071 - .7071i) ...
                  (.7071 - .7071i) (0 - 1i)];
elseif(CONST == 4)
     % 16 QAM constellation table
     Const_Table = [(1 + j) (3 + j) (1 + 3j) (3 + 3j) (1 - j)
                  (1 - 3j) (3 - j) (3 - 3j) ...
                  (-1 + j) (-1 + 3j) (-3 + j) (-3 + 3j)
                  (-1 -j) (-3 - j) (-1 - 3j) (-3 - 3j)];
end

% Generating training from [12] %
% Training data is modulated using BPSK
if L==2
     C = [1 1 -1 1 1 1 1 -1 1 1 -1 1 -1 -1 -1 1];
elseif L == 4
     C = [1 1 -1 1 1 1 1 -1];
elseif L == 8
     C = [1 1 -1 1];
elseif L == 16
     C = [1 1];
end

if L==2
     Sign_Pattern = [1 1];
elseif L == 4
     Sign_Pattern = [-1 1 -1 -1];
elseif L == 8
     Sign_Pattern = [1 1 -1 -1 1 -1 -1 -1];
elseif L == 16
     Sign_Pattern = [1 -1 -1 1 1 1 -1 -1 1 -1 1 1 -1 1 -1 -1];
end

Training = [];
for i=1:L
     Training = [Training Sign_Pattern(i)*C];
end
Length_Training = length(Training);
Length_Frame = Length_Training + DATA_SIZE;

% Modulated frame generation %
Frame_Blocks = [];
Data_Blocks = [];
for n = 1:NO_OF_FRAMES
     Data = (randn(1,DATA_SIZE*CONST)>0);
     % Bits to symbols
     DataSymbols = reshape(Data,CONST,length(Data)/CONST);
     % Symbols to constellation table indices
     Table_Indices = 2.^(CONST-1:-1:0) * DataSymbols + 1;
```

```
    % Indices to constellation pPoint
    Block = Const_Table(1,Table_Indices);
    Data_Blocks = [Data_Blocks Block];
    % Frame = training + modulated data
    Frame_Blocks = [Frame_Blocks Training Block];
end

% Passing the signal through the channel %
Received_Signal = filter(hl,1, Frame_Blocks);

% Adding AWGN noise %
Received_Signal = awgn(Received_Signal, SNR, 'measured');

% Equalizer design at the reciver %
Length_Rx_Signal = length(Received_Signal);
No_Frame = fix(Length_Rx_Signal/Length_Frame);
hn=zeros(MAX_SPREAD,1);
Detected_Blocks = [];
for frame=0:No_Frame-1

% Using training to update filter coefficients using NLMS algorithm %
start_training_index = frame*Length_Frame+1;
end_training_index = start_training_index+Length_Training-1;
Training_Data =
Received_Signal(start_training_index:end_training_index);
for i=MAX_SPREAD:Length_Training
    xn=Training_Data(i:-1:i-MAX_SPREAD+1);
    y(i)=xn*hn;
    d(i)=Training(i);
    e(i)=d(i)-y(i);
    hn = hn + mue*e(i)*xn'/(norm(xn)^2);
end

start_data_index = end_training_index+1;
end_data_index = start_data_index+DATA_SIZE-1;
Received_Data_Frame =
Received_Signal(start_data_index:end_data_index);

% Using the updated coefficients to equalize the received signal %
Equalized_Signal = filter(hn,1,Received_Data_Frame);
for i=1:DATA_SIZE
    % Slicer for decision making
    % Finding const point with min euclidean distance
    [Dist Decimal] = min(abs(Const_Table - Equalized_Signal(i)));
    Detected_Symbols(i) = Const_Table(Decimal);
end
scatterplot(Received_Data_Frame), title('Unqualized Received Signal');
scatterplot(Equalized_Signal), title('Equalized Signal');
Detected_Blocks = [Detected_Blocks Detected_Symbols];
end
[Detected_Blocks' Data_Blocks'];
```

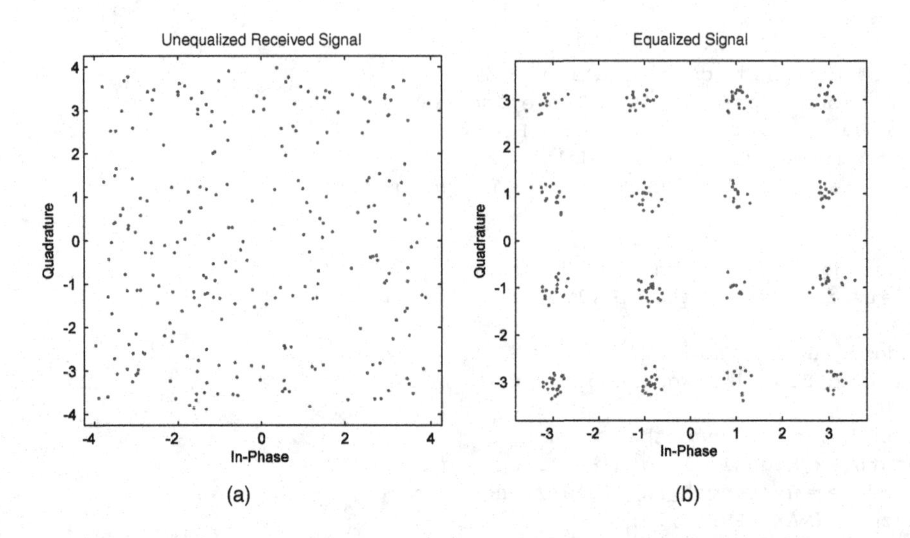

Figure 11.5 Equalizer example. (a) Unequalized received signal. (b) Equalized QAM signal at the receiver

Figure 11.5 shows a comparison between an unequalized QAM signal and an equalized signal using the NLMS algorithm.

Usually divisions in the algorithm are avoided while selecting algorithms for implementation in hardware. For the case of NLMS, The Newton Repson method can be used for computing $1/\mathbf{x}_n\mathbf{x}_n^{\mathrm{T}}$. Modified MATLAB® code of the NLMS loop of the implementation is given here:

```
for i=MAX_SPREAD:Length_Training
xn=Training_Data(i:-1:i-MAX_SPREAD+1);
y(i)=xn*hn;
d(i)=Training(i);
e(i)=d(i)-y(i);
Energy=(norm(xn)^2);
% Find a good initial estimate
if Energy > 10
    invEnergy = 0.01;
elseif Energy > 1
    invEnergy = 0.1;
else
    invEnergy = 1;
end

% Compute inverse of energy using Newton Raphson method
for iteration = 1:5
    invEnergy = invEnergy*(2-Energy*invEnergy);
end
    hn = hn + mue*e(i)*xn'*invEnergy;
end
```

To avoid division a simple LMS algorithm can be used. For the LMS algorithm, the equation for updating the coefficients need not perform any normalization and the MATLAB® code for adaptation is given here:

```
for i=MAX_SPREAD:Length_Training
    xn=Training_Data(i:-1:i-MAX_SPREAD+1);
    y(i)=xn*hn;
    d(i)=Training(i);
    e(i)=d(i)-y(i);
    hn = hn + mue*e(i)*xn';
end
```

11.5 Echo Canceller

An echo canceller is an application where an echo of a signal that mixes with another signal is removed from that signal. There are two types of canceller, acoustic and line echo.

11.5.1 Acoustic Echo Canceller

In a speaker phone or hands-free mobile operation, the far-end speaker voice is played on a speaker. Multiple and delayed echoes of this voice are reflected from the physical surroundings and are picked up by the microphone along with the voice of the near-end speaker. Usually these echo paths are long enough to cause an unpleasant perception to the far-end speaker. An acoustic echo canceller is used to cancel the effects. A model of the echo cancellation problem is shown in Figure 11.6.

Considering all digital signals, let $u[n]$ be the far-end voice signal. The echo signal is $r[n]$ and it gets added in to the near-end voice signal $x[n]$. An echo canceller is an adaptive filter $h[n]$ that ideally takes $u[n]$ and generates a replica of the echo signal, $\hat{r}[n]$. This signal is cancelled from the received signal to get an echo-free signal, $\hat{x}[n]$, where:

$$\hat{x}[n] = x[n] + r[n] - u[n] * \boldsymbol{h}[n]$$
$$\hat{x}[n] = x[n] + r[n] - \hat{r}[n].$$

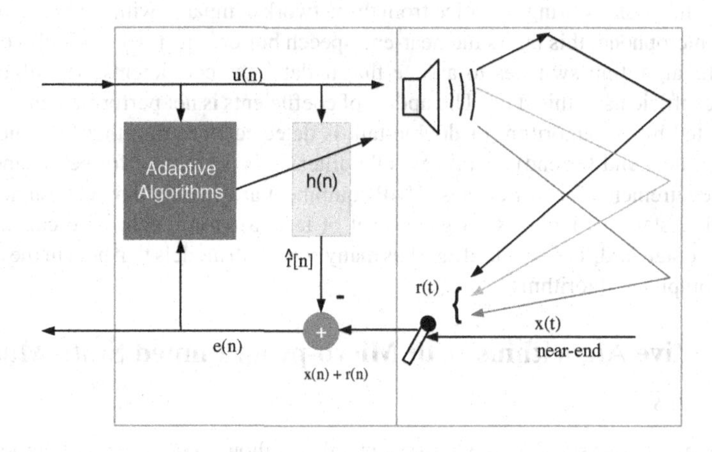

Figure 11.6 Model of an acoustic echo canceller

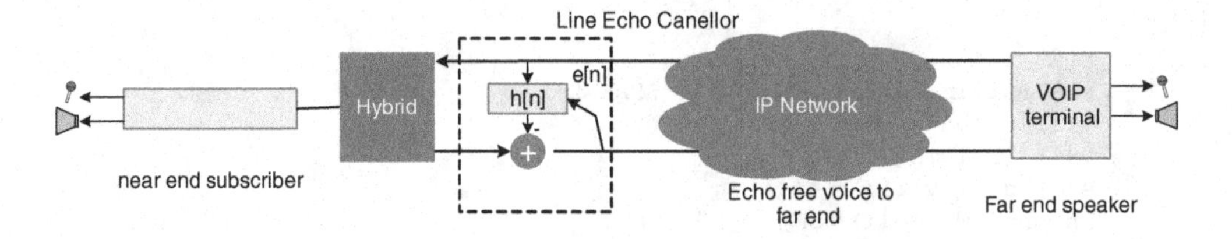

Figure 11.7 Line echo due to impedance mismatch in the 4:2 hybrids at the central office and its cancellation in Media Gateway

To cancel a worst echo path of 500 milliseconds for a voice sampled at 8 kHz requires an adaptive filter with 4000 coefficients. There are several techniques to minimize the computational complexity of an echo canceller [13–15].

11.5.2 Line Echo Cancellation (LEC)

Historically, most of the local loops in telephony use two wires to connect a subscriber to the central office (CO). A hybrid is used at the CO to decouple the headphone and microphone signals to four wires. A hybrid is a balanced transformer and cannot be perfectly matched owing to several physical characteristics of the two wires from the subscriber's home to the CO. Any mismatch in the transformer results in echo of the far-end signal going back with the near-end voice. The phenomenon is shown in Figure 11.7. The perception of echo depends on the amount of round-trip delay in the network.

The echo is more apparent in VoIP as there are some inherent delays in processing the compressed voice signal that are accumulated over buffering and queuing delays at every hop of the IP network. In a VoIP network the delays can build up to 120–150 ms. This makes the round trip delay around 300 ms. It is therefore very critical for VoIP applications to cancel out any echo coming from the far end [16].

A line echo cancellation algorithm detects the state in which only far-end voice is present and the near-end is silent. The algorithm implements a voice-activity detection technique to check voice activity at both ends. This requires calculating energies on the voice received from the network and the voice acquired in a buffer from the microphone of the near end speaker. If there is significantly more energy in the buffer storing the voice from the network compared with the buffer of voice from the near-end microphone, this infers the near-end speech buffer is just stori the echo of the far-end speech and the algorithm switches to a state that updates the coefficients. The algorithm keeps updating the coefficients in this state. The update of coefficients is not performed once the near-end voice is detected by the algorithm or double-talk is detected (meaning that the voice activity is present at both near and far end). In all cases the filtering is performed for echo cancellation.

The LEC requirements for carrier-class VoIP equipment are governed by ITU standards such as G.168 [17]. The standard includes a rigorous set of tests to ensure echo-free call if the LEC is compliant to the standard. The standard also has many echo path models that help in the development of standard compliant algorithms.

11.6 Adaptive Algorithms with Micro-programmed State Machines

11.6.1 Basics

It is evident from the discussion so far in this chapter that, although coefficient adaptation is the main component of any adaptive filtering algorithm, several auxiliary computations are required for complete implementation of the associated application. For example, for echo cancelling the algorithm needs first to detect double-talk to find out whether coefficients should be updated or not. The algorithms for double-talk detection are of a very different nature [18–20] than filtering or coefficient adaptation. In addition the application also requires decision logic to implement one or the other part of the algorithm. All these requirements favor a micro-programmed accelerator to implement. This also augments the requisite flexibility of modifying the algorithm at the implementation stage of the design. The algorithm can also be modified and recoded at an any time in the life cycle of the product.

For LEC-type applications, the computationally intensive part is still the filtering and coefficient adaption as the filter length is quite large. The accelerator design is optimized to perform these operations and the programmability helps in incorporating the auxiliary algorithms like voice activity detection and state machine coding without adding any additional hardware. To illustrate this methodology, the remainder of this section gives a detailed design of one such application.

11.6.2 Example: LEC Micro-coded Accelerator

A micro-coded accelerator is designed to implement adaptive filter applications and specifically to perform LEC on multiple channels in a carrier-class VoIP gateway. The processor is primarily optimized to execute a time-domain LMS-based adaptive LEC algorithm on a number of channels of speech signals. The filter length for each of the channels is programmable and can be extended to 512 taps or more. As the sampling rate of speech is 8 kHz, these taps correspond to (512/8) ms of echo tail length.

11.6.2.1 Top-level Design

The accelerator consists of a datapath and a controller. The datapath has two MAC blocks capable of performing four MAC operations every cycle, a logic unit, a barrel shifter, two address generation units (AGUs) and two blocks of dual-ported data memories (DMs). The MAC block can also be used as two independent multipliers and two adders. The datapath has two sets of register files and a few application-specific registers for maximizing reuse of data samples for convolution operation.

The controller has a program memory (PM), instruction decoder (ID), a subroutine module that supports four nested subroutine calls, and a loop machine that provides zero overhead support for four nested loops. The accelerator also has access to an on-chip DMA module that fills in the data from external memory to DMs. All these features of the controller are standard capabilities that can be designed and coded once and then reused in any design based on a micro-coded state machine.

11.6.2.2 Datapath/Registers

The most intensive part of the algorithm is to perform convolution with a 512-coefficient FIR filter. The coefficients are also updated when voice activity is detected on the far-end speech signal. To perform these operations effectively, the datapath has two MAC blocks with two multipliers and one adder in each block. These blocks support application-specific instructions of filtering and adaptation of coefficients. The two adders of the MAC blocks can also be independently used as general-purpose adders or subtractors. The accelerator also supports logic instruction such as AND, OR and EXOR of two operands. The engine has two register files to store the coefficients and input data. Figure 11.8 shows the complete datapath of the accelerator.

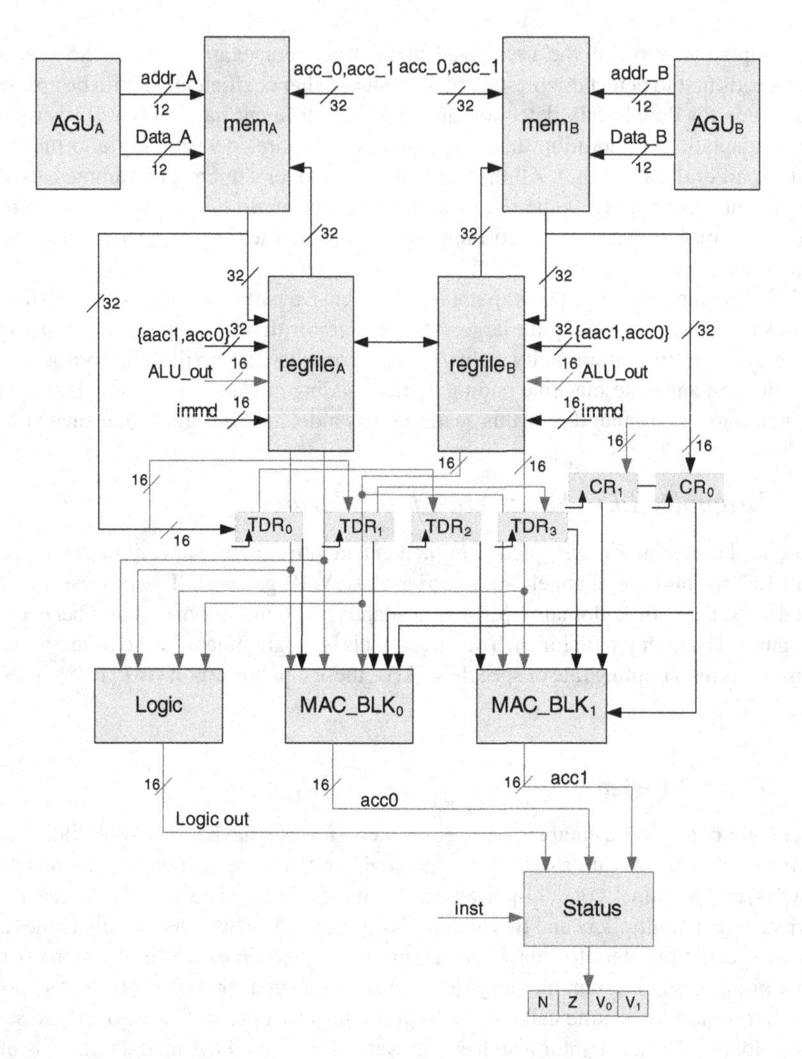

Figure 11.8 Datapath of the multi-channel line echo canceller

Register Files

The datapath has two 16-bit register files, A and B, with 8 registers each. Each register file has one read port to read 32-bit aligned data from two adjacent registers R_i and R_{i+1} ($i=0, 2, 4, 6$), where $R \in \{A, B\}$, or 16-bit data from memory in any of the registers in the file. The 32-bit aligned data from memory *mem_R*[31:0] can be written in the register files at two consecutive registers. A cross-path enables copying of the contents of two aligned registers from one register file to the other. The accelerator also supports writing a 16-bit result from ALU operations to any register, or ACC_0 and ACC_1 in two consecutive registers. A 16-bit immediate value can also be written in any one of the registers. A register transfer level (RTL) design of one of the register files is given in Figure 11.9.

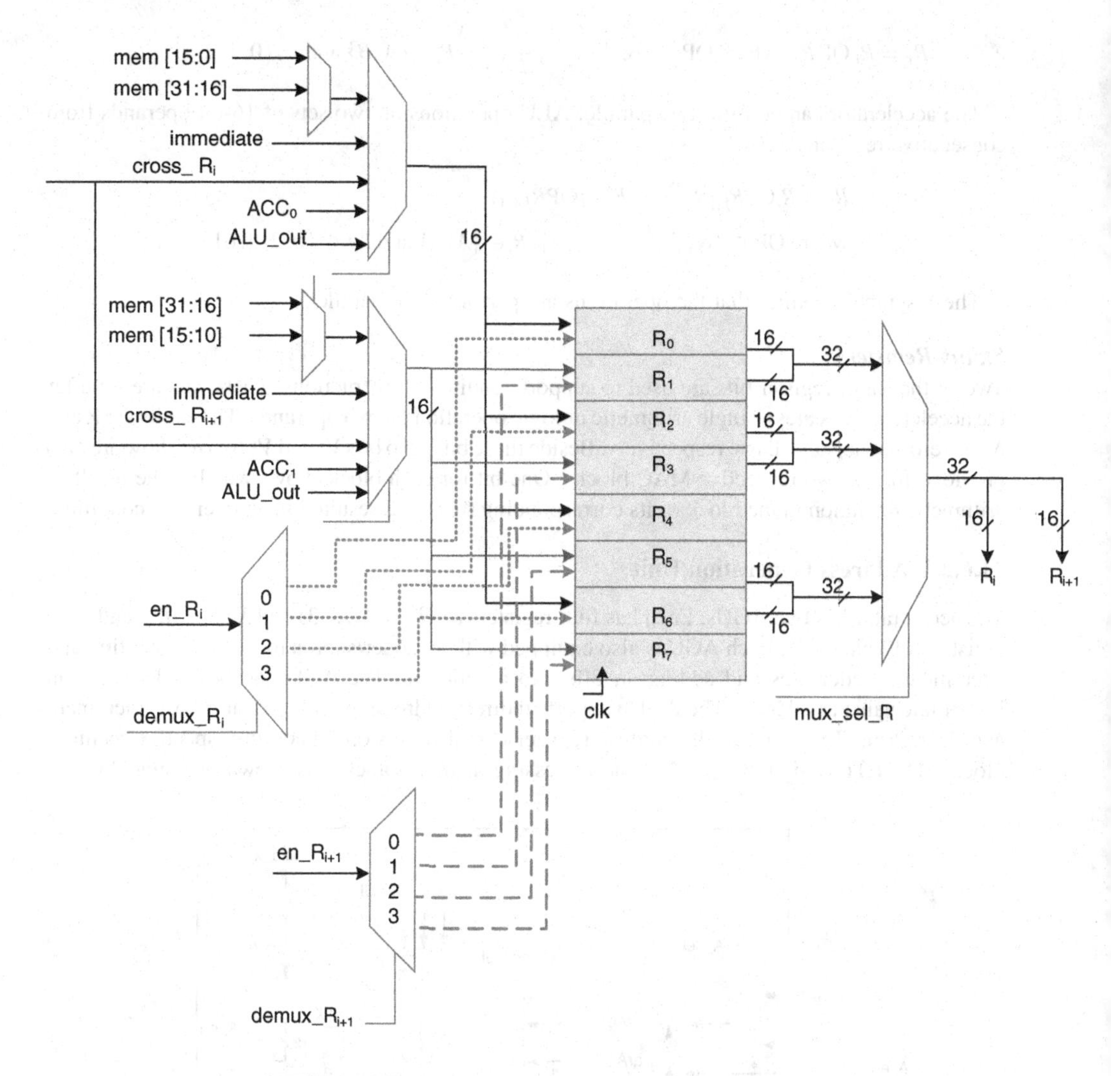

Figure 11.9 RTL design of one of the register files

Special Registers

The datapath has two sets of special registers for application-specific instructions. Four 16-bit registers are provided to support tap-delay line operation of convolution. The data is loaded from *mem_A* to tap-delay line registers TDR_0 and TDR_1, while the contents of these registers are shifted to TDR_2 and TDR_3 in the same cycle. Similarly to store the coefficients, two coefficient registers CR_0 and CR_1 are provided that are used to store the values of coefficients from *mem_B*. These registers are shown in Figure 11.8.

Arithmetic and Logic Operations

The accelerator can perform logic operations of AND, OR and EXOR and arithmetic operations of $\times$, $+$ and $-$ on two 16-bit operands stored in any register files:

$$R_i = R_i \, \mathrm{OP} \, R_j, \text{ where OP} \in \{\&, |, \wedge, \times, +, -\}, R \in \{A, B\} \text{ and } i,j\{0,1,2,\ldots,7\}$$

The accelerator can perform two parallel ALU operations on two sets of 16-bit operands from consecutive registers:

$$R_i = R_i \mathrm{OP} R_j || R_{i+1} = R_{i+1} \mathrm{OP} R_{j+1},$$
$$\text{where OP} \in \{\&, |, \wedge, \times, +, -\}, R \in \{A, B\} \text{ and } i,j \in \{0,2,4,6\}.$$

The || symbol signifies that the operations are performed in parallel.

Status Register

Two of the status register bits are used to support conditional instructions. These bits are set after the accelerator executes a single arithmetic or logic operation on two operands. These bits are Z and N for zero and negative flags, respectively. Beside these bits, two bits V_0 and V_1 for overflow are also provided for the two respective MAC blocks. One bit for each block shows whether the result of arithmetic operation in the block or its corresponding ACC_i has resulted in an overflow condition.

11.6.2.3 Address Generation Unit

The accelerator has two AGUs. Each has four registers $arXi$ ($i = 0 \ldots 3$) and $X \in \{A, B\}$, and each register is 12 bits wide. Each AGU is also equipped with an adder/subtractor for incrementing and decrementing addresses and adding an offset to an address. The AGU can be loaded with an immediate value as address. The AGU supports indirect addressing with pre- and post- increment and decrement. The value of the address register can also be stored back in respective memory blocks. The RTL design of the AGU for addressing memory block A is shown in Figure 11.10.

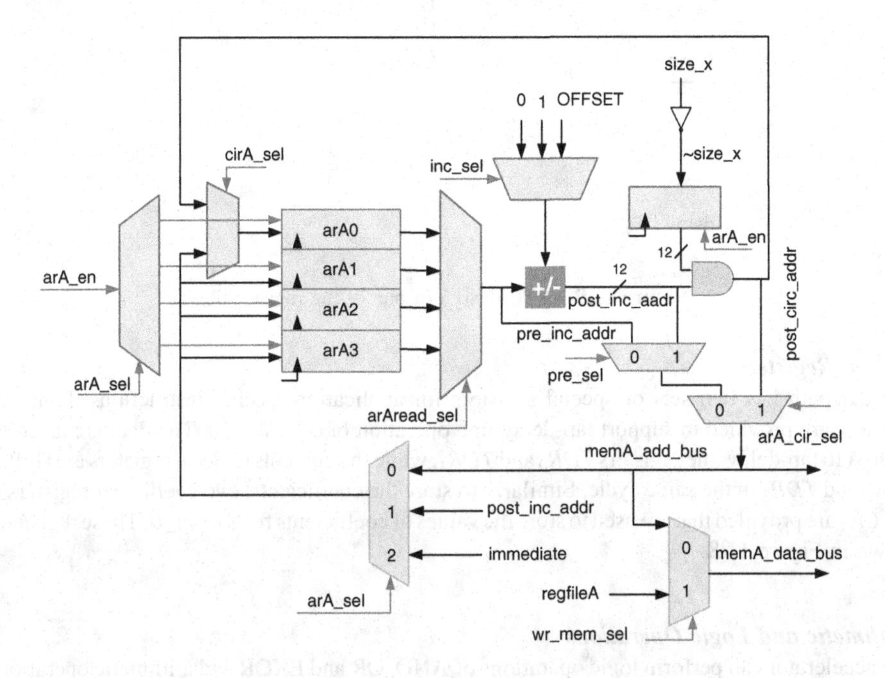

Figure 11.10 Address generation unit A

Address registers *arA*0 and *arB*0 can also be used for circular addressing. For circular addressing the AGU adds *OFFSET* to the content of the respective register, AND operation then performs modulo-by-*SIZE* operation and stores the value in the same register. The size of the circular buffer *SIZE* should always be a power of 2, and the cicular buffer must always be placed at memory address with $\log_2 SIZE$ least significant zeros. The logic to perform modulo operation is given here:

$$arX0 = (arX0 + OFFSET)\&(\sim SIZE);\ \text{where}\ X \in \{A, B\}$$

This simple example illustrates the working of the logic. Let a circular buffer of size 8 start at address 0 with address bus width 6. Assume the values $arA0 = 7$ and $OFFSET = 3$. Ten modulo addressing logic computes the next address as:

$$arA0 = (7 + 3 = 00\_1010)\&(7'b00\_0111) = 7'b00\_0010 = 2$$

MAC Blocks

The datapath has two MAC blocks. Each MAC block has two multipliers and one adder/subtractor. The MAC performs two multiplications on 16-bit signed numbers x_{n-k}, h_k and x_{n-k-1}, h_{k+1} and adds the products to a previously accumulated sum in the accumulator ACC_i:

$$ACCi_i = ACC_i + x_{n-k}h_k + x_{n-k-1}h_{k+1}$$

The accumulator register is 40 bits wide. The accumulator has 8 bits as guard bits for intermediate overflows. The result in the accumulator can be rounded to 16 bits for writing it back to one of the register files. The rounding requires adding 1 to the LSB of the 16-bit value. This is accomplished by placing a 1 at the 16th bit location on accumulator reset instruction, taking the guard bits off, this bit is the LSB of the 16 bits to be saved in memory. The adder and one of the multipliers in the MAC block can also add or multiply, respectively, two 16-bit operands and store the result directly in one of the registers in the register files.

Figure 11.11 shows the configuration of the MAC blocks for maximum data reuse while performing the convolution operation. The application-specific registers are placed in a setting that allows maximum reuse of data. The taped delay line of data and two registers for the coefficients allow four parallel MAC operations for accumulating for two output samples y_n and y_{n-1} as:

$$ACC_0 = ACC_0 + x_n h_0 + x_{n-1}h_1$$
$$ACC_1 = ACC_1 + x_{n-1}h_0 + x_{n-2}h_1$$

This effective reuse of data is shown in Figure 11.12.

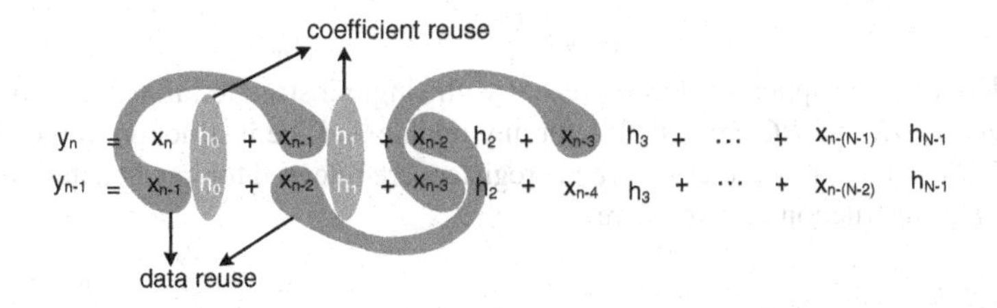

Figure 11.11 Reuse of data for computing two consecutive output values of convolution summation

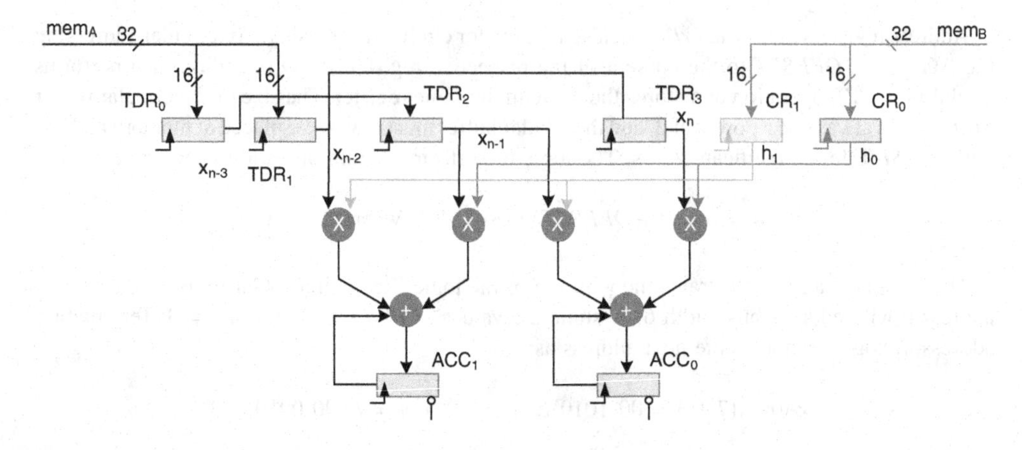

Figure 11.12 Two MAC blocks and specialized registers for maximum data reuse

11.6.2.4 Program Sequencer

The program sequencer handles four types of instruction: next instruction in the sequence, branch instruction, repeat instruction, and subroutine instruction.

11.6.2.5 Repeat Instruction

A detailed design of a loop machine is given in Chapter 10. The machine supports four nested loop instruction. The format of a loop instruction is given here.

repeat LABEL COUTN

The instructions following the loop instruction marks the start of the loop and the address of this instruction is stored in Loop Start Address (*LSA*) register. Address of the *LABEL* marks the address of the last instruction in the loop and the address is stored in Loop End Address (*LEA*) register. The *COUNT* gives the number of times the zero overhead loop needs to be repeated and the value is stored in Loop Counter (*LC*) register. The logic that loop machine implements is given here

```
if (PC == LEA && LC ! = 0)
   next_address = LSA;
   LC --;
else
next_address = PC ++;
```

The accelerator also supports a special repeat with single instruction in the loop body. For this repeat instruction, *IR* and *PC* are disabled for any update and the instruction in the *IR* register is executed *COUNT* number of cycles before the registers are released for any updates. The format of the repeat signal instruction is given here

*repeat*1 *COUNT*
Instruction

11.6.2.6 Conditional Branch Instruction

LEC accelerator supports the following jump instructions:

if (Z) jump LABEL
if (N) jump LABEL
jump LABEL

The first two conditional branch instructions check the specified status flag and if the status flag is set then the machine starts executing micro-code stored in address specified as *LABEL*. The design of the program sequencer for LEC accelerator is given in Figure 11.13.

11.6.2.7 Condition Multiplexer (C_MUX)

Conditional Multiplexer logic checks if the instruction is a conditional branch instruction and accordingly set the conditional flag to the next address generation logic. The C_MUX implements the logic in Table 11.1.

11.6.2.8 Next Address Logic

The Next Address Logic (NALogic) provides the address of the next instruction to be fetched from Program Memory (PM). The block has the following inputs:

1. **PC register:** For normal program execution
2. **Subroutine Return Address (SRA) register:** This register keeps the return address of the last subroutine called and this address is used once a return to subroutine is made.
3. **LSA register:** This register has the loop start address of the current loop and the address is used once the loop count is not zero and the program executes the last instruction in the loop.
4. **Branch Address Fields from IR:** For branch instruction and subroutine calls the IR has address of the next instruction. This address is used for instructions that execute a jump or subroutine call.

The next address block implements the following logic

```
if (branch || call subroutine)
   next_address = IR[branch address fields]
else if (instruction == return)
   next_address = SRA
else if (PC == LEA && LC != 0)
   next_address = LSA
else
   next_address = PC;
```

11.6.2.9 Data Memories

The accelerator has two memories each consisting N kB. For defining rest of the memory related fields in the current design, assume 8 kB of SRAM with 32-bit data bus. This makes 2048 memory locations and 11-bit address is required to access any memory location. These fields accordingly

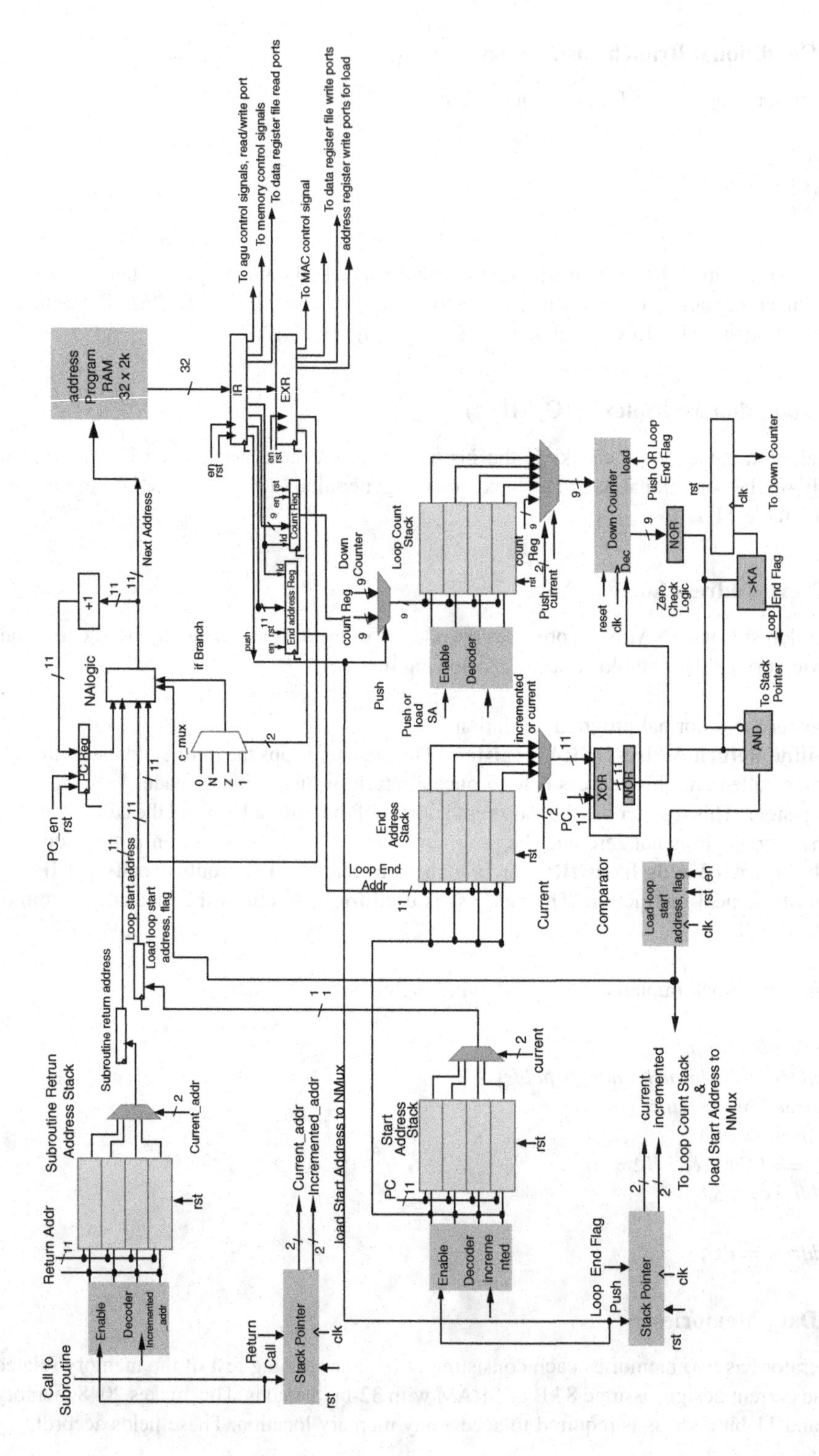

Figure 11.13 Program sequences of the LEC accelerator

Table 11.1 Logic implemented by C_MUX (see text)

Select signals		Output (cond_flag)()
N	Z	
0	0	FALSE
0	1	T/F depends on flag Z
1	0	T/F depends on flag N
1	1	TRUE unconditional

change with different memory sizes. The memories are dual-ported to give access to the DMA for simultaneous read and write in the memories.

11.6.2.10 Instruction Set Design

The instruction set has some application-specific instructions for implementing LMS adaptive algorithm. These instructions perform 4 MAC operations utilizing the tap-delay line registers and coefficients registers. Beside these instructions, the accelerator supports many general-purpose instructions as well.

The accumulator supports load and store instructions in various configurations.

11.6.2.11 Single Load Long Instruction

$$Xi.l = *arXj[++/--] \text{ where } X \in \{A, B\}, i \in \{0, 2, 4\} \text{ and } j \in \{0, 1, 2, 3\}$$

The instruction loads a 32 bit word stored at address location pointed by *arXj* register and the instruction also post increment or decrement the address as defined in the instruction. The terms in the brackets show the additional options in any basic instruction.

Example:

$arA0 = 0x23c$
$A2.l = *arA0 ++$

This instruction stores the 32-bit content at memory location $0x23c$ in $A2$, $A3$ registers.

Similarly the same instructions for B register file are given here:

$arB0 = 0x23c$ // address in mem B
$B2.l = *arB0 ++$

11.6.2.12 Parallel Load Long Instructions

The two load instruction on respective register files A and B can be used in parallel as well. The format of the instruction is given here:

$$Ai.l = *arAj[++/--] || Bm.l = *arBn[++/--]$$

where $i, m \in \{0,2,4\}$ and $j, n \in \{0,1,2,3\}$

11.6.2.13 Single and Parallel Load Short Instruction

$$Xi.s = *arXj[++/--] \text{ where } X \in \{A, B\}, i \in \{0, 1, 2, \dots 7\} \text{ and } j \in \{0, 1, 2, 3\}$$

This instruction loads a 16 bit value pointed by *arXj* in *Xi* register. The address register can optionally be post incremented or decremented.

Two of these instructions on respective register files *A* and *B* can be used in parallel as well. The instruction format is shown here

$$Ai.s = *arAj[++/--]||Bk.s = *arBl[++/--]$$

These instructions are orthogonal and can be used as mix of short and long load instructions in parallel.

$$Ai.s/l = *arAj[++/--]||Bk.s/l = *arBl[++/--]$$

11.6.2.14 Single Long Store Instruction

$$*arXj[++/--] = Xi.l \text{ where } X \in \{A, B\}, i \in \{0, 2, 4\} \text{ and } j \in \{0, 1, 2, 3\}$$

11.6.2.15 Parallel Long Store Instruction

$*arAj[++/--] = Ai.l||*arBn[++/--]Bm.l$ where $i, m \in \{0,2,4\}$ and $j, n \in \{0,1,2,3\}$

11.6.2.16 Single and Parallel Load Short Instruction

$$*arXj[++/--] = Xi.s \text{ where } X \in \{A, B\}, i \in \{0, 1, 2, \dots, 7\} \text{ and } j \in \{0, 1, 2, 3\}$$
$$*arAj[++/--] = Ai.s/l||Bk.s/l = *arBl[++/--] = Bk.s/l$$

11.6.2.17 Pre-increment Decrement Instruction

All the load and store instruction support pre increment/decrement addressing as well. In this case the address is first incremented or decremented as defined and then used for memory access. For example the single load instruction with pre-increment addressing is coded as

$$Xi.l = *[++]arXj \text{ where } X \in \{A, B\}, i \in \{0, 2, 4\} \text{ and } j \in \{0, 1, 2, 3\}$$

11.6.3 Address Registers Arithmetic

Load Immediate Address in an Address Register....

This instruction loads an immediate value *CONST* as address into any one of the registers of an address register file.

$$arXj = CONST \text{ where } X \in \{A, B\} \text{ and } j \in \{0, 1, 2, 3\}$$

A 12-bit constant offset *OFFSET* can be added or subtracted in any register of the address register files and the result can be stored in the same or any other register in the register files.

$$arXj = arYk + OFFSET \text{ where } X, Y \in \{A, B\} \text{ and } j, k \in \{0, 1, 2, 3\}$$

Example:

$$arA0 = arB2 + 0x131$$

The instruction adds $0x131$ in the content of $arB2$ and stores the value back in $arA0$ address register.

11.6.3.1 Circular Addressing

The accelerator supports circular addressing where the size of the circular buffer N should be a power of 2 and needs to be placed in memory address with $\log_2 N$ least significant zeroes in the address.

$$arX = Address\%N$$

11.6.3.2 Arithmetic and Logic Instruction

The accelerator supports one logic and two arithmetic instructions on 16-bit signed or unsigned (S/U) operands. The instruction format is

$$Xi = Xj(S/U)OPYk(S/U) \text{ where } OP \in \{+, -, *, \gg, , \&, \wedge\}, X, Y \in \{A, B\} \text{ and } i, j, k \in \{0, 1, 2, \ldots, 7\}$$

The shift is performed using the multiplier in the MAC block. Two of the arithmetic instructions can be used in parallel, the destination registers of the two parallel instructions should be in different register files.

$$Xi = Xj(S/U)OP1Yk(S/U) || Xm = Xn(S/U)OP2Yo(S/U)$$

$$\text{where } OP1, OP2 \in \{+, -, *, \gg\}, X, Y \in \{A, B\} \text{ and } i, j, k, m, n, o \in \{0, 1, 2, \ldots, 7\}$$

The conditional flags of N and Z are only set if the accelerator executes a single arithmetic and logic instruction.

11.6.3.3 Accumulator Instructions

The value of accumulator is truncated and 16 bits (31:16) of the respective Acc are stored in any register of the register files. In case of single instruction execution the condition flags are also set.

$$Xi = ACCj \text{ where } X \in \{A, B\}, j \in \{0, 1\} \text{ and } i \in \{0, 1, 2, \ldots, 7\}$$
$$ACCj = 0;$$

This instruction resets the ACC regsiter.

$$ACCj = 0x00008000$$

This instruction resets the register and places one at 15th bit location for rounding.

$$ACCj+ = Xi(S/U, F/I)*Yj(S/U, F/I)$$

This instruction performs MAC operation, where F and I are used to specify fraction or integer mode of operations.

$$ACCj = Xi(S/U, F/I)*Yj(S/U, F/I)$$

This instruction multiplies two signed or unsigned operands in fraction or integer mode and store them in the ACC.

Two of these instructions can be used in parallel for both the ACC registgers. For example the instruction here implements two MAC operations on signed operands in parallel.

$$ACC0+ = A0(S)*B1(S)||ACC1+ = A2(S)*B2(S)$$

11.6.3.4 Branch Instruction

The accelerator supports conditional and unconditional branch instructions. Two flags that are set as a result of single arithmetic instructions are used to make the decision. The instructions are

if (Z) jump LABEL
if (N) jump LABEL

The accelerator also supports unconditional instruction

jump LABEL

Where LABEL is the address of the next instruction the accelerator executes if the condition is true, otherwise the accelerator keeps executing instructions in a sequence.

11.6.3.5 Application Specific Instructions

The accelerator supports application specific instructions for accelerating the performance of convolution in adaptive filtering algorithms. In signal processing the memory accesses are the main bottlenecks of processing data. Inherently in many signal processing algorithms especially while implementing convolution summation the data is reused across multiple instructions. This reuse of data is exploited by providing a tap delay line and registers for filter coefficients. The data can be loaded in tap delay line registers $TDR0$ and $TDR1$ and coefficient registers $CR0$ and $CR1$ for optimal filtering operation:

$$TDL = *arA0--||CR = *arB1++||CONV$$

The instruction saves the 32-bit of data from the address pointed by $arA0$ in Tap Delay Line by storing the two 16-bit samples in $TDR0$ and $TDR1$ while respectively shifting the older values into $TDR2$ and $TDR3$. The instruction also decrements the address stored in the address register for accessing the new samples in the next iteration. Similarly, the coefficients stored at address pointed by $arB1$ are also read in coefficient registers $CR0$ and $CR1$ while the address register is incremented by 1. The datapath in the same cycle performs four MAC operations:

$$*arX[++/--] = ACCs \text{ where } X \in \{A, B\}$$

The 16-bit values in two accumulators $ACC0$ and $ACC1$ are stored in memory location pointed by arX register.

Example: The MAC units along with special registers are configured for adaptive filter specific instructions. The MAC units can perform four MAC operations to simultaneously compute two output samples.

$$y_n = x_n h_0 + x_{n-1} h_1 + x_{n-2} h_2 + x_{n-3} h_3 + \ldots + x_{n-(N-1)} h_{N-1}$$
$$y_{n-1} = x_{n-1} h_0 + x_{n-2} h_1 + x_{n-3} h_2 + x_{n-4} h_3 + \ldots + x_{n-(N-2)} h_{N-1}$$

The values of x_n and x_{n-1} are loaded first in the TDL and subsequently x_{n-2} and x_{n-3} are loaded in the TDL while h_0 and h_1 are loaded in CRs. The length of the filter is 40. The convolution operation is performed in a loop $N/2$ number of times to compute the two samples, use circular addressing.

```
arA0 = xBf%128; //point arA0 to the current samples, use circular addressing
arB0 = hbf;
arA1 = yBf%128
accs = 0x00008000; //reset the two acculators for rounding
TDR = *arA0--%//load x[n] and x[n-1] in TDR0 and TDR1
TDR = *arA0--%//load x[n-2] and x[n-3] while shifting old values
repeatc20// repeat the next instruction 20 times
TDR = *arA0--||CR = *arB0++||CONV
*arA1 = ACCs
```

11.6.3.6 Instruction Encoding

Though the instruction can be encoded in a way that saves complex decoding but in this design a simple coding technique is used to avoid complexity. The accelerator supports both regular and special instructions (Figure 11.14). The regular instructions are a set of orthogonal instructions that are identical for both the data paths and AGUs. A 0 at bit location 0 identifies that the sub-instructions in the current instruction packet are orthogonal and a 1 at the bit location specifies that this is a special or non orthogonal instruction.

0	Orthogonal instructions			
0	15	15	10	10
0	Data Path A	Data Path B	AGU-A	AGU-B
1	Special and non-orthogonal instruction			

Figure 11.14 Instruction encoding

11.6.4 Pipelining Options

The pipelining in the datapath is not shown but the accelerator can be easily pipelined. Handling pipeline hazard is an involved area and good coverage of pipelining hazards can be found in [21]. An up to six pipeline stages are suggested. In the first stage an instruction is fetched from PM into an Instruction Register. The second stage of pipeline decodes the instruction and in third stage operands are fetched in the operand registers. A host of other operations related to memory reads and writes and address-register arithmetic can also be performed in this cycle. The third stage of pipelining performs the operation. This stage can be further divided to use pipelined multipliers and adders. The last stage of pipelining writes the result back to the register files. Figure 11.15 proposes a pipelined design of the accelerator. The pipeline stages are Instruction Read (IR), Instruction Decode (ID), Operand Read (OR), and optional pipelining in execution unit as Execution 1 (EX1) and Execution 2 (EX2), and finally Write Back (WB) stage.

11.6.4.1 Delay Slot

Depending upon the number of pipeline stages, the delay slot can be effectively used to avoid pipeline stalls. Once the program sequencer is decoding a branch instruction the next instruction in the program is fetched. The program can fill the delay slot with a valid instruction that logically

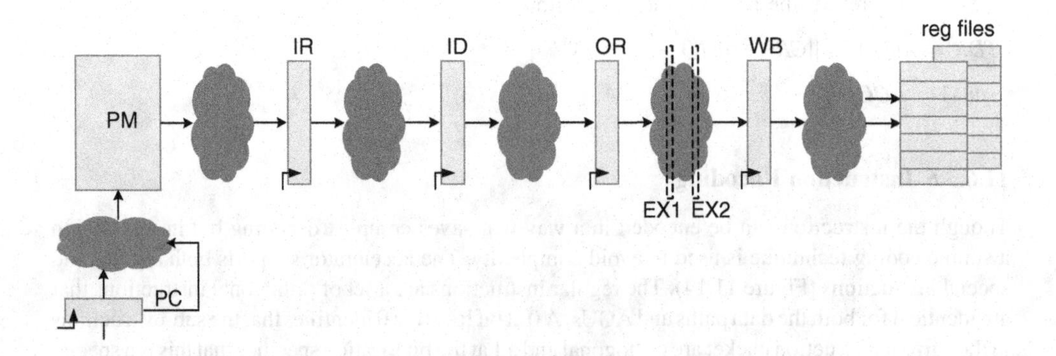

Figure 11.15 Pipeline stages for the LEC accelerator

precedes the branch instruction but is placed after the instruction. The delay slot instruction must not be a branch instruction. In case the programmer could not effectively use the delay slot, it should be filled with NOP instruction.

11.6.5 Optional Support for Coefficient Update

In the design of micro-coded state machine, HW support can also be added to speedup coefficients update operation of (11–3). This requires addition of a special error register *ER* that saves the scaled error value $\mu e[n]$ and few provisions in the datapath for effective use of multipliers, adders and memory for coefficient updates. This additional configuration is given in Figure 11.16. This augmentation in the datapath and AGUs can be simply incorporated in the existing design of the respective units given in Figures 11.8 and 11.10 with additional address register in the AGU, interconnections and multiplexers. The setting assumes arB0 stores the address of the coefficient buffer. The coefficient memory is a dual port memory that allows reading and writing of 32-bit word simultaneously. The current value of the address in arB0 register is latched in the delay address register arB0D for use in the subsequent write in mem_B. The updated coefficients are calculated in ACCs and are stored in the coefficient buffer in the next cycle.

arB0 = hBf
arA0 = xBf %256
ER = A0*B0; // assuming μ in A0 and e[n] is in B0 registers
CR = *arB0++ || TDR=*arA0--%
repeat1 20
    CR=*arB0++||TDR=*arA0--||*UPDATE*

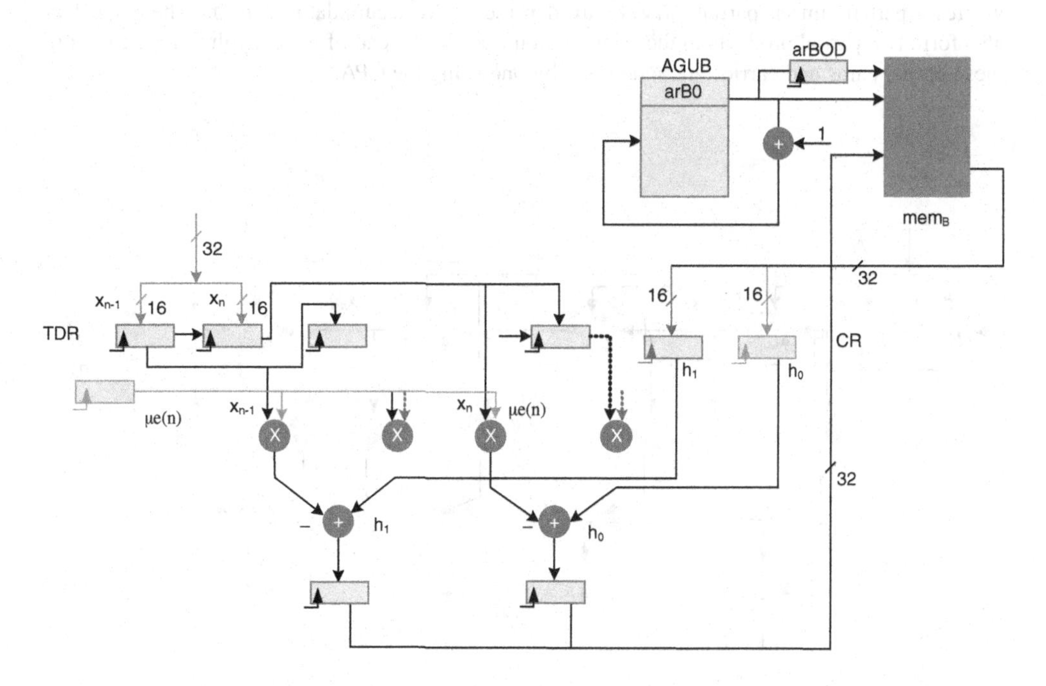

Figure 11.16 Optional additional support for coefficient updating

The UPDATE micro-code performs the following operations in HW in a single cycle:

```
ACC0 = CR0-TR0*ER
ACC1 = CR1 - TR1*ER
*arB0D = truncate{ACC1,ACC0}
```

This setting enables the datapath to update two coefficients in one cycle. The same cycle loads values of two new input samples and two coefficients.

11.6.6 Multi MAC Block Design Option

The datapath can be easily extended to include more MAC blocks to accelerate the execution of filtering operation. A configuration of the accelerator with four MAC blocks is depicted in Figure 11.17.

It is important to highlight that the current design can easily be extended with more number of MAC blocks for enhanced performance. The new design still works with just loading two samples of data and two coefficients of the filter from the two memories respectively

.

11.6.7 Compression Tree and Single CPA-based Design

The design can be further optimized for area by using compression trees with a single CPA. The design is shown in Figure 11.18. Each MAC block is implemented with a single compression tree where the partial sum and partial carry are saved in a set of two accumulator registers. These registers also form two partial products in the compression tree. At the end of the convolution calculation, these partial sums and carries are added one by one using the CPA.

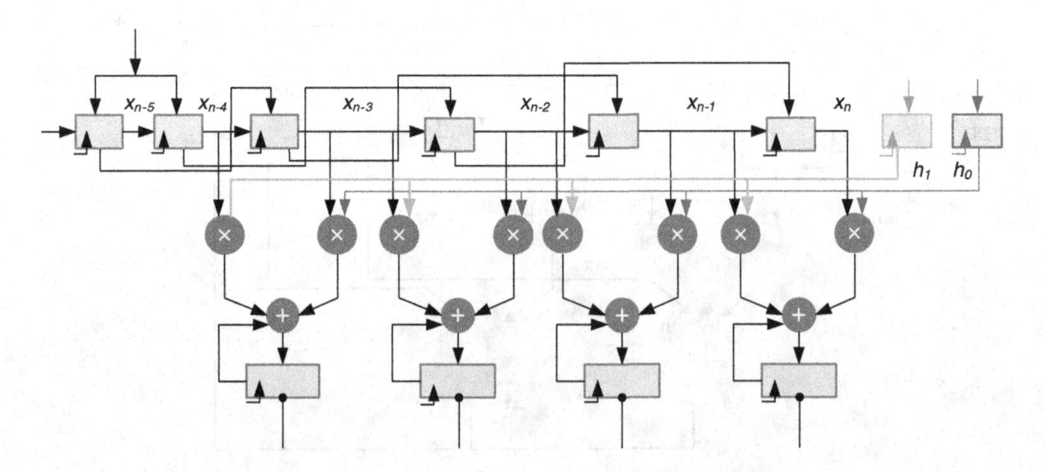

Figure 11.17 Optional additional MAC blocks for accelerating convolution calculation

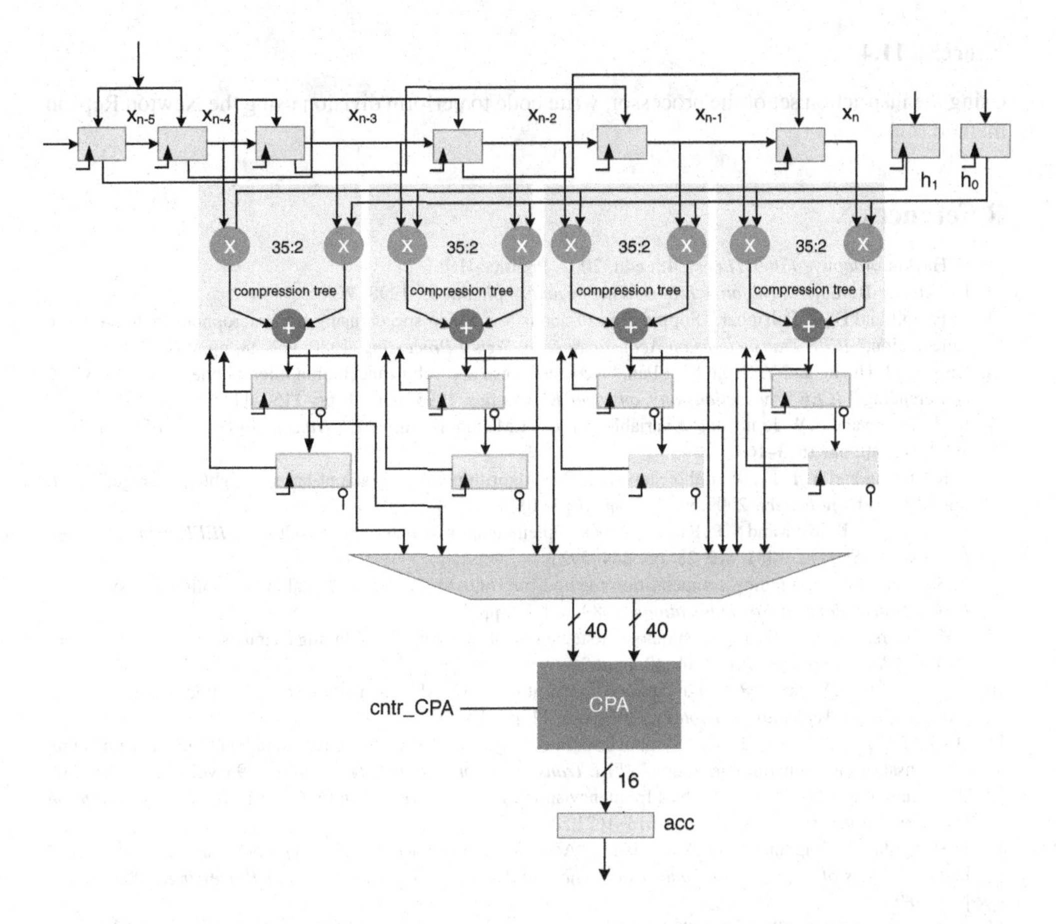

Figure 11.18 Optional use of compression trees and one CPA for optimized hardware

Exercises

Exercise 11.1

Add bit-reverse addressing support in the arA0 register of the AGU. The support increments and decrements addresses in bit-reverse addressing mode. Draw RTL diagram of the design.

Exercise 11.2

Add additional TDL registers and MAC blocks in Figure 11.12 to support IIR filtering operation in adaptive mode.

Exercise 11.3

Add additional hardware support with an associated MAC to compute $\mathbf{x}_n\mathbf{x}_n^{\mathrm{T}}$ while performing a filtering operation. Use this value to implement the NLMS algorithm of (11.4).

Exercise 11.4

Using the instruction set of the processor, write code to perform division using the Newton Repson method.

References

1. S. Haykin, *Adaptive Filter Theory*, 4th edn, 2002, Prentice-Hall.
2. B. Farhang-Boroujeny, *Adaptive Filters: Theory and Applications*, 1998, Wiley.
3. S. F. Boll and D. C. Pulsipher, "Suppression of acoustic noise in speech using two-microphone adaptive noise cancellation," *IEEE Transactions on Acoustic Speech, Signal Processing*, 1980, vol. 28, pp. 752–753.
4. Kuo, S. M. Huang and Y. C. Zhibing Pan, "Acoustic noise and echo-cancellation microphone system for video conferencing," *IEEE Transactions on Consumer Electronics*, 1995, vol. 41, pp. 1150–1158.
5. R. H. Kwong and E. W. Johnston, "A variable step size LMS algorithm," *IEEE Transactions on Signal Processing*, 1992, vol. 40, pp. 1633–1642.
6. J.-K. Hwang and Y.-P. Li, "Variable step-size LMS algorithm with a gradient-based weighted average," *IEEE Signal Processing Letters*, 2009, vol. 16, pp. 1043–1046.
7. G. A. Clark, S. K. Mitra and S. R. Parker, "Block implementation of adaptive digital filters," *IEEE Transactions on Circuits and Systems*, 1981, vol. 28, pp. 584–592.
8. S. Sampei, "Rayleigh fading compensation method for 16QAM modem in digital land mobile radio systems," *IEICE Transactions on Communications*, 1989, vol. 73, pp. 7–15.
9. J. K. Cavers, "An analysis of pilot symbol assisted modulation for Rayleigh fading channels," *IEEE Transactions on Vehicular Technology*, 1991, vol. 40, pp. 686–693.
10. S. Sampei and T. Sunaga, "Rayleigh fading compensation for QAM in land mobile radio communications," *IEEE Transactions on Vehicular Technology*, 1993, vol. 42, pp. 137–147.
11. H.-B. Li, Y. Iwanami and T. Ikeda, "Symbol error rate analysis for MPSK under Rician fading channels with fading compensation based on time correlation," *IEEE Transactions on Vehicular Technology*, 1995, vol. 44, pp. 535–542.
12. T. M. Schmidl and D. C. Cox, "Robust frequency and timing synchronization for OFDM," *IEEE Transactions on Communications*, 1997, vol. 45, pp. 1613–1621.
13. J. P. Costa, A. Lagrange and A. Arliaud, "Acoustic echo cancellation using nonlinear cascade filters," in *Proceedings of IEEE International Conference on Acoustics, Speech and Signal Processing*, 2003, vol. 5, p. V–389–392.
14. A. Stenger and W. Kellermann, "Nonlinear acoustic echo cancellation with fast converging memoryless preprocessor," in *Proceedings of IEEE International Conference on Acoustics, Speech an Signal Processing*, 2000, vol. 2, pp. II-805–808.
15. A. Guerin, G. Faucon and R. L. Bouquin-Jeannes, "Nonlinear acoustic echo cancellation based on Volterra filters," *IEEE Transactions on Speech Audio Processing*, 2003, vol. 11, pp. 672–683.
16. V. Krishna, J. Rayala and B. Slade, "Algorithmic and implementation aspects of echo cancellation in packet voice networks," in *Proceedings of 36th IEEE Asilomar Conference on Signals, Systems and Computers*, 2000, vol. 2, pp. 1252–1257.
17. International Telecommunications Union, (ITU). G.168: "Digital network echo cancellers," 2004.
18. H. Ye and B. X. Wu, "A new double-talk detection based on the orthogonality theorem," *IEEE Transactions on Communications*, 1991, vol. 39, pp. 1542–1545.
19. T. Gansler, M. Hansson, C.-J. Ivarsson and G. Salomonsson, "A double-talk detector based on coherence," *IEEE Transactions on Communications*, 1996, vol. 44, pp. 1421–1427.
20. J. Benesty, D. R. Morgan and J. H. Cho, "A new class of doubletalk detectors based on cross-correlation," *IEEE Transactions on Speech and Audio Processing*, 2000, vol. 8, pp. 168–172.
21. J. L. Hennessy and D. A. Patterson, *Computer Architecture: a Quantitative Approach*, 4th edn, 2007, Morgan Kaufmann Publishers, Elsevier Science, USA.
22. D. G. Manolakis, V. K. Ingle and S. M. Kogon, *Statistical and Adaptive Signal Processing*, 2000, McGraw-Hill.
23. S. G. Sankaran and A. A. Beex, "Convergence behavior of affine projection algorithm," *IEEE Transactions on Signal Processing*, 2000, vol. 48, pp. 1086–1096.

12

CORDIC-based DDFS Architectures

12.1 Introduction

This chapter considers the hardware mapping of an algorithm to demonstrate application of the techniques outlined in this book. Requirement specifications include requisite sampling rate and circuit clock. A folding order is established from the sampling rate and circuit clock. To demonstrate different solutions for HW mapping, these requirements are varied and a set of architectures is designed for these different requirements.

The chapter explores architectures for the digital design of a *direct digital frequency synthesizer* (DDFS). This generates sine and cosine waveforms. The DDFS is based on a CoORDinate DIgital Computer (CORDIC) algorithm. The algorithm, through successive rotations of a unit vector, computes sine and cosine of an input angle θ. Each rotation is implemented by a CORDIC element (CE). An accumulator in the DDFS keeps computing the next angle for the CORDIC to compute the sine and cosine values. An *offset* to the accumulator in relation with the circuit clock controls the frequency of the waveforms produced.

After describing the algorithm and its implementation in MATLAB®, the chapter covers design techniques that can be applied to implement a DDFS in hardware. The selection is based on the system requirements. First, a *fully dedicated architecture* (FDA) is given that puts all the CEs in cascade. Pipelining is employed to get better performance. This architecture computes new sine and cosine values in each cycle. If more circuit clock cycles are available for this computation then a time-shared architecture is more attractive, so the chapter considers time-shared or folding architectures. The folding factor defines the number of CEs in the design. The folded architecture is also pipelined to give better timings.

Several novel architectures have been proposed in the literature. The chapter gives some alternate architecture designs for the CORDIC. A distributed ROM-based CORDIC uses read-only memory for initial iterations. Similarly, a collapsed CORDIC uses look-ahead transformation to merge several iterations to optimize the CORDIC design.

The chapter also presents a novel design that uses a single iteration to compute the values of sine and cosine, to demonstrate the extent to which a designer can creatively optimize a design.

Digital Design of Signal Processing Systems: A Practical Approach, First Edition. Shoab Ahmed Khan.

12.2 Direct Digital Frequency Synthesizer

A DDFS is an integral component of high-performance communication systems [1]. In a linear digital modulation system like QAM, the baseband signal is translated in frequency. This translation is usually done in more than one stage. The first stage of mixing is performed in the digital domain and a complex baseband signal $s_b[n]$ is translated to an intermediate frequency (IF) by mixing it with an exponential of equivalent digital frequency ω_o:

$$s_{IF}[n] = s_b[n]e^{j\omega_o n} \tag{12.1}$$

This IF signal is converted to analog using a D/A converter and then mixed using analog mixers for translation to the desired frequency band.

At the receiver the similar complex mixing is performed. The mixer in the digital domain is best implemented using a DDFS. Usually, due to differences in the crystals for clock and frequency generation at the transmitter and receiver, the mixing leaves an offset:(12.2)

$$s_b[n]e^{j\omega_o n}e^{-j\omega_o' n} \tag{12.3}$$

The receiver needs to compute this frequency error, the offset being $\Delta\omega = \omega_o - \omega_o'$. The computation of correction is made in a frequency correction loop. The frequency adjustment again requires generation of an exponential to make this correction. A DDFS is used to generate the desired exponential:

$$\left(s_b[n]e^{j\omega_o n}e^{-j\omega_o' n}\right)e^{-j\Delta\omega n} \tag{12.3}$$

A DDFS generates a spectrally pure sine and cosine for quadrature mixing and frequency and phase correction in a digital receiver, as shown in Figure 12.1. At the front end, a DDFS mixes with the digitized IF signal. The decimation stage down-converts the signal to baseband. The phase detector computes the phase error and the output of the loop filter generates phase correction for DDFS that generates the correction for quadrature mixing.

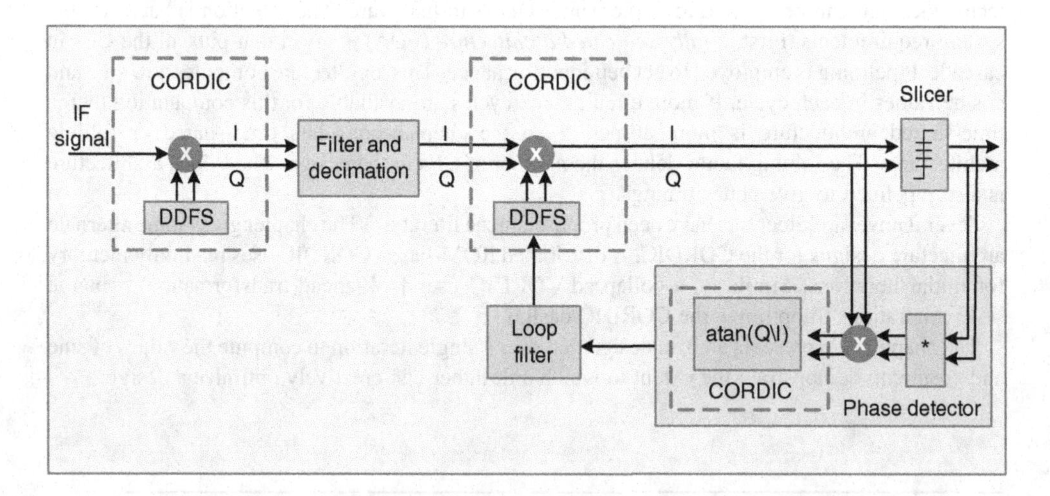

Figure 12.1 Use of DDFS in a digital communication receiver

For a high data-rate communication system, this value needs to be computed in a single cycle. In designs where multiple cycles are available, a folded and time-shared architecture is designed to effectively use area. Similarly, while demodulating an M-ary phase-modulated signal, the slicer computes $\tan^{-1}(Q/I)$, where Q is the quadrature and I is the in-phase component of a demodulated signal. All these computations require an effective design of CORDIC architecture.

A DDFS is also critical in implementing high-speed frequency and phase modulation systems as it can perform fast switching of frequency at good frequency resolution. For example, a GMSK baseband modulating signal computes the expression [2]:

$$s(t) = \sqrt{\frac{2E_b}{T}} \exp\left[j\pi \sum_{n=0}^{k} \beta_n \theta(t-nt)\right] \tag{12.4}$$

for $kT < t < (k+1)T$, where T is the symbol period, E_b is the energy per bit, $\beta_n \in \{1,-1\}$ is the modulating data, and $\theta(t)$ is the phase pulse. A DDFS is very effective in generating a baseband GMSK signal.

A DDFS is characterized by its *spectral purity.* A measure of spectral purity is the 'spurious free dynamic range' (SFDR), defined as the ratio (in dB) of amplitude of the desired frequency to the highest frequency component of undesired frequency.

12.3 Design of a Basic DDFS

A simple design of a DDFS is given in Figure 12.2. A frequency control word W in every clock cycle of frequency f_{clk} is added in an N-bit phase accumulator. If $W = 1$, it takes the clock 2^N cycles to make the accumulator overflow and starts again. The output of the accumulator is used as an address to a ROM/RAM where the memory stores a complete cycle of a sinusoid. The data from the memory generates a sinusoid. The DDFS can generate any frequency f_o by an appropriate selection of W using:

$$f_o = \frac{wf_{CLK}}{2^N} \tag{12.5}$$

The digital signals $\cos(\omega_o n)$ and $\sin(\omega_o n)$ can be input to a D/A converter at sampling rate $T = 1/f_{CLK}$ for generating analog sinusoids of frequency f_o, where:

$$\omega_o = 2\pi f_o T$$

The maximum frequency from the DDFS is constrained by the Nyquist sampling criterion and is equal to $f_{clk}/2$.

The basic design of Figure 12.2 is improved by exploiting the symmetry of sine and cosine waves. The modified design is shown in Figure 12.3. The output of the accumulator is truncated from N to L

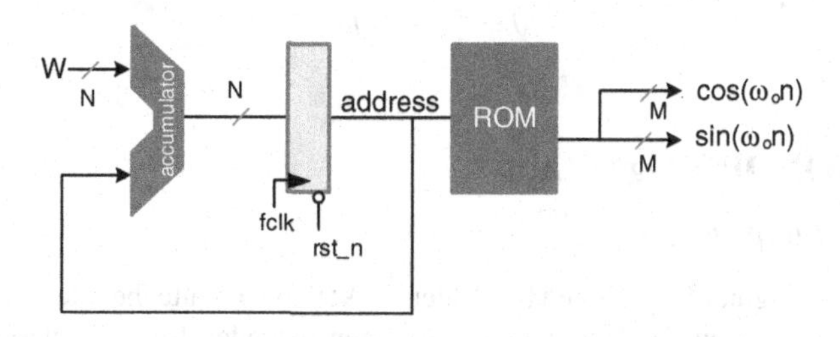

Figure 12.2 Design of a basic DDFS

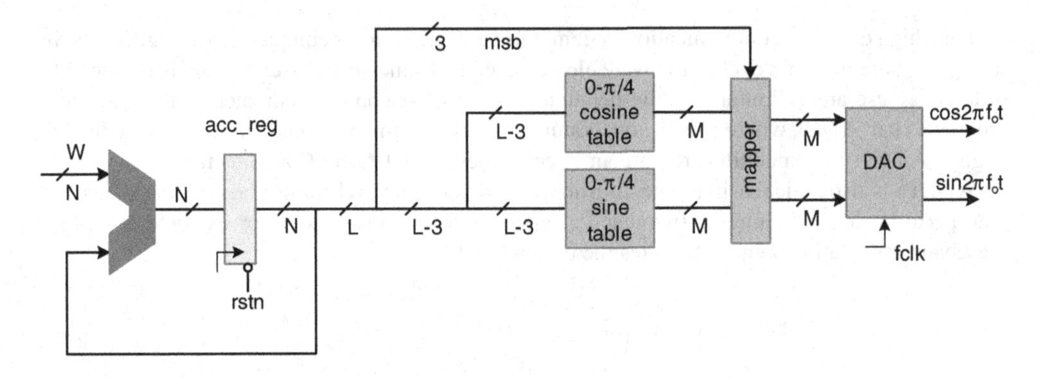

Figure 12.3 Design of a DDFS with reduced ROM size

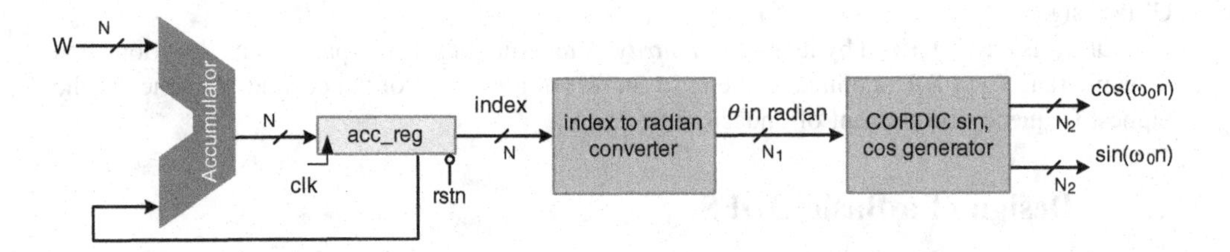

Figure 12.4 DDFS with CORDIC block for generating sine and cosine of an angle θ

bits to reduce the memory requirement. A complete period 0 to 2π of sine and cosine waves can be generated from values of the two signals from 0 to $\pi/4$. The sizes of the two memories are reduced by one-eighth by only storing the values of sine and cosine from 0 to $\pi/4$. The $L-3$ bits are used to address the memories, and then three most signficant bits (MSBs) of the address are used to map the values to generate complete periods of cosine and sine.

A ROM/RAM-based DDFS requries two 2^{L-3} deep memories of width M. The design takes up a large area and dissipates significant power. Several algorithms and techniques have been proposed that reduce or completely eliminate look-up tables in memories. An efficient algorithm is CORDIC, which uses rotation of vectors in Cartesian coordinates to generate values of sine and cosine.

The DDFS with a CORDIC block is shown in Figure 12.4. The CORDIC algorithm takes angle θ in radians, whereas the DDFS accumulator specifies the angle as an index value. To use a CORDIC block in DDFS, a CSD multiplier is required that converts index n to angle θ in radians, where:

$$\theta = \frac{2\pi}{2^N} \times index$$

12.4 The CORDIC Algorithm

12.4.1 Introduction

This algorithm was originally developed by Volder in 1959 to compute the rotation of a vector in the Cartesian coordinate system [3]. The method has been extended for computation of hyperbolic functions, multiplication, division, exponentials and logarithms [4].

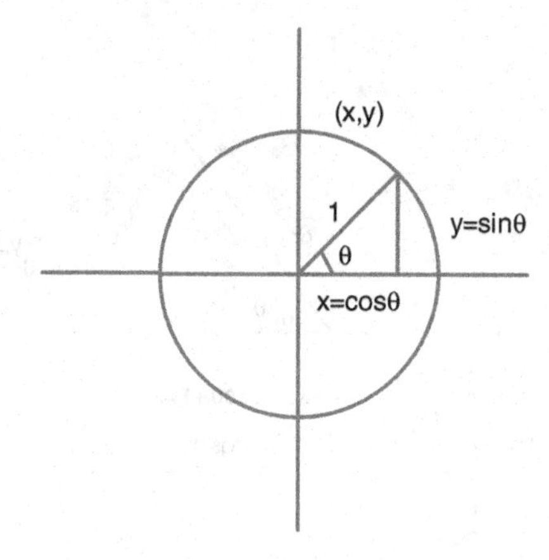

Figure 12.5 The CORDIC algorithm successively rotates a unit vector to desired angle θ

Multiplication by sine and cosine is an integral part of any communication system, so area- and time-efficient techniques for iterative computation of CORDIC are critical. The best application of the CORDIC algorithm is its use in DDFS. Several architectural designs around CORDIC have been reported in the literature [5–8].

To bring a unit vector to desired angle θ, the basic CORDIC algorithm gives known recursive rotations to the vector. Once the vector is at the desired angle, then the x and y coordinates of the vector are equal to $\cos\theta$ and $\sin\theta$, respectively. This basic concept is shown in Figure 12.5.

Mathematically the angle θ is approximated by addition and subtraction of angles of N successive rotations $\Delta\theta_i$. The expression for the approximation is:

$$\theta = \sum_{i=0}^{N-1} \sigma_i \Delta\theta_i \ \text{ for } \sigma_i = \begin{cases} 1 \text{ for positive rotation} \\ -1 \text{ for negative rotation} \end{cases}$$

As depicted in Figure 12.6, the unit vector in iteration i is rotated by some angle θ_i and the next rotation is made by an angle $\Delta\theta_i$ that brings the vector to new angle θ_{i+1}, then:

$$\cos\theta_{i+1} = \cos(\theta_i + \sigma_i\Delta\theta_i) = \cos\theta_i \cos\Delta\theta_i - \sigma_i \sin\theta_i \sin\Delta\theta_i \quad (12.6a)$$

$$\sin\theta_{i+1} = \sin(\theta_i + \sigma_i\Delta\theta_i) = \sin\theta_i \cos\Delta\theta_i - \sigma_i \cos\theta_i \sin\Delta\theta_i \quad (12.6b)$$

From the figure it can be easily established that $x_i = \cos\theta_i$, $y_i = \sin\theta_i$ and similarly $x_{i+1} = \cos\theta_{i+1}$, $y_{i+1} = \sin\theta_{i+1}$. Substituting these in (12.6) we get:

$$x_{i+1} = x_i \cos\Delta\theta_i - \sigma_i y_i \sin\Delta\theta_i \quad (12.7a)$$

$$y_{i+1} = \sigma_i x_i \sin\Delta\theta_i + y_i \cos\Delta\theta_i \quad (12.7b)$$

In matrix forms, the set of equations in (12.7) can be written as:

$$\begin{bmatrix} x_{i+1} \\ y_{i+1} \end{bmatrix} = \begin{bmatrix} \cos\Delta\theta_i & -\sigma_i \sin\Delta\theta_i \\ \sigma_i \sin\Delta\theta_i & \cos\Delta\theta_i \end{bmatrix} \begin{bmatrix} x_i \\ y_i \end{bmatrix}$$

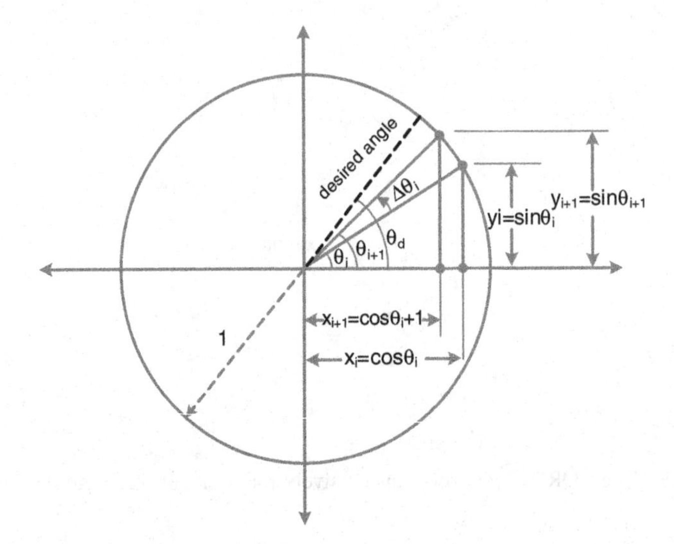

Figure 12.6 CORDIC algorithm incremental rotation by $\Delta\theta_i$

Taking $\cos\Delta\theta_i$ common yields:

$$\begin{bmatrix} x_{i+1} \\ y_{i+1} \end{bmatrix} = \cos\Delta\theta_i \begin{bmatrix} 1 & -\sigma_i \tan\Delta\theta_i \\ \sigma_i \tan\Delta\theta_i & 1 \end{bmatrix} \begin{bmatrix} x_i \\ y_i \end{bmatrix} \tag{12.8}$$

In this expression $\cos\Delta\theta_i$ can also be written in terms of $\tan\Delta\theta_i$ as:

$$\cos\Delta\theta_i = \frac{1}{\sqrt{1+\tan^2\Delta\theta_i}}$$

To avoid multiplication in computing (12.8), the incremental rotation is set as:

$$\tan\Delta\theta_i = 2^{-i} \tag{12.9}$$

which implies $\Delta\theta_i = \tan^{-1}2^{-i}$, and the algorithm applies N such successive micro-rotations of $\pm\Delta\theta_i$ to get to the desired angle θ. This arrangement also poses a limit to the desired angle as:

$$\sum_{i=0}^{N-1}\Delta\theta_i \le \theta \le -\sum_{i=0}^{N-1}\Delta\theta_i$$

Now substituting expression (12.9) in (12.8) gives:

$$\begin{bmatrix} x_{i+1} \\ y_{i+1} \end{bmatrix} = \frac{1}{\sqrt{1+2^{-2i}}} \begin{bmatrix} 1 & -\sigma_i 2^i \\ \sigma_i 2^i & 1 \end{bmatrix} \begin{bmatrix} x_i \\ y_i \end{bmatrix}$$

This can be written as a basic rotation of the CORDIC algorithm:

$$\begin{bmatrix} x_{i+1} \\ y_{i+1} \end{bmatrix} = k_i R_i \begin{bmatrix} x_i \\ y_i \end{bmatrix} \tag{12.10}$$

where

$$k_i = \frac{1}{\sqrt{1+2^{-2i}}} \text{ and } R_i = \begin{bmatrix} 1 & -\sigma_i 2^i \\ \sigma_{i2^i} & 1 \end{bmatrix}$$

Starting from index $i=0$, the expression in (12.10) is written as:

$$\begin{bmatrix} x_1 \\ y_1 \end{bmatrix} = k_0 R_0 \begin{bmatrix} x_0 \\ y_0 \end{bmatrix} \tag{12.11a}$$

and for index $i=1$ the expression becomes:

$$\begin{bmatrix} x_2 \\ y_2 \end{bmatrix} = k_1 R_1 \begin{bmatrix} x_1 \\ y_1 \end{bmatrix} \tag{12.11b}$$

Now substituting the value of $\begin{bmatrix} x_1 \\ y_1 \end{bmatrix}$ from (12.11a) into this expression, we get:

$$\begin{bmatrix} x_2 \\ y_2 \end{bmatrix} = k_0 k_1 R_0 R_1 \begin{bmatrix} x_0 \\ y_0 \end{bmatrix} \tag{12.11c}$$

Writing (12.10) for indices $i=3,4,\ldots,N-1$ and then substituting values for the previous indices, we get:

$$\begin{bmatrix} x_N \\ y_N \end{bmatrix} = k_0 k_1 k_2 \ldots k_{N-1} R_0 R_1 R_2 \ldots R_{N-1} \begin{bmatrix} x_0 \\ y_0 \end{bmatrix} \tag{12.12}$$

All the k_i are constants and their product can be pre-computed as constant k, where:

$$k = k_0 k_1 k_2 \ldots k_{N-1} = \prod_{i=0}^{N-1} \frac{1}{\sqrt{1+2^{-2i}}} \tag{12.13}$$

For each stage or rotation, the algorithm computes σ_i to determine the direction of the ith rotation to be positive or negative. The factor k is incorporated in the first iteration and the expression in (12.12) can then be written as:

$$\begin{bmatrix} x_N = \cos\theta \\ y_N = \sin\theta \end{bmatrix} = R_0 R_1 R_2 \ldots R_{N-1} \begin{bmatrix} k \\ 0 \end{bmatrix}$$

12.4.2 CORDIC Algorithm for Hardware Implementation

For effective HW implementation the CORDIC algorithm is listed as follows:

S0 To simplify the hardware, θ_0 is set to the desired angle θ_d and θ_1 is computed as:

$$\theta_1 = \theta_0 - \sigma_0 \tan^{-1} 2^0$$

where σ_0 is the sign of θ_0. Also initialize $x_0 = k$ and $y_0 = 0$.

S1 The algorithm then performs N iterations for $i = 1, \ldots, N-1$, and performs the following set of computations:

$$\sigma_1 = 1 \text{ when } \theta_i > 0 \text{ else it is } -1 \tag{12.14a}$$

$$x_{i+1} = x_i - \sigma_i 2^{-i} y_i \tag{12.14b}$$

$$y_{i+1} = \sigma_i 2^{-1} x_i + y_i \tag{12.14c}$$

$$\theta_{i+1} = \theta_i - \sigma_i \tan^{-1} 2^{-i} \tag{12.14d}$$

All the values for $\tan^{-1}2^{-i}$ are pre-computed and stored in an array.

S2 The final iteration generates the desired results as:

$$\cos\theta_d = x_N$$

$$\sin\theta_d = y_N$$

The following MATLAB® code implements the CORDIC algorithm:

```
% CORDIC implementation for generating sin and cos of desired angle
% theta_d
close all
clear all
clc
% theta resolution that determines number of rotations of CORDIC
% algorithm
N = 16;
% generating a tan table and value of constant k
tableArcTan = [];
k = 1;
for i=1:N
    k = k * sqrt(1+(2^(-2*i)));
    tableArcTan = [tableArcTan atan(2^(-i))];
end

k = 1/k;
x = zeros(1,N+1);
y = zeros(1,N+1);
theta = zeros(1,N+1);

sine = [];
cosine = [];

% Specify all the values of theta
% Theta value must be within -0.9579 and 0.9579 radians

theta_s = -0.9;
theta_e = 0.9;
for theta_d = theta_s:.1:theta_e
% CORDIC algorithm starts here
theta(1) = theta_d;
x(1) = k;
```

```
y(1) = 0;
for i=1:N,
    if (theta(i) > 0)
        sigma = 1;
    else
        sigma = -1;
end
x(i+1) = x(i) - sigma*(y(i) * (2^(-i)));
y(i+1) = y(i) + sigma*(x(i) * (2^(-i)));
theta(i+1) = theta(i) - sigma*tableArcTan(i);
end
% CORDIC algorithm ends here and computes the values of
% cosine and sine of the desired angle in y(N+1) and
% x(N+1), respectively
cosine = [cosine x(N+1)];
sine = [sine y(N+1)];
end
```

Figure 12.7 shows output of the MATLAB® code that correctly computes values of sine and cosine using the basic CORDIC algorithm. The range of angle θ is from –0.9 radians (−54.88 degrees) to 0.9 radians (54.88 degrees).

One instance of the algorithm for $\theta = 43$ degrees is explained in Figure 12.8. The CORDIC starts with $\theta_0 = 43$. As CORDIC tries to make the resultant angle equal to 0 it applies a negative rotation $\Delta\theta_0 = \tan^{-1}2^{-0}$ to bring the angle $\theta_1 = 16.44$. Two more negative rotations take the angle to the

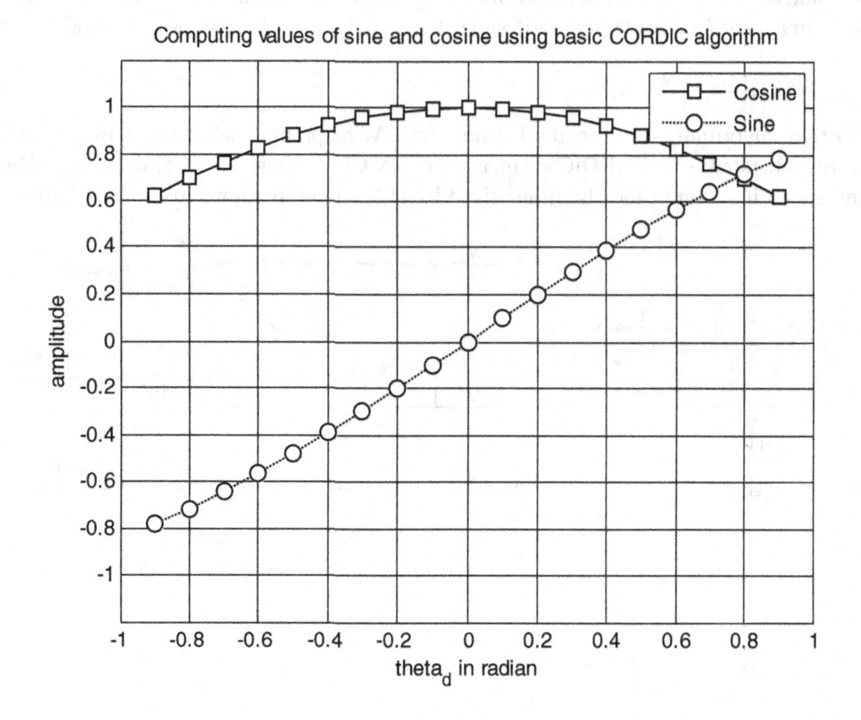

Figure 12.7 Computation of sine and cosine using basic CORDIC algorithm

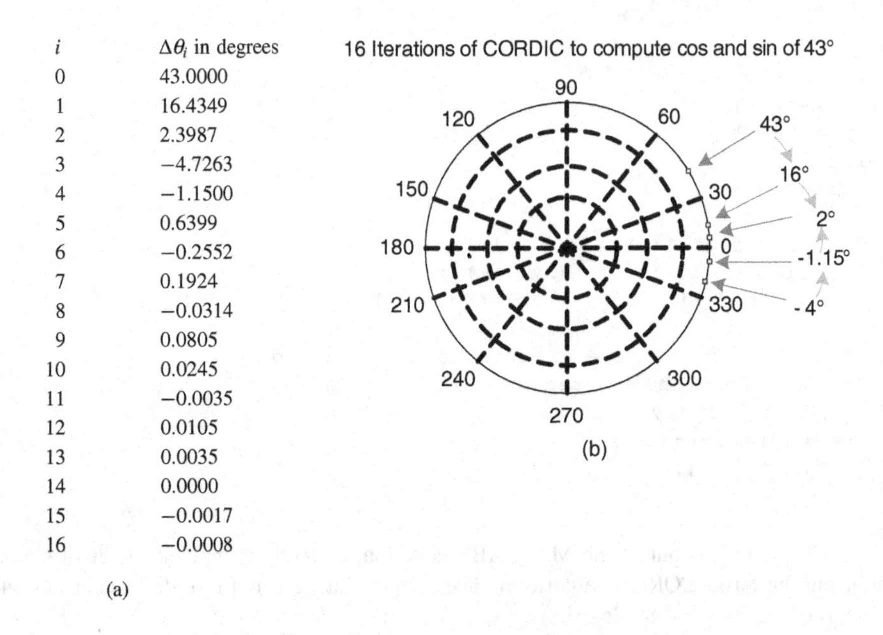

i	$\Delta\theta_i$ in degrees
0	43.0000
1	16.4349
2	2.3987
3	−4.7263
4	−1.1500
5	0.6399
6	−0.2552
7	0.1924
8	−0.0314
9	0.0805
10	0.0245
11	−0.0035
12	0.0105
13	0.0035
14	0.0000
15	−0.0017
16	−0.0008

Figure 12.8 Example of CORDIC computing sine and cosine for $\theta = 43$ degrees. (a) Table of values of θ_i for every iteration of the CORDIC algorithm. (b) Plot showing convergence

negative side with $\theta_3 = -4.73$. The algorithm now gives positive rotation $\Delta\theta_3 = \tan^{-1}2^{-3}$ and keeps working to make the final angle equal to 0, and in the final iteration the angle $\theta_{16} = -0.0008$ degrees.

12.4.3 Hardware Mapping

The CORDIC algorithm exhibits natural affinity for HW mapping. Each iteration of the algorithm can be implemented as a CORDIC element (CE). A CE is shown in Figure 12.9(a). This CE implements ith iteration of the algorithm given by (12.14) and is shown in Figure 12.9(b).

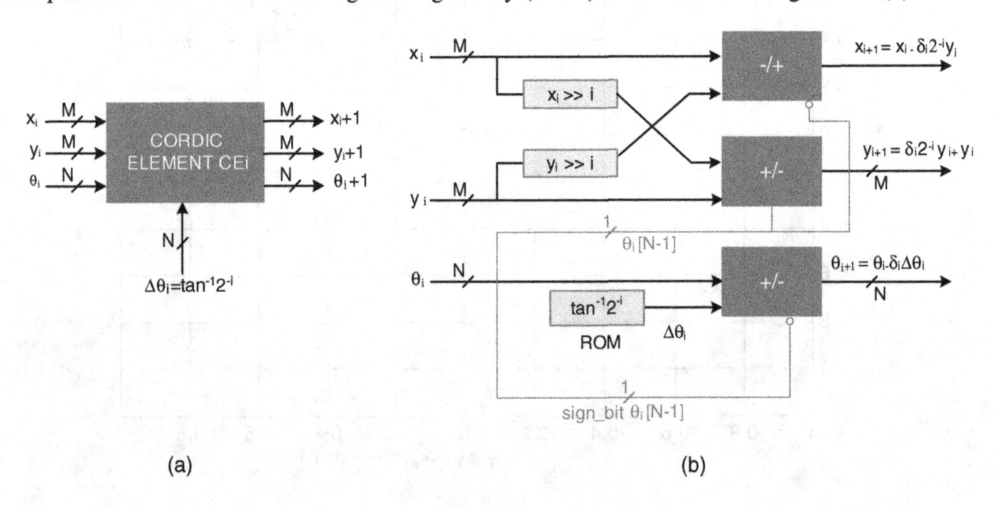

Figure 12.9 Hardware mapping. (a) A CORDIC element. (b) An element implementing the ith iteration of the CORDIC algorithm

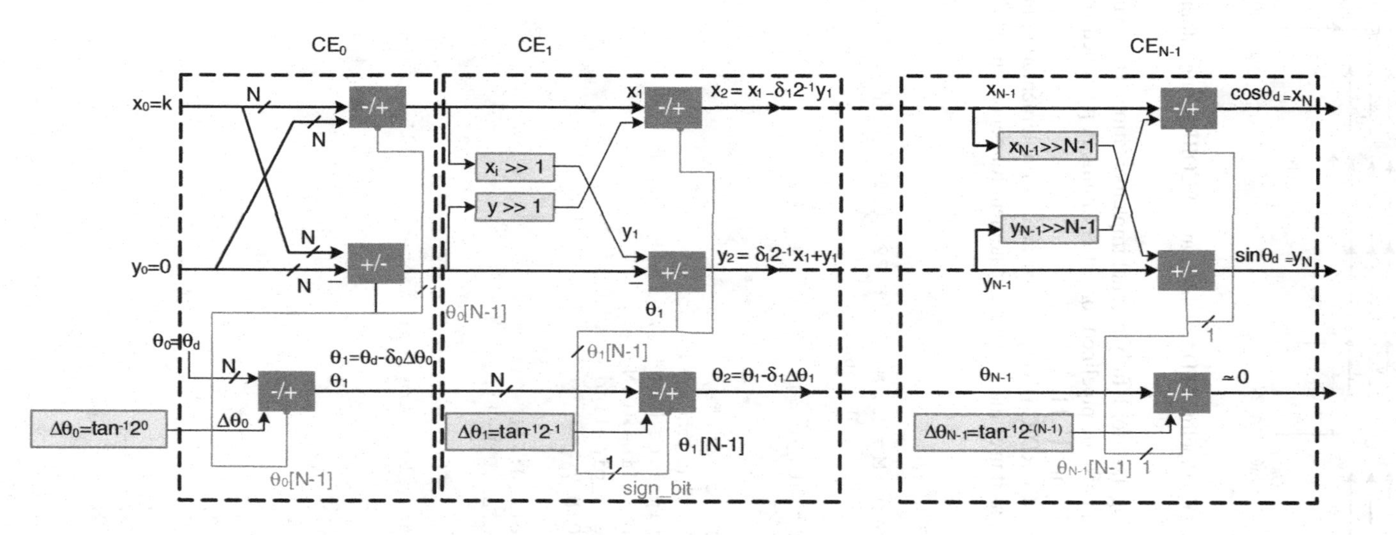

Figure 12.10 FDA architecture of CORDIC as cascade of CEs

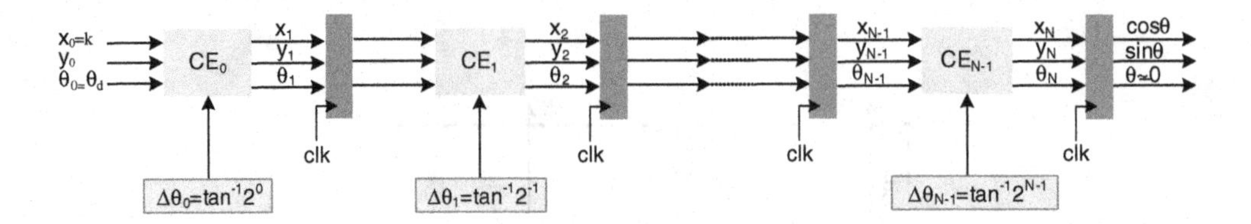

Figure 12.11 Pipelined FDA architecture of CORDIC algorithm

These CEs are cascaded together for a fully parallel implementation. This mapping is shown in Figure 12.10. This design can also be pipelined for better timing. A pipelined version of the fully parallel design is shown in Figure 12.11.

This pipeline structure for CORDIC is very regular and its implementation in Verilog is given below. The code is compact as it uses the `for` loop statement for generating multiple instances of CEs and pipeline registers.

```
module CORDIC #(parameter M = 22, N = 16, K=22'h0DBD96)
// In Q2.20 format value of K = 0.8588
(
input signed [N - 1:0] theta_d,
input clk,
input rst_n,
output signed [M - 1:0] cos_theta,
output signed [M - 1:0] sine_theta);
reg signed [M-1:0] x_pipeline [0:N-1];
reg signed [M-1:0] y_pipeline [0:N-1];
reg signed [N-1:0] theta_pipeline [0:N-1];
reg signed [M-1:0] x[0:N];
reg signed [M-1:0] y[0:N];
reg signed [N-1:0] theta[0:N];
reg signed [N-1:0] arcTan[0:N-1];
integer i;

// Arctan table: radian values are represented in Q1.15 format
always @*
begin
  arcTan[0] = 16'h3B59;
  arcTan[1] = 16'h1F5B;
  arcTan[2] = 16'h0FEB;
  arcTan[3] = 16'h07FD;
  arcTan[4] = 16'h0400;
  arcTan[5] = 16'h0200;
  arcTan[6] = 16'h0100;
  arcTan[7] = 16'h0080;
  arcTan[8] = 16'h0040;
  arcTan[9] = 16'h0020;
  arcTan[10] = 16'h0010;
  arcTan[11] = 16'h0008;
  arcTan[12] = 16'h0004;
  arcTan[13] = 16'h0002;
  arcTan[14] = 16'h0001;
```

```
  arcTan[15] = 16'h0000;
end
always @*
begin
  x[0] = K;
  y[0] = 0;
  theta[0] = theta_d;
  CE_task(x[0], y[0], theta[0], arcTan[0], 4'd1, x[1],y[1], theta[1]);
  for (i=0; i<N-1; i=i+1)
  begin
    CE_task(x_pipeline[i], y_pipeline[i], theta_pipeline[i],
     arcTan[i+1], i+2, x[i+2], y[i+2], theta[i+2]);
  end
end
always @(posedge clk)
begin
 for(i=0; i<N-1; i=i+1)
 begin
   x_pipeline[i] <= x[i+1];
   y_pipeline[i] <= y[i+1];
   theta_pipeline[i] <= theta[i+1];
 end
end

assign cos_theta = x_pipeline[N-2];
assign sine_theta =y_pipeline[N-2];

task CE_task(
input signed [M - 1:0] x_i,
input signed [M - 1:0] y_i,
input signed [N - 1:0] theta_i,
input signed [N - 1:0] Delta_theta,
input [3:0]i,
output reg signed [M - 1:0] x_iP1,
output reg signed [M - 1:0] y_iP1,
output reg signed [N - 1:0] theta_iP1);

reg sigma, sigma_bar;
reg signed [M - 1:0] x_input, y_input;
reg signed [M - 1:0] x_shifted, y_shifted, x_bar_shifted,
 y_bar_shifted;
reg signed [N - 1:0] Delta_theta_input, Delta_theta_bar;
begin
  sigma = theta_i[N-1]; // Sign bit of the angle
  sigma_bar = _sigma;
  x_shifted = x_i >>> i; // Shift by 2^-i
  y_shifted = y_i >>> i; // Shift by 2^-i
  x_bar_shifted = _x_shifted + 1;
  y_bar_shifted = _y_shifted + 1;
  Delta_theta_bar = _Delta_theta + 1;
  if ((sigma) || (theta_i == 0))
 begin
   x_input = x_bar_shifted; // Subtract if sigma is negative
   y_input = y_shifted; // Add if sigma is negative
```

```
 Delta_theta_input = Delta_theta; // Add if sigma is negative
end
else
begin
   x_input = x_shifted;
   y_input = y_bar_shifted;
   Delta_theta_input = Delta_theta_bar;
end
   x_iP1 = x_i + y_input;
   y_iP1 = x_input + y_i;
   theta_iP1 = theta_i + Delta_theta_input;
end
endtask
endmodule
```

The Verilog implementation is tested for multiple values of desired angle θ_d. The output values are stored in a file. The Verilog code for the stimulus is given here:

```
module stimulus;
parameter M = 22;
parameter N = 16;
reg signed [N - 1:0] theta_d;
reg clk, rst_n;
wire signed [M - 1:0] cos_theta;
wire signed [M - 1:0] sine_theta;
integer i;
reg valid;
integer outFile;

CORDIC cordicParallel
(
    theta_d,
    clk,
    rst_n,
    cos_theta,
    sine_theta);
initial
begin
    clk = 0;
    valid = 0;
    outFile = $fopen("monitor.txt","w");
    # 300 valid = 1;
    # 395 valid = 0;
end
initial
begin
    #5
    theta_d = -29491;
    // For theta_d = -0.9:0.1:0.9
    for (i=0; i<20 ; i=i+1)
        #20 theta_d = theta_d + 3276;
        #400 $fclose(outFile);
        $finish;
```

```
end
always
     #10 clk = ~clk;
initial
$monitor($time, " theta = %d, cos_theta = %d, sine_theta = %d",
  theta_d, cos_theta, sine_theta);
// $monitor(" \t%d \t%d \t%d", theta_d, cos_theta, sine_theta);
  always@ (posedge clk)
    if(valid)
    $fwrite(outFile, " %d %d\n", cos_theta, sine_theta);
endmodule
```

The `monitor.txt` file is read in MATLAB® as a 19 × 2 matrix, *results*. The first and second columns list the values of $\cos\theta$ and $\sin\theta$, respectively. The following MATLAB® code compares the results by placing them on adjacent columns and their difference in the next column:

```
a_v =results(1:19,1)/2^20; % Converting from Q2.20 to floating point
a=(cos(-0.9:0.1:0.9))';
b_v=results(1:19,2)/2^20; % Converting from Q2.20 format to floating point
b=(sin(-0.9:0.1:0.9))';
[a a_v a-a_v b b_v b-b_v]
```

The results for values of θ from −0.9 to +0.9 with an increment of 0.1 radians are computed. The output from double-precision MATLAB® code using functions $\cos\theta$ and $\sin\theta$ and from fixed-point Verilog simulation are compared in Table 12.1. The first two columns are values of $\cos\theta$ from

Table 12.1 Comparison of results of double-precision floating-point MATLAB® and fixed-point Verilog implementations

θ	MATLAB® $\cos\theta_M$	Verilog $\cos\theta_V$	Difference $\cos\theta_M - \cos\theta_V$	MATLAB® $\sin\theta_M$	Verilog $\sin\theta_V$	Difference $\sin\theta_M - \sin\theta_V$
−0.9000	0.6216	0.6216	0	−0.7833	−0.7833	0
−0.8000	0.6967	0.6967	0	−0.7174	−0.7173	0
−0.7000	0.7648	0.7648	0	−0.6443	−0.6442	0
−0.6001	0.8253	0.8253	0	−0.5647	−0.5647	0
−0.5001	0.8775	0.8775	0	−0.4795	−0.4795	0
−0.4001	0.9210	0.9210	0	−0.3895	−0.3895	0
−0.3001	0.9553	0.9553	0	−0.2957	−0.2956	−0.0001
−0.2002	0.9800	0.9800	0	−0.1988	−0.1988	0
−0.1002	0.9950	0.9950	0	−0.1000	−0.1000	0
−0.0002	1.0000	1.0000	0	−0.0002	−0.0002	0
0.0998	0.9950	0.9950	0	0.0996	0.0996	0
0.1997	0.9801	0.9801	0	0.1984	0.1984	0
0.2997	0.9554	0.9554	0	0.2952	0.2952	0
0.3997	0.9212	0.9212	0	0.3891	0.3891	0
0.4997	0.8777	0.8777	0	0.4791	0.4792	0
0.5996	0.8255	0.8255	0	0.5643	0.5644	0
0.6996	0.7651	0.7651	0	0.6439	0.6439	0
0.7996	0.6970	0.6970	0	0.7171	0.7171	0
0.8996	0.6219	0.6219	0	0.7831	0.7831	0

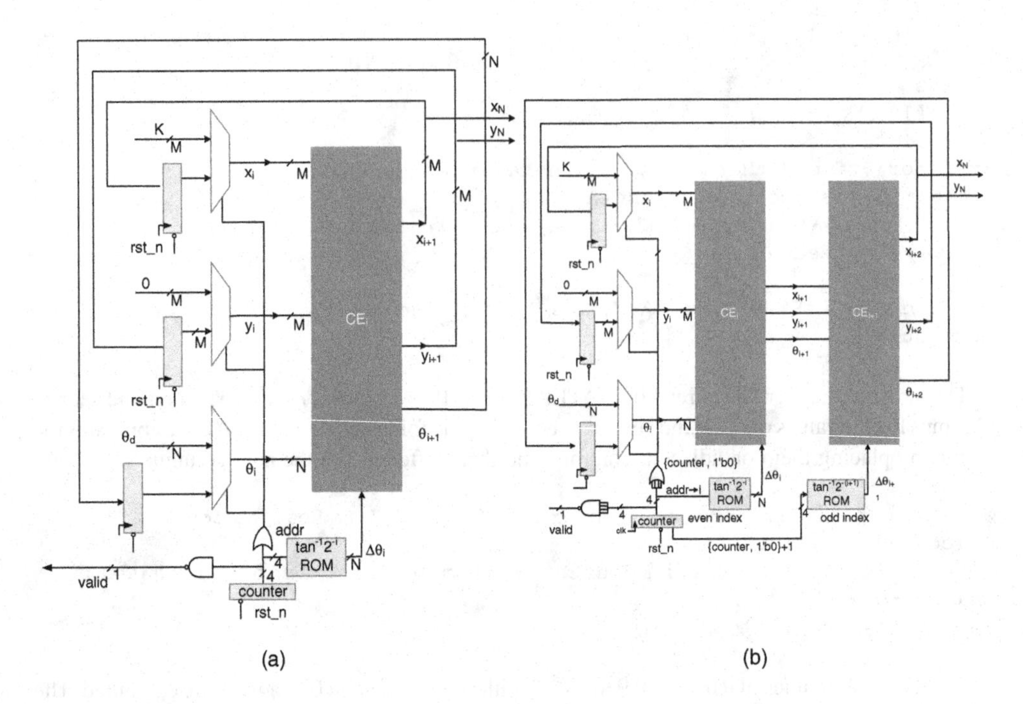

Figure 12.12 Folded CORDIC architecture. (a) Folding factor of 16. (b) Folding factor of 8

MATLAB® and Verilog, respectively, the third column shows the difference in the results. Similarly the last three columns are the same set of values for $\sin\theta$.

12.4.4 Time-shared Architecture

The CORDIC algorithm is very regular and a folded architecture can be easily crafted. Depending on the number of cycles available for computing the sine and cosine values, the CORDIC algorithm can be folded by any folding factor. For $N = 16$, the architecture in Figure 12.12(a) is folded by a folding factor of 16. Similarly Figure 12.13(b) shows a folded architecture by folding factor of 8. The valid signal is asserted after the design computes the N iterations. The RTL Verilog for a folding factor of 16 is given here:

```
module CORDIC_Shared #(parameter M = 22, N = 16, LOGN = 4, K=22'h0DBD96)
// In Q2.20 format, value of K = 0.8588 K=22'h0DBD96)
(
input signed [N - 1:0] theta_d,
input clk,
input rst_n,
output reg signed [M - 1:0] cos_theta,
output reg signed [M - 1:0] sin_theta);
reg signed [M-1:0] x_reg;
reg signed [M-1:0] y_reg;
reg signed [N-1:0] theta_reg;
```

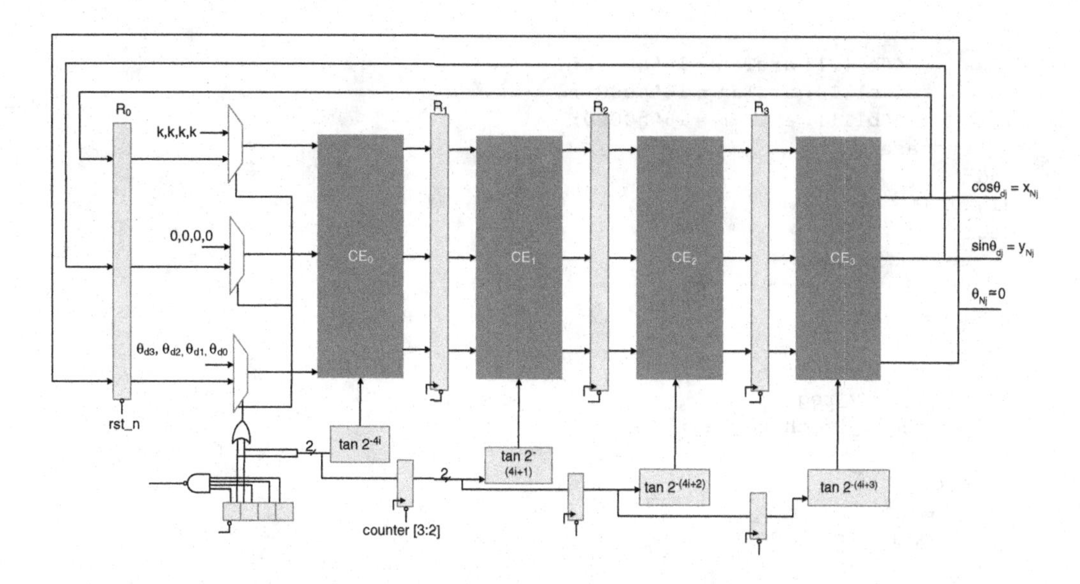

Figure 12.13 Four-slow folded architecture by a folding factor of 4

```
reg signed [M - 1:0] x_o;
reg signed [M - 1:0] y_o;
reg signed [M-1:0] x_i;
reg signed [M-1:0] y_i;
reg signed [N-1:0] theta_i, theta_o;
reg signed [N-1:0] arcTan;
reg [LOGN-1:0] counter;
reg sel, valid;

integer i;
// Arctan table: radian values are represented in Q1.15 format
always@*
begin
   case (counter)
         4'b0000: arcTan = 16'h3B59;
         4'b0001: arcTan = 16'h1F5B;
         4'b0010: arcTan = 16'h0FEB;
         4'b0011: arcTan = 16'h07FD;
         4'b0100: arcTan = 16'h0400;
         4'b0101: arcTan = 16'h0200;
         4'b0110: arcTan = 16'h0100;
         4'b0111: arcTan = 16'h0080;
         4'b1000: arcTan = 16'h0040;
         4'b1001: arcTan = 16'h0020;
         4'b1010: arcTan = 16'h0010;
         4'b1011: arcTan = 16'h0008;
         4'b1100: arcTan = 16'h0004;
```

```
         4'b1101: arcTan = 16'h0002;
         4'b1110: arcTan = 16'h0001;
         4'b1111: arcTan = 16'h0000;
    endcase
end
always@*
begin
    sel = |counter;
    valid = &counter;
    if(sel)
      begin
        x_i = x_reg;
        y_i = y_reg;
        theta_i = theta_reg;
      end
    else
      begin
        x_i = K;
        y_i = 0;
        theta_i = theta_d;
      end
      CE_task(x_i, y_i, theta_i, arcTan, counter
    + 1, x_o, y_o, theta_o );
end
always@(posedge clk)
  begin
  if(!rst_n)
    begin
        x_reg <= 0;
        y_reg <= 0;
        theta_reg <= 0;
        counter <= 0;
      end
else if (!valid)
  begin
      x_reg <= x_o;
      y_reg <= y_o;
      theta_reg <= theta_o;
      counter <= counter + 1;
    end
else
  begin
      cos_theta <= x_o;
      sin_theta <= y_o;
    end
endmodule

module stimulus;
parameter M = 22;
parameter N = 16;
reg signed [N - 1:0] theta_d;
reg clk;
```

```
reg rst_n;
wire signed [M - 1:0] x_o;
wire signed [M - 1:0] y_o;
integer i;

CORDIC_Shared CordicShared(
    theta_d,
    clk,
    rst_n,
    x_o,
    y_o);
initial
begin
    clk = 1; rst_n = 0; theta_d = 0;
    #10
    rst_n = 1;
    theta_d = -29491;
    // For theta_d = -0.9:0.1:0.9
    for (i=0; i<20 ; i=i+1)
begin
    rst_n = 0;
    #2 rst_n = 1;
    #400
     theta_d = theta_d + 3276;
    end
    #10000
    $finish;
end
always
    #10 clk = ~clk;
initial
$monitor($time, " theta = %d, cos_theta = %d, sine_theta = %d,
  reset = %d", theta_d, x_o, y_o, rst_n);
endmodule
```

The recursive nature of the time-shared architecture poses a major limitation in folding the architecture at higher degrees as it increases the critical path of the design. The datapath needs to be pipelined but the registers cannot be added straight away. The look-ahead transformation of Chapter 7 can also not be directly applied as the decision logic of computing σ_i in every stage of the algorithm restricts application of this transform.

12.4.5 C-slowed Time-shared Architecture

The time-shared architecture in its present configuration cannot be pipelined and can only be C-slowed. The technique to C-slow a design for running multiple threads or instances of input sequences is discussed in detail in Chapter 7. The feedback register is replicated C times and then these registers are retimed to reduce the critical path. This enables the user to initiate C computations of the algorithm in parallel in the time-shared architecture. Each iteration takes N cycles, but the design computes C values in these cycles. A design for $N = 16$ and $C = 4$ is shown in Figure 12.13.

A simple counter-based controller is used to appropriately select the input to each CE_i. Two MSBs of the counter are used as select line to three multiplexers. In the first four cycles, four desired angles and associated values of x_0 and y_0 are input to CE_0. All subsequent cycles feed the values from CE_3 to R_0 to CE_0. The successive working of the algorithm for the initial few cycles is elaborated in the timing diagram of Figure 12.14.

The architecture works in lock step calculating N iterations for every input angle, and produces $\cos\theta_{di}$ and $\sin\theta_{di}$ for $i = 0, \ldots, 3$ in four consecutive cycles after the 16th cycle.

12.4.6 Modified CORDIC Algorithm

An algorithm selected for functional simulation in software often poses serious a limitation in achieving HW-specific design objectives. The basic CORDIC algorithm described in Section 12.4.1 is a good example of this limitation. The designer in these cases should explore other HW affine algorithmic options for implementing the same functionality. For the CORDIC algorithm, a modified version is an option of choice that eliminates limitations in exploring parallel and time-shared architecture.

The CORDIC algorithm of Section 12.4.1 requires computation of σ_i and only then it conditionally adds or subtracts one of the operands while implementing (12.14). This conditional logic restricts the HW design to exploit inherent parallelism in the algorithm. A simple modification can eliminate this restriction and efficient parallel architectures can be realized.

The standard CORDIC algorithm assumes θ as a summation of N positive or negative micro-rotations of angles $\Delta\theta_i$ as given by (12.14). A binary representation of a positive value of θ as below can also be considered for micro rotations:

$$\theta = \sum_{i=0}^{N-1} b_i 2^{-i} \quad \text{for } b_i \in \{0, 1\} \tag{12.15}$$

where each term in the summation requires either a positive rotation equal to 2^{-i} or no rotation, depending on the value of the bit b_i at location i. This representation cannot be directly used in HW as the constant k of (12.13) becomes data dependent. A modification in the binary representation of θ of (12.15) is thus required that makes values of k data independent.

This independence can be accomplished by recoding the expression in (12.15) to only use $+1$ or -1. This recoding of the binary representation is explained next.

12.4.7 Recoding of Binary Representation as ±1

An N-bit unsigned number b in Q1.(N-1) format can be represented as:

$$b = \sum_{i=0}^{N-1} b_i 2^{-i} \text{ where } b_i \in \{0, 1\}$$

The bits b_i in the expression can be recoded to $r_i \in \{+1, -1\}$ as:

$$b = \sum_{i=0}^{N-1} b_i 2^{-i} = \sum_{i=0}^{N-1} r_i 2^{-(i+1)} + 2^{-0} - 2^{-N} \tag{12.16}$$

$$r_i = 2b_i - 1 \text{ where } r_i \in \{+1, -1\}$$

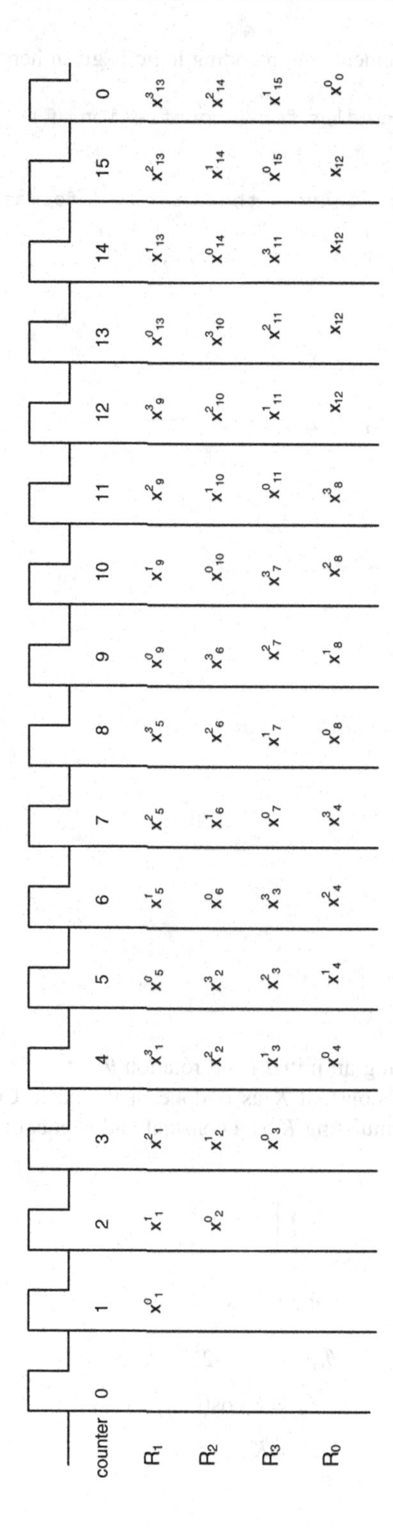

Figure 12.14 Timing diagram of 4-slow folded CORDIC architecture of Figure 12.13

The MATLAB® code that implements the recoding logic is given here:

```
theta_d = 1.394; % An example value for demonstration of the equivalence
theta = theta_d;
N = 16;
theta = fix(theta * 2^(N-1)); % Convert theta to Q1.15 format

% Seperating bits of theta as b=b[1],b[2],....,b[N]
for i=1:N
    b(N+1-i) = rem(theta, 2);
    theta = fix(theta/2);
end

% Recoding 0, 1 bits of b as -1,+1 of r respectively
for i=1:N
    r(i) = 2*b(i) - 1;
end

% Computing value of theta using b
Kb = 0;
for i = 1:N
    Kb = Kb + b(i)*2^(-(i-1));
end

% Computing value of theta using r
Kr=0;
for i = 1:N
    Kr = Kr + r(i)*2^(-(i));
end

% The constant part
Kr = Kr + (2^0) - 2^(-N);

% The three values are the same
[Kr Kb theta_d]
```

This recoding requires first giving an initial fixed rotation θ_{init} to cater for the constant factor $(2^0 - 2^{-N})$ along with computing constant K as is done in the basic CORDIC algorithm. The recoding of b_is as ± 1 helps in formulating K as a constant and is equal to:

$$K = \prod_{i=1}^{N-1} \cos(2^{-(i+1)})$$

The rotation for θ_{init} can then be first applied, where:

$$\theta_{init} = 2^{-0} - 2^{-(N+1)}$$

$$x_0 = K\cos(\theta_{init})$$

$$y_0 = K\sin(\theta_{init})$$

Now to cater for the recoding part, the problem is reformulating to compute the following set of iterations for $i = 1, \ldots, N-1$:

$$x_i = x_{i-1} - r_i \tan 2^{-i} y_{i-1} \tag{12.17a}$$

$$y_i = r_i \tan 2^{-i} x_{i-1} + y_{i-1} \tag{12.17b}$$

where, unlike σ_i, the values of r_i are predetermined, and these iterations do not include any computations of the $\Delta\theta_i$ as are done in the basic CORDIC algorithm.

```
clear all
close all
N=16;

% Computing the constant value K, the recording part
K = 1;
for i = 0:N-1
    K = K * cos(2^(-(i+1)));
end

% The constant initial rotation
theta_init = (2)^0 - (2)^(-N);
x0 = K*cos(theta_init);
y0 = K*sin(theta_init);
cosine = [];
sine = [];
for theta = 0:.1:pi/2
    angle = [0:.1:pi/2];
theta = round(theta * 2^(N-1)); % convert theta in Q1.15 format

% Separating bits of theta as b=b[1],b[2],....,b[N]
for k=1:N
    b(N+1-k) = rem(theta, 2);
    theta = fix(theta/2);
end

% Recoding shifted bits of b as r with +1,-1
for k=1:N
    r(k) = 2*b(k) - 1;
end

% First modified CORDIC rotation
x(1) = x0 - r(1)*(tan(2^(-1)) * y0);
y(1) = y0 + r(1)*(tan(2^(-1)) * x0);
% Remainder of the modified CORDIC rotations
for k=2:N,
    x(k) = x(k-1) - r(k)* tan(2^(-k)) * y(k-1);
    y(k) = y(k-1) + r(k) * tan(2^(-k)) * x(k-1);
end

cosine = [cosine x(k)];
sine = [sine y(k)];

end
```

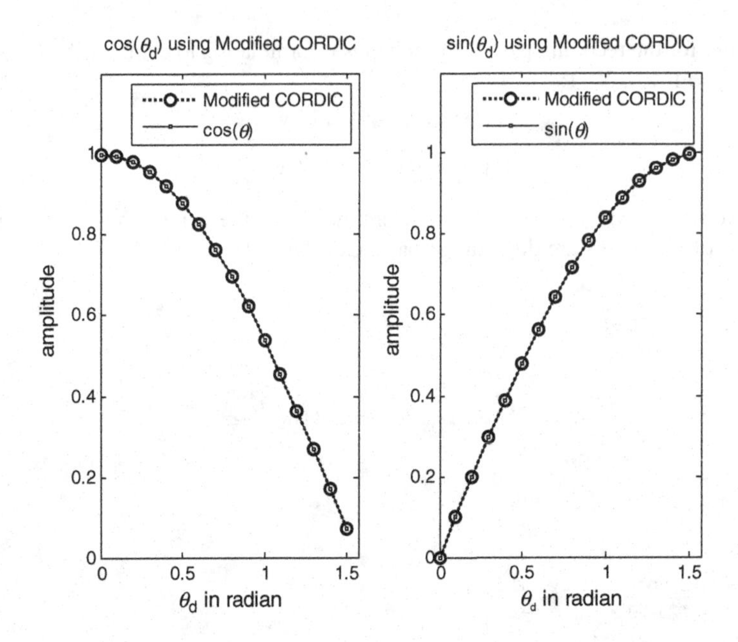

Figure 12.15 Results using CORDIC and modified CORDIC algorithms

Figure 12.15 shows the results of using CORDIC and modified CORDIC algorithms. A mean squared error (MSE) comparison of the two algorithms for different numbers of rotations N is given in Figure 12.16. The error is calculated for P sets of computation by computing the mean squared difference between the value $\cos\theta_i$ in double-precision arithmetic and the value using the CORDIC algorithm for the quantizing value of θ_i as θ_{Qi} in Q1.($N-1$) format. The expression for MSE while considering θ_i as an N-bit number in Q1.($N-1$) format is:

$$MSE_N = \frac{1}{P}\sum_{i=0}^{P-1}(\cos\theta_i - CORDIC(\theta_{Qi}))^2 \text{ where } \theta_{Qi} = \text{round}(\theta_i \times 2^{N-1})$$

It is clear from the two plots that, for $N > 10$, mean squared error for both the algorithm is very small, so $N = 16$ is a good choice for the CORDIC algorithms.

12.5 Hardware Mapping of Modified CORDIC Algorithm

12.5.1 Introduction

One issue in a modified CORDIC algorithm is eliminating $\tan 2^{-i}$ from (12.17) as it requires multiplication in every stage. This multiplication can be avoided for stages for $i > 4$ as the values of $\tan 2^{-i}$ can be approximated as:

$$\tan 2^{-i} \approx 2^{-i} \text{ for } i > 4 \tag{12.18}$$

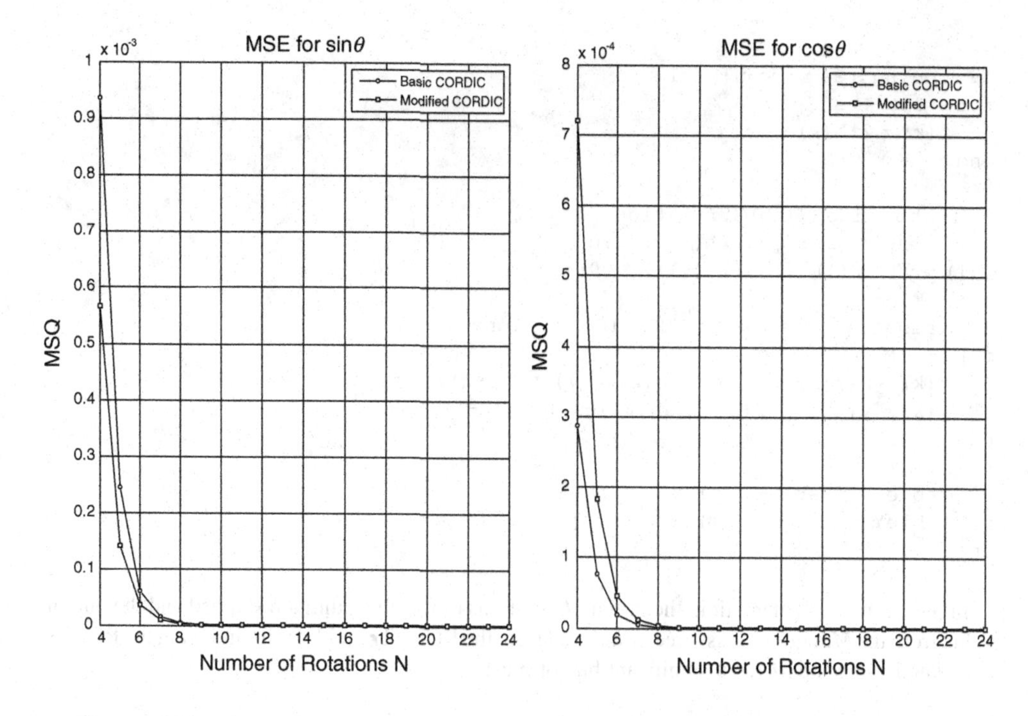

Figure 12.16 Mean square error comparison of CORDIC and modified CORDIC algorithms

This leaves us to pre-compute all possible values of the initial four iterations and store them in a ROM. The MATLAB® code for generating this ROM input is given here:

```
tableX=[];
tableY=[];
N = 16;
K = 1;
for i = 1:N
   K = K * cos(2^(-(i)));
end

% The constant initial rotation
theta_init = (2)^0 - (2)^(-N);
x0 = K*cos(theta_init);
y0 = K*sin(theta_init);
cosine = [];
sine = [];

M = 4;
for index = 0:2^M-1
for k=1:M
   b(M+1-k) = rem(index, 2);
   index = fix(index/2);
  end
```

```
% Recoding b as r with +1,-1
for k=1:M
    r(k) = 2*b(k) - 1;
end

% First modified CORDIC rotation
x(1) = x0 - r(1)*(tan(2^(-1)) * y0);
y(1) = y0 + r(1)*(tan(2^(-1)) * x0);

% Remainder of the modified CORDIC rotations
for k=2:M,
    x(k) = x(k-1) - r(k)* tan(2^(-k)) * y(k-1);
    y(k) = y(k-1) + r(k) * tan(2^(-k)) * x(k-1);
end

    tableX = [tableX x(M)];
    tableY = [tableY y(M)];
end
```

In hardware implementation, the initial M iterations of the algorithm are skipped and the output value from the Mth iteration is directly indexed from the ROM. The address for indexing the ROM is calculated using the M most significant bits of b as:

$$index = b_0 2^{M-1} + b_1 2^{M-2} + \ldots + b_{M-1} 2^0 \tag{12.19}$$

Reading from the ROM directly gives $x[M-1]$ and $y[M-1]$ values. The rest of the values of $x[k]$ and $y[k]$ are then generated by using the approximation of (12.18). This substitution replaces multiplication by $\tan 2^{-i}$with a shift by 2^{-i} operation. The equations implementing simplified iterations for $i = M+1, M+2, \ldots, N$ are:

$$x_i = x_{i-1} - r_i 2^{-i} y_{i-1} \tag{12.20a}$$

$$y_i = r_i 2^{-i} x_{i-1} + y_{i-1} \tag{12.20b}$$

Fixed-point implementation of the modified CORDIC algorithm that uses tables for directly indexing the value for the Mth iteration and implements (12.20) for the rest of the iterations is given here:

```
P = 22;
N = 16;
M = 4;
% Tables are computed for P=22, M=4 and values are in Q2.20 format
tableX = [
      1045848
      1029530
      997147
      949203
      886447
```

```
    809859
    720633
    620162
    510013
    391906
    267683
    139283
    8710
    -122000
    -250805
    -375697];
tableY = [
    65435
    195315
    322148
    443953
    558831
    664988
    760769
    844678
    915406
    971849
    1013127
    1038596
    1047857
    1040767
    1017437
    978229];
cosine = [];
sine = [];
for theta = 0:.01:pi/2
    angle = [0:.01:pi/2];
    theta = round(theta * 2^(N-1)); % convert theta in Q1.N-1 format

% Seperating bits of theta as b=b[1],b[2],....,b[N]
for k=1:N
    b(N+1-k) = rem(theta, 2);
    theta = fix(theta/2);
  end

% Compute index for M = 4;
index = b(1)*2^3 + b(2)*2^2 + b(3)*2^1 + b(4)*2^0+1;
x(4)=tableX(index);
y(4)=tableY(index);

% Recoding b[M+1], b[M+2], ..., b[n] as r with +1,-1
for k=M+1:N
  r(k) = 2*b(k) - 1;
end

% Simplified iterations for modified CORDIC rotations
for k=M+1:N,
```

```
    x(k) = x(k-1) - r(k) * 2^(-k) * y(k-1);
    y(k) = y(k-1) + r(k) * 2^(-k) * x(k-1);
end

% For plotting, convert the values of x and y to
                              floating point format
        cosine = [cosine x(k)/(2^(P-2))];
        sine = [sine y(k)/(2^(P-2))];
end
```

This modification results in simple fully parallel and time-shared HW implementations. The fully parallel architecture is shown in Figure 12.17. The architecture can also be easily pipelined.

These iterations can also be merged together to be reduced using a compression tree for effective HW mapping [9]. A merged cell that takes partial-sum and partial-carry results from state $(i-1)$ and then generates partial results for iteration i while using a 5:2 compression tree is shown in Figure 12.18.

For the time-shared architecture, multiple CEs can be used to fold the architecture for any desired folding factor.

12.5.2 Hardware Optimization

In many applications the designer may wish to explore optimization above what is apparently perceived from the algorithm. There is no established technique that can be generally used, but a deep analysis of the algorithm often reveals novel ways of optimization. This section demonstrates this assertion by presenting a novel algorithm that computes the sine and cosine values in a single stage, thus enabling zero latency with very low area and power consumption [10].

Consider a fixed-point implementation of the modified CORDIC algorithm. As the iterations now do not depend on values of $\Delta\theta_i$, the values of previous iterations can be directly substituted into the current iteration. If we consider $M=4$, then indexing into the tables gives the values of x_4 and y_4. Now these values are used to compute the iteration for $i=5$ as:

$$x_5 = x_4 - r_5 2^{-5} y_4 \tag{12.21a}$$

$$y_5 = r_5 2^{-5} x_4 + y_4 \tag{12.21b}$$

The iteration for $i=6$ calculates:

$$x_6 = x_5 - r_6 2^{-6} y_5$$

$$y_6 = r_6 2^{-6} x_5 + y_5$$

Now substituting expressions for x_5 and y_5 from (12.21) in the above expressions, we get:

$$\begin{aligned} x_6 &= \left(1 - r_5 r_6 2^{-11}\right) x_4 - \left(r_5 2^{-5} + r_6 2^{-6}\right) y_4 \\ y_6 &= \left(r_5 2^{-5} + r_6 2^{-6}\right) x_4 + \left(1 - r_5 r_6 2^{-11}\right) y_4 \end{aligned}$$

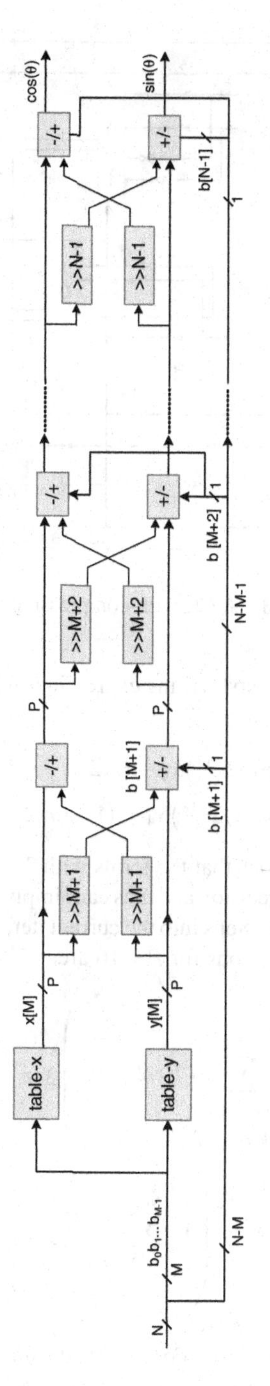

Figure 12.17 FDA of modified CORDIC algorithm

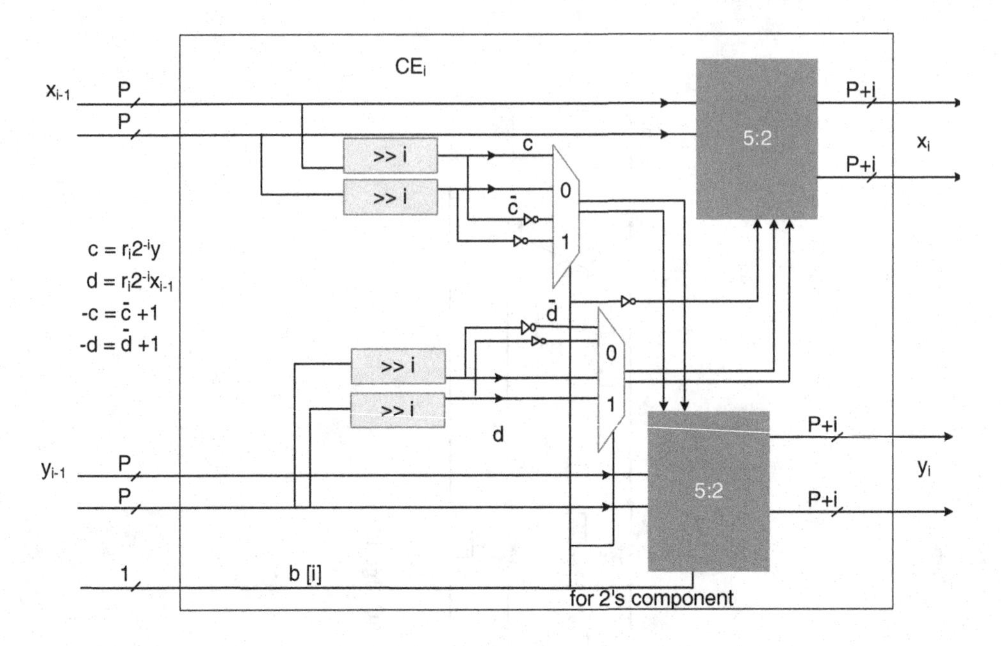

Figure 12.18 A CE with compression tree

Adopting the same procedure – that is, first writing expressions for x_7 and y_7 and then substituting values of x_6 and y_6 – we get:

$$x_7 = \left(1 - r_5 r_6 2^{-11} - r_5 r_7 2^{-12} + r_6 r_7 2^{-13}\right) x_4 - \left(r_5 2^{-5} + r_6 2^{-6} + r_7 2^{-7} - r_5 r_6 r_7 2^{-18}\right) y_4$$

$$y_7 = \left(r_5 2^{-5} + r_6 2^{-6} + r_7 2^{-7} - r_5 r_6 r_7 2^{-18}\right) x_4 + \left(1 - r_5 r_6 2^{-11} - r_5 r_7 2^{-12} + r_6 r_7 2^{-13}\right) y_4$$

It is evident from the above expressions that the terms with 2^{-x} with $x > P$ will shift the entire value outside the range of the required precision and thus can simply be ignored. If all these terms are ignored and we substitute previous expressions into the current iteration, we get the final iteration as a function of x_4 and y_4. The final expressions for $P = 16$ are:

$$\cos\theta = \left(1 - \sum_{i=5}^{N-1} \sum_{\substack{j=i+1 \\ (i+j) \le N}}^{N-1} r_i r_j 2^{-(i+j)} \right) x_4 - \left(\sum_{i=5}^{N-1} r_i 2^{-i} \right) y_4$$

$$\sin\theta = \left(\sum_{i=5}^{N-1} r_i 2^{-i} \right) x_4 + \left(1 - \sum_{i=5}^{N-1} \sum_{\substack{j=i+1 \\ (i+j) \le N}}^{N-1} r_i r_j 2^{-(i+j)} \right) y_4$$

Each term in a bracket is reduced to one term. For $P = 16$, the first term in the bracket results in 12 terms and the second bracket also contains 12 terms. These expressions can be implemented as two compression trees. All the iterations of the CORDIC algorithm are merged into one expression and the expression can be effectively computed in a single cycle.

The MATLAB® code for the algorithm is given below. This uses the same table as generated in MATLAB® code earlier for computing values of x_4 and y_4:

```
cosine = [];
sine = [];
N = 16;
P = 16;
M = 4;

for theta = .00:.05:pi/2
    angle =[0.00:.05:pi/2];
    theta = round(theta * 2^(N-1)); % convert theta in Q1.N format

% Separating bits of theta as b=b[1],b[2],....,b[N]
for k=1:N
    b(N+1-k) = rem(theta, 2);
    theta = fix(theta/2);
end

% Compute index for M = 4;
index = b(1)*2^3 + b(2)*2^2 + b(3)*2^1 + b(4)*2^0+1;
    x4=tableX(index);
    y4=tableY(index);

% Recoding b[M+1], b[M+2], ..., b[n] as r with +1,-1
for k=M+1:N
    r(k) = 2*b(k) - 1;
end

    xK = (1-r(5)*r(6)*2^(-11) -r(5)*r(7)*2^(-12) -r(5)*r(8)*2^(-13)
                -r(5)*r(9)*2^(-14) -r(5)*r(10)*2^(-15)
                -r(5)*r(11)*2^(-16) ... -r(6)*r(7)*2^(-13)
                -r(6)*r(8)*2^(-14)-r(6)*r(9)*2^(-15)
                -r(6)*r(10)*2^(-16) ... -r(7)*r(8)*2^(-15)
                -r(7)*r(9)*2^(-16)*x4 - (r(5)*2^(-5)+r(6)*2^(-6)
                +r(7)*2^(-7)+r(8)*2^(-8)+r(9)*2^(-9)+r(10)*2^(-10)
                ... +r(11)*2^(-11)+r(12)*2^(-12)+r(13)*2^(-13)
                +r(14)*2^(-14)+r(15)*2^(-15)+r(16)*2^(-16))*y4;
    yK = (r(5)*2^(-5)+r(6)*2^(-6)+r(7)*2^(-7)+r(8)*2^(-8)+r(9)*2^(-9)
                +r(10)*2^(-10) ... +r(11)*2^(-11)+r(12)*2^(-12)
                +r(13)*2^(-13)+r(14)*2^(-14)+r(15)*2^(-15)
                +r(16)*2^(-16))*x4+ ... (1-r(5)*r(6)*2^(-11)
                -r(5)*r(7)*2^(-12) -r(5)*r(8)*2^(-13)
                -r(5)*r(9)*2^(-14) -r(5)*r(10)*2^(-15)
                -r(5)*r(11)*2^(-16) ... -r(6)*r(7)*2^(-13)
                -r(6)*r(8)*2^(-14)-r(6)*r(9)*2^(-15)
                -r(6)*r(10)*2^(-16) ... -r(7)*r(8)*2^(-15)
                -r(7)*r(9)*2^(-16))*y4;
% For plotting, convert the values of x and y to floating point format,
% using the original format used for table generation where P=22
cosine = [cosine xK/(2^(20))];
sine = [sine yK/(2^(20))];
end
```

12.5.3 Novel Optimal Hardware Design

Yet another approach to designing optimal hardware of the CORDIC is to map the expressions in r_i into two binary constants using reverse encoding and then use four multipliers and two adders to compute the desired results. This novel technique results in a single-stage CORDIC architecture. The design is shown in Figure 12.19.

The inverse coding is accomplished using (12.16). The inverse coding for *const1* is simply derived as:

$$\sum_{i=M+1}^{N-1} r_i 2^{-(i+1)} = \sum_{i=M+1}^{N-1} b_i 2^{-i} - 2^{-M} + 2^{-N} \tag{12.22}$$

In HW design, as the original b_i are kept intact they are used for computing the two constants. To cater for the 2^{-N} term in (12.22), a 1 is appended to b as the LSB b_N, and for the -2^{-M} factor the corresponding b_{M+1} bit is flipped and is assigned a negative weight. Then *const1* is given as:

$$const1 = -b'_{M+1} 2^{-M} + \sum_{i=M+2}^{N} b_i 2^{-(i-1)} \tag{12.23}$$

where b'_{M+1} is the complement of bit b_{M+1}. Similarly for *const2*, the first r_k are computed for $i = M+1, \ldots, N-1$ as:

$$rr_k = r_i r_j \text{ where } k = i + j \text{ and } k \le P$$

These rr_k are then inverse coded using (12.22) and an equivalent value is computed similar to the *const1* computation by expression in (12.23). For $N = 16$ and $P = 16$, this requires computing the

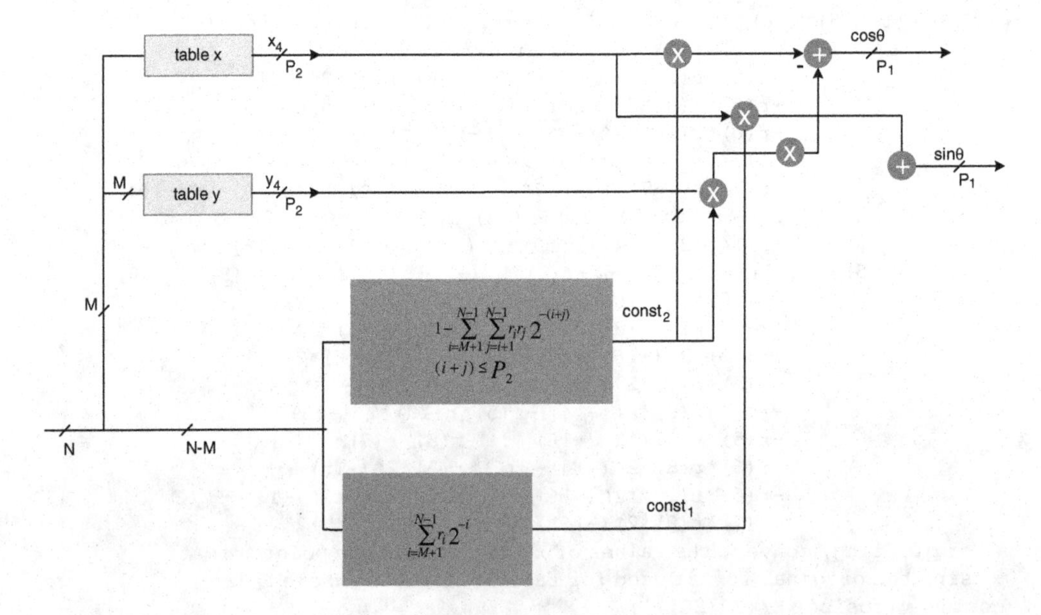

Figure 12.19 Optimal hardware design with single-stage implementation of the CORDIC algorithm

values of rr_k for three values of i. First for $i = 5$, rr_k are computed and inverse coded as a constant value:

$$t_{5,j} = r_5 r_j \quad \text{for} \quad j = 6, 7, \ldots, 11$$

These $t_{5,j}$ are reverse-coded as $b'_k s$ as:

$$\sum_{j=6}^{11} t_{5,j} 2^{-(5+j)} = \sum_{j=6}^{11} \left(b_5 \sim \hat{b}_j \right) - 2^{-2M} + 2^{-N}$$

$$= \sum_{k=2M+1}^{N-1} c_k 2^{-k} - 2^{-2M} + 2^{-N}$$

$$\text{where } c_k = b_5 \sim \hat{b}_j \quad \text{for} \quad j = 6, 7, \ldots, 11 \quad \text{where } k = 5 + j$$

Using t_k after catering for the two terms -2^{-2M} and 2^{-N}, the value of the constant beta is computed as:

$$beta_0 = -c'_{2M+1} 2^{-2M} + \sum_{k=2M+2}^{N} c_k 2^{-(k-1)}$$

Following the same steps, values of $beta_1$ and $beta_2$ are computed for $i = 6$ and $i = 7$, respectively. The details of the computation are given here in MATLAB® code:

```
clear all
close all
% Values for first four iterations in Q2.16 format
table=[65366     4090
       64346    12207
       62322    20134
       59325    27747
       55403    34927
       50616    41562
       45040    47548
       38760    52792
       31876    57213
       24494    60741
       16730    63320
        8705    64912
         544    65491
       -7625    65048
      -15675    63590
      -23481    61139];

N = 16;
P = 18; % 18-bit width for catering for truncation effects
M = 4;
cosSSC = []; sinSSC = []; % arrays of single stage cordic
cosQ = []; sinQ = []; % arrays for 16-bit fixed point precision
cosD = []; sinD = []; % arrays for double precision floating point values

for theta_f = .0:.001:pi/2
% convert theta in Q1.5 format
theta = round(theta_f * 2^(N-1)); % 16-bit fixed-point value of theta
```

```
thetaQ = theta/2^(N-1); % 16-bit quantized theta in floating point format

% separating bits of theta as
% b=b[1],b[2],....,b[N]
    for k=1:N
        b(N+1-k) = rem(theta, 2);
        theta = fix(theta/2);
    end
    %compute index for M = 4;
    index=b(1)*2^3 + b(2)*2^2 + b(3)*2^1 + b(4)+1;
    x4 = table(index,1); y4 = table(index,2);
    % recoding b[M+1],b[M+2],...,b[n] as r-> +1,-1
    for k=5:N
        r(k) = 2*b(k) - 1;
    end
    % computing const2 for all values of i and j
    % i+j<=P
    sum = 0;
    for i=5:7;
        for j=i+1:P-i
            k = i+j;
            sum = sum + r(i)*r(j)*2^(-k);
        end
    end
    const2 = 1 - sum;
    const1 = 0;
    for k=M+1:N
        const1= const1+r(k)*2^(-(k));
    end

    % single stage modified CORDIC butterfly
    xK = const2 * x4 - const1 * y4;
    yK = const1 * x4 + const2 * y4;
    % arrays for recording single state, quantized and double precision
    % results for final comparision
    cosSSC = [cosSSC xK]; sinSSC = [sinSSC yK];
    cosD = [cosD cos(thetaQ)]; sinD = [sinD sin(thetaQ)];
    cosQ = [cosQ (round(cos(thetaQ)*2^14)/2^14)]; sinQ = [sinQ
(round(sin(thetaQ)*2^14)/2^14)];
end
% computing mean square errors, Single Stage CORDIC performs better
MSEcosQ = mean((cosQ-cosD).^2); % MSEcosQ = 3.0942e-010
MSEcosSSC = mean((cosSSC./2^16-cosD).^2) % MSEcosSSC = 1.3242e-010

MSEsinQ = mean((sinQ-sinD).^2); % MSEsinQ = 3.1869e-010
MSEsinSSC = mean((sinSSC./2^16-sinD).^2); % MSEsinSSC = 1.2393e-010
MSEcosQ MSEcosSSC]
MSEsinQ MSEsinSSC]
```

The code is mapped in hardware for optimal implementation of the CORDIC algorithm. Verilog code of the design is given here (the code can be further optimized by using a compression tree for computing *const2*):

```
module CORDIC_Merged #(parameter M = 4, N = 16, P1 = 18) // Q2.16 format values of
x4 and y4 are stored in the ROM)
  (
  input [N-1:0] bin, // value of theta_d in Q1.N-1 format
   // input clk,
  output reg signed  [P1 - 1:0] cos_theta,
  output reg signed  [P1 - 1:0] sin_theta);

  reg [N-1:0] b;
  reg signed [P1-1:0] x4;
  reg signed [P1-1:0] y4;
  reg signed [P1-1:0] beta0, beta1, beta2;
  reg signed [P1-1:0] const1, const2;
  reg signed [2*P1-1:0] mpy0, mpy1, mpy2, mpy3;
  reg signed [P1-1:0] xK, yK;

  wire [5:0] c0;
  wire [3:0] c1;
  wire [1:0] c2;

  reg [M-1:0] index;

  // table to store all possible values of x4 and y4 in Q2.16 format
  always@*
  begin
  case (index)
   4'b0000: begin x4 = 65366;  y4 = 4090; end
   4'b0001: begin x4 = 64346;  y4 = 12207; end
   4'b0010: begin x4 = 62322;  y4 = 20134; end
   4'b0011: begin x4 = 59325;  y4 = 27747; end
   4'b0100: begin x4 = 55403;  y4 = 34927; end
   4'b0101: begin x4 = 50616;  y4 = 41562; end
   4'b0110: begin x4 = 45040;  y4 = 47548; end
   4'b0111: begin x4 = 38760;  y4 = 52792; end
   4'b1000: begin x4 = 31876;  y4 = 57213; end
   4'b1001: begin x4 = 24494;  y4 = 60741; end
   4'b1010: begin x4 = 16730;  y4 = 63320; end
   4'b1011: begin x4 = 8705;   y4 = 64912; end
   4'b1100: begin x4 = 544;    y4 = 65491; end
   4'b1101: begin x4 = -7625;  y4 = 65048; end
   4'b1110: begin x4 = -15675; y4 = 63590; end
   4'b1111: begin x4 = -23481; y4 = 61139; end
  endcase
  end

// the NOVALITY, computing the values of constants for single
// stage CORDIC
assign c0 = {6{b[5]}}~^b[11:6];
assign c1 = {4{b[6]}}~^b[10:7];
assign c2 = {2{b[7]}}~^b[9:8];
  always@*
  begin
```

```
    index = b[15:12];
    beta0 = {{12{~c0[5]}},c0[4:0],1'b1};
    beta1 = {{14{(~c1[3])}},c1[2:0],1'b1};
    beta2 = {{16{~c2[1]}},c2[0], 1'b1};
    const2 = {1'b0,16'h8000,1'b0} - (beta0 + beta1 + beta2);
    const1 = {{6{~b[11]}},b[10:0],1'b1};
  end

  always@*
  begin
    mpy0 = const2 * x4;
    mpy1 = const1 * y4;
    xK = mpy0[34:16] - mpy1[34:16];

    mpy2 = const1 * x4;
    mpy3 = const2 * y4;
    yK = mpy2[34:16] + mpy3[34:16];
  end

  always @* // replace by sequential block with clock for sythesis timing
  begin
   b = bin;
   cos_theta = xK;
   sin_theta = yK;
  end

endmodule

//
module stimulus;
  parameter P = 18;
    parameter N = 16;

  reg signed [N - 1:0] theta_d;
  wire signed  [P - 1:0] x_o;
  wire signed  [P - 1:0] y_o;
  integer i;
  integer outFile;

 CORDIC_Merged merged (
 theta_d,
 x_o,
 y_o);
 initial
    outFile = $fopen("monitor_merge.txt","w");

 always@ (x_o, y_o)
   $fwrite(outFile, " %d %d\n", x_o, y_o);

initial
begin
#5 theta_d = 0;
```

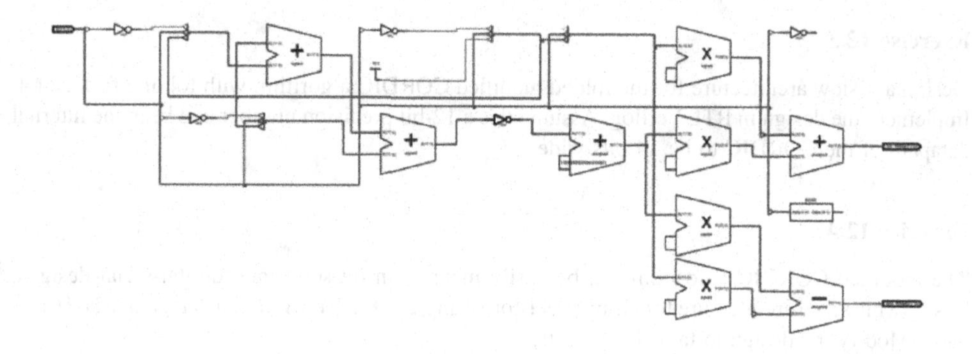

Figure 12.20 Schematic of single-stage CORDIC design

```
   for (i=0; i<16 ; i=i+1)
       #20 theta_d = theta_d + 3276;

   #400 $fclose(outFile);
   $finish;
end

initial
$monitor($time, " theta = %d, cos_theta = %d, sine_theta = %d\n", theta_d, x_o,
y_o); endmodule
```

The code is synthesized using Xilinx ISE 10.1 on a Vertix-2 device. The design computes values of $\sin\theta$ and $\cos\theta$ in a single stage and single cycle. The design can also be easily pipelined for high-speed operation.

The RTL schematic of the synthesized code is given in Figure 12.20. The schematic clearly shows the logic for computing the constants, a ROM for table look-up, four multipliers and an adder and a subtractor for computing the single-stage CORDIC.

Exercises

Exercise 12.1

Pipeline the modified CORDIC algorithm and implement the design in RTL Verilog. Assume θ is a 16-bit precision number and keep the internal datapath of the CORDIC to be 22 bits wide. Pipeline the design at every CE level.

Exercise 12.2

Design a folded architecture for a modified CORDIC algorithm with folding factors of 16, 8 and 4. Assume θ is a 16-bit precision number and keep the internal datapath of the CORDIC to be 16 bits wide.

Exercise 12.3

Design a 4-slow architecture for the folded modified CORDIC algorithm with folding factor of 4. Implement the design in RTL Verilog. Assume θ is a 12-bit precision number and keep the internal datapath of the CORDIC to be 14 bits wide.

Exercise 12.4

The modified CORDIC algorithm can be easily mapped in bit-serial and digital-serial designs. Design a bit-serial architecture for 16-bit precision of angle θ. Keep the internal datapath to be 16 bits wide. Modify the design to take 4-bit digits.

Exercise 12.5

Use a compression tree to compute *const 2* of the novel CORDIC algorithm of Section 12.5.3. Rewrite RTL code and synthesize the design. Compare the design with the one given in the section for any improvement in area and timing.

References

1. J. Valls, T. Sansaloni, A. P. Pascual, V. Torres and V. Almenar, "The use of CORDIC in software defined radios: a tutorial," *IEEE Communications Magazine*, 2006, vol. 44, *no.* 9, pp. 46–50.
2. K. Murota, K. Kinoshita and K. Hirade, "GMSK modulation for digital mobile telephony," *IEEE Transactions on Communications*, 1981, vol. 29, pp. 1044–1050.
3. J. Volder, "The CORDIC computing technique," *IRE Transactions on Computing*, 1959, pp. 330–334.
4. J. S. Walther, "A unified algorithm for elementary functions," in *Proceedings of AFIPS Spring Joint Computer Conference*, 1971, pp. 379–385.
5. S. Wang, V. Piuri and E. E. Swartzlander, "Hybrid CORDIC algorithms," *IEEE Transactions on Computing*, 1997, vol. 46, pp. 1202–1207.
6. D. De Caro, N. Petra and G. M. Strollo, "A 380-MHz direct digital synthesizer/mixer with hybrid CORDIC architecture in 0.25-micron CMOS," *IEEE Journal of Solid-State Circuits*, 2007, vol. 42, pp. 151–160.
7. T. Rodrigues and J. E. Swartzlander, "Adaptive CORDIC: using parallel angle recoding to accelerate rotations," *IEEE Transactions on Computers*, 2010, vol. 59, pp. 522–531.
8. P. K. Meher, J. Valls, T.-B. Juang, K. Sridharan and K. Maharatna, "50 years of CORDIC: algorithms, architectures, and applications," *IEEE Transactions on Circuits and Systems I*, 2009, vol. 56, pp. 1893–1907.
9. T. Zaidi, Q. Chaudry and S. A. Khan, "An area- and time-efficient collapsed modified CORDIC DDFS architecture for high-rate digital receivers," in *Proceedings of INMIC*, 2004, pp. 677–681. S. A. Khan, "A fixed-point single stage CORDIC architecture", Technical report CASE, June 2010.

13

Digital Design of Communication Systems

13.1 Introduction

This chapter covers the methodology for designing a complex digital system, and an example of a communication transmitter is considered.

Major blocks of the transmitter are the source coding block for voice or data compression, forward error correction (FEC) for enabling error correction at the receiver, encryption for data security, multiplexing for adding multiple similar channels in the transmitted bit stream, scrambling to avoid runs of zeros or ones, first stage of modulation that packs multiple bits in a symbol and performing phase, frequency, amplitude or a hybrid of these modulations, and digital up-conversion (DUC) to translate a baseband modulated signal to an intermediate frequency (IF). This digital signal at IF is passed to a digital-to-analog (D/A) converter and then forwarded to an analog front end (AFE) for processing and onward transmission in the air.

The receiver contains the same components, cascaded together in reverse order. The receiver first digitizes the IF signal received from its AFE using an A/D converter. It then sequentially passes the digital signal to a digital down-converter (DDC), demodulator, descrambler, demultiplexer, decryption, FEC decoder and source decoder blocks. All these blocks re-do whatever transformations are performed on the signal at the transmitter.

Receiver design, in general, is the more challenging because it has to counter the noise introduced on the signal on its way from the transmitter. Also, the receiver employs components that are running at its own clock, and frequency synthesizers that are independent of the transmitter clock, so this causes frequency, timing and phase synchronization issues. Multi-path fading also affects the received signal. All these factors create issues of carrier frequency and phase synchronization, and frame and symbol timing synchronization. The multi-path also adds inter-symbol interference.

For high data-rate communication systems most of the blocks are implemented in hardware (HW). The algorithms for synchronizations in the receiver require complex nested feedback loops.

The transmitter example in this chapter uses a component-based approach. This approach is also suitable for software-defined radios (SDRs) using reconfigurable field-programmable gate arrays (FPGAs) where precompiled components can be downloaded at runtime to configure the functionality.

Digital Design of Signal Processing Systems: A Practical Approach, First Edition. Shoab Ahmed Khan.

Using techniques explained in earlier chapters, this chapter develops time-shared, dedicated and parallel architectures for different building blocks in a communication transmitter. MATLAB® code is listed for the design. A critical analysis is usually required for making major design decisions. The communication system design gives an excellent example to illustrate how different design options should be used for different parts of the algorithm for effective design.

A crucial step in designing a high-end digital system is the top-level architecture, so this chapter first gives design options and then covers each of the building blocks in detail. For different applications these blocks implement different algorithms. For example, the source encoding may compress voice, video or data. The chapter selects one algorithm out of many options, and gives architectural design options for that algorithm for illustration.

13.2 Top-Level Design Options

The advent of software-defined radios has increased the significance of top-level design. A high data-rate communication system usually consists of hybrid components comprising ASICs, FPGAs, DSPs and GPPs. Most of these elements are either programmable or reconfigurable at runtime. A signal processing application in general, and a digital communication application in particular, requires sequential processing of a stream of data where the data input to one block is processed and sent to the next block for further processing. The algorithm that runs in each block has different processing requirements.

There are various design paradigms used to interconnect components. These are shown in Figure 13.1.

13.2.1 Bus-based Design

In this design paradigm, all the components are connected to a shared bus. When there are many components and the design is complex, the system is mapped as a system-on-chip (SoC). These components are designed as processing elements (PEs). A shared bus-based design then connects all the PEs in an SoC to a single bus. The system usually performs poorly from the power consumption perspective. In this arrangement, each data transfer is broadcast. The long bus has a very high load

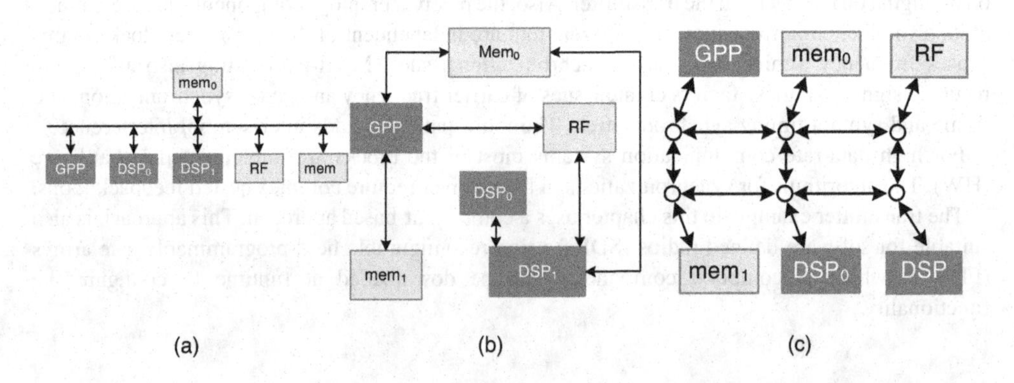

Figure 13.1 Interconnection configurations. (a) Shared bus-based design. (b) Peer-to-peer connections. (c) NoC-based design

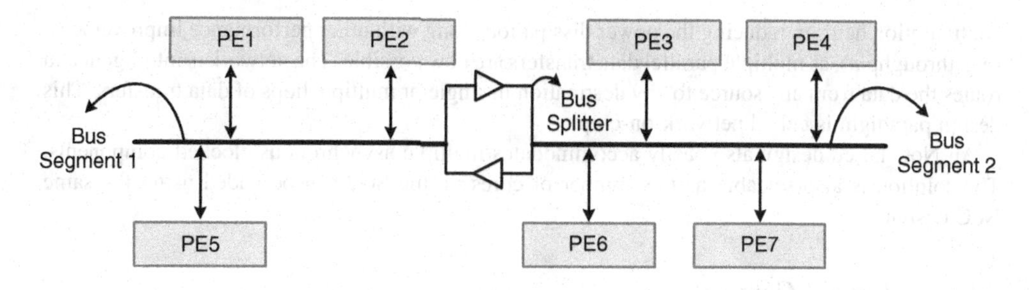

Figure 13.2 Split shared bus-based design for a complex digital system

capacitance C, and this results in high power dissipation as power is directly proportional to C as given by the formula:

$$P = \frac{1}{2} C f V^2$$

where f and V are clock frequency and supply voltage, respectively.

Segmenting the shared bus is commonly used to improve the performance. This segmentation also enables the bus to make parallel transfers in multiple segments. As the length of each segment is small, so is its power dissipation. A segmented bus design is shown in Figure 13.2.

Examples of on-chip shared-bus solutions are AMBA (advanced microcontroller bus architecture) by ARM [1], SiliconBackplane μNetwork by Sonics [2], Core Connect by IBM [3], and Wishbone [4].

A good example of a shared bus-based SoC is the IBMCell processor. In this, an element interconnect bus (EIB) connects all the SPEs for internal communication. The bus provides four rings of connections for providing shared bandwidth to the attached computing resources [5].

Although designs based on a shared bus are very effective, they have an inherent limitation. Resources use the same bus, so bus accesses are granted using an arbitration mechanism. In network-on-chip (NoC) designs, parallel transfers of data among the PEs are possible. This flexibility comes at the cost of additional area and power dissipation.

13.2.2 Point-to-Point Design

In this design paradigm, all the components that need to communicate with each other are connected together using dedicated and exclusive interconnects. This obviously, for a complex design with many components, requires a large area for interconnection logic. In design instances with deterministic inter-PE communication, and especially for communication systems where data is sequentially processed and only two PEs are connected with each other, this design option is very attractive.

13.2.3 Network-based Design

Shrinking geometries reduce gate delays and their dynamic power dissipation, but the decreasing wire dimensions increase latency and long wires require repeaters. These repeaters consume more power. Power consumption is one of the issues with a shared bus-based design in SoC configuration. In a network-based architecture, all the components that need to communicate with other are connected to a network of interconnections. These interconnections require short wires, so this configuration helps in reducing the power dissipation along with other performance improvements (e.g. throughput), as multiple parallel data transfers are now possible. The network is intelligent and routes the data from any source to any destination in single or multiple hops of data transfers. This design paradigm is called network-on-chip.

An NoC-based design also easily accommodates multiple asynchronous clocked components. The solution is also scalable as the number of cores on the SoC can be added using the same NoC design.

13.2.4 Hybrid Connectivity

In a hybrid design, a mix of three architectures can be used for interconnections. A shared bus-based architecture is a good choice for applications that rely heavily on broadcast and multicast. Similarly, point-to-point architecture best suits streaming applications where the data passes through a sequence of PEs. NoC-based interconnections best suit applications where any-to-any communication among PEs is required. A good design instance for the NoC-based paradigm is a platform with multiple programmable homogenous PEs that can execute a host of applications. A multiple-processor SoC is also called an MPSoC. In MPSoC, some of the building blocks are shared memories and external interfaces. The Cisco Systems CRS-1 router is a good example; it uses 188 extensible network processors per packet processing chip.

A hybrid of these three settings can be used in a single device. The connections are made based on the application mapped on the PEs on the platform.

13.2.5 Point-to-Point KPN-based Top-level Design

A KPN-based architecture for top-level design is a good choice for mapping a digital communication system. This is discussed in detail in Chapter 4.

Figure 13.3 lays out a design. Without any loss of generality, this assumes all PEs are either micro-programmed state machine based architecture or time-shared architecture and run at asynchronous independent clocks, or a global clock where input to the system may be at the A/D converter sampling clock and the interface at the output of the system then works at the D/A converter clock. A communication system is usually a multi-rate system. A KPN-based design synchronizes these multi-rate blocks using FIFOs between every two blocks.

13.2.6 KPN with Shared Bus and DMA Controller

A direct memory access (DMA) is used to communicate with external memory and off-chip peripheral devices. The design is augmented with components to support on chip transfers. A representative top-level design is shown in Figure 13.4.

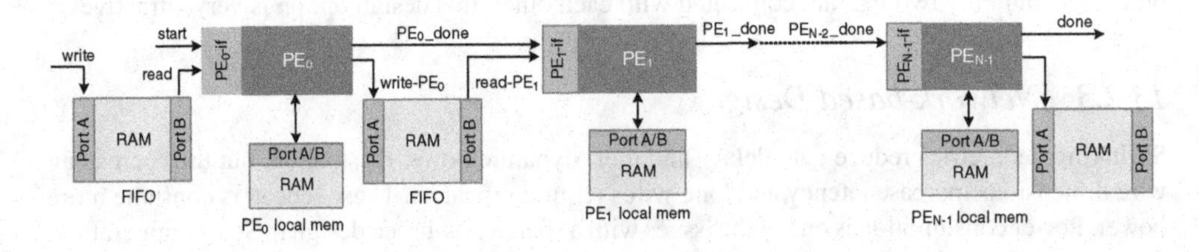

Figure 13.3 KPN-based system for streaming applications

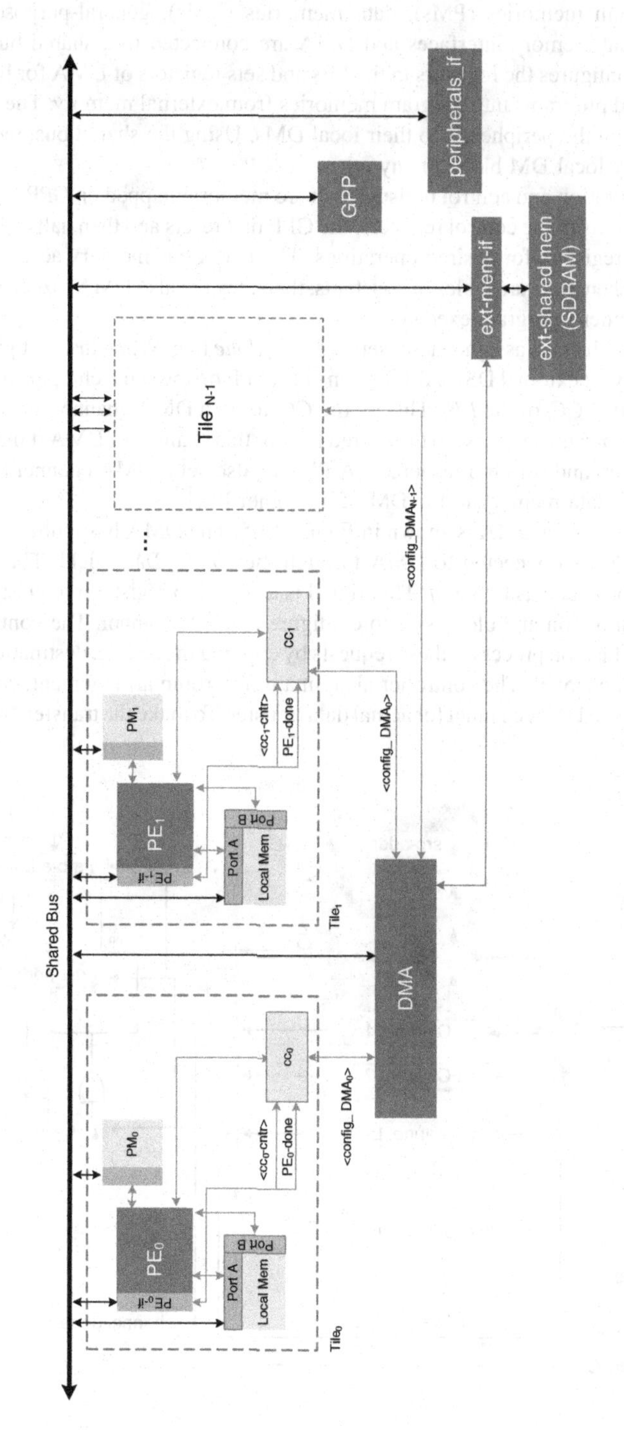

Figure 13.4 Shared-bus and KPN-based top-level design with DMA for memory transfers

All the PEs, program memories (PMs), data memories (DMs), general-purpose processor, peripherals and external memory interfaces and DMA are connected to a shared bus. The GPP acting as bus master configures the registers in the PEs and sets registers of DMA for bringing data into data memories and programs into program memories from external memory. The PEs also set DMA to bring data from the peripherals to their local DMs. Using the shared bus, the DMA also transfers data from any local DM block to any other.

All PEs have configuration and control registers that are memory mapped on GPP address space. By writing appropriate bits in the control registers, the GPP first resets and then halts all PEs. It then sets the configuration registers for desired operations. The DMA has memory access to program memory of each PE. When there are tables of constants, then they are also DMA to data memories of specific PEs before a micro-program executes.

PE_i, after completing the task assigned to it, sets the PE_i*_done* flag. When the next processing on the data is to be done by an external DSP, ASIC, or any other PE on the same chip, the done signal is also sent to the respective CC_i of the PE_i. This let the CC_i to set a DMA channel for requisite data transfer. The DMA when gets the access to the shared bus for this channel of DMA it makes the data transfer between on-chip and off-chip resources. A PE can also set a DMA channel to make data transfers from its local data memory to the DM of any other PE.

A representative design of a DMA is shown in Figure 13.5. The DMA has multiple channels to serve all CCs. Each CC_i is connected to DMA through the *config_DMA* field. This has several configuration-related bits such as CC_i*_req*. This signal is asserted to register a request to DMA by specifying source, destination and block size to configure a DMA channel. The controller of the DMA in a round-robin fashion processes these requests by copying the source, destination and block size to an empty DMA channel. The controller also, in a round-robin arrangement, copies a filled DMA channel to execute a DMA channel for actual data transfer. To make the transfer, the DMA gets

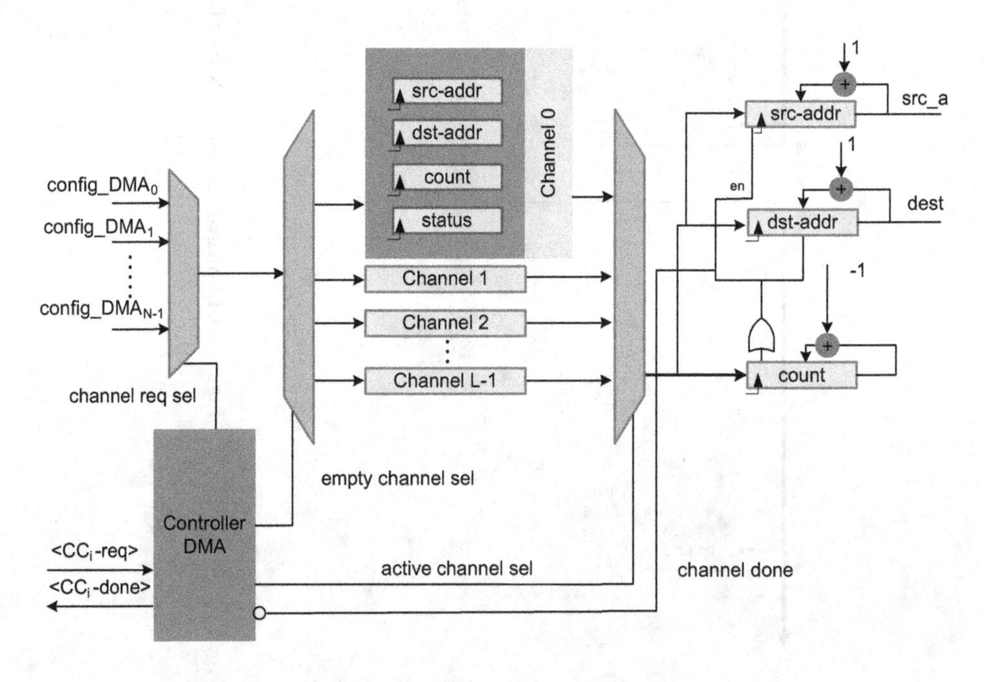

Figure 13.5 Multi-channel DMA design

access to the shared bus and then makes the requisite transfer. After the DMA is done with processing the channel, the DMA assert $CC_i\_req\_done$ flag and also set the DMA channel free and starts processing the next request if there is any.

13.2.7 Network-on-Chip Top-level Design

The NoC paradigm provides an effective and scalable solution to inter-processor communication in MPSoC designs [6, 7]. In these designs the cores of PEs are embedded in an NoC fabric. The network topology and routing algorithms are the two main aspects of NoC designs. The cores of PEs may be homogenous (arranged in a regular grid connected to a grid of routers). In many application-specific designs, the PEs may not be homogenous, so they may or may not be connected in a regular grid. To connect the PEs a number of interconnection topologies are possible.

13.2.7.1 Topologies

Topology refers to the physical structure and connections of nodes in the network. This connectivity dictates the routing ability of the design. Figure 13.6 gives some configurations. The final selection is based on the traffic pattern, the requirements of quality of service (QoS), and budgeted area and power for interconnection logic. For example, grid [8] and torus [6] arrangements of routers are shown in Figures 13.6(a) and (b). In these, a router is connected to its nearest neighbors. Each router is also connected with a PE core that implements a specified functionality. This arrangement is best for homogenous PEs with almost uniform requirements on inter-processor communication, or a generic design that can be used for different applications. It also suits run-time reconfigurable

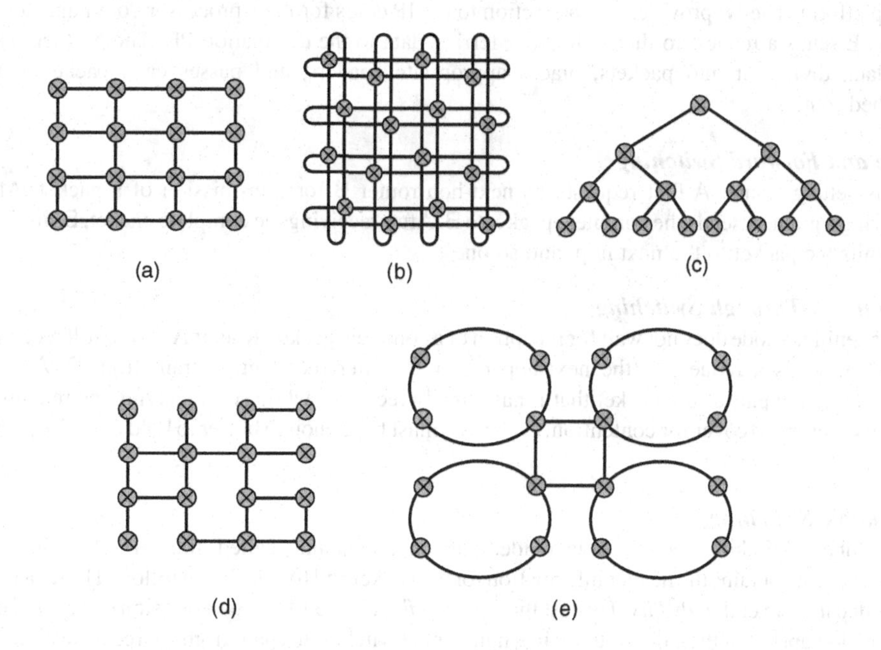

Figure 13.6 Network-on-chip configurations: (a) grid; (b) torus; (c) binary tree; (d) irregular connectivity; (e) mixed topology

architectures. The generic and flexible nets of routers enable any type of communication among the PEs.

In many designs a physically homogenous network and regular tile shape of processing elements are not always possible. The PEs, owing to their varied complexity, may be of different sizes. Application-specific SoCs require different handling of configurations and routing algorithms [9]. To select optimal topologies for these SoCs, irregular structures are usually designed.

13.2.7.2 Switching Options

There are two switching options in NoC-based design: circuit switching and packet switching. The switching primarily deals with transportation of data from a source node to a destination node in a network setting. A virtual path in NoC traverses several intermediate switches while going from source to destination node.

In circuit switching, a dedicated path from source to destination is first established [10]. This requires configuration of intermediate switches to a setting that allows the connection. The switches are shared among the connected nodes but are dedicated to make the circuit-switched transmission possible. The data is then transmitted and, at the end, the shared resource is free to make another transmission.

In contrast, packet switching breaks the data into multiple packets. These are transported from source to destination. A routing scheme determines the route of each packet from one switch to another. A packet-based network design provides several powerful features to the communication infrastructure. It uses very well established protocols and methodologies in IP networks, to enable the nodes to perform unicast, multicast and QoS-based routing. Each PE is connected with a network interface (NI) to the NoC. This interface helps in seamless integration of any PE with the others on an SoC platform. The NI provides an abstraction to the IP cores for inter-processor communications. Each PE sends a request to the NI for transferring data to the destination PE. The NI then takes the data, divides it into packets, places appropriate headers, and passes each packet to the attached routers.

Store and Forward Switching

In this setting, router A first requests its next-hop router B for transmission of a packet. After receiving a grant, A sends the complete packet to B. After receiving the complete packet, B similarly transmits the packet to the next hop, and so on.

Virtual Cut-Through Switching

In this setting, a node does not wait for reception of a complete packet. Router A, while still receiving a packet, sends a request to the next-hop router B. After receiving a grant from B, A starts transmitting the part of the packet that it has already received, while still receiving the remainder from its source. To cover for contention, each node must have enough buffer to hold an entire packet of data.

Wormhole Switching

In wormhole switching, the packet is divided into small data units called ‘flits’. The first flit is the *head flit* and it contains the header information for the packet and for the flits to follow. The data itself is divided into several *body flits*. The last flit is the *tail flit*. Figure 13.7 shows division of a packet into flits and a snapshot of their distribution in a number of switches in a path from source to destination. All routers operate on these smaller data units. In this switching the buffers are allocated for flits rather than for entire packets.

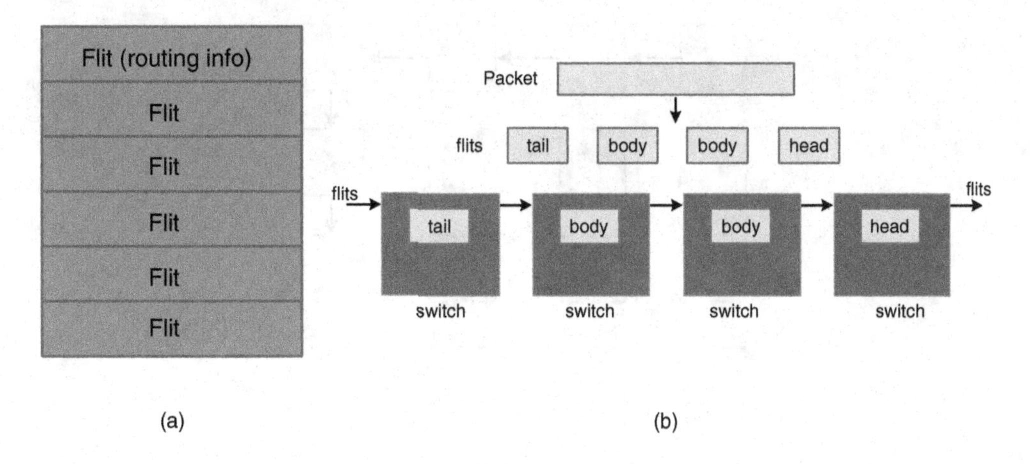

Figure 13.7 Wormhole switching that divides a packet into multiple flits. (a) Packet divided into flits. (b) Wormhole switching of flits

13.2.7.3 Routing Protocols

A router implements options out of a number of routing protocols. All these protocols minimize some measure of delays and packet drops. The delay may be in terms of the number of hops or overall latency. In a router on a typical NoC, a switch takes the packet from one of the inputs to the desired output port.

The system may perform source routing, in which the entire route from source to destination is already established and the information is placed in the packet header. Alternatively the system may implement routing where only the destination and source IDs are appended in the packet header. A router based on the destination ID routes the packets to the next hop. For a more involved design where the PEs are processing voice, video and data simultaneously, a QoS based routing policy may also be implemented that looks into the QoS bits in each received packet and then prioritizes processing of packets at each router. For the simplest designs, pre-scheduled routing may also be implemented. An optimal schedule for inter-processor communication can then be established.

Based on the schedule, the NoC can be optimized for minimum area and power implementation.

13.2.7.4 Routing Algorithms

For an NOC, a routing algorithm determines a path from a set of possible paths that can virtually connect a source node to a destination node. There are three types of routing algorithm: deterministic, oblivious and adaptive.

- The *deterministic algorithm* implements a predefined logic without any consideration of present network conditions [11–19]. This setting for a particular source and destination always selects the same path. This may cause load imbalance and thus affects the performance of the NoC. The routing algorithm is, however, very simple to implement in hardware.
- The *oblivious algorithm* adds a little flexibility to the routing options. A subset of all possible paths from source to destination is selected. The hardware selects a path from this subset in a way that it equally distributes the selection among all the paths. These algorithms do not take network conditions into account.

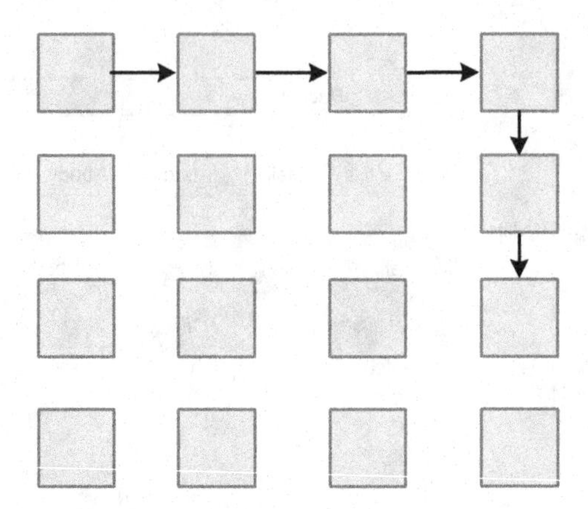

Figure 13.8 XY routing

- The *adaptive algorithm*, while making a routing decision, takes into account the current network condition. The network condition includes the status of links and buffers and channel load condition. A routing decision at every hop can be made using a look-up table. An algorithm implanted as combinational logic may also makes the decision on the next hop.

XY Routing

This is a very simple to design logic for routing a packet to the next hop. The address of the destination router is in terms of (x,y) coordinates. A packet first moves in the x direction and then in the y direction to reach its destination, as depicted in Figure 13.8. For cases that result in no congestion, the transfer latency of this scheme is very low, but its packet throughput decreases sharply as the packet injection rate in the NoC increases. This routing is also deadlock free but results in an unbalanced load on different links.

Toggled XY Routing

This is from the class of oblivious schemes. It splits packets uniformly between XY and YX. If a packet is routed using YX, it first moves in the y direction and then in the x direction. It is also very simple to implement and results in balanced loads for symmetric traffic. Figure 13.9 shows an NoC

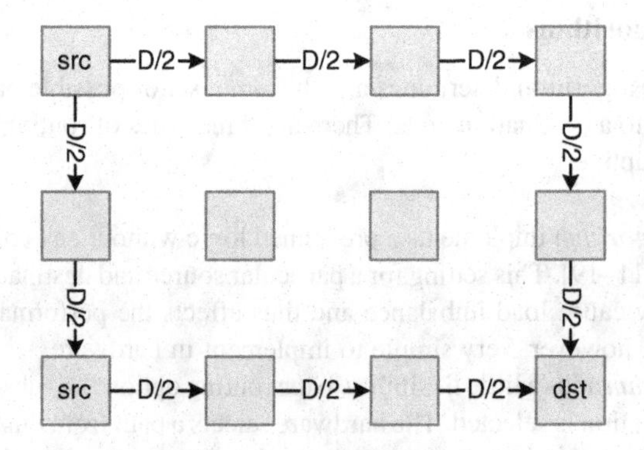

Figure 13.9 Toggled XY routing

performing toggled XY routing by switching D/2 packets in XY and the others in YX while routing a uniform load of D packets.

Weighted Toggled XY Routing

The weighted version of toggled XY routing works on *a priori* knowledge of traffic patterns among nodes of the NoC. The ratios of XY and YX are assigned to perform load balancing. Each router has a field that is set to assign the ratios. To handle out-of-sequence packet delivery, the packets for one set of source and destination are constrained to use the same path.

Source-Toggled XY Routing

In this routing arrangement, the nodes are alternately assigned XY and YX for load balancing. The scheme results in balance loads for symmetric traffic. This is shown in Figure 13.10.

Deflection Routing

In this routing scheme, packets are normally delivered using predefined shortest-path logic. In cases where there is contention, with two packets arriving at a router needing to use the same path, the packet with the higher assigned priority gets the allocation and the other packet is deflected to a non-optimal path. Deflection routing helps in removing buffer requirements on the routers [20].

13.2.7.5 Flow Control

Each router in an NoC is equipped with buffers. These buffers have finite capacity and the flow control has to be defined in a way that minimizes the size of the buffers while achieving the required performance. Each router has a defined throughput that specifies the number of bits of data per second a link on the router transfers to its next hop. The throughput depends on the width of the link and clock rate on each wire. To minimize buffer sizes, each packet to be transferred is divided into a number of flits of fixed size.

The transfer can be implemented as circuit switching, as described in Section 13.2.7.2. This arrangement does not require any buffers in the routers, but the technique results in an overhead for allocation and de-allocation of links for each transfer.

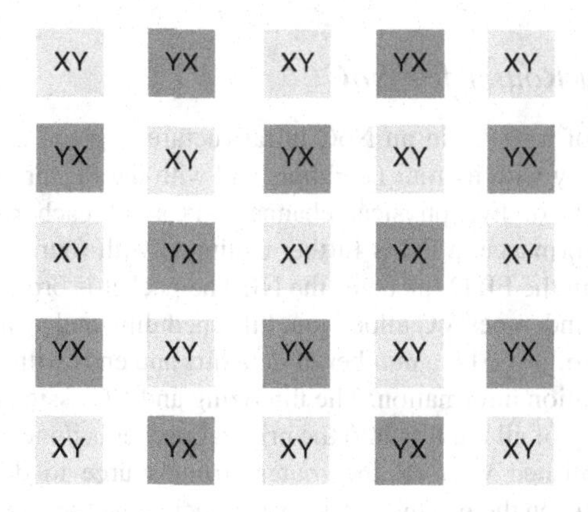

Figure 13.10 Source-toggled XY routing

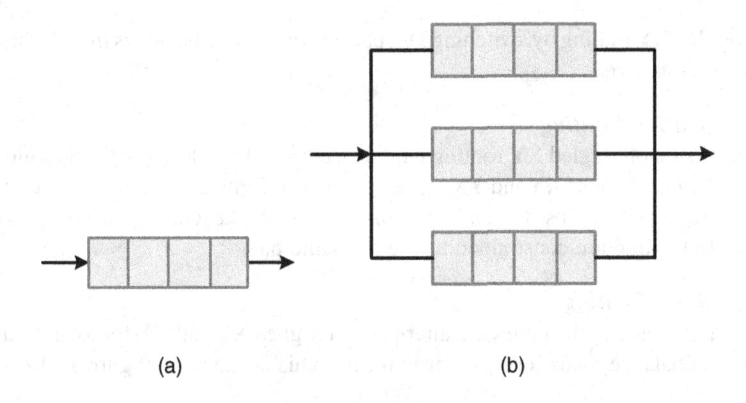

Figure 13.11 Flow control. (a) One channel per link. (b) Multiple virtual channels per link

Packet switching is a more effective for flow control. This requires a buffer on each outgoing or incoming link of each router. For routers supporting wormhole switching, the unit of transfer is a flit. Flits that are competing for the same link are stored in a buffer and wait for their turn for transfer to the next hop.

13.2.7.6 Virtual Channels

For buffer-based flow control, each link of a router is equipped with one buffer to hold a number of flits of the current transfer. A link with one buffer is shown in Figure 13.11(a). If the packet is blocked at some node, its associated flits remain in buffers at different nodes. This results in blockage of all the links in the path for other transfers. To mitigate this blocking, each link can be equipped with multiple parallel buffers, and these are called 'virtual channels' (VCs). A link with three virtual channels is shown in Figure 13.11(b).

The addition of virtual channels in NoC architecture provides flexibility but also adds complexity and additional area. Switching schemes have been proposed that use VCs for better throughput and performance [21–23].

13.2.8 Design of a Router for NoC

A representative design of a router in an NoC infrastructure is given in Figure 13.12 [24]. The router assumes connectivity with its four neighbors and with the PE through a network interface. This connectivity consists of five physical channels (PCs) of each router. To provide more flexibility and better performance, a PC is further equipped with four virtual channels (VCs). A packet of data is written in the FIFO queue of the NI. The packet is broken into a number of flits. The packet transmission undergoes VC allocation, flit scheduling and switch arbitration. A packet is broken into a head flit followed by a number of data flits and ends with a tail flit. The header flit has the source and destination information. The fliterizing and VC assignment is performed at the source PE, and reassembly of all the flits into the original packet is done at the destination PE. An association is also established with all the routers from source to destination to ensure the availability of the same VC on them. This one-to-one association eases the routing of the body flits through the route. Only when this VC to VC association is established is the packet assigned a VC

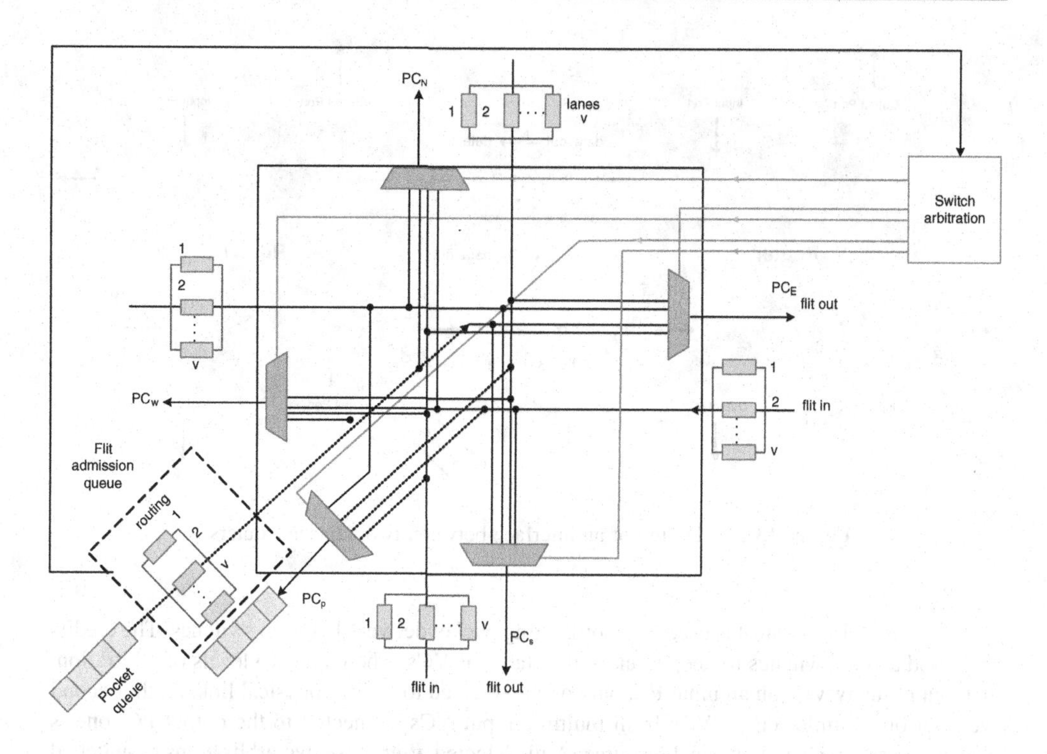

Figure 13.12 Design of a router with multiple VCs for each physical channel

identification (VCid). The process of breaking a packet into flits and assigning a VCid to each flit is given in Figure 13.13.

At the arrival of a header flit, the router runs a routing algorithm to determine the output PC on which the switch needs to route this packet for final destination. Then the VC allocator allocates the assigned VC for the flit on the input PC of the router.

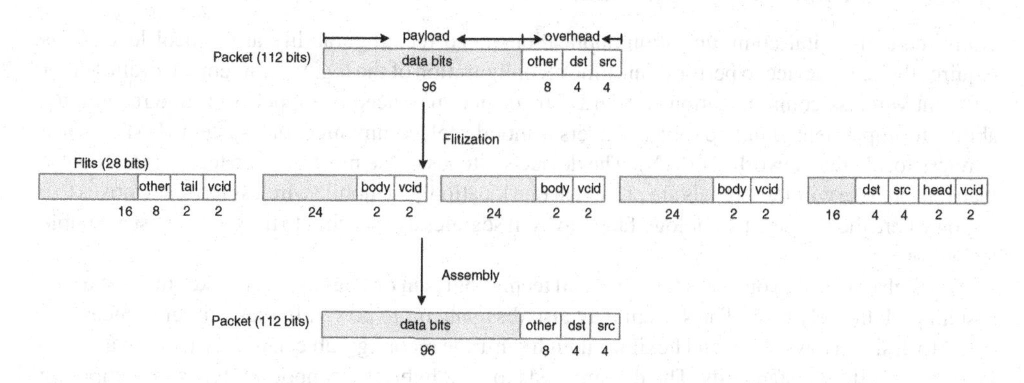

Figure 13.13 Process of breaking a packet into a number of flits and assigning a VCid to each

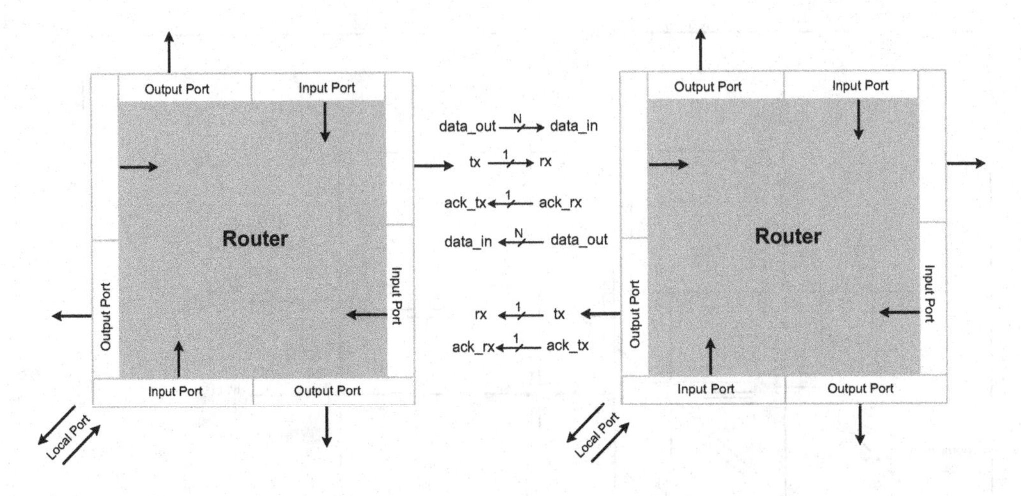

Figure 13.14 Design of an interface between two adjacent routers

Credit-based flow control is used for coordinating packet delivery between switches. The credits are passed across switches to keep track of the status of VCs. There are two levels of arbitration. First, out of many VCs on an input PC, only one is selected to use the physical link. At the second level, out of multiple sets of VCs from multiple input VCs connected to the output PC, one is selected to use the shared physical resource. A flit selected after these two arbitrations is switched out on the VC to the next router. Each body flit passes through the same set of VCs in each router. After the tail flit is switched out from a router, the VC allocated for the packet is released for further use.

A simple handshaking mechanism can be used for transmitting a flit of data on the PC [25]. The transmit-router places the flit on the data-out bus and a *tx* signal is asserted. The receive-router latches the flit in a buffer and asserts the *ack-tx* signal to acknowledge the transfer. Several of these interfaces are provided on each router to provide parallel bidirectional transfers on all the links. Figure 13.14 shows interfacing details of the links between two adjacent routers.

13.2.9 Run-time Reconfiguration

Many modern digital communication applications need reconfigurability at protocol level. This requires the same device to perform run-time reconfiguration of the logic to implement an altogether different wireless communication standard. The device may need to support run-time reconfigurability to implement standards for a universal mobile telecommunications system (UMTS) or a wireless local area network (WLAN). The device switches to the most suitable technology if more than one of these wireless signals are present at one location. If a mobile wireless terminal moves to a region where the selected technology fades away, it seamlessly switches to the second best available technology.

Although one design option is to support all technologies in one device, that makes the cost of the system prohibitively high. The system may also dissipate more power. In these circumstances it is better to design a low-power and flexible enough run-time reconfigurable hardware that can adapt to implement different standards. The design needs to be a hybrid to support all types of computing requirements. The design should have digital signal processors (DSPs), general-purpose processors

(GPPs) and reconfigurable cores, accelerators and PEs. All the computing devices need to be embedded in the NoC grid for flexible inter-processor communication.

13.2.10 NoC for Software-defined Radio

An SDR involves multiple waveforms. A waveform is an executable application that implements one of the standards or proprietary wireless protocols. Each waveform consists of a number of components. These components are connected through interfaces with other components. There are standards to provide the requisite abstraction to SDR development. Software communication architecture (SCA) is a standard that uses an 'object request broker' (ORB) feature of common object request broker architecture (CORBA) to provide the abstraction for inter-component communication [26]. The components in the design may be mapped on different types of hardware resources running different operating systems. The development of ORB functionality in HW is ongoing, the idea being to provide a level of abstraction in the middleware that relieves the developer from low-evel issues of inter-component communication. This requires coding the interfaces in an interface definition language (IDL). The interfaces are then compiled for the specific platform.

Figure 13.15(a) shows a library of components developed for waveform creation. Most of them have been developed for mapping in HW or SW platforms. The interfaces of these components are written in an IDL. While crafting a waveform, the user needs to specify the hardware resource on which a component is to be mapped. The HW resource may be a GPP, a DSP, an ASIC or an IP core. This hardware mapping of components on target technologies is shown in Figure 13.15(b).

The component library lists several developed components for source encoding, FEC, encryption, scrambling, framing and modulation. The designer is provided with a platform to select resources from the library of components and connect them with other components to create a waveform. The designer also specifies the target technologies. In the example, the 16-QAM OFDM modulation is mapped on the DSP, whereas encryption and FEC are mapped on FPGA. The framer is mapped on a GPP.

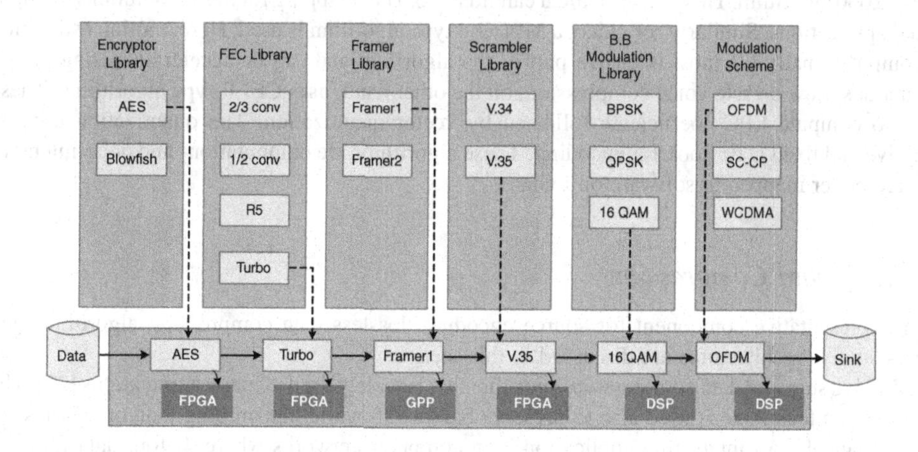

Figure 13.15 Component-based SDR design. (a) Component library. (b) Waveform creaction and mapping on different target technologies in the system

The FPGA for many streaming applications implements MPSoC-based architecture. The components that are mapped on FPGA are already developed. An ideal SDR framework generates all the interfaces and makes an executable waveform mapped on a hybrid platform.

In the case of an SCA (software communications architecture) there may be CORBA-compliant and non-CORBA-compliant components in the design. A CORBA-compliant resource runs ORB as a middleware and provides an abstraction for inter-component communication among CORBA-compliant resources.

13.3 Typical Digital Communication System

A digital communication system is shown in Figure 13.16. A digital transmitter primarily consists of a cascade of sub-systems. Each sub-system performs a different set of computations on an input stream of data. Most of the blocks in the system are optional. Depending on the application, a transmitter may not source encode or encrypt data. Similarly the system may not multiplex data from different users in a single stream for onward transmission. The system at the receiver usually has the same set of blocks but in reverse order. As the receiver needs to correct impairment introduced by the environment and electronic components, its design is more complex. In mobile/cellular applications a base station or an access point may be simultaneously communicating with more than one end terminal, so it does a lot more processing than individual mobile end terminals.

13.3.1 Source Encoding

The input stream consists of digital data or a real-time digitized voice, image or video. The raw data is usually compressed to remove redundancy. The data, image and video processing algorithms are regular in structure and can be implemented in hardware, whereas the voice compression algorithms are irregular and are usually performed on a DSP.

For lossless data compression, LZW or LZ77 standard techniques are used for source encoding. These algorithms for high data rates are ideally mapped in HW. The images are compressed using the JPEG-2000 algorithm. This is also an ideal candidate for HW mapping for high-resolution and high-speed applications. Similarly for video, a MPEG-4 type algorithm is used. Here, motion estimation is computationally the most intensive part of the algorithm and can be accelerated using digital techniques. Low bit rate voice compression, on the other hand, uses CELP-type algorithms. These have to compute LPC coefficients followed by vector quantization. The quantization uses an adaptive and fixed code book for encoding. These algorithms are computation- and code-intensive and are better mapped in software on DSPs.

13.3.2 Data Compression

As a representative component for source encoding, lossless data compression algorithms and representative architectures are discussed in this section.

LZ77 is a standard data compression algorithm that is widely used in many applications [27]. The compression technique is used also to preserve bandwidth while transmitting data on wireless or wired channels. An interesting application is in computer networks where, before data for each session is ported to the network, it is compressed at the source and then decompressed at the destination.

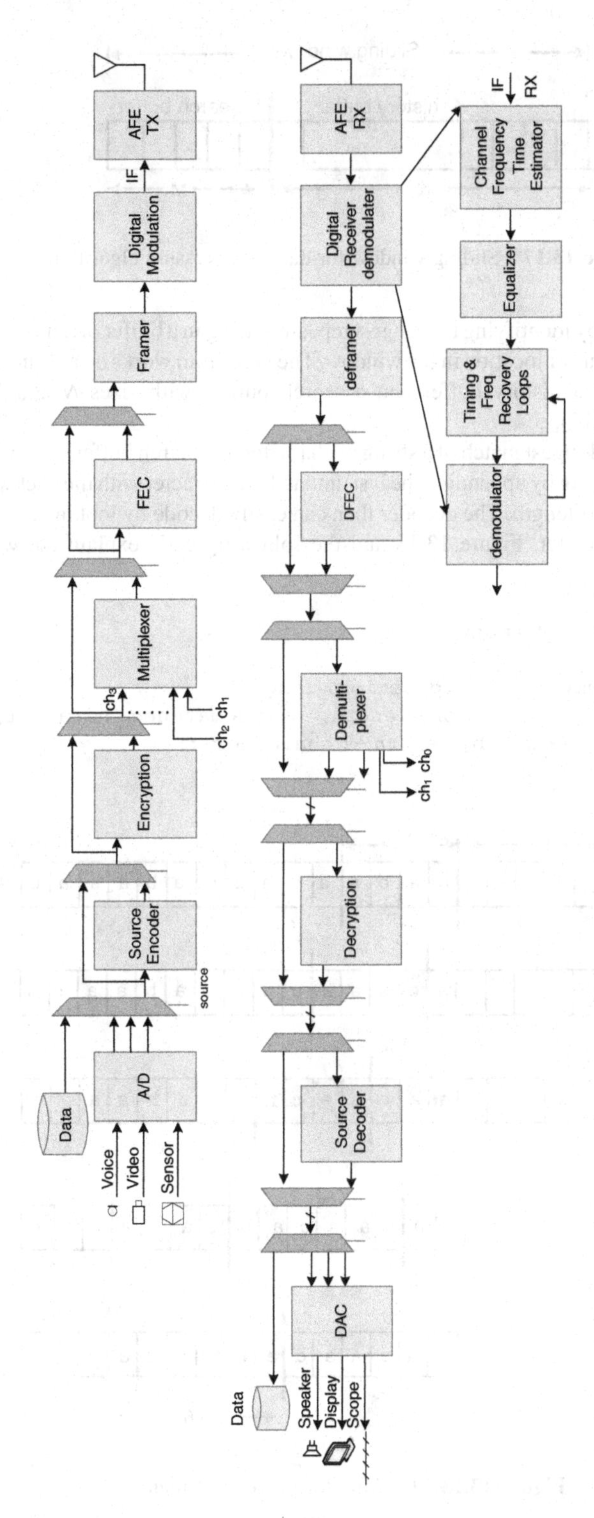

Figure 13.16 Representative communication system based on digital technologies

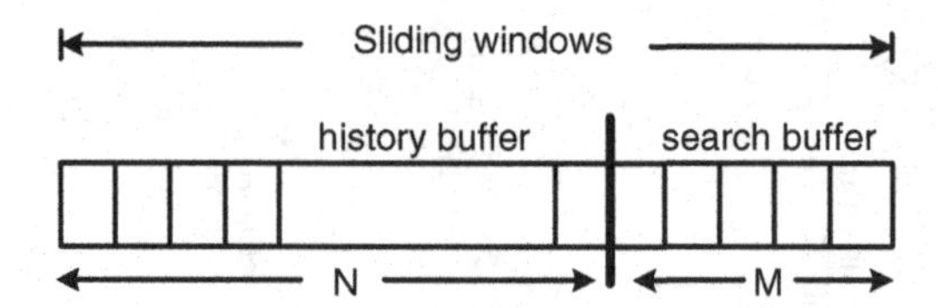

Figure 13.17 Sliding window for data compression algorithm

The algorithm works by identifying the longest repeated string in a buffer of data in a window. The string is coded with its earlier location in the window. The algorithm works on a sliding window. The window has two parts, a history buffer and a search buffer, with sizes N and M characters, respectively (Figure 13.17).

The encoder finds the longest match of a string of character in a search buffer into the entire sliding window and codes the string by appending the last unmatched character with the backward pointer to the matched string and its length. The decoder then can easily decode by looking at the pointer and the last unmatched character. Figure 13.18 and the following code explain the working of the algorithm.

```
while (search buffer not empty)
{
        find_longest_match() → (position, length);
        code_match() → (position, length, next unmatched character);
        shift left the window by length +1 characters;
}
```

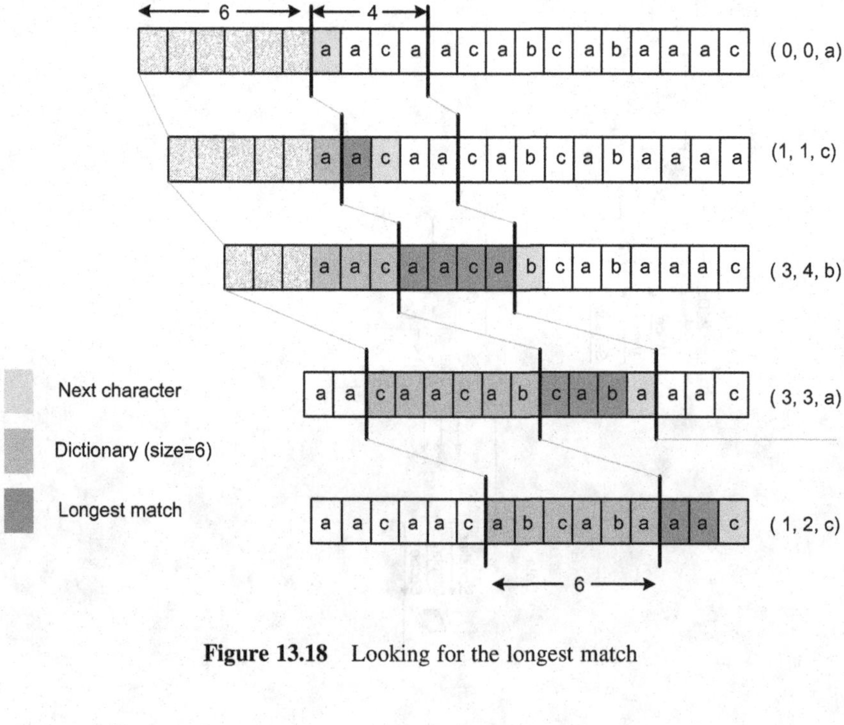

Figure 13.18 Looking for the longest match

h0	h1	h2	h3	h4	h5	h6	h7	s0	s1	s2
s0	s1	s2								
	s0	s1	s2							
		s0	s1	s2						
			s0	s1	s2					
				s0	s1	s2				
					s0	s1	s2			
						s0	s1	s2		
							s0	s1	s2	

Figure 13.19 Parallel comparisons of characters listed in the first row with the three characters in the search buffer

The compression algorithm is sequential, so an iteration depends on the result of the previous iteration. It might appear that a single iteration of the algorithm can be parallelized, but multiple iterations are difficult to parallelize. A technique can be used where the architecture computes all possible results for the subsequent iterations and then finally makes the selection. This technique is effective to design high-speed architectures for applications that have dependencies across iterations. An implementation of the technique to design a massively parallel architecture to perform LZ77 compression at a multi-gigabit rate is presented in [28–30]. A representative architecture is given in this section.

Here the example assumes a history buffer h of size 8 and maximum match of 3 characters that requires a search buffer s of size 3. The architecture performs all the matches in parallel. These matches are shown in the grid of Figure 13.19. The first row shows the values of the history and search buffers that are compared in parallel with the three characters in the search buffer.

The architecture computes the longest match of the search buffer [s0 s1 s2] to the string of characters listed in the first row. When it finds a match it computes the length of the match and also identifies its location. The size of history and the search buffers offer a tradeoff between area and achievable compression ratio, because the larger the buffer the larger the area.

Figure 13.20(a) computes the best match for one iteration of the algorithm. Each row computes a comparison listed in a row of the grid of Figure 13.19. Each comparator compares the two characters and generates 1 if there is a match. A cascaded AND operation in each row carries forward the 1 for successive matches. The output of the AND operation is also used in each column to select the row that results in a successful match. The serial logic is a column is shown in Figure 13.21(a). The logic ORs the results of all rows in each column, and the last row of multiplexers selects the length of the maximum match and the multiplexer `pointer_mux` selects the pointer.

Beside serial realization of the column logic, the logic can be optimized for logarithmic time. The optimized logic groups the serial logic to work in pairs and is depicted in Figure 13.21(b). This results in that identifies the location of the match logarithmic rather than serial complexity.

Architecture can be designed that computes multiple iterations of the algorithm. The architecture is optimized as computations are shared across multiple iterations. Figure 13.22 displays a table showing all the possible comparisons required for performing multiple iterations of the algorithm, highlighting the sharing of comparisons in the first three columns.

The first column of the table shows the first iteration. The second column computes the result if no match is established in the first iteration, and s0 is moved from the search buffer to the history buffer with coding (0 0 s0). Similarly the third, fourth and fifth columns show if the first iteration results in match length of 1, 2 and 3, respectively. If the first iteration matches two characters, than

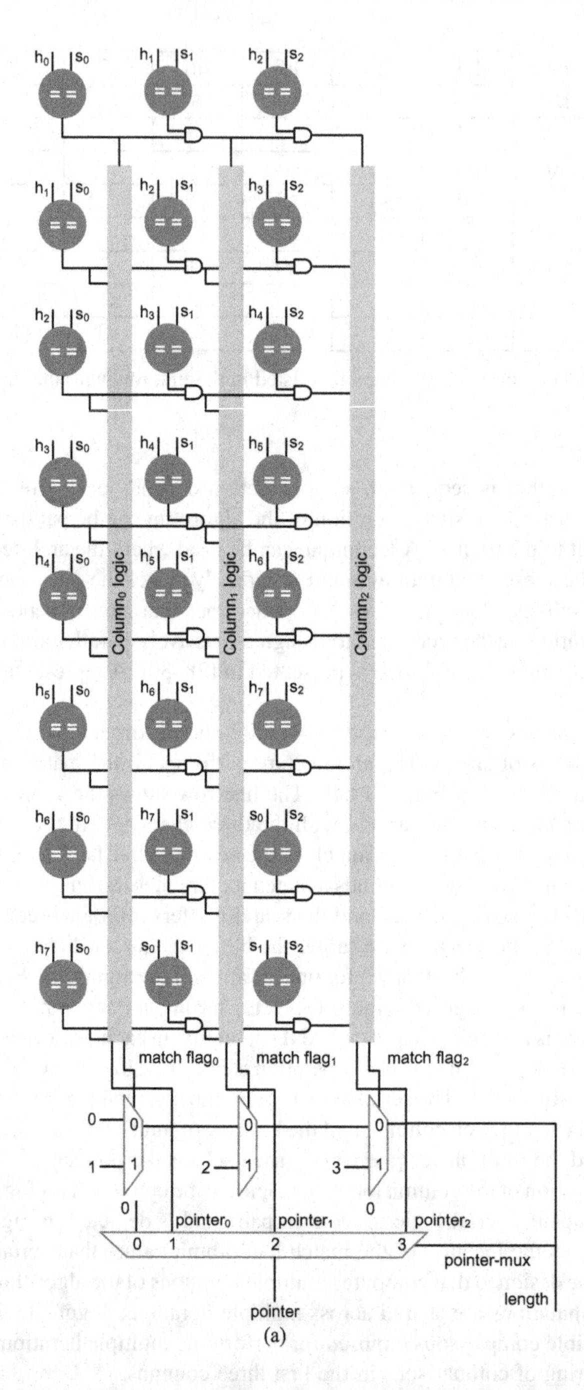

Figure 13.20 Digital design for computing iteration of the LZ77 algorithm in parallel. (a) Identifying logic in each column. (b) A block representing the logic of three columns and pointer and length calculation

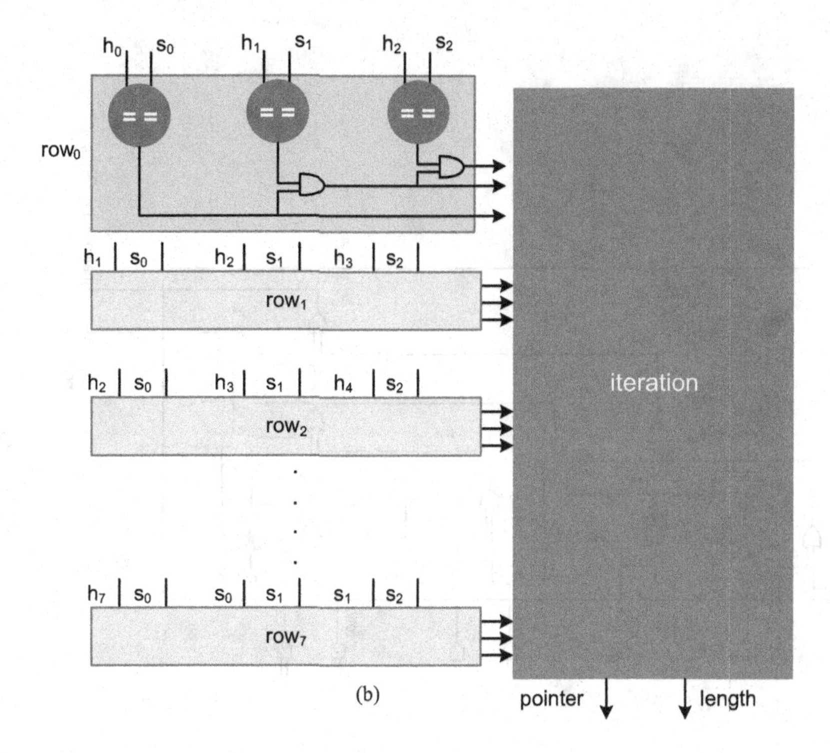

(b)

Figure 13.20 (*Continued*)

columns 4, 5 and 6 are considered for the next iteration. If all three characters are matched, then columns 5 and 6 are considered for the next iteration. If no match is found, the architecture implements six iterations of the algorithm in parallel. The table also highlights the sharing that is performed across iterations.

Figure 13.23 shows the architecture that implements these multiple iterations. The selection block finally selects the appropriate answer based on the results of the multiple iterations.

13.3.3 Encryption

The encryption block takes source encoded data as clear text and ciphers the text using a pre-stored key. AES is the algorithm of choice for most digital communication applications. The algorithm is very regular and designed for hardware mapping. A representative architecture for moderate data rate applications is described below.

13.3.3.1 AES Algorithm

An AES algorithm encrypts a 128-bit block of plain text using one of three sizes of key, 128, 192 or 256 [31]. The encryption is performed in multiple rounds. For the three key sizes the number of rounds for data encryption are $n = 10$, 12 and 14. In the key scheduling stage, the key is expanded to

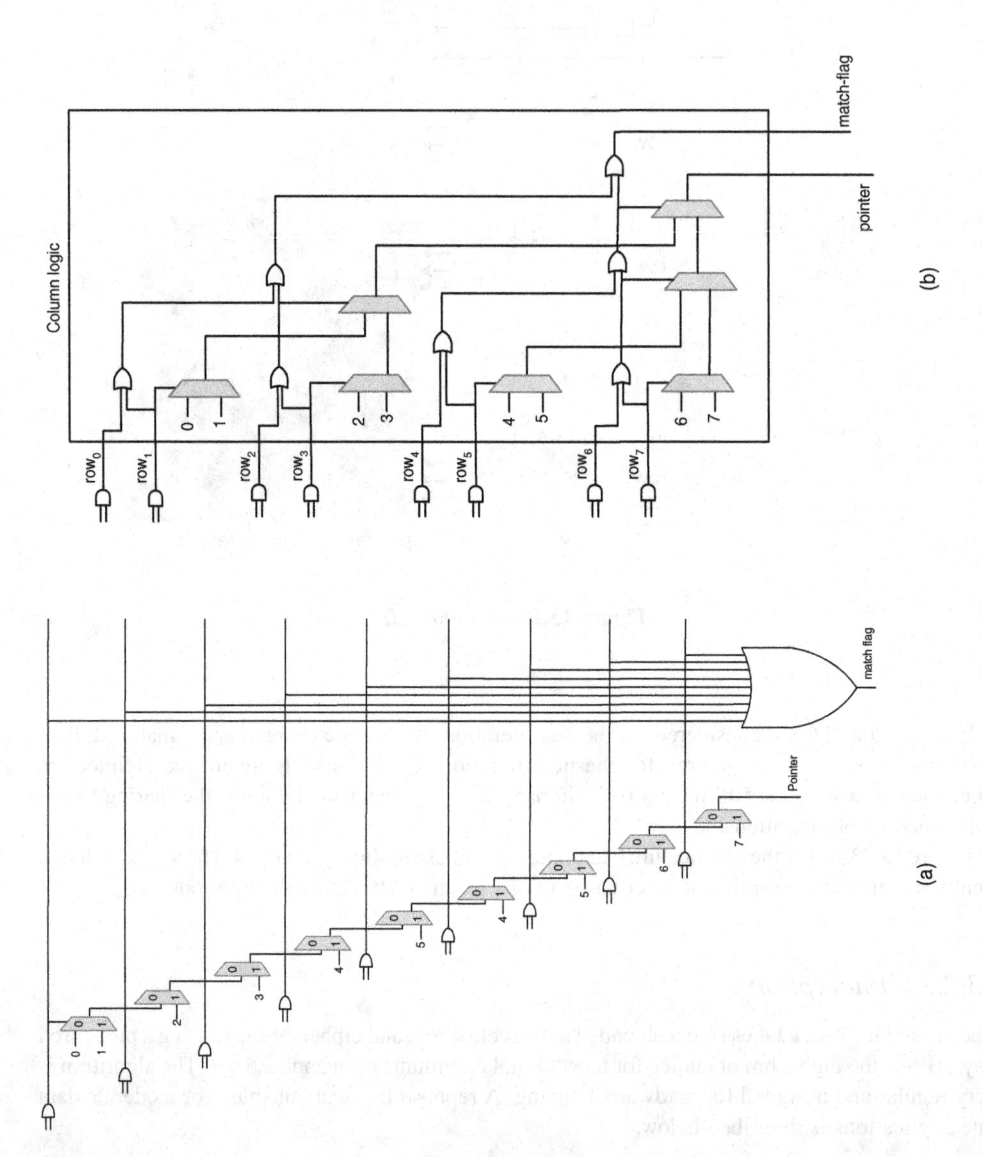

Figure 13.21 Pointer calculation logic for each column: (a) with serial complexity, and (b) with logarithmic complexity

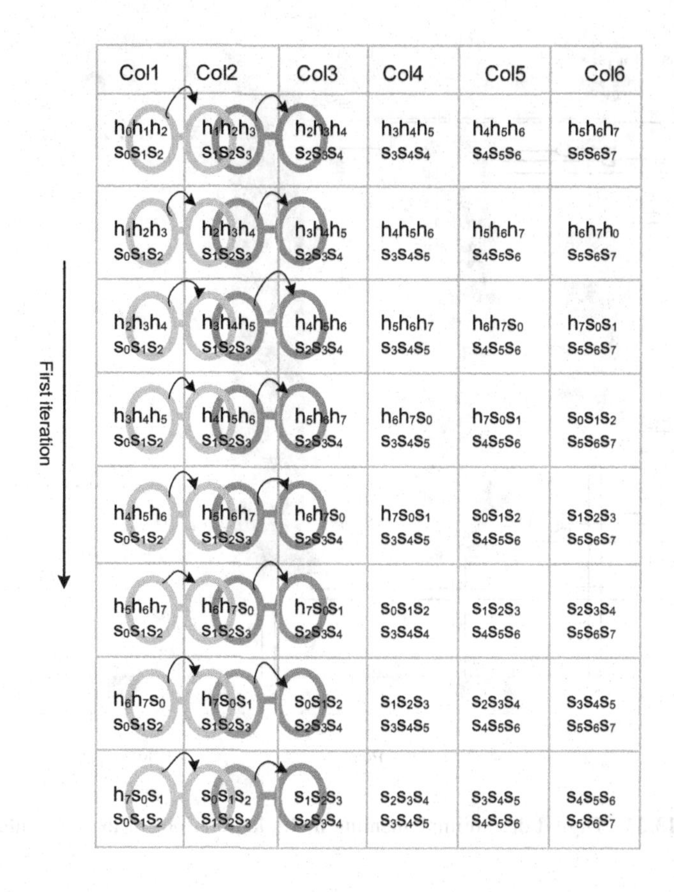

Figure 13.22 Multiple iteration of the algorithm with shared comparisons

$(4n + 4)$ number of 32-bit words. This operation is done once for an established key and can be performed offline to support multiple rounds of operation. The plain text is arranged into four 32-bit words in a state matrix. Most of the operations in the algorithm are performed on 32-bit data. These operations are 'add round key' (ARK), 'byte substitution' (BS), 'shift rows' (SR) and 'mix column' (MC). A block diagram is shown in Figure 13.24(a).

The flow of the algorithm is given in Figure 13.24(b). The first step is filling of the state matrix with 128 bits of plain text. The arrangement is shown in Figure 13.25, where the data is stored column-wise. The first byte takes the first element of the first column.

In the first operation of ARK, 128 bits of states are XORed with 128 bits of the round key. The key is derived from the cipher key in the key scheduling stage. In the next operation of ARK, the 8-bit element of the state matrix is indexed in a pre-stored 256-entry table of S-box and the entry is replaced by the entry in S-box. The substitution is followed by a shift row operation where the rows 0, 1, 2 and 3 are circularly rotated to the left by 0, 1, 2 and 3, respectively. The operation is shown in Figure 13.26(a). After shifting rows, the state matrix is transformed in an MC operation. In this, each column of the state matrix in multiplied in GF(2^8) to the matrix given in Figure 13.26(b). The MC operation is performed in all rounds except the final round of the algorithm. Pseudo-code of the AES Implementation is given here:

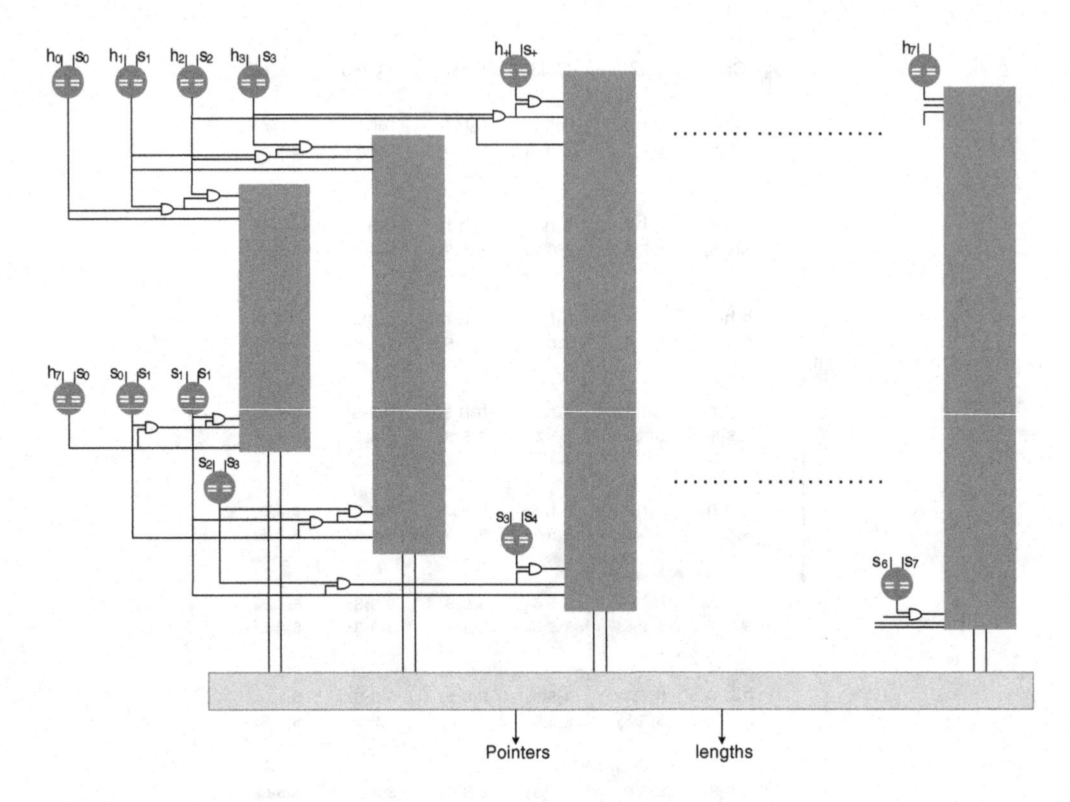

Figure 13.23 Digital design implementing multiple iterations of the LZ77 algorithm

```
// Nb = 4
// Nr = 10, 12, 14 for 128, 192 and 256 bit AES
// Hard parameter, Nk=4,6,8 for 128, 192, 256 AES
// UWord8 and UWord32 are 8-bit and 32-bit unsigned numbers
UWord32 key[Nk];
UWord32 w[Nb*(Nr-1)]; /* Expended Key */
AES_encryption (
UWord8 plain_text[4*Nb],
UWord8 cipher_text[4*Nb],
UWord32 w[Nb*(Nr-1)]);
begin
    UWord8 state[4*Nb];
    state = Plain_text;
           ARK(state, w[0, Nb-1]);
   for(round=1, round<=Nr-1, round++)
    begin
        SB(state);
        SR(state);
        MC(state);
        ARK(state, w[round*Nb, (round+1)*Nb-1)]);
    end
    SB(state);
    SR(state);
    ARK(state, w[Nr*Nb, (Nr+1)*Nb-1)]);
end
```

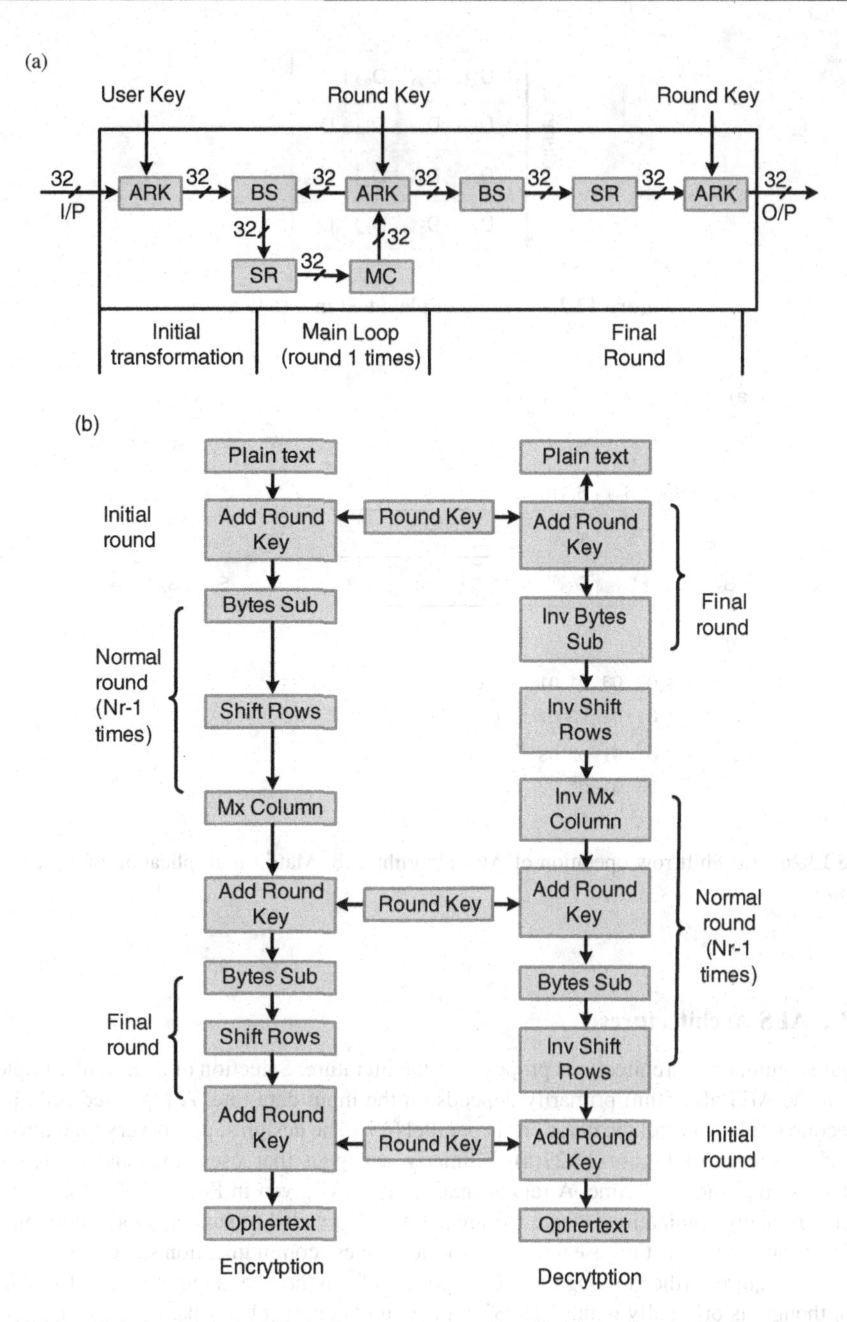

Figure 13.24 AES algorithm. (a) Block diagram of encryption. (b) Flow of algorithm for encryption and decryption

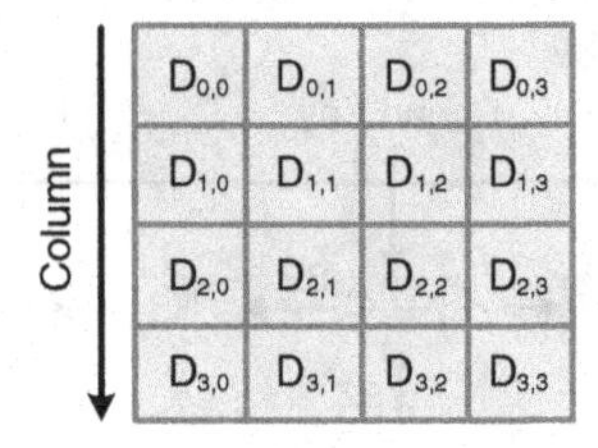

Figure 13.25 Filling of plain text in a state matrix

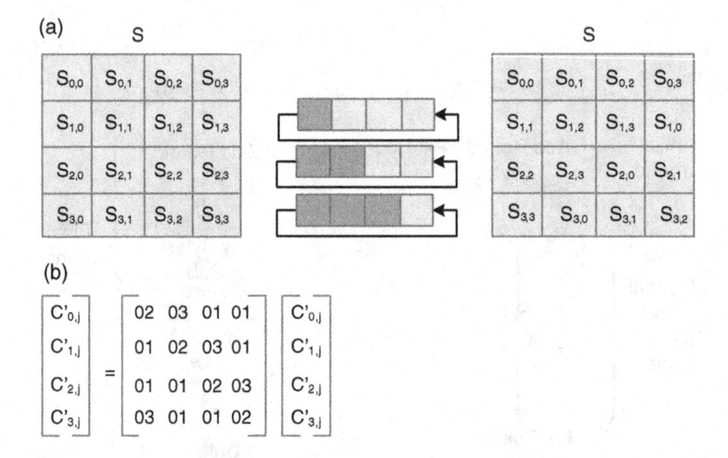

$$\begin{bmatrix} C'_{0,j} \\ C'_{1,j} \\ C'_{2,j} \\ C'_{3,j} \end{bmatrix} = \begin{bmatrix} 02 & 03 & 01 & 01 \\ 01 & 02 & 03 & 01 \\ 01 & 01 & 02 & 03 \\ 03 & 01 & 01 & 02 \end{bmatrix} \begin{bmatrix} C'_{0,j} \\ C'_{1,j} \\ C'_{2,j} \\ C'_{3,j} \end{bmatrix}$$

Figure 13.26 (a) Shift row operation of AES algorithm. (b) Matrix multiplication of mixed column operation

13.3.3.2 AES Architectures

There are a number of architectures proposed in the literature. Selection of a particular implementation of the AES algorithm primarily depends on the input data rate. A pipelined fully parallel architecture implements all the iterations in parallel [35]. The design supports very high throughput rate and is shown in Figure 13.27(a). Similarly a design that uses time-shared architecture implements one round at a time. A representative layout is given in Figure 13.27(b).

There are many applications where the architecture is needed to support moderately high data rates in a small area, as is the case with most of the wireless communication standards. An optimal design may require further folding the AES algorithm from the one given in Figure 13.27(b). The design. though. is originally defined for 32-bit data units but can be broken into 8- or 16-bit data units [32, 33]. The HW mapping then reduces the requirement on buses, registers and memories from 32-bit to 8- or 16-bit while supporting the specified throughput. For an 8-bit design this amounts to almost four-fold reduction in hardware.

An example of an implementation of AES in a KPN network framework is shown in Figure 13.28. The AES processor reads plain text data from an input FIFO. The processor has an internal RAM that is used for local storage of temporary results. The cipher data is then stored into a FIFO and a *done*

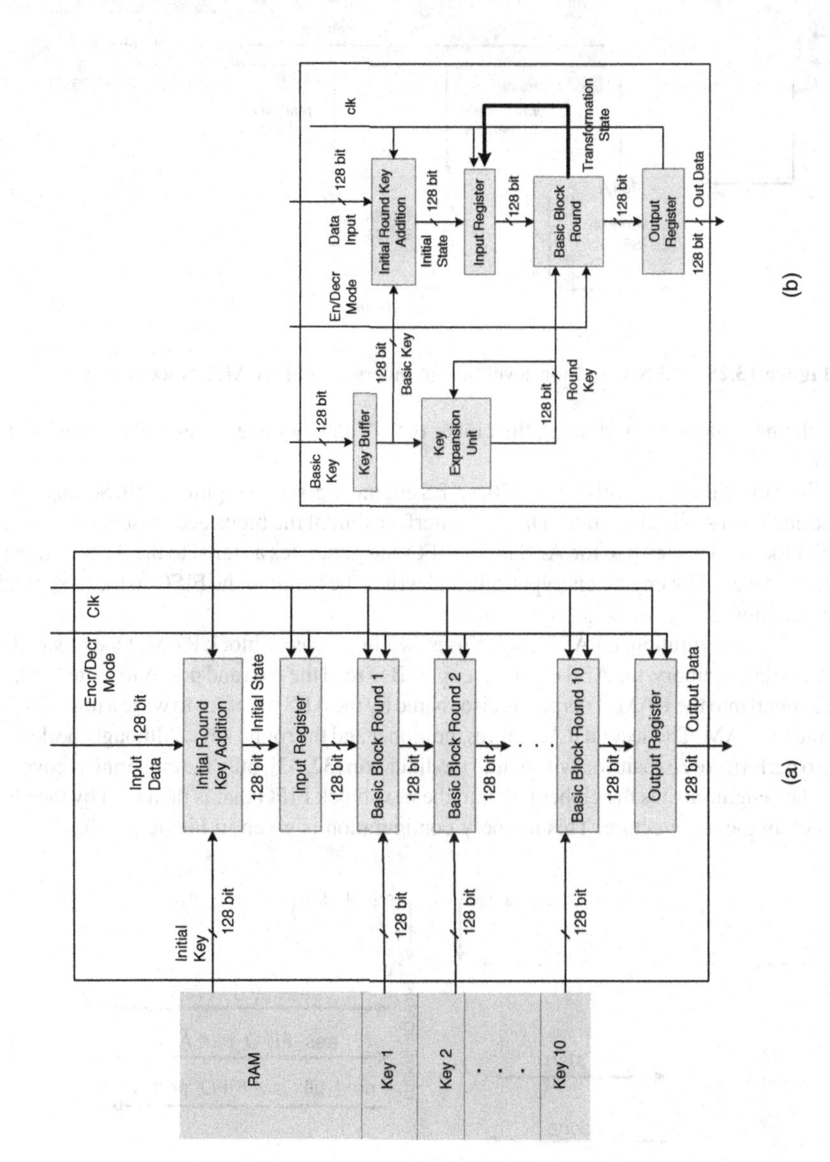

Figure 13.27 AES architectures. (a) Pipelined fully parallel design of AES. (b) Time-shared design

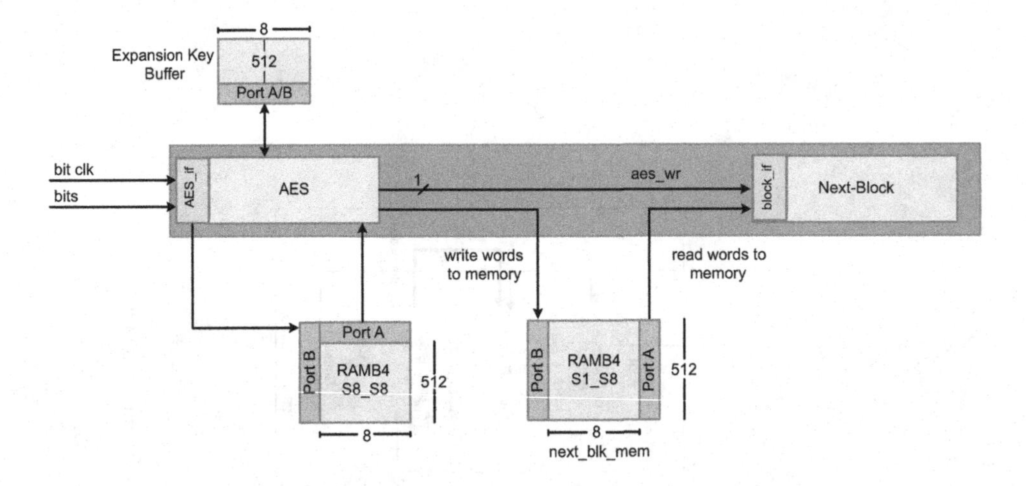

Figure 13.28 KPN-based top-level design incorporating an AES processor

flag is set for the next block to start using the cipher data. The key once expended is stored in an internal RAM.

The block diagram of the top-level design of the AES engine is given in Figure 13.29. Serial data is received to the engine at serial clock rate. The AES interface unit of the block accumulates the bits to make a 128-bit block and writes it to the AES input FIFO and generates a signal to the AES engine to start processing the data. The engine encrypts data and writes the result to the FIFO to the next block for onward processing.

In a typical KPN configuration, an AES engine uses two instances of block RAM. One block also acts as a FIFO and data memory, the AES engine uses port B to read the data and port A to write the data back after each round into the RAM. This port is also shared by the AES interface to write a new 128-bit of plain text into the RAM. The second RAM stores the cipher and the round keys. Although the design of AES can also perform key expansion with a little modification [32, 33], only the ciphering is covered in this design. The engine writes the cipher text into the next block FIFO that is then read by the next processing block in the architecture. This memory configuration is given in Figure 13.30.

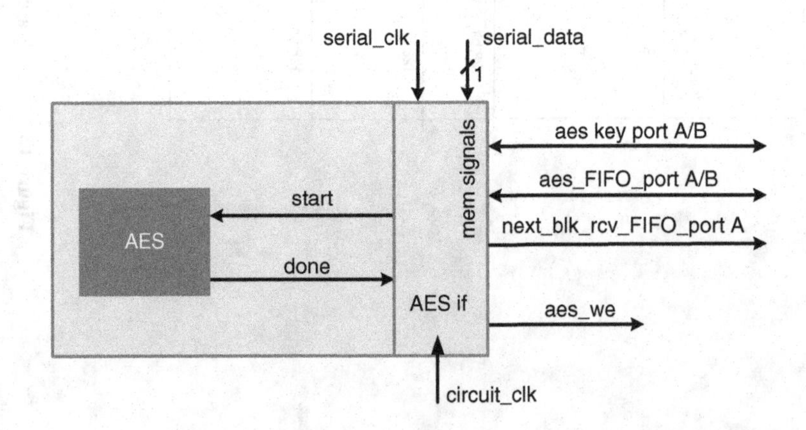

Figure 13.29 Top-level design of AES engine

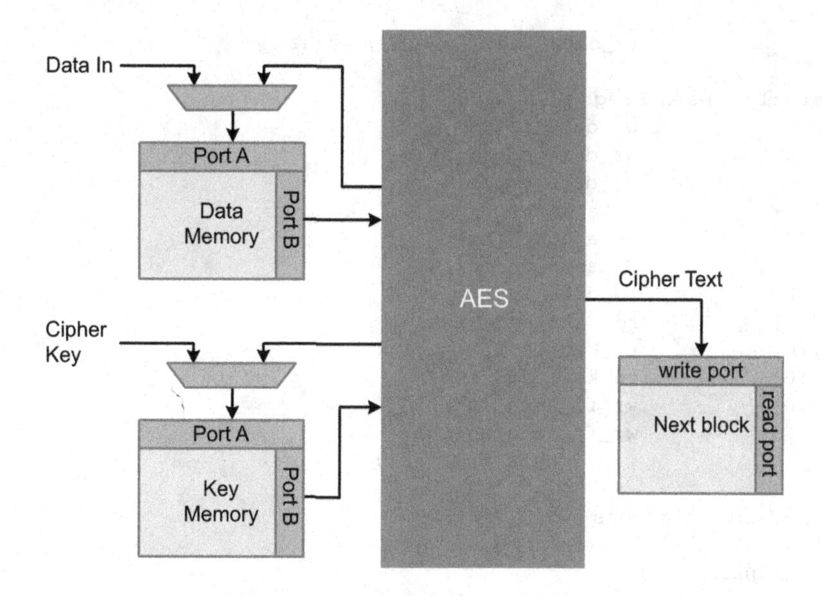

Figure 13.30 Memory configuration of AES engine in KPN settings

There are two types of memory in most of the FPGAs: distributed RAM and block RAM. The LUTs can be configured as distributed RAM whereas block RAM is a dedicated dual-port memory. An FPGA contains several of these blocks. For example, XC3S5000 in the Spartan-3 family of devices has 104 blocks of RAM totaling 1872 kilobits of memory. Similarly the LUT of a CLB in the Spartan-3 family can be optionally used as 16-deep $\times$ 1-bit synchronous RAM. These memories can be cascaded to form deeper and wider units. The distributed RAM should be used only for small memories as it consumes LUTs that are primarily meant for implementing digital logic. The block RAM should be used for large memories. Each block RAM is wrapped with a synchronous interface. The details on instantiating these two types of memory can be found from the user manual of a specific device [36, 37]. The memories and FIFOs are generated using the Xilinx core generation tool. The top-level design using generated modules are given here.

```
module aes_top(
input clk,
input reset,

// Input interface
input [7:0]         wr_data_i,
input [3:0]         wr_data_addr_i,
input               wr_data_en_i,
input [7:0]         wr_key_i,
input [4:0]         wr_key_addr_i,
input               wr_key_en_i,
input               start_i,
input               rd_ff_data_en_i,

// Output interface
output [7:0]        rd_ff_data_empty,
    // Use to indicate 128-bit in FIFO
```

```
output                ff_data_avail_o // FIFO done
);
// Internal wires and regs
wire [7:0]            wr_data_int;
wire [3:0]            wr_data_addr_int;
wire                  wr_data_en_int;
wire [7:0]            data_mux;
wire [3:0]            data_addr_mux;
wire                  data_wr_en_mux;
wire [7:0]            rd_data_int;
wire [3:0]            rd_data_addr_int;
wire [7:0]            rd_key_int;
wire [3:0]            rd_key_addr_int;
wire [7:0]            wr_ff_data_int;
wire                  wr_ff_data_en_int;
wire                  wr_ff_data_full_int;

// Module instantiations

// Bus assignment
/* Initially the data is written in the data_mem from the
           external controller. When signal start_i is asserted
           the in-place AES encryption starts processing /
assign data_mux = (start_i) ? wr_data_int : wr_data_i;
assign data_addr_mux = (start_i) ? wr_data_addr_int : wr_data_addr_i;
assign data_wr_en_mux = (start_i) ? wr_data_en_int : wr_data_en_i;

// Input data memory: for holding 16 bytes, memory width of 8-bit
input_data_mem input_data_mem_inst(
    .clka      (clk),
    .clkb      (clk),
    .wea       (data_wr_en_mux), // Write interface
    .addra     (data_addr_mux),
    .dina      (data_mux),
    .addrb     (rd_data_addr_int), // Read interface to AES
    .doutb     (rd_data_int)
);

// Key sets
key_mem key_mem_inst(
    .clka      (clk),
    .clkb      (clk),
    .wea       (wr_key_en_i), // Write interface
    .addra     (wr_key_addr_i),
    .dina      (wr_key_i),
    .addrb     (rd_key_addr_int), // Read interface to AES engine
    .doutb     (rd_key_int)
);

// AES engine
aes aes_inst(
    .clk      (clk),
    .reset    (reset),
```

```
// Write interface to input data memory for in-place computation
    .wr_data_o    (wr_data_int),
    .wr_data_addr_o    (wr_data_addr_int),
    .wr_data_en_o    (wr_data_en_int),

// Read interface to input data memory
    .rd_data_i           (rd_data_int),
    .rd_data_addr_o      (rd_data_addr_int),
    .rd_key_i            (rd_key_int), // Read interface to keys
    .rd_key_addr_o       (rd_key_addr_int),

// Start signal from the external controller
    .start_i ( start_i ),

// Write interface to data FIFO
    .wr_ff_data_o        (wr_ff_data_int),
    .wr_ff_data_en_o     (wr_ff_data_en_int),
    .wr_ff_data_full     (wr_ff_data_full_int),

// Done and output written in FIFO
    .ff_data_avail_o     (ff_data_avail_o)
);

// Output data FIFO
output_ff output_ff_inst(
    .rst          (reset),
    .wr_clk       (clk),
    .rd_clk       (clk),
    .din          (wr_ff_data_int ), // Write interface
    .wr_en        (wr_ff_data_en_int),
    .full         (wr_ff_data_full_int),

// Read interface for extranl controller
.dout             (rd_ff_data_o),
.rd_en            (rd_ff_data_en_i),
.empty            (rd_ff_data_empty)
);
endmodule
```

13.3.3.3 Time-Shared 8-bit Folding Architecture

There is no straightforward method of further folding AES architecture as the algorithm is iterative and nonlinear while it performs computation for a round. The standard folding techniques covered in Chapter 8 cannot be directly applied. A trace scheduling technique is used in [32, 33, 38]. In this technique all the flow of parallel computations in the algorithm are traced. The interdependencies across these traces are established. All these traces are then folded while taking account of their interdependencies and then are mapped in hardware.

The processor is designed to use 8-bit data. The architecture implements a 256 AES that requires 14 rounds. The processor uses an 8 × 240-bit memory for storing cipher and round key. The memory is managed as $i = 0, 1, 2, \ldots, 14$ sections for 15 ARK operations of the algorithm. Each section i

(0,0) 0	(0,1) 4	(0,2) 8	(0,3) 12
(1,0) 1	(1,1) 5	(1,2) 9	(1,3) 13
(2,0) 2	(2,1) 6	(2,2) 10	(2,3) 14
(3,0) 3	(3,1) 7	(3,2) 11	(3,3) 15

(a)

(0,0) 0	(0,1) 4	(0,2) 8	(0,3) 12
(1,1) 5	(1,2) 9	(1,3) 13	(1,0) 1
(2,2) 10	(2,3) 14	(2,0) 2	(2,1) 6
(3,3) 15	(3,0) 3	(3,1) 7	(3,2) 11

(b)

Figure 13.31 State indexing for shift-row operation: (a) original indexing, and (b) shift-row indexing

stores 16 bytes (128-bit) of key for the *i*th iteration of the algorithm. The ARK block reads corresponding bytes of the key and the state in each cycle and XORs them. The data is read from memories in shift-row format, so logic for the SR operation is not required. The block BS is performed as reading from SBOX ROM.

The technique maps the algorithm on to a byte-systolic architecture. Each iteration of the implementation reads an 8-bit state directly in row-shifted order. The original indexing of the states is shown in Figure 13.31(a) and the row-shifted indices are shown in Figure 13.31(b). To access the values of the states shown column-wise, a simple circular addressing mode is used.

The incremented SR index is generated by implementing a modulo accumulator:

$$addrSR = (addrSR + 5)\%16.$$

The %16 is implemented by a 4-bit free-running accumulator that ignores overflows. The same address is used to index the round key from key memory. A 4-bit counter *roundCount* is appended to the address to identify the round:

$$addrKey = \{roundCount, addrSM\}$$

where *addrKey* is the address to the key memory. The design works on reading 8-bit data from the memory and ciphering the plain text in multiple iterations. Each round of AES requires 16 cycles of 8-bit operations. The design of ARK, SR and BS for 8-bit architecture is shown in Figure 13.32.

The 8-bit result computed from BS is then passed to the mix-column block. The 8-bit design works on a byte by byte input value from the BS block. The design multiplies each byte with four constant values in GF(2^8) as defined in the standard. The values in the constant matrix are such that each row has the same values just shifted by one to the right. This multiplication for the first column is shown here:

$$\begin{bmatrix} C_{00} \\ C_{01} \\ C_{02} \\ C_{03} \end{bmatrix} = \begin{bmatrix} 02 & 03 & 01 & 01 \\ 01 & 02 & 03 & 01 \\ 01 & 01 & 02 & 03 \\ 03 & 01 & 01 & 02 \end{bmatrix} \begin{bmatrix} C_{00} \\ C_{01} \\ C_{02} \\ C_{03} \end{bmatrix}.$$

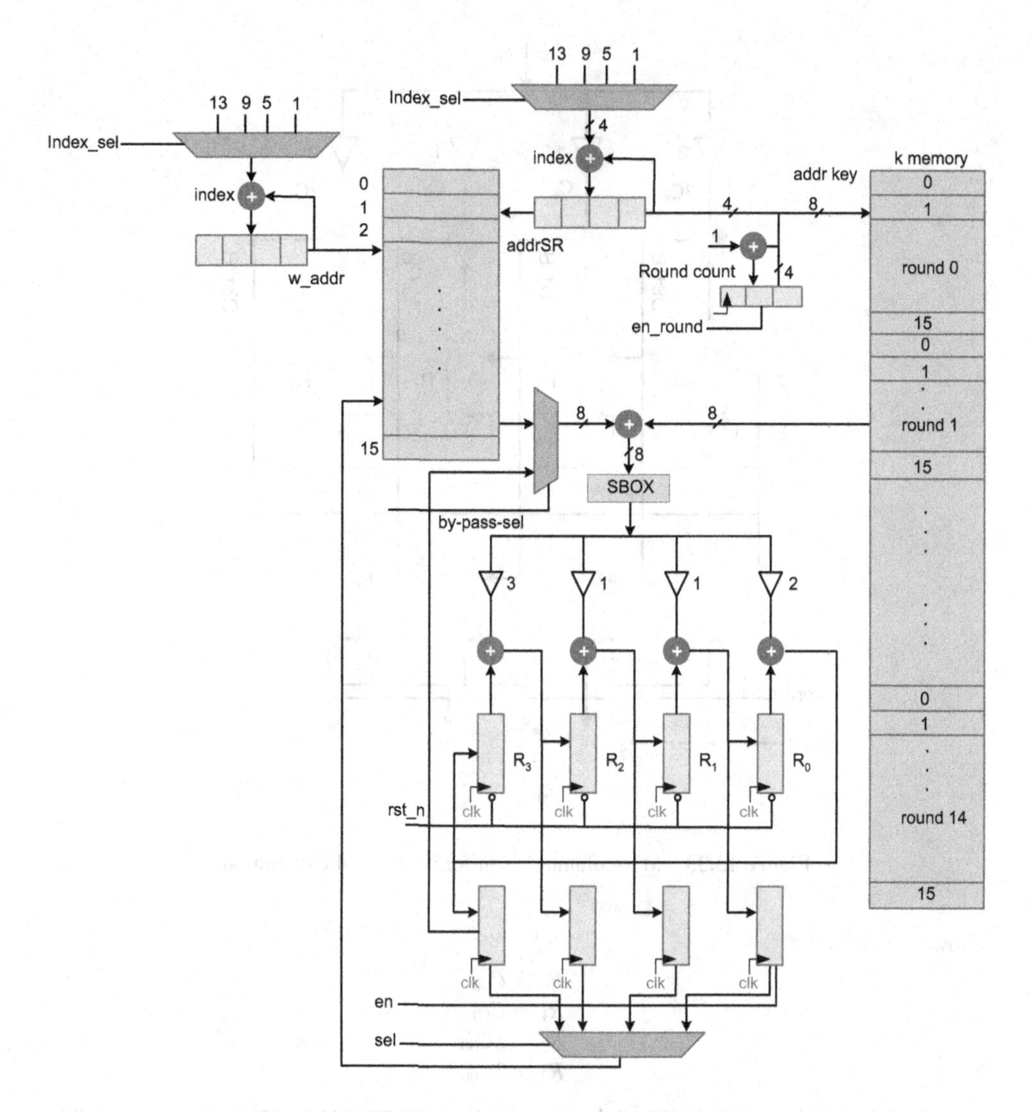

Figure 13.32 Eight-bit time-shared AES architecture

This arrangement helps in implementing the mix-column operation in byte-systolic manner. The architecture of the MC block is shown in Figure 13.33. The results of multiplication and addition are saved in registers shifted to the right by one. In the first cycle, C_{00} is input to the MC module and the output values of the computation are in the registers and the four registers have the

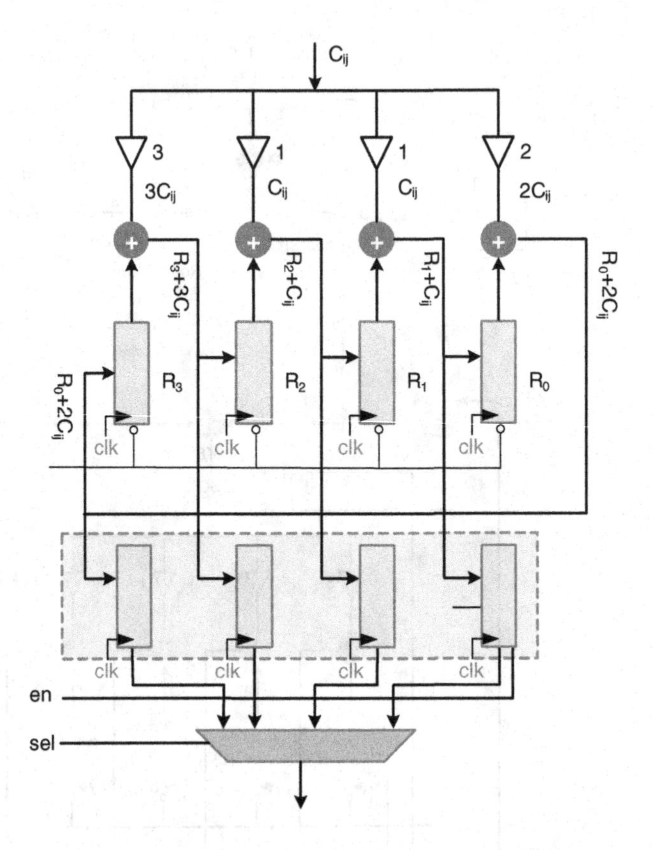

Figure 13.33 Mix-column design for byte systolic operation

following values:

$$
\begin{aligned}
R_0 &= C_{00} \\
R_1 &= C_{00} \\
R_2 &= 3C_{00} \\
R_3 &= 2C_{00}
\end{aligned}
$$

In the next cycle the second byte of states C_{01} is used to compute the next term in the partial sums. At the end of the second cycle the registers have the following values:

$$
\begin{aligned}
R_0 &= R_1 + C_{01} = C_{00} + C_{01} \\
R_1 &= R_2 + C_{01} = 3C_{00} + C_{01} \\
R_2 &= R_3 + 3C_{01} = 2C_{00} + 3C_{01} \\
R_3 &= R_0 + 2C_{01} = C_{00} + 2C_{01}
\end{aligned}
$$

Similarly in third cycle the following values are computed and stored in the registers:

$$
\begin{aligned}
R_0 &= R_1 + C_{02} = 3C_{00} + C_{01} + C_{02} \\
R_1 &= R_2 + C_{02} = 2C_{00} + 3C_{01} + C_{02} \\
R_2 &= R_3 + 3C_{02} = C_{00} + 2C_{01} + 3C_{02} \\
R_3 &= R_0 + 2C_{02} = C_{00} + C_{01} + 2C_{02}
\end{aligned}
$$

And finally the fourth cycle computes a complete column and the values are latched in shadow registers:

$$\begin{aligned} SR_0 &= R_1 + C_{03} = 2C_{00} + 3C_{01} + C_{02} + C_{03} \\ SR_1 &= R_2 + C_{03} = C_{00} + 2C_{01} + 3C_{02} + C_{03} \\ SR_2 &= R_3 + 3C_{03} = C_{00} + C_{01} + 2C_{02} + 3C_{03} \\ SR_3 &= R_0 + 2C_{03} = 3C_{00} + C_{01} + C_{02} + 2C_{03} \end{aligned}$$

For in-place computation, these values from the shadow registers are moved into the locations of byte-addressable memory that are used to calculate these values. This requires generating index values in a particular pattern. The logic that generates addressing patterns for in-place computation is explained in the next section. The MC module multiplies all the columns and saves the values in the data memory.

The first row of registers are also reset every fourth cycle for computing the multiplication of the next column. An *enable* signal latches the values of the final result in the second row of shadow registers. These values are then saved in the data memory by in-place addressing. As the same HW block is time-shared for performing the next round of the AES algorithm, the SR requires reading the fifteenth value in the fourth cycle for the MC computation in the next round. The value is still in the register of MC in this clock cycle. The value is directly passed for ARK computation. The timing diagram of Figure 13.34 illustrates the generation of different signals.

It is important to appreciate that in many applications no established technique can be applied to design an effective architecture. The 8-bit AES architecture described in this section is a good example that demonstrates this assertion. The algorithm is explored by expending all the computations in an iteration. Then, intelligently, all dependencies are resolved by tracing out single-byte operations to make the algorithm work on a single byte of input.

13.3.3.4 Byte-Systolic Fully Parallel Architecture

This section presents a novel byte-systolic fully parallel AES architecture [38]. The architecture works on byte in-place indexing. A byte of plain text is input to the architecture and a byte of cipher text is output in every clock cycle after an initial latency of 16×15 cycles. All the rounds of encryption are implemented by cascading all the stages with pipeline logic. The data is input to the first stage in byte-serial fashion. When the 16 bytes have been written in the first data RAM block, the stage starts executing the first round of the algorithm. At the same time the input data for the second frame is written in the RAM block by employing byte in-place addressing. This scheme writes the input data at locations that are already used in the current cycle of the design. For example, the first stage reads the RAM in row-shifted order reading indices 0, 5, 10 and 15 in the first four cycles. The four bytes of input data for the second frame are written at these locations in the first RAM block.

The four tables in first row of Figure 13.35 show the write addresses for the first four frames in the first RAM block for data. These address patterns repeat after every four frames. The memory locations are given column-wise. The memory location is numbered in the corner of each box, while the indices of the values written in these locations are in the center. The four tables in the second row show the sequential ordering of these indices for reading in row-shifted order. The column-wise numbers show the memory locations that are read in a sequence to input the data in the specified format.

Figure 13.36 shows the systolic architecture. Each stage has its own RAM blocks for storing data for a round and corresponding key. The addressing for reading from the RAM block and writing the data in the same location is performed using the address generation unit shown with each memory block. The addressing is done using an index. For four successive frames the value of the index is

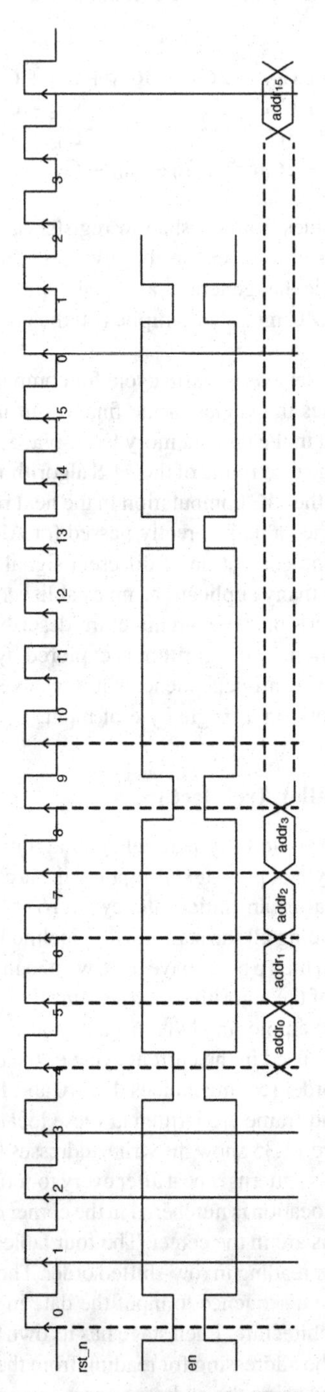

Figure 13.34 Timing diagram for the time-shared AES architecture

rounds 4i

[0] 0	[4] 4	[8] 8	[12] 12
[1] 1	[5] 5	[9] 9	[13] 13
[2] 2	[6] 6	[10] 10	[14] 14
[3] 3	[7] 7	[11] 11	[15] 15

rounds 4i+1

[0] 0_0	[4] 4_4	[8] 8_8	[12] 12_{12}
[1] 13	[5] 1	[9] 5	[13] 9
[2] 10	[6] 14	[10] 2	[14] 6
[3] 7	[7] 11	[11] 15	[15] 3

rounds 4i+2

[0] 0_0	[4] 4_4	[8] 8_8	[12] 12_{12}
[1] 9	[5] 13	[9] 1	[13] 5
[2] 2	[6] 6	[10] 10	[14] 14
[3] 11	[7] 15	[11] 3	[15] 7

rounds 4i+3

[0] 0_0	[4] 4_4	[8] 8_8	[12] 12_{12}
[1] 5	[5] 9	[9] 13	[13] 1
[2] 10	[6] 14	[10] 2	[14] 6
[3] 15	[7] 3	[11] 7	[15] 11

(a)

rounds 4i

0	4	8	12
5	9	13	1
10	14	2	6
15	3	7	11

Index = (index+4)%16=5;
addr = (addr+index)%16

rounds 4i+1

0	4	8	12
9	13	1	5
2	6	10	14
11	15	3	7

Index = (index+4)%16=9;
addr = (addr+index)%16

rounds 4i+2

0	4	8	12
13	1	5	9
10	14	2	6
7	11	15	3

Index = (index+4)%16=13;
addr = (addr+index)%16

rounds 4i+3

0	4	8	12
1	5	9	13
2	6	10	14
3	7	11	15

Index = (index+4)%16=1;
addr = (addr+index)%16

(b)

Figure 13.35 Byte in-place indexing for byte systolic AES architecture. (a) Indices for writing data for first four frames. (b) Memory locations for reading data in shift-row format

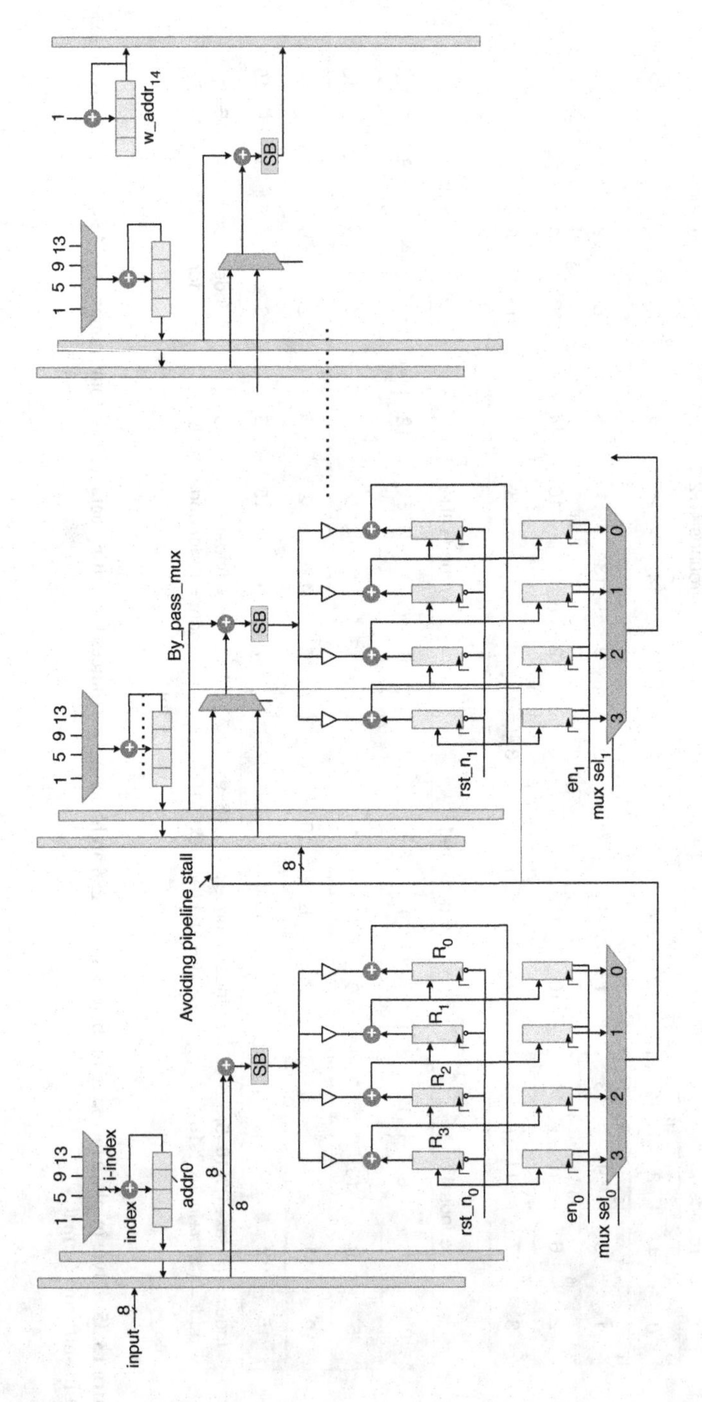

Figure 13.36 Byte systolic architecture implementing AES algorithm

generated by adding 4 into the previous index. For writing into all the RAM blocks, the index is initialized to 1. So the first write in each RAM block is in sequential order. For reading the data written in the RAM and at the same time writing the second frame on a byte by byte basis, the address is incremented as:

$$w\_addr = (w\_addr + index)\%16.$$

It is clear from Figure 13.36 that in-place storing of data into memory as given in (a) and then reading in data in row-shifted form for the second frame requires. The index is updated after every 16 cycles as

$$index = (index + 4)\%16$$

This pattern is repeated every four frames. The index value repeats as 1, 5, 9, and 13 and can be directly selected from a 4:1 MUX as shown in Figure 13.36. The last value (i.e. the fifteenth) computed for each round is directly sent to ARK of the next block for avoiding pipeline stall. The architecture perfectly works in lock step. Once all the pipeline stages are filled, each clock cycle inputs one byte of plain text data and generates one byte of ciphered output.

13.3.4 Channel Coding

In many communication systems, the next block in a transmitter is channel coding. This is performed at the transmitter to automatically correct errors at the receiver. The transmitter performs coding of the input data using convolution, Reed–Solomon or turbo codes. In many applications more than one type of error-correction coding is also performed. All these algorithms can be mapped in HW for high-throughput applications. Description and architecture design of an encoder for block turbo code (BTC) is given in this section.

The BTC serially combines two linear block codes, C_1 and C_2, to form the product code $P = C_1 \times C_2$ [39]. A block code is represented by (n,k), where n and k stand for code word length and number of information bits, respectively. For example, consider (7,4) extended Hamming code. This code takes four information bits, computes three parity bits and appends them to the information bits to create seven bits of code word. This can be represented by *IIIIPPP*, where *I* and *P* represent information and parity bits. A two-dimensional turbo code is constructed by successively applying coding to a 2-D information matrix first for all the rows and then for all the columns.

Figure 13.37 shows application of block turbo coding (7,4) by (7,4) to produce (49,16) block turbo code. The information is arranged in a 4×4 matrix. Each row of the matrix is block coded and three parity bits P_H are attached. This operation results in a 4×7 matrix. Now each column of this matrix is again coded and the resultant matrix is shown in the figure.

A configurable design of a BTC encoder can be incorporated in a communication system that supports a wide range of code rates. For example, by appropriately selecting from $n_1, n_2 = 11,26,57$ the block size of information matrix can range from 121 bits to 4096 bits. For these values a number of options are available to the user, as given in Table 13.1. The 11×11 block of input data is coded as a 16×16 block where each row and column has 4 bits of c data and 1 bit of parity.

The standard Hamming encoder used for row and column operations is shown in Figure 13.38. The registers r_i for $i = 0, \ldots, 5$ compute the redundant bits, whereas r_p computes the parity bit. The encoding polynomial is selected according to the code rate selected for encoding. Some of the registers are not used for low code rates. The *mode-sel* selects one of the modes of operation.

The BTC is best suited to be implemented in hardware in a typical KPN setting of Figure 13.3. This setting assumes three dual-port memories with an interface to the datapath. Figure 13.39 shows

I	I	I	I	P_H	P_H	P_H
I	I	I	I	P_H	P_H	P_H
I	I	I	I	P_H	P_H	P_H
I	I	I	I	P_H	P_H	P_H
P_V	P_V	P_V	P_V	P_{VH}	P_{VH}	P_{VH}
P_V	P_V	P_V	P_V	P_{VH}	P_{VH}	P_{VH}
P_V	P_V	P_V	P_V	P_{VH}	P_{VH}	P_{VH}

Figure 13.37 (49,16) block turbo code formulation by combining (7,4) $\times$ (7,4) codes

the interfaces and the memory blocks. The BTC starts encoding the data once enough bits are available in AES out-memory which is filled by the AES engine of Section 13.3.3. The size is easily calculated using the read and write pointers of the FIFO controller of AES out-memory. The BTC reads the input data and keeps the intermediate results in its internal memory, and finally the output is written to BTC out-memory.

Table 13.1 Options for block turbo code for supporting different code rates

Mode number	Input block dimensions		Input block size	Output block dimensions		Output block size	Code rate
	Rows	*Columns*	*K*	*Rows*	*Columns*	*n*	*k/n*
4	11	11	121	16	16	256	0.47
1	11	26	286	16	32	512	0.55
2	11	57	627	16	64	1024	0.61
9	26	11	286	32	16	512	0.55
5	26	26	676	32	32	1024	0.66
3	26	57	1482	32	64	2048	0.72
10	57	11	627	64	16	1024	0.61
11	57	26	1482	64	32	2048	0.72
6	57	57	3249	64	64	4096	0.79

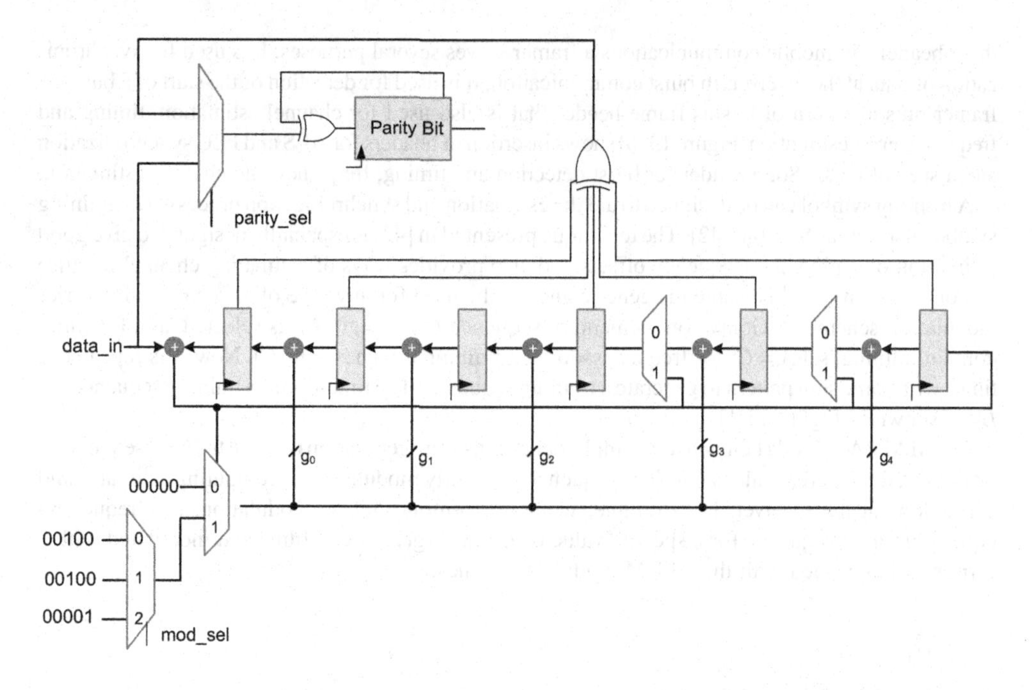

Figure 13.38 Standard Hamming encoder for BTC

The FEC engine reads data written by the AES engine in AES out-memory to perform horizontal encoding. The encoded data is written back into internal memory of the FEC engine. For the vertical coding, the data is read from the internal FEC memory and the encoded data is written into FEC out-memory.

13.3.5 Framing

For synchronization, different delimiters are inserted in data. For example, to indentify and synchronize an AES frame, an 8-bit AES header can be inserted after every defined number of AES buffers. Similarly, for synchronizing an FEC block, a header may also be inserted for synchronization. Beside

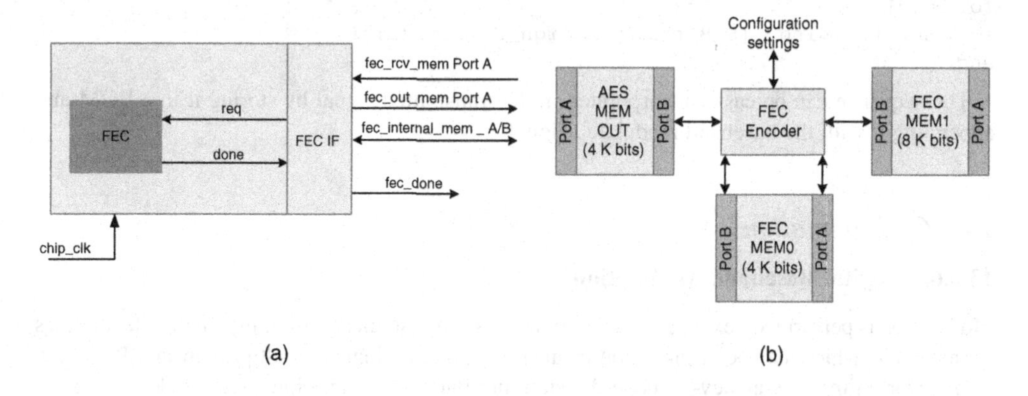

Figure 13.39 BTC system-level design. (a) BTC interfaces. (b) BTC memory block settings

these headers, in mobile communications a framer serves several purposes. It is used for synchronization of data at the receiver. In burst communication, it is used for detection of the start of a burst. A framer puts a pattern of bits as frame header that is also used for channel estimation, timing and frequency error estimation. Figure 13.40 shows insertion of headers for AES and FEC synchronization and a start of burst (SoB) header for burst detection and timing, frequency and channel estimation.

A training symbol can be designed to aid the estimation and synchronization process. The training symbol also acts as SoB [40–42]. The technique presented in [42] is especially designed to give good estimation of timing and frequency offsets and also provides ways of estimating channel impulse response. The method presented is generic and can be used for any type of single or multi carrier modulation scheme. A Golay complementary sequence C of length N_G is selected and, for time-domain training, set $A = C$; for frequency-domain training assign $A = fft(C)$. Now A is repeated L times with some sign pattern to generate a training sequence. The time-domain training sequence for $L = 4$ shown in Figure 13.41.

The MATLAB® code below lists an implementation of the generation of these training sequences, from [42], for different values of L. The sequence is directly modulated for computing estimates and corrections at the receiver. For example, while performing OFDM modulation, the frequency-domain training sequence for a specific value of L can be generated offline and modulated with a carrier and appended with the OFDM modulated frames.

```
if L==2
    Sign_Pattern = [1 1];
    C = [1 1 -1 1 1 1 1 -1 1 1 -1 1 -1 -1 -1 1];
elseif L == 4
    Sign_Pattern = [-1 1 -1 -1];
    C = [1 1 -1 1 1 1 1 -1];
elseif L == 8
    Sign_Pattern = [1 1 -1 -1 1 -1 -1 -1];
    C = [1 1 -1 1];
elseif L == 16
    Sign_Pattern = [1 -1 -1 1 1 1 -1 -1 1 -1 1 1 -1 1 -1 -1];
    C = [1 1];
end

A = fft(C);
SoB_training = [];
for i=1:L
    SoB_training = [SoB_training Sign_Pattern(i)*A];
end
```

The sequence can be easily incorporated in the transmitted signal by storing it in a ROM and appending it with the baseband modulated signal.

13.3.6 Modulation

13.3.6.1 Digital Baseband Modulation

Modulation is performed next in a digital communication system. A host of modulation techniques are available, which may be linear or nonlinear. They may use single or multiple carriers. They may also perform time-, frequency- or code-division multiplexing for multiple users. A description of one of these techniques is given below.

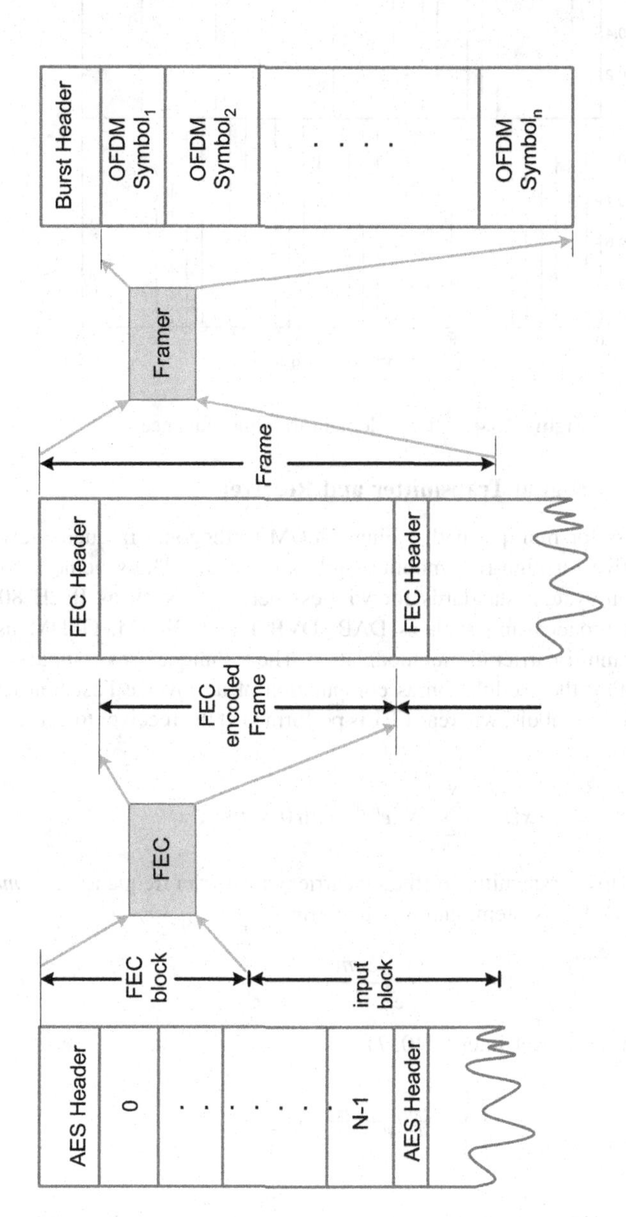

Figure 13.40 Header and frame insertion for synchronization

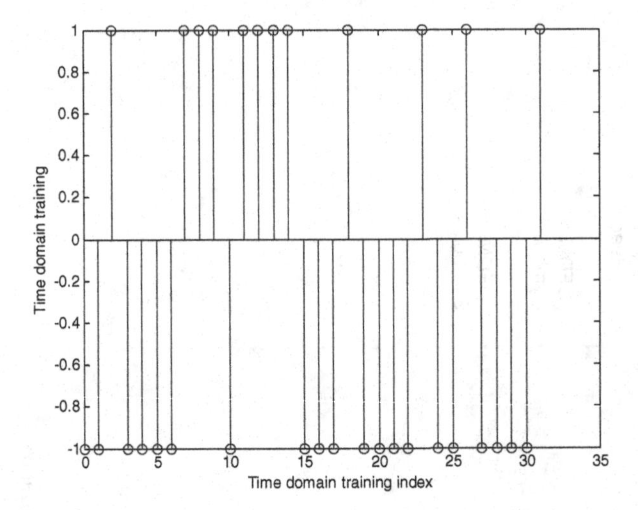

Figure 13.41 Time-domain training sequence

13.3.6.2 OFDM-based Digital Transmitter and Receiver

Owing to its robustness for multiple path fading, OFDM (orthogonal frequency-division multiplexing) is very effective for high-rate mobile wireless terminals. The scheme is used in many modern communication system standards for wireless networks, such as IEEE 802.11(a) and 802.16(a), and digital broadcasting such as DAB, DVB-T and DRAM. OFDM uses multiple orthogonal carriers for multi-carrier digital modulation. The technique uses complex exponentials. This helps in formulating the modulation as computation of an inverse Fast Fourier Transform (IFFT) of a set of parallel symbols, whereas FFT is performed at the receiver to extract the OFDM-modulated symbols:

$$x(t) = \sum_{m=0}^{N-1} X_m e^{\frac{j2\pi m}{T}t} \text{ with } 0 \leq t \leq T$$

where X_m is the symbol to be transmitted on the subcarrier with carrier frequency $f_m = m/T$. There are N multi-carriers in an OFDM system, and the subcarriers

$$\mathrm{e}^{j\frac{2\pi m}{T}t}$$

are orthogonal to each other over interval [0, T] as

$$\frac{1}{T}\int_0^T \mathrm{e}^{j\frac{2\pi l}{T}t} e^{-j\frac{2\pi n}{T}t} dt = \delta_{\mathrm{ln}}$$

where $l, n = 01, 2, ..., N-1$. At the transmitter, the bits are first mapped to symbol X_m, and N of these symbols (for $m = 0, \ldots, N-1$) are placed in parallel for IFFT computation. The output of IFFT is then serially modulated by a carrier frequency. To avoid intersymbol interference, and keeping the periodicity of the IFFT block intact, a cyclic prefix from the end of the IFFT block is copied to the start. The addition of a cyclic prefix is shown in Figure 13.42.

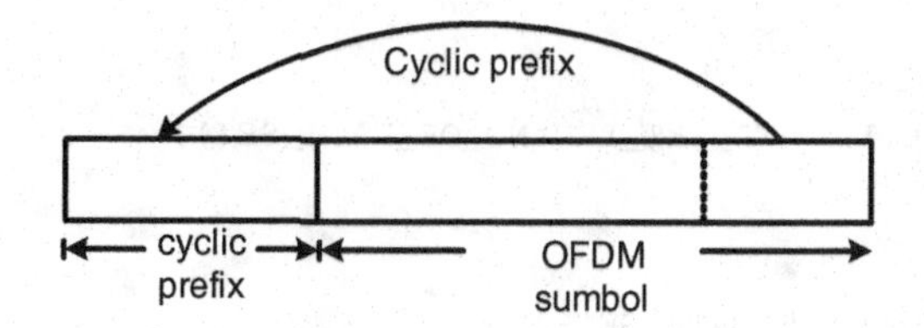

Figure 13.42 Adding cyclic prefix to an OFDM symbol

An OFDM system needs to address several issues that are standard to a communication system. These issues are carrier and sampling clock frequency offsets, OFDM symbol or frame offset, dealing with channel effects, and analog component noise and imperfections.

As the computation of timing and frequency offsets and the channel estimation are very critical for OFDM-based systems, some of the IFFT carriers may also be used as pilot tones. Training-based techniques are also popular that use a training sequence for these estimations. The MATLAB® code below shows the use of a training sequence for estimation and a 32-point IFFT-based OFDM transmitter. In the code the raw data is generated. The data is then block coded using a Reed–Solomon encoder. This encoding is followed by convolution encoding using a 2/3 encoder. The coded data is then mapped to QPSK symbols. These symbols are then grouped in parallel for OFDM coding:

```
clear all
close all
fec_en = 1;

% Each data set consists of 2 sets of 24 8-bit raw symbols,
% thus making the data length equal to 2*8*24=384
ofdm_burst_size = 24; % Number of OFDM frames in a burst of Tx
ofdm_train_per_burst = 1; % Number of trainings per burst
bps = 2;
DATA_SET_LEN = 384;
NO_OF_DATA_SET = 4; % For simulation

% RS code parameters adds 8 extra 8-bit symbols for error correction
RS_CODE_m = 8;
RS_CODE_k = 24;
RS_CODE_n = 32;

% After Reed-Solomn coding, the length of the burst becomes 2*8*32=512
% Convolution code parameters, the length becomes 512*3/2=768
% punct_code = [1 1 0 1];
code_rate = 1/2; %2/3;
trellis = poly2trellis(7, [133 171]);
tb_length = 3;
const_map = [-1-j  -1+j  1-j  1+j];
N = 32; % dft_size
K = 4; % cyclic prefix

% OFDM symbol size with cyclic Prefix
sym_size = N + K;
L = 4; % for training
M = N/L;
```

```
% TRANSMITTER
% Generate raw data
raw_data = randint(1, DATA_SET_LEN*NO_OF_DATA_SET);
frame_out = [];
out = [];
un_coded = [];
for burst_index=0:NO_OF_DATA_SET-1
% process data on frame by frame basis
frame = raw_data(DATA_SET_LEN*burst_index+1:
  DATA_SET_LEN*(burst_index+1));
if (fec_en)
    % Block encoding
    % Put the raw data for RS encoder format
    % The encoder generates n m-bit code symbols, that has k m-bit message
    % symbols
    % First put the bits in groups of m-bits to convert them
    % to RS symbols
    msg_in = reshape(frame,RS_CODE_m,DATA_SET_LEN/RS_CODE_m)';

    % Convert the m-bit symbols from binary to decimal while
    % considering the left bit as MSB
    msg_sm = bi2de(msg_in, 'left-msb');

    % Put the data in k m-bit message symbols
    rs_msg_sm = reshape(msg_sm, length(msg_sm)/RS_CODE_k,RS_CODE_k);

    % Convert the symbols into GF of 2^RS_CODE_m
    msg_tx = gf(rs_msg_sm, RS_CODE_m);

    % Encode using RS (n,k) encoder
    code = rsenc(msg_tx, RS_CODE_n, RS_CODE_k);

        % Convolution encoding
        tx_rs = code.x.';
        tx_rs_st = tx_rs(:);
        bin_str = de2bi(tx_rs_st,RS_CODE_m,'left-msb')';
        bin_str_st = bin_str(:);
        un_coded = [un_coded bin_str_st'];

        % Perform convolution encoding
        % conv_enc_out = convenc(bin_str_st, trellis, punct_code);
             frame = convenc(bin_str_st, trellis)';
end
             frame_out = [frame_out frame];
end

% QPSK modulation
out_symb =reshape(frame_out, bps, length(frame_out)/bps)';
out_symb_dec = bi2de(out_symb,'left-msb');
mod_out = const_map(out_symb_dec+1);
```

```
% OFDM modulation
ofdm_mod_out = [];

% Generate OFDM symbols for one burst of transmission
for index=0:ofdm_burst_size-1,
    ifft_in = mod_out(index*N+1:(index+1)*N);
    ifft_out = ifft(ifft_in, N);
    ifft_out_cp = [ifft_out(end-K+1:end) ifft_out];
    ofdm_mod_out=[ofdm_mod_out ifft_out_cp];
end
```

A burst is formed that appends a training sequence of Figure 13.41 to the OFDM data. A number of OFDM symbols are sent in a burst. This number depends on the coherent bandwidth of the channel that ensures the flatness of the channel on the entire bandwidth of transmission. The training is used to estimate different parameters at the receiver:

```
% Generating training for start of burst, synchronization and estimation
% Generate the training offline and store in memory for
% appending it with input data
C = [1 1 -1 1 1 1 1 -1];
A = fft(C);
Sign_Pattern = [-1 1 -1 -1];
G = [];
for i=1:L
    G = [G Sign_Pattern(i)*A];
end

% Add cyclic prefix to G
train_sym = [G(end-K+1:end) G]/4;

% Appending training to start of burst
ofdm_mod_out = [train_sym ofdm_mod_out];
```

The signal is transmitted after it is up-converted and mixed with a carrier. In a non-line-of-sight mobile environment the transmitted signal gets to the receiver through multiple paths. If there is a relative velocity between transmitter and receiver, the received signal frequency also experiences Doppler shift. The multi-path effects cause frequency fading. The signal also suffers from timing and frequency offsets. The channel impurities are modeled using a Rayleigh fading channel where timing and frequency offsets are also added to test the effectiveness of estimation and recovery techniques in the receiver. An additive white Gaussian noise (AWGN) is also added in the signal to test the design for varying signal to noise ratios (SNRs):

```
% Multi-path fading frequency selective channel model
ts = 1/(186.176e3); % sample time of the input signal
fd = 5; % maximum Doppler shift in Hz
tau = [0 8e-6 15e-6]; % path delays
pdb = [0 -10 -20]; % avg path gain
theta = 0.156; % frequency offset
SNR = 5; & signal to noise ratio in dbs
timing_offset = 18;
chan = rayleighchan(ts,fd,tau,pdb);
```

```
chan.StoreHistory = 1; % To plot the time varying channel

% Frequency selective multi-path fading channel
% Passing the signal through a multipath Rayleigh fading channel
% and adding other channel impurities
Received_Signal = filter(chan, ofdm_mod_out);

% Plot(chan)
Received_Signal = Received_Signal .*
                  exp(j*2*pi*(theta/N)*(0:length(Received_Signal)-1));
Received_Signal = awgn(Received_Signal, SNR, 'measured');
Received_Signal = [zeros(1,timing_offset) Received_Signal];
```

The coarse timing estimation is based on correlation of the training signal that consists of L parts of M size sequence. The algorithm computes the peak of the timing matrix given by the following set of equations:

$$\upsilon_\varepsilon(d) = \left(\frac{L}{L-1}\frac{|P(d)|}{E(d)}\right)^2$$

$$P(d) = \sum_{k=0}^{L-2} b(k) \sum_{m=0}^{M-1} r^*(d+kM+m)\cdot r(d+(k+1)M+m)$$

$$E(d) = \sum_{i=0}^{M-1}\sum_{k=0}^{L-1} |r(d+i+kM)|^2.$$

The MATLAB® code here implements these equations.

```
% RECEIVER
% Timing and frequency estimation
for k = 1:L-1
b(k) = Sign_Pattern(k)*Sign_Pattern(k+1);
end

% Timing estimation
for d=1:N
    E(d) = sum(abs(Received_Signal(d:d+L*M-1)).^2);
    P(d) = 0;
    for k=0:L-2
        index = d+k*M;
        indexP1 = d+(k+1)*M;
        P(d) = P(d) + (b(k+1) * Received_Signal(indexP1:
               indexP1+M-1)*Received_Signal(index:index+M-1)');
    end
    Timing_Metric(d) = ((L/(L-1))*abs(P(d))/E(d))^2;
end

[x y] = max(Timing_Metric);
Coarse_Timing_Point = round(y);
d_max = Coarse_Timing_Point;
```

For the frequency estimation, the training signal is first modified such that all the L parts have the same sign. This modified training symbol is represented by $\{y(k)\text{: } k = 0, \ldots, N-1\}$.The coarse estimate is given by:

$$R_y(m) = \frac{1}{N-mM}\sum_{k=mM}^{N-1} y^*(k-mM)y(k), 0 \leq m \leq H$$

$$\varphi(m) = \left[\arg\{R_y(m)\} - \arg\{R_y(m-1)\}\right]_{2\pi}, 1 \leq m \leq H$$

$$w(m) = 3\frac{(L-m)(L-m+1)-H(L-H)}{H(4H^2-6LH+3L^2-1)}$$

$$\acute{\upsilon} = \frac{L}{2\Pi}\sum_{m=1}^{H} w(m)\varphi(m).$$

The following MATLAB® code implements these equations and computes frequency offset estimation:

```
% Frequency estimation
H = L/2;
H = L/2;
w(1)=0; % not used
for m = 1:H
     w(m+1)=3*L/(2*pi)*((L-m)*(L-m+1)-H*(L-H))/(H*(4*H^2-6*L*H+3*L^2-1));
end
y=[];
for i=1:L
     y=[y Sign_Pattern(i)*Received_Signal(d_max+M*(i-1):d_max+i*M-1)];
end
for m = 0:H
     index = m*M;
     Ry(m+1) = 1/(N-index) * y(index+1:N)*y(1:N-index)';
     arg(m+1) = angle(Ry(m+1));
end
phi(1)=0; % not used
for m = 2:H+1
     phi(m) = arg(m)-arg(m-1);
     if phi(m) > pi
          phi(m) = phi(m) - 2*pi;
     elseif phi(m) < -pi
          phi(m) = phi(m) + 2*pi;
     end
end
Coarse_Frequency = phi*w';
```

These coarse estimates are further refined in [42] and the block diagram of the system is shown in Figure 13.43.

The training signal also aids in estimating the channel impulse response. The channel is assumed to be static for a few OFDM symbols. If gains of K paths are:

$$\mathbf{h} = h_0, h_1, \ldots, h_{K-1}$$

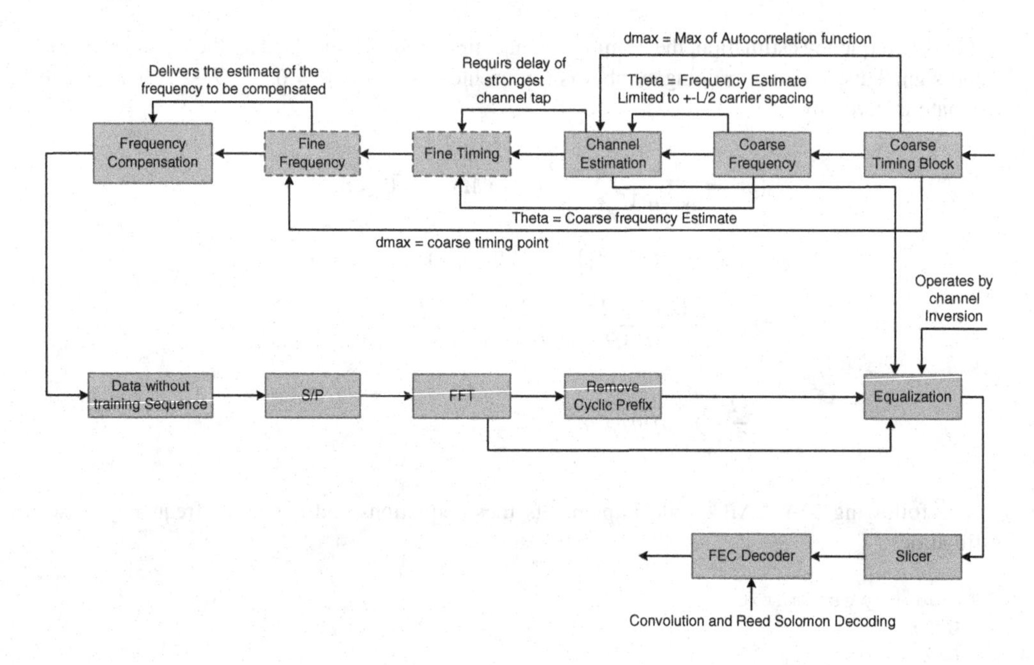

Figure 13.43 Block diagram of complete OFDM receiver

then the received signal can be written as:

$$r = e^{j}\mathbf{W}(v)\cdot\mathbf{S}\cdot\mathrm{h}+\mathbf{n}$$

where

$$W(v)\triangleq\mathrm{diag}\{1, e^{j2\pi v/N}, e^{j2\pi 2v/N}, \ldots, e^{j2\pi(N-1)v/N}\}$$

$$\mathbf{n}\triangleq[n(0)n(1)\ldots n(N-1)]$$

$$S\begin{bmatrix} s(0) & s(-1)\ldots & s(-k+1) \\ s(1) & s(0)\ldots & s(-k+2) \\ & \cdot & \\ & \cdot & \\ & \cdot & \\ s(N-1) & s(N-2)\ldots & s(N-k) \end{bmatrix}$$

Here, $\{s(k)\colon k=-N_g, -N_g+1, \ldots, N-1\}$ are the samples of the training sequence with cyclic prefix, $\mathbf{n}$ is the noise vector, v is the frequency offset, and $\mathbf{r}$ is the received signal. The channel estimation using coarse frequency offset can be written as:

$$\hat{\mathbf{h}} = \left[\mathbf{S}^{H}\cdot\mathbf{S}\right]^{-1}\mathbf{S}^{H}\cdot\mathbf{W}^{H}(\hat{v})\cdot\mathbf{r}$$

where $\hat{\nu}$ is the coarse frequency offset estimate. The following MATLAB® code implements the channel estimation:

```
ct = Coarse_Timing_Point;
cf = Coarse_Frequency;
r = Received_Signal(ct:ct+N-1);
for i = 1:N
     w(i) = exp(j*2*pi*cf*(i+K-1)/N);
end
W = diag(w);
S = zeros(N, K);
for i = 1:N
     S(i,:) = fliplr(train_sym(i+1:K+i));
end
h_cap = inv(S'*S)*S'*W'*r.';
Channel_Estimate = h_cap;
```

The receiver first needs to estimate the frequency and timing offset. It also needs to estimate the channel impulse response. When the timing and frequency errors have been estimated, the receiver makes requisite corrections to the received signal using these estimations. The channel estimation is also used to equalize the received signal. The OFDM, because of its multiple carriers, assumes the channel for each carrier as flat. This flat fading factor is calculated for each carrier and the output is adjusted accordingly. The symbols are extracted by converted output of FFT computation to serial and applying slicer logic to the output. The symbols are then mapped to bits:

```
% Frequency and timing offsets compensation
Received_Signal = Received_Signal(Coarse_Timing_Point-K:end);
Received_Signal = Received_Signal.* exp(-j*2*pi *
            ((Coarse_Frequency)/N)*(0:length(Received_Signal)-1));
Initial_Point = K+N+K+1;
Received_Signal = Received_Signal(Initial_Point:Initial_Point+(N+K) *
            (ofdm_burst_size-1)-1);

% Processing the synchronized data
Received_Signal = reshape(Received_Signal, N+K,ofdm_burst_size-1);
% Serial to parallel
Received_Signal_WOGI = Received_Signal(1:N,:); % remove cyclic prefix
Received_Signal_WOGI = fft(Received_Signal_WOGI,[],1);
Received_Signal_WOZP = zeros(N,ofdm_burst_size-1);
Received_Signal_WOZP(1:N/2,:) = Received_Signal_WOGI(1:N/2,:);
Received_Signal_WOZP((N/2)+1:end,:) = Received_Signal_WOGI(end-(N/2)+1:
end,:);

% Estimation
H = fft(Channel_Estimate,N);
CH = zeros(N,1);
CH(1:N/2) = H(1:N/2);
CH((N/2)+1:end) = H(end-(N/2)+1:end);
H_l_hat_square = abs(CH);
W = 1./CH;
```

```
% Equalization
Estimated_Signal = [];
for Symbol_Number = 1:1:ofdm_burst_size-1
     Present_Symbol = Received_Signal_WOZP(:,Symbol_Number);
     Estimated_Symbol = Present_Symbol.*W;
     Estimated_Signal = [Estimated_Signal Estimated_Symbol];
end
Received_Signal_WZP = reshape(Estimated_Signal, 1, N *
                    (ofdm_burst_size-1)); % Parallel to serial
% Scatterplot(Received_Signal_WZP)
% Title('Equalizer Output')
Recovered_Signal = Received_Signal_WZP;
% Received_Signal_WZP = reshape(Estimated_Signal, 1,
N*(ofdm_burst_size-1)); % Parallel to serial

% Slicer
Bit_Counter = 1;
Bits = zeros(1,length(Recovered_Signal)*bps);
for Symbol_Counter = 1:length(Recovered_Signal)
                         % Slicer for decision making
     [Dist Decimal] = min(abs(const_map -
               Recovered_Signal(Symbol_Counter)));
                    % Checking for mimimum Euclidean distance
     est = const_map(Decimal);
     Bits(1,Bit_Counter:Bit_Counter+bps-1) = de2bi(Decimal-1,bps,
                              'left-msb');
                         % Constellation Point to Bits
     Bit_Counter = Bit_Counter + bps;
end
scatterplot(Recovered_Signal)
title('Received Constellation')
grid
Data = bit_stream(1:N*(ofdm_burst_size-1)*bps);
ErrorCount = sum((Data~=Bits))
scatterplot(Received_Signal_WZP); title('Received Constellation');
grid

store = ErrorCount/length(Data)
```

Figure 13.44 shows the constellation of QPSK-based OFDM receiver of this section for different channel conditions. For deep fades in the channel frequency resource, an OFDM-based system is augmented with error-correction codes and interleaving. The interleaving helps in spreading the consecutive bits in the input sequence to different locations in the transmitted signal.

It is quite evident from the MATLAB® code that simple FFT and IFFT cores can be used for mapping the OFDM Tx/Rx in hardware. The channel estimation should preferably be mapped on a programmable DSP. The timing error and frequency errors can also be calculated in hardware. The system needs to be thought of as a cascade of PEs implementing different parts of the algorithm. Depending on the data rates, the HW mapping may be fully parallel or time-shared.

13.3.7 Digital Up-conversion and Mixing

The modulated signal is to be digitally mixed with an intermediate frequency (IF) carrier. In a digital receiver one of the two frequencies, 70 MHz or 21.4 MHz, is commonly used. The digital carrier is

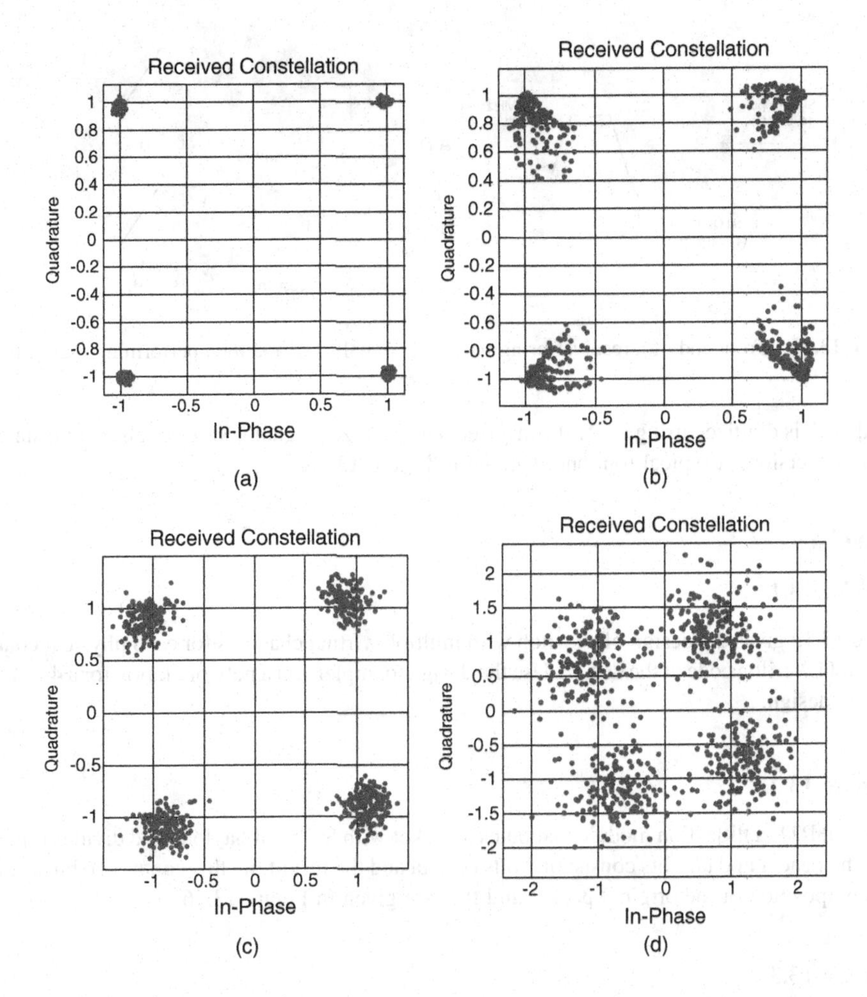

Figure 13.44 Constellation of QPSK-based OFDM received signal for different channel conditions and frequency offsets. (a) 40 dB SNR. (b) 40 dB SNR with 30 Hz Doppler shift. (c) 20 dB SNR with 5 Hz Doppler shift. (d) 10 dB SNR

generated adhering to the Nyquist sampling criterion. To digitally mix the baseband signal with the digitally generated carrier, the baseband signal is also up-converted to the same sampling rate as that of the IF carrier. A digital up-converter performs the conversion in multiple stages using polyphase decomposition. The designs of digital up- and down-converters using polyphase FIR and IIR filters are given in Chapter 8.

In direct communication transmitter and receiver the inphase and quadrature components of a modulated signal are converted to analog at baseband. The signal is then mixed using one or multiple stages of quadrature mixing and finally the signal is transmitted after passing it through a power amplifier.

13.3.8 Front End of the Receiver

The receiver front end usually performs processing in the analog domain. The received signal is amplified and goes through one or multiple stages of mixing and is brought down to IF. The analog

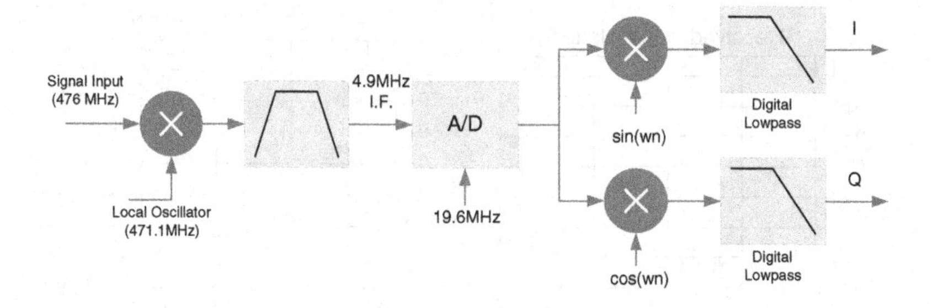

Figure 13.45 Front end of a receiver bringing an RF signal to IF and then performing digital mixing

signal at IF is digitized using an A/D converter. The digitized signal goes through several stages of down-conversion. A typical font end is given in Figure 13.45.

Exercises

Exercise 13.1

Figure 13.12 gives the design of a switch with multiple virtual channels for each physical channel. Write RTL Verilog code of the design. Use the design to implement a network fabric for a 4×4 array MPSoC design.

Exercise 13.2

Design at RTL a flitization module that takes a packet with 80-bit of payload and divides it into six flits. The header and tail flits consist of 8 bits of data and the rest of the flits contain 16 bits for data. The composition of the original packet and flits are given in Figure 13.46.

Exercise 13.3

Use the CORE Generator utility of ISE Xilinx and generate memories and FIFOs for the design of Figure 13.47. All data buses are 8 bits wide, the FIFOs are 16 deep and local memories are 512 deep. Select an FPGA from the Virtex-5 family. Write RTL Verilog code of the top-level module.

Exercise 13.4

Design and code in RTL Verilog a block turbo encoder to perform encoding on a 2-D 11×11-bit block. The encoder first performs coding row-wise, adding four coded bits and one parity bit in each row, and then encoding on each resultant column to add four coded bits and one parity bit. This makes the size of the coded block 16×16 as shown in Figure 13.48. The encoder is given in Figure 13.38. For 11×11-bit encoding, registers r_4 and r_5 are not used. After 11 cycles the registers r_0, r_1, r_2 and r_3 have the coded bits and the parity bit comes from the parity-bit register r_p. The parity max uses data bits for the first 11 cycles and then it uses the coded bits to calculate the parity bit. The same encoding is used to code rows and columns.

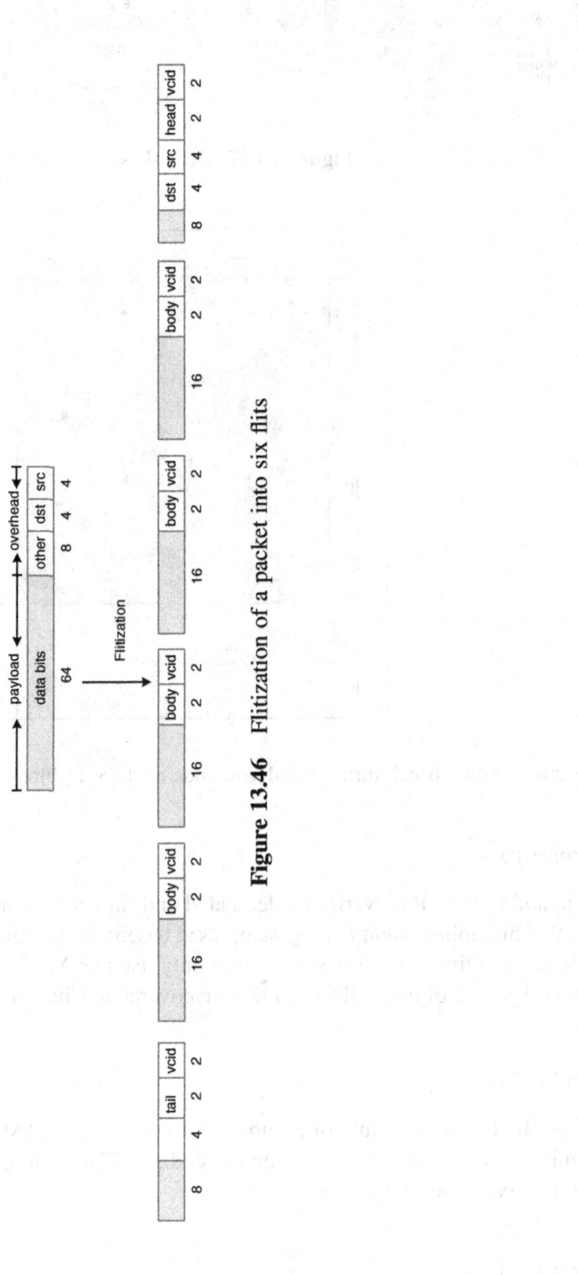

Figure 13.46 Flitization of a packet into six flits

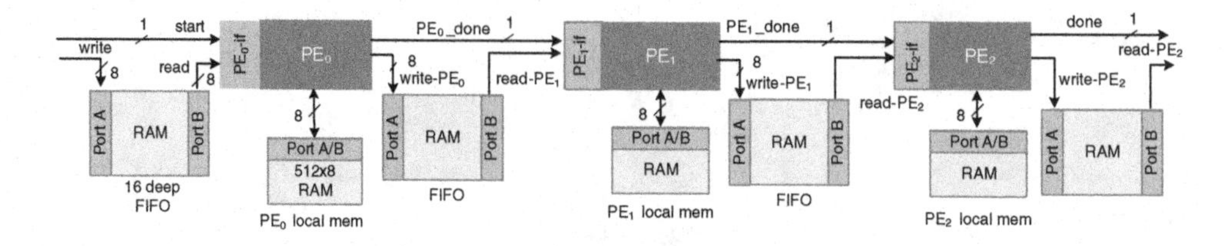

Figure 13.47 KPN-based top-level design

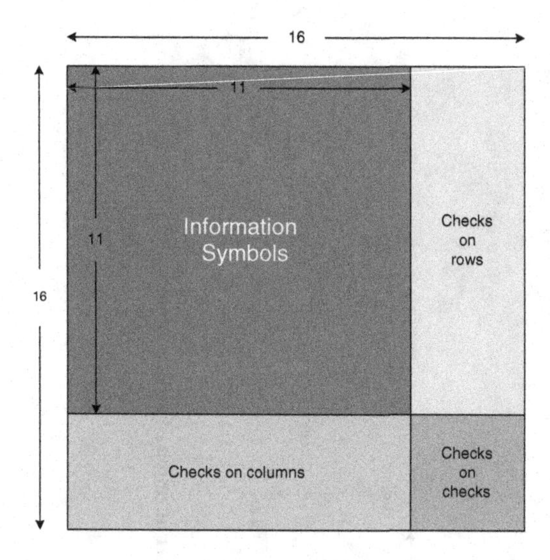

Figure 13.48 Block turbo encoder to code an 11×11 block to make a 16×16 coded block

Exercise 13.5

Design and code in RTL Verilog a dedicated and time-shared architecture of a digital quadrature mixer that multiplies 16-bit I and Q samples at baseband to digitally generated 16-bit sin $(\omega_0 n)$ and cos $(\omega_0 n)$. The time-shared design should only use one MAC unit. Truncate the output to 16-bit numbers. Use one of the DDFS architecture covered in Chapter 12 for sine and cosine generation.

Exercise 13.6

Use systolic FFT architecture of Figure 8.14 to design 16-QAM-based 8 sub-carrier based OFDM transmitter and receiver blocks. Write the code in RTL Verilog and test your design with Matlab generated fixed-point implementation.

Exercise 13.7

Implement a 16-bit folded and parallel systolic AES architecture that uses the in-place indexing technique of Sections 13.3.3.3 and 13.3.3.4, respectively.

Exercise 13.8

Convert the code of the OFDM transmitter and receiver of Section 13.3.6.2 in fixed-point format. Use the fi tool of MATLAB® and keep the datapath to 16-bit numbers.

Exercise 13.9

Design a fully parallel architecture to implement the LZ-77 algorithm using architecture presented in Section 13.3.1. Keep the lengths of the search buffer and history buffer to 4 and 8 characters wide.

References

1. www.arm.com/products/system-ip/amba/
2. D. Wingard, "Micronetwork-based integration for SoCs," in *OpenCores Organization, Revision B.3,* 2002, available at sen.enst.fr/filemanager/active?fid=662.
3. IBM. "CoreConnect bus cores," product brief.
4. "WISHBONE system-on-chip (SoC) interconnection architecture for portable IP cores," *OpenCores Organization, Revision B.3*, 2002, available at sen.enst.fr/filemanager/active?fid=662.
5. D. Pham *et al*, "Design and implementation of a first-generation CELL processor," *Digest of Technical Papers for ISSCC*, 2005, pp. 184–185.
6. W. J. Dally and B. Towles, "Route packet, not wires: on-chip interconnection networks," in *Proceedings of DAC*, 2001, pp. 684–689.
7. L. Benini and G. De Micheli, "Networks on chips: a new SoC paradigm," *IEEE Computer Magazine*, 2002, vol. 35, pp. 70–78.
8. S. Kumar *et al.*, "A network-on-chip architecture and design methodology.," in *Proceedings of IEEE Computer Society Annual Symposium on VLSI*, 2002, pp. 105–112.
9. M. Palesi, R. Holsmark, S. Kumar and V. Catania, "*Application specific routing algorithms for network on chip*," *IEEE Transactions on Parallel and Distributed Systems*, 2009, vol. 20, pp. 316–330.
10. D. Wiklund and D. Liu, "SoCBUS: Switched network on chip for hard real-time embedded systems," in *Proceedings of International Symposium on Parallel and Distributed Processing*, 2003, pp. 78–85, IEEE Computer Society, USA.
11. H. Sullivan, T. R. Bashkow and D. Klappholz, "A large-scale homogeneous, fully distributed parallel machine," in *Proceedings of 4th ACM Symposium on Parallel Algorithms and Architectures*, 1977, pp. 105–117.
12. T. Nesson and S. L. Johnsson, "ROM routing on mesh and torus networks," in *Proceedings of 7th ACM Symposium on Parallel Algorithms and Architectures*, 1995, pp. 275–287.
13. D. Seo, A. Ali, W. T. Lim and N. Rafique, "Near-optimal worst-case throughput routing for two-dimensional mesh networks," in *Proceedings of ISCA*, 2005, pp. 432–443.
14. J. Hu and R. Marculescu, "DyAD: smart routing for networks-on-chip," in *IEEE Proceedings of 41st Conference on Design and Automation*, 2004, pp. 260–263, ACM, NY, USA.
15. W. J. Dally and C. L. Seitz, "Deadlock-free message routing in multiprocessor interconnection networks," *IEEE Transactions on Computers*, 1987, vol. 36, pp. 547–553.
16. J. Duato, "A new theory of deadlock-free adaptive routing in wormhole networks," *IEEE Transactions on Parallel and Distributed Systems*, 1993, vol. 4, pp. 1320–1331.
17. G. M. Chiu, "The odd–even turn model for adaptive routing," *IEEE Transactions on Parallel and Distributed Systems*, 2000, vol. 11, pp. 729–728.
18. C. J. Glass and L. M. Ni, "The turn model for adaptive routing," *Journal of ACM*, 1994, vol. 31, pp. 874–902.
19. C. J. Glass and L. M. Ni, "Maximally fully adaptive routing in 2D meshes," in *Proceedings of International Conference on Parallel Processing*, 1992, vol. I, pp. 101–104, ACM, NY, USA.
20. M. Millberg, E. Nilsson, R. Thid and A. Jantsch, "Guaranteed bandwidth using looped containers in temporally disjoint networks within the Nostrum network on chip," in *Proceedings of Design Automation and Test in Europe Conference*, 2004, vol. 2, pp. 890–895, IEEE Computer Society, USA.
21. J. Hu and R. Marculescu, "Exploiting the routing flexibility for energy/performance aware mapping of regular NoC architectures," in *Proceedings of Design Automation and Test in Europe Conference*, 2003, vol. 1, pp. 10688–10693, IEEE Computer Society, USA.

22. Y. H. Kang, T.-J. Kwon and J. Draper, "Dynamic packet fragmentation for increased virtual channel utilization in on-chip routers," in *Proceedings of 3rd ACM/IEEE International Symposium on Networks-on-Chip*, 2009, pp. 250–255.
23. M. Zhang and C.-S. Choy, "Low-cost VC allocator design for virtual channel wormhole routers in networks-on-chip," in *Proceedings of 2nd ACM/IEEE International Symposium on Networks-on-Chip*, 2008, pp. 207–208.
24. Z. Lu, "Design and analysis of on-chip communication for network-on-chip platforms," PhD thesis, Royal Institute of Technology, Stockholm, 2007.
25. S. A. Asghari, H. Pedram, M. Khademi and P. Yaghini, "Designing and implementation of a network-on-chip router based on handshaking communication mechanism," *World Applied Sciences Journal*, 2009, vol. 6, no. 1, pp. 88–93.
26. C. R. A. Gonzalez, C. B. Dietrich and J. H. Reed, "Understanding the software communications architecture," *IEEE Communications Magazine*, 2009, vol. 47, no. 9, pp. 50–57.
27. J. Ziv and A. Lempel, "A universal algorithm for sequential data compression," *IEEE Transactions on Information Theory*, 1977, vol. 23, pp. 337–343.
28. R. Mehboob, S. A. Khan, Z. Ahmed, H. Jamal, and M. Shahbaz, "Multigig Lossless Data Compression Device" *IEEE Transactions on Consumer Electronics*, 2010, vol. 56, no. 3.
29. S.-A. Hwang and C.-W. Wu, "Unified VLSI systolic array design for LZ data compression," *IEEE Transactions on Very Large Scale Integration Systems*, 2001, vol. 9, pp. 489–499.
30. J. l. Nunez and S. Jones, "Gbits/s lossless data compression hardware," *IEEE Transactions on Very Large Scale Integration Systems*, 2003, vol. 11, pp. 499–510.
31. www.esat.kuleuven.ac.be/~rijmen/rijndael
32. S. M. Farhan, S. A. Khan and H. Jamal, "An 8-bit systolic AES architecture for moderate data rate applications," *Microprocessors & Microsystems*, 2009, vol. 33, pp. 221–231.
33. S. M. Farhan, S. A. Khan and H. Jamal, "Low-power area-efficient high data-rate 16-bit AES crypto processor," in *Proceedings of 18th ICM International Conference on Microelectronics, Dhahran*, 2006, pp. 186–189.
34. N. Sklavos and O. Koufopavlou, "Architectures and VLSI implementations of the AES-proposal Rijndael," *IEEE Transactions on Computers*, 2002, vol. 51, pp. 1454–1459.
35. A. Hodjat and I. Verbauwhede, "Area–throughput trade-offs for fully pipelined 30 to 70 Gbits/s AES processors," *IEEE Transactions on Computers*, 2006, vol. 55, pp. 366–372.
36. "Using look-up tables as distributed RAM in Spartan-3 generation FPGAs," in application note for XAPP464 (v2.0), March 2005.
37. "Using block RAM in Spartan-3 generation FPGAs," in application note for XAPP463 (v2.0), March 2005.
38. S. A. Khan, "8-bit time-shared and systolic architecture for AES," Technical report, CASE May 2010.
39. R. Pyndiah, "Near-optimum decoding of product codes: block turbo codes," *IEEE Transactions on Communications*, 1998, vol. 46, pp. 1003–1010.
40. T. M. Schmidl and D. C. Cox, "Robust frequency and timing synchronization for OFDM," *IEEE Transactions on Communications*, 1997, vol. 45, pp. 1613–1621.
41. H. Minn, M. Zeng and V. K. Bhargava, "On timing offset estimation for OFDM systems," *IEEE Commununications Letters*, 2000, vol. 4, pp. 242–244.
42. H. Minn, V. K. Bhargava and K. B. Letaief, "A robust timing and frequency synchronization for OFDMsystems," *IEEE Transactions on Wireless Communication*, 2003, vol. 2, pp. 822–839.

Index

Digital Design of Signal Processing Systems: A Practical Approach, First Edition. Shoab Ahmed Khan.